Cytochrome P-450

Structure, Mechanism, and Biochemistry

Cytochrome P-450
Structure, Mechanism, and Biochemistry

Edited by
Paul R. Ortiz de Montellano
University of California, San Francisco
San Francisco, California

Plenum Press • New York and London

Library of Congress Cataloging in Publication Data

Main entry under title:

Cytochrome P-450.

Includes bibliographical references and index.
1. Cytochrome P-450. 2. Metalloenzymes. I. Ortiz de Montellano, Paul R.
QP671.C83C98 1986   574.19′25   85-30103
ISBN 0-306-42147-X

QP
671
.C83
C98
1986

© 1986 Plenum Press, New York
A Division of Plenum Publishing Corporation
233 Spring Street, New York, N.Y. 10013

All rights reserved

No part of this book may be reproduced, stored in a retrieval system, or transmitted in any form or by any means, electronic, mechanical, photocopying, microfilming, recording, or otherwise, without written permission from the Publisher

Printed in the United States of America

Contributors

JOHN R. BATTISTA • Department of Chemistry, Wayne State University, Detroit, Michigan 48202

SHAUN D. BLACK • Department of Biological Chemistry, Medical School, The University of Michigan, Ann Arbor, Michigan 48109

MINOR J. COON • Department of Biological Chemistry, Medical School, The University of Michigan, Ann Arbor, Michigan 48109

HOWARD J. EISEN • Laboratory of Developmental Pharmacology, National Institute of Child Health and Human Development, National Institutes of Health, Bethesda, Maryland 20205

JOHN T. GROVES • Department of Chemistry, The University of Michigan, Ann Arbor, Michigan 48109

MAGNUS INGELMAN-SUNDBERG • Department of Physiological Chemistry, Karolinska Institute, S-104 01 Stockholm, Sweden

COLIN R. JEFCOATE • Department of Pharmacology, University of Wisconsin Medical School, Madison, Wisconsin 53706

MALIYAKAL E. JOHN • Departments of Biochemistry and Obstetrics and Gynecology, University of Texas Health Science Center, Dallas, Texas 75235

ANTHONY Y. H. LU • Department of Animal Drug Metabolism, Merck Sharp and Dohme Research Laboratories, Rahway, New Jersey 07065

THOMAS J. McMURRY • Department of Chemistry, The University of Michigan, Ann Arbor, Michigan 48109

LAWRENCE J. MARNETT • Department of Chemistry, Wayne State University, Detroit, Michigan 48202

GERALD T. MIWA • Department of Animal Drug Metabolism, Merck Sharp and Dohme Research Laboratories, Rahway, New Jersey 07065

RALPH I. MURRAY • Department of Biochemistry, University of Illinois, Urbana, Illinois 61801

PAUL R. ORTIZ DE MONTELLANO • Department of Pharmaceutical Chemistry, School of Pharmacy, University of California, San Francisco, California 94143

JULIAN A. PETERSON • Department of Biochemistry, University of Texas Health Sciences Center, Dallas, Texas 75235

THOMAS L. POULOS • Genex Corporation, Gaithersburg, Maryland 20877

RUSSELL A. PROUGH • Department of Biochemistry, University of Texas Health Sciences Center, Dallas, Texas 75235

NORBERT O. REICH • Department of Pharmaceutical Chemistry, School of Pharmacy, University of California, San Francisco, California 94143

EVAN R. SIMPSON • Departments of Biochemistry and Obstetrics and Gynecology, and Cecil H. and Ida Green Center for Reproductive Biology Sciences, University of Texas Health Science Center, Dallas, Texas 75235

STEPHEN G. SLIGAR • Department of Biochemistry, University of Illinois, Urbana, Illinois 61801

MICHAEL R. WATERMAN • Departments of Biochemistry and Obstetrics and Gynecology, University of Texas Health Science Center, Dallas, Texas 75235

DAVID J. WAXMAN • Department of Biological Chemistry and Dana–Farber Cancer Institute, Harvard Medical School, Boston, Massachusetts 02115

PAUL WELLER • Department of Chemistry, Wayne State University, Detroit, Michigan 48202

Preface

Major advances have been made in recent years in clarifying the molecular properties of the cytochrome P-450 system. These advances stem, in practical terms, from the generally recognized importance of cytochrome P-450 in the metabolism of drugs and in the bioactivation of xenobiotics to toxic products. The fascinating multiplicity and differential regulation of cytochrome P-450 isozymes, and their ability to catalyze extraordinarily difficult chemical transformations, have independently drawn many chemists and biochemists into the P-450 circle. Progress in the field, from a technical point of view, has been propelled by the development of reliable procedures for the purification of membrane-bound enzymes, by the growing repertoire of molecular biological techniques, and by the development of chemical models that mimic the catalytic action of P-450. As a result, our understanding of the P-450 system is moving from the descriptive, pharmacological level into the tangible realm of atomic detail.

The rapid progress and multidisciplinary character of the cytochrome P-450 field, which cuts across the lines that traditionally divide disciplines as diverse as inorganic chemistry and genetics, have created a need for an up-to-date evaluation of the advances that have been made. It is hoped that this book, with its molecular focus on the cytochrome P-450 system, will alleviate this need. The authors of the individual chapters have strived to emphasize recent results without sacrificing the background required to make their chapters comprehensible to informed nonspecialists. Some overlap exists in the coverage of certain topics, but I have deemed this overlap desirable because it both preserves the integrity of individual chapters and offers the reader different perspectives on questions that are still open to debate. The success of this book, in my view, will be measured by the extent to which it helps to consolidate progress in the field, instigates new experimental approaches, and engenders productive discussion.

The book is informally divided into three sections. The first section

(Chapters 1 and 2) reviews the metalloporphyrin and non-P-450 hemeprotein models that have provided the molecular background for our current understanding of the catalytic mechanisms of P-450 enzymes. The second section (Chapters 3–9) is devoted to the structure, mechanism, and regulation of the hepatic P-450 isozymes. The chapters in this second section cover (1) the active-site topologies of P-450 enzymes (Chapter 3), (2) the interactions of P-450 enzymes with the proteins that provide them with electrons and with the membrane in which these protein–protein interactions occur (Chapters 4 and 5), (3) the primary and secondary structures of P-450 enzymes (Chapter 6), (4) the mechanisms of oxygen activation and transfer (Chapter 7), (5) the mechanisms by which P-450 enzymes are inhibited (Chapter 8), and (6) the mechanisms by which hepatic P-450 enzymes are regulated (Chapter 9). The third and last section of the book deals with the P-450 enzymes involved in sterol biosynthesis and catabolism (Chapters 10 and 11) and the bacterial P-450 enzymes, particularly P-450$_{cam}$ (Chapters 12 and 13). The book closes with an appendix intended to help readers determine the identity of the hepatic isozymes isolated by different laboratories.

No task of the magnitude of this volume is made without a price, frequently paid by the innocent. I can only acknowledge with gratitude the forbearance of my family, who saw less of me than they might have, and of my students, for whom I did less than I might have, had I not been immersed in the preparation of this volume. I also gratefully acknowledge the help of Elizabeth Komives, whose criticism of the chapters hopefully helped to make them more intelligible to her fellow graduate students.

Paul R. Ortiz de Montellano

San Francisco

Contents

1. *Metalloporphyrin Models for Cytochrome P-450* 1
   Thomas J. McMurry and John T. Groves

2. *Comparison of the Peroxidase Activity of Hemeproteins and Cytochrome P-450* 29
   Lawrence J. Marnett, Paul Weller, and John R. Battista

3. *The Topology of the Mammalian Cytochrome P-450 Active Site* 77
   Gerald T. Miwa and Anthony Y. H. Lu

4. *Cytochrome P-450 Reductase and Cytochrome b_5 in Cytochrome P-450 Catalysis* 89
   Julian A. Peterson and Russell A. Prough

5. *Cytochrome P-450 Organization and Membrane Interactions* 119
   Magnus Ingelman-Sundberg

6. *Comparative Structures of P-450 Cytochromes* 161
   Shaun D. Black and Minor J. Coon

7. *Oxygen Activation and Transfer* 217
   Paul R. Ortiz de Montellano

8. *Inhibition of Cytochrome P-450 Enzymes* 273
   Paul R. Ortiz de Montellano and Norbert O. Reich

9. *Induction of Hepatic P-450 Isozymes: Evidence for Specific Receptors* 315
   Howard J. Eisen

10. *Regulation of Synthesis and Activity of Cytochrome P-450 Enzymes in Physiological Pathways* 345
    Michael R. Waterman, Maliyakal E. John, and Evan R. Simpson

11. *Cytochrome P-450 Enzymes in Sterol Biosynthesis and Metabolism* 387
    Colin R. Jefcoate

12. *Cytochrome P-450$_{cam}$ and Other Bacterial P-450 Enzymes* 429
    Stephen G. Sligar and Ralph I. Murray

13. *The Crystal Structure of Cytochrome P-450$_{cam}$* 505
    Thomas L. Poulos

Appendix. Rat Hepatic Cytochrome P-450: Comparative Study of Multiple Isozymic Forms 525
    David J. Waxman

Index 541

CHAPTER 1

Metalloporphyrin Models for Cytochrome P-450

THOMAS J. McMURRY and JOHN T. GROVES

1. Introduction

The elucidation of the molecular mechanisms of biological oxygen activation has been the focus of sustained attention for over a decade.[1] In this time, cytochrome P-450 has become a Rosetta stone among the heme-containing monooxygenases.[2,3] The wide variety of oxygenations mediated by P-450 and the significance of these processes in steroid metabolism, drug detoxification, and the carcinogenic activation of polycyclic aromatic hydrocarbons have stimulated an effort to understand these processes. The selective hydroxylation of unactivated alkanes, in particular, lacks a classical paradigm in organic chemistry. Accordingly, there has been an effort to develop synthetic models of P-450 which might be used for the oxyfunctionalization of hydrocarbons. An understanding of the mechanism of these simple cases has begun to provide a conceptual base for the understanding of the enzymatic pathway.

In this chapter, pertinent elements of the mechanism of action of P-450 are considered in chemical terms. The developing chemistry of synthetic metalloporphyrins is then presented in this light.

$$\text{C}_6\text{H}_{11}\text{-H} + \text{O}_2 \xrightarrow[2\text{H}^+]{2e^-} \text{C}_6\text{H}_{11}\text{-OH} + \text{H}_2\text{O}$$

The active site of P-450 has long been known to contain a single iron

---

THOMAS J. McMURRY and JOHN T. GROVES • Department of Chemistry, The University of Michigan, Ann Arbor, Michigan 48109.

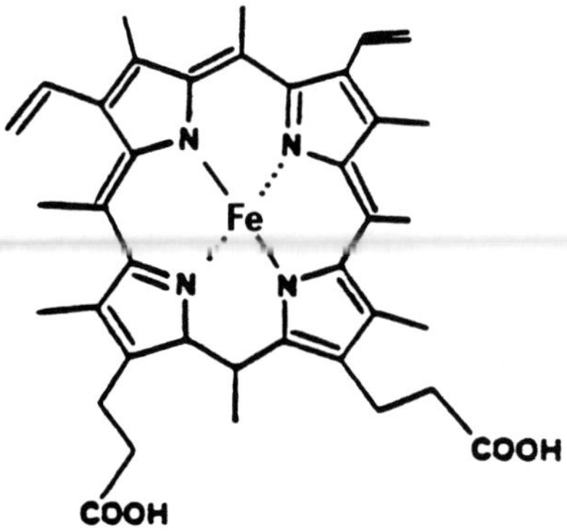

FIGURE 1. Structure of protoporphyrin IX.

protoporphyrin IX prosthetic group (Fig. 1). Dioxygen is bound, reduced, and activated at this site. An understanding of the events leading up to the substrate oxidation has resulted in general acceptance of the catalytic cycle for P-450[3] (Fig. 2). The salient features of the catalytic cycle include:

1. Binding of the substrate to give a high-spin ferric complex
2. One-electron reduction of the iron to the iron(II) state
3. Binding of dioxygen to generate the oxy form $SFe^{3+}O_2^-$
4. A second, one-electron reduction to yield the iron peroxo species $Fe^{3+}O_2^{2-}$

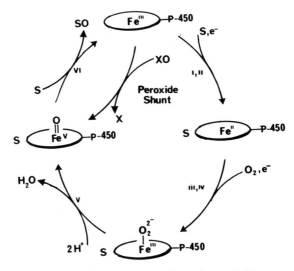

FIGURE 2. Catalytic cycle of cytochrome P-450.

5. Formal heterolysis of the O–O bond with concomitant generation of the reactive oxidant [FeO]$^{3+}$ and a molecule of water
6. A two-electron oxidation of substrate to produce SO and regenerate the ferric resting state of the enzyme

The exact nature of the active oxidant, depicted as [Fe^V=O] in Fig. 2, remains unknown. If the oxidation equivalents are localized on the heme, the oxidant can be represented as an oxoiron(V) species **1** or as an oxoiron(IV) porphyrin radical **2**. In the latter case, the second oxidation equivalent resides on the macrocycle in the form of a delocalized π radical instead of at the metal center. Thus, if one considers the porphyrin ligand to be a dianion, the oxidized porphyrin radical is a monoanion. Alternatives to these formulations include condensed protein peroxide compounds such as a peroxyacid **3** or peroxyimidic acid **4**.³

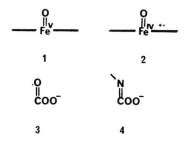

An observation of signal importance with regard to the chemistry of oxygen activation by P-450 was that the binding and reduction of dioxygen could be circumvented by the addition of exogenous oxygen sources such as ROOH,⁴ PhIO,⁵ and NaIO$_4$.⁶ We have referred to this step as the peroxide shunt pathway (Fig. 2). Evidence has accumulated which suggests that the oxidant derived via the peroxide shunt pathway is similar to that formed by the reduction of dioxygen.⁷ Furthermore, the ability of the two-electron oxotransfer agents iodosylbenzene and periodate to function in the enzyme suggests that an oxidized iron center with a single oxygen atom is a viable active intermediate.[5,7b,7c]

2. Molecular Probes of the P-450 Mechanism

The mechanism of P-450-catalyzed reactions has been studied using substrates designed to probe the nature of the ultimate oxidant. Stereoselectivity, molecular rearrangements, and hydrogen isotope effects characteristic of the oxygen transfer reaction have been the most useful probes. Surprisingly, the hydroxylation of a saturated methylene (CH$_2$)

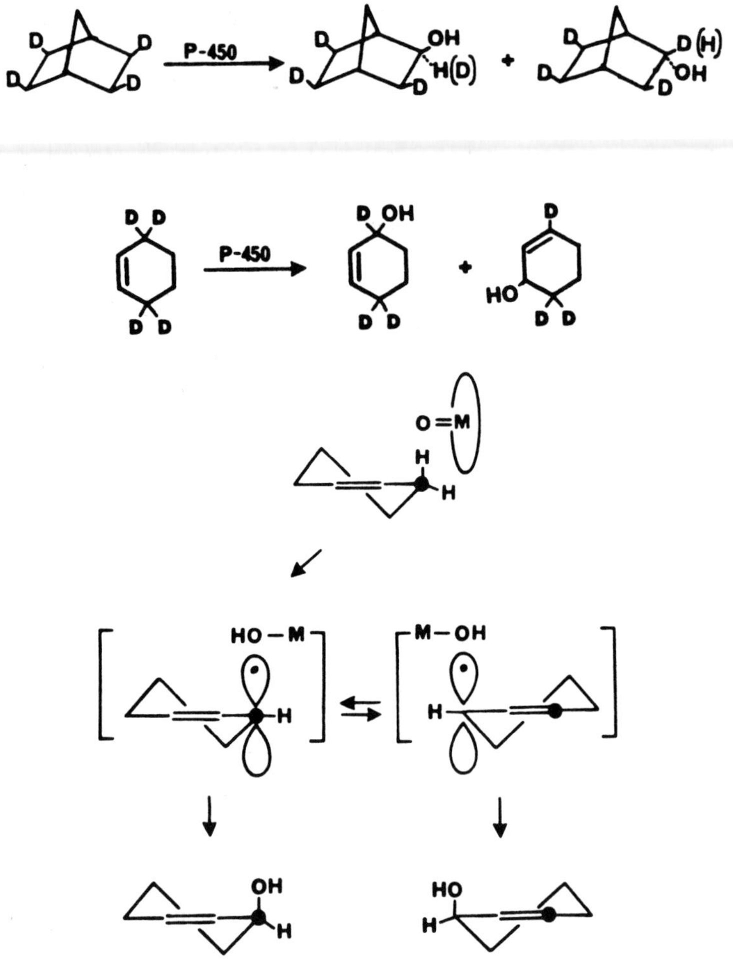

FIGURE 3. Epimerization and allylic scrambling observed for P-450-catalyzed hydroxylations.

is accompanied by a significant amount of epimerization at the carbon center and a large isotope effect (k_H/k_D 10–12). These results were first observed for the hydroxylation of *exo-exo-exo-exo*-tetradeuterionorbornane[8] by P-450$_{LM}$ (Fig. 3) and subsequently for camphor hydroxylation by P-450$_{cam}$.[9] More recently, it has been established that the hydroxylation of selectively deuterated cyclohexenes proceeds with substantial allylic scrambling (Fig. 3).[7a] Collectively, these results provide compelling evidence for a hydrogen atom abstraction, radical recombination mechanism for the hydroxylation reaction. Consistent with model studies (*vide infra*), the reactive intermediate can be inferred from these results to be a ferryl species ($[FeO]^{3+}$). Hydrogen atom abstraction from the substrate by such a species to form a caged radical hydroxoiron(IV) complex and partial scrambling or epimerization of the caged alkyl radical before collapse with the hydroxoiron(IV) is indicated by the observation of rearranged products (Fig. 3).

Results reported by Miwa *et al.*[9] have indicated that the *O*-dealkylation reaction catalyzed by P-450 also proceeds by a hydrogen atom abstraction, radical recombination mechanism. Although a substantial intrinsic isotope effect (k_H/k_D 12.8–14) was observed for the deethylation of 7-ethoxycoumarin, the decrease in the rate resulting from the deuteration did not affect the rate of O_2 consumption, the rate of disappearance of 7-ethoxycoumarin, or the rate of H_2O_2 production (H_2O_2 is a product derived from the uncoupling of the intermediate complexes of P-450 and O_2). The rate decrease *was* accounted for largely by the production of another metabolite derived from hydroxylation of the aromatic ring.[10] The authors concluded that the formation of the active oxidant was irreversible and that the high commitment to catalysis was consistent with the formation of a ferryl oxidant, $[FeO]^{3+}$.

Due to the high reactivity of the active oxidant of P-450, no physical characterization of this species is presently available. However, the related peroxidases[11] and catalases[12] form stable oxidized intermediates which have been extensively studied. Horseradish peroxidase (HRP) reacts with hydrogen peroxide to generate a complex which is two electrons oxidized above the ferric resting state. This complex, designated as Compound I (HRP-I), reacts with a variety of oxidizable organic substrates in consecutive, one-electron steps (equations 1–3).

$$HRP + H_2O_2 = HRP\text{-}I + H_2O \qquad (1)$$

$$HRP\text{-}I + AH_2 = HRP\text{-}II + AH\cdot \qquad (2)$$

$$HRP\text{-}II + AH\cdot = HRP + A + H_2O \qquad (3)$$

Compound I has been definitively characterized as an oxoiron(IV) porphyrin radical species on the basis of Mössbauer,[13] EPR,[10] ENDOR,[14] EXAFS,[15] NMR,[16] and MCD[17] studies. The intermediate HRP-II is best formulated as a neutral oxoiron(IV) complex[18] resulting from a one-electron reduction of the porphyrin π radical of HRP-I. The axial coordination site in HRP is occupied by a histidine ligand,[19] whereas P-450 has unusual thiolate ligation. This difference has been suggested to account for the fact that HRP does not catalyze the oxygenation of hydrocarbons as does P-450.[20]

3. Oxidations Catalyzed by Synthetic Iron Metalloporphyrins

Insights into the mechanistic problems presented by the putative catalytic cycle for P-450 have been gained through the use of simple chemical models for the biological systems.[7c,21] Synthetic tetraarylporphyrins (Fig. 4) are readily prepared, bind a variety of metals (M = Fe, Co, Cr, Mn, Ru, and so on), and have been used extensively to model the natural heme prosthetic group protoporphyrin IX. The bulky aryl substituents reduce the tendency of the natural heme to agglomerate in solution and help protect the *meso* carbons from oxidation. In addition, the phenyl groups can be substituted with a variety of substituents as exemplified by the celebrated "picket fence" porphyrin[22] (Fig. 4b), which incorporates a pocket to facilitate oxygen binding.[23] The tetramesitylporphyrin (Fig. 4c) was found to be well suited for solution NMR studies (M. Nakamura, unpublished results) because of its relatively high solubility and symmetrical structure. The *ortho* methyl groups also inhibited the formation of μ-oxo dimers[24] and enhanced the stability of oxidized iron porphyrinates[25] by simultaneous protection of all four *meso* carbons.

The first priority in developing a relevant model system for P-450 was to establish that synthetic ferric porphyrinates could catalyze hydrocarbon oxidations in a manner consistent with P-450. This was first accomplished by the demonstration in our laboratories[26] that iron tetra-

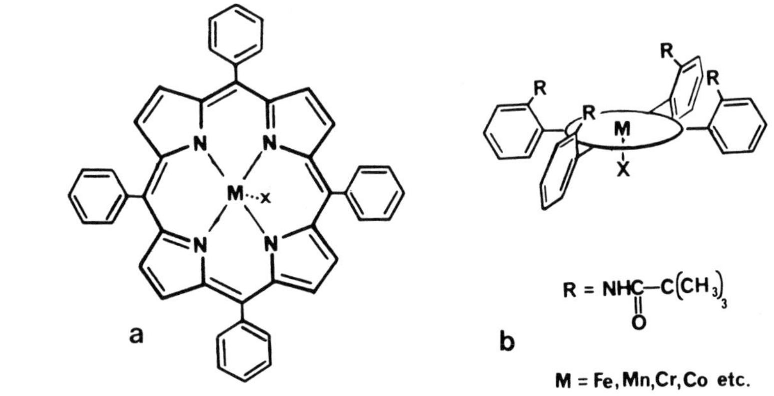

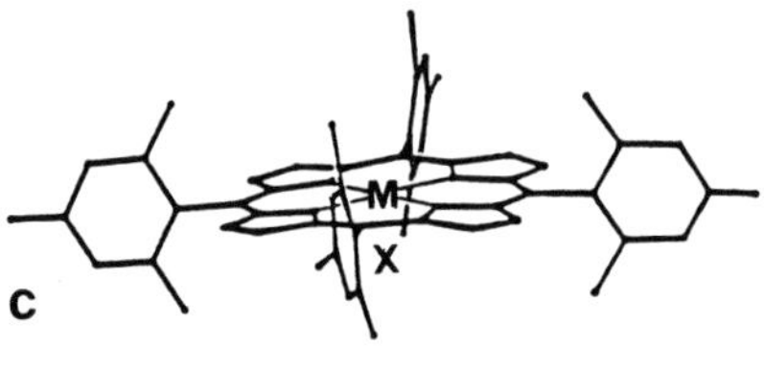

FIGURE 4. Examples of synthetic tetraarylporphyrinates. (a) Tetraphenylporphyrin (TPP); (b) "picket fence" porphyrin; (c) tetramesitylporphyrin (TMP) (M = Fe, Mn, Cr, Co, and so on).

phenylporphyrins catalyzed the epoxidation of alkenes and the hydroxylation of hydrocarbons using iodosylbenzene as the oxygen source.[27]

Remarkable shape selectivities resulted from relatively small variations in the structure of the catalyst, particularly at the *ortho*-phenyl position.[28] For example, the epoxidation of *cis*-stilbene with tetraphenylporphyrinatoiron(III) chloride (FeTPPCl) and iodosylbenzene was very much more facile than that of *trans*-stilbene. The preference for *cis*-epoxidation was further demonstrated by the selective epoxidation of *cis*-cyclododecatriene. With FeTPPCl, a 1.35:1 mixture of *cis* and *trans* epoxides was obtained but the more hindered tetramesitylporphyrinatoiron(III) chloride (FeTMPCl) catalyst increased the selectivity for *cis*-epoxidation to about 9:1. Applied to the epoxidation of *trans,trans,cis*-1,5,9-cyclododecatriene, the selective formation of the *cis*-monoepoxide was observed.

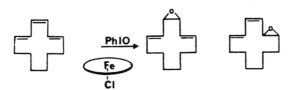

Asymmetric induction has been observed for the catalytic epoxidation of styrene (48% ee) with a chiral porphyrin catalyst, 5α,10β,15α,20β-tetrakis(*o* - [(*s*) - 2' - carboxymethyl - 1,1' - binaphthyl - 2 - carboxamido]phenyl)porphyrin **5**.[28]

5

The asymmetric induction resulted from purely nonbonded interactions between the catalyst and substrate and represents the highest degree of catalytic asymmetric induction yet achieved for unfunctionalized olefins with a nonenzymatic catalyst.

The hydroxylation of unactivated alkanes, the most difficult P-450-mediated oxidation, was also effected using iodosylbenzene and ferric porphyrin catalysts.[29] Significantly, the reaction was found to be dependent on the nature of the porphyrin substituents, as expected for a metal-centered oxidation. This was illustrated by comparing the ratio of 3°/2° hydrogen abstraction for the oxidation of adamantane.

In the presence of $BrCCl_3$, the oxidation of cycloheptane produced cycloheptanol, cycloheptanone, and substantial amounts of alkyl bromide. This result indicated the reaction had substantial radical character and that the radicals could be captured by bromine atom abstraction from the solvent.

The mechanism of the ferric porphyrin-catalyzed iodosylbenzene oxidation was postulated to involve the oxoiron(IV) porphyrin radical **6**. The electronic structure expected for **6** has two singly occupied d_{xz} and d_{yz} Fe=O(π) antibonding orbitals which are expected to have considerable oxy radical character, allowing the abstraction of a hydrogen atom from a carbon–hydrogen bond approaching parallel to the porphyrin plane (Fig. 5). The shape-selective hydroxylations and the sensitivity of these reactions to relatively small changes in the steric environment of the porphyrin periphery were consistent with such an approach. Recombination of the Fe(IV)OH with the caged alkyl radical generated the product alcohol, while occasional escape of the radical rationalized the brominated products in the presence of $BrCCl_3$. We have referred to this process as

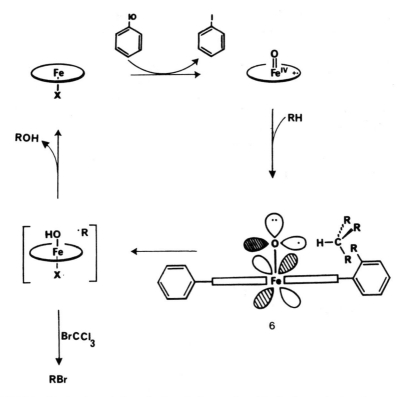

FIGURE 5. Mechanism of alkane hydroxylation catalyzed by ferric porphyrins: the oxygen rebound mechanism.

the oxygen rebound mechanism based upon analogies to our earlier results with iron–peroxide systems in acetonitrile.[7b,c]

Interestingly, the recent crystal structure of the P-450$_{cam}$ camphor complex shows that the 5-*exo* hydrogen of the substrate is displayed in precisely the arrangement shown in Fig. 5[30] (see Chapter 13).

It has been demonstrated by Bruice and co-workers that soluble, monomeric dimethylaniline *N*-oxides can function as oxygen donors in the ferric porphyrin-catalyzed oxidation of hydrocarbons.[31] When used as an exogenous oxygen source for P-450, the reduced tertiary amines were demethylated by the enzyme in a process which was suggested to involve hydrogen atom abstraction based on an intramolecular isotope effect (k_H/k_D 2–4.3 depending on the oxidant).[32] It should be noted that similar numbers obtained by other workers for the P-450-catalyzed *N*-

demethylation of N-methyl-N-trideuteriomethylaniline were interpreted as evidence for an electron transfer, deprotonation mechanism.[33]

A series of papers investigating the hydroxylation of aromatic compounds catalyzed by an iron heme in the presence of thiol and oxygen have been published.[34] The reactions were suggested to mimic P-450 activity based on the observation of an NIH shift during the oxidation of *para*-methylanisole catalyzed by a heme thioglycolic ester complex.[35] Superoxide ion was shown to be generated by these systems.[36]

Traylor *et al.*[37] have recently succeeded in preparing two chlorinated porphyrins, *meso*-tetra(2,6-dichlorophenyl)porphyrinatoiron(III) and *meso*-tetra(pentachloro)phenylporphyrinatoiron(III), which are efficient catalysts and are remarkably resistant to degradation under oxidizing conditions. When used in conjunction with pentafluoroiodosylbenzene, high catalyst turnovers and good yields of epoxides and alcohols were obtained.

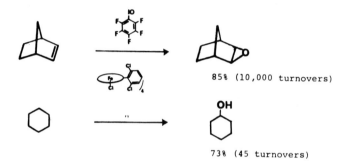

85% (10,000 turnovers)

73% (45 turnovers)

4. Preparation and Characterization of Oxidized Iron Porphyrinates

The development of iron(IV) porphyrin models for the oxidized heme proteins has been difficult to achieve due to the instability of such complexes.[38] As alluded to earlier, metalloporphyrins can be oxidized at either the metal, the macrocycle, or both. An iron(III)tetraphenylporphyrin complex which has been oxidized by one electron can therefore be represented either as an iron(IV) species **7** or as an iron(III) porphyrin radical **8**.

Early formulations of one-electron oxidized ferric porphyrins as iron(IV)[39] have been revised in favor of the π radical structure **8**.[40,41] This assignment was based on the insensitivity of the oxidation potentials of FeTPPX (X = Cl$^-$, F$^-$, N$_3^-$, PhO$^-$, and so on) to variation of the axial ligand.[42] Further characterization by NMR spectroscopy[39,42] demonstrated the existence of large isotropic shifts of the phenyl protons resulting from the delocalization of spin density onto the phenyl ring from the macrocycle π radical.

Crystal structures and detailed Mössbauer and magnetic moment measurements were performed on FeTPP·(ClO$_4$)$_2$ **9**, FeTPP·(CL)(SbF$_6$) **10**, and FeTPP·(Cl)(ClO$_4$) **11**.[44,45] These studies verified the iron(III) π radical assignment and addressed the potential coupling interactions which can occur when a paramagnetic $S = \frac{5}{2}$ iron(III) is in close proximity to an $S = \frac{1}{2}$ free radical. Structures **10** and **11** were structurally related 5-coordinate iron (III) chlorides differing only in the anion (SbCl$_6^-$ or ClO$_4^-$). While FeTPP·(Cl)(SbF$_6$) **10** exhibited properties consistent with an antiferromagnetically coupled $S = 2$ system, FeTPP·(ClO$_4$)$_2$ (was described as a noninteracting $S = \frac{5}{2}$, $S = \frac{1}{2}$ system. The presence of spin coupling was suggested to be predominately controlled by the symmetry of the iron d-orbitals and the molecular orbital in which the radical is localized (A$_{2u}$).[46] In the centrosymmetric complex **9**, overlap of the iron d-orbitals and the porphyrin A$_{2u}$ is symmetry forbidden, while the noncentrosymmetric environment of **10** (and **11**) provides a route for spin coupling to occur.

The only ligands which have been reported to stabilize the iron(IV) oxidation state relative to the porphyrin π cation radical in a monomeric* complex are the oxo group and the alkoxy group.[43] A neutral oxo-iron(IV)TPP(N-MeIM) complex **12** has been identified as the product

---

* An oxidized dimeric μ-nitrido complex, [(FeTPP)$_2$N]·ClO$_4^-$, exhibits Mössbauer properties more in accordance with iron(IV) than with iron(III): English, D. R., Hendrickson, D. N., and Suslick, K. S., 1983, Mössbauer spectra of oxidized iron porphyrins, *Inorg. Chem.* 22:367–368.

resulting from cleavage of an iron(III) peroxo species by *N*-methylimidazole.[47,48]

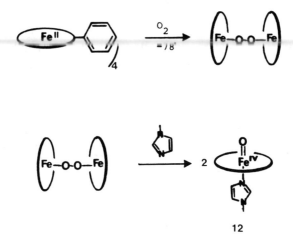

12

Complete formation of the oxoiron(IV) complex **12** required 2 moles of imidazole per peroxo dimer. In contrast to the dimer, the temperature dependence of the proton NMR of **12** strictly obeyed the Curie law between −80 and −15°C, indicating that **12** is a monomeric species. The visible spectrum of **12** is unusual compared to the electronic spectra of iron(III) porphyrins and exhibits absorbances at 420, 560, and 590 nm. **12** was capable of oxidizing triphenylphosphine but not hydrocarbons.

A similar oxoiron(IV) has been inferred to exist as an intermediate in the pyridine-induced conversion of $[FeTPP \cdot (H_2O)_2]^{2+}$ to the oxidized dimer $[TPPFeOFeTPP \cdot]^+ ClO_4^-$.[49] Bajdor and Nakamoto[50] have recently succeeded in identifying the iron–oxygen stretching frequency in FeOTPP formed by laser photolysis of $Fe(O_2)TPP$ at 15°K. The resonance Raman spectrum of this species revealed a signal at 852 cm^{-1} which shifted to 818 cm^{-1} upon substitution with ^{18}O. The authors noted that the force constant calculated for the ferryl group, 5.32 mdyn/Å, was significantly larger than that of the Fe–O bonds in $(FeTPP)_2$–O (3.8 mdyn/Å),[51] indicating formulas such as $Fe(IV)=O^{2-}$ are preferable to $Fe(III)-O^-$.

The sterically hindered 5,10,15,20-tetramesitylporphyrinatoiron complex has been shown to produce an unusual oxidized species upon reaction with a variety of oxidants. Thus, the oxidation of this complex with *m*-chloroperoxybenzoic acid at −78°C has been shown to produce a green species **6**. Several lines of evidence now support the formulation of **6** as an oxoiron(IV) porphyrin radical formally equivalent to the celebrated Compound I of horseradish peroxidase.[25]

6

The visible spectrum of **6** shows the characteristic long-wavelength absorption of a porphyrin radical. Further, large downfield shifts for the aryl hydrogens in the proton NMR spectrum of **6** [δ 68 (*m* H); 24 and 26 (*o*-methyl); 11.1 (*p*-methyl)] are also indicative of substantial spin and charge density on the porphyrin ring. The isomer shift in the iron Mössbauer spectrum of **6** (δ 0.06) is in the range of other iron(IV) complexes.[43,52] The complicated magnetic field and temperature dependence of the Mössbauer data are closely modeled by an $S = 1$ iron(IV) coupled strongly and isotropically to the $S = \frac{1}{2}$ porphyrin.[52] Magnetic moment measurements and the EPR spectrum were also consistent with a $\frac{3}{2}$ spin system for **6**. Recent EXAFS data have provided evidence that the iron–oxygen distance in **6** is 1.6 Å,[15] characteristic of a bond order greater than one.

The high chemical reactivity of **6** toward hydrocarbons has indicated that it is kinetically competent to be the reactive species in the iron porphyrin catalytic systems. That the O–O bond is broken in the reactive complex is supported by the efficient incorporation of ^{18}O into the product epoxide in the presence of olefinic substrates. Peroxyacids do not exchange the peroxidic oxygen with water whereas the oxo ligand of metalloxo complexes do. Thus, a mechanism for oxygen transfer to olefins via **6** is:

Traylor et al.[53,54] have examined the kinetics of the oxidation of a

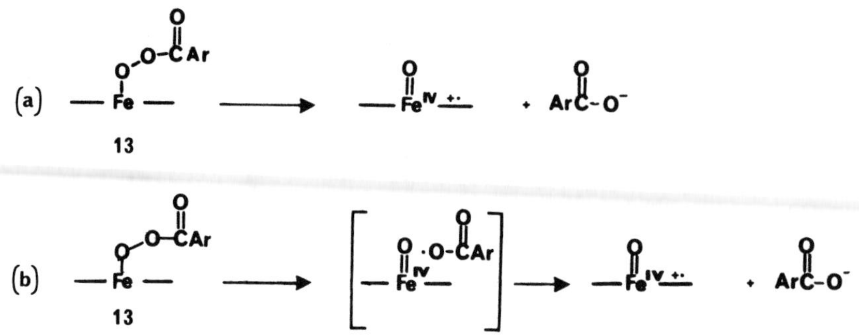

FIGURE 6. Heterolysis versus homolysis of an acylperoxyiron (III) porphyrinate.

reactive substrate (2,4,6-tri-*tert*-butylphenol) by an iron porphyrin–peroxyacid system. Heme destruction was avoided at high phenol concentrations and the reaction could be studied at room temperature. These results indicated that heterolytic cleavage of the O–O bond of the peroxyacid had occurred since CO_2 production was not observed when phenylperoxyacetic acid was employed as the oxidant. Homolytic cleavage of the O–O bond would generate an alkyl carboxyl radical, which is known to decarboxylate even in the presence of phenol. Thus, a heterolytic, oxygen atom transfer reaction (path a, Fig. 6) was favored over the more traditional one-electron cleavage to give a peroxy radical (path b) on this basis.

Consistent with the work of Traylor, McCarthy and White[55] have reported that peroxide O–O bond heterolysis was operative in the HRP-, chloroperoxidase-, catalase-, and metmyoglobin-catalyzed decomposition of peroxyphenylacetic acid. However, significant *homolysis* of the O–O bond was promoted by P-450. It was later shown that the peroxide homolysis reaction was not on a pathway that resulted in product formation. (For further discussion see Chapter 7.)

The existence of a peroxyacyl iron porphyrin complex (**13**) in the conversion of iron(III) to a reactive oxoiron(IV) porphyrin radical **6** was inferred from the kinetic studies discussed above, and was postulated to be formed by the acylation of a peroxoiron(III) tetraphenylporphyrin complex $[Fe(III)O_2]^-$ with acetic anhydride.[56] The acylperoxy complex decomposed at $-50°C$ to form an EPR-silent green complex **14** which was capable of oxygenating hydrocarbons and was suggested to be equivalent to **6** above.

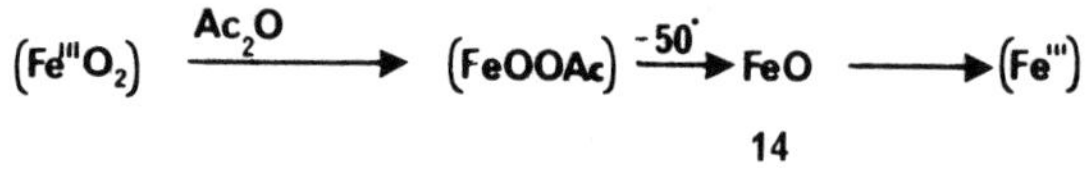

$$(Fe^{III}O_2) \xrightarrow{Ac_2O} (FeOOAc) \xrightarrow{-50°} FeO \longrightarrow (Fe^{III})$$

14

The formation of an analogous acylperoxymanganese(III) porphyrin has been documented.[57] The acylation of a manganese(II) superoxide [or manganese(III)peroxo] complex[58] **15** with benzoyl chloride afforded the peroxyacylmanganese(III) species **16**. This species decomposed in a process which was first order in hydroxide to generate an oxomanganese(V) porphyrinate **17** which was capable of epoxidizing olefins. Since the binding of dioxygen by manganese(II) porphyrins has been established,[59] this cycle constitutes a complete model of the oxygen activation by P-450. The role of hydroxide ion in facilitating O–O bond cleavage in the peroxyacylmanganese(III) complex **16** is significant and suggests that the thiolate ligand in P-450 performs a similar function.

5. Oxygenations Catalyzed by Synthetic Manganese Porphyrins

Manganese porphyrins have been found to be more efficient catalysts for the oxygenation of hydrocarbons than iron porphyrins.[60,61]

The reactive intermediate in the manganese-catalyzed reactions has been postulated to be an oxomanganese(V) complex **17**, although such a

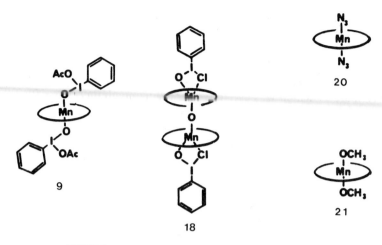

FIGURE 7. Oxidized manganese porphyrin complexes.

species has not been definitively characterized. The characterization of the oxidized manganese species has resulted in the isolation of several iodosylbenzene complexes of manganese porphyrins[62] as well as some monomeric manganese(IV) species (Fig. 7).[63] Although structural identification of **17** remains elusive, complexes formulated as manganese(V) have been characterized in solution.[57,64]

The hydroxylation of substrates catalyzed by manganese porphyrins has been shown to involve the formation of alkyl radicals which can be trapped by the axial ligand X of MnTPPX.[65] Decomposition of **18** in the presence of *tert*-butylbenzene was shown to involve formation of manganese(IV) by EPR.[66] A mechanism consistent with this observation was presented (Fig. 8).

Axial ligand transfer other than oxygen has been achieved by supplying the system with an excess of NaX (X = halogen, acetate, azide).[67] By employing a phase transfer system with saturated NaX in the aqueous phase and MnTPPX catalyst and substrate in the organic phase, yields of up to 55% of alkyl azides were obtained when iodosylbenzene was used as the oxidant. Oxidants such as RCO_3H and $NaIO_4$ were found to be ineffective under these reaction conditions.

Phase transfer catalysis was found to provide a convenient and efficient source of oxidant for the manganese porphyrin-catalyzed oxidation of hydrocarbons.[68] Especially noteworthy was the alkali metal hydrochlorite system developed by Meunier and Collman.[69] This system is re-

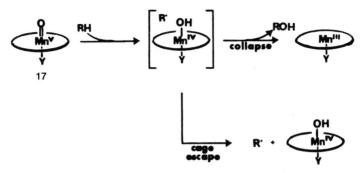

FIGURE 8. Mechanism of manganese porphyrin-catalyzed iodosylbenzene oxidation of hydrocarbons.

markably efficient (up to 2.5 catalyst turnovers/sec) and highly stereoselective for the epoxidation of *cis* olefins. It has also been shown that the immobilization of the porphyrin on an isocyanide-containing polymer resulted in a more active catalyst using hypochlorite as the oxidant.[70] The reactive intermediate involved in the hypochlorite oxidations was suggested to be an oxomanganese(V) complex,[71] although other formulations have not been ruled out.

Two differing views have been advanced regarding the kinetics of this reaction. Rozenberg *et al.*[72] have investigated the MnTPPCl-catalyzed epoxidation of cyclohexene using NaOCl as the oxidant and concluded that the rate-determining step involved the breakdown of the Mn(III)TPPOCl complex to form Mn(V)OTPPCl. This hypothesis was based on the fact that the rate of the reaction was independent of the olefin concentration. Other kinetic investigations[71,73] of the phase transfer, MnTPPCl-catalyzed oxidation system using LiOCl as the oxidant and 4'-imidazolylacetophenone as the ligand revealed that the epoxidation of cyclooctene and *trans*-β-methylstyrene proceeded at different rates and that the rate was independent of olefin concentration. A competitive oxidation of the two olefins showed a *reversal in the reactivity* of the olefins. Consistent with these results, a mechanism (Fig. 9) was proposed which incorporates a reversibly formed metallaoxetane 22 similar to that proposed by Sharpless *et al.* for the chromyl chloride epoxidation of olefins.[74] The reverse process, conversion of an epoxide to a metallaoxetane, has been observed for the addition of *low-valent* transition metal complexes to tetracyanoethylene oxide[75] and has been suggested as a viable path for the oxidative addition of Ir(I) to a variety of epoxides.[76]

Rappé and Goddard[77] have evaluated this 2 + 2 cycloaddition path and have concluded that the formation of the metallacycle in the chromyl

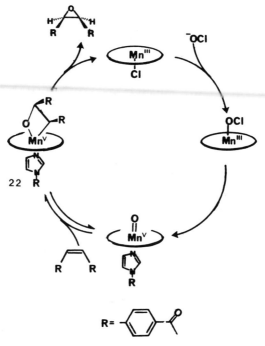

FIGURE 9. Proposed catalytic cycle for the manganese porphyrin-catalyzed epoxidation of olefins with alkali hypochlorites.

chloride epoxidation was favored primarily due to the stability of the second "spectator" oxo group. Further experimental support for this hypothesis was provided by Walba et al.[78] who investigated the oxidation of ethylene by the cationic chromyl complex **23** in the gas phase (Fig. 10). The production of the carbon-containing ionic product **24** in addition to **25**, **26**, and **27** was suggested to arise from the oxametallacyclobutane **28** via a retro "2 + 2" process to afford **29** which lost CH_2O to give **24**.

Two examples of olefin oxygenation catalyzed by manganese porphyrins in the presence of dioxygen and a reducing agent have been reported. A two-phase (H_2O/C_6H_6) system devised by Mansuy et al.[79] utilized ascorbate as the reducing agent, whereas Tabushi[80] has introduced the use of collodial platinum to catalyze electron transfer from hydrogen to Mn(III)TPPX to form Mn(II)TPP. When stirred under an atmosphere of H_2 and O_2 in the presence of colloidal platinum, cyclohexene was epoxidized in a yield translating into 64 catalyst turnovers. The relative reactivities of several olefins compared to cyclohexene were nearly identical to that observed with iodosylbenzene as the oxidant, suggesting an

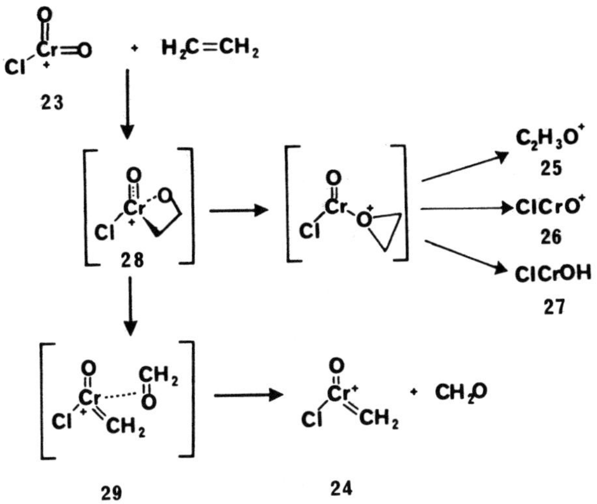

FIGURE 10. Gas phase oxidation of ethylene by [Cro$_2$Cl]$^+$.

oxidized manganese complex was also the reactive oxidant in this aerobic system.

6. Cobalt and Molybdenum Porphyrins as Oxidation Catalysts

It has recently been suggested that epoxidation by these metal/reductant/O$_2$ systems can involve *olefin activation* and not oxygen activation.[81] Olefin activation has been suggested for the CoTPP-catalyzed oxidation of substituted styrenes to substituted 1-phenylethanols in the presence of NaBH$_4$ and oxygen.

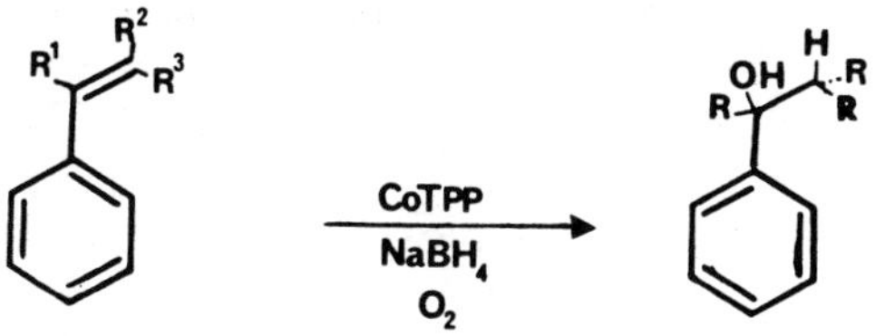

When sodium borodeuteride was employed as the reducing agent, deuterium incorporation was observed in the R^1 and R^2 positions of the 1-phenylethanol derived from styrene. At higher conversions, more complex deuteration was observed and deuterium was observed in the re-

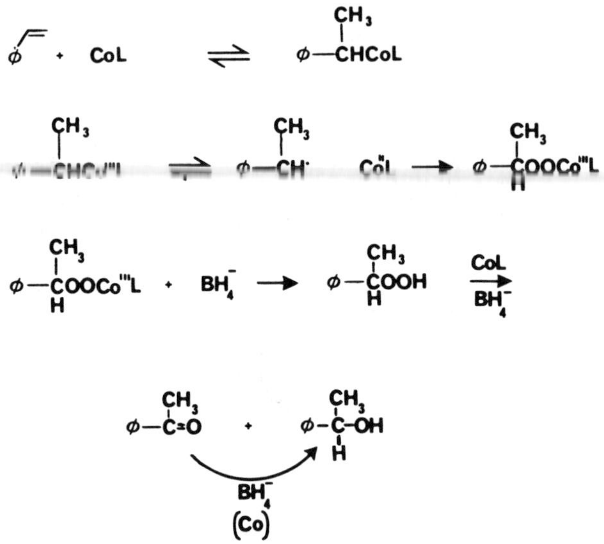

FIGURE 11. Olefin activation by CoTPP/NaBH$_4$.

covered styrene. The distribution of products at various temperatures was strikingly similar to the reported values of deuterium incorporation in 1-arylethanols produced by the reaction of substituted (1-arylethyl)peroxy cobaloxime with NaBD$_4$ in methanol.[82] The following mechanism was proposed to account for the CoTPP-catalyzed results (Fig. 11).

It was suggested that the manganese porphyrin-catalyzed oxygenation of hydrocarbons employing dioxygen and a reducing agent followed a similar mechanism, and that the high yields of epoxides obtained were the result of a manganese-catalyzed hydroperoxide epoxidation. Ledon et al.[83] have reported that molybdenum porphyrins activate hydroperoxides toward nucleophilic attack by olefins. However, it should be noted that MnTPPCl does not catalyze the epoxidation of cyclohexene in the presence of cumene hydroperoxide.[84]

The Ledon catalytic system incorporates an oxomolybdenum(V) tetraphenylporphyrinmethoxide catalyst and tertiary alkyl hydroperoxides to epoxidize olefins with high stereoselectivity[83,85] (tBHP = *tert*-butyl hydroperoxide).

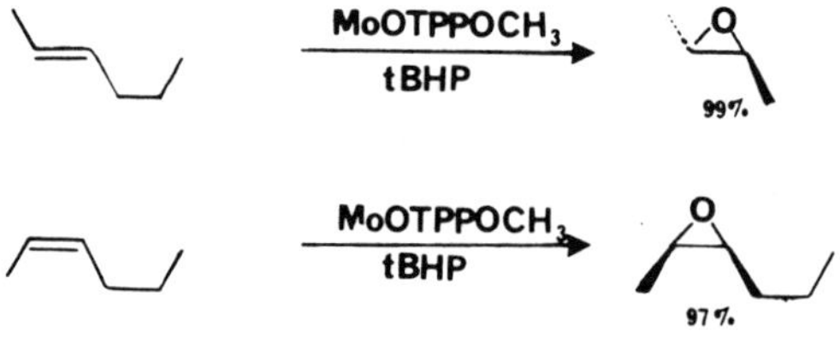

Since the selectivity of the reaction varied as a function of the hydroperoxide used, the active porphyrin oxidant was envisioned to be the hydroxo-peroxo molybdenum(IV) complex **30**. The mechanism was suggested to involve nucleophilic attack of the olefin on the coordinated peroxide with resulting peroxy bond heterolysis, in analogy to the mechanism of olefin epoxidation by molybdenum peroxo complexes.[86,87] Accordingly, reactions of iron-coordinated peroxides with substrate molecules may be considered for the peroxide shunt pathway in P-450 as well.

30

The *cis*-dioxomolybdenum(IV) porphyrinate **31** was found to undergo reduction either by a one-electron pathway in the oxidation of 2-propanol to acetone or by a two-electron oxidation, ligand transfer reaction as in the case of phosphine oxidations.[88] The oxo bond of this complex was reportedly activated by the electrochemical oxidation of the porphyrin macrocycle. Although [MoOTPP]$^+$ was observed, the oxidation of substrate was not demonstrated.[89]

31

7. Conclusion

Suggestions that the oxidized forms of horseradish peroxidase, HRP-I and HRP-II, contained the oxoiron moiety and that such an intermediate could be the active oxygen species in the oxygenation cycle of P-450 have now received strong support from model systems. The characterization of simple oxometalloporphyrin complexes strengthens the view that porphyrinato *ferrates* are accessible biological intermediates with oxidizing properties related to inorganic reagents such as chromates.

Several aspects of the P-450 cycle still lack a chemical analogue. The role of the unusual thiolate ligand is a matter of speculation and no oxidized iron porphyrin complexes have been prepared with sulfur ligation. The mechanism of O–O bond scission and, in particular, the chemistry of hydroperoxy iron porphyrins have not been extensively explored.

These and related issues form the current focus of the chemistry of P-450.

References

1. Hayaishi, O. (ed.), 1974, *The Molecular Mechanisms of Oxygen Activation*, Academic Press, New York.
2. Groves, J. T., 1979, Cytochrome P-450 and other hemecontaining oxygenases, *Adv. Inorg. Biochem.* **1**:119–145.
3. (a) White, R. E., and Coon, M. J., 1980, Oxygen activation by cytochrome P-450, *Annu. Rev. Biochem.* **49**:315–356. (b) Guengerich, F. P., and Macdonald, T. L., 1984, Chemical mechanisms of catalysis by cytochromes P-450: A unified view, *Acc. Chem. Res.* **17**:9–16. (c) Ullrich, V., 1979, Cytochrome P-450 and biological hydroxylation reactions, *Top. Curr. Chem.* **83**:67–104.
4. Hrycay, E. G., Gustafsson, J.-A., Ingelman-Sundberg, M., and Ernster, L., 1975, Sodium periodate, sodium chlorite, and organic hydroperoxides as hydroxylating agents in hepatic microsomal steroid hydroxylation reactions by cytochrome P-450, *FEBS Lett.* **56**:161–165.
5. Lichtenberger, F., Nastainczyk, W., and Ullrich, V., 1976, Cytochrome P-450 as an oxene transferase, *Biochem. Biophys. Res. Commun.* **70**:939–946.
6. Hrycay, E. G., Gustafsson, J., Ingelman-Sundberg, M., and Ernster, L., 1975, Sodium periodate, sodium chlorite, organic hydroperoxides, and H_2O_2 as hydroxylating agents in steroid hydroxylation reactions catalyzed by partially purified cytochrome P-450, *Biochem. Biophys. Res. Commun.* **66**:209–216.
7. (a) Groves, J. T., and Subramanian, D. V., 1984, Hydroxylation by cytochrome P-450 and metalloporphyrin models: Evidence for allylic rearrangement, *J. Am. Chem. Soc.* **106**:2177–2181. (b) Groves, J. T., and Van Der Puy, M., 1974, Stereospecific aliphatic hydroxylation by an iron-based oxidant, *J. Am. Chem. Soc.* **96**:5274–5275. (c) Groves, J. T., and McClusky, G. A., 1976, Aliphatic hydroxylation via oxygen rebound: Oxygen transfer catalyzed by iron, *J. Am. Chem. Soc.* **98**:859–861.
8. Groves, J. T., McClusky, G. A., White, R. E., and Coon, M. J., 1978, Aliphatic hydroxylation by highly purified liver microsomal cytochrome P-450: Evidence for a carbon radical intermediate, *Biochem. Biophys. Res. Commun.* **81**:154–160.
9. Miwa, G. T., Walsh, J. S., and Lu, A. Y. H., 1984, Kinetic isotope effects on cytochrome P-450-catalyzed oxidation reactions: The oxidative O-dealkylation of 7-ethoxycoumarin, *J. Biol. Chem.* **259**:3000–3004.
10. Harada, N., Miwa, G. T., Walsh, J. S., and Lu, A. Y. H., 1984, Kinetic isotope effects on cytochrome P-450-catalyzed oxidation reactions: Evidence for the irreversible formation of an activated oxygen intermediate of cytochrome P-448, *J. Biol. Chem.* **259**:3005–3010.
11. (a) Dunford, H. B., and Stillman, J. S., 1976, On the function and mechanism of action of peroxidases, *Coord. Chem. Rev.* **19**:187–251. (b) Hewson, W. D., and Hager, L. P., 1979, Peroxides, catalases, and chloroperoxidase in: *The Porphyrins*, Volume VII (D. Dolphin, ed.), Academic Press, New York, pp. 295–332.
12. Jones, P., and Wilson, I., 1978, in: *Metal Ions in Biological Systems*, Volume 17 (H. Sigel, ed.), Dekker, New York, p. 187.
13. Schultz, C. E., Devaney, P. W., Winkler, H., DeBrunner, P. G., Doan, N., Chiang, R., Rutter, R., and Hager, L. P., 1979, Horseradish peroxidase compound I: Evidence for spin coupling between the heme iron and the "free" radical, *FEBS Lett.* **103**:102–105.

14. (a) Roberts, J. E., Hoffman, B. M., Rutter, R., and Hager, L. P., 1981, Electron–nuclear double resonance of horseradish peroxidase compound I: Detection of the porphyrin π-cation radical, *J. Biol. Chem.* **256**:2118–2121. (b) Roberts, J E., Hoffman, B. M., Rutter, R., and Hager, L. P., 1981, ^{17}O ENDOR of horseradish peroxidase compound I, *J. Am. Chem. Soc.* **103**:7656–7659.
15. Penner-Hahn, J. E., McMurry, T. J., Renner, M., Latos-Grazynsky, L., Eble, K. S., Davis, I. M., Balch, A. L., Groves, J. T., Dawson, J. R., and Hodgson, K. O., 1983, X-ray absorption spectroscopic studies of high valent iron porphyrins: Horseradish peroxidase compounds I and II and synthetic models, *J. Biol. Chem.* **258**:12761–12764.
16. LaMar, G. N., deRopp, J. S., Smith, K. M., and Langry, K. C., 1981, Proton nuclear magnetic resonance investigation of the electronic structure of compound I of horseradish peroxidase, *J. Biol. Chem.* **256**:237–243.
17. Browlett, W. R., Gasyna, Z., and Stillman, M. J., 1983, The temperature dependence of the MCD spectrum of horseradish peroxidase of compound I, *Biochem. Biophys. Res. Commun.* **112**:515–520.
18. Schultz, C., Chiang, R., and DeBrunner, P. G., 1979, Mössbauer parameters of Fe^{4+} heme proteins of spin $S = 1$, *J. Phys. (Paris) Suppl.* **40**(C2):534–536.
19. Yonetani, T., Yamamoto, H., Erman, J. E., Leigh, J. S., and Reed, G. H., 1972, Electromagnetic properties of hemoproteins. V. Optical and electron paramagnetic resonance characteristics of nitric oxide derivatives of metalloporphyrin–apohemoprotein complexes, *J. Biol. Chem.* **247**:2447–2455.
20. Ullrich, V., 1980, Dioxygen activation by heme–sulfur proteins, *J. Mol. Catal.* **7**:158–167.
21. Ochai, E.-I., 1977, *Bioinorganic Chemistry*, Allyn & Bacon, Rockleigh, N.J.
22. Sorrell, T. N., 1980, 37. (Dioxygen) (N-methylimidazole) [(all-cis)-5,10,15,20-tetrakis[2-(2,2-dimethylpropionamido)-phenyl]porphyrinato(2-)]iron(II), in: *Inorganic Syntheses*, Volume XX (D. H. Busch, ed.), Wiley, New York, pp. 161–169.
23. Collman, J. P., Gagne, R. R., Reed, C. A., Halbert, T. R., Lang, G., and Robinson, W. T., 1975, "Picket fence porphyrins." Synthetic models for oxygen binding hemoproteins, *J. Am. Chem. Soc.* **97**:1427–1439.
24. (a) Nemo, T. E., 1980, Epoxidation and hydroxylation catalyzed by ferric porphyrins, Ph.D. Thesis, The University of Michigan. (b) Cheng, R. J., Grazynsky, L. L., and Balch, A. L., 1982, Preparation and characterization of some hydroxy complexes of iron(III) porphyrins, *Inorg. Chem.* **21**:2412–2418.
25. Groves, J. T., Haushalter, R. C., Nakamura, M., Nemo, T., and Evans, B. J., 1981, High-valent iron-porphyrin complexes related to peroxidase and cytochrome P-450, *J. Am. Chem. Soc.* **103**:2884–2886.
26. Groves, J. T., Nemo, T. E., and Myers, R. S., 1979, Hydroxylation and epoxidation catalyzed by iron-porphine complexes: Oxygen transfer from iodosylbenzene, *J. Am. Chem. Soc.* **101**:1032–1033.
27. (a) Lindsay-Smith, J. R., and Sleath, P. R., 1982, Model systems for cytochrome P-450 dependent mono-oxygenases. Part 1. Oxidation of alkenes and aromatic compounds by tetraphenylporphinatoiron(III) chloride and iodosylbenzene, *J. Chem. Soc. Perkin Trans. 2* **1982**:1009–1015. (b) Lindsay-Smith, J. R., Nee, M. W., Noar, J. B., and Bruice, T. C., 1984, Oxidation of N-nitrosodibenzylamine and related compounds by metalloporphyrin-catalyzed model systems for the cytochrome P-450 dependent mono-oxygenases, *J. Chem. Soc. Perkin Trans. 2* **1984**:255–260. (c) Chang, C. K., and Kuo, M.-S., 1979, Reaction of iron(III) porphyrins and iodosoxylene: The active oxene complex of cytochrome P-450, *J. Am. Chem. Soc.* **101**:3413–3415.
28. Groves, J. T., and Myers, R. S., 1983, Catalytic asymmetric epoxidations with chiral iron porphyrins, *J. Am. Chem. Soc.* **105**:5791–5796.

29. Groves, J. T., and Nemo, T. E., 1983, Aliphatic hydroxylation catalyzed by iron porphyrin complexes, *J. Am. Chem. Soc.* **105**:6243–6248.
30. Poulos, T., 1984, The crystal structure of cytochrome P-450$_{CAM}$, SE Regional ACS Meeting, Raleigh, N.C., Oct. 1984.
31. (a) Nee, M. W., and Bruice, T. C., 1982, Use of the N-oxide of p-cyano-N,N-dimethylaniline as an "oxygen" donor in a cytochrome P-450 model system, *J. Am. Chem. Soc.* **104**:6123–6125. (b) Shannon, P., and Bruice, T. C., 1981, A novel P-450 model system for the N-dealkylation reaction, *J. Am. Chem. Soc.* **103**:4580–4582.
32. Heimbrook, D. C., Murray, R. I., Egeberg, K. D., Sligar, S. G., Nee, M. W., and Bruice, T. C., 1984, Demethylation of N,N-dimethylaniline and p-cyano-N,N-dimethylaniline and their N-oxides by cytochromes P-450$_{LM2}$ and P-450$_{CAM}$, *J. Am. Chem. Soc.* **106**:1514–1515.
33. Miwa, G. T., Walsh, J. S., Kedderis, G. L., and Hollenberg, P. F., 1983, The use of intramolecular isotope effects to distinguish between deprotonation and hydrogen atom abstraction mechanisms in cytochrome P-450- and peroxidase-catalyzed N-demethylation reactions, *J. Biol. Chem.* **258**:14445–14449.
34. (a) Sakurai, H., Hatayama, E., and Nishida, M., 1983, Aromatic hydroxylation of acetanilide and aniline by hemin–thiolate complex as a cytochrome P-450 model, *Inorg. Chim. Acta* **80**:7–12. (b) Sakurai, H., and Ogawa, S., 1979, A model system of cytochrome P-450: Hydroxylation of aniline by iron– or hemin–thiol compound systems, *Chem. Pharm. Bull.* **1979**:2171–2176.
35. Sakurai, H., Hatayama, E., Fujitani, K., and Kata, H., 1982, Occurrence of aromatic methyl migration (NIH-shift) during oxidation of p-methylanisole by hemin–thiolester complex as a cytochrome P-450 model, *Biochem. Biophys. Res. Commun.* **35**:1649–1654.
36. (a) Sakurai, H., Ishizu, K., and Okada, K., 1984, Superoxide generation by an iron-tetraphenylporphyrin-thiolate-oxygen system and its significance in relation to the coordination site of cytochrome P-450, *Inorg. Chim. Acta* **91**:L9–L11. (b) Sakurai, H., and Ishizu, K., 1982, Generation of superoxide in a cobalt(II) tetraphenylporphyrin-thiolate-oxygen system, *J. Am. Chem. Soc.* **104**:4960–4962.
37. Traylor, P. S., Dolphin, D., and Traylor, T. G., 1984, Sterically protected hemins with electronegative substituents: Efficient catalysts for hydroxylation and epoxidation, *J. Chem. Soc. Chem. Commun.* **1984**:279–280.
38. Reed, C. A., 1982, Iron(I) and Iron(IV) porphyrins, *Adv. Chem. Ser.* **201**:333–356.
39. (a) Felton, R. H., Owen, G. S., Dolphin, D., and Fajer, J., 1971, Iron(IV) porphyrins, *J. Am. Chem. Soc.* **93**:6332–6334. (b) Felton, R. H., Owen, G. S., Dolphin, D., Forman, A., Borg, D. C., and Fajer, J., 1973, Oxidation of ferric porphyrins, *Ann. N.Y. Acad. Sci.* **206**:504–514.
40. Phillippi, M. A., and Goff, H. M., 1982, Electrochemical synthesis and characterization of the single-electron oxidation products of ferric porphyrins, *J. Am. Chem. Soc.* **104**:6026–6034.
41. Gans, P., Marchon, J.-C., Reed, C. A., and Regnard, J. R., 1981, One-electron oxidation of chloroiron(III) tetraphenylporphyrin: Evidence for porphyrin cation radical in the oxidized product, *Nouv. J. Chim.* **5**:203–204.
42. (a) Phillippi, M. A., Shimomura, E. T., and Goff, H. M., 1981, Investigation of axial anionic ligand and porphyrin substituent effects on the oxidation of iron(III) porphyrins: Porphyrin-centered vs. metal-centered oxidation, *Inorg. Chem.* **20**:1322–1325. (b) Goff, H. M., and Goff, H. M., and Phillippi, M. A., 1983, Imidazole complexes of low-spin iron(III) porphyrin π-cation radical species: Models for the compound I π-cation radical state of peroxidases, *J. Am. Chem. Soc.* **105**:7567–7571.
43. Groves, J. T., Quinn, R., McMurry, T. J., Lang, G., and Boso, B., 1984, Iron(IV)

porphyrins from iron(III) porphyrin cation radicals, *J. Chem. Soc. Chem. Commun.* **1984**:1455–1456.
44. Schlotz, W. F., Reed, C. A., Lee, Y. J., Scheidt, W. R., and Lang, G., 1982, Magnetic interactions in metalloporphyrin π-radical cations of copper and iron, *J. Am. Chem. Soc.* **104**:6791–6793.
45. Buisson, G., Deronzier, A., Duee, E., Gans, P., Marchon, J.-C., and Regnard, J.-R., 1982, Iron(III)-porphyrin π-cation radical complexes: Molecular structures and magnetic properties, *J. Am. Chem. Soc.* **104**:6793–6796.
46. (a) Hanson, L. K., Chang, C. K., Davis, M. S., and Fajer, J., 1981, Electron pathways in catalyase and peroxidase enzymic catalysis: Metal and macrocycle oxidations of iron porphyrins and chlorins, *J. Am. Chem. Soc.* **103**:663–670. (b) Loew, G. H., Kert, C. J., Hjelmeland, L. M., and Kirchner, R. F., 1977, Active site models of horseradish peroxidase compound I and a cytochrome P-450 analogue: Electronic structure and electric field gradients. (c) Loew, G. H., and Herman, Z. S., 1980, Calculated spin densities and quadrupole splitting for model horseradish peroxidase compound I: Evidence for iron(IV) porphyrin (S = 1) π-cation radical electronic structure, *J. Am. Chem. Soc.* **102**:6173–6174.
47. (a) Chin, D. H., Balch, A. L., and LaMar, G. N., 1980, Formation of porphyrin ferryl (FeO^{2+}) complexes through the addition of nitrogen bases to peroxo-bridged iron(III) porphyrins, *J. Am. Chem. Soc.* **102**:1446–1448. (b) Chin, D. H., LaMar, G. N., and Balch, A. L., 1980, On the mechanism of autoxidation of iron(II) porphyrins: Detection of a peroxo-bridged iron(III) porphyrin dimer and the mechanism of its thermal decomposition to the oxobridged iron(III) porphyrin dimer, *J. Am. Chem. Soc.* **102**:4344–4350.
48. Simonneaux, W. F., Schlotz, W. F., and Reed, C. A., 1982, Mössbauer spectra of unstable iron porphyrins: Models for compound I of peroxidase, *Biochim. Biophys. Acta* **716**:1–7.
49. Arena, F., Gans, P., and Marchon, J.-C., 1984, Dual electron-transfer reactivity of a high-spin iron(III) porphyrin cation radical complex, *J. Chem. Soc. Chem. Commun.* **1984**:196–197.
50. Bajdor, K., and Nakamoto, K., 1984, Formation of ferryltetraphenylporphyrin by laser irradiation, *J. Am. Chem. Soc.* **106**:3045–3046.
51. Burke, J. M., Kincaid, J. R., and Spiro, T. G., 1978, Resonance Raman spectra and vibrational modes of iron(III) tetraphenylporphine μ-oxo dimer: Evidence for phenyl interaction and lack of dimer splitting, *J. Am. Chem. Soc.* **100**:6077–6083.
52. Boso, B., Lang, G., McMurry, T. J., and Groves, J. T., 1983, Mössbauer effect study of tight spin coupling in oxidized chloro-5,10,15,20-tetra(mesityl)porphyrinato-iron(III), *J. Chem. Phys.* **79**:1122–1126.
53. Traylor, T. G., Lee, W. A., and Stynes, D. V., 1984, Model compound studies related to peroxidases: Mechanisms of reactions of hemins with peracids, *J. Am. Chem. Soc.* **106**:755–764.
54. Traylor, T. G., Lee, W. A., and Stynes, D. V., 1984, Model compound studies related to peroxidases. II. The chemical reactivity of a high valent protohemin compound, *Tetrahedron* **40**:553–568.
55. McCarthy, M.-B., and White, R. E., 1983, Functional differences between peroxidase compound I and the cytochrome P-450 reactive oxygen intermediate, *J. Biol. Chem.* **258**:9153–9158.
56. (a) Khenkin, A. M., and Shteinman, A. A., 1982, *Izv. Akad. Nauk SSSR Ser. Khim.* **7**:1668. (b) Khenkin, A. M., and Shteinman, A. A., 1984, The mechanism of oxidation of alkanes by peroxo complexes of iron porphyrins in the presence of acylating agents: A model for activation of O_2 by cytochrome P-450, *J. Chem. Soc. Chem. Commun.* **1984**:1219–1220.

57. Groves, J. T., Watanabe, Y., and McMurry, T. J., 1983, Oxygen activation by metalloporphyrins: Formation and decomposition of an acylperoxymanganese(III) complex, *J. Am. Chem. Soc.* **105**:4489–4490.
58. Shirazi, A., and Goff, H. M., 1982, Characterization of superoxide–metalloporphyrin reaction products: Effective use of deuterium NMR spectroscopy, *J. Am. Chem. Soc.* **104**:6318–6322.
59. (a) Hoffman, B. M., Szymanski, T., Brown, T. G., and Basolo, F., 1978, The dioxygen adducts of several manganese(II) porphyrins: Electron paramagnetic resonance studies, *J. Am. Chem. Soc.* **100**:7253–7254. (b) Jones, R. D., Summerville, D. A., and Basolo, F., 1978, Manganese(II) porphyrin oxygen carriers: Equilibrium constants for the reaction of dioxygen with para-substituted meso-tetraphenylporphinatomanganese(II) complexes, *J. Am. Chem. Soc.* **100**:4416–4424. (c) Hanson, L. K., and Hoffman, B. M., 1980, Griffith model bonding in dioxygen complexes of manganese porphyrins, *J. Am. Chem. Soc.* **102**:4602–4609.
60. Groves, J. T., Kruper, W. J., Jr., and Haushalter, R. C., 1980, Hydrocarbon oxidations with oxometalloporphinates: Isolation and reactions of a (porphinato)manganese(V) complex, *J. Am. Chem. Soc.* **102**:6375–6377.
61. Hill, C. L., and Schardt, B. C., 1983, Alkane activation and functionalization under mild conditions by a homogeneous manganese(III) porphyrin–iodosylbenzene oxidizing system, *J. Am. Chem. Soc.* **102**:6374–6375.
62. (a) Smegal, J. A., Schardt, B. C., and Hill, C. L., 1983, Isolation, purification and characterization of intermediate (iodosylbenzene)metalloporphyrin complexes from the (tetraphenylporphinato)manganese(III)–iodosylbenzene catalytic hydrocarbon functionalization system, *J. Am. Chem. Soc.* **105**:3510–3515. (b) Smegal, J. A., and Hill, C. L., 1983, Synthesis, characterization and reaction chemistry of a bis(iodosylbenzene)–metalloporphyrin complex, [PhI(OAc)O]$_2$Mn^{IV}TPP: A complex possessing a five-electron oxidation capability, *J. Am. Chem. Soc.* **105**:2920–2922.
63. (a) Camenzind, M. J., Hollander, F. J., and Hill, C. L., 1982, Synthesis, ground electronic state, and crystal and molecular structure of the monomeric manganese(IV) porphyrin complex dimethoxy(5,10,15,20-tetraphenylporphinato)manganese(IV), *Inorg. Chem.* **21**:4301–4308. (b) Camenzind, M. J., Hollander, F. J., and Hill, C. L., 1983, Synthesis, characterization and ground electronic state of the unstable monomeric manganese(IV) porphyrin complexes diazo- and bis(isocyanato) (5,10,15,20-tetraphenylporphinato)manganese(IV): Crystal and molecular structure of the bis(isocyanato) complex, *Inorg. Chem.* **22**:3776–3785.
64. (a) Carnieri, N., Harriman, A., and Porter, G., 1982, Photochemistry of manganese porphyrins. Part 6. Oxidation–reduction equilibria of manganese(III) porphyrins in aqueous solution, *J. Chem. Soc. Dalton Trans.* **1982**:931–938. (b) Carnieri, N., Harriman, A., Porter, G., and Kalyanasundraram, K., 1982, Photochemistry of manganese porphyrins. Part 7. Characterization of manganese porphyrins in organic and aqueous/organic microheterogeneous systems, *J. Chem. Soc. Dalton Trans.* **1982**:1231–1238.
65. Kruper, W. J., Jr., 1982, The isolation, characterization and reactivity of high valent oxometalloporphyrinates of chromium and manganese, Parts 1 and 2, Ph.D. Thesis, The University of Michigan.
66. Smegal, J. A., and Hill, C. L., 1983, Hydrocarbon functionalization by the (iodosylbenzene)manganese(IV)porphyrin complexes from the (tetraphenylporphinato)manganese(III)–iodosylbenzene catalytic hydrocarbon oxidation system: Mechanism and reaction chemistry, *J. Am. Chem. Soc.* **105**:3515–3521.
67. (a) Hill, C. L., Smegal, J. A., and Henly, T. J., 1983, Catalytic replacement of unactivated alkane carbon–hydrogen bonds with carbon–X bonds (X = nitrogen, oxygen, chlorine, bromine, or iodine): Coupling of intermolecular hydrocarbon activation by

Mn^{III}TPPX complexes with phasetransfer catalysis, *J. Org. Chem.* **48**:3277-3281. (b) Hill, C. L., and Smegal, J. A., 1982, Catalytic replacement of unactivated alkane carbon–hydrogen bonds with carbon–nitrogen bonds, *Nouv. J. Chim.* **6**:287-289.
68. Epoxidation of Olefins by Alkali Metal Hypochlorites, Fr. Demande FR 2,518,545, Guilmet, E., and Meunier, B., June 24, 1983.
69. (a) Collman, J. P., Kodadek, T., Raybuck, S. A., and Meunier, B., 1983, Oxygenation of hydrocarbons by cytochrome P-450 model compounds: Modification of reactivity by axial ligands, *Proc. Natl. Acad. Sci. USA* **80**:7039-7041. (b) Caravalho, M.-E., and Meunier, B., 1983, Stereochemical arguments against a possible chlorohydrin route in the catalytic epoxidation of olefins with NaOCl/Mn-porphyrins, *Tetrahedron Lett.* **24**:3621-3624. (c) Guilmet, E., and Meunier, B., 1982, Unexpected modification of selectivity with pyridine in the NaOCl/Mn(TPP)OAc catalytic epoxidation, *Nouv. J. Chim.* **6**:511-513. (d) Guilmet, E., and Meunier, B., 1982, Role of pyridine in the catalytic activation of sodium hypochlorite in the presence of manganese porphyrin, *Tetrahedron Lett.* **23**:2449-2452. (e) Guilmet, E., and Meunier, B., 1980, A new catalytic route for the epoxidation of styrene with sodium hypochlorite activated by transition metal complexes, *Tetrahedron Lett.* **21**:4449-4450. (f) Tabushi, I., and Koga, N., 1979, Synergetic combination of catalysis of the phase transfer–electron transfer type for the oxidation of alcohols or hydrocarbons, *Tetrahedron Lett.* **20**:3681-3684.
70. van der Made, A., Smeets, J. W. H., Nolte, R. J. M., and Drenth, W., 1983, Olefin epoxidation by a mono-oxygenase model: Effect of site isolation, *J. Chem. Soc. Chem. Commun.* **1983**:1204-1206.
71. Bortolini, O., and Meunier, B., 1983, Isolation of a high-valent "oxo-like" manganese porphyrin complex obtained from NaOCl oxidation, *J. Chem. Soc. Chem. Commun.* **1983**:1364-1366.
72. Rosenberg, J. A. S., Nolte, R. J. M., and Drenth, W., 1984, Mechanism of olefin epoxidation by a mono-oxygenase model, *Tetrahedron Lett.* **25**:789-792.
73. Collman, J. P., Brauman, J. I., Meunier, B., Raybuck, S. A., and Kodadek, T., 1984, Epoxidation of olefins by cytochrome P-450 model compounds: Mechanism of oxygen atom transfer, *Proc. Natl. Acad. Sci. USA* **81**:3245-3248.
74. Sharpless, K. B., Teranishi, A. Y., and Bäckvall, J.-E., 1977, Chromyl chloride oxidations of olefins: Possible role of organometallic intermediates in the oxidations of olefins by oxo transition metal species, *J. Am. Chem. Soc.* **99**:3120-3128.
75. (a) Schlodder, R., Ibers, J. A., Lenarda, M., and Graziani, M., 1974, Structure and mechanism of formation of the metallooxacyclobutane complex Pt[C$_2$(CN)$_4$O)][As(C$_6$H$_5$)$_3$]$_2$, the product of the reaction between tetracyanooxirane and Pt[As(C$_6$H$_5$)$_3$]$_4$, *J. Am. Chem. Soc.* **96**:6893-6900. (b) Lenarda, M., Ros, R., Traverso, O., Pitts, W. D., Baddley, W. H., and Graziani, M., 1977, Reactions of tetracyanoethylene oxide with some noble metal complexes, *Inorg. Chem.* **16**:3178-3182.
76. Milstein, D., and Calabrese, J. C., 1982, Oxidative addition of unactivated epoxides to iridium(I) complexes: Formation of stable cis-hydridoformylemethyl and -acylmethyl complexes, *J. Am. Chem. Soc.* **104**:3773-3774.
77. Rappé, A. K., and Goddard, W. A., III, 1982, Hydrocarbon oxidation by high-valent group 6 oxides, *J. Am. Chem. Soc.* **104**:3287-3294.
78. Walba, D. M., DePuy, C. H., Grabowski, J. J., and Bierbaum, V. M., 1984, Oxidation of alkenes by d^0 transition-metal oxo species: A mechanism for the oxidation of ethylene by a dioxochromium(VI) complex in the gas phase, *Organometallics* **3**:498-499.
79. Mansuy, D., Fontecave, M., and Bartoli, J.-F., 1983, Monooxygenase-like dioxygen activation leading to alkane hydroxylation and olefin epoxidation by an Mn^{III} (porphyrin)–ascorbate biphasic system, *J. Chem. Soc. Chem. Commun.* **1983**:253-254.
80. (a) Tabushi, I., and Koga, N., 1979, P-450 type oxygen activation by porphyrin–man-

ganese complex, *J. Am. Chem. Soc.* **101**:6456-6458. (b) Tabushi, I., and Yazaki, A., 1981, P-450-type dioxygen activation using H_2/colloidal Pt as an effective electron donor, *J. Am. Chem. Soc.* **103**:7371-7373. (c) Tabushi, I., and Nishiya, T., 1983, Colloidal platinum as an efficient and selective catalyst for reduction of metalloenzymes and metallocoenzymes, *Tetrahedron Lett.* **24**:5005-5008.
(d) Tabushi, I., and Morimitsu, K., 1984, Stereospecific, regioselective and catalytic monoepoxidation of polyolefins by the use of a P-450 model, H_2-O_2-TPP·Mn-colloidal platinum, *J. Am. Chem. Soc.* **106**:6871-6872.
81. Okamoto, T., and Oka, S., 1984, Oxygenation of olefins under reductive conditions: Cobalt-catalyzed selective conversion of aromatic olefins to benzylic alcohols by molecular oxygen and tetrahydroborate, *J. Org. Chem.* **49**:1589-1594.
82. Bied-Charreton, C., and Gaudemer, A., 1976, Carbonyl compounds as primary products in the reduction of alkyldioxycobaloximes by sodium borohydride, *J. Am. Chem. Soc.* **98**:3997-3998.
83. Ledon, H. J., Durbut, P., and Varescon, F., 1981, Selective epoxidation of olefins by molybdenum porphyrin catalyzed peroxy-bound heterolysis, *J. Am. Chem. Soc.* **103**:3601-3603.
84. Mansuy, D., Bartoli, J.-F., and Momenteau, M., 1982, Alkane hydroxylation catalyzed by metalloporphyrins: Evidence for different active oxygen species with alkylhydroperoxides and iodosobenzene as oxidants, *Tetrahedron Lett.* **23**:2781-2784.
85. Ledon, H. J., Durbut, P., and Varescon, F., 1982, *Proceedings of the Climax Fourth International Conference on the Chemistry and Uses of Molybdenum* (H. F. Barry and P. C. H. Mitchell, eds.), Climax Molybdenum Co., Ann Arbor, Mich., pp. 319-322.
86. Chong, A., and Sharpless, K. B., 1977, On the mechanism of the molybdenum and vanadium catalyzed epoxidation of olefins by alkyl hydroperoxides, *J. Org. Chem.* **42**:1587-1590.
87. Sheldon, R. A., 1980, Synthetic and mechanistic aspects of metal-catalyzed epoxidations with hydroperoxides, *J. Mol. Catal.* **7**:107-126.
88. Ledon, H., Varescon, F., Malinski, T., and Kadish, K. M., 1984, Reduction of cis-dioxo(tetraphenylporphinate) molybdenum(VI): One- or two-electron-transfer pathway, *Inorg. Chem.* **23**:261-263.
89. Malinski, T., Ledon, H., and Kadish, K. M., 1983, Electrochemical activation of a metal oxo bond: Oxidation of cis-dioxomolybdenum(VI)tetraphenylporphyrin, *J. Chem. Soc. Chem. Commun.* **1983**:1077-1079.

CHAPTER 2

Comparison of the Peroxidase Activity of Hemeproteins and Cytochrome P-450

LAWRENCE J. MARNETT, PAUL WELLER, and JOHN R. BATTISTA

1. Introduction

Peroxidases are enzymes that catalyze the oxidation of inorganic and organic substrates at the expense of a hydroperoxide—H_2O_2, alkyl hydroperoxides, or acyl hydroperoxides[1]. The actual function of the peroxidase may be to

$$ROOH + AH_2 \xrightarrow{Enz} ROH + A + H_2O \tag{1}$$

reduce a hydroperoxide or oxidize a particular substrate and, therefore, they are quite versatile and widespread. Most peroxidases contain ferriprotoporphyrin IX as their prosthetic group and "peroxidase activity" is a common characteristic of many hemeproteins and simple heme complexes.[2-4] Table I lists some of the better-characterized peroxidases, their properties, and functions. The discovery that cytochrome P-450 exhibits peroxidase activity and utilizes hydroperoxides to catalyze aliphatic hydroxylation and olefin epoxidation in the absence of NADPH and NADPH-cytochrome P-450 reductase provided a new perspective on its mechanism.[5-7] Much of the present understanding of the catalytic cycle of P-450 evolved from experiments designed to compare its peroxidase activity to those of classical peroxidases such as horseradish peroxidase (HRP). This chapter attempts to compare and contrast the principal fea-

LAWRENCE J. MARNETT, PAUL WELLER, and JOHN R. BATTISTA • Department of Chemistry, Wayne State University, Detroit, Michigan 48202.

TABLE I
Comparison of Peroxidases

Enzyme	Source	Subunit M.W. (oligomer)	Heme environment			Rates of Compound I formation (M^{-1} sec^{-1})	Proposed functions
			Type	Heme per subunit	Fifth ligand		
Horseradish peroxidase	Plant root	40,000 (monomer)[1]	b	1	His[1,2]	2.0×10^{7}[43]	Biosynthesis of plant auxins
Catalase	Beef liver	60,000 (tetramer)[13]	b	1	Tyr[154]	1.7×10^{7}[13]	Consumption of H_2O_2 in peroxisomes
Myeloperoxidase	Polymorphonuclear leukocytes	142,000–149,000[163]	Chlorin[54,69,151–153]	2[163]	—	—	Antimicrobial: via production of HCCl during phagocytosis
Lactoperoxidase	Saliva, milk, eosinophils, tears	77,000[32]	c[2,32]	1[32]	—	1.0×10^{2}[43]	Antimicrobial: via production of

Enzyme	Source	MW	Heme type	Heme/subunit	Axial ligand	Rate constant	Function
							hypothiocyanate during phagocytosis
Chloroperoxidase	*Caldariomyces fumago*	40,000–42,000[31] (monomer)	b^{31}	1^{31}	CySH[165]	—	Biosynthesis of caldariomycin
Thyroid peroxidase	Thyroid	62,000	—	—	—	—	Biosynthesis of thyroxine and triiodothyronine
Cytochrome c peroxidase	Yeast mitochondria[164]	34,000 (monomer)[15]	b	1	His[164]	2.0×10^{7} [43]	Reduction of H_2O_2 oxidation of ferrocytochrome c
	Pseudomonas[50,70–74]	43,000 (monomer)[70]	c^{50}	2	His[74]	1.2×10^{8} [74]	Reduction of H_2O_2 oxidation of ferrocytochrome c
PGH Synthase	Bovine seminal vesicle[35]	71,000	b	1	—	—	Biosynthesis of PGH
	Ovine seminal vesicle[65]	69,000 (dimer)[65]	b	1	—	—	Biosynthesis of PGH

tures of the interaction of peroxidases with hydroperoxides. It is not intended to be comprehensive because, in the words of the author of a previous review "it is unlikely that any reviewer would profess an adequate expertise in the disciplines, ranging from genetics to chemical physics, which a comprehensive discussion would require."[8] The reader can consult any of a number of excellent reviews[2,8-18] or leading references cited herein.

Extensive studies have established that many peroxidase-catalyzed oxidations proceed by the sequence:

$$\text{Enz} + \text{ROOH} \rightarrow \text{Compound I} + \text{ROH} \quad (2)$$

$$\text{Compound I} + \text{AH}_2 \rightarrow \text{Compound II} + \text{AH}\cdot \quad (3)$$

$$\text{Compound II} + \text{AH}_2 \rightarrow \text{Enz} + \text{AH}\cdot \quad (4)$$

$$\text{AH}\cdot \rightarrow \text{Nonradical products} \quad (5)$$

in which Enz is native enzyme, AH_2 is a reducing substrate, $AH\cdot$ is a free radical, and ROH is an alcohol or H_2O. ROH is released from the enzyme concomitant with Compound I formation.[19] H_2O is released from the enzyme concomitant with Compound II reduction.[2] The fate of the reducing substrate-derived radicals is characteristic of each compound. They may combine or disproportionate or undergo other reactions. For example, the reaction of HRP and H_2O_2 with p-cresol results in dimerization to Pummerer's ketone[20]:

$$\text{(6)}$$

Reaction with N,N-dimethylaniline results in N-demethylation[21,22]:

$$\text{(7)}$$

and oxidation of phenylbutazone results in hydroxylation at the 4-position[23,24]:

$$\underset{\text{Bu}}{\underset{\text{HO}}{\text{Ph}}\diagdown\underset{}{\text{N-N}}\diagup\underset{}{\text{Ph}}} \longrightarrow \underset{\text{OH Bu}}{\underset{\text{O}}{\text{Ph}}\diagdown\underset{}{\text{N-N}}\diagup\underset{}{\text{Ph}}\diagdown\text{O}} \qquad (8)$$

Oxidations catalyzed by peroxidases have been extensively reviewed.[11]

Compound I contains two oxidizing equivalents from the hydroperoxide whereas Compound II retains one equivalent. However, not all peroxidase oxidations proceed by the sequential one-electron pathway outlined in reactions (2)–(5). For HRP, iodide[25,26] or bisulfite[27,28] reduces Compound I without forming Compound II.

Catalase is related to other peroxidases in that it is capable of reducing H_2O_2 by mechanisms involving higher oxidation states of the enzyme.[13] It functions to dismutate 2 moles of H_2O_2 to O_2 and H_2O via the following sequence of reactions:

$$\text{Cat} + H_2O_2 \rightarrow \text{Compound I} + H_2O \qquad (9)$$

$$\text{Compound I} + H_2O_2 \rightarrow \text{Cat} + O_2 + H_2O \qquad (10)$$

This "catalatic" activity is among the most efficient enzymatic reactions known.[13] Catalase also effects peroxidatic reactions albeit at comparatively low rates relative to its catalatic reaction.[13] In incubations with phenolic reducing substrates, a transient catalase Compound II is observed.[13,29] Catalase Compound I also exhibits an oxidase activity that stereospecifically oxidizes small organic alcohols to aldehydes while reducing H_2O_2.[13,30]

Myeloperoxidase,[163] chloroperoxidase,[31] lactoperoxidase,[32] and thyroid peroxidase[33] catalyze peroxidatic oxidations and, in addition, halogenate organic substrates via an intermediate formally equivalent to hypohalite ion.[17] Halogenation efficiencies and halide specificities differ. Myeloperoxidase and chloroperoxidase also have appreciable catalatic activity[17]. Thyroid peroxidase[33] and lactoperoxidase[32] catalyze iodination reactions and coupling reactions in the biosynthesis of the thyroid hormones thyroxine and triiodothyronine. HRP can iodinate tyrosine[34] and "haloperoxidase" activities are common to most peroxidases.[17]

Cytochrome c peroxidase functions to oxidize ferrous cytochrome c to ferric cytochrome c.[15] Prostaglandin H (PGH) synthase oxygenates polyunsaturated fatty acids to hydroperoxides and peroxidatically reduces the hydroperoxides to alcohols.[35]

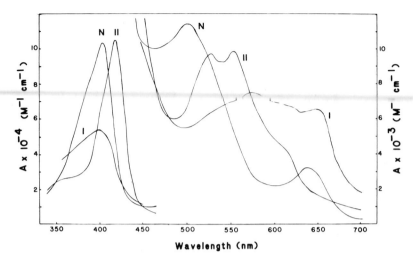

FIGURE 1. Comparison of Soret and visible spectra for native horseradish peroxidase (N), Compound I (I), and Compound II (II). Redrawn from Ref. 14 and references therein with permission.

2. Spectral Changes Associated with Peroxidase Catalysis

Addition of one equivalent of H_2O_2 to a solution of HRP (which is brown in color) causes the appearance of a green color and dramatic visible absorption changes (Fig. 1).[14,36,37] The Soret band is significantly reduced in intensity and multiple bands appear in the long-wavelength visible region (500–650 nm). The species responsible for this spectrum is a covalently modified enzyme that contains both oxidizing equivalents of the peroxide but only one of its oxygens; it is called Compound I.[38] Green compounds with identical visible spectra are produced by reaction of HRP with H_2O_2, peroxy acids, and alkyl hydroperoxides.[18,19,39–41] This implies that Compound I retains the terminal peroxide oxygen but not the residual organic functionality.

A considerable amount of experimental information suggests that the metal atom of the heme prosthetic group is in the +4 (ferryl) oxidation state and that the porphyrin is present as a radical cation.[19] This accounts for transfer of the two redox equivalents of H_2O_2 to the enzyme. The remaining peroxide oxygen is bound as the sixth ligand to iron with an Fe–O distance of 1.64 Å indicative of a double bond.[18] The fifth ligand is an imidazole nitrogen (from histidine) that is the proximal ligand in resting HRP (high-spin ferric) and all derivatives formed from it.[42] For many years, it was believed that an EPR signal of the radical cation could not be detected because of strong coupling to the paramagnetic metal. However, the signal has recently been recorded as a band in the $g = 2$ region with extremely broad wings, which integrates for 0.8 spin/heme.[41] Compound I is a strong oxidizing agent that decomposes spontaneously at a moderate rate on standing at room temperature. It rapidly consumes reducing impurities that might be present in biochemical preparations, which can present difficulties in its preparation and introduce artifacts into experimental results.[43]

Addition of one equivalent of a one-electron reductant such as ferrocyanide quantitatively converts green Compound I to red Compound II (Fig. 1).[44] The intensity of the Soret band is approximately equivalent to that of resting HRP and the long-wavelength visible absorption is considerably narrowed and intensified. Compound II retains the peroxide oxygen, which is probably present as a hydroxyl group and one redox equivalent of H_2O_2.[45,46] Mössbauer spectroscopy indicates that the heme iron remains ferryl following reduction of Compound I to Compound II, whereas EPR spectroscopy indicates that the porphyrin cation radical is reduced.[41] Thus, reduction of the porphyrin cation radical occurs in preference to the metal. The half-cell potential for Compound I → Compound II is +0.96 V.[47] The Fe–O bond distance in Compound I is 1.93 Å, considerably longer than in Compound I.[18]

Reduction of Compound II with one equivalent of ferrocyanide quantitatively regenerates resting enzyme (ferric).[45] The half-cell potential for Compound II → resting enzyme is +0.99 V.[47] The sequence, resting enzyme → Compound I → Compound II → resting enzyme can be repeated numerous times with minimal loss of activity. Although Compounds I and II are potent oxidizing agents, their reactions are well controlled by the protein. This is not true of all hemeproteins with "peroxidase" activity. The spectral changes summarized in Fig. 1 provide a sensitive method for tracking the course of the catalytic cycle, an advantage not often bestowed on enzymologists. Consequently, peroxidase reactions, especially those of HRP, have been intensively studied by biophysical techniques for nearly 50 years. The absorption spectra of HRP-catalyzed reactions operating in the steady state usually correspond to the spectrum of Compound II.[2]

Several other hemeproteins react with peroxides to form colored intermediates. Among these are catalase (bacterial and animal),[48,49] cytochrome c peroxidase (yeast and bacterial),[15,50] lactoperoxidase,[51,52] chloroperoxidase,[53] myeloperoxidase,[54] myoglobin,[55,56] and P-450.[57,58] In addition, transient reactive intermediates can be detected in the reaction of peroxides with the isolated prosthetic group, iron(III) protoporphyrin IX, and other heme complexes[3]. Figure 2 compares the visible absorption spectra of the initial H_2O_2 compounds of HRP, bacterial catalase, chlo-

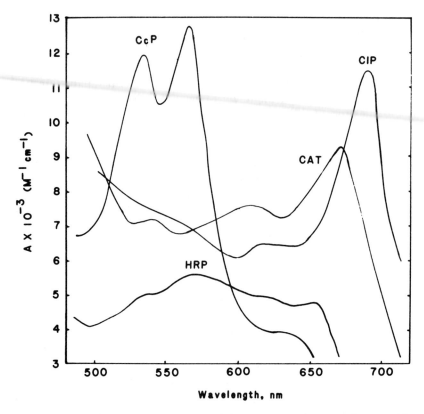

FIGURE 2. Comparison of visible spectra of Compounds I of horseradish peroxidase (HRP),[59] bacterial catalase (CAT),[154] chloroperoxidase (CIP),[53] and cytochrone c peroxidase (CcP).[61] Redrawn from Ref. 59 with permission.

roperoxidase, and yeast cytochrome c peroxidase.[59] It is obvious that there is considerable variation in the spectra. The spectra in the Soret region of the derivatives of HRP, catalase, and chloroperoxidase are similar, exhibiting the low molar absorptivities typical of porphyrin radical cations. The fifth ligands to iron in these hemeproteins are nitrogen (histidine), oxygen (tyrosine), and sulfur (cysteine), respectively. The initial oxidation products of each protein contain both redox equivalents of H_2O_2 and one of its oxygens. The general formulation of their electronic structure is the same—a ferryl-oxo complex surrounded by porphyrin radical cation. If they are electronically so similar, why are their visible spectra different? One hypothesis states that it is due to the symmetry of the highest occupied molecular orbitals of the porphyrin radical cations.[39] The a_{2u} orbital ($^2A_{2u}$ state) exhibits maximal electron density at the *meso* carbons and pyrrole nitrogens whereas the a_{1u} orbital ($^2A_{1u}$ state) exhibits maximal electron density at the pyrrole α-carbons.[60] Indeed, the visible spectrum of HRP Compound I is qualitatively similar to that of [Co(III)-octaethylporphyrin]$^{2+}$·perchlorate, a typical $^2A_{2u}$ ground state, whereas the spectrum of bacterial catalase Compound I is similar to that of [Co(III)-octaethylporphyrin]$^{2+}$·bromide, an $^2A_{1u}$ ground state.[39] By comparison of the visible spectra in Fig. 2, chloroperoxidase Compound I would be classified as a $^2A_{1u}$ ground state. The magnetic circular dichroism spectra of HRP Compound I and beef liver catalase Compound I are somewhat more similar to each other than are the magnetic circular dichroism spectra of the cobalt porphyrin models.[49] However, the visible spectrum of beef liver catalase Compound I is also more similar to the visible spectrum of HRP Compound I than is bacterial catalase Compound I.[49] Recent EPR studies have revealed what appears to be significant electron density at the *meso* carbons of chloroperoxidase Compound I.[59] This is inconsistent with the predictions of a $^2A_{1u}$ ground state and suggests that other factors in addition to or besides orbital symmetry may contribute to the different visible spectra in Fig. 2.

The absorption spectrum of yeast cytochrome *c* peroxidase Compound I is very different from those of HRP, catalase, and chloroperoxidase Compounds I in both the Soret and visible regions (Fig. 2).[15] In fact, it resembles the spectrum of HRP Compound II quite closely, yet cytochrome *c* peroxidase Compound I retains both redox equivalents of H_2O_2.[15,61-64] Furthermore, it contains a stable free radical (~1 spin/molecule) that is not strongly coupled to the ferryl center. Compound I of yeast cytochrome *c* peroxidase appears to contain a ferryl–oxo complex and a free radical derived from an amino acid side chain, possibly tryptophan.[15] The free radical is stable, does not react with O_2, and can be reduced with the ferryl–oxo center to resting enzyme.

HRP, catalase, chloroperoxidase, and yeast cytochrome *c* peroxidase are the most extensively investigated peroxidases. Their higher oxidation states are sufficiently stable for preparation and study. As a result, they have provided the experimental framework for classification of the reaction products of H_2O_2 with other hemeproteins. However, one must proceed with caution. The Compound I-type derivatives of many hemeproteins are very reactive and will oxidize even trace amounts of reducing agents. Endogenous reducing agents are always present in crude preparations such as microsomal fractions so it can be quite difficult to record the spectrum of the initial oxidation product. As a result, data generated using crude preparations are often not very informative. Even quite pure hemeproteins may contain varying levels of reducing agents present as preservatives. For example, PGH synthase usually contains diethyldithiocarbamic acid or phenol as preservatives, both of which are

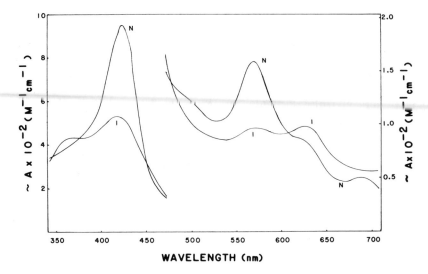

FIGURE 3. Soret and visible spectra of native canine myeloperoxidase (N) and its Compound I (I). Redrawn from Ref. 54 with permission. The absorptivity scalers are approximate. See the original reference for exact values.

peroxidase reducing substrates.[65,66] These agents must be removed before studying the peroxidase activity but after doing so the enzyme is unstable ($t_{1/2} \sim 30$ min for the resting enzyme). Given these difficulties, rapid kinetic spectrophotometry may be very important for accurate biophysical characterization of peroxide-induced derivatives.

Lactoperoxidase forms an H_2O_2 reaction product that resembles HRP Compound I spectroscopically.[51,52] It decays rapidly to a species with a Compound II-like spectrum in the absence of reducing substrates.[67] Recent titrations of the second intermediate indicate that it contains two redox equivalents.[67] The spectral changes suggest the first product is a ferryl-porphyrin cation radical that transfers an electron from an amino acid side chain to the cation radical. The product (lactoperoxidase Compound II) is analogous electronically to yeast cytochrome c peroxidase Compound I. The conversion of the ferryl-cation radical to the ferryl-free radical (Compound I → II) is irreversible and it appears that the two species react with reducing substrates differently.[67]

Hemes of the chlorin and c type also react with H_2O_2 to form covalently modified proteins. Canine myeloperoxidase contains two hemes that are distinct from protoheme and optically similar to chlorin.[68,69,151–153] It reacts with H_2O_2 to form a species spectrally similar to HRP Compound I in the Soret and visible regions (Fig. 3).[54] In contrast to the peroxidases discussed so far, a 40-fold excess of H_2O_2 is required for

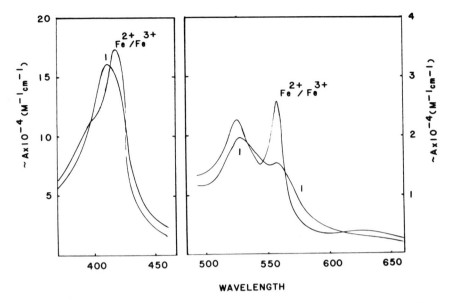

FIGURE 4. Soret and visible spectra of *Pseudomonas* cytochrome c peroxidase. Half-reduced form (Fe^{2+}/Fe^{3+}) and its Compound I (I). Redrawn from Ref. 74 with permission. Absorptivity scaler is approximate.

complete formation of myeloperoxidase Compound I.[54] In the absence of reducing substrates, decay of Compound I to a species spectrally similar to HRP Compound II is apparent within 30 msec.[54] The number of redox equivalents in myeloperoxidase Compound II is unknown.

Cytochrome c peroxidase of *Pseudomonas aeruginosa* is very different from yeast cytochrome c peroxidase. It contains two heme c groups covalently bound to a single polypeptide chain and is isolated as an Fe^{3+}–Fe^{3+} enzyme that is inactive as a peroxidase.[70-72] Reduction of one heme with ascorbate or NADH generates an Fe^{2+}–Fe^{3+} enzyme that reacts rapidly with H_2O_2 to form a Compound I.[50,73] The Soret absorbance is shifted but not significantly decreased in intensity and the visible region does not contain a band above 600 nm (Fig. 4). This rules out the presence of a porphyrin radical cation and it has been suggested that the peroxide derivative is an Fe^{4+}–Fe^{3+}. Compound I is reduced by one electron by its substrate azurin to a species formulated as Fe^{3+}–Fe^{3+}. Curiously, this Fe^{3+}–Fe^{3+} form is different than the Fe^{3+}–Fe^{3+} state of the isolated enzyme.[74] Addition of a second electron from azurin regenerates the half-reduced resting enzyme.

Investigations of the spectral changes associated with the reaction of peroxidases with H_2O_2 provide evidence for great diversity. The model

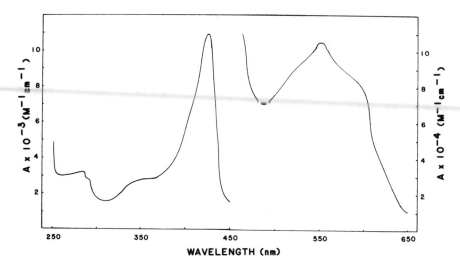

FIGURE 5. Ultraviolet and visible spectra of oxidized metmyoglobin. Redrawn from Ref. 56 with permission.

described for HRP Compound I of a ferryl-porphyrin radical cation does not adequately describe the initial spectroscopically detectable oxidation product of all peroxidases. Rather, the picture that emerges is one of enzymes containing one of the peroxide redox equivalents in a ferryl porphyrin and the other as a porphyrin cation radical, amino acid radical, or metal.

Metmyoglobin reacts slowly with H_2O_2 (10^5 times slower than HRP).[43] An oxidized protein is formed that has the absorption spectrum displayed in Fig. 5.[56] The absorption spectrum is inconsistent with the presence of a porphyrin radical cation. Furthermore, the oxidized protein contains only a single redox equivalent of the peroxide.[75] The other redox equivalent is only detectable transiently during formation of the initial spectral intermediate and seems related to irreversible oxidation of an amino acid residue.[75,76] Amino acid oxidation results in destruction of the protein and limits the number of peroxidatic turnovers to one or two. George first proposed that metmyoglobin reduces H_2O_2 *by a single electron* generating a ferryl porphyrin (Fig. 5) and hydroxyl radical:

$$Fe^{3+} + H_2O_2 \rightarrow Fe^{4+}-OH + \cdot OH \tag{11}$$

The hydroxyl radical presumably reacts with a protein residue (e.g., tyrosine; *vide infra*). This mechanism accounts for the formation of the spectrally detectable oxidized heme that contains a single redox equivalent and the generation of a second, transient redox equivalent (·OH). Overall, both redox equivalents of H_2O_2 are consumed. George's proposal represents a radical departure (excuse the pun) from the heterolytic chemistry discussed for other peroxidases. One cannot rule out the possibility that metmyoglobin reacts heterolytically with H_2O_2 to form an HRP Compound I-like species that abstracts an electron or H atom from an amino acid as yeast cytochrome *c* peroxidase and lactoperoxidase Compounds I do.[77] The difference between metmyoglobin and the latter peroxidases could be that the free radical on the amino acid of metmyoglobin Compound I is exposed to solution and can react with O_2, thereby irreversibly oxidizing it. However, George's mechanism offers an attractive hypothesis to explain several disparate experimental observations that will be considered in other sections of this chapter. It also provides an important precedent for the consideration of P-450.

Rabbit liver P-450 (LM_2) reacts with a variety of organic hydroperoxides and peracids to produce two different absorbing species, termed C and D[57]:

$$P\text{-}450 + XOOH \rightarrow C \rightarrow D \qquad (12)$$

The dependence of the spectral changes on hydroperoxide concentration is not stoichiometric but resembles a binding isotherm. The reactions that generate C and D are reversible, an observation unprecedented in the peroxidase literature.[57] Furthermore, the absorption bands of C and D in the Soret region *differ for each hydroperoxide* (Fig. 6). This finding contrasts sharply with the peroxidase literature and suggests that C and D contain organic functionality in the vicinity of the heme. Addition of a hydroxylatable substrate increases the rate of formation of C, decreases the rate of formation of D, and lowers the steady-state concentration of both.[78] This is consistent with a model in which C is on the pathway to substrate hydroxylation whereas D is a catalytically inconsequential side product. Hydroxylation of substituted toluenes by a series of substituted cumene hydroperoxides is retarded by electron-withdrawing groups in the *para* position of the toluene but accelerated by electron-withdrawing groups in the *para* position of the hydroperoxide.[57] The latter observation suggests that the organic functionality of the hydroperoxide is present in the oxidizing agent that plays a role in hydroxylation. None of these findings are consistent with the paradigm for oxidations catalyzed by peroxidases. Rather, they suggest that P-450 reduces hydroperoxides by one electron to form an alkoxyl radical and a ferryl-hydroxo complex, both

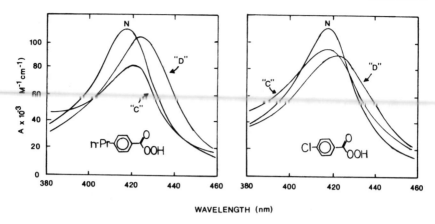

FIGURE 6. Comparison of Soret spectra for rabbit liver cytochrome P-450 native (N), complex C; and complex D with *p-n*-propylperbenzoic acid and *p*-chloroperbenzoic acid. Redrawn from Ref. 57 with permission.

of which are required for hydroxylation. In that regard, P-450 appears more closely related to metmyoglobin than classical peroxidases.

Reaction of P-450$_{cam}$ with peroxy acids generates two transient absorbing species.[58] They exhibit hypsochromically shifted Soret bands that have significantly reduced molar absorptivities. These features are reminiscent of ferryl–cation radicals such as Compound I of HRP and are quite different from the spectral changes recorded for rabbit liver LM$_2$. One similarity is the finding that large excesses of peroxide are required to drive the absorption shifts to completion.

3. Kinetics of Peroxidase Catalysis

Peroxidases exhibit different degrees of sensitivity to irreversible inactivation by peroxides, which limits the applicability of kinetic methods of analysis. HRP is remarkably stable to peroxides, which is one reason why it has been extensively studied.[2] Steady-state kinetic analysis has provided saturation parameters (K_m and k_{cat}) and comparative rate data but cannot be used to dissect each step in the catalytic cycle. Compound I formation can be made rate-determining by limiting the concentration of H_2O_2 and using an excess of reducing substrate. Reduction of Compound I is so rapid that it can rarely be made rate-limiting, except at low pH with a few reducing substrates (e.g., ferrocyanide[79,80] or *p*-aminobenzoic acid[81]). In general, this step is kinetically silent to steady-state analysis. Under conditions of excess peroxide and reducing substrate, reduction of Compound II is usually rate-limiting. Thus, steady-state methods are not applicable to each step in HRP catalysis. Oxidations of ferrocytochrome c[82] or N,N-dimethylaniline[22] by H_2O_2 and HRP exhibit Ping-Pong kinetics as expected from reactions (2)–(5). This implies binding of reducing substrate to the enzyme subsequent to the formation of Compound I. However, spectroscopic evidence is available that reveals low affinity but specific binding of aromatic substrates to resting HRP.[83-87] A rapid equilibrium of reducing substrate and enzyme would also be consistent with Ping-Pong kinetics.

It is possible to adjust conditions so that the formation or reaction of each intermediate of HRP catalysis can be monitored by transient kinetic methods.[88] Determination of pH–rate profiles provides information about the chemical events and potentially important residues at each stage of turnover. However, HRP oxidizes a wide variety of reducing substrates and not all oxidations proceed with the same pH–rate profile.

Compound I formation varies only slightly with pH.[2] The unprotonated form of the enzyme reacts with H_2O_2 with a rate coefficient of 1.8×10^7 M^{-1} sec^{-1} (Fig. 7).[96] An acidic residue with a pK_a of ~ 3.0 influences the reaction. This has been attributed to Asp-43 for HRP isoenzyme C.[89] The overall rate of Compound I formation is diffusion-controlled and limited by formation of an outer-sphere complex that decays to an inner-sphere complex. The inner-sphere complex undergoes chemical reaction to form Compound I.[90] No uptake or release of protons occurs.[19,91] Coordination of HRP with ligands that cannot undergo reduction is also rapid and tight.[2] Cyanide binds with a rate coefficient of 2×10^5 M^{-1} sec^{-1} and a K_d of 1.2×10^{-6} M.[92] The monoprotonated and unprotonated forms of the enzyme react with fluoride with rate coefficients of 6×10^6 and 4.6×10^7 M^{-1} sec^{-1}, respectively.[93] Both forms of the enzyme have K_d's for fluoride of 1×10^{-3} M.[93] The same ionizable groups that are important for Compound I formation appear to be involved in fluoride ligation and outer-sphere complex formation appears rate-limiting. These data suggest that reduction of H_2O_2 ligated to HRP is extremely fast. The rate coefficients for formation of Compound I from CH_3OOH and C_2H_5OOH are lower than HOOH but still large (1.5×10^6 and 3.6×10^6 M^{-1} sec^{-1}, respectively).[94] As mentioned earlier, the Compounds I generated from organic hydroperoxides are spectroscopically indistinguishable from Compound I formed from H_2O_2. Interestingly, t-butyl-hydroperoxide does not react with HRP to form a Compound I.[95] This may indicate a sterically restricted active site although t-butyl-hydroperoxide is less easily reduced chemically than the others.

Reduction of HRP Compound I by reducing substrates has been monitored by stopped-flow spectrophotometry under pseudo-first-order con-

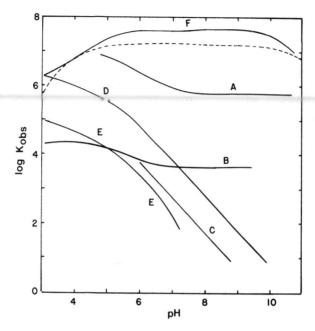

FIGURE 7. pH dependence for the formation of horseradish peroxidase compound I and its reaction with reducing substrates. Dashed line represents formation of HRP-I by reaction of native HRP with H_2O_2. Solid lines show rates of reaction for HRP-I with: (A) ferrocyanide ion; (B) p-aminobenzoic acid; (C) nitrite ion or nitrous acid; (D) iodide ion; (E) bisulfite ion; (F) p-cresol. K_{obs} expressed in units of M^{-1} sec^{-1}. Redrawn from Refs. 2, 16, and 96 with permission.

ditions with an excess of H_2O_2. Formation of Compound I is faster than the dead time of the instrument; reduction of Compound I is then monitored. A protein residue of $pK_a \sim 5.1$–5.4 is important for Compound I reduction, and has been shown to be the distal His-42.[14] The log k_{obs} versus pH plots for ferrocyanide,[79,80] p-aminobenzoic acid,[81,97] p-cresol,[20] nitrite,[98] iodide,[25] and bisulfite[98] are summarized in Fig. 7. All of these substrates show the influence of the distal histidine on the observed rates. In the case of L-tyrosine oxidation (data not shown), the distal histidine appears to participate in base catalysis.[99] For p-aminobenzoic acid and ferrocyanide oxidation, it participates to some degree in acid catalysis.[97]

The reaction of Compound II with reducing substrates is usually rate-limiting. Consequently, reaction of native HRP with one equivalent of H_2O_2 is required to prevent cycling of the enzyme following its reduction.[20] Compound II reduction requires the uptake of an electron and

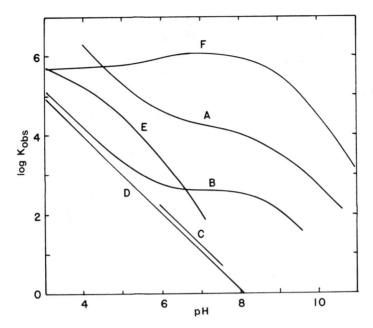

FIGURE 8. pH dependence for the reaction of horseradish peroxidase Compound II with reducing substrates. (A) Ferrocyanide ion; (B) p-aminobenzoic acid; (C) nitrite ion or nitrous acid; (D) iodide ion; (E) bisulfite ion; (F) p-cresol. K_{obs} expressed in units of $M^{-1}\,sec^{-1}$. Redrawn from Ref. 2 with permission.

proton.[91] Figure 8 illustrates k_{obs} versus pH plots for HRP Compound II reduction by some substrates that were previously used for Compound I reduction. At low pH, ferrocyanide[79,80] and p-aminobenzoic acid[81,97] have larger rate constants for the reduction of Compound II than for Compound I. The reduction of p-cresol is most strikingly influenced by an ionizable group with a pK_a of 8.6.[83] The influence of this group is seen in the rate curves for p-aminobenzoic acid[81,97] and ferrocyanide.[80] Another ionizable group of $pK_a \sim 0$ appears to be mechanistically important for iodide oxidation. From these data, both acid and base catalysis appear to function in Compound II reduction.

Table II summarizes the rates of reaction of various reducing substrates with HRP Compound I and Compound II. The rates of reaction of one-electron reducing substrates are approximately 40 times faster for Compound I reduction than for Compound II.

The application of transient-state kinetics to HRP greatly simplifies the interpretation of the HRP catalytic cycle.[2] Important ionizable protein

TABLE II
Summary of Rates of Reduction for HRP Compounds I and II with Reducing Substrates

Reducing substrate	pH	$k_{obs}{}^a$ (M^{-1} sec^{-1}) I → II	II → N	Ref.
p-Aminobenzoic acid	5.4	5×10^4	1.3×10^3	156
Ascorbic acid	4.3	2.3×10^6	2.6×10^5	157
p-Cresol	6.8	—	1.1×10^6	157
	7.0	4.2×10^7	—	158
Ferrocyanide	5.6	7.5×10^6	2.1×10^5	80
	5.9	1.9×10^6	4.4×10^4	80
	6.9	9.1×10^5	1.8×10^4	80
Iodide	5.4	1.8×10^5	4.5×10^2	159
	6.0	5.6×10^4	1.2×10^2	25
Guaiacol	7.0	9×10^6	3×10^5	160
Luminol	8.0	2.3×10^6	7.2×10^4	161
Nitrite	5.4	2×10^5	2.4×10^3	156
Resorcinol	7.0	8.0×10^6	3×10^6	158, 161

residues have been identified and related to potential mechanisms for heterolysis of the peroxide bond as well as the reduction of the higher oxidation states of the enzyme.[14,83,90] In the case of peroxide heterolysis, the requirements for ionizable groups in a precise stereochemical arrangement for participation in proton donation, neutralization of developing charge density, and stabilization of higher valence states of the iron-oxo complexes have been incorporated into mechanistic hypotheses.[2]

Steady-state kinetic analysis of pyrogallol oxidation by H_2O_2 in the presence of several hemeproteins is quite revealing (Table III).[100] Whereas "classical peroxidases" such as HRP, chloroperoxidase, and catalase have predictably high turnover numbers, those of metmyoglobin and P-450 are very low and are actually lower than the turnover number of hemin, the prosthetic group containing no protein. P-420 has a higher k_{cat} than P-450 as does P-450 supported by NADPH and O_2 rather than H_2O_2. The demarcation between classical peroxidases and other hemeproteins is reminiscent of the differences in spectral changes observed on reaction of the proteins with H_2O_2. Taken together, the results suggest a fundamentally different mechanism or at least drastic differences in the rate of reaction of the hemeproteins with H_2O_2. These differences can only be explained by structural differences in the proteins because the prosthetic group is the same.

Pyrogallol is a substrate that is oxidized by electron transfer.[1] When the same group of enzymes was tested for their ability to hydroxylate

TABLE III
Peroxidation of Pyrogallol by Various Hemeproteins[a]

Hemeprotein	$K_m{}^b$	$V_{max}{}^c$
Horseradish peroxidase	1.1	150,000
Catalase	91	295,000
Chloroperoxidase	4.3	137,000
Metmyoglobin		5[d]
P-450$_{LM2}$	20	7.5
P-450$_{LM2}$ (Fpt/NADPH/O$_2{}^e$)		110
P-450$_{LM4}$	No measurable rate	
P-420$_{LM2}$	900	95
Hemin	37	28

[a] From Ref. 100 with permission.
[b] K_m concentration expressed in mM.
[c] V_{max} expressed as moles of purpurogallin produced per mole of hemeprotein per min.
[d] No difference in rate observed with H_2O_2 concentrations from 25 to 150 mM.
[e] V_{max} observed with the complete reconstituted system containing reductase (Fpt), NADPH, and O_2 without peroxide.

aliphatic substrates, only P-450 was able to do so.[100] Cumene hydroperoxide-dependent hydroxylations catalyzed by P-450 result from transfer of the hydroperoxide oxygen to the aliphatic substrate and occur with hydroperoxide/alkane stoichiometry of 1:1.[101] Furthermore, P-450 is the only hemeprotein capable of converting peroxyphenylacetic acid to benzyl alcohol, a reaction that results from one-electron reduction of the peroxide[102]:

$$\text{PhCH}_2\text{C(=O)-OOH} \longrightarrow \text{PhCH}_2\text{OH} + CO_2 \qquad (13)$$

4. Hydroperoxide Transformations

We have compared a series of hemeproteins with respect to their ability to catalytically transform 5-phenyl-4-pentenyl-1-hydroperoxide (PPHP). HRP, catalase, cytochrome c peroxidase (yeast), lactoperoxidase, and PGH synthase effect smooth reduction to the alcohol when they are incubated with PPHP (100 μM) and phenol (200 μM):

$$\text{Ph-CH=CH-CH}_2\text{-CH}_2\text{-OOH} \longrightarrow \text{Ph-CH=CH-CH}_2\text{-CH}_2\text{-OH} \qquad (14)$$

Reduction does not occur if phenol is omitted or the enzyme is denatured before incubation. These results are consistent with the catalytic cycle of peroxidases outlined in reactions (2)–(5). Simultaneous two-electron reduction of hydroperoxide (reaction 2) generates alcohol and Compound I. The catalytic cycle is only completed when a reducing substrate is present (reactions 3 and 4). It is noteworthy that the alcohol is the only product detected with catalase because this enzyme is well known to transform small, primary hydroperoxides to aldehydes (e.g., ethyl hydroperoxide to acetaldehyde).[13]

When similar incubations of PPHP are conducted in the presence of metmyoglobin, methemoglobin, or hematin, a different profile results:

$$Ph\text{-CH=CH-CH}_2\text{-CH}_2\text{-OOH} \longrightarrow Ph\text{-CH=CH-CH}_2\text{-CHO} \qquad (15)$$

Alcohol and aldehyde are produced in a ratio of 1:5. Excess reducing substrate (500 μM) has no effect on the product ratio or the extent of reaction. In addition, metmyoglobin and methemoglobin only turn over 30–40 times before apparently inactivating. Complete reaction of 100 μM PPHP is catalyzed by 0.5 μM hematin in 15 min. The mechanism of aldehyde formation is unknown but intriguing. One possibility is that the alcohol is produced quantitatively and then oxidized. As noted above, catalase, an enzyme for which alcohol oxidation is *precedented*,[13] produces alcohol from PPHP without a trace of aldehyde. In addition, incubation of the alcohol with hematin and H_2O_2 or cumene hydroperoxide produces no aldehyde. This suggests that alcohol oxidation, although possible, is not a major pathway of aldehyde formation. Protic[103] and Lewis acids[104] catalyze dehydration of hydroperoxides to carbonyl compounds but exposure of PPHP to 1 N H_2SO_4 for 15 min generates no aldehyde. Thus, the heme complexes and hemeproteins do not appear to be acting as Lewis acids. Another possibility is that the heme center reduces PPHP by one electron to an alkoxyl radical and forms a ferryl-hydroxo complex:

$$Fe^{3+} + R\text{-CH(H)-OOH} \longrightarrow Fe^{4+}\text{-OH} + R\text{-CH(H)-O}\cdot \longrightarrow Fe^{3+} + R\text{-CH(H)=O} \qquad (16)$$

Oxidation of the alkoxyl radical by the ferryl-hydroxo complex would yield aldehyde and regenerate the heme. We are designing experiments to test this hypothesis.

Reconstituted rabbit liver P-450 $(LM_2)^{67}$ catalyzed the decomposition of PPHP in a manner remarkably similar to metmyoglobin or methemoglobin in that alcohol and aldehyde were formed in the same ratio and reducing substrates had no effect on the PPHP conversion process. A reconstituted preparation consisting of P-450 and NADPH-cytochrome P-450 reductase in dilauroylphosphatidylcholine liposomes, demonstrated active NADPH oxidase activity with dimethylaniline as substrate. The reconstituted enzyme in the presence of NADPH, converts PPHP to alcohol and aldehyde in the same ratio as P-450 alone.

Spectrophotometric, kinetic, and metabolic criteria exist that distinguish hemeproteins in their reactions with hydroperoxides. "Classical peroxidases" such as HRP, catalase, and cytochrome c peroxidase exhibit high turnover numbers toward oxidizable substrates and spectral and metabolic properties consistent with the catalytic cycle outlined in reactions (2)–(5). Other hemeproteins such as metmyoglobin and methemoglobin are poorly efficient at oxidation, do not generate oxidation states analogous to Compound I, and do not reduce all peroxides by two electrons. P-450 clearly belongs in the latter category with the distinction that it will catalyze hydroperoxide-dependent epoxidation and aliphatic hydroxylation. One might wonder, then, why the peroxidase activity of P-450 has been much-heralded. This appears due to two factors. First, many investigators use P-450 preparations from rat liver that contain very large amounts of enzyme (1–3 nmole/mg protein). The concentrations of P-450 present in many experiments are in the range 1–5 μM.[5-7] In contrast, HRP is an effective catalyst at 1–5 nM. Second, many of the studies employed cumene hydroperoxide as oxidant, which is an excellent substrate for P-450 but a poor substrate for many commonly used peroxidases. Consequently, comparisons of peroxidase activity between P-450 and other hemeproteins that use cumene hydroperoxide as substrate will imply that P-450 is more effective.

5. Peroxidase-Catalyzed Oxidations

The types of hydroperoxide-dependent oxidations catalyzed by peroxidases and P-450 provide another means to differentiate them. Peroxidases catalyze electron transfer reactions quite efficiently but are at best very poorly efficient at aliphatic hydroxylation or olefin epoxidation. Electron transfer occurs on the enzyme but the subsequent reactions of the electron-deficient derivative of the reducing substrate occur in solution. Consequently, the products of peroxidase-catalyzed reactions are largely determined by the solution chemistry of the initial product of electron transfer. For example, many aromatic amines are oxidized by peroxidases to radical cations. The radical cations, which are often stable enough to be detected by electron paramagnetic resonance, frequently decompose by bimolecular reactions. Phenidone is oxidized to a radical cation and two molecules of radical cation disproportionate to phenidone and 3-hydroxy-1-phenyl-pyrazole[105,106]:

$$\text{(17)}$$

The net reaction is dehydrogenation but it occurs as a result of electron transfer and disproportionation. Aminopyrine is dehydrogenated by a similar mechanism but the product, an iminium salt, hydrolyzes to formaldehyde and N-methyl-4-aminoantipyrine[107]:

$$\longrightarrow \quad + \text{ HCHO} + \text{H}^+ \quad \text{(18)}$$

N-Demethylation of aromatic amines appears to occur by the same sequence. Phenols donate electrons to peroxidases and form phenoxyl radicals. The chemistry of phenoxyl radicals is complex but coupling of two radicals is common. This accounts for the formation of Pummerer's ketone from p-cresol by HRP[20]:

$$\text{(19)}$$

β-Dicarbonyls are oxidized by peroxidases to carbon-centered radicals that react with O_2 to yield peroxyl radicals:

$$\xrightarrow{O_2} \quad \text{(20)}$$

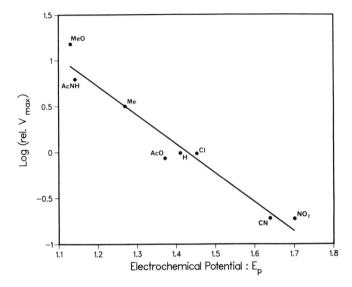

FIGURE 9. Correlation of rate of oxidation of *p*-substituted aryl sulfides to oxidation potential. Unpublished data kindly supplied by Professors A. P. Schaap and T. Kimura.

Phenylbutazone is oxidized to 4-hydroperoxy-phenylbutazone, which is reduced to 4-hydroxy-phenylbutazone.[24] The overall hydroxylation is a result of the chemistry of the electron-deficient derivative of phenylbutazone.

Peroxidases catalyze oxygen transfers to heteroatoms. This may result from electron transfer followed by coupling of the electron-deficient derivative to the ferryl-oxo oxygen:

$$ROOH + X \rightarrow X-O + ROH \tag{21}$$

PGH synthase oxygenates sulfides to sulfoxides and the sulfoxide oxygen is derived quantitatively from the hydroperoxide group.[108,109] The rate of oxygenation of a series of *p*-substituted sulfides by HRP correlates inversely to the oxidation potential of the sulfides (Fig. 9) and exhibits a ρ value of -1.21 versus σ^+ when subjected to Hammett analysis. Similar results have been reported for the NADPH-dependent oxidation of sulfoxides to sulfones by P-450.[110] Oxygen transfer to nitrogen is not a characteristic reaction of peroxidases. Chloroperoxidase oxidizes 2-aminofluorene to nitrosofluorene and a peroxidase in pea seed microsomes oxygenates aniline to phenylhydroxylamine.[21,111] However, recent results suggest the plant peroxidase may be a P-450.[112] Halide oxidation by per-

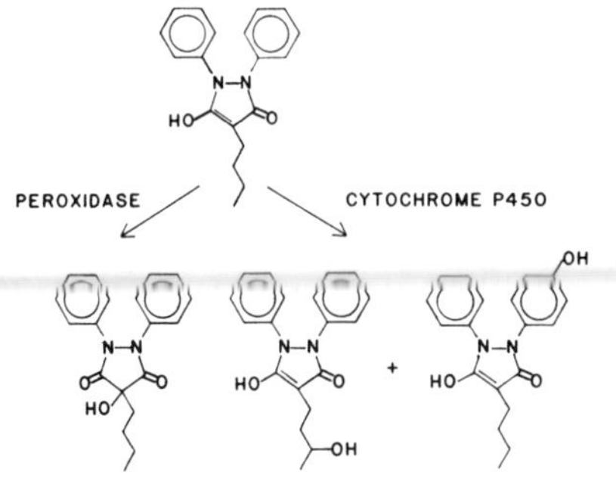

FIGURE 10. Peroxidase- and P-450-catalyzed phenylbutazone oxidation.

oxidases is quite common.[11] Iodide is a reducing substrate for most peroxidases but bromide and chloride are only oxidized by peroxidases such as chloroperoxidase and myeloperoxidase that generate a very strong oxidant. The product of halide oxidation is hypohalous acid.

P-450 catalyzes hydroperoxide-dependent hydroxylation and epoxidation in which the peroxide oxygen is transferred to the substrate.[101,113] The initial step in aliphatic hydroxylation is electron transfer to a potent enzyme-bound oxidant, estimated to have a redox potential of $+1.5$–2.0 V.[114] This is significantly higher than the half-cell potentials for the one-electron reduction of HRP Compound I and HRP Compound II ($+0.96$ and $+0.99$ V, respectively) and may explain the differential ability of P-450 and HRP to catalyze hydroxylation.[18] Despite the high redox potential of its Compound I, P-450 is an inefficient catalyst of electron transfer from electron-rich molecules.[100] Consider, for example, phenylbutazone. As discussed above, peroxidases oxidize it exclusively at the α-carbon of the β-dicarbonyl group to form 4-hydroperoxy-phenylbutazone. In contrast, NADPH supports the P-450-catalyzed hydroxylation of phenylbutazone to oxyphenbutazone and γ-hydroxy-phenylbutazone (Fig. 10).[115] One might wonder why the β-dicarbonyl group, the most electron-rich site in the molecule, is not oxidized by P-450. It appears that steric interactions that orient substrates at the active site play an important role in P-450-catalyzed oxidation.

6. Peroxyl Radicals as Oxidizing Agents in Heme–Hydroperoxide Reactions

Investigation of fatty acid hydroperoxide-dependent oxidations during prostaglandin biosynthesis provides evidence for the production of oxidizing agents that are unrelated to the iron-oxo intermediates of peroxide catalysis.[116] Fatty acid hydroperoxides are the only organic hydro-

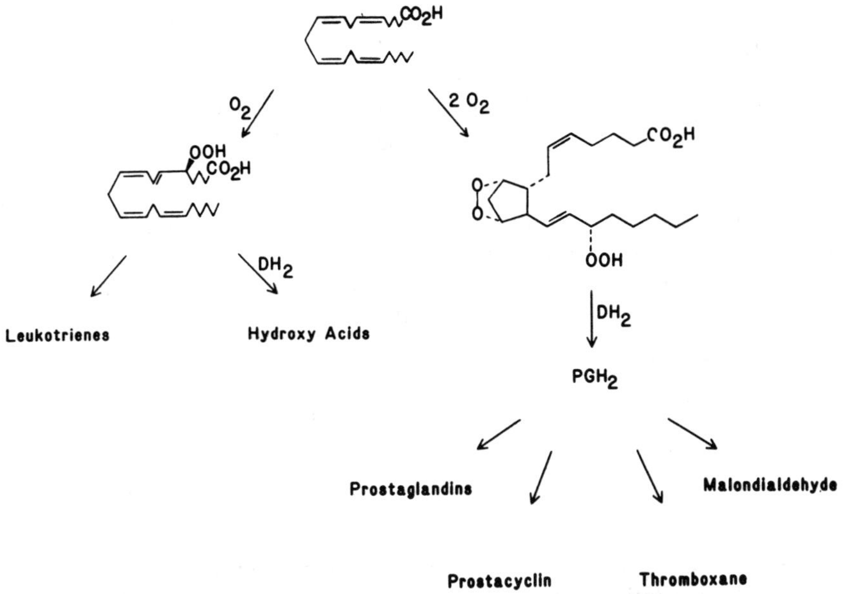

FIGURE 11. Enzymatic pathways of hydroperoxide formation from arachidonic acid.

peroxides commonly found in biological systems.[117] They are generated from unsaturated fatty acids by PGH synthase and a family of lipoxygenases (Fig. 11). These hydroperoxides are efficiently reduced by the peroxidase activity of PGH synthase but are poor substrates for other heme-containing peroxidases. Like H_2O_2, alkyl hydroperoxides, or acyl hydroperoxides, fatty acid hydroperoxides support the oxidation of peroxidase reducing substrates by converting the heme iron of PGH synthase to higher oxidation states.[118] Typical peroxidase reducing substrates are oxidized by the iron-oxo intermediates of PGH synthase.[119]

Fatty acid hydroperoxides also oxidize compounds that are not PGH synthase reducing substrates. Diphenylisobenzofuran,[120] benzo[a]pyrene,[117] and 7,8-dihydroxy-7,8-dihydrobenzo[a]pyrene (BP-7,8-diol)[121] are cooxidized during arachidonic acid oxygenation to dibenzoylbenzene, benzo[a]pyrene quinones, and benzo[a]pyrene diolepoxide, respectively. Because these compounds do not stimulate hydroperoxide reduction, they are not oxidized by higher oxidation states of the enzyme. Therefore, there must be another oxidizing agent generated during PGH synthase turnover.

Oxidation of these three compounds shares the following features:

1. The oxygen incorporated into diphenylisobenzofuran and BP-7,8-

diol is derived from molecular oxygen and not from the hydroperoxide.[120,122]
2. Oxidation is potently inhibited by antioxidants.[123]
3. The stoichiometry of compound oxidized to hydroperoxide added is indicative of a free radical process.[117,120]
4. Only fatty acid hydroperoxides support oxidation.

Taken together, the observations suggest that the oxidizing agent in these reactions is a free radical rather than a peroxidase ferryl-oxo complex.

Hematin, the prosthetic group of peroxidases and P-450, catalyzes epoxidation of BP-7,8-diol by fatty acid hydroperoxides.[124]

$$\text{(22)}$$

Detergent, in excess of its critical micellar concentration, is required for maximum epoxidation. The products, oxygen source, stereochemistry, and sensitivity to antioxidant inhibition are analogous to the oxidation of BP-7,8-diol by microsomal PGH synthase preparations.[124] We have proposed that peroxyl radicals generated by reaction of fatty acid hydroperoxides and hematin are the epoxidizing agents.[124] The products and oxygen labeling pattern of the reaction of 13-hydroperoxy-9,11-octadecadienoic acid and hematin suggest that hematin reduces the hydroperoxide by one electron to an alkoxyl radical that cyclizes to the adjacent double bond[125–127]:

$$\text{(23)}$$

The resultant epoxyallylic radical either couples to the hydroxyl group of the ferryl-hydroxo complex or escapes the solvent cage and couples

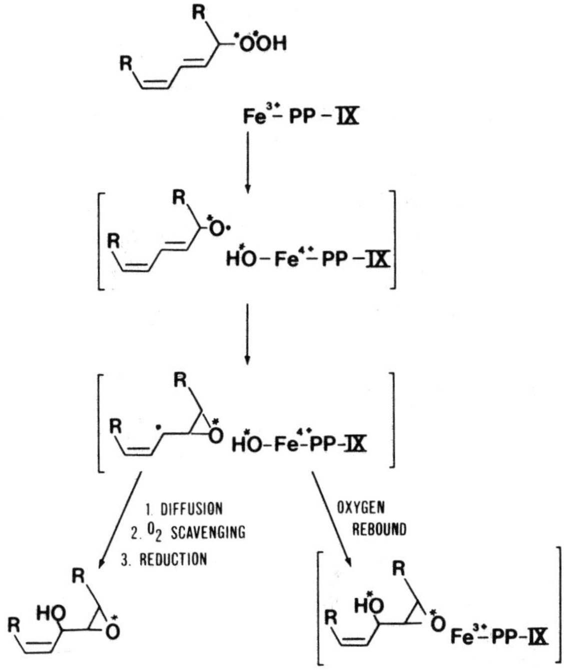

FIGURE 12. Proposed mechanism for the hematin-catalyzed rearrangement of hydroperoxy fatty acids to epoxy alcohols.

to O_2 forming a peroxyl radical (Fig. 12). The peroxyl radical epoxidizes BP-7,8-diol. Quantitation of peroxyl radical-derived products of 13-hydroperoxy-octadecadienoic acid and diolepoxide-derived products of BP-7,8-diol indicates that BP-7,8-diol traps approximately 75% of the peroxyl radicals generated.[126] It is, therefore, a sensitive chemical trapping agent for the detection of peroxyl radicals.

The formation of epoxy alcohols in which the hydroxyl group originates from O_2 provides direct evidence for carbon radical intermediates generated by cyclization of alkoxyl radicals (Fig. 12). Ferrous-cysteine complexes react with 13-hydroperoxy-octadecadienoic acid to form epoxy alcohols in high yield.[128] ^{18}O-labeling studies establish that the hydroxyl oxygens are derived quantitatively from O_2. These results and the mechanism proposed to explain them were important precedents for the mechanism depicted in reaction (23) and Fig. 12. Interestingly, hemoglobin has been reported to convert 13-hydroperoxy-octadecadienoic acid to epoxy alcohols in high yield.[129] This is most likely to occur by hemoglobin reduction of the hydroperoxide by one electron to alkoxyl radicals that cyclize to the 11,12-double bond. This is consistent with George's hypothesis that metmyoglobin reduces H_2O_2 by one electron to form the hydroxyl radical and a ferryl-hydroxo complex.[76]

The results described above clearly indicate that peroxyl radicals generated by coupling of O_2 to carbon radicals epoxidize olefins. This predicts that any biochemical reaction that generates peroxyl radicals will lead to olefin epoxidation. Epoxidation of BP-7,8-diol has been detected during lipid peroxidation,[130] hydrocarbon autoxidation,[131] and phenylbutazone oxidation by peroxidases.[132] Phenylbutazone oxidation proceeds by oxidation of the enolate of phenylbutazone to a carbon radical that couples with O_2.[24] Inclusion of BP-7,8-diol traps the peroxyl radicals with concomitant epoxidation of the hydrocarbon.

H_2O_2, in the presence of methemoglobin or metmyoglobin, epoxidizes styrene.[133] In contrast to NADPH-dependent epoxidations by P-450, both enantiomers of styrene oxide are produced and cis- and trans-[1-^{2}H]-styrene oxides are produced from trans-[1-^{2}H]-styrene:

$$\text{Ph}\diagup\!\!=\!\!\diagdown\text{D} \longrightarrow \text{Ph}\diagup\!\!\triangle\!\!\diagdown\text{D} + \text{Ph}\diagup\!\!\triangle\!\!\diagdown_D \quad (24)$$

The loss of stereochemical integrity of [1-^{2}H]-styrene oxide is particularly significant because it implies the existence of an intermediate that is capable of rotation about the C1–C2 bond. ^{18}O-labeling investigations indicate that the epoxide oxygen of styrene oxide is derived primarily from O_2 and that epoxidation is inhibitable by antioxidants.[133] These findings suggest that a peroxyl radical is the oxidizing agent. A mechanism has been proposed in which H_2O_2 reacts with metmyoglobin to generate a ferryl-hydroxo complex and a tyrosyl radical (from a protein residue).[133] The tyrosyl radical couples with O_2 to form a peroxyl radical that epoxidizes styrene. A tyrosine residue that is a likely candidate for the source of the peroxyl radical is present in metmyoglobin in the vicinity of the heme center. This mechanism implies that a limited number of oxidations can be catalyzed by each heme group and, in fact, it appears that only one or two oxidations occur per heme.

It is very attractive to suggest that metmyoglobin reacts with H_2O_2 to form hydroxyl radical and a ferryl-hydroxo complex and that the hydroxyl radical oxidizes the tyrosine. This would be consistent with George's original hypothesis and would explain why a transient oxidizing agent is generated along with the more stable ferryl-hydroxo complex. It would also be consistent with the much slower rate of reduction of H_2O_2 by metmyoglobin than by HRP (10^5-fold slower). However, it is also pos-

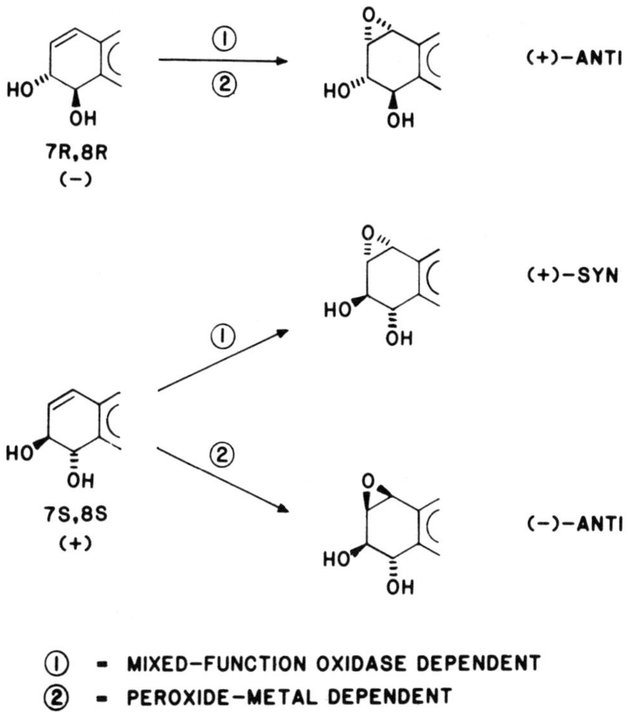

FIGURE 13. The stereochemistry of (±)-BP-7,8-diol epoxidation by peroxyl radicals (peroxide–metal-dependent) and cytochrome P-450 (mixed-function oxidase-dependent).

sible that metmyoglobin reduces H_2O_2 by two electrons forming a ferryl-oxo complex analogous to HRP Compound I, which then oxidizes the tyrosine residue.[44] At present, it is not possible to choose between these two alternatives.

The stereochemistry of epoxidation of BP-7,8-diol by peroxyl radicals is distinct from that of P-450 when the latter is supported by NADPH (Fig. 13).[134,135] The (−)-enantiomer of BP-7,8-diol is epoxidized to (+)-*anti*-diolepoxide by peroxyl radicals and P-450 whereas the (+)-enantiomer of BP-7,8-diol is oxidized to (−)-*anti*-diolepoxide by peroxyl radicals and to (+)-*syn*-diolepoxide by P-450. Tetraol hydrolysis products of the diolepoxides are readily separable by HPLC, which provides a convenient method for distinguishing epoxidation by peroxyl radicals or the iron-oxo complex of P-450.[136] Starting with racemic BP-7,8-diol, *anti*/*syn* ratios of approximately 2.5 or greater are indicative of a peroxyl

TABLE IV
Epoxidation of (±)-BP-7,8-diol by Uninduced Rat Liver Microsomes[a]

Treatment[b]	% tetraol formation	% unknown 1 formation	% unknown 2 formation	TBA reactive materia	Anti/syn ratio
NADPH, ADP/Fe^{3+}, EDTA/Fe^{2+}	19 ± 4.4	N.D.[c]	N.D.	14 ± 1.3	2.5 ± 0.4
NADPH	0.6 ± 0.2	1.2 ± 0.3	1.9 ± 0.7	0.3 ± 0.1	1.1 ± 0.3
NADPH + metyrapone	N.D.	N.D.	N.D.	N.D.	—
NADPH (boiled microsomes)[d]	N.D.	N.D.	N.D.	N.D.	—
Cumene hydroperoxide	15 ± 2.7	N.D.	N.D.	18 ± 1.9	5.0 ± 1.4
Cumene hydroperoxide + butylated hydroxyanisole	2.3 ± 0.5	1.0 ± 0.8	1.3 ± 0.7	3.6 ± 0.6	1.1 ± 0.4
Cumene hydroperoxide + metyrapone	0.1 ± 0.1	N.D.	N.D.	1.0 ± 0.2	—
Cumene hydroperoxide (boiled microsomes)	N.D.	N.D.	N.D.	0.6 ± 0.3	—

[a] Data from Ref. 137.
[b] Experimental conditions: 0.5 mg microsomal protein was combined with 40 μM BP-7,8-diol. Oxidation was initiated by either 1 mM NADPH, 100 μM cumene hydroperoxide, or by the addition of 1 mM NADPH, 4.0 mM ADP, 15 μM Fe^{3+}, 100 μM EDTA, and 110 μM Fe^{2+}. Other additions included 10 μM butylated hydroxyanisole and 1 mM metyrapone.
[c] N.D., not detected.
[d] Microsomes heated to 90°C for 3 min before incubation.

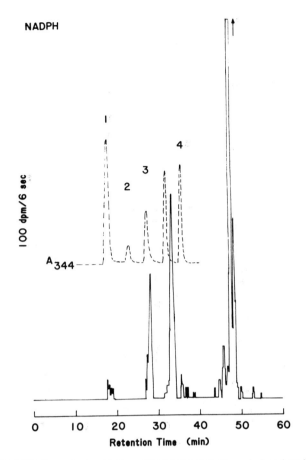

FIGURE 14. NADPH-dependent oxidation of (±)-BP-7,8-diol in uninduced rat liver microsomes. Profile of radioactive metabolites eluting during reverse-phase HPLC. (Inset) UV profile of tetraol standards: (1) *trans-anti,* (2) *trans-syn*, (3) *cis-anti*, (4) *cis-syn*. Redrawn from Ref. 137.

radical-mediated process whereas *anti/syn* ratios of 1 are characteristic of a P-450.[136] More dramatic differences can be detected using (+)-BP-7,8-diol.[136]

Microsomal preparations containing P-450 utilize cumene hydroperoxide to support substrate oxidation.[5-7] Cumene hydroperoxide also initiates lipid peroxidation in liver microsomes in a P-450-dependent process. There exists, therefore, the possibility that peroxyl radicals generated during lipid peroxidation contribute to P-450-mediated hydroperoxide-dependent oxidations in liver microsomes. This possibility was

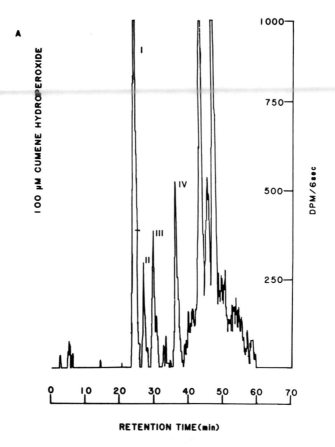

FIGURE 15. Cumene hydroperoxide-dependent oxidation of BP-7,8-diol in uninduced rat liver microsomes. Profiles of radioactive metabolites eluting during reverse-phase HPLC. (A) Effect of 100 μM butylated hydroxyanisole on cumene hydroperoxide-dependent oxidation. Peaks identified by Roman numerals are *trans-anti*-tetraol (I), *trans-syn*-tetraol (II), *cis-anti*-tetraol (III), and *cis-syn*-tetraol (IV). Position of tetraol elution is indicated by arrows in (B).

assessed in uninduced rat liver microsomes using the stereochemistry of BP-7,8-diol epoxidation as a probe for peroxyl radical generation.[137] The results are presented in Table IV. As a control, microsomes were treated with a combination of NADPH and chelated iron, a system shown to initiate lipid peroxidation.[138] The extent of lipid peroxidation was measured by the thiobarbituric acid (TBA) assay.[139] Oxidation was monitored by following conversion of (±)-[^{14}C]-BP-7,8-diol to oxygenated products that were separated by reverse-phase HPLC. The *anti/syn* ratio showed

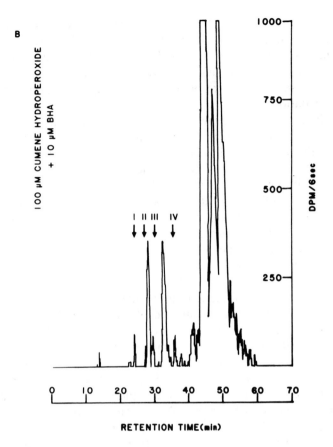

FIGURE 15. (*continued*)

a predominance of *anti*-diolepoxide-derived tetraols, indicating peroxyl radical generation. NADPH in the absence of iron chelates does not trigger lipid peroxidation in microsomes and diolepoxide formation is 30-fold lower than when lipid peroxidation is induced. Further, tetraol formation accounts for only 10% of total metabolism. Two unidentified products (unknowns 1 and 2) appear (Fig. 14) as the major metabolites. Although characterization of these products has not been completed, UV spectra are suggestive of metabolism in the pyrene moiety of the molecule. Formation of a phenol or K-region epoxide seems likely. Formation of nondiolepoxide-derived products from BP-7,8-diol by uninduced rat liver microsomes has been reported elsewhere.[134] The production of both unknowns is an NADPH-dependent P-450-catalyzed reaction since it is inhibited by metyrapone, boiling, or carbon monoxide treatment. Unknowns 1 and 2 are not detected in peroxyl radical-dependent reactions. Thus, in addition to the stereochemistry of BP-7,8-diol epoxidation, the formation of these unidentified products provides a complementary method to assess the nature of the oxidizing agent responsible for BP-7,8-diol oxidation.

Addition of 100 μM cumene hydroperoxide initiates a pattern of oxidation that parallels that of NADPH plus chelated iron apparently due to the fact that cumene hydroperoxide initiates lipid peroxidation. An *anti/syn* ratio of 5.0 correlates to the high level of lipid peroxidation and is consistent with peroxyl radical formation. Failure to detect unknowns 1 and 2 provides additional evidence that P-450 is not the oxidant. Addition of 10 μM butylated hydroxyanisole to cumene hydroperoxide-supplemented microsomes eliminates lipid peroxidation but only reduces the level of BP-7,8-diol oxidation by 50%. However, the metabolite profile changes dramatically (Fig. 15).

These results reveal a dichotomy in the action of P-450 on cumene hydroperoxide. In the absence of antioxidants, the major products arise as a result of one-electron reduction of cumene hydroperoxide to the cumyloxyl radical which initiates lipid peroxidation. In this case, peroxyl radicals derived from unsaturated lipid are the epoxidizing agents. In the presence of the antioxidant butylated hydroxyanisole, the products suggest the enzyme reduces cumene hydroperoxide by two electrons to cumenol while generating a ferryl-oxo complex analogous to the oxidizing agent produced from NADPH and O_2. It seems unlikely that the chemistry of the reaction of P-450 with cumene hydroperoxide changes under the influence of an antioxidant, although there is such a precedent for the effect of imidazole on hematin-dependent oxygenations by cumene hydroperoxide.[140,141] Rather, butylated hydroxyanisole may be altering the balance of two competing reactions. Lipid peroxidation triggered by one-electron reduction of cumene hydroperoxide could rapidly inactivate P-450 and mask ferryl-oxo-dependent epoxidation with huge concentrations of peroxyl radicals. Inhibiting lipid peroxidation by adding butylated hydroxyanisole enables numerous turnovers of the ferryl-oxo pathway of P-450 to occur while lowering the background of epoxidation by peroxyl radicals. Nevertheless, one is left with the dilemma that one- and two-electron reduction of cumene hydroperoxide occurs simultaneously in rat liver microsomes and is dependent on P-450. Homolytic and heterolytic scission of hydroperoxides has been reported for simple heme complexes but is dependent on the nature of the axial ligand or the leaving group ability of the alcohol product of hydroperoxide heterolysis.[142-144] Of course, in the present case, we cannot ignore the possibility that different

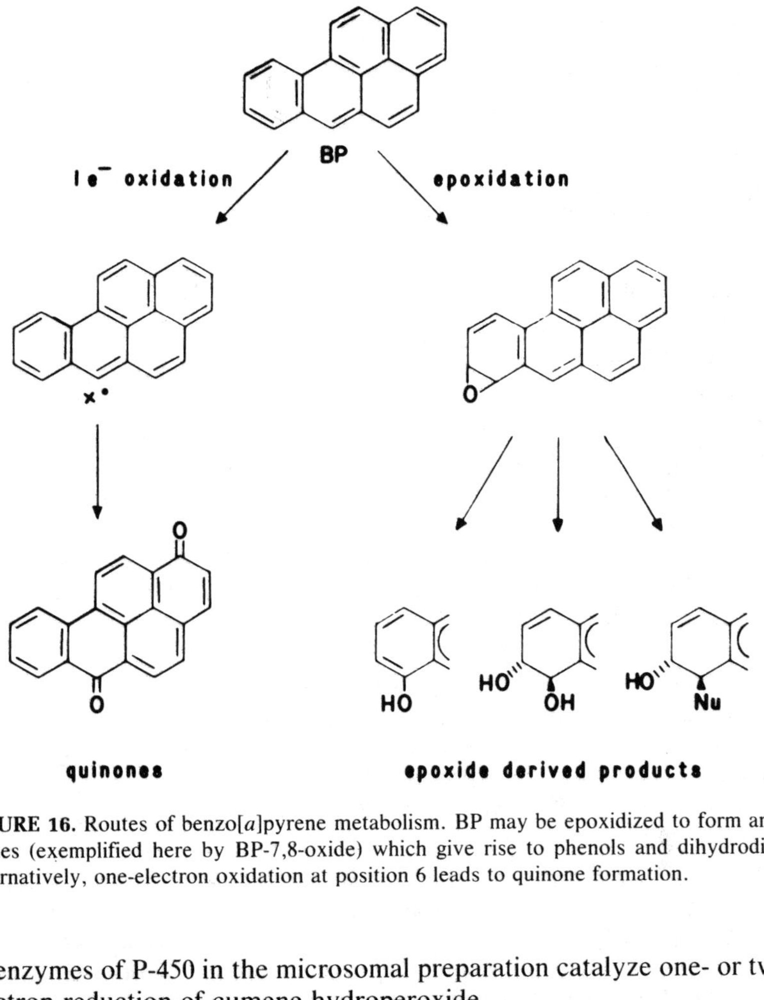

FIGURE 16. Routes of benzo[a]pyrene metabolism. BP may be epoxidized to form arene oxides (exemplified here by BP-7,8-oxide) which give rise to phenols and dihydrodiols. Alternatively, one-electron oxidation at position 6 leads to quinone formation.

isoenzymes of P-450 in the microsomal preparation catalyze one- or two-electron reduction of cumene hydroperoxide.

Our observations on the cumene hydroperoxide-dependent epoxidation of BP-7,8-diol by rat liver microsomes are reminiscent of the differences observed in benzo[a]pyrene (BP) oxygenation by rat liver microsomes supported by NADPH or cumene hydroperoxide.[145] The major products of NADPH-dependent BP oxidation are phenols and dihydrodiols (70%), both of which are secondary metabolites of arene oxides (Fig. 16).[146] Quinones, nonenzymatic oxidation products of 6-hydroxy-BP, are also detected (30%). 6-Hydroxy-BP is a product of one-electron oxidation or radical reactions and is not derived from an arene oxide.[147] When BP oxidation is supported by cumene hydroperoxide, quinones account for 93% of the products. Furthermore, the quinone oxygens are derived primarily from O_2, not cumene hydroperoxide.[148] It seems likely that cumene hydroperoxide-initiated lipid peroxidation is responsible for the dramatic change in BP oxygenation. Indeed, ascorbate-induced lipid peroxidation causes BP oxidation to quinones.[149]

7. Mechanistic Speculation

The previous sections have compared some of the major features of hydroperoxide metabolism by peroxidases and P-450. To recapitulate, peroxidases (HRP, cytochrome c peroxidase, chloroperoxidase, catalase) react rapidly and stoichiometrically with hydroperoxides to produce alcohol and a higher oxidation state of the peroxidase. The higher oxidation states can be characterized spectrally as containing a ferryl-oxo derivative of the heme and an oxidizing equivalent in the porphyrin or in a protein residue. The initial spectral intermediate is the same regardless of the hydroperoxide. Reduction of the higher oxidation state to resting enzyme requires two electrons and is usually, but not always, stepwise. In the case of HRP, reduction of Compound II is the rate-limiting step in the overall catalytic cycle. In contrast, P-450 and some other hemeproteins react slowly with excess hydroperoxide to generate higher oxidation states that do not resemble spectrally those of peroxidases. The initial oxidation product of liver microsomal P-450 appears to contain the organic functionality of the hydroperoxide. In addition, alcohol is not always the sole product of hydroperoxide reduction as it is with peroxidases. P-450 oxidizes typical peroxidase substrates but also catalyzes aliphatic hydroxylation and aromatic epoxidation, reactions that reveal its strength as an oxidant. The turnover numbers for all these oxidations are similar and are several orders of magnitude lower than the turnover numbers of peroxidases for their substrates. It appears that formation of the higher oxidation state rather than its reduction is rate-limiting for P-450.

It is difficult to overestimate the importance that the peroxidase analogy has had on the evolution of the concepts of the "active oxygen" of P-450. Yet most of the data summarized in this chapter highlight the weakness of the analogy. In fact, were it not for the fact that P-450 catalyzes hydroperoxide-dependent aliphatic hydroxylation and olefin epoxidation, it would be little different from hemeproteins such as metmyoglobin that are not considered true peroxidases. Since most of the peroxidases and hemeproteins discussed in this chapter have the same prosthetic group, ferriprotoporphyrin IX, can we understand how the proteins alter reaction of heme with hydroperoxides?

Recent studies of the reaction of (tetraphenylporphyrin)iron chloride with hydroperoxides suggest that two mechanisms are operative.[143] With hydroperoxides containing good leaving groups (e.g., peracids), two-electron reduction predominates:

$$P-Fe^{III} + ROOH \rightarrow P-Fe^{V}=O + ROH \tag{25}$$

However, when the leaving group is poor ($pK_a \geq 10$), one-electron reduction seems to be more important:

$$P-Fe^{III} + ROOH \rightarrow P-Fe^{IV}-OH + RO^\bullet \tag{26}$$

The key question to answer then is how do the protein components of peroxidases make two-electron reduction (heterolysis) of H_2O_2 and simple alkyl hydroperoxides ($pK_a \sim 15-16$) so facile?

An answer to this question has been proposed by Poulos and Kraut[150] and is based on the identification of groups at the active sites of cytochrome c peroxidase. The distal histidine (His-52) is well positioned to participate in general base–general acid catalysis, thereby transferring the hydroperoxide proton to the internal peroxide oxygen (leaving group). In addition, Arg-48 is perfectly located to stabilize the negative charge that develops in the transition state for heterolysis (Fig. 17). Juxtaposition of the pendant His and Arg residues greatly facilitates heterolysis resulting in two-electron heme oxidation. His and Arg residues corresponding to those of cytochrome c peroxidase are also found in HRP and turnip peroxidase although the crystal structures of these proteins are not yet available. Alternatively, the role of the distal His may be played by a carboxylate ion (Asp) in HRP. The Arg residue is proposed to play a critical role in stabilizing the transition state for heterolysis and such a suitably positioned residue is absent in metmyoglobin and P-450$_{cam}$, proteins that exhibit poor peroxidase activity. In fact, P-450$_{cam}$ also appears to lack the distal His and the environment of its heme is distinctly nonpolar. Thus, one might anticipate that heterolytic peroxide cleavage would be an inefficient reaction for the latter proteins. However, beef liver catalase, one of the most efficient of all enzymes and a heterolytic peroxide reductant, does not contain an Arg on the distal side of the heme pocket. It does have a His located similarly to those in cytochrome c peroxidase and metmyoglobin as well as an Asn that may H-bond to the internal peroxide oxygen. Furthermore, catalase possesses a Tyr residue as the fifth ligand to heme which, if ionized, could substantially increase the reducing ability of the metal center. The latter point suggests, as noted by Poulos and Kraut, that one must also consider the effect of ligands coordinated to the fifth position of the heme on its chemistry. Indeed, simple heme complexes are more efficient catalysts of hydroperoxide-dependent olefin

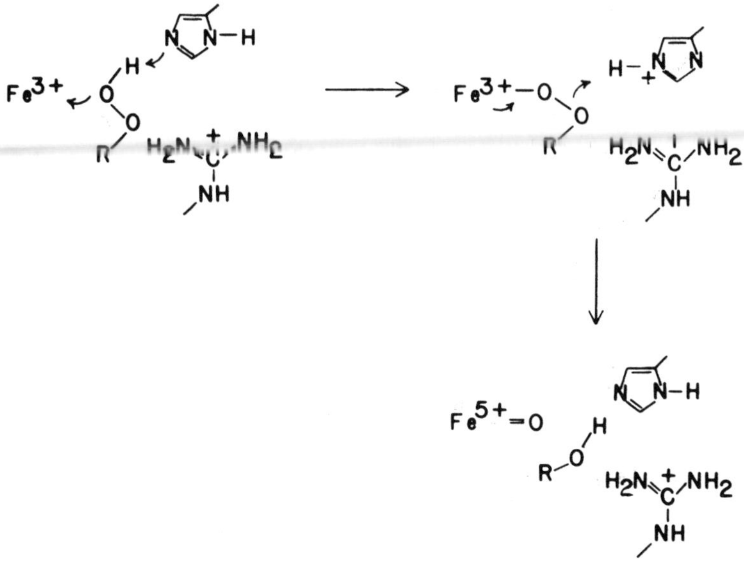

FIGURE 17. Steps in the catalytic reduction of hydroperoxides by cytochrome c peroxidase. According to Ref. 150.

epoxidation when they contain coordinated imidazole.[141] P-450 contains thiolate as the fifth ligand which one anticipates would significantly stabilize higher oxidation states of the heme and possibly enhance heterolytic peroxide reduction. However, it is possible that this ligand would also enhance homolytic cleavage as suggested by White and Coon[9]:

$$\text{CyS}^--\text{Fe}^{3+} + \text{H--O--OR} \rightarrow \text{CyS·--Fe}^{3+}-\text{OH} + \cdot\text{OR} \qquad (27)$$

It is tempting to categorize peroxidases and P-450 as two-electron and one-electron peroxide reductants, respectively. However, such an oversimplification is not supported by all of the data. Under certain conditions, P-450 catalyzes hydroperoxide-dependent reactions that are clearly distinct from one-electron chemistry, that are analogous to NADPH-supported reactions, and that are most easily explained as "oxene-type" chemistry. For example, although P-450 catalyzes reactions initiated by homolysis of cumene hydroperoxide, it is possible to detect aromatic epoxidation when the reactions are carried out in the presence of antioxidants.[137] It is difficult to rationalize olefin epoxidation by one-electron hydroperoxide reduction intermediates. Theoretically, an alkoxyl or hydroxyl radical could abstract an electron from the π-bond to form a radical cation that could be trapped by the ferryl–oxo complex and decompose to epoxide. However, the transfer of an electron from a simple olefin to an alkoxyl radical seems quite unfavorable and it is difficult to understand how the overall epoxidation could be stereospecific. Olefin epoxidations are more readily explained by invoking oxidizing agents—ferryl-oxo complexes—that contain both oxidizing equivalents of the peroxide.

At present then, it seems there is evidence implicating P-450 as a one- and a two-electron reductant of peroxide. The electronic factors that govern this dichotomy are unknown and await experimental elucidation. In the meantime, the frustrated investigator, whose results are inconsistent with the peroxidase paradigm, may take solace in macromolecular anthropomorphism. DNA may be selfish, but P-450 appears to be schizophrenic.

ACKNOWLEDGMENTS. We are indebted to Brian Dunford and Paul Ortiz de Montellano for helpful discussions and to Paul Schaap and Tokuji Kimura for providing data prior to publication. Work in the principal investigator's laboratory has been performed by a number of excellent graduate students and postdoctoral associates and supported by the American Cancer Society and National Institute of General Medical Sciences. L.J.M. is a recipient of a Faculty Research Award from the American Cancer Society (FRA 243).

References

1. Saunders, B. C., Holmes-Siedel, A. G., and Stark, B. P., 1964, *Peroxidases*, Butterworths, London.
2. Dunford, H. B., and Stillman, J. S., 1976, On the function and mechanism of action of peroxidases, *Coord. Chem. Rev.* **19**:187–251.
3. Jones, P., Mantle, D., Davies, D. M., and Kelly, H. C., 1977, Hydroperoxidase activities of ferrihemes: Heme analogues of peroxidase intermediates, *Biochemistry* **16**:3974–3978.
4. Portsmouth, D., and Beal, E. A., 1971, The peroxidase activity of deuterohemin, *Eur. J. Biochem.* **19**:479–487.
5. Hrycay, E. G., and O'Brien, P. J., 1972, Cytochrome P-450 as a microsomal peroxidase in steroid hydroperoxide reduction, *Arch. Biochem. Biophys.* **153**:480–494.
6. Kadlubar, F. F., Morton, K. C., and Ziegler, D. M., 1973, Microsomal-catalyzed hydroperoxide-dependent C-oxidation of amines, *Biochem. Biophys. Res. Commun.* **54**:1255–1260.
7. Rahimtula, A. D., and O'Brien, P. J., 1974, Hydroperoxide catalyzed liver microsomal aromatic hydroxylation reactions involving cytochrome P-450, *Biochem. Biophys. Res. Commun.* **60**:440–447.
8. Jones, P., and Wilson, I., 1978, Catalase and iron complexes with catalase-like prop-

erties, in: *Metal Ions in Biological Systems* (H. Segel, ed.), Dekker, New York, pp. 185–240.
9. White, R. E., and Coon, M. J., 1980, Oxygen activation by cytochrome P-450, *Annu. Rev. Biochem.* **49**:315–356.
10. Coon, M. J., and White, R. E., 1980, Cytochrome P-450, a versatile catalyst in monooxygenation reactions, in: *Dioxygen Binding and Activation by Metal Complexes* (T. G. Spiro, ed.), Wiley, New York, pp. 73–123.
11. Saunders, B. C., 1973, Peroxidase and catalase, in: *Inorganic Biochemistry*, Volumes 1 and 2 (G. L. Eichorn, ed.), Elsevier, Amsterdam, pp. 988–1021.
12. Keilin, D., 1966, *The History of Cell Respiration and Cytochrome*, Cambridge University Press, London.
13. Schonbaum, G. R., and Chance, B., 1976, Catalase, in: *The Enzymes*, Volume 13 (P. Boyer, ed.), Academic Press, New York, pp. 363–408.
14. Dunford, H. B., 1982, Peroxidases, *Adv. Inorg. Biochem.* **4**:41–68.
15. Yonetani, T., 1976, Cytochrome c peroxidase, in: *The Enzymes*, Volume 13 (P. Boyer, ed.), Academic Press, New York, pp. 345–362.
16. Yamazaki, I., 1974, Peroxidase, in: *Molecular Mechanisms of Oxygen Activation* (O. Hayaishi, ed.), Academic Press, New York, pp. 535–558.
17. Morrison, M., and Schonbaum, G. R., 1976, Peroxidase-catalyzed halogenation, *Annu. Rev. Biochem* **45**:861–888.
18. Chance, B., Powers, L., Ching, Y., Poulos, T., Schonbaum, G. R., Yamazaki, I., and Paul, K. G., 1984, X-ray absorption studies of intermediates in peroxidase activity, *Arch. Biochem. Biophys.* **235**:596–611.
19. Schonbaum, G. R., and Lo, S., 1972, Interaction of peroxidases with aromatic peracids and alkyl peroxides: Product analyses, *J. Biol. Chem.* **247**:3353–3360.
20. Hewson, W. D., and Dunford, H. B., 1976, Stoichiometry of the reaction between horseradish peroxidase and p-cresol, *J. Biol. Chem.* **251**:6043–6052.
21. Kedderis, G. L., Koop, D. R., and Hollenberg, P. F., 1980, N-Demethylation reactions catalyzed by chloroperoxidase, *J. Biol. Chem.* **255**:10174–10182.
22. Kedderis, G. L., and Hollenberg, P. F., 1983, Characterization of the N-demethylations catalyzed by horseradish peroxidase, *J. Biol. Chem.* **258**:8129–8138.
23. Portoghese, P. S., Svanborg, K., and Samuelsson, B., 1975, Oxidation of oxyphenylbutazone by sheep vesicular gland microsomes and lipoxygenase, *Biochem. Biophys. Res. Commun.* **63**:748–755.
24. Marnett, L. J., Bienkowski, M. J., Pagels, W. R., and Reed, G. A., 1980, Mechanism of xenobiotic cooxygenation coupled to prostaglandin H_2 biosynthesis, in: *Advances in Prostaglandin and Thromboxane Research, Volume 6* (B. Samuelsson, P. W. Ramwell, and R. Paoletti, eds.), Raven Press, New York, pp. 149–151.
25. Roman, R., and Dunford, H. B., 1972, pH dependence on the oxidation of iodide by compound I of horseradish peroxidase, *Biochemistry* **11**:2076–2082.
26. Bjorksten, F., 1970, The horseradish peroxidase-catalyzed oxidation of iodide: Outline of the mechanism, *Biochim. Biophys. Acta* **212**:396–406.
27. Roman, R., and Dunford, H. B., 1973, Studies on horseradish peroxidase. XII. A kinetic study on the oxidation of sulfite and nitrite by compounds I and II, *Can. J. Chem.* **51**:588–596.
28. Arcaiso, T., Miyoshi, K., and Yamazaki, I., 1976, Mechanism of electron transport from sulfite to horseradish peroxidase compounds, *Biochemistry* **15**:3059–3063.
29. George, P., 1952, Redox reactions of catalase intermediate compounds and a new "peroxidase" role for catalase, *Biochem. J.* **52**:XIX.
30. Stern, K. G., 1936, On the mechanism of enzyme action: A study of the decomposition of monoethyl hydrogen peroxide by catalase and of an intermediate enzyme substrate compound, *J. Biol. Chem.* **114**:473–494.

31. Thomas, J. A., Morris, D. A., and Hager, L. P., 1970, Chloroperoxidase. VIII. Formation of peroxide and halide complexes and their relation to the mechanism of the halogenation reaction, *J. Biol. Chem.* **245**:3135–3142.
32. Courtin, F., Deme, D., Verion, A., Michot, J. L., Pommier, J., and Nunez, J., 1982, The role of lactoperoxidase–H_2O_2 compounds in the catalysis of thyroglobulin iodination and thyroid hormone synthesis, *Eur. J. Biochem.* **124**:603–609.
33. Deme, D., Pommier, J., and Nunez, J., 1978, Specificity of thyroid hormone synthesis: The role of thyroid peroxidase, *Biochim. Biophys. Acta* **540**:73–82.
34. Dunford, H. B., and Ralston, I. M., 1983, On the mechanism of iodination of tyrosine, *Biochem. Biophys. Res. Commun.* **116**:639–643.
35. Ohki, S., Ogino, N., Yamamato, S., and Hayaishi, O., 1979, Prostaglandin hydroperoxidase, an integral part of prostaglandin endoperoxide synthetase from bovine vesicular gland microsomes, *J. Biol. Chem.* **254**:829–836.
36. Theorell, H., 1941, Crystalline peroxidase, *Enzymologia* **10**:250–252.
37. Chance, B., 1943, The kinetics of the enzyme–substrate compound of peroxidase, *J. Biol. Chem.* **151**:553–577.
38. George, P., 1953, Intermediate compound formation with peroxidase and strong oxidizing agents, *J. Biol. Chem.* **201**:413–426.
39. Dolphin, D., and Felton, R. H., 1974, The biochemical significance of porphyrin π cation radicals, *Acc. Chem. Res.* **7**:26–32.
40. LaMar, G. N., de Ropp, J. S., Smith, K. M., and Langry, K. C., 1981, Proton nuclear magnetic resonance investigation of the electronic structure of compound I of horseradish peroxidase, *J. Biol. Chem.* **256**:237–243.
41. Schulz, C. E., Rutter, R., Sage, J. T., DeBrunner, P. G., and Hager, L. P., 1984, Mossbauer and electron paramagnetic resonance studies of horseradish peroxidase and its catalytic intermediates, *Biochemistry* **23**:4743–4754.
42. Felton, R. H., Romans, A. Y., Yu, N. T., and Schonbaum, G. R., 1976, Laser Raman spectra of oxidized hydroperoxidases, *Biochim. Biophys. Acta* **434**:82–89.
43. Dunford, H. B., and Nadezhdin, A. D., 1982, On the past eight years of peroxidase research, in: *Oxidases and Related Redox Systems* (T. E. King, H. S. Mason, and M. Morrison, eds.), Pergamon Press, Elmsford, N.Y., pp. 653–670.
44. George, P., 1952, The specific reactions of iron in some hemoproteins, *Adv. Catal.* **4**:367–428.
45. George, P., 1953, The chemical nature of the second hydrogen peroxide compound formed by cytochrome c peroxidase and horseradish peroxidase. 1. Titration with reducing agents, *Biochem. J.* **54**:267–276.
46. George, P., 1953, The chemical nature of the second hydrogen peroxide compound formed by cytochrome c peroxidase and horseradish peroxidase. 2. Formation and decomposition, *Biochem. J.* **55**:220–230.
47. Hayashi, Y., and Yamazaki, I., 1979, The oxidation–reduction potentials of compound I/compound II and compound II/ferric couples of horseradish peroxidases A_2 and C, *J. Biol. Chem.* **254**:9101–9106.
48. Schonbaum, G. R., and Chance, B., 1976, Catalase, in: *The Enzymes*, Volume 13 (P. Boyer, ed.), Academic Press, New York, pp. 363–408.
49. Browett, W. R., and Stillman, M. J., 1981, Evidence for heme π cation radical species in compound I of horseradish peroxidase and catalase, *Biochim. Biophys. Acta* **660**:1–7.
50. Araiso, T., Ronnenberg, M., Dunford, H. B., and Ellfolk, N., 1980, The formation of the primary compound from hydrogen peroxide and *Pseudomonas* cytochrome c peroxidase, *FEBS Lett.* **118**:99–102.
51. Chance, B., 1949, The properties of the enzyme substrate compounds of horseradish and lactoperoxidase, *Science* **109**:204–208.

52. Maguire, R. J., Dunford, H. B., and Morrison, M., 1971, The kinetics of the formation of the primary lactoperoxidase-hydrogen peroxide compound, *Can. J. Biochem.* **49**:1165–1171.
53. Palcic, M. M., Rutter, R., Araiso, T., Hager, L. P., and Dunford, H. B., 1980, Spectrum of chloroperoxidase compound I, *Biochem. Biophys. Res. Commun.* **94**:1123–1127.
54. Harrison, J. R., Araiso, T., Palcic, M. M., and Dunford, H. B., 1980, Compound I of myeloperoxidase, *Biochem. Biophys. Res Commun.* **94**:34–40.
55. George, P., and Irvine, D. H., 1954, Reaction of metmyoglobin with strong oxidizing agents, *Biochem. J.* **58**:188–195.
56. King, N. K., and Winfield, M. E., 1963, The mechanism of myoglobin oxidation, *J. Biol. Chem.* **238**:1520–1528.
57. Blake, R. C., II, and Coon, M. J., 1980, On the mechanism of action of cytochrome P-450: Spectral intermediates in the reaction of P-450LM$_2$ with peroxy compounds, *J. Biol. Chem.* **255**:4100–4111.
58. Wagner, G. C., Palcic, M. M., and Dunford, H. B., 1983, Absorption spectra of cytochrome P-450cam in the reaction with peroxy acids, *FEBS Lett.* **156**:244–248.
59. Rutter, R., Valentine, M., Hendrich, M. P., Hager, L. P., and Debrunner, P. G., 1983, Chemical nature of the porphyrin π cation radical in horseradish peroxidase compound I, *Biochemistry* **22**:4769–4774.
60. Gouterman, M., 1961, Spectra of porphyrins, *J. Mol. Spectrosc.* **6**:138–163.
61. Yonetani, T., 1965, Stoichiometry between enzyme, H_2O_2, and ferrocytochrome c, and enzymic determination of extinction coefficients of cytochrome c, *J. Biol. Chem.* **240**:4509–4514.
62. Yonetani, T., 1966, Cytochrome c peroxidase. IV. A comparison of peroxide-induced complexes of horseradish and cytochrome c peroxidases, *J. Biol. Chem.* **241**:2562–2571.
63. Yonetani, T., Schleyer, H., and Ehraenberg, A., 1966, Cytochrome c peroxidase. VII. Electron paramagnetic resonance absorptions of the enzyme and complex ES in dissolved and crystalline forms, *J. Biol. Chem.* **241**:3240–3243.
64. Coulson, A. F. W., Erman, J. E., and Yonetani, T., 1971, Cytochrome c peroxidase, XVII. Stoichiometry and mechanism of the reaction of compound ES with donors, *J. Biol. Chem.* **246**:9117–9124.
65. Van der Ouderaa, F. J., Buytenhek, M., Nugteren, D. H., and Van Dorp, D. A., 1977, Purification and characterization of prostaglandin endoperoxide synthetase from sheep vesicular glands, *Biochim. Biophys. Acta* **487**:315–331.
66. Hemler, M. E., and Lands, W. E. M., 1980, Protection of cyclooxygenase activity during heme-induced destabilization, *Arch. Biochem. Biophys.* **201**:586–593.
67. Weller, P., Hollenberg, P., and Marnett, L. J., manuscript in preparation.
68. Schultz, J., and Schmuckler, H. W., 1964, Myeloperoxidase of the leukocyte of normal human blood. II. Isolation, spectrophotometry, and amino acid analysis, *Biochemistry* **3**:1234–1238.
69. Harrison, J. E., and Schultz, J., 1978, Myeloperoxidase: Confirmation and nature of heme-binding inequivalence. Resolution of a carbonyl-substituted heme, *Biochim. Biophys. Acta* **536**:341–349.
70. Ellfolk, N., and Soininen, R., 1971, *Pseudomonas* cytochrome c peroxidase. III. The size and shape of the enzyme molecule, *Acta Chem. Scand.* **25**:1535–1540.
71. Soininen, R., Ellfolk, N., and Kalkkinen, N., 1973, *Pseudomonas* cytochrome c peroxidase. IX. Molecular weight of the enzyme in dodecyl sulfate–polyacrylamide gel electrophoresis, *Acta Chem. Scand.* **27**:1106–1107.

72. Ellfolk, N., Ronnberg, M., Aasa, R., Andreasson, L. E., and Vanngard, T., 1983, Properties and function of the two hemes in *Pseudomonas* cytochrome c peroxidase, *Biochim. Biophys. Acta* **743**:23–30.
73. Ronnberg, M., Araiso, T., Ellfolk, N., and Dunford, H. B., 1981, The catalytic mechanism of *Pseudomonas* cytochrome c peroxidase, *Arch. Biochem. Biophys.* **207**:197–204.
74. Ronnberg, M., Lambeir, A.-M., Ellfolk, N., and Dunford, H. B., 1985, A rapid-scan spectrometric and stopped-flow study of compound I and compound II of *Pseudomonas* cytochrome c peroxidase, *Arch. Biochem. Biophys.* **236**:714–719.
75. George, P., and Irvine, D. H., 1952, Reaction between metmyoglobin and hydrogen peroxide, *Biochem. J.* **52**:511–517.
76. George, P., and Irvine, D. H., 1955, A possible structure for the higher oxidation state of metmyoglobin, *Biochem. J.* **60**:596–604.
77. George, P., and Irvine, D. H., 1956, A kinetic study of the reaction between ferrimyoglobin and hydrogen peroxide, *J. Colloid Sci.* **11**:327–339.
78. Blake, R. C., II, and Coon, M. J., 1981, On the mechanism of action of cytochrome P-450. Role of peroxy spectral intermediates in substrate hydroxylation, *J. Biol. Chem.* **256**:5755–5763.
79. Hasinoff, B. B., and Dunford, H. B., 1970, The kinetics of oxidation of ferrocyanide by horseradish peroxidase compounds I and II, *Biochemistry* **9**:4930–4939.
80. Cotton, M. L., and Dunford, H. B., 1973, Studies on horseradish peroxidase. XI. On the nature of compounds I and II as determined from the kinetics of the oxidation of ferrocyanide, *Can. J. Chem.* **51**:582–587.
81. Dunford, H. B., and Cotton, M. L., 1975, Kinetics of the oxidation of p-aminobenzoic acid catalyzed by horseradish peroxidase compounds I and II, *J. Biol. Chem.* **250**:2920–2932.
82. Santimone, M., 1975, The mechanism of ferrocytochrome c oxidation by horseradish isoperoxidase, *Biochimie* **57**:91–96.
83. Critchlow, J. E., and Dunford, H. B., 1972, Studies on horseradish peroxidase. IX. Kinetics of the oxidation of p-cresol by compound II, *J. Biol. Chem.* **247**:8703–8713.
84. Morishima, I., and Ogawa, S., 1979, Nuclear magnetic resonance studies on hemoproteins, *J. Biol. Chem.* **254**:2814–2820.
85. Leigh, J. S., Maltempo, M. M., Ohlsson, P. I., and Paul, K. G., 1975, Optical, NMR, and EPR properties of horseradish peroxidase and its donor complexes, *FEBS Lett.* **51**:304–308.
86. Schejter, A., Laner, A., and Epstein, N., 1976, Binding of hydrogen donors to horseradish peroxidase: A spectroscopic study, *Arch. Biochem. Biophys.* **174**:36–44.
87. Burns, P. S., Williams, R. J. P., and Wright, P. E., 1975, Conformational studies of peroxidase–substrate complexes: Structure of indolepropionic acid–horseradish peroxidase complex, *J. Chem. Soc. Chem. Commun.* **1975**:795–7967.
88. Critchlow, J. E., and Dunford, H. B., 1972, The use of transition state acid dissociation constants in pH-dependent enzyme kinetics, *J. Theor. Biol.* **37**:307–320.
89. Dunford, H. B., and Araiso, T., 1979, Horseradish peroxidase. XXXVI. On the difference between peroxidase and metmyoglobin, *Biochem. Biophys. Res. Commun.* **89**:764–768.
90. Jones, P., and Dunford, H. B., 1977, On the mechanism of compound I formation from peroxidases and catalases, *J. Theor. Biol.* **69**:457–470.
91. Yamada, H., and Yamazaki, I., 1974, Proton balance in conversions between five oxidation–reduction states of horseradish peroxidase, *Arch. Biochem. Biophys.* **165**:728–738.

92. Dolman, D., Newell, G. A., Thurlow, M. D., and Dunford, H. B., 1975, A kinetic study of the reaction of horseradish peroxidase with hydrogen peroxide, *Can. J. Biochem.* **53**:495–501.
93. Dunford, H. B., and Alberty, R. A., 1967, The kinetics of fluoride binding by ferric horseradish peroxidase, *Biochemistry* **6**:447–451.
94. Chance, B., 1949, The enzyme–substrate compounds of horseradish peroxidase and peroxides. II. Kinetics of formation and decomposition of the primary and secondary complexes, *Arch. Biochem. Biophys.* **22**:224–252.
95. Brill, A. S., 1966, Peroxidases and catalase, in: *Comprehensive Biochemistry*, Volume 14 (M. Florkin and E. H. Stotz, eds.), Elsevier, Amsterdam, pp. 447–479.
96. Dunford, H. B., Hewson, W. D., and Steiner, H., 1978, Horseradish peroxidase. XIX. Reactions in water and deuterium oxide: Cyanide binding, compound I formation and reactions of compound I and II with ferrocyanide, *Can. J. Chem.* **56**:2844–2852.
97. Hubbard, C. D., Dunford, H. B., and Hewson, W. D., 1975, Horseradish peroxidase. XVII. Reactions of compounds I and II with p-aminobenzoic acid in deuterium oxide, *Can. J. Chem.* **53**:1563–1569.
98. Roman, R., and Dunford, H. B., 1973, Studies on horseradish peroxidase. XII. A kinetic study of the oxidation of sulfite and nitrite by compounds I and II, *Can J. Chem.* **51**:588–596.
99. Ralston, I., and Dunford, H. B., 1978, Horseradish peroxidase. XXXII. pH dependence of the oxidation of L-(−)-tyrosine by compound I, *Can. J. Biochem.* **56**:1115–1119.
100. McCarthy, M.-B., and White, R. E., 1983, Functional differences between peroxidase compound I and the cytochrome P-450 reactive oxygen intermediate, *J. Biol. Chem.* **258**:9153–9158.
101. Nordblom, G. D., White, R. E., and Coon, M. J., 1976, Studies on hydroperoxide-dependent substrate hydroxylation by purified liver microsomal cytochrome P-450, *Arch. Biochem. Biophys.* **175**:524–533.
102. White, R. E., Sligar, S. G., and Coon, M. J., 1980, Evidence for a homolytic mechanism of peroxide oxygen–oxygen bond cleavage during substrate hydroxylation by cytochrome P-450, *J. Biol. Chem.* **255**:11108–11111.
103. Hiatt, R., 1971, Hydroperoxides, in: *Organic Peroxides*, Volume 2 (D. Swern, ed.), Wiley–Interscience, New York, pp. 1–152.
104. Gardner, H. W., and Plattner, R. D., 1984, Linoleate hydroperoxides are cleaved heterolytically into aldehydes by Lewis acid aprotic solvents, *Lipids* **19**:294–299.
105. Marnett, L. J., Siedlik, P. H., and Fung, L. W-M., 1982, Oxidation of phenidone and BW755c by prostaglandin endoperoxide synthetase, *J. Biol. Chem.* **257**:6957–6964.
106. Lee, W. E., and Miller, D. W., 1966, The oxidation of pyrazolidone developing agents, *Photogr. Sci. Eng.* **10**:192–201.
107. Lasker, J. M., Sivarajah, R., Mason, R. P., Kalyanaraman, B., Abou-Donia, M. B., and Eling, T. E., 1981, A free radical mechanism of prostaglandin synthase-dependent aminopyrine demethylation, *J. Biol. Chem.* **256**:7764–7767.
108. Egan, R. W., Gale, P. H., Vanden Heuvel, W. J. A., Baptista, E. M., and Kuehl, F. A., 1980, Mechanism of oxygen transfer by prostaglandin hydroperoxidase, *J. Biol. Chem.* **255**:323–326.
109. Egan, R. W., Gale, P. H., Baptista, E. M., Kennicott, K. L., Vanden Heuvel, W. J. A., Walker, R. W., Fagerness, P. E., and Kuehl, F. A., 1981, Oxidation reactions by prostaglandin cyclooxygenase-hydroperoxidase, *J. Biol Chem.* **256**:7352–7361.
110. Watanabe, Y., Iyanagi, T., and Oae, S., 1982, One electron transfer mechanism in the enzymatic oxygenation of sulfoxide to sulfone promoted by a reconstituted system with purified cytochrome P-450, *Tetrahedron Lett.* **23**:533–536.
111. Ishimaru, A., and Yamazaki, I., 1977, Hydroperoxide-dependent hydroxylation in-

volving "H_2O_2-reducible hemoprotein" in microsomes of pea seeds, *J. Biol. Chem.* **252**:6118–6124.
112. Blee, E., Casida, J. E., and Durst, F., 1984, Oxidation metabolism of xenobiotics in higher plants: Sulfoxidation of mesurol by soybean cotyledon microsomes, in: *Ninth European Workshop on Drug Metabolism, Abstracts*, p. 208.
113. Rahimtula, A. D., O'Brien, P. J., Seifried, H. E., and Jerina, D. M., 1978, The mechanism of action of cytochrome P-450: Occurrence of the 'NIH shift' during hydroperoxide-dependent aromatic hydroxylations, *Eur. J. Biochem.* **89**:133–141.
114. Guengerich, F. P., and McDonald, T. A., 1984, Chemical mechanisms of catalysis by cytochromes P-450: A unified view, *Acc. Chem. Res.* **17**:9–16.
115. Burns, J. J., Rose, R. K., Goodwin, S., Reichtal, J., Horning, E. C., and Brodie, B. B., 1955, The metabolic fate of phenylbutazone (Butazolidine) in man, *J. Pharmacol.* **113**:481–489.
116. Marnett, L. J., 1984, Hydroperoxide-dependent oxidations during prostaglandin biosynthesis, in: *Free Radicals in Biology*, Volume 6 (W. A. Pryor, ed.), Academic Press, New York, pp. 63–94.
117. Marnett, L. J., and Reed, G. A., 1979, Peroxidatic oxidation of benzo[a]pyrene during prostaglandin biosynthesis, *Biochemistry* **18**:2923–2929.
118. Nastainczyk, W., Schuhn, D., and Ullrich, V., 1984, Spectral intermediates of prostaglandin hydroperoxidase, *Eur. J. Biochem.* **144**:381–385.
119. Marnett, L. J., and Eling, T. E., 1983, Cooxidation during prostaglandin biosynthesis: A pathway for the metabolic activation of xenobiotics, in: *Reviews in Biochemical Toxicology*, Volume 5 (E. Hodgson, J. R. Bend, and R. M. Philpot, eds.), Elsevier/North-Holland, Amsterdam, pp. 135–172.
120. Marnett, L. J., Bienkowski, M. J., and Pagels, W. R., 1979, Oxygen 18 investigation of prostaglandin synthetase-dependent co-oxidation of diphenylisobenzofuran, *J. Biol. Chem.* **254**:5077–5082.
121. Marnett, L. J., Johnson, J. T., and Bienkowski, M. J., 1979, Arachidonic acid-dependent metabolism of 7,8-dihydroxy-7,8-dihydrobenzo[a]pyrene by ram seminal vesicles. *FEBS Lett.* **106**:13–16.
122. Marnett, L. J., and Bienkowski, M. J., 1980, Hydroperoxide-dependent oxygenation of 7,8-dihydroxy-7,8-dihydrobenzo[a]pyrene by ram seminal vesicle microsomes: Source of the oxygen, *Biochem. Biophys. Res. Commun.* **96**:639–647.
123. Marnett, L. J., Reed, G. A., and Johnson, J. T., 1977, Prostaglandin synthetase dependent benzo[a]pyrene oxidation: Products of the oxidation and inhibition of their formation by antioxidants, *Biochem. Biophys. Res. Commun.* **79**:569–576.
124. Dix, T. A., and Marnett, L. J., 1981, Free radical epoxidation of 7,8-dihydroxy-7,8-dihydrobenzo[a]pyrene by hematin and polyunsaturated fatty acid hydroperoxides, *J. Am. Chem. Soc.* **103**:6744–6746.
125. Dix, T. A., and Marnett, L. J., 1983, Hematin-catalyzed rearrangement of hydroperoxy-linoleic acid to epoxy alcohols via an oxygen-rebound, *J. Am. Chem. Soc.* **105**:7001–7002.
126. Dix, T. A., Fontana, R., Panthani, A., and Marnett, L. J., 1985, Hematin catalyzed epoxidation of 7,8-dihydroxy-7,8-dihydrobenzo[a]pyrene by polyunsaturated fatty acid hydroperoxide, *J. Biol. Chem.* **260**:5358–5365.
127. Dix, T. A., and Marnett, L. J., 1985, Conversion of linoleic acid hydroperoxide to hydroxy, keto, epoxy hydroxy, and trihydroxy fatty acids by hematin, *J. Biol. Chem.* **260**:5351–5357.
128. Gardner, H. W., Weisleder, D., and Kleinman, R., 1978, Formation of trans-12,13-epoxy-9-hydroperoxy-trans-10-octadecadienoic acid from 13-L-hydroperoxy-cis-9-

trans-11-octadecadienoic acid catalyzed by either soybean extract or cysteine-FeCl$_3$, *Lipids* **13**:246–252.
129. Hamberg, M., 1975, Decomposition of unsaturated fatty acid hydroperoxides by hemoglobin-structures of major products of 13-L-hydroperoxy-9,11-octadecadienoic acid, *Lipids* **10**:87–92.
130. Dix, T. A., and Marnett, L. J., 1983, Metabolism of polycyclic aromatic hydrocarbon derivatives to ultimate carcinogens during lipid peroxidation, *Science* **221**:77–79.
131. Mahoney, L. R., Johnson, M. D., Korcek, S., Marnett, L. J., and Reed, G. A., 1982, Inhibition of aldehyde oxidation by polycyclic aromatic hydrocarbons, in: *Abstracts of Papers, 184th American Chemical Society Meeting, Division of Organic Chemistry*, No. 30, American Chemical Society, Washington, D.C., p. 30.
132. Reed, C. A., Brooks, E. A., and Eling, T. A., 1984, Phenylbutazone-dependent epoxidation of 7,8-dihydroxy-7,8-dihydrobenzo[a]pyrene—A new mechanism of prostaglandin H synthase-catalyzed oxidations, *J. Biol. Chem.* **259**:5591–5595.
133. Ortiz de Montellano, P. R., and Catalano, C. E., 1985, Epoxidation of styrene by hemoglobin and myoglobin: Transfer of oxidizing equivalents to the protein surface, *J. Biol. Chem.* **260**:9265–9271.
134. Thakker, D. R., Yagi, H., Akagi, H., Koreeda, M., Lu, A. Y. H., Levin, W., Wood, A. W., Conney, A. H., and Jerina, D. M., 1977, Metabolism of benzo[a]pyrene VI: Stereo-selective metabolism of benzo[a]pyrene 7,8 dihydrodiol to diol epoxides, *Chem. Biol. Interact.* **16**:281–300.
135. Panthananickal, A., and Marnett, L. J., 1981, Arachidonic acid-dependent metabolism of 7,8-dihydroxy-7,8-dihydrobenzo[a]pyrene to polyguanylic acid-binding derivatives, *Chem. Biol. Interact.* **33**:239–252.
136. Dix, T. A., and Marnett, L. J., 1984, Detection of the metabolism of polycyclic aromatic hydrocarbon derivatives to ultimate carcinogens during lipid peroxidation, *Methods Enzymol.* **105**:347–352.
137. Dix, T. A., 1983, The mechanism of the fatty acid hydroperoxide dependent epoxidation of 7,8-dihydroxy-7,8-dihydrobenzo[a]pyrene, *Ph.D. dissertation*, Wayne State University,
138. Wills, E. D., 1969, Lipid peroxide formation in microsomes—General considerations, *Biochem. J.* **113**:315–324.
139. Buege, J. A., and Aust, S. D., 1978, Microsomal lipid peroxidation, *Methods Enzymol.* **52**:302–310.
140. Mansuy, D., Leclaire, J., Fontecave, M., and Momenteau, M., 1984, Oxidation of monosubstituted olefins by cytochromes P-450 and heme models: Evidence for the formation of aldehydes in addition to epoxides and allylic alcohols, *Biochem. Biophys. Res. Commun.* **119**:319–325.
141. Mansuy, D., Battioni, P., and Renaud, J.-P., 1984, In the presence of imidazole, iron- and manganese-porphyrins catalyze the epoxidation of alkenes by alkyl hydroperoxides, *J. Chem. Soc. Chem. Commun.* **1984**:1255–1257.
142. Groves, J. T., 1980, Mechanisms of metal-catalyzed oxygen insertion, in: *Metal Ion Activation of Dioxygen* (T. G. Spiro, ed.), Wiley–Interscience, New York, pp. 125–162.
143. Lee, W. A., and Bruice, T. C., 1985, Homolytic and heterolytic oxygen–oxygen bond scissions accompanying oxygen transfer to iron (III) porphyrins by percarboxylic acids and hydroperoxides: A mechanistic criterion for peroxidase and cytochrome P-450, *J. Am. Chem. Soc.* **107**:513–514.
144. Traylor, T. G., Lee, W. A., and Stynes, D. V., 1984, Model compound studies related to peroxidases. II. The chemical reactivity of a high valent protohemin compound, *Tetrahedron* **40**:553–568.

145. Capdevila, J., Estabrook, R. W., and Prough, R. A., 1980, Differences in the mechanism of NADPH- and cumene hydroperoxide-supported reactions of cytochrome P-450, *Arch. Biochem. Biophys.* **200**:186–195.
146. Holdec, G., Yagi, H., Dansette, P., Jerina, D. M., Levin, W., Lu, A. Y. H., and Conney, A. H., 1974, Effects of inducers and epoxide hydrase on the metabolism of benzo[a]pyrene by liver microsomes and a reconstituted system–Analysis by high pressure liquid chromatography, *Proc. Natl. Acad. Sci. USA* **71**:4356–4360.
147. Nagata, C., Tagashia, Y., and Kodama, M., 1974, Metabolic activation of benzo[a]pyrene: Significance of its free radical in: *The Biochemistry of Disease: Chemical Carcinogenesis*, Volume 4 (P.O.P. T'so and J. A. Di Paolo, eds.), Dekker, New York, pp. 87–111.
148. Capdevila, J., Saeki, Y., and Falck, J. R., 1984, The mechanistic plurality of cytochrome P-450 and its biological ramifications, *Xenobiotica* **14**:109–118.
149. Morgenstern, R., DePierre, J. W., Lind, C., Guthenberg, C., Mannervik, B., and Ernster, L., 1981, Benzo[a]pyrene quinones can be generated by lipid peroxidation and are conjugated with glutathione by glutathione S-transferase B from rat liver, *Biochem. Biophys. Res. Commun.* **99**:682–690.
150. Poulos, T. L., and Kraut, J., 1980, The sterochemistry of peroxidase catalysis, *J. Biol. Chem.* **255**:8199–8205.
151. Eglinton, D. G., Barber, D., Thomson, A. J., Greenwood, C., and Segal, A. W., 1982, Studies of cyanide binding to myeloperoxidase by electron paramagnetic resonance and magnetic circular dichroism spectroscopies, *Biochim. Biophys. Acta* **703**:187–195.
152. Ikeda-Saito, M., Prince, R. C., Argade, P. V., and Rousseau, D. L., 1984, Spectroscopic studies of myeloperoxidase, *Fed. Proc.* **43**:1561.
153. Sibbett, S. S., and Hurst, J. K., 1984, Structural analysis of myeloperoxidase by resonance spectroscopy, *Biochemistry* **23**:3007–3013.
154. Murthy, M. R. N., Reid, T. J., III, Sicignano, A., Tanaka, N., and Rossmann, M. G., 1981, Structure of beef liver catalase, *J. Mol. Biol.* **152**:465–499.
155. Di Nello, R. K., and Dolphin, D. H., 1981, Substituted hemins as probes for structure–function relationships in horseradish peroxidase, *J. Biol. Chem.* **256**:6903–6912.
156. Chance, B., 1952, The kinetics and stoichiometry of the transition from primary to secondary peroxidase peroxide complexes, *Arch. Biochem. Biophys.* **41**:416–424.
157. Critchlow, J. E., and Dunford, H. B., 1972, Studies on horseradish peroxidase. X. The mechanism of oxidation of p-cresol, ferrocyanide, and iodide by compound II, *J. Biol. Chem.* **247**:3714–3725.
158. Job, D., and Dunford, H. B., 1975, Substituent effect on oxidation of phenols and aromatic amines by horseradish peroxidase compound I, *Eur. J. Biochem.* **66**:607–614.
159. Roman, P., Dunford, H. B., and Evell, M., 1971, Studies on horseradish peroxidase. VII. A kinetic study of the oxidation of iodide by horseradish peroxidase compound II, *Can. J. Chem.* **49**:3059–3063.
160. Yamazaki, I., and Yakota, I., 1973, Oxidation states of peroxidase, *Mol. Cell. Biochem.* **2**:39–52.
161. Cormier, M. J., and Prichard, J., 1968, An investigation of the mechanism of the luminescent peroxidation of luminol by stopped flow techniques, *J. Biol. Chem.* **243**:4706–4714.
162. Chance, B., 1951, Enzyme–substrate compounds, in: *Advances in Enzymology*, Volume 12 (F. F. Nord, ed.), Interscience, New York, pp. 153–188.
163. Harrison, J. E., 1982, The role of peroxide in the functional mechanism of myeloperoxidase, in: *Oxidases and Related Redox Systems* (T. E. King, H. S. Mason, and M. Morrison, eds.), Pergamon Press, Elmsford, N. Y., pp. 717–732.
164. Poulos, T. L., Freer, S. T. Alden, R. A., Edwards, S. L., Skogland, U., Tokio, K.,

Eriksson, B., Xuong, N., Yonetani, T., and Kraut, J., 1980, The crystal structure of cytochrome c peroxidase, *J. Biol. Chem.* **255**:575–580.
165. Dawson, J. H., Trudell, J. R., Barth, G., Linder, R. E., Bunnenberg, E., Djerassi, C., Chiang, R., and Hager, L. P., 1976, Chloroperoxidase: Evidence for P-450 type heme environment from magnetic circular dichroism spectroscopy, *J. Am. Chem. Soc.* **98**:3709–3710.

CHAPTER 3

The Topology of the Mammalian Cytochrome P-450 Active Site

GERALD T. MIWA and ANTHONY Y. H. LU

1. Introduction

Currently, the only direct method for visualizing the active site of an enzyme is through high-resolution X-ray crystallography. Indeed, for P-450_{cam}, a soluble enzyme crystallized from *Pseudomonas putida*, this has yielded the most complete three-dimensional information to date on a P-450 active site (Chapter 13). Similar data are not available for any of the mammalian P-450 isozymes, however, since no crystals have yet been successfully produced for these hydrophobic proteins. Consequently, investigations have relied on much more indirect means to obtain information about the topology of the active site of these enzymes.

There are a great number of P-450 isozymes in the mammalian hepatic P-450 family.[1,2] Chemical induction can cause a marked synthesis of specific forms of these proteins. For example, phenobarbital and 3-methylcholanthrene cause greater than 15- to 40-fold increases in isozymes P-450_b and P-450_c, respectively.[3-5] In the last decade, considerable progress has been made on the purification of various isozymes induced by chemical agents and, more recently, on the constitutive forms in untreated animals.[6,7] Further progress on the purification and crystallization of these isozymes will provide the basis for additional studies that offer promise in defining the detailed topology of the mammalian membrane-bound enzymes.

GERALD T. MIWA and ANTHONY Y. H. LU • Department of Animal Drug Metabolism, Merck Sharp and Dohme Research Laboratories, Rahway, New Jersey 07065.

2. Historical

Even before the discovery and isolation of multiple forms of hepatic cytochrome P-450, the broad substrate specificity of these enzymes was recognized in microsomal preparations.[1] The correlation between substrate-induced spectral perturbations[8-11] or catalytic activity[12-15] and the hydrophobicity of the substrate suggested that a hydrophobic barrier restricted substrate access to the active site of microsomal P-450. The spectral binding studies of Backes and Canady[16] provided evidence for sufficient accessibility of small solvent molecules to cause perturbations in the heme spectrum for cytochromes P-450 in microsomal preparations from phenobarbital-induced rats but not in the microsomes from untreated rats. These studies provided the earliest hints that differences in active site architecture existed among various P-450 isozymes. Still unanswered, however, was the question of whether the correlation with hydrophobicity reflected a barrier to substrate partition, as through the microsomal membrane, or the tighter binding of the substrate with a hydrophobic protein domain constituting part of the active site of the enzyme.

Subsequent to the isolation of some of these P-450 isozymes, evidence was obtained for the broad but overlapping substrate specificities of individual isozymes.[1,2] White et al.[9] also demonstrated the correlation between spectral binding and hydrophobic properties for a number of small phenylalkane molecules with purified LM_2, an isozyme obtained from phenobarbital-induced rabbit livers. In addition, 1-phenyloctane was observed to deviate from the binding predicted by its hydrophobicity, a fact the authors attributed to restrictions imposed by the enzyme binding site. These data suggested a large active site, for this purified, single isozyme, but not so large as to permit indiscriminate binding, with random orientation, of all substrates.

The stereoselective oxidation of a number of substrates has suggested the chiral nature of the P-450 active site. For example, selectivity in the metabolism of one enantiomer of a racemic mixture such as with warfarin[17] or amphetamine[18]; the predominant attack on one face of a planar molecule, such as the epoxidation of various polycyclic aromatic hydrocarbons,[19] pyrrole nitrogen alkylation by terminal olefins,[20] and the α-face hydroxylation of steroids[21]; the generation of chiral S-oxides,[22] or the selective removal of a prochiral hydrogen from a substrate[23] all suggested that constraints existed for substrate orientation in the catalytic site of these enzymes. Experiments designed to probe the high stereoselectivity and regioselectivity for mammalian P-450 isozymes have been a powerful means for giving some dimension to the P-450 active site but this indirect evidence has provided topological information largely on only the two-dimensional plane parallel to the face of the porphyrin.

3. Topology of Selected P-450 Isozymes

3.1. Rabbit Liver P-450 Isozymes

The P-450$_{LM_2}$ isozyme from the livers of rabbits induced with phenobarbital was the first P-450 isozyme to be purified to apparent homogeneity from a mammalian source[24,25] and to have the complete amino acid sequence reported.[26,27] Identification of the cysteine residue, coordinated to the heme iron, provided a marker for demonstrating the high hydrophobic character of the amino acids adjacent to this site.

Since the native, oxidized protein is hexacoordinate and low spin, the substrate-induced dissociation of the sixth ligand, presumed to be water, to form a characteristic type I binding spectrum provided a means for White *et al.*[9] to demonstrate the correlation between the binding constant for the purified enzyme and the hydrophobicity of the substrate. A linear correlation was obtained with a homologous series of *n*-alkylbenzenes suggesting that substrate binding was stabilized by hydrophobic interactions of the substrate with the protein in the vicinity of the heme. Some evidence was also obtained on the total dimension of this binding site since 1-phenyl octane deviated from the linear correlation observed for the smaller alkylbenzenes. Proton NMR studies by Novak and Vatsis[28] suggested some selectivity for the binding of specific regions of acetanilide and 2,6-dimethylaniline[29] to the LM$_2$ iron which correlated with the observed regioselectivity in product formation.[30]

Recent studies by White *et al.*[31] demonstrated the almost complete absence of regiospecificity in the hydroxylation of rigid, alicyclic compounds by rabbit P-450$_{LM_2}$. This observation is in marked contrast to the high regio- and stereoselectivity observed by these investigators for P-450$_{cam}$ and suggested that the topology of the LM$_2$ active site is voluminous enough to permit complete mobility of the alicyclic substrates examined.

Studies on the inhibition of binding of ethyl isocyanide, carbon monoxide, and imidazole by polycyclic aromatic hydrocarbons have been used by Imai[32] to define the topology of another rabbit isozyme, LM$_4$, which is inducible by 3-methylcholanthrene but is also known to be present in untreated and phenobarbital-treated rabbits. Structure–activity studies revealed a minimum size and geometry of a fused polycyclic nucleus which inhibited binding of these ligands. Moreover, a rigid polycyclic nucleus was more effective in this inhibition than a nonfused polycyclic system. These data could be explained by an active site model in which a portion of the polycyclic nucleus was anchored on a hydrophobic site on the protein such that a portion of the fused ring system was oriented rigidly over the sixth coordination site of the iron.

3.2. Rat Liver P-450$_c$

The major hepatic P-450 isozyme (P-450$_c$) induced in the rat by 3-methylcholanthrene has been isolated and biochemically characterized (Chapter 6). During the elegant work on the activation of polycyclic aromatic hydrocarbons to carcinogenic metabolites by Jerina, Lovin, and co-workers, these authors observed pronounced stereoselectivity in the 4,5- and 7,8-dihydrodiol products formed from benzo[a]pyrene by liver microsomes from 3-methylcholanthrene-induced rats.[33] Subsequent studies have shown that a (+)-[4S,5R]-oxide was formed with high stereochemical fidelity (> 97%) by purified P-450$_c$.[34] High stereoselectivity (96%) in the formation of benzo[a]pyrene [7R,8S]-oxide was also deduced from the 96% (−)-[7R,8R]-dihydrodiol produced on hydration by purified epoxide hydrolase.[35,36] The regioselectivity with high stereochemical fidelity by a purified single P-450 isozyme suggested a catalytic site capacious enough to accommodate the reorientation of the substrate necessary for the observed positional isomeric metabolites and broad substrate specificity of the enzyme while restrictive enough to permit stereoselective addition of oxygen to only one face of the benzo[a]pyrene molecule.

Since oxygen transfer from the porphyrin to the substrate occurs on only one face of the substrate, the superimposition of the 4,5- and 7,8-double bonds of benzo[a]pyrene that underwent epoxidation permitted definition of the steric constraints imposed by the protein on substrate binding. Elaboration of these studies to other polycyclic aromatic hydrocarbons such as benzo[a]anthracene,[37,38] chrysene,[39] and phenanthrene[40] provided the stereochemical data summarized in Table I. The high stereochemical fidelity observed in all the isomeric products was the rationale for superimposing the double bonds that have undergone epoxidation. This treatment defined the minimum boundary of the P-450$_c$ catalytic site (Fig. 1).

It is noteworthy that the binding site is asymmetric with respect to the porphyrin iron. In fact, as depicted in Fig. 1, evidence suggests that the protein covers a quadrant of the porphyrin accounting for the stereospecificity observed. Consequently, these data provide some measure of the proximity of the protein to the iron-bound oxygen and suggest that this distance is within the dimension of a benzene ring although the detailed relationship of each of the nonidentical pyrrole rings to the binding domain was not revealed from these studies.

An estimation of the energy required to immobilize a substrate on this site, presumably due to hydrophobic interaction of the substrate with the protein forming the perimeter of the binding site, could be deduced from isotope effect studies on the O-deethylation of 7-ethoxycoumarin.[42]

TABLE I
Absolute Configuration of the Dihydrodiol Products Produced in 3-Methylcholanthrene-Induced Rats[a]

Dihydrodiol	Enantiomeric composition (%)		Enantiomeric purity (%)
	R, R	S, S	
Benzo[a]pyrene			
4, 5-	96	4	92
7, 8-	96	4	92
9, 10-	96	4	92
Benzo[a]anthracene			
5, 6-	81	19	62
8, 9-	98	2	96
10, 11-	98	2	96
Phenanthrene			
1, 2-	96.6	3.5	93
3, 4-	98.5	1.5	97
9, 10-	42	58	16
Chrysene			
1, 2-	90	10	80
3, 4-	98.5	1.5	97

[a] From Ref. 41 with permission.

The substitution of deuterium for the hydrogen on the α-carbon which underwent oxidation resulted in a pronounced intrinsic isotope effect ($^Dk = 14$) during O-deethylation corresponding to a bond energy difference of approximately 1.6 kcal/mole. This isotope effect was accompanied by a change in the regioselectivity of oxidation from the O-ethyl group to the C5–C6 double bond on the aromatic ring of the labeled substrate.[43,44] The free energy change required for reorientation of the substrate (from position A to position B in Fig. 2) to produce the observed products could be estimated from the ratio of the O-deethylated/ring hydroxylated prod-

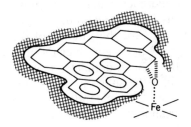

FIGURE 1. Postulated topology of the catalytic site of P-450$_c$. Benzo[a]pyrene is depicted lying over a plane parallel to the porphyrin plane. The protein perimeter in the catalytic site permits only one of two possible faces of the substrate to be oriented in a fashion allowing epoxidation of the C9–C10 double bond. The depicted topology introduces the epoxide in the observed 9R, 10S geometry. From Ref. 49 with permission.

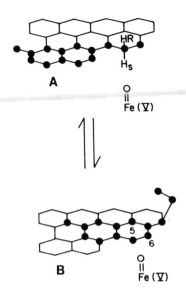

FIGURE 2. Stereochemistry of α-carbon oxidation and alteration in regioselectivity by P-450$_c$ during the oxidation of 1,1-[^{2}H$_2$]-ethoxycoumarin. (A) The substrate is depicted in the active site of P-450$_c$ in the only fashion that does not require the extension of coumarin or the side chain terminal methyl groups into the perimeter of the protein domain shown in Fig. 1. The pro-S hydrogen abstraction predicted by Jerina's model[40] was observed. (B) A second favorable orientation of the substrate places the C5–C6 double bond in a position susceptible to the oxygen addition observed with 1,1-[^{2}H$_2$]-ethoxycoumarin.

uct ratios for the unlabeled and labeled substrates. This value was calculated to be approximately 1.3 kcal/mole.

The similarity between the intrinsic isotope effect and the isotope effect calculated from the free energy change demonstrates that an approximately 1.3–1.6 kcal/mole energy barrier normally limits orientation of the substrate to position A in the binding site depicted in Fig. 2. This is within the range of the binding energy (0.3–0.7 kcal/methylene unit) that could be achieved through hydrophobic interactions of the substrate with this enzyme.[8,9] Moreover, the reorientation of the substrate in the active site can be completely explained in terms of the difference in the energy of cleaving the C–H and C–D bonds; a conformational change of the protein is not required to account for the change in regioselectivity.

The energy barrier and the approximate distance between the binding site and the methylene carbon of 7-ethoxycoumarin could also be estimated from studies employing R- and S-7[1-^{3}H]-ethoxycoumarin. Superimposition of this substrate on the P-450$_c$ active site proposed by Jerina et al.[40] so that the methylene carbon, which undergoes oxidation, is above the iron (Fig. 2, position A) places the pro-S hydrogen toward the oxygen when the terminal methyl group is oriented toward the interior of the active site. In contrast, orientation of the pro-R hydrogen over the iron requires directing the methyl group toward the protein domain surrounding the substrate. The conformation depicted for the substrate is also the lowest energy state for the planar conformation. Consequently, both the conformational stability and the steric constraints of the active site contribute to the high stereoselectivity (> 95%) observed[23] for S-hydrogen abstraction. The more stable substrate conformation is, however, insufficient to always dictate the stereochemical course of hydrogen abstraction as P-450$_b$ exhibited a preference for the opposite hydrogen.[23]

3.3. Rat Liver P-450$_b$

The major P-450 isozyme (P-450$_b$) in the livers of rats induced with phenobarbital has been isolated, sequenced by cDNA analysis,[45] and biochemically characterized.[3] Phenobarbital causes the marked synthesis of this and a closely related isozyme in rats. It has been estimated that approximately 50–60% of the total microsomal P-450 content is composed of these two isozymes after phenobarbital induction.[3]

Ortiz de Montellano and co-workers have recently exploited the heme alkylation, which occurs during the catalytic oxidation of terminal olefins and acetylenes, as a means of defining the topology of the active site of this isozyme.[46] The systematic elucidation of the regiochemistry and absolute stereochemistry of heme addition to various terminal olefins and acetylenes has yielded unique information on the topology of the P-450$_b$ active site. This approach has also defined the orientation of the heme, in relation to the fifth iron ligand, as being identical to the orientation in hemoglobin.[47]

Only a single porphyrin adduct, of the eight adduct isomers possible by addition of one of the four nonidentical pyrrole nitrogens to either end of the asymmetric π-bond, was observed for octene,[46] octyne,[46] propene,[46] ethylene, and propyne.[48] Pyrrole nitrogen addition always occurred at the terminal carbon of the π-bond and molecular oxygen was traced as the source of the hydroxyl group in the internal carbon. The olefins exclusively alkylated the pyrrole nitrogen of ring D while the substituted acetylenes alkylated pyrrole ring A. Acetylene was exceptional in being able to alkylate multiple nitrogens.

Unification of the high regio- and stereospecificity observed for N-alkylation by these terminal π-bonded compounds suggested the active site model depicted in Fig. 3 in which the substrate binding domain is elongated and asymmetrically oriented over the porphyrin iron. The asymmetric shape of the substrate binding domain and the relative arrangement of the binding site over the porphyrin, as depicted with masking of pyrrole ring B, is uniquely defined by this approach because of the bridge created between the substrate binding domain and the porphyrin during heme alkylation. Moreover, the similar masking of ring B has also been observed in P-450$_{cam}$ by X-ray analysis (Chapter 13). Thus, the orientation of the terminal π-bond of the suicide substrate within this active site crevice would permit the observed alkylation of the pyrrole nitrogens on

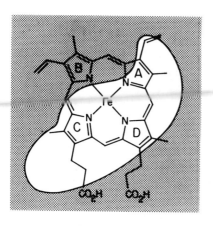

FIGURE 3. Topology of P-450$_b$. The elongated perimeter of the protein domain is not centered over the porphyrin and covers pyrrole ring B. Pyrrole rings A, C, and D are susceptible to alkylation by suitable terminal olefins or acetylenes. From Ref. 2 with permission.

rings A, C, and D but not ring B. The absolute stereochemistry of the carbon bearing the hydroxylfunction, introduced during the pyrrole N-alkylation, was also correctly predicted by this model.[20]

4. Summary

This chapter has reviewed the current information about the topology of various mammalian P-450 isozymes gleaned from substrate binding studies and from the stereo- and regio-selectivity of the oxidations catalyzed by these enzymes. These indirect methods have revealed the orientation of the porphyrin with respect to the fifth ligand and provided a two-dimensional skeleton description of the perimeter and nature of the protein domain constituting the substrate binding site. Evidence is sufficiently strong to further define a common feature in the topology of three forms of P-450. P-450$_b$, P-450$_c$, and P-450$_{cam}$ (Chapter 13) all share the apparent masking by the protein of one of the heme pyrrole rings. For P-450$_b$ and P-450$_{cam}$, ring B appears to be covered while the specific pyrrole ring has not been defined for P-450$_c$. The generality of this "pyrrole masking" and the purpose of this structural feature will require further investigations. In addition, the integration of the known amino acid sequence of some of these isozymes into a three-dimensional picture of the active site must await X-ray crystallographic examination. The organization of the protein within the membrane is discussed in Chapter 6 and the mode of association with other electron transport proteins is discussed in Chapters 4 and 5.

References

1. Lu, A. Y. H., and West, S. B., 1980, Multiplicity of mammalian microsomal cytochromes P-450, *Pharmacol. Rev.* **31:**277–291.
2. Johnson, E. F., 1979, Multiple forms of cytochrome P-450: Criteria and significance, in: *Reviews in Biochemical Toxicology* (E. Hodgson, J. R. Bend, and R. M. Philpot, eds.) Elsevier/North Holland, Amsterdam, pp. 1–26.
3. Ryan, D. E., Thomas, P. E. Reik, L. M., and Levin, W., 1982, Purification, characterization and regulation of five rat hepatic microsomal cytochrome P-450 isozymes, *Xenobiotica* **12:**727–744.
4. Guengerich, F. P., Wang, P., and Davidson, N. K., 1982, Estimation of isozymes of microsomal cytochrome P-450 in rats, rabbits, and humans using immunochemical staining coupled with sodium dodecyl sulfate–polyacrylamide gel electrophoresis, *Biochemistry* **21:**1698–1706.
5. Pickett, C. B., Jeter, R. L., Morin, J., and Lu, A. Y. H., 1981, Electroimmunochemical quantitation of cytochrome P-450, cytochrome P-448, and epoxide hydrolase in rat liver microsomes, *J. Biol. Chem.* **256:**8815–8820.
6. Koop, D. R., Persson, A. V., and Coon, M. J., 1981, Properties of electrophoretically homogeneous constitutive forms of liver microsomal cytochrome P-450, *J. Biol. Chem.* **256:**10704–10711.
7. Cheng, K.-C., and Schenkman, J. B., 1982, Purification and characterization of two constitutive forms of rat liver microsomal cytochrome P-450, *J. Biol. Chem.* **257:**2378–2385.
8. Canady, W. J., Robinson, D. A., and Colby, H. D., 1974, A partition model for hepatic cytochrome P-450–hydrocarbon complex formation, *Biochem. Pharmacol.* **23:**3075–3078.
9. White, R. E., Oprian, D. D., and Coon, M. J., 1980, Resolution of multiple equilibria in binding of small molecules to cytochrome P-450$_{LM}$, in: *Microsomes, Drug Oxidations, and Chemical Carcinogenesis*, Volume I (M. J. Coon, A. H. Conney, R. W. Estabrook, H. V. Gelboin, J. R. Gillette, and P. J. O'Brien, eds.), Academic Press, New York, pp. 243–251.
10. Jefcoate, C. R., Gaylor, J. L., and Calabrese, R. L., 1969, Ligand interactions with cytochrome P-450. I. Binding of primary amines, *Biochemistry* **8:**3455–3463.
11. Jansson, I., Orrenius, S., Ernster, L., and Schenkman, J. B., 1972, A study of the interactions of a series of substituted barbituric acids with the hepatic microsomal monooxygenase, *Arch. Biochem. Biophys.* **151:**391–400.
12. McMahon, R. E., 1961, Demethylation studies. I. The effect of chemical structure and lipid solubility, *J. Med. Pharm. Chem.* **4:**67–78.
13. Martin, Y. C., and Hansch, C., 1971, Influence of hydrophobic character on the relative rate of oxidation of drugs by rat liver microsomes, *J. Med. Chem* **14:**777–779.
14. Cohen, G. M., and Mannering, G. J., 1973, Involvement of a hydrophobic site in the inhibition of the microsomal *p*-hydroxylation of aniline by alcohols, *Mol. Pharmacol.* **9:**383–397.
15. Cho, A. K., and Miwa, G. T., 1973, The role of ionization in the N-demethylation of some N,N-dimethylamines, *Drug Metab. Dispos.* **2:**477–483.
16. Backes, W. L., and Canady, W. J., 1981, The interaction of hepatic cytochrome P-450 with organic solvents; The effect of organic solvents on apparent spectral binding constants for hydrocarbon substrates, *J. Biol. Chem.* **256:**7213–7227.
17. Kaminsky, L. S., Fasco, M. J., and Guengerich, F. P., 1980, Comparison of different forms of purified cytochrome P-450 from rat liver by immunological inhibition of regio- and stereoselective metabolism of warfarin, *J. Biol. Chem.* **255:**85–91.

18. Cho, A. K., and Wright, J., 1978, Minireview: Pathways of metabolism of amphetamine and related compounds, *Life Sci.* **22**:363–372.
19. Jerina, D. M., Michaud, D. P., Feldmann, R. J., Armstrong, R. N., Vyas, K. P., Thakker, D. R., Yagi, H., Thomas, P. E., Ryan, D. E., and Levin, W., 1982, Stereochemical modeling of the catalytic site of cytochrome P-450$_c$, in: *Microsomes, Drug Oxidations and Drug Toxicity* (R. Sato and R. Kato, eds.), Japan Scientific Societies Press, Tokyo, pp. 195–201.
20. Ortiz de Montellano, P. R., Mangold, B. L. K., Wheeler, C., Kunze, K. L., and Reich, N.O., 1983, Stereochemistry of cytochrome P-450-catalyzed epoxidation and prosthetic heme alkylation, *J. Biol. Chem.* **258**:4208–4213.
21. Waxman, D. J., Ko, A., and Walsh, C., 1983, Regioselectivity and stereoselectivity of androgen hydroxylations catalyzed by cytochrome P-450 isozymes purified from phenobarbital-induced rat liver, *J. Biol. Chem.* **258**:11937–11947.
22. Waxman, D. J., Light, D. R., and Walsh, C., 1982, Chiral sulfoxidations catalyzed by rat liver cytochrome P-450, *Biochemistry* **21**:2499–2507.
23. Tullman, R. H., Walsh, J. S., and Miwa, G. T., 1984, The stereochemistry of P-450 and P-448 catalyzed O-dealkylation of 7-ethoxycoumarin, *Fed. Proc.* **43**:346.
24. van der Hoeven, T. A., Haugen, D. A., and Coon, M. J., 1974, Cytochrome P-450 purified to apparent homogeneity from phenobarbital-induced rabbit liver microsomes: Catalytic activity and other properties, *Biochem. Biophys. Res. Commun.* **60**:569–575.
25. Imai, Y., and Sato, R., 1974, A gel-electrophoretically homogeneous preparation of cytochrome P-450 from liver microsomes of phenobarbital-pretreated rabbits, *Biochem. Biophys. Res. Commun.* **60**:8–14.
26. Heinemann, F. S., and Ozols, J., 1983, The complete amino acid sequence of rabbit phenobarbital-induced liver microsomal cytochrome P-450, *J. Biol. Chem.* **258**:4195–4201.
27. Tarr, G. E., Black, S. D., Fujita, V. S., and Coon, M. J., 1983, Complete amino acid sequence and predicted membrane topology of phenobarbital-induced cytochrome P-450 (isozyme 2) from rabbit liver microsomes, *Proc. Natl. Acad. Sci. USA* **80**:6552–6556.
28. Novak, R. F., and Vatsis, K. P., 1982, ^{1}H Fourier transform nuclear magnetic resonance relaxation rate studies on the interaction of acetanilide with purified isozymes of rabbit liver microsomal cytochrome P-450 and with cytochrome b$_5$, *Mol. Pharmacol.* **21**:701–709.
29. Novak, R. F., Kapetanovic, I. M., and Mieyal, J. J., 1977, Nuclear magnetic resonance studies of substrate–hemeprotein complexes in solution, *Mol. Pharmacol.* **13**:15–30.
30. Coon, M. J., and Vatsis, K. P., 1978, Biochemical studies on chemical carcinogenesis: Role of multiple forms of liver microsomal cytochrome P-450 in the metabolism of benzo[a]pyrene and other foreign compounds, in: *Polycyclic Hydrocarbons and Cancer: Environment, Chemistry and Metabolism*, Volume I, (H. V. Gelboin and P.O.P. Ts'o, eds.), Academic Press, New York, pp. 335–360.
31. White, R. E., McCarthy, M.-B., Egeberg, K. D., and Sligar, S. G., 1984, Regioselectivity in the cytochromes P-450: Control by protein constraints and by chemical reactivities, *Arch. Biochem. Biophys.* **228**:493–502.
32. Imai, Y., 1982, Interaction of polycyclic hydrocarbons with cytochrome P-450. III. Effects of hydrocarbon binding on the interaction of some ligands with P-448$_1$ heme, *J. Biochem.* **92**:77–88.
33. Thakker, D. R., Yagi, H., Akagi, H., Koreeda, M., Lu, A. Y. H., Levin, W., Wood, A. W., Conney, A. H., and Jerina, D. M., 1977, Metabolism of benzo[a]pyrene. VI. Stereoselective metabolism of benzo[a]pyrene and benzo[a]pyrene 7,8-dihydrodiol to diol epoxides, *Chem. Biol. Interact.* **16**:281–300.

34. Armstrong, R. N., Levin, W., Ryan, D. E., Thomas, P. E., Mah, H. D., and Jerina, D. M., 1981, Stereoselectivity of rat liver cytochrome P-450c in formation of benzo[a]pyrene 4,5-oxide, *Biochem. Biophys. Res. Commun.* **100**:1077–1084.
35. Boyd, D. R., Gadaginamath, G. S., Kher, A., Malone, J. F., Yagi, H., and Jerina, D. M., 1980, (+)- and (−)-benzo[a]pyrene 7,8-oxide: Synthesis, absolute stereochemistry, and stereochemical correlation with other mammalian metabolites of benzo[a]pyrene, *J. Chem. Soc. Perkin Trans. I* **1980**:2112–2116.
36. Levin, W., Buening, M. K., Wood, A. W., Chang, R. L., Kezierski, B., Thakker, D. R., Boyd, D. R., Gadaginamath, G. S., Armstrong, R. N., Yagi, H., Karle, J. M., Slaga, T. J., Jerina, D. M., and Conney, A. H., 1980, An enantiomeric interaction in the metabolism and tumorigenicity of (+)- and (−)-benzo[a]pyrene 7,8-oxide, *J. Biol. Chem.* **255**:9067–9074.
37. Thakker, D. R., Levin, W., Yagi, H., Turujman, S., Kapadia, D., Conney, A. H., and Jerina, D. M., 1979, Absolute stereochemistry of the trans-dihydrodiols formed from benzo[a]anthracene by liver microsomes, *Chem. Biol. Interact.* **27**:145–161.
38. van Bladeren, P. J., Armstrong, R. N., Cobb, D., Thakker, D. R., Ryan, D. E., Thomas, P. E., Sharma, N. D., Boyd, D. R., Levin, W., and Jerina, D. M., 1982, Stereoselective formation of benz[a]anthracene (+)-(5S,6R)-oxide and (+)-(8R,9S)-oxide by a highly purified and reconstituted system containing cytochrome P-450$_c$, *Biochem. Biophys. Res. Commun.* **106**:602–609.
39. Nordqvist, M., Thakker, D. R., Vyas, K. P., Yagi, H., Levin, W., Ryan, D. E., Thomas, P. E., Conney, A. H., and Jerina, D. M., 1981, Metabolism of chrysene and phenanthrene to bay-region diol epoxides by rat liver enzymes, *Mol. Pharmacol.* **19**:168–178.
40. Jerina, D. M., Michaud, D. P., Feldmann, R. J., Armstrong, R. N., Vyas, K. P., Thakker, D. R., Yagi, H., Thomas, P. E., Ryan, D. E., and Levin, W., 1982, Stereochemical modeling of the catalytic site of cytochrome P-450$_c$, in: *Microsomes, Drug Oxidations and Drug Toxicity* (R. Sato and R. Kato, eds.), Japan Scientific Societies Press, Tokyo, pp. 195–201.
41. Thakker, D. R., Levin, W., Yagi, H., Conney, A. H., and Jerina, D. M., 1982, Regio- and stereoselectivity of hepatic cytochrome P-450 toward polycyclic aromatic hydrocarbon substrates, in: *Biological Reactive Intermediates*, Volume IIA (R. Snyder, D. V. Parke, D. J. Jollow, C. G. Gibson, and C. M. Witmer, eds.), Plenum Press, New York, pp. 525–539.
42. Miwa, G. T., Walsh, J. S., and Lu, A. Y. H., 1984, Kinetic isotope effects on cytochrome P-450-catalyzed oxidation reactions: The oxidative O-dealkylation of 7-ethoxycoumarin, *J. Biol. Chem.* **259**:3000–3004.
43. Harada, N., Miwa, G. T., Walsh, J. S., and Lu, A. Y. H., 1984, Kinetic isotope effects on cytochrome P-450-catalyzed oxidation reactions: Evidence for the irreversible formation of activated oxygen intermediate of cytochrome P-450, *J. Biol. Chem.* **259**:3005–3010.
44. Walsh, J. S., and Miwa, G. T., 1984, The mechanism of the cytochrome P-448 mediated 6-hydroxylation of 7-ethoxycoumarin, *Biochem. Biophys. Res. Commun.* **121**:960–965.
45. Fujii-Kuriyama, Y., Mizukami, Y., Kawajiri, K., Sogawa, K., and Muramatsu, M., 1982, Primary structure of a cytochrome P-450: Coding nucleotide sequence of phenobarbital-inducible cytochrome P-450 cDNA from rat liver, *Proc. Natl. Acad. Sci. USA* **79**:2793–2797.
46. Kunze, K. L., Mangold, B. L. K., Wheeler, C., Beilan, H. S., and Ortiz de Montellano, P. R., 1983, The cytochrome P-450 Active Site: Regiospecificity of prosthetic heme alkylation by olefins and acetylenes, *J. Biol. Chem.* **258**:4202–4207.

47. Ortiz de Montellano, P. R., Kunze, K. L., and Beilan, H. S., 1983, Chiral orientation of prosthetic heme in the cytochrome P-450 active site, *J. Biol. Chem.* **258**:45–47.
48. Ortiz de Montellano, P. R., and Kunze, K. L., 1981, Cytochrome P-450 inactivation: Structure of the prosthetic heme adduct with propyne, *Biochemistry* **20**:7266–7271.
49. Yagi, H., and Jerina, D. M., 1982, Absolute configuration of the *trans*-9,10-dihydrodiol metabolite of the carcinogen benzo[a]pyrene, *J. Am. Chem. Soc.* **104**:4026–4027.

CHAPTER 4

Cytochrome P-450 Reductase and Cytochrome b_5 in Cytochrome P-450 Catalysis

JULIAN A. PETERSON and RUSSELL A. PROUGH

1. Introduction to the Reaction Cycle of Cytochrome P-450 with Particular Emphasis on the Reduction Reactions

As the readers of this monograph should be aware, the monooxygenase reaction catalyzed by cytochrome P-450 requires the input of two electrons[1,2]:

$$AH + O_2 + 2e^- + 2H^+ \rightarrow AOH + H_2O \qquad (1)$$

In mammalian systems these two electrons are derived from NADPH; in the soluble system isolated from the bacterium *Pseudomonas putida*, NADH is used as the external source of electrons. A schematic representation of the two different types of electron transfer chains which deliver electrons from the reduced pyridine nucleotides to P-450 is shown in Fig. 1. The system, here referred to as Type I, is found embedded in the membranous endoplasmic reticulum of most eukaryotic cell types, while the second general class, Type II, is found in mitochondria and bacteria. The most completely described P-450 system, which is not membrane-bound, is associated with camphor metabolism in *P. putida*.[3] The Type I electron transport chain is composed of a complex flavoprotein which has both FAD and FMN as prosthetic groups.[4] FAD serves as the initial electron acceptor from NADPH, while the FMN serves to reduce

---

JULIAN A. PETERSON and RUSSELL A. PROUGH • Department of Biochemistry, University of Texas Health Sciences Center, Dallas, Texas 75235.

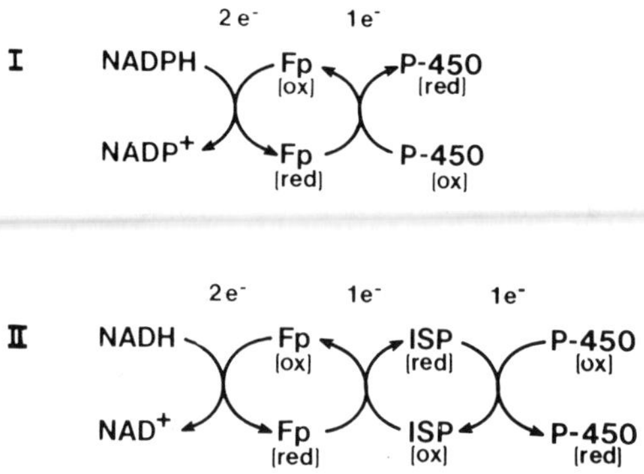

FIGURE 1. Electron transport pathways to cytochrome P-450. See text for details.

the P-450.[5] Later in this review, we will discuss the involvement of cytochrome b_5 as a possible component of this electron transport chain. In Type II systems, the reduced pyridine nucleotide first reduces an FAD-containing reductase which subsequently transfers electrons one at a time to a 2Fe,2S iron–sulfur protein. The iron–sulfur protein serves as an electron "shuttle" between the reductase and P-450.[6]

Estabrook et al.[7] proposed in 1968 that P-450 functioned in a cyclic manner beginning with a substrate binding step which may result in the conversion of the heme iron of P-450 from a low-spin ferric to a high-spin ferric form.[8] Although the monooxygenase reaction requires two electrons to complete the cycle, early work with P-450$_{cam}$ clearly demonstrated that these electrons were introduced in two sequential, one-electron steps.[9] This observation has also been confirmed with purified hepatic, microsomal P-450.[10] The observation of a spectrally distinguishable species present during the steady state of drug oxidation led Estabrook's group to propose the existence of a stable oxygenated form of the enzyme.[11] The identity of this intermediate was confirmed by the characterization of the oxy form of P-450$_{cam}$ with spectrophotometric techniques.[12,13] The "final" step of the reaction cycle is probably a series of reactions which includes electron transfer, oxygen activation, and oxygen insertion steps. Product release, after oxygen insertion into an organic substrate, results in the completion of the cycle returning the enzyme to the ferric, substrate-free, low-spin form.

2. P-450 Reductases: Distribution and Properties

As should be obvious from the preceding section, some enzymatic machinery must be present for the transfer of electrons from the reduced pyridine nucleotides to P-450. A conceptual problem arises with the fact that pyridine nucleotides are two-electron donors, while P-450 can only accept one electron at a time. Thus, enzymes which serve as the electron transfer agent(s) from reduced pyridine nucleotide must be two-electron acceptors as well as one-electron donors in these electron transfer systems. As a consequence of this duality of function, the reductase must be able to store the second electron (a radical species) in a form which is rather insensitive to molecular oxygen, so that electrons are not directly transferred to molecular oxygen resulting in the production of superoxide anion.

2.1. Microsomal

The endoplasmic reticulum is a physically complex environment in which the enzymes of the P-450-dependent monooxygenase system are embedded. This complexity is a consequence of the fact that the bilayer membranes are in essence two-dimensional surfaces, and that the surface of most membranes is heterogeneous with respect to lipid and protein distribution.[14,15] While bilayers do have a third dimension, most investigators acknowledge that the P-450-dependent enzymes are asymmetrically distributed between the two surfaces of the microsomal bilayer with the reductase and P-450 facing the cytoplasm rather than the lumen.[16]

The liver microsomal NADPH-dependent cytochrome P-450 reductase was initially identified on the basis of its ability to oxidize NADPH and to reduce cytochrome c.[17] This nonphysiological reaction has served as a convenient tool for monitoring the content of this enzyme. The original preparations of the NADPH-cytochrome c reductase were unable to reduce P-450 because they had been freed from the endoplasmic reticulum by proteolytic cleavage.[18] During the late 1960s, several groups were involved in the purification of detergent-solubilized forms of the enzymes which would catalyze drug metabolism.[3] However, they could not directly compare their impure reductase preparations to the purified NADPH-cytochrome c reductase because the reductase prepared by proteolytic cleavage would not reduce P-450. It was not until about 1970 when antibodies to the protease-solubilized reductase were prepared by Orrenius',[19] Masters'[20] and Omura's[21] groups that good evidence was obtained that NADPH-cytochrome c reductase actually was the initial electron acceptor from NADPH and subsequently transferred electrons to P-450. Not only would antibodies to the protease-solubilized NADPH-cytochrome c reductase inhibit drug metabolism in microsomal preparations, but they also inhibited the reaction in the purified, reconstituted enzyme system. The difference between the protease-solubilized and detergent-solubilized reductases is the presence of a hydrophobic segment on the N-terminal portion of the protein which attaches the reductase to the membrane surface.[22-24] Without this hydrophobic piece, the enzyme is unable to reduce P-450. The detergent-solubilized NADPH-cytochrome P-450 reductase has a molecular weight of 78,000,[22] while the hydrophobic portion has a molecular weight of 6100.[22-24]

One of the interesting features of this enzyme is the extreme stability of its one-electron reduced form.[23] It is not unusual for the enzyme to remain in this state during the typical purification procedure.[22] However, the air-stable one-electron reduced form is usually oxidized by $K_3Fe(CN)_6$ in the final isolation steps resulting in the purified enzyme being fully oxidized.[22] It is not widely recognized that when the activity of the reconstituted (fully oxidized) enzyme is studied, the first step is a "priming" reaction in which the enzyme is initially converted to the two-electron reduced form.[25] The relationship between the kinetics of interaction with and reduction of P-450 by this two-electron reduced form of the reductase and the same reaction by the three-electron reduced form is not readily apparent. Coon's group was the first to postulate that the FAD served as the entrance point for electron transfer into the reductase, while the FMN served as the exit to P-450.[5] The redox potentials of the detergent-solubilized enzyme have been determined and the most interesting feature of these results is that the potential for the one- to two-electron reduced forms (-270 mV) is very nearly the same as that of the two- to three-electron reduced forms (-290 mV).[26] That is to say, the putative forms which reduce P-450 are equipotential.

2.2. Mitochondrial

The transfer of electrons from NADPH to $P-450_{scc}$ in adrenal cortex mitochondria is mediated by an FAD-containing flavoprotein, adrenodoxin reductase,[27,28] which has a minimum molecular weight of 54,000.[28] The reductase seems to be rather weakly associated with the mitochondrial inner membrane. The second protein involved in the electron transfer process is the 2Fe,2S iron–sulfur protein, adrenodoxin,[27] which has a minimum molecular weight of 12,500.[29] Both of these proteins have been isolated in homogeneous form and their properties described. The elegant work in Waterman's laboratory has demonstrated that both of these proteins, as well as $P-450_{scc}$ and $P-450_{11\beta}$, are synthesized in the cytosol as larger preprotein forms, transported into mitochondria, and processed to the final mature forms (Waterman *et al.*, this volume). The ratio of adrenodoxin reductase to adrenodoxin to total cytochrome P-450 in adrenal mitochondria is approximately $1:8:8$.[30]

The properties of adrenodoxin reductase, adrenodoxin, and $P-450_{scc}$ as they interact during partial reactions of the electron transfer system, as well as during turnover, have been described in detail by Kamen's group and more recently by Lambeth. Since most of these results have been discussed in review articles, we will only present recent findings as they relate to our overall understanding of how P-450 can/may be reduced during turnover. The studies of the Kamen/Lambeth group have focused either on the interaction of adrenodoxin reductase and adrenodoxin or on a kinetic analysis of the turnover of the complete enzyme system. An important finding from Kamen's group which will be discussed in more detail later is that during catalytic turnover, adrenodoxin reductase does not appear to return to the completely oxidized state, but remains instead as a one-electron reduced form.[31] The results of these kinetic studies are most readily interpreted as indicating that the enzyme cycles between the one-electron reduced starting state and the three-electron fully reduced form. During reoxidation by adrenodoxin, the reductase gives up electrons one at a time and traverses the semiquinone state.[6]

Adrenodoxin appears to function as a shuttle between adrenodoxin reductase and P-450$_{scc}$ with the association reactions with both of these proteins being primarily electrostatic in nature.[6] Recently, Millett's group has identified three carboxyl groups on adrenodoxin which are apparently involved in the interaction with adrenodoxin reductase and cytochrome c.[32] It is interesting to note that the reduction of cytochrome c by adrenodoxin does not require its dissociation from the reductase. Thus, it is not clear whether the carboxyl groups on adrenodoxin involved in the reduction of cytochrome c are the same ones which are involved in protein–protein interaction during electron transfer to P-450$_{scc}$.

2.3. Microbial

The bacterial reductases which transfer electrons to P-450 can be like either the microsomal or the mitochondrial enzymes. The most completely described enzyme system is the one found in *P. putida* which is involved in camphor catabolism.[33] This enzyme is a simple FAD-containing protein with a molecular weight of 48,000 which accepts electrons from NADH and sequentially transfers these electrons[34] to the appropriate iron–sulfur protein, putidaredoxin (M_r 12,500).[35] It is interesting to note that during induction of this catabolite-degrading enzyme system, the ratio of putidaredoxin reductase to putidaredoxin to P-450$_{cam}$ is approximately 1:8:8.[33] The activities measured with the purified reductase and putidaredoxin are sufficient to account for the observed turnover of P-450$_{cam}$ both in whole cells and in reconstituted systems.

Recent results have shown that the reaction cycle of putidaredoxin reductase (Fig. 2)[34] in the transfer of electrons to putidaredoxin is very similar to that of adrenodoxin reductase.[31] The "resting" enzyme must

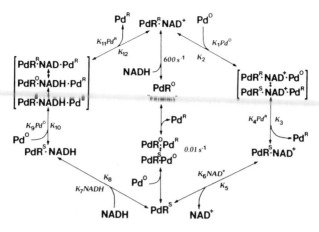

FIGURE 2. Reaction cycle of putidaredoxin reductase. The "resting" enzyme is pictured in the center of this cycle as PdR^0 which reacts with NADH in a very fast priming reaction to give the two-electron reduced enzyme $PdR^R \cdot NAD^+$. This form of the enzyme proceeds clockwise around the cycle first reacting with a molecule of putidaredoxin which removes an electron from the enzyme leaving the semiquinone form. Steady-state kinetic analysis indicates that the next step is dissociation of the NAD^+ followed by binding of a second molecule of NADH giving what is formally a three-electron reduced form. The final step of this cycle is the binding of a second molecule of putidaredoxin which removes one electron leaving the enzyme in the two-electron reduced-NAD^+ charge transfer complex.

undergo a priming reaction of the initial reduction of the flavin by NADH followed by binding of a single molecule of putidaredoxin and subsequent transfer of an electron. Kinetic studies have shown that NAD^+ dissociates from the semiquinone form of the enzyme, followed by the binding of a second molecule of NADH. It is this semiquinone–NADH complex of putidaredoxin reductase which is catalytically competent to bind a second molecule of putidaredoxin and to transfer the second electron from the initial NADH. With the dissociation of the reduced putidaredoxin, the catalytic cycle is complete, leaving the two-electron reduced putidaredoxin reductase–NAD^+ charge transfer complex. As is true in the case of adrenodoxin reductase–adrenodoxin, putidaredoxin serves as a mobile electron carrier between the reductase and P-450. In subsequent sections of this chapter, we will discuss the reduction of P-450 by putidaredoxin.

One of the striking differences between the mitochondrial and bacterial reductase–redoxin systems is the lack of ionic strength effect on the reduction of putidaredoxin by putidaredoxin reductase.[33] The effect of various salts on the bacterial system follows the Hoffmeister series for chaotrophic effect, rather than ionic strength. In contrast, the adrenodoxin reductase–adrenodoxin system seems to follow a simple ionic strength dependency,[36] indicating that in the latter case the protein–protein recognition and interaction is probably primarily electrostatic.

3. Reduction of P-450

As should be apparent from the reaction cycle for P-450, transfer of electrons to P-450 must be a very complex process involving the introduction of electrons at two chemically distinct reaction steps. The reduction of these forms of P-450 does not preclude the reduction of other forms of the enzyme. For example, the reduction potential of the low-spin, substrate-free form of P-450$_{cam}$[9,37] would seem to preclude reduction because its potential is about -270 mV. However, the potential of the high-spin, substrate-bound form of this enzyme is about -170 mV. Thus, this dramatic shift in the reduction potential is usually presumed to order the reaction cycle in the case of the bacterial enzyme. Recent studies of the reduction of camphor-free P-450$_{cam}$ have shown that the rate of this reaction is about 30% of the rate of reduction of the comparable camphor-bound enzyme (C. Newton-Bieker and J. A. Peterson, unpublished results). Similar shifts in the redox potential of mammalian microsomal P-450 have not been definitively demonstrated.[40,41] Initially, Waterman and Mason measured the redox potential of microsomal P-450 associated with the low-spin EPR signal of this enzyme. The value was shown to be about -360 mV.[38] More recent studies with purified and membrane-bound forms of P-450 have also shown the potential to be about -350 mV and not to vary systematically with the addition of substrate.[39,40]

Although most investigators have focused on the entry of the first electron into P-450 as an important process to study, this approach is a matter of convenience due to the difficulty of studying the introduction of the second electron. This second step is complicated by competing reactions and enzyme turnover which cloud the results obtained. Recently, several groups have begun to establish the factors governing the introduction of the second electron; an important electron transfer step in oxygen activation.[41,42]

3.1. Characteristics of the Reduction of P-450$_{cam}$

The transfer of an electron from reduced putidaredoxin to P-450$_{cam}$ occurs following the formation of a binary complex between these two physiologically functional partners.[43,44] The intracomplex electron transfer obeys first-order kinetics; however, the actual kinetics are dependent on the salt concentration, temperature, and pH of the reaction mixture. The simplest interpretation of the data is that following the rapid, reversible formation of the complex, there is an intracomplex "conformational" change which is temperature-dependent. This process is followed by a temperature-independent reaction which we have interpreted to be electron tunneling between the two redox active centers in this complex—the iron–sulfur center and the heme iron.

3.2. Characteristics of the Reduction in Microsomal Vesicles

In this part of the discussion of the reduction of microsomal P-450, we will focus on the introduction of the first electron and will leave the discussion of the properties of the second electron transfer reaction to the presentation of the role of cytochrome b_5 in this enzyme system.

Many years ago, it was observed that the reduction of P-450 could be readily followed by monitoring the formation of the carbon monoxide complex at 450 nm.[45] Early studies were hampered by the fact that the reaction was extremely sensitive to oxygen contamination and that under even the best of conditions used, the progress curve was clearly not a simple first-order plot. Gillette's group first noticed that the shape of the curve was markedly affected by the amount of oxygen present in the reaction solution. Another group measuring the rate of reduction of microsomal P-450, prepared from animals which had been pretreated with phenobarbital, found that the reaction could be resolved into two simple first order processes.[46] However, due to the nature of the equipment available, a substantial fraction of the reaction progress curve was missed because the components were mixed by hand and measured with spectrophotometers which had a response time of 0.5 sec^{-1}. With the advent of commercial stopped-flow spectrophotometers which could be utilized with turbid suspensions,[47,48] we were able to examine the time course of the complete reduction reaction from 5% to greater than 95% and to rapidly analyze many different sets of data.[49] We found that the reaction could be resolved into two first-order processes which for analytical purposes can be expressed as either two concurrent or consecutive reactions as shown in equations (2) and (3):

$$A \rightleftharpoons C$$
$$B \rightleftharpoons C' \qquad (2)$$

where A and B are separate, distinguishable reactants which are concurrently converted to the products C and C' which are spectrally *indistinguishable*.

$$A \rightleftharpoons B \rightleftharpoons C \qquad (3)$$

This type of reaction system involves the consecutive conversion of A to B and B to C. In the case where the conversion of B to C is fast compared to the conversion of A to B, this system will exhibit "burst" kinetics.

It is well recognized that typical first-order analyses are extremely sensitive to the analytical procedure utilized as well as the choice of the end point of the reaction.[50] Initially, we chose to use a nonlinear estimation procedure which involved a rather elaborate statistical analysis to determine the rate constants and the fraction of enzyme which was in each phase.[49,51] In each case, when our data had a correlation coefficient less than 0.9985, we subsequently found that not all of the P-450 had been reduced enzymatically during the time course of data collection. We used this procedure to calculate the fraction of enzyme which was reduced in either the fast or slow phases. Using this analytical procedure, the measurement of the reduction became routine and quite reproducible.[48] We typically use this procedure to teach stopped-flow spectrophotometry to graduate students and fellows.

Because of the nature of microsome-bound P-450, we chose to fit the data to equation (2) which is for two concurrent reactions. We noted that several precautions must be observed in the collection and analysis of similar kinetic data: (1) the reaction must go to completion, (2) oxygen must be rigorously excluded from the system, (3) the temperature must be accurately controlled, and (4) the microsomal fractions must be prepared in a highly reproducible fashion.[49] Matsubara et al.[51] reported that the apparent rate of reduction of P-450 in the fast phase at 25°C was a function of the particular drug substrate which was present in the reaction mixture and that for some substrates, the rate of reduction of P-450 was greater than the turnover number for the metabolism of the substrate. For other compounds, the rate was the same as the turnover number. It is no mere coincidence that the spectral species which we identify as the oxy form of P-450 accumulates during the steady state of metabolism of certain substrates, i.e., when the rate of introduction of the first electron is greater than the turnover of the enzyme. For similar reasons, we would not expect to observe the oxy form during the steady state of metabolism of drugs if the rate of introduction of the first electron is equal to the rate of drug metabolism.

The actual process for the interaction of NADPH-cytochrome P-450 reductase and P-450 has interested investigators for many years. In early observations by Franklin and Estabrook,[52] P-450 reductase was shown to be very sensitive to the mercurial, mersalyl. In microsomal preparations which had been treated with mersalyl, the fraction of P-450 which was reduced in the fast phase was decreased in proportion to the amount of reductase which had been inhibited by the mercurial. They interpreted this result to indicate that there were two populations of P-450 which were reduced by the reductase and that if a reductase molecule was inhibited, then "its" cytochromes P-450 apparently had to diffuse through the microsomal membrane to a different reductase to be reduced. The term which was used to express the segregation of cytochromes P-450 into these two populations was the "cluster" model.

More recently, Peterson et al.[49] conducted a rather detailed study of the temperature dependence of the reduction of microsomal P-450 from phenobarbital-pretreated rats. They concluded that the simplest explanation of their data for the temperature dependence of the biphasic kinetics was that the reduction proceeded via two concurrent first-order processes. It is interesting to note that it is mathematically impossible to differentiate between models consisting of two concurrent first-order processes or two consecutive first-order reactions if they have the rate constants measured for the P-450 system.[49] Two consecutive reactions have been referred to as exhibiting "burst" kinetics in the literature.[52,53] Several important lines of evidence led us to the conclusion which was expressed in that publication: (1) the data fit extremely well (correlation coefficient > 0.998) to the mathematical model; (2) the fraction of enzyme reduced in the fast phase did *not* correlate with the fraction of enzyme which was in the high-spin state; (3) as shown in Fig. 3, the majority (> 85%) of the enzyme was reduced in the fast phase at temperatures above 25°C while greater than 50% was reduced in the fast phase at all temperatures above 4°C; and (4) there was no break in the Arrhenius plot for the temperature dependence of the fast-phase rate constant, in contrast to the slow phase. This break in the plot seemed to be a function of the particular lipophilic substrate added to the reaction mixture. We felt that our data were best interpreted as indicating that there were two populations of P-450 which were reduced by *random* interactions with the reductase. Each population had unique rate constants which were dependent on the particular temperature, salt, and substrate concentration. Our naivete at the time led us to propose a "structural" model to account for this observation; however, we should have simply stated that the cluster or segregation of P-450 into two populations was due to the known microheterogeneity of the microsomal membrane and that the kinetics of reduction of P-450 reflected this segregation. Clearly, the cartoon which was drawn in that paper in 1976 is a vast oversimplification of the state of our *current* knowledge of the microsomal membrane, but was reasonable for the time it was proposed.[49]

Other hypotheses have been put forward to account for the nonlinearity of the kinetics of reduction of microsomal P-450. Most groups have fit the initial fast phase to a first-order curve, but Ullrich's group initially proposed that the reduction of P-450 in the slow phase was a second-order process reflecting the bimolecular collision between P-450 and the

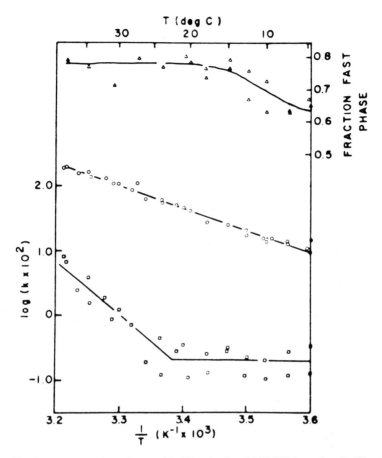

FIGURE 3. Temperature dependence of P-450 reduction. NADPH-dependent P-450 reductase activity was measured at different temperatures with the dual-wavelength stopped-flow apparatus.[48,49] ○, □: the values obtained for k_f and k_s, the rate constants for the fast and slow phases, respectively. △: the fraction of the P-450 which is reduced, in the fast phase.

reductase.[54] Sligar and Schenkman's group proposed that the biphasic nature reflected the rapid first-order reduction of high-spin (substrate bound?) enzyme followed by the slower reduction of the remaining enzyme as it was converted to the more readily reduced high-spin form.[55] This hypothesis has gone through several modifications including an initial hypothesis that the slow phase of reduction was in reality "substrate binding." However, transient-state measurements have clearly shown that substrate binding to microsomal P-450 is quite rapid on the time scale of reduction.[56]

It was the nonlinear analytical procedure discussed above as well as the knowledge of the rate of substrate binding to microsomal P-450 which led us away from explanations which included spin-state control of reduction. In addition, there is a smooth temperature-dependent effect on the spin state of microsomal P-450 as measured by the "binding spectrum."[37] Figure 3 shows the temperature dependence of the fast phase, slow phase, and the fraction of enzyme which was reduced in the fast phase in the presence of this substrate. Clearly, this temperature-dependent effect does not correlate with the spin state transition. It is interesting to note that the break in the temperature dependence of the slow-phase rate constant coincides with the shift in the fraction of enzyme which is reduced in the fast phase. Whether this results from a change in the fluidity of the membrane has not been determined to date for a membrane system which is as heterogeneous as microsomes.

Other experimental techniques have been utilized to determine whether P-450 and P-450 reductase formed stable aggregates in microsomal membranes. Analysis of absorption anisotropy of eosin-labeled P-450 in rat liver microsomes has shown that there are both mobile and immobile populations of P-450.[58] In fact, these results indicate that at 20°C, greater than 50% of the enzyme is immobile. However, a direct correlation cannot be made between the enzyme which is immobile and that which is reduced in either the fast or slow phase. Cross-linking studies of the interaction of the proteins of liver microsomes have demonstrated that complexes exist between proteins of 52,000 and 79,000 daltons.[59] In microsomes prepared from the livers of either phenobarbital-pretreated or control rats, oligomeric complexes were formed which contained three proteins, two of about 52,000 daltons and one about 79,000 daltons. The analytical methods employed did not permit determining the precise identity of the proteins concerned.

3.3. Characteristics of the Reaction in Various Reconstituted Systems

As biochemists, we often feel compelled to study a system by reducing it to the simplest possible components and learning about the individual parts, and reconstructing the system to learn if the whole system is simply the sum of the parts. While this experimental approach served our discipline admirably in the early study of soluble enzymes, we must continually keep in mind when studying membrane-bound enzymes that the intact membrane may be contributing to the interactions between the components. This is most readily apparent in some of the studies of the various membrane-bound electron transport systems.

Most groups will readily acknowledge that the ratio of P-450 reductase to P-450 is approximately 1:20 in microsomes prepared from the livers of phenobarbital-pretreated rats. As shown in Fig. 3, greater than 85% of the P-450 is reduced in a rapid, first-order process under temperature and salt conditions which are physiological. In contrast, purified P-450$_{LM_2}$ and P-450 reductase must be at a 1:1 ratio to achieve comparable activity in reconstituted systems.[60,61] An additional complication is that there is still a fast and a slow phase in this reconstituted system. Coon's group has proposed that the slow phase in the reconstituted system is a function of the reductase rather than some particular interaction between the reductase and P-450.[62] Others have extended their results obtained with the microsomal enzymes and have concluded that the fast phase and slow phase of reduction in purified enzyme systems are due to a slow spin state change of the enzyme.[63,64] Due to the complexity of the reaction catalyzed by P-450 reductase, we are not certain that the investigators who are studying the purified enzymes have taken into account the state of the reductase in the initial reaction mixture. Early studies of the reductase clearly demonstrated that the resting form was the air-stable, one-electron reduced species. It is not immediately apparent to us exactly what the effect on the overall kinetics of the reduction would be if the enzyme had to first be "primed" via the conversion to the one-electron reduced species. Given the complicated nature of the reaction, it is not difficult to envision problems in deducing which of the nine possible reduced forms of the reductase reacts with P-450.[4] It is certain that not all of the reduced forms will be equally reactive. It must always be remembered that the microsome-bound enzymes can achieve comparable activity to the purified enzymes with a mole ratio of reductase to P-450 of 1:20.

4. What Is the Role of b_5 in the Reduction of P-450?

4.1. The Physiological Function of b_5

The existence of another cytochrome in the microsomal membrane of most mammalian tissues, namely b_5, stimulated a series of studies to biochemically characterize this protein in the 1960s and 1970s. An excellent review on b_5 and its physiological significance was written by Oshino in 1980[65] and we will only briefly review this aspect of its function. To date, the physiological functions include those listed in Table I. The focus of the present review will be the possible role of b_5 in the electron transfer reactions involving P-450.

b_5 is a protohemeprotein which was first discovered by Sanborn and Williams as a component of cecropia silkworm larvae[79] and by Chance

TABLE I
Physiological Reactions Involving Cytochrome b_5

Reaction	Reference
Acyl CoA Δ^9-desaturase	66
Acyl CoA Δ^6 desaturase	65
Acyl CoA Δ^5-desaturase	65
Fatty acid (C2) elongation	67
Δ^7-Sterol Δ^5-desaturase	68
Phospholipid desaturase	69
1-Alkyl-2-acyl-sn-glycero-3-phosphoryl ethanolamine desaturase	70
β-Ketostearoyl-CoA reductase	71
4-Methylsterol oxidase	72
Methemoglobin reductase	73, 74
Phenol oxidase	75
Prostaglandin synthase reductase	76
N-Hydroxylamine reductase	77, 78
Cytochrome P-450 reductase	This review

and Williams as a component of rat liver microsomes.[80] Two research groups (Omura and Sato in Japan and Strittmatter in the United States) have performed many of the biochemical studies to elucidate the nature of this cytochrome. It exists as an amphipathic protein of 16,000 daltons and is composed of at least two domains: a hydrophobic portion which interacts with biological membranes and a hydrophilic portion containing the redox active heme moiety.[81,82] These two domains are connected by a flexible region of approximately 15 amino acid residues. The oxidized hemeprotein displays a wavelength maximum at 413 nm and the reduced cytochrome has wavelength maxima at 423, 526, and 557 nm.[83] Alterations in the redox status of the cytochrome can be followed kinetically by measuring absorbance changes utilizing the wavelength pairs of either 413–423 nm ($\Delta\epsilon = 185,000$ M^{-1} cm^{-1}), 424–500 nm ($\Delta\epsilon = 130,000$ M^{-1} cm^{-1}), or 557–567 nm ($\Delta\epsilon = 19,000$ M^{-1} cm^{-1}), respectively. Early preparations of the hemeprotein were obtained after proteolytic digestion of microsomal fractions to yield the hydrophilic domain of approximately 12,000 daltons.[83] Two microsomal flavoproteins are known to reduce the hemeprotein: NADH-cytochrome b_5 reductase[84] and NADPH-cytochrome c (P-450) reductase.[85] This fact explains why the membrane-bound cytochrome can be reduced with either NADH or NADPH and, therefore, either reduced pyridine nucleotide can serve as an electron donor for electron transfer reactions involving b_5.[86]

b_5 is ubiquitously distributed within mammalian tissues and is localized in several cellular compartments. The hemeprotein has been shown to exist in the endoplasmic reticulum of most cells and also on the outer aspect of the outer mitochondrial membrane.[65,87] These two locations may contain different gene products of b_5, as the proteins do not appear to be immunochemically identical. However, the b_5 reductases purified from either the microsomal or the outer mitochondrial membranes do appear to be nearly identical gene products.[88] In addition, erythrocytes possess a soluble form of the hemeprotein[73] which is immunochemically similar to the hepatic microsomal cytochrome.[74] A number of cell types, including lymphocytes and leukocytes, contain measurable quantities of b_5, possibly for the synthesis of certain necessary phospholipids.[66,89]

4.2. The Synergistic Effect of NADH and NADPH on P-450 Function

In the early 1970s, Cohen and Estabrook described the phenomenon of synergism of NADPH by NADH in promoting the function of P-450.[90-92] The K_m values for NADPH and NADH were measured independently for the demethylation of an N-methyl drug and the C-hydroxylation of an aromatic hydrocarbon and were shown to be approximately 0.01 and 5 mM, respectively.[90,93] The rate of metabolism in the presence of 0.25 mM NADH alone is only 5–10% as large as that seen with 0.25 mM NADPH. However, the rate of metabolism of aminopyrine (N-demethylation) increases nearly twofold (Fig. 4) when equimolar amounts of both pyridine nucleotides were used. This synergism, as described by Cohen and Estabrook, also included the fact that total product formation (HCHO from aminopyrine) increased nearly twofold as well (Fig. 4). Cohen and Estabrook demonstrated that while microsomal oxygen and NADPH utilization were nearly stoichiometric, HCHO produced during metabolism of aminopyrine was only 50–60% as large. A portion of the NADPH utilized was shown to result in H_2O_2 production.[94] In the presence of both NADH and NADPH, the stoichiometry of formaldehyde formation from aminopyrine and NADPH oxidation approached a value of 0.9. However, the utilization of NADH was also increased twofold, and the extra reducing equivalents provided by NADH could not be accounted for by Cohen and Estabrook.[90,91] It was postulated that NADH saturated electron transfer to an endogenous substrate-metabolizing pathway(s), thereby allowing NADPH to be preferentially utilized in the P-450-dependent reaction. In a subsequent study, Hildebrandt and Estabrook[95] provided partial evidence that b_5 may participate in electron transfer to P-450. It was hypothesized that the first electron donated to reduce P-450 most likely is provided by P-450 reductase, but that the second electron donated to reduce oxycytochrome P-450 (formally $Fe^{2+}-O_2-S$) may be donated by another component, presumably b_5 (Fig. 5). While a number of other substrates have been studied besides aminopyrine, no additional

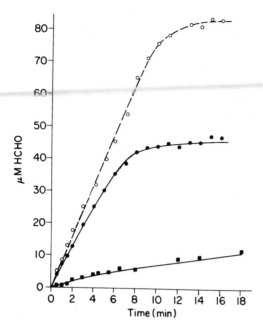

FIGURE 4. Aminopyrine metabolism with NADH and/or NADPH as sources of reducing equivalents. Microsomes (1.5 mg/ml) were incubated in the presence of 8 mM aminopyrine and either 0.11 mM NADH (■), 0.09 mM NADPH (●), or 0.11 NADH plus 0.09 mM NADPH (○). NADPH-regenerating system was included and the product of aminopyrine metabolism, HCHO, was measured as a function of time.[26]

insight was provided to explain the original phenomenon described by Cohen and Estabrook.[90–92]

4.3. Subsequent Studies on the Role of b_5 in Microsomal Electron Transport Reactions

Further understanding of the role of b_5 in the function of P-450 required the development of other experimental techniques to test the in-

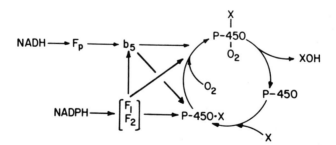

FIGURE 5. A scheme for microsomal electron transport reactions of P-450. Fp, NADH-cytochrome b_5 reductase; F_1F_2, NADPH-cytochrome P-450 reductase; X, hydroxylatable substrate. The ferrous oxycytochrome can also decompose to yield H_2O_2 (not shown).

teractions between the two cytochromes. Two approaches have led to our current understanding of this problem: (1) development of specific inhibitory antibodies against the various microsomal proteins, and (2) the purification (and reconstitution) of the native microsomal proteins.

Several research groups developed inhibitory antibodies elicited toward b_5 and the two flavoprotein reductases, NADH-cytochrome b_5 reductase and NADPH-cytochrome c (P-450) reductase purified after proteolytic treatment of microsomal fractions. Studies with these inhibitory antibodies confirmed the role of b_5 and b_5 reductase in electron transfer from NADH to the fatty acid desaturase[21,96] and of P-450 reductase to P-450.[19,20] The anti-b_5 and anti-b_5 reductase immunoglobulins had high affinities toward the liver microsome-bound proteins and inhibited the NADH-dependent reduction of c by rat liver microsomes by more than 90%. Similarly, the anti-c (P-450) reductase immunoglobulin potently inhibited the NADPH-dependent reduction of both c and P-450. Mannering et al. utilized the anti-b_5 globulin to establish whether the NADH synergism of the NADPH-dependent demethylation of ethylmorphine was dependent upon electron transfer by b_5.[97] In the presence of both NADH and NADPH, the anti-b_5 globulin prevented the synergism observed by Cohen and Estabrook. However, the antibody did not affect the reaction performed in the presence of NADPH alone. The effect of the inhibitory antibody to b_5 on the NADH-synergism could be reversed by preincubating the antibody with purified b_5. These results suggested that, indeed, b_5 appeared to be involved as an electron donor in the synergism phenomenon, but in the presence of NADPH alone, some other electron donor, such as P-450 reductase, must provide the second electron.

Prough and Burke[93] utilized the anti-P-450 reductase immunoglobulin to study the NADH-dependent 2- and 4-hydroxylation of biphenyl by rat liver and lung microsomes. At all concentrations of NADH studied (0.25–10 mM), the antibody to P-450 reductase potently inhibited the hydroxylation of biphenyl. 2'-AMP, a potent inhibitor of c (P-450) reductase, also inhibited the NADH-dependent hydroxylation of biphenyl. In addition, Prough and Masters showed that NADH could reduce purified c (P-450) reductase.[98] The apparent K_m was noted to be approximately 5 mM and the turnover numbers for both c and flavin reduction were identical to those obtained with NADPH. These results indicated that at least one electron from NADH must be transferred to P-450 by P-450 reductase during the reaction.

Noshiro and Omura[99] used the anti-rat liver immunoglobulins to b_5, b_5 reductase, and c (P-450) reductase to study the role of these proteins in electron transfer to P-450 in mouse liver microsomes. Anti-c (P-450) reductase potently inhibited both the NADH- and NADPH-dependent metabolism of a number of compounds: 7-ethoxycoumarin, benzo[a]pyrene, benzphetamine, and aniline. This observation supported the contention of Prough and Burke[93] that P-450 reductase must be involved in the transfer of at least one electron to P-450. The anti-b_5 and b_5 reductase globulins had no effect on the NADPH-dependent metabolism of the substrates studied. However, the anti-b_5 globulin inhibited the NADH-dependent metabolism of all of the substrates, except aniline. Anti-b_5 reductase only inhibited the NADH-dependent metabolism of 7-ethoxycoumarin and benzphetamine. Taken together, these results provide additional evidence that the first electron donated to P-450 must be transferred by c(P-450) reductase, but that depending on the rate-limiting step of the reaction (transfer of the second electron versus some other process), b_5 may provide the second electron when NADH is present. Other reactions which compete with P-450 for reducing equivalents may affect the rate of reaction and the stoichiometry of product formation relative to NADPH or oxygen consumption.

It is interesting to note that Sasame et al.[100] have observed that the NADH- and NADPH-dependent hydroxylation reactions of laurate catalyzed by kidney cortex and liver microsomes were inhibited by the antibody to b_5. Burke and Prough also noted that anti-b_5 inhibited the NADPH-dependent 4-hydroxylation of biphenyl by lung microsomes (unpublished results). However, unlike the work of Sasame et al., anti-P-450 reductase globulin inhibited both the NADH- and NADPH-supported reactions in lung microsomes. While the antibody used by Sasame et al. displayed titers weaker than those of Noshiro and Omura,[99] the results suggest that under certain conditions b_5 may provide reducing equivalents necessary for the function of P-450.

4.4. Studies Involving the Reconstitution of Purified Microsomal Enzymes

An initial study on the involvement of b_5 in P-450-mediated reactions using reconstituted systems containing various purified enzymes from rat liver microsomes was performed by Lu and Levin's groups.[101,102] Their results suggested that only a limited number of NADPH-dependent reactions, notably the C-hydroxylation of chlorobenzene, catalyzed by reconstituted P-450 were potentiated by the addition of b_5 to the reaction mixture. They also demonstrated that the b_5/b_5 reductase system could provide electrons to mediate the metabolism of benzo[a]pyrene in the presence of NADH.[102] However, addition of P-450 reductase with the other flavoprotein and b_5 enhanced metabolism, but this effect was dependent on the concentration of NADH added to the reaction mixtures. These results were interpreted to indicate that only specific forms of the monooxygenase may accept electrons from b_5 and that the metabolism of certain compounds (chlorobenzene) was enhanced by b_5 when compared to substrates like benzphetamine or benzo[a]pyrene. Electron transfer at high concentrations of NADH involved both b_5 and P-450 reductase as suggested by the work of Prough and Burke[93] and Noshiro and Omura.[99] Similar studies by Imai and Sato[103,104] confirmed a role for b_5 from rabbit in the NADPH-dependent reactions of reconstituted P-450 with two other substrates, benzphetamine and N,N-dimethylaniline. They observed that the stoichiometry of product formation from benzphetamine relative to NADPH oxidized changed from approximately 0.6 to 0.9 in the presence of b_5. This result was similar to that seen by Cohen and Estabrook.[90] In addition, the presence of b_5/b_5 reductase in the reaction mixture allowed the utilization of electrons from NADH to provide reducing equivalents to the monooxygenase. Based on these results, they suggested that b_5 may function in two possible ways: (1) increasing the stoichiometry (or coupling) of the reaction, and (2) supplying the second electron to P-450.

Subsequently, Yamano's group[105-107] immobilized trypsin-solubilized b_5 on a Sepharose 6B matrix in an attempt to isolate forms of P-450 from rabbit liver which have a high affinity for b_5. The isozyme isolated demonstrated an obligatory requirement for b_5 when reconstituted with the two flavoprotein reductases (b_5 reductase and P-450 reductase) and phospholipid. Both flavoproteins were required for maximal activity, but P-450 reductase alone could partially support turnover of the P-450 ($\sim$ 20%). However, maximal activity required a 1:1 molar ratio of the two cytochromes. The synergistic effect of NADH was most notable when lower reduction levels of b_5 existed in the steady state of the reaction. When various heme- or metal-substituted b_5 derivatives were utilized, substitution of the protoheme by meso- or deuteroheme leads to similar turnover numbers when a 1:1 ratio of b_5: P-450 was obtained. However, Co-protoporphyrin-substituted b_5 did not effectively replace the native hemeprotein and apocytochrome b_5 was not capable of electron transport. Similar results were shown by Morgan and Coon[108] for the Mn-substituted b_5 with rabbit liver microsomal P-450$_{LM2}$, except the required molar ratio for the two hemeproteins was 2:1. The existence of molecular complexes between the two cytochromes has also been shown by Chiang[109] and Bosterling and Trudell[110] in reconstituted vesicles.

Recently, Werringloer et al. noted that there is a pH dependence of the synergistic effect of NADH on NADPH-dependent monooxygenase reactions of rat liver.[111] Their results indicated that the pH profile for the NADH synergism was similar to the differences in the pH-dependent steady-state levels of b_5 reduced by NADPH plus NADH relative to NADPH. Both parameters were maximal at low pH (6.5–7.2) and were minimal at high pH (7.5–9.0). However, the amounts of oxycytochrome

P-450 were very different, i.e., the steady-state amount of oxy-P-450 was minimal at low pH (pH 6.5–7.2) and maximal at high pH (above 7.5). Noshiro et al.[112] observed that the steady-state levels of oxy-P-450 were increased by high ionic strength or pH. Decreased pH, low ionic strength, or added b_5 all lowered the steady-state level of oxy-P-450. In support of these observations, Noshiro et al. used an inhibitory anti-b_5 globulin to demonstrate that the antibody was inhibitory to monooxygenase activity at low pH, but not at high pH. These studies support the contention that b_5 (Fe^{2+}) can reduce oxy-P-450, but indicate the wide number of variables which must be controlled to demonstrate the role of b_5.

Werringloer[113,114] has proposed a "counterpoise-regulation" of the steady-state concentration of oxy-P-450 which defines the rate-limiting steps of the reactions catalyzed by P-450. Based on his studies of the pH profiles of the various reactions measured (monooxygenase and H_2O_2 production) and the steady-state levels of the various forms of the two cytochromes [b_5 (Fe^{2+}/Fe^{3+}), P-450 (high-/low-spin, and oxy-P-450], he has predicted that two rate-limiting steps exist for monooxygenase function. At low pH, the rate-limiting step is reduction of the ferric high-spin hemeprotein and at high pH, the rate-limiting step is reduction of oxy-P-450. As the pH of the cell is 6.8–7.2, it would appear that the physiological rate-limiting step most likely is the reduction of ferric, high-spin P-450. However, this hypothesis requires more study and the reconstituted cytochromes must be examined to eliminate artifacts due to other components of the microsomal membrane. More recently, Bonfils et al.[115] and Pompon and Coon[41] have directly demonstrated that ferrous b_5 can reduce the ferrous dioxygen form of certain isozymes of P-450. The reactions required 1:1 molar ratios of the two cytochromes for maximal reaction and the rate constants for reduction were sufficient to account for the turnover of oxy-P-450 during catalysis. Pompon and Coon have shown that this reaction is dependent on the isozyme of P-450 used and that under certain circumstances, oxy-P-450 may reduce ferric b_5. They also indicated that the various isozymes studied may possess different rate-limiting steps and differing electron-accepting properties toward b_5 or P-450 reductase.

5. Problems for the Future

While the study of the enzyme system which has become known as "cytochrome P-450" is completing its first quarter century, the future bodes well for a more complete understanding of how this fascinating complex of enzymes functions. Other chapters of this review have dealt with many of the ongoing studies related to the synthesis and genetic regulation of this system so we will confine our projections to studies of the reductase–P-450 interaction and its definition.

5.1. Electron Transfer Mechanism

As we obtain a more complete picture of the structure of these proteins, we will be able to describe the specific residues involved in electron transfer from the input of electron equivalents to their transfer to the heme iron of P-450. One of the central questions in the study of electron transfer reactions is the mechanism/pathway of transfer between two redox active centers. Does the reaction involve tunneling through space without the intervention of bonds or must there be a pathway which includes overlapping π orbitals of aromatic amino acid residues? As the three-dimensional structure of the enzymes is determined, the residues, which might be involved in protein–protein recognition, binding, or electron transfer, will be identified. These residues could be modified by site-specific mutagenesis which can more clearly define the role of these residues in electron transfer.

5.2. Protein–Protein Interaction during Electron Transfer

Work on the interaction of the various reductases and their electron acceptor molecules has convincingly demonstrated that there are specific electrostatic recognition sites on the protein surfaces. Using the combined techniques of chemical modifications, protein structural analysis, and site-specific mutagenesis, we should be able to clearly describe the determinants which control the specific interactions between these proteins.

5.3. Metabolic Control of P-450

While we have dwelled on the events of electron transfer between the various proteins of this monoxygenase system, it should be clear that the availability of reducing equivalents in the form of reduced pyridine nucleotides will also radically affect the balance of which reactions of this system predominate. The availability of extremely powerful microcomputers and associated software and graphics should enable us to explore the mathematical modeling of the metabolic control of the P-450 system. The goal of the modeling studies should be to first replicate the *in vitro* results, but investigators involved in such studies must bear in mind that the reaction conditions in a test tube may not mimic the cellular conditions very well. The perturbations of protein concentration, $NADP^+/NADPH$, $NAD^+/NADH$ ratios, and oxygen and substrate availability will certainly result in alterations of the activity of these and competing reactions.

Another area of needed future study is the characterization of the reactions of oxy-P-450. As seen in the previous sections, little is known about the chemical and biochemical properties of this species of P-450. In terms of the reaction mechanism of the hemeprotein, an understanding of the processes involved in the formation and decomposition of this species of the cytochrome is required for the final description of its biochemical mechanism.

To date, several possible species of reduced oxygen have been postulated to be formed during the decomposition of the oxy-P-450 in the absence of hydroxylatable substrate; they include superoxide anion radical, hydrogen peroxide, and water. A thorough investigation of the form-specific decomposition of this penultimate species either substrate-bound or free must be performed to fully understand the chemical mechanism of the reaction cycle of P-450. In addition, we will need to more clearly define the chemical reactions required for the activation of molecular oxygen and substrate in the steps prior to and during oxygen insertion. A final area related to the metabolic control of the cytochrome will involve study of the enzyme at the molecular level. A number of groups have been involved in understanding the DNA sequence and mechanism of gene expression for the family of cytochromes P-450 (see Chapter 10). This study will provide exciting information about the number and evolution of this unique family of hemeproteins. The complete details of the role of the Ah receptor in the induction of a certain class of the cytochromes are also an area of intense study. The tools of modern molecular biology will aid in this quest and provide complete information about the control of the Ah receptor levels *in vivo*. Further work will be necessary to understand how other inducers which do not utilize the Ah receptor bring about the increase in the tissue content of P-450. These areas remain an important future area of study due to their extreme importance in cancer research, pharmacology, and toxicology.

5.4. The Use of Site-Specific Mutagenesis—Will It Help Us to Understand the Mechanism of This Enzyme System?

With the advent of the exciting possibilities of molecular biology, new vistas will open up to those investigators who can avail themselves of the tools which will be made available. The use of site-specific mutagenesis to explore the determinants of substrate binding and recognition as well as electron transfer and protein–protein interaction is very attractive. However, we must always remember that the tools afforded by molecular biology are like any other new experimental technology: it will be fraught with false starts and wrong answers if we rely on it solely for our answers. We, as explorers of this fascinating enzyme system, will obtain lasting answers only through the complementary use of these and other biophysical, biochemical, pharmacological, and physiological tools.

ACKNOWLEDGMENTS. After completing a survey of the recent literature, it was with some trepidation that we began to actually formulate this review. The level of active interest and research on this field has grown beyond belief in the past six years. Over 1300 papers have appeared which have cytochrome P-450 and either reduction or reductase as keywords. Thus, it was inevitable that we would unintentionally omit citations which might have been more appropriately included than the ones which were selected. In addition, we have chosen to omit some topics which should have been covered but for which there was not adequate space. We apologize to those in both of these classes.

This work was supported in part by grants from the USPHS (GM-19036, JAP; CA-32511, RAP), the American Cancer Society (BC-336, RAP), and the R. A. Welch Research Foundation (I-405, JAP; I-616, RAP).

REFERENCES

1. Mason, H. S., Fowlks, W. L., and Peterson, E., 1955, Oxygen transfer and electron transport by the phenolase complex, *J. Am. Chem. Soc.* **77**:2914–2915.
2. Hayaishi, O., Katagiri, M., and Rothberg, S., 1955, Mechanism of the pyrocatechase reaction, *J. Am. Chem. Soc.* **77**:5450–5451.
3. White, R. E., and Coon, M. J., 1980, Oxygen activation by cytochrome P-450, *Annu. Rev. Biochem.* **49**:315–356.
4. Iyanagi, T., and Mason, H. S., 1973, Some properties of hepatic reduced nicotinamide adenine dinucleotide phosphate-cytochrome *c* reductase, *Biochemistry* **12**:2297–2308.
5. Vermilion, J. L., and Coon, M. J., 1978, Identification of the high and low potential flavins of liver microsomal NADPH-cytochrome P-450 reductase, *J. Biol. Chem.* **253**:8812–8819.
6. Lambeth, J. D., and Kamin, H., 1976, Properties of the complexes of reduced enzyme with $NADP^+$ and NADPH, *J. Biol. Chem.* **251**:4299–4306.
7. Estabrook, R. W., Hildebrandt, A., Remmer, H., Schenkman, J. B., Rosenthal, O., and Cooper, D. Y., 1968, Role of cytochrome P-450 in microsomal mixed-function oxidation, in: *Biochem. Sauerst., Colloq. Ges. Biol. Chem., 19th* (B. Hess and H. Staudinger, eds.), Springer-Verlag, Berlin, pp. 142–177.
8. Tsai, R., Yu, C.-A., Gunsalus, I. C., Peisach, J., Blumberg, W., Orme-Johnson, W. H., and Beinert, H., 1970, Spin-state changes in cytochrome $P-450_{cam}$ on binding of specific substrates, *Proc. Natl. Acad. Sci. USA* **66**:1157–1163.
9. Peterson, J. A., 1971, Camphor binding by *Pseudomonas putida* cytochrome P-450, *Arch. Biochem. Biophys.* **144**:678–693.
10. Peterson, J. A., White, R. E., Yasukochi, Y., Coomes, M. L., O'Keeffe, D. H., Ebel, R. E., Masters, B. S. S., Ballou, D. P., and Coon, M. J., 1976, Evidence that purified cytochrome $P-450_{LM}$ is a one electron acceptor, *J. Biol. Chem.* **251**:4010–4016.
11. Estabrook, R. W., Hildebrandt, A. G., Baron, J., Netter, K. J., and Leibman, K.,

1971, A new spectral intermediate associated with cytochrome P-450 function in liver microsomes, *Biochem. Biophys. Res. Commun.* **42:**132–139.
12. Ishimura, Y., Ullrich, V., and Peterson, J. A., 1971, Oxygenated cytochrome P-450 and its possible role in enzymic hydroxylation, *Biochem. Biophys. Res. Commun.* **42:**140–146.
13. Peterson, J. A., Ishimura, Y., and Griffin, B. W., 1972, *Pseudomonas putida* cytochrome P-450: Characterization of an oxygenated form of the hemoprotein, *Arch. Biochem. Biophys.* **149:**197–208.
14. DePierre, J. W., and Ernster, L., 1977, Enzyme topology of intracellular membranes, *Annu. Rev. Biochem.* **46:**201–262.
15. Pearse, B. M. F., and Bretscher, M. S., 1981, Membrane recycling by coated vesicles, *Annu. Rev. Biochem.* **50:**85–101.
16. Peterson, J. A., O'Keeffe, D. H., Werringloer, J., Ebel, R. E., and Estabrook, R. W., 1978, Patches and pockets: The microenvironments of a membrane bound hemeprotein, in: *Microenvironments and Metabolic Compartmentation* (P. A. Srere and R. W. Estabrook, eds.), Academic Press, New York, pp. 433–450.
17. Horecker, B. C., 1950, Triphosphopyridine nucleotide-cytochrome c reductase in liver, *J. Biol. Chem.* **183:**593–605.
18. Coon, M. J., Strobel, H. W., and Boyer, R. F., 1973, On the mechanism of hydroxylation reactions catalyzed by cytochrome P-450, *Drug Metab. Dispos.* **1:**92–97.
19. Raftell, M., and Orrenius, S., 1970, Preparation of antisera against cytochrome b_5 and NADPH-cytochrome c reductase from rat liver microsomes, *Biochim. Biophys. Acta* **233:** 358–365.
20. Masters, B. S. S., Baron, J., Taylor, W. E., Isaacson, E. L., and LoSpalluto, J., 1971, Immunochemical studies on electron transport chains involving cytochrome P-450. I. Effects of antibodies to pig liver microsomal reduced triphosphopyridine nucleotide-cytochrome c reductase and the non-heme iron protein from bovine adrenocortical mitochrondria, *J. Biol. Chem.* **246:**4143–4150.
21. Oshino, N., and Omura, T., 1973, Immunochemical evidence for the participation of cytochrome b_5 in microsomal stearyl-CoA desaturation reaction, *Arch. Biochem. Biophys.* **157:**395–404.
22. Yasukochi, Y., and Masters, B. S. S., 1976, Some properties of a detergent-solubilized NADPH-cytochrome c (cytochrome P-450) reductase purified by biospecific affinity chromatography, *J. Biol. Chem.* **251:**5337–5344.
23. Gum, J. R., and Strobel, H. W., 1979, Isolation of the membrane-binding peptide of NADPH-cytochrome P-450 reductase: Characterization of the peptide and its role in the interaction of reductase with cytochrome P-450, *J. Biol. Chem.* **254:**4177–4185.
24. Black, S. D., French, J. S., Williams, C. H., Jr., and Coon, M. J., 1979, Role of a hydrophobic polypeptide in the N-terminal region of NADPH-cytochrome P-450 reductase in complex formation with P-450$_{LM}$, *Biochem. Biophys. Res. Commun.* **91:**1528–1535.
25. Yasukochi, Y., Peterson, J. A., and Masters, B. S. S., 1979, NADPH-cytochrome c(P-450 reductase): Spectrophotometric and stopped-flow kinetic studies on the formation of reduced flavoprotein intermediates, *J. Biol. Chem.* **254:**7079–7104.
26. Vermilion, J. L., and Coon, M. J., 1978, Purified liver microsomal NADPH-cytochrome P-450 reductase: Spectral characterization of oxidation-reduction states, *J. Biol. Chem.* **253:**2694–2704.
27. Omura, T., Sanders, E., Estabrook, R. W., Cooper, D. Y., and Rosenthal, O., 1966, Isolation from adrenal cortex of a nonheme iron protein and a flavoprotein functional as a reduced triphosphopyridine nucleotide-cytochrome P-450 reductase, *Arch. Biochem. Biophys.* **177:** 660–673.

28. Chu, J.-W. and Kimura, T., 1973, Adrenal steroid hydroxylases: Complex formation of the hydroxylase components, *J. Biol. Chem.* **248**:2089–2094.
29. Huang, J. J., and Kimura, T., 1973, Studies on adrenal steroid hydroxylases: Oxidation-reduction properties of adrenal iron-sulfur protein (adrenodoxin), *Biochemistry* **12**:406–409.
30. Estabrook, R. W., Baron, J., Peterson, J., and Ishimura, Y., 1972, Oxygenated cytochrome P-450 as an intermediate in hydroxylation reactions, in: *Biological Hydroxylation Mechanisms* (G. S. Boyd and R. M. S. Smellie, eds.), Academic Press, New York, pp. 159–185.
31. Lambeth, J. D., and Kamin, H., 1977, Adrenodoxin reductase and adrenodoxin: Mechanisms of reduction of ferricyanide and cytochrome c, *J. Biol. Chem.* **252**:2908–2917.
32. Lambeth, J. D., Geren, L. M., and Millett, F. 1984, Adrenodoxin interaction with adrenodoxin reductase and cytochrome P-450$_{scc}$: Cross-linking of protein complexes and effects of adrenodoxin modification by EDC, *J. Biol. Chem.* **259**:10025–10029.
33. Roome, P. W., Jr., Philley, J. C., and Peterson, J. A., 1983, Purification and properties of putidaredoxin reductase, *J. Biol. Chem.* **258**:2593–2598.
34. Roome, P. W., Jr., 1985, Mechanism of and physical properties of putidaredoxin reductase, Ph.D. dissertation, The University of Texas Health Science Center at Dallas, Southwestern Graduate School of Biomedical Sciences.
35. Tanaka, M., Hanu, M., Yasunobu, K. T., Dus, K., and Gunsalus, I. C., 1974, Amino acid sequence of putidaredoxin, an iron-sulfur protein from *Pseudomonas putida*, *J. Biol. Chem.* **249**:3689–3701.
36. Lambeth, J. D., Seybert, D. W., and Kamin, H., 1979, Ionic effects on adrenal steroidogenic electron transport: The role of adrenodoxin as an electron shuttle, *J. Biol. Chem.* **254**:7255–7264.
37. Sligar, S., 1976, Coupling of spin, substrate, and redox equilibria in cytochrome P-450, *Biochemistry* **15**:5399–5406.
38. Waterman, M. R., and Mason, H. S., 1972, Redox properties of liver cytochrome P-450, *Arch. Biochem. Biophys.* **150**:57–63.
39. Guengerich, F. P., Ballou, D. P., and Coon, M. J., 1975, Purified liver microsomal cytochrome P-450: Electron-accepting properties and oxidation-reduction potential, *J. Biol. Chem.* **250**:7405–7414.
40. Backstrom, D., Ingelman-Sundberg, M., and Ehrenberg, A., 1983, Oxidation-reduction potential of soluble and membrane-bound rabbit liver microsomal cytochrome P-450$_{LM2}$, *Acta Chem. Scand.* **37**:891–894.
41. Pompon, D., and Coon, M. J., 1984, On the mechanism of action of cytochrome P-450: Oxidation and reduction of the ferrous dioxygen complex of liver microsomal cytochrome P-450 by cytochrome b_5, *J. Biol. Chem.* **259**:15377–15385.
42. Brewer, C. B., and Peterson, J. A., 1985, Single turnover studies with oxy-P-450$_{cam}$, *Fed. Proc.* **44**:1609.
43. Hintz, M. J., and Peterson, J. A., 1981, The kinetics of reduction of cytochrome P-450$_{cam}$ by reduced putidaredoxin, *J. Biol. Chem.* **256**:6721–6728.
44. Hintz, M. J., Mock, D. M., Peterson, L. L., Tuttle, K., and Peterson, J. A., 1982, Equilibrium and kinetic studies of the interaction of cytochrome P-450$_{cam}$ and putidaredoxin, *J. Biol. Chem.* **257**:14324–14332.
45. Gigon, P. L., Gram, T. E., and Gillette, J. R., 1968, Effect of drug substrates on the reduction of hepatic microsomal cytochrome P-450 by NADPH, *Biochem. Biophys. Res. Commun.* **31**:558–562.
46. Holtzman, J. L., and Carr, M. L., 1972, The temperature dependence of the components of the hepatic microsomal mixed-function oxidases, *Arch. Biochem. Biophys.* **150**:227–234.

47. Peterson, J. A., and Mock, D. M., 1975, Dual wavelength stopped-flow spectrophotometry: computer acquisition and analysis, *Anal. Biochem.* **68**:545–553.
48. Peterson, J. A., Ebel, R. E., and O'Keeffe, D. H., 1978, Computerized stopped-flow spectrophotometric measurement of cytochrome P-450 reductase, *Methods Enzymol.* **52**:221–226.
49. Peterson, J. A., Ebel, R. E., O'Keeffe, D. H., Matsubara, T., and Estabrook, R. W., 1976, Temperature dependence of cytochrome P-450 Reduction: A model for NADPH-cytochrome P-450 reductase: cytochrome P-450 interaction, *J. Biol. Chem.* **251**:4010–4016.
50. Hiromi, K., 1979, in: *Kinetics of Fast Enzyme Reactions*, Halsted Press, New York, p. 247.
51. Matsubara, T., Baron, J., Peterson, L. L., and Peterson, J. A., 1976, NADPH-cytochrome P-450 reductase, *Arch. Biochem. Biophys.* **172**:463–469.
52. Franklin, M. R., and Estabrook, R. W., 1971, On the inhibitory action of mersalyl on microsomal drug oxidation: A rigid organization of the electron transport chain, *Arch. Biochem. Biophys.* **143**:318–329.
53. Backes, W. L., Sligar, S. G., and Schenkman, J. B., 1980, Cytochrome P-450 reduction exhibits burst kinetics, *Biochem. Biophys. Res. Commun.* **97**:860–867.
54. Diehl, H., Schadelin, J., and Ullrich, V., 1970, Studies on the kinetics of cytochrome P-450 reduction in rat liver microsomes, *Hoppe-Seylers Z. Physiol. Chem.* **351**:1359–1371.
55. Cinti, D. L., Sligar, S. G., Gibson, G. G., and Schenkman, J. B., 1979, Temperature-dependent spin equilibrium of microsomal and solubilized cytochrome P-450 from rat liver, *Biochemistry* **18**:36–42.
56. Ristau, O., Rein, H., Janig, G.-R., and Ruckpaul, K., 1978, Quantitative analysis of the spin equilibrium of cytochrome P-450$_{LM2}$ fraction from rabbit liver microsomes, *Biochim. Biophys. Acta* **536**:226–234.
57. Ebel, R. E., O'Keeffe, D. H., and Peterson, J. A., 1978, Substrate binding to hepatic microsomal cytochrome P-450: Influence of the microsomal membrane, *J. Biol. Chem.* **253**:3888–3897.
58. Kawato, S., Gut, J., Cherry, R. J., Winterhalter, K. H., and Richter, C., 1982, Rotation of cytochrome P-450. I. Investigations of protein–protein interactions of cytochrome P-450 in phospholipid vesicles and liver microsomes, *J. Biol. Chem.* **257**:7023–7029.
59. Baskin, L. S., and Yang, C. S., 1982, Cross-linking studies of the protein topography of rat liver microsomes, *Biochim. Biophys. Acta* **684**:263–271.
60. Miwa, G. T., West, S. B., Huang, M.-T., and Lu, A. Y. H., 1979, Studies on the association of cytochrome P-450 and NADPH-cytochrome c reductase during catalysis in a reconstituted hydroxylating system, *J. Biol. Chem.* **254**:5695–5700.
61. French, J. S., Guengerich, F. P., and Coon, M. J., 1980, Interactions of cytochrome P-450, NADPH-cytochrome P-450 reductase, phospholipid, and substrate in the reconstituted liver microsomal enzyme system, *J. Biol. Chem.* **255**:4112–4119.
62. Oprian, D. D., Vatsis, K. P., and Coon, M. J., 1974, Kinetics of reduction of cytochrome P-450$_{LM4}$ in a reconstituted liver microsomal enzyme system, *J. Biol. Chem.* **254**:8895–8902.
63. Cinti, D. L., Sligar, S. G., Gibson, G. G., and Schenkman, J. B., 1979, Temperature-dependent spin equilibrium of microsomal and solubilized cytochrome P-450 from rat liver, *Biochemistry* **18**:36–42.
64. Taniguchi, H., Imai, Y., Iyanagi, T., and Sato, R., 1979, Interaction between NADPH-cytochrome P-450 reductase and cytochrome P-450 in the membrane of phosphatidylcholine vesicles, *Biochim. Biophys. Acta* **550**:341–356.
65. Oshino, N., 1980, Cytochrome b$_5$ and its physiological significance, in: *Hepatic Cy-*

tochrome P-450 Monooxygenase System (J. B. Schenkman and D. Kupfer, eds.), Pergamon Press, Elmsford, N. Y., pp. 407–447.

66. Oshino, N., Imai, Y., and Sato, R., 1971, Function of cytochrome b_5 in fatty acid desaturation by rat liver microsomes, *J. Biochem.* **69:**155–167.
67. Keyes, S. R., Alfano, J. A., Jansson, I., and Cinti, D. L., 1979, Rat liver microsomal elongation of fatty acids: Possible involvement of cytochrome b_5, *J. Biol. Chem.* **254:**7778–7784.
68. Reddy, V. V., Kupfer, D., and Caspi, E., 1977, Mechanism of C-5 double bond introduction in the biosynthesis of cholesterol by rat liver microsomes: Evidence for the participation of cytochrome b_5, *J. Biol. Chem.* **252:**2797–2801.
69. Pugh, E. L., and Kates, M., 1977, Direct desaturation of eicosatrienoyl lecithin to arachidonyl lecithin by rat liver microsomes, *J. Biol. Chem.* **252:**68–73.
70. Paltauf, F., Prough, R. A., Masters, B. S. S., and Johnston, J. M., 1974, Evidence for the participation of cytochrome b_5 in plasmalogen synthesis, *J. Biol. Chem.* **249:**2661–2662.
71. Nagao, M., Ishibishi, T., Okayasu, T., and Imai, Y., 1983, Possible involvement of NADPH-cytochrome P450 reductase and cytochrome b_5 on β-ketostearoyl-CoA reduction in microsomal fatty acid chain elongation supported by NADPH, *FEBS Lett.* **155:**11–14.
72. Fukushima, H., Grinstead, G. F., and Gaylor, J. L., 1981, Total enzymic synthesis of cholesterol from lanosterol: Cytochrome b_5-dependence of 4-methyl sterol oxidase, *J. Biol. Chem.* **256:**4822–4826.
73. Passon, P. G., Reed, D. W., and Hultquist, D. E., 1972, Soluble cytochrome b_5 from human erythrocytes, *Biochim. Biophys. Acta* **275:**51–61.
74. Kuma, F., Prough, R. A., and Masters, B. S. S., 1976, Studies on methemoglobin reductase: Immunochemical similarity of soluble methemoglobin reductase and cytochrome b_5 of human erythrocytes with NADH-cytochrome b_5 reductase and cytochrome b_5 of rat liver microsomes, *Arch. Biochem. Biophys.* **172:**600–607.
75. Oshino, N., and Sato, R., 1971, Stimulation by phenols of the reoxidation of microsomal bound cytochrome b_5 and its implication to fatty acid desaturation, *J. Biochem.* **69:**169–180.
76. Strittmatter, P., Machuga, E. T., and Roth, G. J., 1982, Reduced pyridine nucleotides and cytochrome b_5 as electron donors for prostaglandin synthetase reconstituted in dimyristyl phosphatidylcholine vesicles, *J. Biol. Chem.* **257:**11883–11886.
77. Kadlubar, F. F., McKee, E. M., and Ziegler, D. M., 1973, Reduced pyridine nucleotide-dependent N-hydroxyamine oxidase and reductase activities of hepatic microsomes, *Arch. Biochem. Biophys.* **156:**46–57.
78. Kadlubar, F. F., and Ziegler, D. M., 1974, Properties of a NADH-dependent N-hydroxyamine reductase isolated from pig liver microsomes, *Arch. Biochem. Biophys.* **169:**83–92.
79. Sanborn, R. C., and Williams, C. M., 1950, The cytochrome system in the cecropia silkworm with special reference to the properties of a new component, *J. Gen. Physiol.* **33:**579–588.
80. Chance, B., and Williams, G. R., 1955, Kinetics of cytochrome b_5 in rat liver microsomes, *J. Biol. Chem.* **209:**945–951.
81. Ito, A., and Sato, R., 1968, Purification by means of detergents and properties of cytochrome b_5 from liver microsomes, *J. Biol. Chem.* **243:**4922–4923.
82. Spatz, L., and Strittmatter, P., 1971, A form of cytochrome b_5 that contains an additional hydrophobic sequence of 40 amino acid residues, *Proc. Natl. Acad. Sci. USA* **68:**1042–1046.
83. Strittmatter, P., and Velick, S. F., 1956, The isolation and properties of microsomal cytochrome, *J. Biol. Chem.* **221:**253–264.

84. Strittmatter, P., and Velick, S. F., 1957, The purification and properties of microsomal cytochrome reductase, *J. Biol. Chem.* **228**:785–799.
85. Enoch, H. G., and Strittmatter, P., 1979, Cytochrome b_5 reduction by NADPH-cytochrome P-450 reductase, *J. Biol. Chem.* **254**:8976–8981.
86. Oshino, N., Imai, Y., and Sato, R., 1966, Electron-transfer mechanism associated with fatty acid desaturation catalyzed by liver microsomes, *Biochim. Biophys. Acta* **128**:13–28.
87. Fukushima, K., and Sato, R., 1973, Purification and characterization of cytochrome b_5-like hemoprotein associated with outer mitochondrial membrane of rat liver, *J. Biochem.* **74**:161–173.
88. Kuwahara, S., Okada, Y., and Omura, T., 1978, Evidence for molecular identity of microsomal and mitochondrial NADH-cytochrome b_5 reductase of rat liver, *J. Biochem.* **83**:1049–1059.
89. Prough, R. A., Imblum, R. L., and Kouri, R. A., 1976, NADH-cytochrome c reductase activity in cultured human lymphocytes: Similarity to the liver microsomal NADH-cytochrome b_5 and cytochrome b_5 enzyme system, *Arch. Biochem. Biophys.* **176**:119–126.
90. Cohen, B. S., and Estabrook, R. W., 1971, Microsomal electron transport reactions. I. Interaction of reduced triphosphopyridine nucleotide during the oxidative demethylation of aminopyrine and cytochrome b_5 reduction, *Arch. Biochem. Biophys.* **143**:37–45.
91. Cohen, B. S., and Estabrook, R. W., 1971, Microsomal electron transport reactions. II. The use of reduced triphosphopyridine nucleotide and/or reduced diphosphopyridine nucleotide for the oxidative N-demethylation of aminopyrine and other drug substrates, *Arch. Biochem. Biophys.* **143**:46–53.
92. Cohen, B. S., and Estabrook, R. W., 1971, Microsomal electron transport reactions. III. Cooperative interactions between reduced diphosphopyridine nucleotide and reduced triphosphopyridine nucleotide linked reactions, *Arch. Biochem. Biophys.* **143**:54–65.
93. Prough, R. A., and Burke, M. D., 1975, The role of NADPH-cytochrome P-450 reductase in microsomal hydroxylation reactions, *Arch. Biochem. Biophys.* **170**:160–168.
94. Werringloer, J., 1976, The formation of hydrogen peroxide during hepatic microsomal electron transport reactions, in: *Microsomes and Drug Oxidation* (V. Ullrich, I. Roots, A. Hildebrandt, R. W. Estabrook, and A. H. Conney, eds.), Pergamon Press, Elmsford, N.Y., pp. 261–268.
95. Hildebrandt, A., and Estabrook, R. W., 1971, Evidence for the participation of cytochrome b_5 in hepatic microsomal mixed-function oxidation reactions, *Arch. Biochem. Biophys.* **143**,66–79.
96. Kuriyama, Y., Omura, T., Siekevitz, P., and Palade, G. E., 1969, Effects of phenobarbital on the synthesis and degradation of the protein components of rat liver microsomal membranes, *J. Biol. Chem.* **244**:2017–2023.
97. Mannering, G. J., Kuwahara, S., and Omura, T., 1974, Immunochemical evidence for the participation of cytochrome b_5 in the NADH synergism of the NADPH-dependent mono-oxidase system of hepatic microsomes, *Biochem. Biophys. Res. Commun.* **57**:476–481.
98. Prough, R. A., and Masters, B. S. S., 1976, Kinetic and spectral studies on the reduction of liver microsomal NADPH-cytochrome c reductase by NADH, in: *Flavins and Flavoproteins* (T. P. Singer, ed.), Elsevier, Amsterdam, pp. 668–673.
99. Noshiro, M., and Omura, T., 1978, Immunochemical study on the electron pathway from NADH to cytochrome P-450 of liver microsomes, *J. Biochem.* **83**:61–77.
100. Sasame, H. A., Thorgeirsson, S. S., Mitchell, J. R., Gillette, J. R., 1974, The possible

involvement of cytochrome b_5 in the oxidation of lauric acid by microsomes from kidney cortex and liver of rats, *Life Sci.* **14**:35–46.
101. Lu, A. Y. H., and Levin, W., 1974, Liver microsomal electron transport systems. III. Involvement of cytochrome b_5 in the NADPH-supported cytochrome P-450-dependent hydroxylation of chlorobenzene, *Biochem. Biophys. Res. Commun.* **61**:1348–1355.
102. West, S. B., and Lu, A. Y. H., 1977, Liver microsomal electron transport systems: Properties of a reconstituted, NADH-mediated benzo(a)pyrene hydroxylation system, *Arch. Biochem. Biophys.* **182**:369–378.
103. Imai, Y., and Sato, R., 1977, The roles of cytochrome b_5 in a reconstituted N-demethylase system containing cytochrome P-450, *Biochem. Biophys. Res. Commun.* **75**:420–426.
104. Imai, Y., 1981, The roles of cytochrome b_5 in reconstituted monooxygenase systems containing various forms of hepatic microsomal cytochrome P-450, *J. Biochem.* **89**:351–362.
105. Miki, N., Sugiyama, T., and Yamano, T., 1980, Purification and characterization of cytochrome P-450 with high affinity for cytochrome b_5, *J. Biochem.* **88**:307–316.
106. Sugiyama, T., Miki, N., and Yamano, T., 1980, NADH- and NADPH-dependent reconstituted p-nitroanisole O-demethylation system containing cytochrome P-450 with high affinity for cytochrome b_5, *J. Biochem.* **87**:1457–1467.
107. Sugiyama, T., Miki, N., Miyake, Y., and Yamano, T., 1982, Interaction and electron transfer between cytochrome P-450 in the reconstituted p-nitroanisole O-demethylase system, *J. Biochem.* **92**:1793–1803.
108. Morgan, E. T., and Coon, M. J., 1984, Effects of cytochrome b_5 on cytochrome P-450-catalyzed reactions: Studies with manganese-substituted cytochrome b_5, *Drug Metab. Dispos.* **2**:358–364.
109. Chiang, J. Y. L., 1981, Interaction of purified microsomal cytochrome P-450 with cytochrome b_5, *Arch. Biochem. Biophys.* **211**:662–673.
110. Bosterling, B., and Trudell, J. R., 1982, Association of cytochrome b_5 and cytochrome P-450 reductase with cytochrome P-450 in the membrane of reconstituted vesicles, *J. Biol. Chem.* **257**:4783–4787.
111. Werringloer, J., Kawano, S., and Kuthan, H., 1982, Regulation of the cyclic function of liver microsomal cytochrome P-450: On the role of cytochrome b_5, in: *Cytochrome P-450: Biochemistry, Biophysics and Environmental Implications* (E. Hietanen, E. Laitinen, and O. Hanninen, eds.), Elsevier, Amsterdam, pp. 509–512.
112. Noshiro, M., Ullrich, V., and Omura, T., 1981, Cytochrome b_5 as electron donor for oxy-cytochrome P-450, *Eur. J. Biochem.* **116**:521–526.
113. Werringloer, J., and Kawano, S., 1980, The control of the cyclic function of liver microsomal cytochrome P-450: "Counterpoise"-regulation of the electron transfer reactions required for the activation of molecular oxygen, in: *Biochemistry, Biophysics and Regulation of Cytochrome P-450* (J.-A. Gustafsson, J. Carlstedt-Duke, A. Mode, and J. Rafter, eds.), pp. 359–362.
114. Werringloer, J., 1982, "Counterpoise"-regulation of the steady-state concentration of oxy-cytochrome P-450: A definition of the rate-limiting step in the catalytic cycle of liver microsomal cytochrome P-450, in: *Microsomes, Drug Oxidations, and Drug Toxicity* (R. Sato and R. Kato, eds.), Japan Scientific Societies Press, Tokyo, pp. 171–178.
115. Bonfils, C., Balny, C., and Maurel, P., 1981, Direct evidence for electron transfer from ferrous cytochrome b_5 to the oxyferrous intermediate of liver microsomal cytochrome P-450 LM2, *J. Biol. Chem.* **256**:9457–9465.

CHAPTER 5

Cytochrome P-450 Organization and Membrane Interactions

MAGNUS INGELMAN-SUNDBERG

1. Introduction

The components of the microsomal hydroxylase system, as well as mitochondrial cytochromes P-450, are integral membrane proteins deeply embedded in the membrane matrix. Consequently, the properties of these protein components and the rates and specificities of the reactions they catalyze, will be influenced by the nature of the other membrane constituents, in particular the phospholipids. The major physical properties of the membrane matrix of importance in this respect are the fluidity, which influences the lateral and rotational mobilities of the protein components, and the membrane charge, which determines the interaction with ionic groups on the proteins. The membrane matrix also provides a hydrophobic environment for the P-450 enzymes, which mostly utilize lipophilic substrates. The membrane thus constitutes a reservoir for the substrates of P-450 and the membrane composition may therefore influence the type of substrate–P-450 interactions that occur.

Phospholipids constitute about 30–40% of the dry weight of the hepatic endoplasmic reticulum.[1,2] This corresponds roughly to a molar ratio between phospholipid and protein of 35:1. The phospholipid composition, 55% phosphatidylcholine (PC), 20–25% phosphatidylethanolamine (PE), 8–10% phosphatidylserine, 5–10% phosphatidylinositol, and 4–7% sphingomyelin,[1,3–5] reveals that the membrane is negatively charged at a neutral pH. The fatty acid composition of the microsomal phospholipids, about 25% arachidonic acid and numerous other unsaturated fatty acids,[4] man-

---

MAGNUS INGELMAN-SUNDBERG • Department of Physiological Chemistry, Karolinska Institute, S-104 01 Stockholm, Sweden.

ifests its fluid nature, but the fluidity is, of course, subject to modifications by changes in the composition of the dietary lipids.[6,7]

The resolution and purification to homogeneity of the components of the P-450 systems and the subsequent reconstitution of the protein components into functional units, in particular into artificial membranous vesicles of various compositions, have allowed detailed investigations concerning the influence of the membrane structure and membrane constituents on the properties of the monooxygenase systems. The purpose of this chapter is to review studies intended to elucidate the membrane interactions of the enzyme systems and to discuss the results of these studies with respect to the influence of the membrane on the dynamics and functional properties of the P-450 systems. Special emphasis will be placed on the liver microsomal hydroxylase system.

2. Methods for Reconstitution of P-450 Systems

Resolution and reconstitution is the classical biochemical approach to the study of multienzyme systems. Following the initial resolution in 1968 of the liver microsomal hydroxylase system into P-450, NADPH-cytochrome P-450 reductase, and the heat-stable factor,[8] the latter later identified as a lipid[9–11] and more specifically as PC,[12] a variety of techniques have been applied for the reconstitution of membrane-bound P-450-dependent hydroxylase systems (for previous reviews see Refs. 13–15). In early studies, reconstitution was achieved by mixing the lipid fraction with P-450 and P-450 reductase.[9–12] Subsequent to the identification of PC as the active lipid factor, different types of this phospholipid were investigated for stimulatory properties of P-450-dependent NADPH-oxidase activity.[12] A mixture of mono- and dilauroylphosphatidylcholine (DLPC) turned out to be most efficient, and DLPC was thereafter chosen by many investigators for studies of reconstituted P-450-dependent activities.[16–24] Optimal reconstitution is reached following incubation of concentrated protein solutions with the phospholipid in a small volume at room temperature, before dilution with buffer.[25] In this manner, effective complexes between P-450 and the reductase are formed. The lipid appears to facilitate complex formation between the two types of protein components.

The advantage of using a phospholipid such as DLPC for reconstitution lies in its content of solely saturated phospholipids, which prevents side effects in the reaction mixtures caused by lipid peroxidative processes. DLPC is too short a phospholipid to be able to form bilayers.[26] Micelles of the phospholipid are formed at a critical micellar concentration (CMC) of about 45 μM.[27] The major part of the reconstitution studies with DLPC have been carried out at a phospholipid concentration much below the CMC. Such a system consists of small aggregates of a nonmembranous and nonmicellar structure.[28] The disadvantage of this type of reconstituted system is that it does not permit studies of protein–lipid and protein–protein interactions, including such phenomena as lateral and rotational diffusion or transient cluster formation, that occur in a phospholipid membrane.

In 1974 it was found that nonionic detergents could replace the phospholipid in the reconstitution of P-450-dependent monooxygenase systems.[29] Detergents like Emulgen 911, Triton X-100, and Triton N-101 stimulated N-demethylation of benzphetamine almost as effectively as DLPC.[29] At high concentrations of the detergents, the activities were usually inhibited[30–32] due to the formation of such an amount of detergent micelles that the system disintegrated. Other types of detergents successfully used for reconstitution include n-octylglucoside[32,33] and a zwitterionic detergent, 3-(3-cholamido-propyl)-dimethylammonio-1-propanesulfonate (CHAPS).[30]

Reconstitution with nonionic detergents has been used preferentially by Japanese workers in studies of the liver microsomal hydroxylase system,[34–36] the adrenal microsomal P-450-dependent 21-hydroxylase system,[37] and mitochondrial P-450 specific for side chain cleavage of cholesterol.[38–40]

Mixed micelles of cholate and DLPC are reported to be more effective for reconstitution of rabbit liver microsomal P-450-dependent activities than either of the two amphipatic compounds alone.[41] This combination has constituted a good alternative for reconstitution by some investigators.[42–45]

As discussed above, one of the roles of the lipid and/or detergent in the reconstitution process is to bring the electron transfer components together. Recently, it was shown that effective reconstitution can be attained even in the absence of lipid or detergent, provided that high enough concentrations (10–20 μM) of the proteins are used and that long incubation times, allowing complex formation, are used.[46]

2.1. Preparation of Membranous P-450 Systems

Upon hydration, phospholipids assume a bilayer structure. Mechanical shaking induces them to form a concentric multilayer particle.[47] Such vesicles can be transformed into single-shelled liposomes by sonic disruption.[48] Unilamellar liposomes can also be formed by removal of cholate from a phospholipid mixture either by dialysis[49] or by gel filtration.[50] The protein to be incorporated can either be present at the beginning together with phospholipids and cholate or subsequently added to preformed phospholipid vesicles.[51]

The cholate dialysis technique was developed by Racker and collaborators originally for the reconstitution of oxidative phosphorylation.[52–54] Sonicated phospholipids are treated with 1–3% of cholate, followed by slow removal of the detergent by dialysis for about 20 hr. A much faster detergent removal occurs by gel filtration of a similar mixture.[34] When first applied, the success of this method was limited[54] but later experiments revealed the unidirectional incorporation of small intestinal sucrase isomaltase into PC vesicles.[55] This method was applied in 1977 for incorporation of the liver microsomal hydroxylase system into a membrane bilayer.[56] P-450$_{LM_{3c}}$ and P-450 reductase were introduced into unilamellar vesicles of egg yolk phosphatidylcholine (EYPC). Upon insertion, the activity of androstenedione hydroxylation catalyzed by this isozyme was enhanced ten-fold in comparison to a nonmembranous reconstituted system based upon the addition of phospholipid to the purified electron transfer components.[56] This method was subsequently used for incorporation of mitochondrial P-450$_{scc}$ into PC vesicles[57,58] and for reconstitution of the liver microsomal hydroxylase system.[59–62] The cholate dialysis technique, however, is the most frequently used method for vesicle reconstitution of P-450 systems.[63–74] The composition of phospholipids in the vesicles is usually chosen to resemble the situation in intact microsomes (see Section 1). Most commonly, a mixture between PC:PE:PA (phosphatidic acid) (mole ratio 18:8:1) that yields a negatively charged reconstituted membrane is employed.[63–65,67–70,73]

If cholate is replaced by the nonionic detergent octylglucoside in the dialysis technique, much larger vesicles are obtained.[76,77] This method has been applied for the introduction of P-450$_{LM_2}$ together with P-450 reductase into unilamellar vesicles of the PC–PE–PA mixture.[69] The vesicles, having diameters of 200–300 nm, are thus more suitable for studies of protein dynamics since vesicle tumbling is minimized.

Attempts to incorporate the liver microsomal P-450 system into preformed phospholipid vesicles have been unsuccessful.[56,59,78] It appears that complexes formed between liver microsomal P-450 and P-450 reductase prevent their incorporation into vesicles in the absence of detergents.[78] The binding kinetics of cytochrome b_5 to preformed vesicles are consistent with a model whereby the protein micelle formed by this amphipathic protein is in equilibrium with its monomeric form, which in turn is incorporated into the vesicles.[79] P-450 reductase and P-450 form large aggregates at small concentrations of the proteins.[27] It thus seems plausible that this will prevent their subsequent incorporation into preformed vesicles.

In contrast to the situation when using components of the liver mi-

TABLE I
Properties of Vesicular P-450-Containing Systems

Method of preparation	Lipid:protein (mole:mole)	Phospholipid	Size φ (nm)	Protein component	Ref.
Cholate gel filtration	2400:1	EYPC	60	LM_{3c}	56
	1200:1	EYPC	56	LM_2	78
	1200:1	EYPC	34	LM_4	78
Cholate dialysis	230:1	EYPC	30–200	LM_2 and reductase	66
	300:1	PC:PE:PA 2:1:0.06 (w/w)	40–60	LM_2 and reductase	70
Octylglucoside dialysis	1200:1	PC:PE:PA	200–300	LM_2 and reductase	69
Cholate gel filtration	500:1	DPPC:DPPE 1:1 (mole:mole)	30–40	$P\text{-}450_{scc}$	57
Spontaneous	50–100:1	EYPC	100	$P\text{-}450_{scc}$	87, 88
Cholate gel filtration	200:1	DOPC	20–60	$P\text{-}450_{scc}$	58
	450:1	DMPC	100	LM_2	59
	420:1	DMPC	138	LM_2 and reductase	59

crosomal hydroxylase system, incorporation into preformed phospholipid vesicles appears to be a very convenient method for the insertion of mitochondrial $P\text{-}450_{scc}$ into membranes.[80–82] The incorporation is facilitated by decreasing vesicle size and increasing extent of unsaturation of the fatty acyl chains of the PC, but is slowed down by a decrease of the fluidity of the membrane caused by the presence of cholesterol or stearic acid in the membrane.[75]

Incorporation of $P\text{-}450_{LM_2}$ into preformed vesicles composed of microsomal phospholipids is more facile than the corresponding insertion into neutral EYPC liposomes.[59,78] This is possibly due to the presence of nonlamellar structures in membranes composed of microsomal phospholipids (see Ref. 83). The incorporation is accompanied by a fusion of the vesicles induced by addition of the protein.[78] A similar increase in vesicle size is observed upon incorporation of $P\text{-}450_{LM_2}$ into dimyristoylphosphatidylcholine (DMPC) liposomes by the cholate dilution technique.[59]

2.2. Properties of P-450-Containing Vesicles

Some properties of P-450-containing liposomes prepared by different techniques are summarized in Table I. With the exception of the octylglucoside dialysis method, vesicles with diameters of 50–100 nm are usually obtained irrespective of the method used. The stability of the reconstituted membranes is remarkable: 2–3 weeks when stored at 4°C without any loss of P-450.[69,70]

Upon incorporation of P-450 into preformed vesicles, the protein is, as expected, inserted in a unidirectional manner.[80] However, the cholate dilution technique also yields almost unidirectional orientation of the enzymes toward the outside of the vesicles.[58,70,78]

In general, it appears that the functional properties of the reconstituted vesicular P-450-containing systems are similar to the more native microsomal system with respect to substrate turnover rate and rate of electron transfer. For example, at a molar ratio of P-450 reductase to P-450 of 0.05–0.1:1, which is similar to the situation in microsomes,[84] the rate constant of the fast phase reduction of P-450 in reconstituted vesicles (0.5 sec^{-1})[85] is similar to the corresponding rate in microsomes.[86]

3. Transverse Topology of the Liver Microsomal P-450 System

All the electron transport components of the liver microsomal P-450 system are localized toward the cytoplasmic side of the endoplasmic reticulum and are, for the most part, distributed heterogeneously along the lateral plane of the membrane (see Refs. 89–92). However, the degree of their integration into the membrane varies. Treatment of microsomes with proteases,[93–95] non-membrane-penetrating denaturing reagents such as p-diazobenzene sulfonate,[94] or antibodies against P-450[96] has revealed that all enzymes in the electron transport chain are exposed with at least some parts accessible to the water environment. The addition to microsomes of small amounts of different proteases, however, causes the release of catalytically active parts of NADH-cytochrome b_5 reductase, NADPH-cytochrome P-450 reductase, and b_5 whereas P-450, when affected at all by this treatment, is destroyed (see Ref. 89). These studies are compatible with the assumption of a more pronounced integration of P-450 than the other electron transfer components into the membrane. This is supported by the finding with purified reconstituted systems that the membrane protects rabbit liver microsomal $P-450_{LM_{3c}}$ from destruction by p-diazobenzene sulfonate[78] and $P-450_{LM_2}$ from phosphorylation in the presence of ATP by the catalytic subunit of cAMP-dependent protein kinase.[72] However, different forms of rabbit liver microsomal P-450 appear to differ in their membrane interactions according to studies of their topology in reconstituted vesicular P-450 systems.[78] Under similar preparative conditions, vesicles formed in the presence of $P-450_{LM_2}$ were large with a mean diameter of 54 nm, vesicles containing $P-450_{LM_{3c}}$ were of intermediate size, and vesicles containing $P-450_{LM_4}$ were the smallest (34 nm).

This difference was interpreted in terms of an inherently different membrane geometry among the cytochromes; LM_2 being the most and LM_4 the least deeply embedded into the membrane matrix. This proposal was supported by the findings on the susceptibility of the proteins to treatment with *p*-diazobenzene sulfonate in the membrane-bound and soluble forms.[78]

The principal differences in the membranous topology of the components of the microsomal P-450-dependent hydroxylase system discussed above are verified by the primary and secondary structures of the proteins. b_5 is anchored to the membrane matrix with a hydrophobic tail of about 40 amino acid residues[97,98] at its C-terminus.[99-102] In microsomes and in vesicles prepared in the presence of deoxycholate, this tail spans the membrane,[99] making intervesicle transfer of b_5 impossible,[99] whereas in liposomes, where the hemeprotein has been incorporated into preformed vesicles, the C-terminus remains on the same side of the membrane as the catalytic, heme-containing hydrophilic part and thus is more loosely integrated.[99] b_5 without its C-terminus is not able to interact with P-450, either in the presence or in the absence of phospholipid.[44,103,104] Cleavage from the C-terminus of a peptide as small as 1000 daltons completely blocks the ability of b_5 to stimulate P-450-catalyzed reactions.[44] The membrane-binding properties of b_5 are lost when 27 but not 18 amino acid residues are digested from the C-terminus,[105] i.e., at least half of the hydrophobic peptide is required to anchor the protein.

Generally similar requirements for membrane incorporation and interaction with P-450 are evident for P-450 reductase. Steapsin or trypsin treatment of the enzyme yields a hydrophilic part of the protein that is unable to bind either to microsomes[106,107] or to phospholipid vesicles,[78,108] and that cannot interact in a functional manner with P-450 or b_5 (see Refs. 109–112). The membrane-binding part has been isolated from steapsin-treated rat reductase[113] and from trypsin-treated rabbit reductase[114,115] and has been characterized as a peptide of about 6000 daltons.[113,114] Analysis of the primary structure of the rabbit peptide revealed the presence of a very hydrophobic sequence at the N-terminal region from residue 10 (Val) to 32 (Phe) that is probably the membrane-spanning section.[115] The membrane-binding moiety has been reported to inhibit the interaction between reductase and P-450 by binding competitively to the hemeprotein,[114] but other workers have not observed this effect.[113]

The primary structural analysis of P-450 has revealed the presence of eight regions of hydrophobic sequences that probably represent four loops of the enzyme that span the membrane (see Chapter 6).[116-120] The general distribution of the hydrophobic parts in the cytochromes P-450 sequenced so far are very similar (see Refs. 119, 120). The data thus confirm the results discussed above suggesting an ample interaction of P-450 with the hydrophobic regions of the membrane matrix.

The heme orientation of P-450 in relation to the plane of the membrane appears at present not to be conclusively established. Studies of the decay of absorption anisotropy after laser flash photolysis of the heme–CO complex of purified phenobarbital-inducible rat liver P-450 in reconstituted membrane vesicles are consistent with a heme plane tilted by $\theta_N = 40°$ from the plane of the membrane,[63] By the same method, the angle of the P-450 heme in microsomes relative to the plane of the membrane was determined to be 55°.[64] In contrast, EPR measurements of intact rat liver microsomes suggest a heme orientation parallel to the plane of the membrane surface.[121] In the latter study, the heme plane of b_5 was determined to be randomly organized in the microsomes. A heme orientation parallel to the plane of the membrane has also been determined by angular EPR studies for bovine cortex P-450 in submitochondrial particle multilayers.[122] The contradictory results might result from difficulties inherent in the interpretations of the data but might also derive from different relative heme orientations among various forms of P-450.

4. Lateral Organization and Mobility of P-450 in Membranes

The mobility of membrane proteins, first demonstrated in the early 1970s, from the start involved studies of the extent to which proteins that are initially randomly distributed in the plane of the membrane become aggregated when treated with antibodies, lectins, or changes in temperature. More sophisticated techniques, including measurements of rotational diffusion using either flash photolysis or saturation transfer EPR and of lateral diffusion using fluorescence photobleaching recovery studies (see Refs. 123–125 for reviews), have recently been used to study rotational and lateral mobility of P-450 in liver microsomes and reconstituted membranous systems.

During the last decade, there has been an intensive debate concerning to what extent the protein components of the liver microsomal electron transport chains are clustered (see Refs. 89, 126 for reviews). One important point to be clarified in this respect is the diffusion rate of the components of the hydroxylase system in the lateral plane and if this rate is so slow that it might govern the rate of electron transfer in the system.

Diffusion coefficients of soluble proteins in water solution are in the range of 10^{-3} to 10^{-6} cm^2/sec, whereas diffusion of membrane proteins in the lateral plane of the membrane takes place at a rate 10^2 to 10^5 times slower.[123] The rate of lateral diffusion of rhodopsin in rod outer segments is about $1-6 \times 10^{-9}$ cm^2/sec,[127,128] but larger proteins and protein complexes such as the concanavalin A–receptor complex,[129] acetylcholine receptors,[130] and Fc receptors in mast cells[131] diffuse much slower. Consequently, when considering a membrane-bound electron transport chain like the microsomal hydroxylase system, it might be argued that complexation of the protein components would greatly facilitate the rate of electron transfer. This should also be evident from the fact that a fraction of the collisions due to random lateral diffusion of the enzyme molecules would be ineffective due to inappropriate relative orientation of the proteins or disturbance of the electron transport component interactions by other membrane-bound proteins. Enzymes that catalyze consecutive reactions linked by common intermediates, such as fatty acid synthetase, are generally found as organized multienzyme clusters.

Calorimetric studies of the microsomal membrane have revealed that no phase transition of the lipid phase occurs in the temperature range 0–45°C,[132] i.e., the bulk of the membrane is entirely in the fluid phase in this temperature range. The rate of lateral diffusion of microsomal lipids has been determined by ESR measurements of the spin exchange rate of a spin-labeled fatty acid,[133] to equal 11×10^{-8} cm^2/sec at 30°C. This is of similar magnitude as the rate of diffusion of the small hydrophobic peptide gramicidin S in EYPC multilayers[134] (3.5×10^{-8} cm^2/sec at 24°C) and to what is observed in general for the diffusion of lipids in cell membranes.[123,124] Below the crystalline phase transition temperature, a decrease of at least two orders of magnitude in the rate of lipid diffusion is usually seen.[136,137]

The rate of lateral diffusion of P-450 has been determined either indirectly, from calculations of the rate of rotational diffusion, or by the method of fluorescence recovery after photobleaching (FRAP). The latter method has been used to determine the rate of lateral diffusion in phospholipid multilayers of fluorescein maleimide-labeled liver microsomal P-450 from phenobarbital-treated rats.[138] The rate of diffusion of P-450 in egg PC and DMPC multilayers at 25°C was surprisingly fast: 2×10^{-8} cm^2/sec, i.e., about ten times faster than the rate of lateral diffusion of apolipoprotein (ApoC-III) in similar membranes.[139]

Lateral diffusion of P-450 from phenobarbital-treated rats has also been estimated from the rotational relaxation time ($\phi_\parallel$) of the protein in reconstituted vesicles from the expression

$$D_L = (\ln \eta/\eta' - \gamma)a^2/\phi_\parallel$$

where η is the membrane viscosity, η' the viscosity of the aqueous phase, γ Euler's constant, and a the radius of the protein.[140] In this manner, a local diffusion constant of 10^{-9} cm^2/sec was calculated.[64] This value should also be valid for small aggregates of P-450 since the local lateral diffusion coefficient is insensitive to the protein size and, in addition, should also be applicable to the situation of the mobile fraction of P-450 in microsomes, since the rotational relaxation time in these structures is about twice that obtained in the vesicles.[64] Incorporation of P-450 reductase into vesicles also containing P-450 did not affect the rate of lateral diffusion of P-450 as based on data from measurements of rotational diffusion, i.e., a local lateral diffusion rate constant of about 10^{-9} cm^2/sec was also calculated for the complex between the flavoprotein and P-450.[65]

The rate of electron transfer between P-450 reductase and P-450 is in the range of 1 to 10 per sec in microsomes[86] and reconstituted membranes.[66,85,141,142] The electron transfer between b_5 and P-450 takes place at a similar rate according to measurements performed in microsomes[143] and in reconstituted micelles[20]; a somewhat smaller rate constant for the electron transport between the proteins in reconstituted membrane vesicles has been reported.[142] Calculations based upon the rate of lateral diffusion of P-450 in microsomes and reconstituted vesicles are consistent with a rate of collision between P-450 and its electron donors at least three orders of magnitude higher than the rate of electron transfer.[54,138] It may thus be concluded that the lateral diffusion *per se* cannot be the sole rate-determining factor in the hydroxylation reactions but rather that molecular shape, orientation of the proteins, and rate of rotation of the electron transport components will be important factors in determining the electron transfer rate.

From the reasons given above, the probability for electron transfer to P-450 will be determined by the number of appropriately oriented collisions between P-450 reductase, b_5, and P-450. According to this view, the rate of electron transfer will increase in parallel with increases in the concentration of the electron transport components in the membrane. Incorporation of P-450 reductase into liver microsomes thus enhances the rate of various P-450-dependent hydroxylation reactions.[106,107,144,146] Exogenously added P-450 is similarly able to couple with microsomal P-450 reductase, as evidenced by the enhanced rate of benzo[a]pyrene hydroxylation observed after incorporation of P-448[147,148] and the increased rate of N-demethylation observed after incorporation of P-450.[144,147] A similar approach was taken in previous studies to show the free lateral diffusion of b_5 and b_5 reductase in liver microsomes.[150,151]

Inactivation of the major part of either b_5 reductase[151] or P-450 reductase[149] does not prevent the reduction, although at a much slower rate, of almost all of b_5 and P-450, respectively (see Ref. 152).

Another approach to investigating the extent of lateral diffusion in the P-450-dependent hydroxylase system has been to change the fluidity of the membrane either by incorporating excess cholesterol or by removing cholesterol from the microsomal membrane.[153] The conclusion that emerged from these studies is that the NADPH-dependent reduction of P-450 is insensitive to changes in the membrane viscosity, whereas the

NADH/NADPH-supported reduction of b_5 is affected by the amount of cholesterol in the membrane. However, the rate constants for P-450 reduction in the microsomes obtained in these studies were about one order of magnitude lower than measured elsewhere.[86]

In conclusion, calculations based upon the apparent rate of lateral diffusion in the microsomal P-450 system reveal that the number of collisions exceeds by far the rate of electron transport during substrate hydroxylation. It therefore appears that other factors, such as the redox potentials of the proteins and steric factors, are more important determinants of the rate of substrate hydroxylation.

5. Rotational Diffusion of P-450 and P-450 Complexes in Membranes

Surprisingly long rotational relaxation times have been estimated for P-450 in liver microsomal membranes from measurements of the decay of absorption anisotropy after laser photolysis of the heme–CO complex. In rat liver microsomes, isolated from phenobarbital-treated rats, relaxation times of 120–270 μsec were estimated at 20°C.[64,154] A substantial fraction (> 50%) of the P-450 was considered as immobile in these membranes. About one-third the P-450 in liver microsomes from phenobarbital-treated rabbits was immobile, whereas in membranes from β-naphthoflavone-treated animals, no rotation of P-450 was detected at 22°C. Preferential cross-linking of rabbit liver microsomal P-450 forms 4 and 6 with cupric ions and orthophenanthroline in liver microsomes from β-naphthoflavone-treated rabbits converted about one-fourth of the total P-450 to dimers, suggesting the presence of oligomers of P-450 in the membrane as an explanation for the long rotational relaxation times.[155]

Similar results have been reached using saturation transfer EPR.[156] Sulfhydryl groups on P-450 were spin labeled with maleimide and the rotational correlation time, measured at 20°C, was 480 μsec. Cross-linking the microsomal proteins with glutardialdehyde resulted in complete immobilization of the P-450, whereas the rotational correlation time calculated for spin-labeled LM_2 in buffer solution was 0.22 nsec.[156] The rate of rotation of monomeric P-450 was calculated, assuming a value of 10 poise for the viscosity of the phospholipid membrane, as 21 μsec, a value considerably lower than experimentally determined in the microsomes.

The conclusions from the rotational studies performed with microsomes are consistent with the proposal that a P-450 molecule has strong intermolecular interactions with other proteins. An important question is whether the clusters in the microsomes are made up solely from the components of the electron transport chain or whether they also include other proteins that specifically or randomly interact with P-450. Studies performed in reconstituted membrane vesicles seem to support the conclusion that P-450 itself forms molecular complexes. By saturation transfer EPR, a rotational correlation time of 180 μsec was obtained for LM_2 when incorporated into vesicles of microsomal phospholipids with a molar ratio between phospholipid and protein of about 750:1.[62] This corresponds to the formation of clusters of 8–12 molecules of P-450 if one assumes a membrane viscosity of 10 poise.[62] A similar value of the rotational correlation time (111 μsec) was determined by measuring the delayed fluorescence of eosin-labeled LM_2[73,157] in negatively charged vesicles having a molar protein/lipid ratio of 115:1. The rotational relaxation time of rat P-450 isolated from phenobarbital-treated animals, determined from the decay of absorption anisotropy after photolysis of the heme–CO complex, was 95 μsec despite the presence of a 700-fold or 20-fold molar excess of phospholipid in the reconstituted membrane.[64] However, in the latter case, about 35% of the P-450 molecules were immobile. The composition of the phospholipids in the membrane did not affect the rate of rotation,[64] i.e., the membrane charge was without effect. In contrast to what might be expected, incorporation of P-450 reductase into vesicles together with P-450 results in a pronounced decrease of the rotational relaxation time of P-450 (40 μsec at 20°C) indicating the disruption of intermolecular P-450 clusters and the subsequent formation of smaller complexes between P-450 and the flavoprotein.[65]

The binding of a substrate by LM_2, when present alone in reconstituted vesicles, markedly enhances its rotational correlation time, probably as a consequence of a conformational change of the enzyme, whereas reduction of the hemeprotein–substrate complex accelerates rotation in the membrane.[73,158]

In conclusion, the rotation of P-450 in the reconstituted membrane is uniaxial in character[73] and takes place at a rate similar to that of other membrane proteins (see Ref. 123). The relatively slow rate of rotation of P-450 in reconstituted membrane vesicles, as determined by many investigators, is consistent with the presence of P-450 in clusters of 6–10 molecules in the membrane,[62,73,158] forming a rotamer with a diameter of about 64 Å.[158] Such clusters are, however, not formed in the presence of P-450 reductase. In the microsomal membrane, it appears that P-450 interacts intimately with other proteins, although the exact nature of the P-450–protein interactions remains open for future investigations.

6. Influence of Membrane Lipid on the Properties of P-450 and P-450-Dependent Reactions

The activities of a large number of membrane-bound enzymes depend on the type of phospholipids immediately surrounding them (see, e.g.,

Refs. 159–161). The interactions between phospholipids and the enzymes are of relatively low specificity in the sense that detergents often can mimic the activating effect of the phospholipids and that approximately a 20-fold molar excess of the phospholipid is required for maximum effect. In situations where somewhat more specific enzyme–phospholipid interactions are observed, the selectivity is based mainly upon the charge of the phospholipid, as for example evidenced from results with a hydrophobic protein,[162] Na^+,K^+-ATPase,[163] oligomycin-sensitive mitochondrial ATPase,[164–166] α-1,2-mannosidase,[167] and cytochrome oxidase.[168] In contrast, very specific phospholipid–enzyme interactions have been reported in some cases. The best documented example is D-β-hydroxybutyrate dehydrogenase which has an absolute and specific requirement for PC for catalytic turnover.[169–171] Addition of PC to the enzyme enables the protein to bind its coenzyme NAD(H) and, furthermore, facilitates the formation of the active enzyme complex. A specific phospholipid requirement has also been reported for delipidated beef heart cytochrome oxidase.[172,173] Optimum activity is reached in the presence of a particular cardiolipin at a stoichiometric relationship of 2–3 moles of phospholipid per mole of enzyme.[172–174]

As discussed in previous sections, the components of the microsomal hydroxylase systems are integrated into the membrane in two different manners: (1) deeply buried in the hydrophobic interior with many parts of the amino acid chain in tight contact with the fatty acyl chains of the phospholipids, as exemplified by LM_2 and LM_4, or (2) with a hydrophobic tail anchoring the protein to the membrane, which results in a relatively limited number of interactions with the hydrophobic part of the membrane, as exemplified by b_5 and P-450 reductase. It is therefore reasonable to expect a more pronounced dependency on the physical state of the lipids and on the phospholipid structure for the function of P-450 than might be expected for the electron donor components.

The types and specificities of the interactions of phospholipids with mitochondrial and microsomal P-450 have been the subject of many investigations during the last decade. The major questions have concerned the extent to which the phospholipids influence substrate binding, substrate specificity, protein conformation, and electron transport to P-450.

6.1. Spin State Control by Membrane Phospholipids

In the absence of exogenous substrate, up to half of the P-450 present in microsomes from untreated rats is in its high-spin form.[175,176] A similar amount of P-450 undergoes a temperature-dependent spin transition in these membranes.[175] Solubilization of the microsomes results in the conversion of a substantial portion of high-spin liver microsomal P-450 to the low-spin form.[175,177] Soluble and purified P-450 are also in temperature-dependent spin-equilibrium.[175,177,178] Investigations considering what components of the microsomal membrane are responsible for the high-spin conversion of P-450 have revealed that free fatty acids[177] as well as P-450 reductase,[179] b_5[10,61,103,104,180] phospholipids,[27,178,179,181] and perhaps also epoxide hydrolase[181] may contribute to this spectral effect.

Titration of purified forms of P-450 with DLPC has revealed that the transition of the hemeprotein into its high-spin form occurs in the presence of phospholipids with apparent spectral dissociation constants (K_s) in the range of 3 μM[179] to 450 μM.[181] LM_2 apparently yields much stronger type I difference spectra than does LM_4.[181] In the absence of substrate, LM_2 exhibits biphasic binding kinetics: binding of the phospholipid to the cytochrome in the high-affinity phase occurs almost stoichiometrically with an apparent dissociation constant of 3–6 μM and in the low-affinity phase with a dissociation constant of 50–70 μM.[179]

Incorporation of LM_2 into preformed phospholipid vesicles converts the protein into its high-spin form.[178] The extent of temperature-dependent high-spin conversion has been reported to depend upon the phospholipid composition of the membranes. At maximum, the high-spin content was 50%, when the enzyme interacted with liposomes composed of PA and when the measurements were performed at 37°C.[178] A pronounced relationship between the negative charge of the membranes and the degree of high-spin conversion was established in which acidic phospholipids favor the high-spin transition. In contrast, nonionic detergents like Triton N-101 and Tween 20 are not capable of bringing about the spin conversion.

In conclusion, the reports presented are consistent with the proposal that direct interactions between phospholipids and P-450 occur that result in conversion of the enzyme into its high-spin form. Whether this effect on the enzyme is a consequence of interactions of the phospholipids with the substrate-binding domains of the proteins, is partly due to the presence of impurities in the lipid preparations, or is dependent on interactions of the phospholipids with specific amphipathic sites on the enzymes, remains to be established.

6.2. Phospholipid Binding and Effects on P-450 Conformation

About a decade ago, it was recognized that negligible P-450-dependent hydroxylase activities remain after extraction of microsomes with organic solvents such as butanol and acetone that remove all neutral lipids and about 80% of the phospholipids.[183] The activity of the monooxygenase system as well as the fast-phase NADPH-dependent reduction of P-450 could be restored by the addition of synthetic PC to the delipidated preparations. These studies suggested that the phospholipids have functions in the monooxygenase system other than solely facilitating the association of P-450 reductase with P-450. In particular, it appeared that the phospholipid could renature P-450 partially destroyed by organic solvent extraction[184] and increase the stability and the conformational rigidity of microsomal[185,186] and mitochondrial P-450.[57]

The finding that certain mitochondrial enzymes have a preference for certain types of phospholipids, so-called boundary lipids, stimulated investigations into whether this was also true for P-450. Partially purified P-450$_{scc}$ was reported to contain PC and PE[187,188]; LM$_2$ was shown to have a preference for PA relative to PC and PE.[67] The latter conclusion was based on the fact that spin-labeled PA did not exhibit a phase transition in membrane vesicles when LM$_2$ was present. This was in contrast to the situation in empty phospholipid vesicles where a pronounced phase transition was observed at about 30°C.

The preference of LM$_2$ for negatively charged phospholipids has also been demonstrated when the catalytic activities of P-450-catalyzed reactions or the reducibility of the protein were examined.[85,141,189] In the presence of, e.g., PS as the only phospholipid, LM$_2$ is readily denatured,[189] providing a further indication of the special relationship between this enzyme and negatively charged phospholipids. Rat liver microsomal P-450 isolated from phenobarbital-treated animals, when bound to DMPC liposomes, is protected to some extent against thermal denaturation when small amounts of PI are present in the membranes.[74]

The interactions between LM$_2$ and phospholipids have been studied by means of second derivative spectroscopy. This method gives an indication of changes in the environment of aromatic amino acids in the protein, in particular tyrosine. Upon the interaction of phospholipid with LM$_2$, the amplitude of the tyrosine band was increased, which was taken as an indication of a decrease in the polarity in the immediate environment of the tyrosine residues, or a reduced extent of tyrosine ionization in the presence of the phospholipid.[190] It was later shown that detergents also could bring about the increase in the tyrosine signal and it was therefore suggested that this phenomenon was due to environmental effects on the exposed tyrosine residues rather than to functionally linked conformational changes.[178]

According to gel filtration studies, LM$_2$ binds more or less specifically about 20 moles of PC per mole of enzyme.[27] About the same amount of phospholipid was required to transform phenobarbital-inducible rat liver microsomal P-450 into a form that readily bound benzphetamine.[135] In contrast, b_5, with only a small membrane-spanning segment, interacts specifically with only 2–4 molecules of phospholipid per enzyme molecule.[191] P-450 from phenobarbital-treated rats also affected the phase transition of DMPC when present in reconstituted liposomes.[74] By using differential scanning microcalorimetry, it was calculated, on the basis of changes in the phase transition enthalpy, that one rat liver microsomal P-450 influences about 350 DMPC molecules.[74]

The phospholipid exchange between the outer and the inner layer of reconstituted membranes, as well as the exchange of phospholipids between different vesicles, are enhanced by the presence of P-450 in the liposomes.[192,193] The rate of phospholipid vesicle transfer was also reported to depend on the phospholipid composition of the vesicles.[192] It might be suggested that the presence of P-450 in the membrane affects the phospholipid packing and facilitates the phospholipid exchange in the transverse plane.

In contrast to the situation with microsomal P-450, relatively specific phospholipid–P-450 interactions are reported for mitochondrial P-450$_{scc}$, the enzyme responsible for side chain cleavage of cholesterol. The rate of this reaction is mainly regulated by the availability of cholesterol for the enzyme (see Chapter 11). Cardiolipin has a remarkable stimulatory effect on the substrate-binding reaction in reconstituted membranes[194] and, on the basis of EPR and other studies, it was concluded that the enzyme has a specific effector site for cardiolipin.[195] The cardiolipin effect was competitively inhibited by α-glycerophosphate and was half-maximal at 4 mole% of the phospholipid in the membrane, a concentration which is in the physiological range.[195]

Only a limited number of studies have considered whether the secondary or the tertiary structure of P-450 is influenced by phospholipids. Interactions of LM$_2$ and, in particular, LM$_4$ with DLPC enhanced the circular dichroism spectra in the UV region, indicating an increase in the amount of α-helical content.[181] Other workers, however, were not able to detect the small changes of the CD spectra of LM$_2$ due to interactions with phospholipids.[185,186,189] Other evidence for the conformational phospholipid dependency of P-450 comes from studies of the effect of DMPC on the binding of a substrate by rat liver phenobarbital-inducible microsomal P-450.[135] In the presence of PC, the relaxation time for the benzphetamine-induced high-spin transition was decreased by a factor of 10, whereas the phospholipid had no influence on the extent of substrate binding at equilibrium. This indicated that the phospholipid provided an environment that facilitated conformational transitions of the enzyme.[135]

In summary, P-450 seems to bind phospholipids in the amount and with a specificity that is typical of integral membrane proteins. The negative charge of the phospholipid appears to be the physical factor of major importance in this respect, at least for some forms of microsomal P-450. The function of the phospholipid as a regulator of side chain cleavage activity, based on the finding of an effector site on the enzyme, constitutes a point of considerable interest.

6.3. Membrane Lipid as Modulator of Substrate Binding and Catalysis

The importance of membrane phospholipids for substrate binding and catalysis in the liver microsomal hydroxylase system was early recognized. Treatment of microsomes with phospholipase A converted P-450 into P-420.[196] P-420 was also formed subsequent to the incubation of liver microsomes with phospholipase C.[197] The same treatment destroyed the type I binding sites[197] and decreased the rate of drug metabolism.[197,198] Isooctane extraction of liver microsomes, which removes PC and PE, also destroyed the type I binding sites of the microsomes[198] as well as their ability to metabolize drugs.[199] The mitochondrial P-450-catalyzed 11β-hydroxylation of deoxycorticosterone was inhibited by phospholipase C treatment.[187] Although there are no simple explanations for all of these effects, the results emphasize the intimate relationship between the membrane phospholipids and the ability of P-450 to bind and oxidize substrates.

One of the major roles of the membrane matrix in P-450 catalysis is to provide a hydrophobic milieu suitable for the bulk of hydrophobic compounds that are substrates for P-450. A relatively low affinity of the substrate for P-450 is compensated for by the partition effect[200] that concentrates the hydrophobic substrates in the membrane interior. An excellent relationship exists between the logarithm of the octanol–water partition coefficients of various aliphatic, aromatic, and alicyclic compounds and the apparent K_s values determined from their abilities to produce type I spectral shifts in microsomes from control, phenobarbital-, and 3-methylcholanthrene-treated hamsters.[201] This effect depends in part on the membrane partitioning phenomenon but also, of course, on the properties of the substrate-binding site of the various P-450. However, when the spectral constants (K_s) obtained for various substrates were corrected for their partition coefficients, the extent of binding was mostly related to their solubilities in the membrane matrix.[176]

Recent results suggest that the substrate-binding site of P-450 is in intimate contact with the membrane lipid phase. The van't Hoff plot of the spectral change (K_s) induced by the binding of benzphetamine to DMPC liposomal liver LM_2 showed a marked break at the transition temperature of the phospholipid, whereas no such break was observed in liver microsomes[206] that have lipids that are entirely in the fluid phase above 0°C.[132] This was taken as an indication of substrate exclusion from the membrane hydrophobic region below the phase transition temperature in the DMPC liposomes (see Refs. 202, 203) rather than as an expression of conformational changes in the P-450 molecule.[206] The substrate-binding site of mitochondrial $P-450_{scc}$ has also been proposed to face the hydrophobic interior of the membrane. This proposal is based on the facts that (1) dilution of cholesterol-containing vesicles with phospholipids results in a decreased rate of side chain cleavage, (2) the hydroxylation rate is sensitive to the fatty acid composition of the membrane, and (3) P-450$_{scc}$, incorporated into a cholesterol-free vesicle, cannot metabolize cholesterol present in a separate vesicle.[80,87]

The rate of the substrate-induced spectral shift for liver microsomal P-450 isolated from phenobarbital-treated rats was enhanced ten-fold by DMPC.[135] The phospholipid had no effect below its transition temperature and the phospholipid did not influence the equilibrium distribution of the substrate (benzphetamine). These findings indicate that the phospholipid in this case provides a better environment for the enzyme to undergo substrate-induced conformational changes.

Studies of the temperature dependence of substrate binding and catalytic turnover have revealed that changes in the free energy and activation energy, respectively, occur at distinct temperatures. The breaks in the van't Hoff plot of K_s for the binding of 17-hydroxyprogesterone to adrenocortical microsomes at 21 and 31°C, according to measurements of fluorescence polarization with 1,6-diphenyl-1,3,5-hexatriene as a probe, correlate with the lipid phase transition at these temperatures.[204] These temperatures for fluidity changes of the microsomal lipids also agree with results obtained using spin-labeled lipophilic nitroxide radicals[205]; the higher temperature also agrees with a break in the Arrhenius plot for P-450 reductase-dependent reduction of a lipophilic spin label.[133] In adrenocortical microsomes depleted of 80% of their phospholipids, similar breaks in the van't Hoff plot for substrate binding were observed that were interpreted in terms of a "boundary" role for the phospholipids.[207]

Discontinuities in the Arrhenius plots for P-450-catalyzed reactions using various substrates and in the activity of P-450 reductase are generally seen at about 20[86,204,208,209] and 30°C.[133,204,208] This has generally been taken as an indication of a temperature- and lipid-dependent rate of electron flow to P-450 (see Ref. 126 for further discussion). The break, at 20°C in the Arrhenius plot, is eliminated on treatment of the microsomes with glycerol or deoxycholate, a finding attributed to interactions of the reductase with P-450 because the activities of the two separate proteins were not temperature dependent in a similar manner.[209] However, discontinuities in the Arrhenius plots can also result from conformational changes in the enzyme molecule itself.[210]

The lipid requirement for the catalytic function of P-450 apparently does not originate from an altered substrate–enzyme interaction. The phospholipid does not influence the K_d or K_m values for common substrates like benzphetamine[135,211] or 7-ethoxycoumarin,[46] although preliminary experiments showed that the K_d of benzphetamine for LM_2 is decreased in response to added phospholipid.[27] Furthermore, (1) the apparent activation energy for P-450-dependent O-deethylation of 7-ethoxycoumarin is not influenced by phospholipids,[46] (2) the enhanced rate of LM_2-dependent hydroxylation reactions in negatively charged membranes is not attributable to changes in the activation energy of the reaction, and (3) the charge effect does not reside in an altered K_m for the substrate.[189] In addition, the phospholipid does not influence the isotope effect seen for the O-deethylation of 7-ethoxycoumarin,[16] indicating that no change occurs in the kinetic role of the carbon–hydrogen bond cleavage.

The kinetics for the hydroxylation of androstenedione by a reconstituted membranous system containing P-450 reductase and LM_{3c} were suggestive of positive cooperativity, whereas no such indication of an effector role for the substrate was obtained in a nonmembranous system reconstituted with small amounts of DLPC.[212] The effector role is probably a consequence of the substrate-concentrating effect of the membrane matrix that results in substrate-induced alterations in the catalytic function of P-450. An allosteric effect by pregnane derivatives on the progesterone 16α-hydroxylase activity of LM_{3b} has also been described.[213]

The regulatory role of the membrane in P-450-dependent catalysis, resulting from a control of substrate availability, is nicely illustrated in mitochondria. The rate of the cholesterol side chain cleavage reaction catalyzed by P-450$_{scc}$ is governed by the amount of cholesterol in the inner mitochondrial membrane (see Refs. 40, 81, 214 and Chapters 10, 11 for reviews). ACTH is believed to regulate the transport of cholesterol from adrenal lipid inclusion droplets to P-450$_{scc}$ in a still not clearly established manner. The effect is manifested as an increase in the extent of high-spin P-450$_{scc}$ after stimulation with ACTH.[215–218] The stimulatory process includes the action of a labile protein with a short half-life. The synthesis of this protein and, in addition, the cholesterol transfer are blocked by cycloheximide.[219] The regulatory protein appears to participate in the intermitochondrial membrane cholesterol transfer.[220,221] These findings have stimulated research into how the membrane composition might influence cholesterol interactions with the cytochrome and into the extent to which ACTH acts by changing the membrane lipid composition.

P-450$_{scc}$, when isolated, is in its high-spin form due to the presence of bound cholesterol.[39,75,222,223] Incorporation of the protein into phospholipid vesicles causes the formation of the low-spin form of the protein due to equilibration of cholesterol into the membrane matrix.[57,75] Binding of cholesterol to the enzyme, which triggers the side chain cleavage activity in part because of the higher affinity of adrenodoxin for the substrate-bound form of the enzyme,[87,224–226] is stimulated by an increase in the unsaturated phospholipid content[80,88,226] but is also affected by the nature of the phospholipid head groups. Substrate binding is facilitated the most by cardiolipin, less by phosphatidylglycerol and phosphatidylserine, and least by phosphatidylinositol and its phosphorylated derivatives.[194,195] Poly-L-lysine can also stimulate substrate binding.[227] The binding of hydroxycholesterols, which have a 100-fold higher affinity than cholesterol for the enzyme, is not influenced by the phospholipid composition.[228] A different ranking of phospholipids is obtained when incorporation of P-450$_{scc}$ into cholesterol-containing vesicles is the rate-determining step.[82] From these results, the conclusion has emerged that a specific phospholipid effector site exists on P-450$_{scc}$ (see above) and that its occupation by one cardiolipin or two phosphatidylcholines stimulates the binding of cholesterol to the enzyme. This leads in turn to the enhanced binding of adrenodoxin and a concomitant increase in the side chain cleavage activity (see Ref. 195).

6.4. Roles of Phospholipids in Enzyme Complex Formation and Electron Transport

The accumulating evidence suggests that, in many microsomal P-450-catalyzed hydroxylation reactions, the introduction of the second electron to P-450 is the overall rate-determining step in the process (see Chapter 4).[43,152,229] Consequently, phospholipid effects on the extent and the nature of the interactions between b_5, P-450 reductase, and P-450 in a functional hydroxylating system, are of major importance.

In early studies, it was recognized that omission of phospholipid from the reconstituted P-450 system diminished the extent of the fast phase of NADPH-dependent P-450 reduction and it was concluded that the phospholipid was necessary for electron transport from NADPH to P-450.[12] Depletion of phospholipids by acetone extraction of the microsomes greatly decreased the fast-phase reduction of P-450.[230] These studies, together with results obtained in the first reconstituted systems,[8-12] suggested that the phospholipid was of great importance in bringing the electron transport components together rather than functioning itself as an electron transport component. It has turned out, however, that complex formation *per se* is not sufficient to permit electron transport between the participating proteins.

6.4.1. Enzyme Complex Formation in Soluble Systems

In solution, both P-450 reductase and P-450 self-aggregate. Molecular complexes with apparent molecular weights of 300,000–500,000, representing species of 6–10 monomers, are formed from liver microsomal P-

450 in solution.[27,28,31,41,179,231] The same type of aggregation is observed with P-450$_{scc}$ in solution.[232] P-450 reductase aggregates under similar conditions to a complex with an apparent molecular weight of about 450,000.[179,233] The clustering of P-450 and P-450 reductase can also be detected by SDS–polyacrylamide gel electrophoresis after cross-linking of the molecular complexes.[155] The aggregation state of P-450 is not markedly influenced by phospholipids.[231] However, addition of detergents such as Triton X-100 or n-octylglucoside, provided that enough detergent is added, causes disintegration of the aggregates and formation of monomolecular species in detergent micelles.[31–33,234]

P-450 and P-450 reductase, as evidenced by gel filtration studies, form large aggregates when present together in solution.[46,179] Similar complexes are detected by gel filtration in the presence of detergents,[37] but this may to a certain extent result from inclusion in detergent micelles. The reductase–P-450 complexes have been detected after cross-linking with dimethyl-3,3'-dithiobis(propionimidate)[235] by the aqueous two-phase partition technique[104] and by measurements of the altered CD spectra that accompany complex formation.[179] The phospholipid does not influence the pattern of cross-linking between the reductase and P-450.[235] This is to be compared with the ability of the phospholipid to stimulate the catalytic activity by a factor of 6 in a similar system,[235] which emphasizes the difference between functional and nonfunctional enzyme complexes. It therefore appears plausible that the phospholipid may influence the relative orientation of the proteins.

In the presence of a large amount of DLPC, P-450 reductase and P-450 coelute during gel filtration as a complex distinct from the phospholipid micelles.[27] The formation of such molecular complexes is favored by the presence of phospholipid. The apparent dissociation constant of P-450 reductase from P-450 is decreased by a factor of 4 on addition of DLPC to the system.[16,27,46] This effect of the phospholipid may not be attributed to the dispersing effect of the amphiphilic lipid[46] because the size of the protein complexes formed under these conditions is only influenced to a limited extent by phospholipid addition.[231]

Molecular complexes between b_5 and P-450, as demonstrated by aqueous two-phase partition of the enzymes, are formed even in the absence of phospholipid.[104] Complex formation between the two proteins can also be followed spectrophotometrically by the high-spin transformation of P-450 that occurs on interaction with b_5 in the presence or absence of phospholipid,[20,61,103,104,180] or by the associated changes in the CD spectra of the proteins.[61] An apparent dissociation constant of 275 nM for the interaction of b_5 with rat liver microsomal P-450 in the presence of phospholipid has been determined.[180] It should be pointed out in this context that the evidence is overwhelming that the stimulatory properties of b_5 in P-450-catalyzed hydroxylation reactions require the presence of phospholipid.[20,44,104,180,236] Here also, it appears that the phospholipid affects the nature and the quality of the enzyme complex.

6.4.2. Enzyme Complexes in Membranous Systems

Liver microsomal P-450 interacts intimately with itself, with P-450 reductase, with b_5, and perhaps also with epoxide hydrolase in membranous systems. As discussed in Section 5, the slow rotation of P-450 in liver microsomes is consistent with a high extent of immobilization and numerous protein–protein interactions. Cross-linking of P-450 monomers with cupric phenanthroline in liver microsomes does not cause any appreciable increase in the rotational relaxation time.[155] Treatment of rat liver microsomes with the cross-linking reagent dimethylsuberimidate, followed by detergent solubilization, enables antibodies against P-450 or P-450 reductase to precipitate epoxide hydrolase. An interaction between these proteins in the microsomes is indicated by this result.[182] P-450, according to rotational diffusion studies,[63] is immobilized by antibodies against the reductase, when present together with P-450 reductase in reconstituted membrane vesicles, whereas the reductase itself, when incorporated into P-450-containing vesicles, mobilizes P-450, by promoting the formation of reductase–P-450 complexes of smaller size than the membranous P-450 complexes themselves.[65]

Enzyme complex formation in membranous systems has been detected by comparing the magnetic CD spectra of P-450 when present alone or together with its electron donors in reconstituted negatively charged membranes. b_5 and P-450 reductase both caused, when incorporated together or separately into P-450-containing vesicles, a decrease in the absolute intensity of the Soret band of the magnetic CD spectrum of P-450.[237] The apparent binding constant of the b_5–P-450 complex decreased on incorporation of the proteins into the vesicles to a value of 5 μM.[103]

6.4.3. Enzyme Complex Formation as Evidenced from Studies of Catalysis and Electron Transport

Kinetic evidence has been presented for the formation of a 1:1 molar complex between P-450 reductase and P-450 in nonmembranous[16,211] and vesicular[71] reconstituted systems. When the total concentration of the two enzymes was held constant, but the relative concentration of each protein varied, a dependence in the apparent rate of P-450-catalyzed reactions directly proportional to the concentration of a proposed 1:1 molar complex of the enzymes was observed.[16,71,211] As discussed above, the phospholipid facilitates the formation of such catalytic complexes, but they can also be formed in the absence of lipid provided that high enough protein concentrations are used and that the proteins are incubated for 2 hr at 25°C.[46] It appears that one of the major roles of the phospholipid in the system is to facilitate complex formation. In particular, the redox potential of P-450 is totally unaffected by phospholipid[145,238] and is not altered by incorporation of the enzyme into phospholipid vesicles.[239]

The NADPH-dependent reduction of P-450 proceeds in at least two phases (see Chapter 4). The question of whether this phenomenon has its origin in the type of lateral organization of the electron transport chain in the membrane has been actively debated during the last decade (see Refs. 86, 126, 141). The fast phase of the reduction was attributed to reduction within a cluster of reductase and P-450 molecules, whereas the slow phase was identified with the reduction of P-450 molecules not directly associated with these clusters.[86] However, the biphasicity of the reduction process is not abolished when the proteins are reconstituted in micelles with small amounts of Emulgen 913[37,66,240] or in the protein aggregates obtained with concentrations of DLPC below the CMC.[18] The biphasicity is influenced, however, by the amount of NADPH in the system[18] and by the concentration of the electron transport components in the vesicles/aggregates.[66,85,141,142] Thus, dilution of reconstituted membrane vesicles[66,85,141] or liver microsomes[241] with phospholipids diminishes the extent and/or rate of the fast reduction phase and decreases the rate of P-450-catalyzed hydroxylation reactions.[66,71,141] In particular, it is noteworthy that a very good correlation exists between the hydroxylation rate and the reciprocal value of the PC content with respect to the P-450 content in reconstituted vesicles.[71] These and other[66,142,212] results are consistent with the proposal that the rate of NADPH-dependent reduction of P-450 and the rate of P-450-catalyzed reactions in membranous systems are proportional to the fraction of active reductase–P-450 complexes in the membrane. The concentration of these complexes, in turn, may be proportional to the concentration of the enzyme relative to the lipid in the vesicles, i.e., long-lived, nondissociable reductase–P-450 complexes are not formed in such PC vesicles.

The stimulatory effects of b_5 on P-450-catalyzed reactions depend on the type of substrate and the type of P-450 in question (see Chapter 4). The stimulation in many cases appears to be connected with a more effective transfer of the second electron to P-450 in the presence of b_5,[43,45,236,242–244] but an additional role of this hemeprotein as an effector of P-450 cannot be excluded.[61,245] Initial methods for the reconstitution of b_5 into P-450-dependent hydroxylation systems involved mixing the phospholipid fraction with the P-450 system and gave nonoptimal results with regard to the stimulatory properties of b_5.[246] It appears that the effects of b_5 in nonmembranous systems depend critically on the phospholipid/protein ratio.[17] In membranous reconstituted systems, where the phospholipid usually is present in large excess, this problem is of less importance.[142,236]

Kinetic experiments have shown that maximal effects are seen at a 1:1 molar ratio of the proteins when reconstituted systems are titrated with b_5, indicating the formation of 1:1 catalytically active complexes between the proteins.[20,42,44,104,142,243,247,248] This is in accord with the types of complexes between these proteins shown to be formed by physical methods (see above). When the relative content of all three electron components in reconstituted membrane vesicles was changed, the rate of LM_2-catalyzed hydroxylation reactions was determined by the component present in deficit.[243,247] That is, kinetic evidence was obtained for the existence of a catalytically active ternary complex. Evidence has also been presented in the three-component reconstituted membrane system for the presence of diffusible protein complexes[142] rather than long-lived protein clusters. The system in this respect is similar to the reconstituted b_5–b_5 reductase membranous system.[151,249]

6.4.4. The Membrane Charge as Effector of P-450 Reduction and Catalysis

When the catalytic activities of various forms of rabbit liver microsomal P-450 reconstituted into membrane vesicles composed of PC or microsomal lipids were compared, it was found that the rates of LM_2- and LM_4-dependent hydroxylation reactions were lower in the PC vesicles than in vesicles composed of more "native" phospholipids.[212] A detailed analysis of the reason for this difference shows that the charge on the phospholipids is the most important physical parameter.[189] A good correlation was obtained when the net negative charge of various types of vesicles[165] was compared with the rate of LM_2-catalyzed oxidation of *para*-nitroanisole.[189] The critical step in the P-450 reaction cycle affected by the membrane charge was the rate of NADPH-dependent reduction of the hemeprotein.[141,189] It was observed that dilution of the membranes with phospholipid caused an extensive decrease in the rate of reduction in neutral PC vesicles, whereas the activity of LM_2-dependent reactions was not affected by dilution.[141] Titration of neutral and negatively charged LM_2-containing vesicles with reductase and subsequent measurement of the reduction process revealed that in both types of membranes, increasing amounts of reductase enhanced the extent of the fast reduction phase and increased the rate constant of the fast reduction phase. However, this increase in reduction rate occurred at much lower reductase concentrations in negative than neutral vesicles.[85] Calculation of the apparent dissociation constants for the formation of a reactive 1:1 molar complex of reductase and P-450 in the two types of vesicles revealed a difference of a factor of 10 in favor of negatively charged membranes: 0.05 μM in negative membranes and 0.5 μM in neutral membranes.[85]

It appears from these data that the membrane charge rather than fluidity may determine the affinity of P-450 reductase for at least some forms of P-450.[153] The lower affinity of the reductase for P-450 in neutral vesicles can be compensated for by increasing the protein versus lipid concentration in the reconstituted membranes; a molar ratio of 1:150 yields a high rate of electron transport in neutral vesicles.[46,66,141] The interaction between these two proteins appears in some cases to be influenced by cellular effectors such as flavonoids[250,251] and polyamines[252] and may thus represent a point for regulation of the hydroxylase system *in vivo*. A membrane charge-dependent regulation, viz by hormone-induced changes in the phospholipid composition through increased incorporation of, e.g., phosphorylated phosphatidylinositols into the microsomal membrane, or by decarboxylation of membranous PS, can be considered in this context.

The effect of the negative membranes on the affinity between reductase and P-450 implies that structures on one or both of these enzymes are altered by ionic interactions with the membrane matrix. Important ion pair interactions have been documented between c and c reductase,[253,254] b_5 and b_5 reductase,[255,256] and P-450 reductase and b_5[257]. Modification of the ε-amino group of lysine-382 of LM_2 inhibits hemeprotein-catalyzed N-demethylation of benzphetamine as well as impairs the rate of electron transfer to P-450 from the reductase, thus indicating the existence of an important ion pair in this interaction.[258] It has furthermore been reported that the interenzyme electron transport is inhibited by high ionic strength,[237] which further strengthens the evidence for the importance of ion pairs in the interaction. It therefore seems plausible that the membrane charge, through ionic bonds, may alter the relative orientation of the ion pair, and thereby determine the overall reduction rate in the system (see Fig. 1).

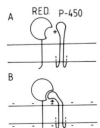

FIGURE 1. Proposed model for the action of the negative membrane in the interaction between P-450 reductase and P-450. (A) neutral membrane; (B) negative membrane.

7. Conclusions

In the sequence of partial reactions that constitute the hydroxylation mechanism of P-450, one may envisage an influence of the membrane in a direct or indirect manner on all the steps. However, it appears that only a few of the steps are sufficiently affected for the interaction to have an impact on the overall rate of hydroxylation. Kinetic analysis of the influence of the membrane on the rate of binding of oxygen or carbon monoxide to P-450$_{scc}$ revealed that the membrane has no effect.[259] It appears, instead, that the major influence of the membrane is on substrate binding, in particular in the mitochondrial side chain cleavage system, and on transfer of the first, and even more importantly, the second, electron to P-450. Recent results suggest, in addition, that the membrane phospholipids may, in some cases, directly participate in the hydroxylation mechanism (see Chapter 2).[260]

The 1:20 stoichiometry of P-450 reductase to P-450 in liver microsomes requires the existence of factors that determine the specificity of the flow of electrons during hydroxylation. The redox potential of P-450 is influenced by the substrate and extensive evidence indicates that substrate binding is connected with conformational alterations in the structure of the protein (see Chapter 3). The presence of a substrate accelerates the rate of reduction of P-450 in reconstituted systems 5- to 20-fold.[66,141,261] In the perfused liver, the addition of a substrate to the perfusate markedly enhances the steady-state level of reduced P-450.[262] One may therefore argue that the specificity of the electron transfer is governed by substrate–P-450 interactions; the selectivity being determined by the affinity of the substrate for the type of P-450 in question. P-450 reductase, as it diffuses laterally, preferentially interacts in a productive manner with substrate-bound P-450, a fact that partially explains the discrepancy between the rate of enzyme collisions determined from lateral diffusion studies and the functional rate of electron transfer.

The data presented in this chapter suggest that complex formation between the protein components of the liver microsomal hydroxylase system can be functional or nonfunctional. This is illustrated schematically in Fig. 2. The rate of P-450-catalyzed hydroxylation reactions will be proportional to the concentration of the functional 1:1 complexes between the reductase and P-450 in the membrane. This concentration should be governed by essentially three factors: the concentration of the various protein components in the membrane, the charge of the membrane, and the presence of a P-450 substrate. The system is in a fast equilibrium, so that long-lived enzyme complexes do not exist. The interaction of b_5 with the reductase–P-450 complex appears to be determined by similar factors.

Although research in the field of P-450–membrane interactions during

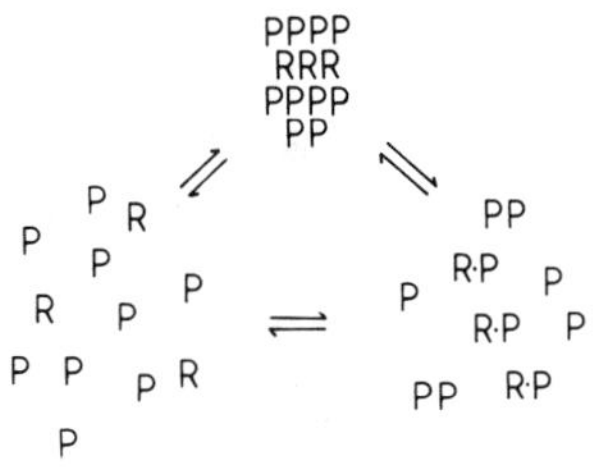

FIGURE 2. Proposed model for the equilibrium between functional and nonfunctional interactions of P-450 reductase and P-450 in membranous systems. P, P-450; R, P-450 reductase.

the last decade has indeed been productive, a need exists for further studies of: (1) the exact dependence of the conformational integrity of P-450 on phospholipids; (2) the P-450 structures that preferentially interact with different parts of the phospholipids; (3) the mechanism behind the facilitating role of the phospholipids on enzyme complex formation; (4) the exact nature of the lateral organization of the hydroxylase system during electron transport and catalytic turnover in microsomes and reconstituted membranous systems; (5) the transverse topology of various types of P-450 including the orientation of the heme and the substrate-binding site in relation to the membrane; (6) the possible role of phospholipids as allosteric effectors of P-450-catalyzed reactions; (7) the possible hormonal mechanisms for the indirect regulation of the hydroxylase system via changes in the membrane phospholipid composition; and (8) the sites of action of compounds like flavonoids and polyamines that acutely regulate the activity of the system.

ACKNOWLEDGMENTS. I am grateful to Mrs. Ylva Ekendahl for generous help in the preparation of this chapter. The work presented from the author's laboratory was supported by grants from the Swedish Medical Research Council.

References

1. Glaumann, H., and Dallner, G., 1968. Lipid composition and turnover of rough and smooth microsomal membranes in rat liver, *J. Lipid Res.* **9**:720–729.
2. Blackburn, G. R., Bornens, M., and Kasper, C. B., 1976. Characterization of the membrane matrix derived from the microsomal fraction of rat hepatocytes. *Biochim. Biophys. Acta* **436**:387–398.
3. Manganiello, V. C., and Phillips, A. H., 1965, The relationship between ribosomes and the endoplasmic reticulum during protein synthesis, *J. Biol. Chem.* **240**:3951–3959.
4. Lee, T.-C., and Snyder, F., 1973. Phospholipid metabolism in rat liver endoplasmic reticulum: Structural analyses, turnover studies and enzymic activities, *Biochim. Biophys. Acta* **291**:71–82.
5. Morin, F., Tay, S., and Simpkins, H., 1972. A comparative study of the molecular

structures of the plasma membranes and the smooth and the rough endoplasmic-reticulum membranes from rat liver, *Biochem. J.* **129**:781–788.
6. Dallner, G., Siekevitz, P., and Palade, G. E., 1966. Biogenesis of endoplasmic reticulum membranes. II. Synthesis of constitutive microsomal enzymes in developing rat hepatocyte, *J. Cell Biol.* **30**:97–117.
7. Wade, A. E., and Norred, W. P., 1976, Effect of dietary lipid on drug-metabolizing enzymes, *Fed. Proc.* **35**:2475–2479.
8. Lu, A. Y. H., and Coon, M. J., 1968, Role of hemoprotein P-450 in fatty acid ω-hydroxylation in a soluble enzyme system from liver microsomes, *J. Biol. Chem.* **243**:1331–1332.
9. Lu, A. Y. H., Junk, K. W., and Coon, M. J., 1969, Resolution of the cytochrome P-450-containing ω-hydroxylation system of liver microsomes into three components, *J. Biol. Chem.* **244**:3714–3721.
10. Lu, A. Y. H., Strobel, H. W., and Coon, M. J., 1969, Hydroxylation of benzphetamine and other drugs by a solubilized form of cytochrome P-450 from liver microsomes: Lipid requirement for drug demethylation, *Biochem. Biophys. Res. Commun.* **36**:545–551.
11. Lu, A. Y. H., Strobel, H. W., and Coon, M. J., 1970, Properties of a solubilized form of the cytochrome P-450 containing mixed-function oxidase of liver microsomes, *Mol. Pharmacol.* **6**:213–220.
12. Strobel, H. W., Lu, A. Y. H., Heidema, J., and Coon, M. J., 1970, Phosphatidylcholine requirement in the enzymatic reduction of hemoprotein P-450 and in fatty acid, hydrocarbon, and drug hydroxylation, *J. Biol. Chem.* **245**:4851–4854.
13. Lu, A. Y. H., and West, S. B., 1978, Reconstituted mammalian mixed function oxidases: Requirements, specificities and other properties, *Pharmacol. Ther.* **2**:337–358.
14. Trudell, J. R., and Bösterling, B., 1983, Interactions of cytochrome P-450 with phospholipids and proteins in the endoplasmic reticulum, in: *Membrane Fluidity in Biology*, Volume 1 (R. C. Aloia, ed.), Academic Press, New York, pp. 201–233.
15. Lu, A. Y. H., and Levin, W., 1974, The resolution and reconstitution of the liver microsomal hydroxylation system, *Biochim. Biophys. Acta* **344**:205–240.
16. Miwa, G. T., and Lu, A. Y. H., 1981, Studies on the stimulation of cytochrome P-450-dependent monooxygenase activity by dilauroylphosphatidylcholine, *Arch. Biochem. Biophys.* **211**:454–458.
17. Bösterling, B., Trudell, J. R., Trevor, A. J., and Bendix, M., 1982, Lipid–protein interactions as determinants of activation or inhibition by cytochrome b_5 of cytochrome P-450-mediated oxidations, *J. Biol. Chem.* **257**:4375–4380.
18. Oprian, D. D., Vatsis, K. P., and Coon, M. J., 1979, Kinetics of reduction of cytochrome P-450 LM_4 in a reconstituted liver microsomal enzyme system, *J. Biol. Chem.* **254**:8895–8902.
19. Dieter, H. H., and Johnson, E. F., 1982, Functional and structural polymorphism of rabbit microsomal cytochrome P-450 form 3b, *J. Biol. Chem.* **257**:9315–9323.
20. Bonfils, C., Balny, C., and Maurel, P., 1981, Direct evidence for electron transfer from ferrous cytochrome b_5 to the oxyferrous intermediate of liver microsomal cytochrome P-450 LM_2, *J. Biol. Chem.* **256**:9457–9465.
21. Nebert, D. W., Heidema, J. K., Strobel, H. W., and Coon, M. J., 1973, Genetic expression of aryl hydrocarbon hydroxylase induction. Genetic specificity resides in the fraction containing cytochrome P_{448} and P_{450}, *J. Biol. Chem.* **248**:7631–7636.
22. Duppel, W., Lebeault, J.-M., and Coon, M. J., 1973, Properties of a yeast cytochrome P-450-containing enzyme system which catalyzes the hydroxylation of fatty acids, alkanes, and drugs, *Eur. J. Biochem.* **36**:583–592.
23. Ryan, D. E., Iida, S., Wood, A. W., Thomas, P. E., Lieber, C. S., and Levin, W.,

1984, Characterization of three highly purified cytochromes P-450 from hepatic microsomes of adult male rats, *J. Biol. Chem.* **259:**1239–1250.
24. Saito, T., and Strobel, H. W., 1981, Purification to homogeneity and characterization of a form of cytochrome P-450 with high specificity for benzo(α)pyrene from β-naphthoflavone-pretreated rat liver microsomes, *J. Biol. Chem.* **256:**984–988.
25. Coon, M. J., 1978, Reconstitution of the cytochrome P-450-containing mixed function oxidase system of liver microsomes, *Methods Enzymol.* **52:**200–206.
26. Mabrey, S., and Sturtevant, J. M., 1978, High sensitivity differential scanning calorimetry in study of biomembranes and related model systems, *Methods Membr. Biol.* **9:**237–274.
27. Coon, M. J., Haugen, D. A., Guengerich, F. P., Vermilion, J. L., and Dean, W. L., 1976, Liver microsomal membranes: Reconstitution of the hydroxylation system containing cytochrome P-450, in: *The Structural Basis of Membrane Function* (Y. Hatefi and L. Djavadi-Ohaniance, eds.), Academic Press, New York, pp. 409–427.
28. Autor, A. P., Kaschnitz, R. M., Heidema, J. K., and Coon, M. J., 1973, Sedimentation and other properties of the reconstituted liver microsomal mixed-function oxidase system containing cytochrome P-450, reduced triphosphopyridine nucleotide-cytochrome P-450 reductase, and phosphatidylcholine, *Mol. Pharmacol.* **9:**93–104.
29. Lu, A. Y. H., Levin, W., and Kuntzman, R., 1974, Reconstituted liver microsomal enzyme system that hydroxylates drugs, other foreign compounds and endogenous substrates. VII. Stimulation of benzphetamine N-demethylation by lipid and detergent, *Biochem. Biophys. Res. Commun.* **60:**266–272.
30. Wagner, S. L., Dean, W. L., and Gray, R. D., 1984, Effect of a zwitterionic detergent on the state of aggregation and catalytic activity of cytochrome P-450 LM2 and NADPH-cytochrome P-450 reductase, *J. Biol. Chem.* **259:**2390–2395.
31. Ingelman-Sundberg, M., 1977, Protein–lipid interactions in the liver microsomal hydroxylase system, in: *Microsomes and Drug Oxidations* (V. Ullrich, I. Roots, A. Hildebrandt, and R. W. Estabrook, eds.), Pergamon Press, Elmsford, N.Y., pp. 67–75.
32. Dean, W. L., and Gray, R. D., 1982, Relationship between state of aggregation and catalytic activity for cytochrome P-450$_{LM2}$ and NADPH-cytochrome P-450 reductase, *J. Biol. Chem.* **257:**14679–14685.
33. Dean, W. L., and Gray, R. D., 1982, Hydrodynamic properties of monomeric cytochromes P-450$_{LM2}$ and P-450$_{LM4}$ in *n*-octylglucoside solution, *Biochem. Biophys. Res. Commun.* **107:**265–271.
34. Imai, Y., 1976, The use of 8-aminooctyl Sepharose for the separation of some components of the hepatic microsomal electron transfer system, *J. Biochem.* **80:**267–276.
35. Sugiyama, T., Miki, N., and Yamano, T., 1979, The obligatory requirement of cytochrome b_5 in the *p*-nitroanisole O-demethylation reaction catalyzed by cytochrome P-450 with a high affinity for cytochrome b_5, *Biochem. Biophys. Res. Commun.* **90:**715–720.
36. Kuwahara, S.-I., and Omura, T., 1980, Different requirement for cytochrome b_5 in NADPH-supported O-deethylation of *p*-nitrophenetole catalyzed by two types of microsomal cytochrome P-450, *Biochem. Biophys. Res. Commun.* **96:**1562–1568.
37. Kominami, S., Hara, H., Ogishima, T., and Takemori, S., 1984, Interaction between cytochrome P-450 (P-450$_{C21}$) and NADPH-cytochrome P-450 reductase from adrenocortical microsomes in a reconstituted system, *J. Biol. Chem.* **259:**2991–2999.
38. Nakajin, S., Ishii, Y., Shinoda, M., and Shikita, M., 1979, Binding of Triton X-100 to purified cytochrome P-450$_{scc}$ and enhancement of the cholesterol side chain cleavage activity, *Biochem. Biophys. Res. Commun.* **87:**524–531.
39. Takikawa, O., Gomi, T., Suhara, K., Itagaki, E., Takemori, S., and Katagiri, M., 1978, Properties of adrenal cytochrome P-450 (P-450$_{scc}$) for the side chain cleavage of cholesterol, *Arch. Biochem. Biophys.* **190:**300–306.

40. Kimura, T., 1981, ACTH stimulation on cholesterol side chain cleavage activity of adrenocortical mitochondria: Transfer of stimulus from plasma membrane to mitochondria, *Mol. Cell. Biochem.* **36**:105–122.
41. van der Hoeven, T. A., and Coon, M. J., 1974, Preparation and properties of partially purified cytochrome P-450 and reduced nicotinamide adenine dinucleotide phosphate-cytochrome P-450 reductase from rabbit liver microsomes, *J. Biol. Chem.* **249**:6302–6310.
42. Imai, Y., 1979, Reconstituted O-dealkylase systems containing various forms of liver microsomal cytochrome P-450, *J. Biochem.* **86**:1697–1707.
43. Imai, Y., 1981, The roles of cytochrome b_5 in reconstituted monooxygenase systems containing various forms of hepatic microsomal cytochrome P-450, *J. Biochem.* **89**:351–362.
44. Waxman, D. J., and Walsh, C., 1983, Cytochrome P-450 isozyme 1 from phenobarbital-induced rat liver: Purification, characterization and interactions with metyrapone and cytochrome b_5, *Biochemistry* **22**:4846–4855.
45. Imai, Y., and Sato, R., 1977, The roles of cytochrome b_5 in a reconstituted N-demethylase system containing cytochrome P-450, *Biochem. Biophys. Res. Commun.* **75**:420–426.
46. Müller-Enoch, D., Churchill, P., Fleischer, S., and Guengerich, F. P., 1984, Interaction of liver microsomal cytochrome P-450 and NADPH-cytochrome P-450 reductase in the presence and absence of lipid, *J. Biol. Chem.*. **259**:8174–8182.
47. Hauser, H., Phillips, M. C., and Stubbss, M., 1972, Ion permeability of phospholipid bilayers, *Nature* **239**:342–344.
48. Bangham, A. D., Hill, M. W., and Miller, N. G. A., 1974, Preparation and use of liposomes as models of biological membranes, *Methods Membr. Biol.* **1**:1–68.
49. Racker, E., 1972, Reconstitution of a calcium pump with phospholipids and a purified Ca^{++}-adenosine triphosphatase from sarcoplasmic reticulum, *J. Biol. Chem.* **247**:8198–8200.
50. Brunner, J., Skrabal, P., and Hauser, H., 1976, Single bilayer vesicles prepared without sonication: Physico-chemical properties, *Biochim. Biophys. Acta* **455**:322–331.
51. Eytan, G. D., Matheson, M. J., and Racker, E., 1976, Incorporation of mitochondrial membrane proteins into liposomes containing acidic phospholipids, *J. Biol. Chem.* **251**:6831–6837.
52. Kagawa, Y., and Racker, E., 1971, Partial resolution of the enzymes catalyzing oxidative phosphorylation. XXV. Reconstitution of vesicles catalyzing ^{32}P-adenosine triphosphate exchange, *J. Biol. Chem.* **246**:5477–5487.
53. Racker, E., and Kandrach, A., 1973, Partial resolution of the enzymes catalyzing oxidative phosphorylation. XXXIX. Reconstitution of the third segment of oxidative phosphorylation, *J. Biol. Chem.* **248**:5841–5847.
54. Kagawa, Y., Kandrach, A., and Racker, E., 1973, Partial resolution of the enzymes catalyzing oxidative phosphorylation. XXVI. Specificity of phospholipids required for energy transfer reactions, *J. Biol. Chem.* **248**:676–684.
55. Brunner, J., Hauser, H., and Semenza, G., 1978, Single bilayer lipid–protein vesicles formed from phosphatidylcholine and small intestinal sucrase isomaltase, *J. Biol. Chem.* **253**:7538–7546.
56. Ingelman-Sundberg, M., and Glaumann, H., 1977, Reconstitution of the liver microsomal hydroxylase system into liposomes, *FEBS Lett.* **78**:72–76.
57. Hall, P. F., Watanuki, M., and Hamkalo, B. A., 1979, Adrenocortical cytochrome P-450 side chain cleavage: Preparation of membrane-bound side chain cleavage system from purified components, *J. Biol. Chem.* **254**:547–552.
58. Yamakura, F., Kido, T., and Kimura, T., 1981, Characterization of cytochrome P-450_{scc}-containing liposomes, *Biochim. Biophys. Acta* **649**:343–354.

59. Kisselev, P. A., Smettan, G., Kissel, M. A., Elbe, B., Zirwer, D., Gast, K., Ruckpaul, K., and Akhrem, A. A., 1984, Reconstitution of the liver microsomal monooxygenase system in liposomes from dimyristoylphosphatidylcholine, *Biomed. Biochim. Acta* **43**:281–293.
60. Noshiro, M., Ruf, H. H., and Ullrich, V., 1980, The role of NADPH-cytochrome P-450 reductase and cytochrome b_5 in the transfer of electrons from NADPH and NADH to cytochrome P-450, in: *Biochemistry, Biophysics and Regulation of Cytochrome P-450* (J.-Å. Gustafsson, J. Carlstedt-Duke, A. Mode, and J. Rafter, eds.), Elsevier/North-Holland, Amsterdam, pp. 351–354.
61. Hlavica, P., 1984, On the function of cytochrome b_5 in the cytochrome P-450-dependent oxygenase system, *Arch. Biochem. Biophys.* **228**:600–608.
62. Schwarz, D., Pirrwitz, J., Coon, M. J., and Ruckpaul, K., 1982, Mobility and clusterlike organization of liposomal cytochrome P-450 LM2: Saturation transfer EPR studies, *Acta Biol. Med. Ger.* **41**:425–430.
63. Gut, J., Richter, C., Cherry, R. J., Winterhalter, K. H., and Kawato, S., 1983, Rotation of cytochrome P-450: Complex formation of cytochrome P-450 with NADPH-cytochrome P-450 reductase in liposomes demonstrated by combining protein rotation with antibody-induced crosslinking, *J. Biol. Chem.* **258**:8588–8594.
64. Kawato, S. Gut, J., Cherry, R. J., Winterhalter, K. H., and Richter, C., 1982, Rotation of cytochrome P-450. I. Investigations of protein–protein interactions of cytochrome P-450 in phospholipid vesicles and liver microsomes, *J. Biol. Chem.* **257**:7023–7029.
65. Gut, J., Richter, C., Cherry, R. J., Winterhalter, K. H., and Kawato, S., 1982, Rotation of cytochrome P-450. II. Specific interactions of cytochrome P-450 with NADPH-cytochrome P-450 reductase in phospholipid vesicles, *J. Biol. Chem.* **257**:7030–7036.
66. Taniguchi, H., Imai, Y., Iyanagi, T., and Sato, R., 1979, Interaction between NADPH-cytochrome P-450 reductase and cytochrome P-450 in the membrane of phosphatidylcholine vesicles, *Biochim. Biophys. Acta* **550**:341–356.
67. Bösterling, B., Trudell, J. R., and Galla, H. J., 1981, Phospholipid interactions with cytochrome P-450 in reconstituted vesicles: Preference for negatively-charged phosphatidic acid, *Biochim. Biophys. Acta* **643**:547–556.
68. Nisimoto, Y., Kinosita, K., Jr., Ikegami, A., Kawai, N., Ichihara, I., and Shibata, Y., 1983, Possible association of NADPH-cytochrome P-450 reductase and cytochrome P-450 in reconstituted phospholipid vesicles, *Biochemistry* **22**:3586–3594.
69. Schwartz, D., Gast, K., Meyer, H. W., Lachmann, U., Coon, M. J., and Ruckpaul, K., 1984, Incorporation of the cytochrome P-450 monooxygenase system into large unilamellar liposomes using octylglucoside, especially for measurements of protein diffusion in membranes, *Biochem. Biophys. Res. Commun.* **121**:118–125.
70. Bösterling, B., Stier, A., Hildebrandt, A. G., Dawson, J. H., and Trudell, J. R., 1979, Reconstitution of cytochrome P-450 and cytochrome P-450 reductase into phosphatidylcholine–phosphatidylethanolamine bilayers: Characterization of structure and metabolic activity, *Mol. Pharmacol.* **16**:332–342.
71. Miwa, G. T., and Lu, A. Y. H., 1984, The association of cytochrome P-450 and NADPH-cytochrome P-450 reductase in phospholipid membranes, *Arch. Biochem. Biophys.* **234**:161–166.
72. Pyerin, W., Taniguchi, H., Stier, A., Oesch, F., and Wolf, C. R., 1984, Phosphorylation of rabbit liver cytochrome P-450 LM_2 and its effect on monooxygenase activity, *Biochem. Biophys. Res. Commun.* **122**:620–626.
73. Greinert, R., Finch, S. A. E., and Stier, A., 1982, Cytochrome P-450 rotamers control mixed-function oxygenation in reconstituted membranes: Rotational diffusion studied by delayed fluorescence depolarization, *Xenobiotica* **12**:717–726.
74. Akhrem, A. A., Andrianov, V. T., Bokut, S. B., Luka, Z. A., Kissel, M. A., Skor-

nyakova, T. G., and Kisselev, P. A., 1982, Thermotropic behaviour of phospholipid vesicles reconstituted with rat liver microsomal cytochrome P-450, *Biochim. Biophys. Acta* **692**:287–295.
75. Tuckey, R. C., and Kamin, H., 1982, Kinetics of the incorporation of adrenal cytochrome P-450$\times_{cc}$ into phosphatidylcholine vesicles, *J. Biol. Chem.* **257**:2887–2893.
76. Baron, C., and Thompson, T. E., 1975, Solubilization of bacterial membrane proteins using alkyl glucosides and dioctanoyl phosphatidylcholine, *Biochim. Biophys. Acta* **382**:276–285.
77. Helenius, A., Fries, E., and Kartenbeck, J., 1977, Reconstitution of Semliki forest virus membrane, *J. Cell Biol.* **75**:866–880.
78. Ingelman-Sundberg, M., and Glaumann, H., 1980, Incorporation of purified components of the rabbit liver microsomal hydroxylase system into phospholipid vesicles, *Biochim. Biophys. Acta* **599**:417–435.
79. Leto, T. L., and Holloway, P. W., 1979, Mechanism of cytochrome b_5 binding to phosphatidylcholine vesicles, *J. Biol. Chem.* **254**:5015–5019.
80. Seybert, D. W., Lancaster, J. R., Jr., Lambeth, J. D., and Kamin, H., 1979, Participation of the membrane in the side chain cleavage of cholesterol: Reconstitution of cytochrome P-450$_{scc}$ into phospholipid vesicles, *J. Biol. Chem.* **254**:12088–12098.
81. Lambeth, J. D., Seybert, D. W., Lancaster, J. R., Jr., Salerno, J. C., and Kamin, H., 1982, Steroidogenic electron transport in adrenal cortex mitochondria, *Mol. Cell. Biochem.* **45**:13–31.
82. Kowluru, R. A., George, R., and Jefcoate, C. R., 1983, Polyphosphoinositide activation of cholesterol side chain cleavage with purified cytochrome P-450$_{scc}$, *J. Biol. Chem.* **258**:8053–8059.
83. Stier, A., Finch, S. A. E., and Bösterling, B., 1978, Non-lamellar structure in rabbit liver micrososmal membranes: A ^{31}P-NMR study, *FEBS Lett.* **91**:109–112.
* 84. Estabrook, R. W., Franklin, M. R., Cohen, B., Shigamatzu, A., and Hildebrandt, A. G., 1971, Biochemical and genetic factors influencing drug metabolism: Influence of hepatic microsomal mixed function oxidation reactions on cellular metabolic control, *Metabolism* **20**:187–199.
85. Blanck, J., Smettan, G., Ristau, O., Ingelman-Sundberg, M., and Ruckpaul, K., 1984, Mechanism of rate control of the NADPH-dependent reduction of cytochrome P-450 by lipids in reconstituted phospholipid vesicles, *Eur. J. Biochem.* **144**:509–513.
86. Peterson, J. A., Ebel, R. E., O'Keeffe, D. H., Matsubara, T., and Estabrook, R. W., 1976, Temperature dependence of cytochrome P-450 reduction: A model for NADPH-cytochrome P-450 reductase:cytochrome P-450 interaction, *J. Biol. Chem.* **251**:4010–4016.
87. Lambeth, J. D., Seybert, D. W., and Kamin, H., 1980, Phospholipid vesicle-reconstituted cytochrome P-450$_{scc}$: Mutually facilitated binding of cholesterol and adrenodoxin, *J. Biol. Chem.* **255**:138–143.
88. Lambeth, J. D., Kamin, H., and Seybert, D. W., 1980, Phosphatidylcholine vesicle reconstituted cytochrome P-450$_{scc}$: Role of the membrane in control of activity and spin state of the cytochrome, *J. Biol. Chem.* **255**:8282–8288.
89. DePierre, J. W., and Ernster, L., 1977, Enzyme topology of intracellular membranes, *Annu. Rev. Biochem.* **46**:201–262.
90. Seidegård, J., Moron, M. S., Eriksson, L. C., and DePierre, J. W., 1978, The topology of expoxide hydratase and benzpyrene monooxygenase in the endoplasmic reticulum of rat liver, *Biochim. Biophys. Acta* **543**:29–40.
91. Morimoto, T., Matsuura, S., Sasaki, S., Tashiro, Y., and Omura, T., 1976, Immunochemical and immunoelectron microscope studies on localization of NADPH-cytochrome c reductase on rat liver microsomes, *J. Cell Biol.* **68**:189–201.

92. Matsuura, S., Fujii-Kuriyama, Y., and Tashiro, Y., 1978, Immunoelectron microscope localization of cytochrome P-450 on microsomes and other membrane structures of rat hepatocytes, *J. Cell Biol.* **78:**504–519.
93. Cooper, M. B., Craft, J. A., Estall, M. R., and Rabin, B. R., 1980, Asymmetric distribution of cytochrome P-450 and NADPH-cytochrome P-450 (cytochrome c) reductase in vesicles from smooth endoplasmic reticulum of rat liver, *Biochem. J.* **190:**737–746.
94. Nilsson, O. S., DePierre, J. W., and Dallner, G., 1978, Investigation of the transverse topology of the microsomal membrane using combinations of proteases and the nonpenetrating reagent diazobenzene sulfonate, *Biochim. Biophys. Acta* **511:**93–104.
95. Nilsson, O. S., and Dallner, G., 1977, Enzyme and phospholipid asymmetry in liver microsomal membranes, *J. Cell. Biol.* **72:**568–583.
96. Thomas, P. E., Lu, A. Y. H., West, S. B., Ryan, D., Miwa, G. T., and Levin, W., 1977, Accessibility of cytochrome P450 in microsomal membranes: Inhibition of metabolism by antibodies to cytochrome P450, *Mol. Pharmacol.* **13:**819–831.
97. Fleming, P. J., Dailey, H. A., Corcoran, D., and Strittmatter, P., 1978, The primary structure of the nonpolar segment of bovine cytochrome b_5, *J. Biol. Chem.* **253:**5369–5372.
98. Ozols, J., and Gerard, C., 1977, Covalent structure of the membranous segment of horse cytochrome b_5, *J. Biol. Chem.* **252:**8549–8553.
99. Enoch, H. G., Fleming, P. J., and Strittmatter, P., 1979, The binding of cytochrome b_5 to phospholipid vesicles and biological membranes: Effect of orientation on intermembrane transfer and digestion by carboxypeptidase Y, *J. Biol. Chem.* **254:**6483–6488.
100. Takagaki, Y., Gerber, G., Nikei, K., and Khorana, H. G., 1980, Amino acid sequence of the membranous segment of rabbit liver cytochrome b_5: Methodology for separation of hydrophobic peptides, *J. Biol. Chem.* **255:**1536–1541.
101. Kondo, K., Takjima, S., Sato, R., and Narita, K., 1979, Primary structure of the membrane-binding segment of rabbit cytochrome b_5, *J. Biochem.* **86:**1119–1128.
102. Ozols, J., and Gerard, C., 1977, Primary structure of the membranous segment of cytochrome b_5, *Proc. Natl. Acad. Sci. USA* **74:**3725–3729.
103. Bendzko, P., Usanov, S. A., Pfeil, W., and Ruckpaul, K., 1982, Role of the hydrophobic tail of cytochrome b_5 in the interaction with cytochrome P-450 LM2, *Acta Biol. Med. Ger.* **41:**K1–K8.
104. Chiang, J. Y. L., 1981, Interaction of purified microsomal cytochrome P-450 with cytochrome b_5, *Arch. Biochem. Biophys.* **211:**662–673.
105. Dailey, H. A., and Strittmatter, P., 1978, Structural and functional properties of the membrane binding segment of cytochrome b_5, *J. Biol. Chem.* **253:**8203–8209.
106. Prkrovsky, A., Mishin, V., Rivkind, N., and Lyakhovich, V., 1977, The binding of NADPH-cytochrome *c* reductase to rat liver microsomes, *Biochem. Biophys. Res. Commun.* **77:**912–917.
107. Yang, C. S., Strickhart, F. S., and Kicha, L. P., 1978, Interaction between NADPH-cytochrome P-450 reductase and hepatic microsomes, *Biochim, Biophys. Acta* **509:**326–337.
108. Gum, J. R., and Strobel, H. W., 1979, Purified NADPH cytochrome P-450 reductase, *J. Biol. Chem.* **254:**4177–4185.
109. Vermilion, J. L., and Coon, M. J., 1978, Purified liver microsomal NADPH-cytochrome P-450 reductase: Spectral characterization of oxidation–reduction states, *J. Biol. Chem.* **253:**2694–2704.
110. Coon, M. J., Strobel, H. W., and Boyer, R. F., 1973, On the mechanism of hydroxylation reactions catalyzed by cytochrome P-450, *Drug Metab. Dispos.* **1:**92–97.

111. Masters, B. S. S., Prough, R. A., and Kamin, H., 1975, Properties of the stable aerobic and anaerobic half-reduced states of NADPH cytochrome c reductase, *Biochemistry* **14:**607–613.
112. Mayer, R. T., and Durrant, J. L., 1979, Preparation of homogenous NADPH cytochrome c (P-450) reductase from house flies using affinity chromatography techniques, *J. Biol. Chem.* **254:**756–761.
113. Gum, J. R., and Strobel, H. W., 1981, Isolation of the membrane binding peptide of NADPH-cytochrome P-450 reductase, *J. Biol. Chem.* **256:**7478–7486.
114. Black, S. D., French, J. S., Williams, C. H., Jr., and Coon, M. J., 1979, Role of a hydrophobic polypeptide in the N-terminal region of NADPH cytochrome P-450 reductase in complex formation with P-450LM, *Biochem. Biophys. Res. Commun.* **91:**1528–1535.
115. Black, S. D., and Coon, M. J., 1982, Structural features of liver microsomal NADPH-cytochrome P-450 reductase: Hydrophobic domain, hydrophilic domain and connecting region, *J. Biol. Chem.* **257:**5929–5938.
116. Heinemann, F. S., and Ozols, J., 1983, The complete amino acid sequence of rabbit phenobarbital-induced liver microsomal cytochrome P-450, *J. Biol. Chem.* **258:**4195–4201.
117. Tarr, G. E., Black, S. D., Fujita, V. S., and Coon, M. J., 1983, Complete amino acid sequence and predicted membrane topology of phenobarbital-induced cytochrome P-450 (isozyme 2) from rabbit liver microsomes, *Proc. Natl. Acad. Sci. USA* **80:**6552–6556.
118. Heinemann, F. S., and Ozols, J., 1982, The covalent structure of rabbit phenobarbital-induced cytochrome P-450, *J. Biol. Chem.* **257:**14988–14999.
119. Kimura, S., Gonzalez, F. J., and Nebert, D. W., 1984, The murine Ah locus: Comparison of the complete cytochrome P_1-450 and P_3-450 cDNA nucleotide and aminoacid sequences, *J. Biol. Chem.* **259:**10705–10713.
120. Leighton, J. K., DeBrunner-Vossbrinck, B. A., and Kemper, B., 1984, Isolation and sequence analysis of three cloned cDNAs for rabbit liver proteins that are related to rabbit cytochrome P-450 (form 2), the major phenobarbital-inducible form, *Biochemistry* **23:**204–210.
121. Rich, P. R., Tiede, D. M., and Bonner, W. D., Jr., 1979, Studies on the molecular organization of cytochromes P-450 and b_5 in the microsomal membrane, *Biochim. Biophys. Acta* **546:**307–315.
122. Blum, H., Leigh, J. S., Salerno, J. C., and Ohnishi, T., 1978, The orientation of bovine adrenal cortex cytochrome P-450 in submitochondrial particle mutlilayers, *Arch. Biochem. Biophys.* **187:**153–157.
123. Cherry, R. J., 1979, Rotational and lateral diffusion of membrane proteins, *Biochim. Biophys. Acta* **559:**289–327.
124. Vaz, W. L. C., Goodsaid-Zalduondo, F., and Jacobson, K., 1984, Lateral diffusion of lipids and proteins in bilayer membranes, *FEBS Lett.* **174:**199–207.
125. Vaz, W. L. C., Derzko, Z. I., and Jacobson, K. A., 1982, Photobleaching measurements of the lateral diffusion of lipids and proteins in artificial phospholipid bilayer membranes, in: *Membrane Reconstitution* (G. Poste and G. I. Nicholson, eds.), Elsevier, Amsterdam, pp. 83–136.
126. Yang, C. S., 1977, Minireview: The organization and interaction of monooxygenase enzymes in the microsomal membrane, *Life Sci.* **21:**1047–1058.
127. Poo, M., and Cone, R. A., 1974, Lateral diffusion of rhodopsin in the photoreceptor membrane, *Nature* **247:**438–441.
128. Liebman, P. A., and Entine, G., 1974, Lateral diffusion of visual pigment in photoreceptor disk membranes, *Science* **185:**457–459.

129. Schlessinger, J., Koppel, D. E., Axelrod, D., Jacobson, K., Webb, W. W., and Elson, E. L., 1976, Lateral transport on cell membranes: Mobility of concanavalin A receptors on myoblasts, *Proc. Natl. Acad. Sci. USA* **73**:2409–2413.
130. Axelrod, D., Ravdin, P., Koppel, D. E., Schlessinger, J., Webb, W. W., Elson, E. L., and Podleski, T. R., 1976, Lateral motion of fluorescently labeled acetylcholine receptors in membranes of developing muscle fibers, *Proc. Natl. Acad. Sci. USA* **73**:4594–4598.
131. Schlessinger, J., Webb, W. W., and Elson, E. L., 1976, Lateral motion and valence of Fc receptors on rat peritoneal mast cells, *Nature* **264**:550–552.
132. Mabrey, S., Powis, G., Schenkman, J. B., and Tritton, T. R., 1977, Calorimetric study of microsomal membrane, *J. Biol. Chem.* **252**:2929–2933.
133. Stier, A., and Sackmann, E., 1973, Spin labels as enzyme substrates: Heterogenous lipid distribution in liver microsomal membranes, *Biochim. Biophys. Acta* **311**:400–408.
134. Wu, E.-S., Jacobson, K., Szoka, F., and Portis, A., Jr., 1978, Lateral diffusion of a hydrophobic peptide, N-4-nitrobenz-2-oxa-1, 3-diazole gramicidin S, in phospholipid multibilayers, *Biochemistry* **17**:5543–5550.
135. Tsong, T. Y., and Yang, C. S., 1978, Rapid conformational changes of cytochrome P-450: Effector of dimyristoyl lecithin, *Proc. Natl. Acad. Sci. USA* **75**:5955–5959.
136. Rich, P. R., Tiede, D. M., and Bonner, W. D., Jr., 1979, Studies on the molecular organization of cytochromes P-450 and b_5 in the microsomal membrane, *Biochim. Biophys. Acta* **546**:307–315.
137. Fahey, P. F., and Webb, W. W., 1978, Lateral diffusion in phospholipid bilayer membranes and multilamellar liquid crystals, *Biochemistry* **17**:3046–3053.
138. Wu, E.-S., and Yang, C. S., 1984, Lateral diffusion of cytochrome P-450 in phospholipid bilayers, *Biochemistry* **23**:28–33.
139. Vaz, W. L. C., Jacobson, K., Wu, E.-S., and Derzko, Z., 1979, Lateral mobility of an amphipathic apolipoprotein, ApoC-III, bound to phosphatidylcholine bilayers with and without cholesterol, *Proc. Natl. Acad. Sci. USA* **76**:5645–5649.
140. Saffman, P. G., and Delbrüch, M., 1975, Brownian motion in biological membranes, *Proc. Natl. Acad. Sci. USA.* **72**:3111–3113.
141. Ingelman-Sundberg, M., Blanck, J., Smettan, G., and Ruckpaul, K., 1983, Reduction of cytochrome P-450 LM_2 by NADPH in reconstituted phospholipid vesicles is dependent on membrane charge, *Eur. J. Biochem.* **134**:157–162.
142. Taniguchi, H., Imai, Y., and Sato, R., 1984, Role of the electron transfer system in microsomal drug monooxygenase reaction catalyzed by cytochrome P-450, *Arch. Biochem. Biophys.* **232**:585–596.
143. Werringloer, J., and Kawano, S., 1980, Cytochrome b_5 and the integrated microsomal electron transport system, in: *Microsomes, Drug Oxidations and Chemical Carcinogenesis*, Volume 1 (M. J. Coon, A. H. Conney, R. W. Estabrook, H. V. Gelboin, J. R. Gillette, and P. J. O'Brien, eds.), Academic Press, New York, pp. 469–478.
144. Miwa, G. T., West, S. B., and Lu, A. Y. H., 1978, Studies on the rate limiting enzyme component in the microsomal monooxygenase system: Incorporation of purified NADPH-cytochrome c reductase and cytochrome P-450 into rat liver microsomes, *J. Biol. Chem.* **253**:1921–1929.
145. Guengerich, F. P., 1983, Oxidation–reduction properties of rat liver cytochromes P-450 and NADPH-cytochrome P-450 reductase related to catalysis in reconstituted systems, *Biochemistry* **22**:2811–2820.
146. Miwa, G. T., and Cho, A. K., 1976, Stimulation of microsomal N-demethylation by solubilized NADPH-cytochrome c reductase, *Life Sci.* **18**:983–988.
147. Yang, C. S., 1977, Interactions between solubilized cytochrome P-450 and hepatic

microsomes: Characterizations of the binding and enhanced catalytic activities, *J. Biol. Chem.* **252**:293–298.
148. Yang, C. S., and Strickhart, S., 1975, Interactions between solubilized cytochrome P-450 and hepatic microsomes, *J. Biol. Chem.* **250**:7968–7972.
149. Yang, C. S., 1975, The association between cytochrome P-450 and NADPH-cytochrome P-450 reductase in microsomal membrane, *FEBS Lett.* **54**:61–64.
150. Rogers, M. J., and Strittmatter, P., 1974, The binding of reduced nicotinamide adenine dinucleotide-cytochrome b_5 reductase to hepatic microsomes, *J. Biol. Chem.* **249**:5565–5569.
151. Rogers, M. J., and Strittmatter, P., 1974, Evidence for random distribution and translational movement of cytochrome b_5 in endoplasmic reticulum, *J. Biol. Chem.* **249**:895–900.
152. Ito, A., 1974, Evidence obtained by cathepsin digestion of microsomes for the assembly of cytochrome b_5 and its reductase in the membrane, *J. Biochem.* **75**:787–793.
153. Archakov, A. I., Borodin, E. A., Dobretsov, G. E., Karasevich, E. I., and Karyakin, A. V., 1983, The influence of cholesterol incorporation and removal on lipid-bilayer viscosity and electron transfer in rat-liver microsomes, *Eur. J. Biochem.* **134**:89–95.
154. Richter, C., Winterhalter, K. H., and Cherry, R. J., 1979, Rotational diffusion of cytochrome P-450 in rat liver microsomes, *FEBS Lett.* **102**:151–154.
155. McIntosh, P. R., Kawato, S., Freedman, R. B., and Cherry, R. J., 1980, Evidence from cross-linking and rotational diffusion studies that cytochrome P-450 can form molecular aggregates in rabbit-liver microsomal membranes, *FEBS Lett.* **122**:54–58.
156. Schwarz, D., Pirrwitz, J., and Ruckpaul, K., 1982, Rotational diffusion of cytochrome P-450 in the microsomal membrane—Evidence for a clusterlike organization from saturation transfer electron paramagnetic resonance spectroscopy, *Arch. Biochem. Biophys.* **216**:322–328.
157. Greinert, R., Staerk, H., Stier, A., and Weller, A., 1979, E-type delayed fluorescence depolarization: A technique to probe rotational motion in the microsecond range, *J. Biochem. Biophys. Methods* **1**:77–83.
158. Greinert, R., Finch, S. A. E., and Stier, A., 1982, Conformation and rotational diffusion of cytochrome P-450 changed by substrate binding, *Biosci. Rep.* **2**:991–994.
159. Coleman, R., 1973, Membrane-bound enzymes and membrane ultrastructure, *Biochim. Biophys. Acta* **300**:1–30.
160. Farias, R. N., Bloj, B., Morero, R. D., Sineriz, F., and Trucco, R. E., 1975, Regulation of allosteric membrane-bound enzymes through changes in membrane lipid composition, *Biochim. Biophys. Acta* **415**:231–251.
161. Gazzotti, P., and Peterson, S. W., 1977, Lipid requirement of membrane-bound enzymes, *J. Bioenerg. Biomembr.* **9**:373–386.
162. Boggs, J. M., Wood, D. D., Moscarello, M. A., and Papahadjopoulos, D., 1977, Lipid phase separation induced by a hydrophobic protein in phosphatidylserine–phosphatidylcholine vesicles, *Biochemistry* **16**:2325–2329.
163. Brotherus, J. R., Jost, P. C., Griffith, O. H., Keana, J. F. W., and Hokin, L. E., 1980, Charge selectivity at the lipid–protein interface of membranous Na,K-ATPase, *Proc. Natl. Acad. Sci. USA* **77**:272–276.
164. Brown, R. E., and Cunningham, C. C., 1982, Negatively charged phospholipid requirement of the oligomycin-sensitive mitochondrial ATPase, *Biochim. Biophys. Acta* **684**:141–145.
165. Cunningham, C. C., and Sinthusek, G., 1979, Ionic charge on phospholipids and their interaction with the mitochondrial adenosine triphosphatase, *Biochim. Biophys. Acta* **550**:150–153.
166. Pitotti, A., Contessa, A. R., Dabbeni-Sala, F., and Bruni, A., 1972, Activation by

phospholipids of particulate mitochrondrial ATPase from rat liver, *Biochim. Biophys. Acta* **274:**528–535.
167. Forsee, W. T., and Schutzbach, J. S., 1983, Interaction of α-1,2-mannosidase with anionic phospholipids, *Eur. J. Biochem.* **136:**577–582.
168. Griffith, O. H., and Jost, P. C., 1979, in: *Proceedings of the Japanese–American Seminar on Cytochrome Oxidase* (B. Chance, T. E. King, K. Okunuki, and Y. Oril, eds.), Elsevier, Amsterdam, pp. 207–218.
169. McIntyre, J. O., Holladay, L. A., Smigel, M., Puett, D., and Fleischer, S., 1978, Hydrodynamic properties of D-β-hydroxybutyrate dehydrogenase, a lipid-requiring enzyme, *Biochemistry* **17:**4169–4177.
170. Gazzotti, P., Bock, H.-G., and Fleischer, S., 1975, Interaction of D-β-hydroxybutyrate apodehydrogenase with phospholipids, *J. Biol. Chem.* **250:**5782–5790.
171. Fleischer, S., McIntyre, J. O., Churchill, P., Fleer, E., and Mauver, A., 1983, in: *Structure and Functions of Membrane Proteins* (E. Quagliariello and F. Palmieri, eds.), Elsevier, Amsterdam, pp. 283–290.
172. Robinson, N. C., 1982, Specificity and binding affinity of phospholipids to the high-affinity cardiolipin sites of beef heart cytochrome c oxidase, *Biochemistry* **21:**184–188.
173. Robinson, N. C., Strey, F., and Talbert, L., 1980, Investigation of the essential boundary layer phospholipids of cytochrome c oxidase using Triton X-100 delipidation, *Biochemistry* **19:**3656–3661.
174. Fry, M., and Green, D. E., 1980, Cardiolipin requirement by cytochrome oxidases and the catalytic role of phospholipid, *Biochem. Biophys. Res. Commun.* **93:**1238–1246.
175. Cinti, D. L., Sligar, S. G., Gibson, G. G., and Schenkman, J. B., 1979, Temperature-dependent spin equilibrium of microsomal and solubilized cytochrome P-450 from rat liver, *Biochemistry* **18:**36–42.
176. Ebel, R. E., O'Keeffe, D. H., and Peterson, J. A., 1978, Substrate binding to hepatic microsomal cytochrome P-450: Influence of the microsomal membrane, *J. Biol. Chem.* **253:**3888–3897.
177. Gibson, G. G., Cinti, D. L., Sligar, S. G., and Schenkman, J. B., 1980, The effect of microsomal fatty acids and other lipids on the spin state of partially purified cytochrome P-450, *J. Biol. Chem.* **255:**1867–1873.
178. Ruckpaul, K., Rein, H., Blanck, J., and Coon, M. J., 1982, Molecular mechanisms of interactions between phospholipids and liver microsomal cytochrome P-450 LM2, *Acta Biol. Med. Ger.* **41:**193–203.
179. French, J. S., Guengerich, F. P., and Coon, M. J., 1980, Interactions of cytochrome P-450, NADPH-cytochrome P-450 reductase, phospholipid, and substrate in the reconstituted liver microsomal enzyme system, *J. Biol, Chem.* **255:**4112–4119.
180. Tamburini, P. P., and Gibson, G. G., 1983, Thermodynamic studies of the protein–protein interactions between cytochrome P-450 and cytochrome b_5: Evidence for a central role of the cytochrome P-450 spin state in the coupling of substrate and cytochrome b_5 binding to the terminal hemoprotein, *J. Biol. Chem.* **258:**13444–13452.
181. Chiang, Y.-L., and Coon, M. J., 1979, Comparative study of two highly purified forms of liver microsomal cytochrome P-450: Circular dichroism and other properties, *Arch. Biochem. Biophys.* **195:**178–187.
182. Guengerich, F. P., and Davison, N. K., 1982, Interaction of epoxide hydrolase with itself and other microsomal proteins, *Arch. Biochem. Biophys.* **215:**462–477.
183. Vore, M., Hamilton, J. G., and Lu, A. Y. H., 1974, Organic solvent extraction of liver microsomal lipid. I. The requirement of lipid for 3,4-benzpyrene hydroxylase, *Biochem. Biophys. Res. Commun.* **56:**1038–1044.
184. Ingelman-Sundberg, M., 1977, Phospholipids and detergents as effectors in the liver microsomal hydroxylase system, *Biochim. Biophys. Acta* **488:**225–234.

185. Uvarov, V. Y., Backmanova, G. I., Archakov, A. I., Sukhomudrenko, A. G., and Myasoedova, K. N., 1980, Conformation and thermostability of soluble cytochrome P-450 and cytochrome P-450 incorporated into liposomal membrane, *Biokhimiya* **45:**1463–1469.
186. Archakov, A. I., Bachmanova, G. I., and Uvarov, V. Y., 1980, Interactions of cytochrome P-450 with phospholipid bilayer, in: *Biochemistry, Biophysics and Regulation of Cytochrome P-450* (J.-Å. Gustafsson, J. Carlstedt-Duke, A. Mode, and J. Rafter, eds.), Elsevier/North-Holland, Amsterdam, pp. 551–558.
187. Wang, H.-P., Pfeiffer, D. R., Kimura, T., and Tchen, T. T., 1974, Phospholipids of adrenal cortex mitochondria and the steroid hydroxylases: The lipid environment of cytochrome P-450, *Biochem. Biophys. Res. Commun.* **57:**93–99.
188. Hall, P. F., Watanuki, M., DeGroot, J., and Rouser, G., 1978, Composition of lipids bound to pure cytochrome P-450 of cholesterol side-chain cleavage enzyme from bovine adrenocortical mitochondria, *Lipids* **14:**148–155.
189. Ingelman-Sundberg, M., Haaparanta, T., and Rydström, J., 1981, Membrane charge as effector of cytochrome P-450$_{LM2}$ catalyzed reactions in reconstituted liposomes, *Biochemistry* **20:**4100–4106.
190. Ruckpaul, K., Rein, H., Ballou, D. P., and Coon, M. J., 1980, Analysis of interactions among purified components of the liver microsomal cytochrome P-450-containing monooxygenase system by second derivative spectroscopy, *Biochim. Biophys. Acta* **626:**41–56.
191. Dehlinger, P. J., Jost, P. C., and Griffith, O. H., 1974, Lipid binding to the amphipathic membrane protein cytochrome b$_5$, *Proc. Natl. Acad. Sci. USA* **71:**2280–2284.
192. Bösterling, B., and Trudell, J. R., 1982, Phospholipid transfer between vesicles: Dependence on presence of cytochrome P-450 and phosphatidylcholine–phosphatidylethanolamine ratio, *Biochim. Biophys. Acta* **689:**155–160.
193. Barsukov, L. I., Kulikov, V. I., Bachmanova, G. I., Archakov, A. I., and Bergelson, L. D., 1982, Cytochrome P-450 facilitates phosphatidylcholine flip-flop in proteoliposomes, *FEBS Lett.* **144:**337–340.
194. Lambeth, J. D., 1981, Cytochrome P-450$_{scc}$: Cardiolipin as an effector of activity of a mitochondrial cytochrome P-450, *J. Biol. Chem.* **256:**4757–4762.
195. Pember, S. O., Powell, G. L., and Lambeth, J. D., 1983, Cytochrome P-450$_{scc}$–phospholipid interactions: Evidence for a cardiolipin binding site and thermodynamics of enzyme interactions with cardiolipin, cholesterol, and adrenodoxin, *J. Biol. Chem.* **258:**3198–3206.
196. Omura, T., and Sato, R., 1964, The carbon monoxide-binding pigment of liver microsomes. I. Evidence for its hemoprotein nature, *J. Biol. Chem.* **239:**2370–2378.
197. Chaplin, M. D., and Mannering, G. J., 1970, Role of phospholipids in the hepatic microsomal drug-metabolizing system, *Mol. Pharmacol.* **6:**631–640.
198. Eling, T. E., and DiAugustine, R. P., 1971, A role for phospholipids in the binding and metabolism of drugs by hepatic microsomes: Use of the fluorescent hydrophobic probe 1-anilinonaphthalene-8-sulphonate, *Biochem. J.* **123:**539–549.
199. Tagg, J., and Mitoma, C., 1968, Studies on the microsomal drug-metabolizing enzyme system—Effect of isooctane and pyridine nucleotides, *Biochem. Pharmacol.* **17:**2471–2479.
200. Parry, G., Palmer, D. N., and Williams, D. J., 1976, Ligand partitioning into membranes: Its significance in determining K_M and K_S values for cytochrome P-450 and other membrane bound receptors and enzymes, *FEBS Lett.* **67:**123–129.
201. Al-Gailany, K. A. S., Houston, J. B., and Bridges, J. W., 1978, The role of substrate lipophilicity in determining type 1 microsomal P-450 binding characteristics, *Biochem. Pharmacol.* **27:**783–788.

202. McConnell, H. M., Wright, K. L., and McFarland, B. G., 1972, The fraction of the lipid in a biological membrane that is in a fluid state: A spin label assay, *Biochem. Biophys. Res. Commun.* **47**:273–281.
203. Shimshick, E. J., and McConnell, H. M., 1973, Lateral phase separation in phospholipid membranes, *Biochemistry* **12**:2351–2360.
204. Narasimhulu, S., 1977, Thermotropic transitions in fluidity of bovine adrenocortical microsomal membrane and substrate–cytochrome P-450 binding reaction, *Biochim. Biophys. Acta* **487**:378–387.
205. Eletr, S., Zakim, D., and Vessey, D. A., 1973, A spin-label study of the role of phospholipids in the regulation of membrane-bound microsomal enzymes, *J. Mol. Biol.* **78**:351–362.
206. Taniguchi, H., Imai, Y., and Sato, R., 1984, Substrate binding site of microsomal cytochrome P-450 directly faces membrane lipids, *Biochem. Biophys. Res. Commun.* **118**:916–922.
207. Narasimhulu, S., 1979, Constraint on the substrate cytochrome P-450 binding reaction in bovine adrenocortical microsomes at physiological temperature, *Biochim. Biophys. Acta* **556**:457–468.
208. Becker, J. F., Meehan, T., and Bartholomew, J. C., 1978, Fatty acid requirements and temperature dependence of monooxygenase activity in rat liver microsomes, *Biochim. Biophys. Acta* **512**:136–146.
209. Duppel, W., and Ullrich, V., 1976, Membrane effects on drug monooxygenation activity in hepatic microsomes, *Biochim. Biophys. Acta* **426**:399–407.
210. Bador, H., Morelis, R., and Louisot, P., 1984, Breaks in Arrhenius plots of reactions involving membrane-bound and solubilized sialyltransferases, due to temperature dependence of kinetic parameters, *Biochim. Biophys. Acta* **800**:75–86.
211. Miwa, G. T., West, S. B., Huang, M.-T., and Lu, A. Y. H., 1979, Studies on the association of cytochrome P-450 and NADPH-cytochrome c reductase during catalysis in a reconstituted hydroxylating system, *J. Biol. Chem.* **254**:5695–5700.
212. Ingelman-Sundberg, M., and Johansson, I., 1980, Catalytic properties of purified forms of rabbit liver microsomal cytochrome P-450 in reconstituted phospholipid vesicles, *Biochemistry* **19**:4004–4011.
213. Johnson, E. F., Schwab, G. E., and Dieter, H. H., 1983, Allosteric regulation of the 16α-hydroxylation of progesterone as catalyzed by rabbit microsomal cytochrome P-450 3b, *J. Biol. Chem.* **258**:2785–2788.
214. Takemori, S., and Kominami, S., 1984, The role of cytochromes P-450 in adrenal steroidogenesis, *Trends Biochem. Sci.* **9**:393–396.
215. Paul, D. P., Gallant, S., Orme-Johnson, N. R., Orme-Johnson, W. H., and Brownie, A. C., 1976, Temperature dependence of cholesterol binding to cytochrome P-450$_{scc}$ of the rat adrenal: Effect of adrenocorticotropic hormone and cycloheximide, *J. Biol. Chem.* **251**:7120–7126.
216. Brownie, A. C., Simpson, E. R., Jefcoate, C. R., and Boyd, G. S., 1972, Effect of ACTH on cholesterol side-chain cleavage in rat adrenal mitochondria, *Biochem. Biophys. Res. Commun.* **46**:483–490.
217. Jefcoate, C. R., Orme-Johnson, W., and Beinert, H., 1973, Effect of ACTH on adrenal mitochondrial cytochrome P-450 in the rat, *Ann. N. Y. Acad. Sci.* **212**:344–360.
218. Simpson, E. R., Jefcoate, C. R., Brownie, A. C., and Boyd, G. S., 1972, The effect of ether anaesthesia stress on cholesterol-side chain cleavage and cytochrome P-450 in rat-adrenal mitochondria, *Eur. J. Biochem.* **28**:442–450.
219. Privalle, C. T., Crivello, J. F., and Jefcoate, C. R., 1983, Regulation of intramitochondrial cholesterol transfer to side-chain cleavage cytochrome P-450 in rat adrenal gland, *Proc. Natl. Acad. Sci. USA* **80**:702–706.

220. Stevens, V. L., Aw, T. Y., Jones, D. P., and Lambeth, J. D., 1984, Oxygen dependence of adrenal cortex cholesterol side chain cleavage, *J. Biol. Chem.* **259:**1174–1179.
221. Vahouny, G. V., Dennis, P., Chanderbhan, R., Fiskum, G., Noland, B. J., and Scallen, T. J., 1984, Sterol carrier protein$_2$ (SCP$_2$)-mediated transfer of cholesterol to mitochondrial inner membranes, *Biochem. Biophys. Res. Commun.* **122:**509–515.
222. Hsu, D. K., Huang, Y. Y., and Kimura, T., 1984, Thermodynamic properties of the cholesterol transfer reaction from liposomes to cytochrome P450$_{scc}$: An enthalpy–entropy compensation effect, *Biochem. Biophys. Res. Commun.* **118:**877–884.
223. Jefcoate, C. R., 1977, Cytochrome P-450 of adrenal mitochondria: Steroid binding sites on two distinguishable forms of rat adrenal mitochondrial cytochrome P-450$_{scc}$, *J. Biol. Chem.* **252:**8788–8796.
224. Lambeth, J. D., Lancaster, J. R., Jr., Seybert, D. W., and Kamin, H., 1980, Binding of cholesterol and adrenodoxin to phospholipid vesicle reconstituted P-450$_{scc}$, in: *Microsomes, Drug Oxidations, and Chemical Carcinogenesis* (M. J. Coon, A. H. Conney, R. W. Estabrook, H. V. Gelboin, J. R. Gillette, and P. J. O'Brien, eds.), Academic Press, New York, pp. 553–557.
225. Kido, T., Arakawa, M., and Kimura, T., 1979, Adrenal cortex mitochondrial cytochrome P-450 specific to cholesterol side chain cleavage reaction: Spectral changes induced by detergents, alcohols, amines, phospholipids, steroid hydroxylase inhibitors, and steroid substrates, and conditions for adrenodoxin binding to the cytochrome, *J. Biol. Chem.* **254:**8377–8385.
226. Hanukoglu, I., Spitsberg, V., Bumpus, J. A., Dus, K. M., and Jefcoate, C. R., 1981, Adrenal mitochondrial cytochrome P-450$_{scc}$: Cholesterol and adrenodoxin interactions at equilibrium and during turnover, *J. Biol. Chem.* **256:**4321–4328.
227. Kido, T., Kimura, T., 1981, Stimulation of cholesterol binding to steroid-free cytochrome P-450$_{scc}$ by poly(L-lysine): The implication in functions of labile protein factor for adrenocortical steroidogenesis, *J. Biol. Chem.* **256:**8561–8568.
228. Lambeth, J. D., Kitchen, S. E., Farooqui, A. A., Tuckey, R., and Kamin, H., 1982, Cytochrome P-450$_{scc}$ substrate interactions: Studies of binding and catalytic activity using hydroxycholesterols, *J. Biol. Chem.* **257:**1876–1884.
229. Sato, R., and Omura, T., 1978, *Cytochrome P-450*, Academic Press, New York.
230. Vore, M., Lu, A. Y. H., Kuntzman, R., and Conney, A. H., 1974, Organic solvent extraction of liver microsomal lipid. II. Effect on the metabolism of substrates and binding spectra of cytochrome P-450, *Mol. Pharmacol.* **10:**963–974.
231. Guengerich, P. F., and Holladay, L. A., 1979, Hydrodynamic characterization of highly purified and functionally active liver microsomal cytochrome P-450, *Biochemistry* **18:**5442–5449.
232. Takagi, Y., Shikita, M., and Hall, P. F., 1975, The active form of cytochrome P-450 from bovine adrenocortical mitochondria, *J. Biol. Chem.* **250:**8445–8448.
233. Knapp, J. A., Dignam, J. D., and Strobel, H. W., 1977, NADPH-cytochrome P-450 reductase: Circular dichroism and physical studies, *J. Biol. Chem.* **252:**437–443.
234. Ingelman-Sundberg, M., Montelius, J., Rydström, J., and Gustafsson, J.-Å., 1978, The active form of cytochrome P-450$_{11\beta}$ from adrenal cortex mitochondria, *J. Biol. Chem.* **253:**5042–5047.
235. Baskin, L. S., and Yang, C. S., 1980, Cross-linking studies of cytochrome P-450 and reduced nicotinamide adenine dinucleotide phosphate-cytochrome P-450 reductase, *Biochemistry* **19:**2260–2264.
236. Ingelman-Sundberg, M., and Johansson, I., 1980, Cytochrome b$_5$ as electron donor to rabbit liver cytochrome P-450$_{LM2}$ in reconstituted phospholipid vesicles, *Biochem. Biophys. Res. Commun.* **97:**582–589.
237. Bösterling, B., and Trudell, J. R., 1982, Association of cytochrome b$_5$ and cytochrome

P-450 reductase with cytochrome P-450 in the membrane of reconstituted vesicles, *J. Biol. Chem.* **257**:4783–4787.
238. Guengerich, F. P., Ballou, D. P., and Coon, M. J., 1975, Purified liver microsomal cytochrome P-450: Electron-accepting properties and oxidation–reduction potential, *J. Biol. Chem.* **250**:7405–7414.
239. Bäckström, D., Ingelman-Sundberg, M., and Ehrenberg, A., 1983, Oxidation–reduction potential of soluble and membrane-bound rabbit liver microsomal cytochrome P-450$_{LM2}$ *Acta Chem. Scand. Ser. B* **37**:891–894.
240. Kominami, S., and Takemori, S., 1982, Effect of spin state on reduction of cytochrome P-450 (P-450$_{C21}$) from bovine adrenocortical microsomes, *Biochim. Biophys. Acta* **709**:147–153.
241. Archakov, A. I., Borondin, E. A., Davydov, D. R., Karyakin, A. I., and Borovyagin, V. L., 1982, Random distribution of NADPH-specific flavoprotein and cytochrome P-450 in liver microsomes, *Biochem. Biophys. Res. Commun.* **109**:832–840.
242. Hildebrandt, A., and Estabrook, R. W., 1971, Evidence for the participation of cytochrome b$_5$ in hepatic microsomal mixed-function oxidation reactions, *Arch. Biochem. Biophys.* **143**:66–79.
243. Ingelman-Sundberg, M., and Johansson, I., 1984, Electron flow and complex formation during cytochrome P-450-catalyzed hydroxylation reactions in reconstituted membrane vesicles, *Acta Chem. Scand.* **B38**:845–851.
244. Ingelman-Sundberg, M., Edvardsson, A.-L., and Johansson, I., 1981, Electron transport and cytochrome P-450-dependent oxygenations mediated by hydroxyl radicals in reconstituted membrane vesicles containing the rabbit liver microsomal monooxygenase system, in: *Microsomes, Drug Oxidations, and Drug Toxicity* (R. Sato and R. Kato, eds.), Japan Scientific Press Societies Wiley–Interscience, New York, pp. 187–194.
245. Morgan, E. T., and Coon, M. J., 1984, Effects of cytochrome b$_5$ on cytochrome P-450-catalyzed reactions: Studies with manganese-substituted cytochrome b$_5$, *Drug Metab. Dispos.* **12**:358–365.
246. Lu, A. Y. H., West, S. B., Vore, M., Ryan, D., and Levin, W., 1974, Role of cytochrome b$_5$ in hydroxylation by a reconstituted cytochrome P-450-containing system, *J. Biol. Chem.* **249**:6701–6709.
247. Ingelman-Sundberg, M., Johansson, I., Brunström, A., Ekström, G., Haaparanta, T., and Rydström, J., 1980, The importance of cytochrome b-5 and negatively charged phospholipids in electron transport to different types of liver microsomal cytochrome P-450 in reconstituted phospholipid vesicles, in: *Biochemistry, Biophysics and Regulation of Cytochrome P-450* (J.-Å. Gustafsson, J. Carlstedt-Duke, A. Mode, and J. Rafter, eds.), Elsevier/North-Holland, Amsterdam, pp. 299–306.
248. Vatsis, K. P., Theoharides, A. D., Kupfer, D., and Coon, M. J., 1982, Hydroxylation of prostaglandins by inducible ioszymes of rabbit liver microsomal cytochrome P-450, *J. Biol. Chem.* **257**:11221–11229.
249. Strittmatter, P., and Rogers, M. J., 1975, Apparent dependence of interactions between cytochrome b$_5$ and cytochrome b$_5$ reductase upon translational diffusion in dimyristoyl lecithin liposomes, *Proc. Natl. Acad. Sci. USA* **72**:2658–2661.
250. Huang, M.-T., Johnson, E. F., Muller-Eberhard, U., Koop, D. R., Coon, M. J., and Conney, A. H., 1981, Specificity in the activation and inhibition by flavonoids of benzo(a)pyrene hydroxylation by cytochrome P-450 isozymes from rabbit liver microsomes, *J. Biol. Chem.* **256**:10897–10901.
251. Huang, M.-T., Chang, R. L., Fortner, J. G., and Conney, A. H., 1981, Studies on the mechanism of activation of microsomal benzo(α)pyrene hydroxylation by flavonoids, *J. Biol. Chem.* **256**:6829–6836.
252. Dalet, M. C., Andersson, K. K., Dalet-Beluche, I., Bonfils, C., and Maurel, P., 1983,

Polyamines as modulators of drug oxidation reactions catalyzed by cytochrome P-450 from liver microsomes, *Biochem. Pharmacol.* **32**:593–601.
253. König, B. W., Osheroff, N., Wilms, J., Juijsers, A. O., Dekker, H. L., and Margoliash, E., 1980, Mapping of the interaction domain for purified cytochrome c_1 on cytochrome c, *FEBS Lett.* **111**:395–398.
254. Koppenol, W. H., and Margoliash, E., 1982, The asymmetric distribution of charges on the surface of horse cytochrome c: Functional implications, *J. Biol. Chem.* **257**:4426–4437.
255. Dailey, H. A., and Strittmatter, P., 1979, Modification and identification of cytochrome b_5 carboxyl groups involved in protein–protein interaction with cytochrome b_5 reductase, *J. Biol. Chem.* **254**:5388–5396.
256. Loverde, A., and Strittmatter, P., 1968, The role of lysyl residues in the structure and reactivity of cytochrome b_5 reductase, *J. Biol. Chem.* **243**:5779–5787.
257. Dailey, H. A., and Strittmatter, P., 1980, Characterization of the interaction of amphipathic cytochrome b_5 with stearyl coenzyme A desaturase and NADPH: cytochrome P-450 reductase, *J. Biol. Chem.* **255**:5184–5189.
258. Bernhardt, R., Makower, A., Jänig, G.-R., and Ruckpaul, K., 1984, Selective chemical modification of a functionally linked lysine in cytochrome P-450$_{LM2}$, *Biochim. Biophys. Acta* **785**:186–190.
259. Tuckey, R. C., and Kamin, H., 1983, Kinetics of O_2 and CO binding to adrenal cytochrome P-450$_{scc}$: Effect of cholesterol, intermediates and phosphatidylcholine residues, *J. Biol. Chem.* **258**:4232–4237.
260. Dix, T. A., and Marnett, L. J., 1983, Metabolism of polycyclic aromatic hydrocarbon derivatives to ultimate carcinogens during lipid peroxidation, *Science* **221**:77–79.
261. Imai, Y., Sato, R., and Iyanagi, T., 1977, Rate-limiting step in the reconstituted microsomal drug hydroxylase system, *J. Biochem.* **82**:1237–1246.
262. Iyanagi, T., Suzaki, T., and Kobayashi, S., 1981, Oxidation–reduction states of pyridine nucleotide and cytochrome P-450 during mixed-function oxidation in perfused rat liver, *J. Biol. Chem.* **256**:12933–12939.

CHAPTER 6

Comparative Structures of P-450 Cytochromes

SHAUN D. BLACK and MINOR J. COON

1. Introduction

Rapid research progress during the past decade has led to the isolation and characterization of many apparently distinct forms of cytochrome P-450 and has shown that the inducers, substrates, and inhibitors of this versatile enzyme are almost unlimited in number. Only in the past 3 years, however, have complete primary structures become known through protein and DNA sequencing, thus providing firm evidence that these cytochromes are distinct gene products. This chapter is concerned with comparative structures of the P-450 cytochromes as revealed by general properties as well as by partial or complete sequence information on about 30 such proteins.

Although it would be highly desirable to name all P-450's by their catalytic functions, this is presently feasible in only a few cases, and even then may be subject to change when additional xenobiotics or, especially, naturally occurring lipids are tested as possible substrates. Particularly with hepatic microsomal P-450, the overlap in substrate specificity among the isozymes and the use of different inducers and substrates have led to much confusion over the possible identity of the enzymes isolated by different laboratories. For simplicity in the present review, we are employing the designations by number and letter most generally used in the literature: 1, 2, 3a, 3b, 3c, 4, 5, 6, and so on for rabbit,[1] a, b, c, d, e, f, g, h, and so on for rat,[2] and 1, 2, 3, and so on for mouse.[3] Investigators in this field are urged to exchange newly isolated P-450 cytochromes for

---

SHAUN D. BLACK and MINOR J. COON • Department of Biological Chemistry, Medical School, The University of Michigan, Ann Arbor, Michigan 48109.

comparison as to spectral and catalytic properties and N-terminal sequence, the usefulness of which is described below. In publications, investigators are urged to designate each form of P-450 by the generally used nomenclature, along with that laboratory's own designation, if necessary, and such essential information as the species, tissue, organelle, sex of the animal, and inducer used if any.*

Although much remains to be learned about the structure of P-450's, the sequence information presently available permits a number of generalizations to be made, if only as a stimulus to further research. Perhaps of most importance, it is now clear that the various cytochromes so far examined in detail are discrete gene products; that is, they differ in amino acid sequence rather than in some detail such as posttranslational modification. Furthermore, the evidence presently available provides no support for the view that P-450's possess constant and variable regions as do the immunoglobulins.

2. Comparison of P-450 Cytochromes by Methods Other Than Sequencing

2.1. Spectral Properties and Molecular Weight Estimates

Characterization of highly purified P-450 cytochromes has generally included assessment of the ferric spin state and ferrous–carbonyl absorption wavelength maximum, as well as electrophoretic determination of purity and minimal molecular weight. These data, with corresponding biological sources and typical inducers, are summarized in Table I for about 50 cytochromes that have been isolated in an electrophoretically homogeneous state and with high specific content from a total of 13 species. Although these proteins are generally low spin in the oxidized state,

---

* Rabbit liver microsomal P-450's, designated on the basis of relative electrophoretic mobility as LM_1, LM_2, and so on, or simply as forms or isozymes 1, 2, and so on, have also been referred to by the abbreviations given in parentheses: 2 ($P-450_1$[4]), 3b ($P-450_5$[5]), 3c ($B1$[6], or $P-450_9$[5]), 4 ($P-448$[6] or $P-448_1$[4]), and 6 ($P-448_2$[5]); some additional forms described briefly by Aoyama et al.[5] may be distinct from those studied by other laboratories. Rabbit lung P-450's are now designated by Philpot and colleagues in the same manner as the corresponding liver cytochromes: 2 (I[7]), 5 (II[7]), and 5a ($P-450_{PG-\omega}$ from pregnant animals[8]). Rabbit kidney P-450's: 2 (b[9]), 4 (tentatively a[9]), and 6 (P-448[10]). Rabbit intestinal mucosa: 4 (tentatively a[11]). Rat liver P-450's: a (UT-F,[12] 1,[13] or PB-3[14]), b (PB-B,[12] PB-4,[14] or PB-1[15]), c (BNF-B,[12] 5,[13] or MC-1[15]), d (BNF/ISF-G,[12] 4,[13] or MC-2[15]), e (PB-D[12] or PB-5[16]), f (possibly, PB-2[15]), g (tentatively RLM3[17]), h (male-specific P-450[18] or tentatively UT-A[12] or tentatively PB-2c[14]), and i (female-specific P-450[18]); isozymes not given a letter designation include PCN[19] (PB/PCN-E[12] or PB-2a[20]), CLO (P-452[21]), PB-C[12] (PB-1[20]), and RLM5[17] (tentatively CC-25[22]). Mouse liver P-450's: 1 (P_1-450[23]), 2 (P_2-450[3]), and 3 (P_3-450[24] or P-448[23]).

TABLE I
Sources, Spectral Properties, and Molecular Weight Estimates of Highly Purified P-450 Cytochromes

Species	Source[a]	Typical inducer[b]	Isozyme	Fe^{II}CO λ max (nm)	Fe^{III} spin state	M_r (av.) SDS–PAGE[c]	References
Rabbit	L, Mc	PB	1	450	Low	48,000	25
	L & Lg & K, Mc		2	451	Low	49,500	4, 7, 9, 26, 27
	L, Mc	ALC	3a	452	High	51,000	28
	L, Mc		3b	450	Low	52,000	29–31
	L, Mc	TAO	3c	449	Low	53,000	6, 30
	L & K & I, Mc	BF, ISF	4	448	High	54,000	4, 9, 11, 26
	L, Mc	CHM	4 (7α)	447	High	53,000	32
	L, Mc		4 (12α)	447	High	53,000	32
	L, Mt		4 (26)	450	Low	53,000	33
	L & Lg, Mc	PB	5	449	Low	57,000	7
	L & Lg, Mc	Pregnancy	5a	450	Low	56,000	8
	L & K, Mc	BF, ISF	6	448	Low	57,500	10, 34, 35
Rat	L, Mc	PB, MC	a	452	Low	47,800	2, 12, 13
	L, Mc	PB	b	450	Low	51,300	2, 12, 15, 27, 36, 37
	L, Mc	MC, ISF	c	447	Low	54,300	2, 12, 13, 15, 36
	L, Mc	ISF, MC	d	447	High	53,000	12, 13, 15, 38
	L, Mc	PB	e	450	Low	51,900	12, 36, 37, 39
	L, Mc		f	448	Mixed	51,000	15, 40
	L, Mc		g	448	Low	50,000	17, 40
	L, Mc		h	451	Low	52,800	12, 18, 40
	L, Mc		i	449	Low	50,500	18, 40

(continued)

TABLE I (continued)

Species	Source[a]	Typical inducer[b]	Isozyme	$Fe^{II}CO$ λ max (nm)	Fe^{III} spin state	M_r (av.) SDS-PAGE[c]	References
	L, Mc		A	450	Low	52,200	41
	L, Mc		B	451	Low	52,400	41
	L, Mc	PCN	PCN	450	Low	51,000	12, 19
	L, Mc	CLO	CLO	452	Mixed	52,000	21
	L, Mc	PB	PB-C	450	Low	49,200	12, 20
	L, Mc		UT-H	449	Low	52,000	42
	L, Mc		RLM5	451	Low	50,000	17
Mouse	L, Mc	MC	1	449	Low	55,000	3, 23
	L, Mc	ISF	2	448	Low	55,000	3
	L, Mc	MC	3	448	Low	55,000	3, 23
	L, Mc	PB	C_2	450	Low	56,000	43
Bovine	Ad & Cl, Mt		scc	448	Low	51,700	44–49
	Ad, Mt		11β	448	Low	46,000	45, 50
	Ad, Mc		C-21	450	Low	49,000	51–53
Pig	Ad, Mc		C-21	450	Mixed	54,000	54
	Ad & T, Mc		17α	448	Low	54,000	55–57
	Ao, Mc		PGI_2Synth	451	Low	49,000	58
Guinea pig	Ad, Mc		17α	448	High[d]	52,000	59
Scup	L, Mc		E	447	Low	54,300	60

COMPARATIVE STRUCTURES OF P-450

Tulip	B, Mc		450	Low	52,500[d]	61
Saccharomyces cerevisiae	Mc	14DM	447	Low	58,000	62
Pseudomonas putida		cam	446	Low	46,000	63, 64
Bacillus megabacterium		meg	450	Low	52,000	65
Rhizobium japonicum		c	447	Low	46,000	66, 67
Human	L, Mc	1	450	Low	53,000	68, 69
	L, Mc	2	447	Low	53,200	69
	L, Mc	3	450	Low	54,300	69
	L, Mc	4	450	Low	54,300	69
	L, Mc	5	450	Low	53,700	69
	L, Mc	7	450	Low	53,200	69
	L, Mc	8	450	Low	49,600	69

[a] L, liver; Lg, lung; K, kidney cortex; I, intestinal mucosa; Ad, adrenal; Cl, corpus luteum; T, testis; Ao, aorta; B, bulb; Mc, microsomes; Mt, mitochondria.
[b] PB, phenobarbital; ALC, ethanol; TAO, triacetyloleandomycin; BF, 5,6-benzoflavone; ISF, isosafrole; CHM, cholestyramine; MC, 3-methylcholanthrene; PCN, pregnenolone 16α-carbonitrile; CLO, clofibrate.
[c] Each value represents the mean of up to 11 determinations.
[d] This cytochrome, as other steroidogenic P-450's, is isolated in a substrate-bound high-spin form.

isozymes 3a and 4 from rabbit and isozyme d from rat are high spin, and rat isozymes f and CLO exhibit mixed high- and low-spin character. The adrenal steroidogenic cytochromes are usually isolated in the presence of substrate to enhance stability and thus show high-spin spectra, but are apparently low spin in the absence of bound substrate. Ferrous–carbonyl wavelength maxima for all of the P-450's are found in the range 446–452 nm. The location of the exact maximum for a given cytochrome does not appear to be correlated with the species, tissue, organelle, inducer, spin state, or molecular weight, let alone the function, and therefore is evidently a property of the unique heme-binding site in each polypeptide. Thus, there is presently no basis for a general classification of these proteins as "P-450" and "P-448" types.

Minimal molecular weight estimates by SDS–PAGE in the presence of denaturants and reducing agents fall in the range 46,000–57,000, and thus define a "P-450 region" of the electrophoretogram for a given source. However, because other microsomal polypeptides such as epoxide hydrolase[70] ($M_r \sim$ 48,000) and the FAD-containing monooxygenase[71] $M_r \sim$ 54,000) are known to exhibit mobilities in this region, the "P-450 region" is clearly not unique. Nonetheless, all presently characterized P-450 cytochromes appear to have similar minimal molecular weights. Molecular weight estimates for a given cytochrome show considerable dispersion, apparently due to the particular electrophoretic technique used. The work of Guengerich[36] in which six different methods were applied to rat isozymes b, c, and e showed differences in molecular weight of 8200, 7000, and 4700, respectively; use of cetyltrimethylammonium bromide as denaturant consistently resulted in the lowest estimates. Substitution of lithium dodecyl sulfate for SDS caused pronounced and differential effects on rat isozyme molecular weight estimates.[12] Electrode buffer concentration has also been shown to dramatically alter the molecular weight estimate; higher ionic strength resulted in higher, and more likely correct, estimates for a number of rat isozymes.[14,20] This is in accord with the findings of Black et al.,[27] who noted that electrophoresis-derived molecular weights appeared to be uniformly low for these hydrophobic polypeptides. (See Section 3.4.2 for a further discussion of this point.) Thus, caution should be exercised in the comparison of molecular weights estimated by different laboratories; critical electrophoretic comparisons necessitate the exchange of samples between the pertinent investigators.

Molecular weights have also been determined under nondenaturing conditions for a limited number of P-450's. Rabbit liver microsomal isozymes 2 and 4, rat liver microsomal isozyme b, and bovine adrenal mitochondrial P-450$_{scc}$ are each associated in the absence of detergents to the extent of approximately 6,[4,72–74] 9,[75] 7,[36] and 8 molecules per aggregate,[47] respectively. In contrast to these membranous eukaryotic cyto-

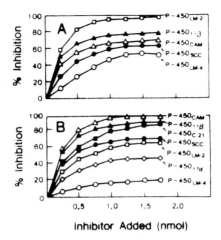

FIGURE 1. Radioimmunoassay based on competitive binding. (A) Competition for anti-P-450$_{LM2}$ IgG between ^{125}I-labeled P-450$_{LM2}$ and various unlabeled cytochromes. (B) Competition for anti-P-450$_{cam}$ IgG between ^{125}I-labeled P-450$_{cam}$ and various unlabeled cytochromes. From Ref. 77 with permission.

chromes, prokaryotic bacterial P-450$_{cam}$ and rhizobial P-450c appear to be monomeric under all conditions used. This is one of the most striking differences observed among the purified P-450's.

2.2. Immunochemical Properties

Polyclonal antibodies have been used by many laboratories as structural probes to assess the extent of relatedness of various forms of P-450. Radioimmunoassays,[76,77] as shown in Fig. 1, indicated cross-reaction of rabbit isozyme 2 with rabbit isozyme 4, P-450$_{cam}$, and steroidogenic P-450$_{11\beta}$ and P-450$_{scc}$; anti-P-450$_{cam}$ antibody showed cross-reaction with rabbit isozymes 2 and 4, as well as with steroidogenic P-450$_{11\beta}$, P-450$_{C-21}$, P-450$_{scc}$, and P-450$_{17\alpha}$. However, the Ouchterlony double-diffusion analysis has been by far the most widely applied technique. With use of this method, Thomas et al.[78] showed earlier that rat isozyme b antibody cross-reacted weakly with rabbit isozyme 2, thus suggesting some common structural features. Later, immunochemical relatedness was shown among rat isozymes b, e, and f[39,40] and between rat isozymes c and d.[79] The cross-reactivity of microsomal P-450's from mouse, rat, rabbit, and human has also been noted.[68,80,81] In contrast, many reports of immunochemical distinctness are available. Rabbit isozymes 1, 2, 3b, 3c, and 4 appeared to be immunochemically distinct[6,25,31,82] as judged by the Ouchterlony method, as did rat isozymes a, b, c, d, g, and h,[2,13,15,38,40] and bovine steroidogenic P-450$_{scc}$, P-450$_{11\beta}$, and P-450$_{C-21}$.[51] Although these results would appear to contradict certain of those given above, they depend upon the specificity of the particular antibody preparation and the technique used. One may conclude from the immunochemical comparisons in the literature that most of the purified P-450's have some structural relatedness. This could be due, of course, to common amino acid sequences or to similar three-dimensional features.

2.3. Peptide Mapping

2.3.1. Electrophoretic Peptide Maps

Peptide maps generated from electrophoretic analysis of limited proteolytic digestion mixtures by the method of Cleveland *et al.*[83] have proven to be a rapid and useful means to examine primary structural similarities and differences of purified forms of P-450. *Staphylococcus aureus* V$_8$ protease, chymotrypsin, papain, trypsin, endoproteinase Lys-C, and CNBr have been used to produce peptide fragments of appropriate size. Samples are analyzed in parallel, and comparison of various forms is readily achieved as can be seen in Fig. 2 for five rabbit isozymes and Fig. 3 for eight rat isozymes. A unique map is produced from each cytochrome, although some equivalent-sized fragments can be seen among the various forms. Furthermore, the patterns of rat isozymes b and e are clearly related, as are those of isozymes c and d, in accord with immunochemical findings stated in the previous section. Electrophoretic peptide maps have also been compared for rabbit isozymes 1, 2, 4, and 6,[34] isozymes 2 and 5,[7] isozymes 2, 3b, and 4,[31] isozymes 2, 3b, 3c, and 4,[30] and different isozyme 4 preparations[32]; with rat isozymes a, b, c, d, e, h, PCN, and PB-C,[12] isozymes a, b, c, and d,[38] isozymes a, c, and d,[2,13] isozymes b, c, and PCN,[19] isozymes b, c, and CLO,[21] isozymes b, e, and PB-1,[20] and isozymes g and RLM5[17]; with mouse isozymes 1 and 3[23]; and with seven human isozymes.[69] Caution is warranted, however, in the comparison of results from different laboratories due to variations in conditions used for proteolysis and electrophoresis.

2.3.2. High-Performance Liquid Chromatographic Peptide Maps

The advent of HPLC with efficient reversed-phase columns has permitted high-resolution peptide mapping of P-450 cytochromes. This has been useful for analytical purposes as well as in preparation of peptides for sequence investigations. Analytical applications have generally utilized limited tryptic digestion, with far-ultraviolet detection of small peptides in HPLC eluates. Although serial processing of samples is necessary,

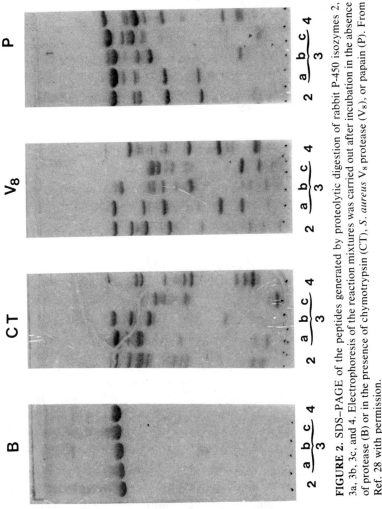

FIGURE 2. SDS–PAGE of the peptides generated by proteolytic digestion of rabbit P-450 isozymes 2, 3a, 3b, 3c, and 4. Electrophoresis of the reaction mixtures was carried out after incubation in the absence of protease (B) or in the presence of chymotrypsin (CT), *S. aureus* V_8 protease (V_8), or papain (P). From Ref. 28 with permission.

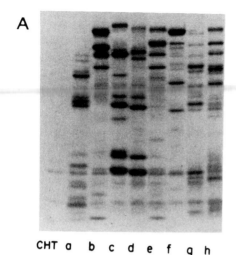

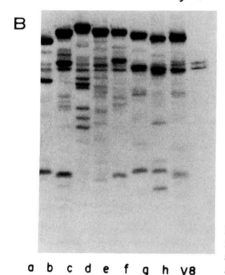

FIGURE 3. SDS–PAGE of peptide fragments of purified rat cytochromes generated by digestion with chymotrypsin (A) or *S. aureus* V_8 protease (B). From Ref. 40 with permission.

this has presented no problem because the technique appears to be highly reproducible. A chromatographic comparison of tryptic digests of five rabbit isozymes with dual absorbance/fluorescence detection is shown in Fig. 4, and a comparison of rat isozymes b and e is shown in Fig. 5. Approximately 50 peptides were visualized in each case, which is two to five times the number usually seen in electrophoretic analyses. This resolution permitted ready discrimination of the rabbit isozymes, whereas immunochemically identical rat isozymes b and e showed remarkably

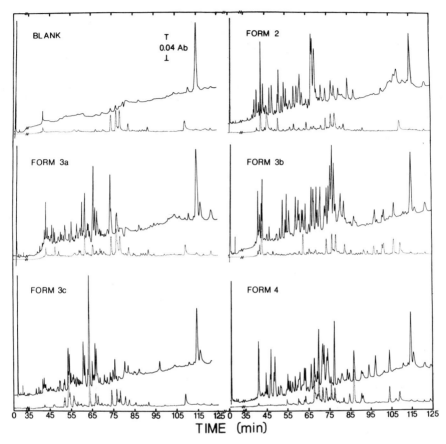

FIGURE 4. Resolution of the tryptic peptides of rabbit P-450 isozymes 2, 3a, 3b, 3c, and 4 by HPLC. The solid line represents the absorbance of the effluent at 214 nm, and the dotted line represents the fluorescence of the effluent with an excitation wavelength of 290 nm and emission measured with a standard 370-nm filter. The fluorescence is in arbitrary units that are the same for each chromatogram. From Ref. 28 with permission.

similar maps. In the latter case, sequence studies showed that the two rat proteins differ by only 13 substitutions.[84] Similarly, Ozols *et al.*[85] used HPLC mapping to identify an analogous peptide from rabbit isozymes 2 and 3b. Another application of HPLC peptide mapping was in the identification of strain variants. Ryan *et al.*[86] used reversed phase analysis of tryptic peptides to detect minor primary structural differences in isozymes b and c, each purified from both Holtzman (H) and Long–Evans (LE) rats; $P-450b_H$ and $P-450b_{LE}$, and $P-450c_H$ and $P-450c_{LE}$ are each pairs of allozymes. Dieter and Johnson[87] used HPLC mapping to identify variants

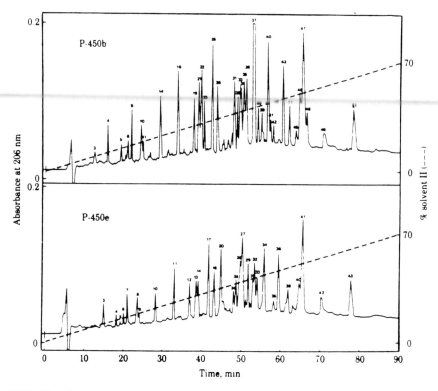

FIGURE 5. Comparative HPLC tryptic maps of rat cytochromes b (upper) and e (lower). From Ref. 84 with permission.

of rabbit isozyme 3b from inbred and outbred populations. HPLC mapping has also been used in the demonstration that preparations of an isozyme from animals treated with different inducing agents are identical. Thus, Koop and Coon[35] used HPLC analyses of trypsin-, *S. aureus* V_8 protease-, and endoproteinase Lys-C-digested isozyme 3a purified from rabbits treated with ethanol or imidazole to provide evidence for identity. In a similar fashion, Yoshioka *et al.*[88] showed that rat isozyme b purified from phenobarbital- and from 1,1-di(*p*-chlorophenyl)-2,2-dichloroethylene-treated animals was indistinguishable.

2.4. Amino Acid Analysis

Representative amino acid analyses of 28 purified P-450 cytochromes submitted to acid hydrolysis are shown in Table II. The proteins contain about 400 to 500 total residues, with exceptionally high leucine content;

COMPARATIVE STRUCTURES OF P-450

TABLE II
Representative Amino Acid Analyses of Highly Purified P-450 Cytochromes

Amino acid residue	Rabbit isozymes							Rat isozymes												Bovine isozymes				Human	Tulip	Prokaryotic			Variation
	2	3a	3b	3c	4	5	6	a	b	c	d	e	f	g	h	i	A	B	PCN	scc	11β	C-21	17α	1		cam	meg	c	$\bar{X} \pm$ S.D.
A	24	17	15	19	29	40	32	25	29	27	26	17	16	18	17	21	40	37	25	23	28	32	26	27	36	34	25	54	27.1 ± 8.9
C	6	5	10	6	7	8	8	4	7	8	5	7	10	8	7	11	11	5	2	3	3	7	5	5	5	6	10	2	6.5 ± 2.5
D+N	39	44	43	47	44	47	48	40	42	47	44	41	43	42	47	39	30	32	27	42	28	31	36	47	48	36	49	33	40.6 ± 6.6
E+Q	42	42	50	44	46	54	45	41	52	47	49	52	38	40	42	41	46	44	52	51	47	54	51	45	42	55	55	54	47.2 ± 5.2
F	35	27	32	30	28	24	31	29	32	35	31	27	28	35	47	36	19	22	34	28	16	20	26	23	29	17	17	17	27.7 ± 7.2
G	32	29	29	29	38	37	36	33	33	36	37	54	26	30	28	33	83	68	33	23	21	35	25	45	42	26	27	33	35.8 ± 13.3
H	15	12	13	8	11	12	16	9	13	15	13	10	14	15	8	13	10	10	15	15	11	15	13	1	7	12	9	12	11.7 ± 3.3
I	23	23	30	34	22	25	20	25	27	25	25	25	30	30	22	31	23	23	31	30	15	19	25	28	19	24	27	14	25.0 ± 5.0
K	21	29	34	40	26	34	22	30	27	28	32	38	31	31	24	31	26	27	31	34	12	18	29	46	39	13	28	14	28.9 ± 8.7
L	57	50	55	56	58	66	54	52	59	55	49	40	54	56	50	57	46	51	46	50	46	76	51	53	45	40	44	46	52.6 ± 7.5
M	7	10	11	18	10	11	8	13	11	8	6	15	15	13	11	12	9	11	12	13	23	11	10	15	7	9	7	8	10.8 ± 2.9
P	31	31	31	24	28	25	24	21	27	27	29	30	32	29	23	29	30	37	45	28	23	31	26	28	27	27	25	25	28.4 ± 4.7
R	36	27	18	28	30	37	36	20	25	26	21	22	19	15	23	17	15	17	18	28	30	30	25	19	20	24	15	29	23.8 ± 6.5
S	31	24	26	28	31	36	39	28	30	36	31	34	29	20	26	26	57	47	24	23	20	26	20	36	37	21	34	20	30.0 ± 8.5
T	25	26	31	29	31	27	28	26	29	28	24	25	33	31	24	24	27	26	25	26	21	23	18	25	27	19	24	24	25.5 ± 4.0
V	30	29	29	32	32	34	40	24	27	33	32	20	25	33	27	28	26	28	37	28	25	36	27	22	34	24	27	25	29.1 ± 4.7
W	1	3	5	5	7	—	6	1	2	3	4	1	—	—	—	—	1	2	1	7	6	9	3	4	—	1	1	1	3.4 ± 2.5
Y	12	15	13	13	11	11	14	14	14	15	10	14	14	6	12	12	11	12	13	18	10	10	17	5	16	9	11	6	12.1 ± 3.1
Total	467	443	475	492	482	528	507	435	486	500	468	472	457	452	438	468	510	499	497	470	373	483	433	474	480	397	435	407	
Literature reference	89	28	30	30	26	7	35	90	36	90	91	36	92	92	92	92	92	41	19	93	94	93	53	68	61	63	65	66	

TABLE III
Comparison of Compositions of P-450 Isozymes Determined by Amino Acid Analysis and from Complete Sequences

Amino acid residue	Rabbit isozyme 2 Analytical data						$\bar{X}$ ± S.D.	Sequence data	Z-score[a]	Rat isozyme b Analytical data			$\bar{X}$ ± S.D.	Sequence data	Z-score	Rat isozyme c Analytical data			$\bar{X}$ ± S.D.	Sequence data	Z-score
A	23	21	23	22	24	26	23.2 ± 1.7	24	−0.47	29	27	26	26.7 ± 2.5	23	+1.48	26	27	27	26.7 ± 0.6	25	+2.93
C	6	3	5	5	6	4	4.5 ± 1.4	4	+0.36	7	8	9	8.0 ± 1.0	6	+2.00	6	8	13	9.0 ± 3.6	8	+0.28
D+N	38	35	37	37	39	39	37.5 ± 1.5	39	−1.00	42	47	38	42.3 ± 4.5	37	+1.18	43	47	47	45.7 ± 2.3	47	−0.56
E+Q	42	39	44	45	42	42	42.3 ± 2.1	44	−0.45	52	47	53	50.7 ± 3.2	49	+0.53	43	47	52	47.3 ± 4.5	50	−0.60
F	31	36	36	37	35	34	34.8 ± 2.1	41	−2.95	32	35	36	34.3 ± 2.1	39	−2.24	28	35	33	32.0 ± 3.6	35	+0.83
G	32	31	34	34	32	34	32.8 ± 1.3	35	−1.69	33	36	33	34.0 ± 1.7	33	+0.59	32	36	39	35.7 ± 3.5	34	+0.49
H	11	12	12	12	15	13	12.5 ± 1.4	14	−1.07	13	15	19	15.7 ± 3.1	17	−0.42	13	15	17	15.0 ± 2.0	16	−0.50
I	19	22	23	23	23	24	22.3 ± 1.8	27	−2.60	27	29	27	27.7 ± 1.2	30	−1.92	24	29	28	27.0 ± 2.7	31	−1.48
K	19	19	20	20	21	20	19.8 ± 0.8	20	−0.27	27	28	28	27.7 ± 0.6	24	+6.17	31	28	31	30.0 ± 1.7	28	+1.18
L	54	52	57	61	57	63	57.3 ± 4.1	64	−1.63	59	55	58	57.3 ± 2.1	63	−2.71	54	55	55	54.7 ± 0.6	59	−7.41
M	7	8	7	8	7	8	7.5 ± 0.6	8	−0.91	11	8	9	9.3 ± 1.5	11	−1.13	10	8	7	8.3 ± 1.5	8	+0.22
P	24	30	29	29	31	28	28.5 ± 2.4	32	−1.46	27	24	31	27.3 ± 3.5	30	−0.77	37	24	26	29.0 ± 7.0	29	0
R	29	32	33	30	36	36	32.7 ± 2.9	38	−1.83	25	26	28	26.3 ± 1.5	31	−3.13	25	26	25	25.3 ± 0.6	28	−4.66
S	30	27	34	30	31	23	29.2 ± 3.8	32	−0.74	30	36	34	33.3 ± 3.1	33	+0.09	38	36	41	38.3 ± 2.5	39	−0.28
T	23	23	25	26	25	24	24.3 ± 1.2	26	−1.67	29	28	28	28.3 ± 0.6	27	+2.17	29	28	36	31.0 ± 4.4	31	0
V	26	26	29	28	30	28	27.8 ± 1.6	30	−1.38	27	33	21	27.0 ± 6.0	23	+0.67	28	33	32	31.0 ± 2.7	34	−1.11
W	1	2	1	1	—	1	1.2 ± 0.5	1	+0.44	2	3	—	2.5 ± 0.7	1	+2.11	6	3	—	4.5 ± 2.1	7	−1.19
Y	9	12	10	12	12	13	11.3 ± 1.5	12	−0.47	14	15	13	14.0 ± 1.0	14	0	14	15	13	14.0 ± 1.0	15	−1.00
Total	424	430	459	458	467	459		491		486	500	489		491		487	500	522		524	
Literature reference	26	4	7	95	89	89		89		36	90	15		96		36	90	15		97	

[a] Z-score is calculated as follows: [($\bar{X}$) − (Sequence value)]/S.D.

for example, P-450$_{C-21}$ was estimated to have nearly 16% of this residue in its composition. The prokaryotic P-450's appear to have relatively fewer total residues and lower hydrophobicity than do the eukaryotic cytochromes, but all of the compositions are rather similar with an average variation of ±6 per residue. The amino acids with the highest percent variation over all the polypeptides were Ala, Cys, Gly, Lys, and Trp, and those with the lowest were Asx, Glx, Leu, Pro, and Val. Individual variations of interest are the high Gly and Ser contents of rat isozymes a and b, the high Pro content of rat P-450$_{PCN}$, and the single His reported for human isozyme 1.

Because differences in amino acid composition have been used as a criterion of structural distinctness among P-450 preparations, and also in the validation of cDNA-derived protein sequences, we have assessed the utility of various residues for such comparisons as shown in Table III. The results of multiple amino acid analyses of rabbit isozyme 2 and rat isozymes b and c are given and compared with the true compositions known from the amino acid sequences. The data show that only Tyr is quantified both accurately and precisely; the average values for Glx and Ser were also accurate, but the individual values, along with those for Pro and Phe, had high scatter and thus poor precision. Met and Lys values showed good precision and reasonable accuracy, while those for Ala, Asx, Gly, His, Thr, Val, and Trp were of moderate accuracy and precision. Cys was generally overestimated, while Leu, Arg, Phe, and Ile were almost uniformly underestimated. Thus, in the comparison of acid hydrolysate-derived amino acid compositions of any two P-450 preparations, Tyr, Met, and Lys can be used as key residues where a range of no more than seven total residue differences is suggestive of identical proteins, taken together with a value of four or less for the average of all residue differences. Close agreement of Pro, Glx, Ser, and Phe values is not necessarily expected. Essentially the same conclusions apply to the comparison of a hydrolysate-derived amino acid composition with that of a primary structure deduced, for example, from cDNA. However, in this case the hydrolysate values for Leu, Arg, Phe, and Ile are expected to be low by averages of 6, 5, 5, and 4 residues, respectively. The above conclusions are valid as drawn from the present data set, but modification of these guidelines will be necessary as HPLC-based phenylthiocarbamyl amino acid analysis[28,98] and computer analysis of the analytical data[27] become more widespread in application.

2.5. Posttranslational Modifications

Glycosylation is the principal modification for which P-450 cytochromes have been examined. Haugen and Coon[26] reported that rabbit isozyme 2 contained 1.7 mannose and 0.7 glucosamine, and isozyme 4 contained 2.0 mannose and 0.8 glucosamine per polypeptide. Later, Imai et al.[4] reported that these two isozymes each contained about 3 hexoses per subunit, in accord with the above findings. Rat isozymes a through e were reported to have 0.9, 0.6–1.5, 0.3, 0.3, and 1.7 equivalents of mannose, respectively, but only traces of glucosamine, galactose, and galactosamine.[99] Guengerich et al.[12] found less than 0.2 equivalent of any detectable sugar in isozymes b and c. Negishi et al.[100] showed that [^{3}H]glucosamine was incorporated into mouse P_3-450 but not P_1-450, although stoichiometry was not assessed. Bacterial P-450$_{cam}$ was first thought to have 0.7 glucosamine per polypeptide,[63] but later studies detected none of this sugar although 0.5 equivalent of hexose was found.[101] In contrast to the above cases, the steroidogenic P-450's appear to be more extensively glycosylated. Bovine adrenal microsomal P-450$_{17\alpha}$ was found to have 2–3 equivalents of glucosamine[55] and bovine liver vitamin D_3 25-hydroxylase to have about 9 carbohydrate residues, including 2.7 glucosamine and 0.9 sialic acid.[102] Bovine adrenal mitochondrial P-450$_{scc}$ was shown to have 2.8 mannose, 1.7 galactose, 1.9 glucosamine, and 0.9 sialic acid with an absence of glucose, fucose, and galactosamine.[47] Treatment of this cytochrome with neuraminidase to remove sialic acid resulted in complete loss of activity; the resulting P-450 preparation had lost the ability to be reduced by adrenodoxin, thereby implicating glycosyl residues in some way in the mediation of protein–protein interactions.

Although phosphorylation has been suggested as a control mechanism in steroid 7α-hydroxylase activity,[103] phosphate groups could not be incorporated into mouse isozymes 1 and 3,[100] and rabbit isozyme 2 as isolated is not phosphorylated (D. Pompon and M. J. Coon, unpublished results). Thus, the role of phosphorylation as a control mechanism for P-450 reactions remains conjectural. Beyond glycosylation and phosphorylation, data on other possible secondary modifications, such as acylation by fatty acids or γ-carboxylation of glutamyl residues, are not currently available.

3. Comparison of P-450 Cytochromes by Sequencing Methods

3.1. N-Terminal Sequences by Edman Degradation

N-terminal sequence analysis has revealed some interesting structural features and has proven to be a very useful preliminary means to assess the equivalence or distinctness of various purified forms of P-450. In Table IV, the N-terminal sequences of 27 cytochromes are detailed. Rabbit isozyme 2 was the first example described in which a mature pro-

TABLE IV
N-Terminal Amino Acid Sequences of P-450 Cytochromes

Species	Isozyme	Residues identified by Edman degradation (1–30)	Reference
Rabbit	2	M E F S L L L L A F L A G L L L L F R G H P K	7, 27, 104
	3a	A V L G I T V A L L G W M V I L L F I S V W K Q I	28
	3b	M D L L I L G I C L S C V V L L S L W K	28, 85, 105
	3c	M D L I F S L E T W V L L A A S L V L L Y L Y G T	6, 28
	4	A M S P A A P L S V T E L L V S A V F C L V F W A V R A S	106
	5	M L G F L	7
	6	M V S D F G L P T F I S A T E L L L A S A V F C L	105
Rat	a	M L D T G L L L V V I L A S L S V M L L V S	90, 92
	b	M E P S I L L L L A L L V G F L L L V R G H P K S R G N F	15, 37, 84, 90, 107
	c	P S V Y G F P A F T S A T E L L L A V T T F X L G F X V	15, 90, 92
	d	A F S Q Y I S L A P E L L L A T A I F C L V F W V L R G X K	15, 91
	e	M E P S I L L L L A L L V G F L L L V R G H P K S R G N F	37, 84
	f	M D L V T F L V L T L S S L I L L S L W	92
	g	M D P V V V L L L S L F F L L	17, 92
	h	M D P V L V L V L T L S S L L L L S L W	92
	i	M D P F V V L V L S L S F L L L L Y X W	92
	PCN	M D L I F M L E T S S L L A	108
	PB-1	M D L V M L L V L T L T X L I L L S I W X Q S S G X G X E K	20
	RLM5	M D P V L V L V L T L L L L	17, 109, 110
Bovine	scc	I S T K T P R P Y S E I P S P G D X G N L N L Y X F	93, 111
	11β	G T S G A V A P K E V L P F E	93
	C-21	M V L A G L L L L T L K A G	93
Pig	C-21	M V L V W L L L L T T L K A G A R L L W G Q W K L R N L H L	54
	17α	M W V L L V F F L L T L T Y L F	57
Guinea pig	17α	M W E L V T L L G L I L A Y L	93
Scup	E	V L M I L P V I G	60
P. putida	cam	T T E T I Q S N A N L A P L P P H V P E H L V F D F D M Y N	112, 113

TABLE V
Comparison of C-Terminal Regions

Species	Isozyme	Method of determination[a]	C-terminal sequence	Accuracy In composition[b]	Accuracy In sequence	Reference
Rabbit	2	CPA/B	-F Y R-COOH	2 (67%)	1 (33%)	26
		CPA/B	-Y(I, V)A(L, F)R-COOH	6 (86%)	3 (43%)	120
		CPY	-L Y F(R, Q)I R F L A R-COOH	8 (73%)	7 (64%)	28
		SEQ	-N V P̲ P̲ S Y Q I R F L A R-COOH	—	—	89
	3a	CPY	-(F, T)(K, Y, L)R I V P̲ L-COOH			28
	3b	CPY	-(V, A, T)(Q, Y, I, L)L F P̲ V V-COOH	6 (50%)	4 (33%)	28
		SEQ	-S V P̲ P̲ S Y E L C F V̲ P̲ A-COOH	—	—	121
	3c	CPY	-(K, Q)(T, F, Y)P̲ V(L, I)A-COOH			28
	4	CPA/B	-S A(R, V)L K-COOH	4 (67%)	1 (17%)	26
		SEQ	-R C E H V Q A R P̲ S F X K-COOH	—	—	106
	6	CPY	-L(V, L)F A-COOH			35
Rat	a	CPY	-(K, G, A)L F I M-COOH			90

COMPARATIVE STRUCTURES OF P-450

		Sequence			Ref.
b	CPA/B	-G A-COOH	1 (50%)	0	36
	CPY	-(R, Y, C)F I A S-COOH	6 (86%)	3 (43%)	90
	SEQ	-K I P P̲ T Y Q I C F S A R-COOH	—	—	84, 96
c	CPA/B	-V-COOH	0	0	36
	CPA/B	-(V, F, K)R A H L-COOH	3 (43%)	0	90
	SEQ	-V Q M R S S G P̲ Q H L Q A-COOH	—	—	97, 122
d	CPA/B	-Y L R F K-COOH	3 (60%)	1 (20%)	91
	SEQ	-T C E H V Q A W P̲ R F S K-COOH	—	—	123
Bovine scc	CPY	-(F, L)A G S-COOH	2 (40%)	0	93
	SEQ	-L V F R P̲ F N Q D P P̲ Q A-COOH	—	—	124
11β	CPY	-(S, A, I, L)G T-COOH			93
C-21	CPY	-(L, V)A G S-COOH			93
P. putida cam	CPA/B	-K T T A V-COOH	5 (100%)	3 (60%)	112
	CPA	-(D, L, M, I, E, K)A L T V-COOH	5 (50%)	1 (10%)	63
	SEQ	-L P̲ L V W N P̲ A T T K A V-COOH	—	—	101

[a] CPA/B, a mixture of carboxypeptidases A and B; CPY, carboxypeptidase Y; SEQ, results from known sequence as determined by other methods.
[b] An amino acid is counted toward accuracy in composition if placed within four residues of the same residue in the known sequence.

tein was found to have retained a membrane-insertion signal peptide[104] similar to those normally removed from secreted proteins by biological processing.[114] The first 20 residues, beginning with Met, are extremely hydrophobic and include two pentaleucine sequences. Beyond residue 20, a number of basic amino acids occur that likely comprise a halt transfer signal.[114] This same pattern of a signal sequence–halt transfer signal appears to be common to all of the P-450's except for mitochondrial P-450_{scc} and P-$450_{11\beta}$ and bacterial P-450_{cam}; the N-terminal regions of these latter cytochromes are quite polar, in clear contrast to the apolar signal peptide in the microsomal cytochromes. The lack of extensive hydrophobicity in the N-terminal regions of P-450_{scc} and P-$450_{11\beta}$ is surprising in view of the fact that these are membrane-bound proteins. The signal sequences of the microsomal cytochromes usually begin with Met, as already noted for isozyme 2, followed by Ser or an acidic residue at position 2 or 3. Beyond this point, the N-terminal sequences are generally hydrophobic and exhibit pairwise identifies in the range of 15 to 60% when optimally aligned. Such alignments suggest that rabbit isozymes 4 and 6 and rat isozymes c and d are all related, as are rabbit isozyme 2 and rat isozymes b and e, rabbit isozyme 3b and rat isozymes f and PB-1, and rabbit isozyme 3c and rat P-450_{PCN}. It is of interest to note that the possibly variable position 10 of rabbit isozyme 3b[30,85] has now been shown to be a Cys[105]; another Cys residue is found at position 13.[105] The occurrence of Cys in the first 25 residues of four of the known N-terminal sequences and the known lability of this residue during Edman degradation indicate the general need to perform such analyses on thiol-alkylated P-450 polypeptides.

The N-terminus of all known P-450's is free to Edman degradation, as has also been found for microsomal epoxide hydrolase.[35,70,115,116] However, this is in contrast to microsomal NADH-cytochrome b_5 reductase,[117] cytochrome b_5,[118] and NADPH-cytochrome P-450 reductase,[119] which are all blocked at the N-terminus, the latter two by acetylation. While acylation of the N-terminus of P-450's does not appear to occur, post-translational modification by terminal cleavage has been observed in four cases. Preparations of rabbit isozyme 3b,[85] rat isozyme b,[84] and bovine P-450_{C-21}[54] have been described that lack the N-terminal residue. Furthermore, rabbit isozyme 4 was earlier found to have several residues in the terminal position and has recently been shown to exist as a mixture of n, n-1, and n-2 forms,[106] and preliminary results with rabbit isozyme 6[35] suggest that this protein may be processed in a similar fashion.

N-terminal sequence analysis has proven to be a highly useful means to compare various purified forms of P-450, but, although differences in N-terminal sequences clearly distinguish two cytochromes, identical sequences do not necessarily imply complete identity of the proteins. This latter point is supported by the example of rat isozymes b and e, which do not differ for at least 30 residues at the N-terminus but show 13 substitutions in the C-terminal half of the polypeptide.[84] Nevertheless, with this exception, N-terminal sequence analysis has proven to be the most definitive means to establish the uniqueness of a given form of P-450.

3.2. C-Terminal Sequences by Treatment with Carboxypeptidase

The results of carboxypeptidase digestions of P-450's are presented in Table V, together with the known C-terminal sequences determined by the Edman procedure, when available. A striking feature of these results is the high error; the mean values for composition and sequence correctness are 59 and 28%, respectively. Although certain of the analyses, such as the carboxypeptidase Y digestion of rabbit isozyme 2, proved to be quite accurate, the results in general suggest a need for caution in the use of such data. Highly accurate C-terminal sequence determinations by this method are clearly not possible, but comparisons between different preparations might be useful.

Two major factors contribute to the observed error. The first is contaminating endopeptidase activity in most carboxypeptidase preparations; electrophoretic experiments are needed to help assess the importance of this factor with each purified cytochrome. The second factor relates to an apparently common structural feature of the C-terminus of the P-450's, which is the presence of Pro residues (underlined in Table V) within the ten C-terminal residues of each sequence so far determined. This residue cannot be removed by carboxypeptidase A or B[125] and is hydrolyzed quite slowly by carboxypeptidase Y.[126] Furthermore, Pro–Pro sequences occur in the C-terminal region of rabbit isozymes 2 and 3b, rat isozymes b and e, and bovine P-450$_{scc}$. This dipeptide is not hydrolyzed by carboxypeptidases A, B, or Y, thereby limiting the amount of sequence information obtainable by this method; in the case of P-450$_{scc}$, only two terminal residues can be obtained through the use of carboxypeptidases. Cys and the amides Asn and Gln, which are also commonly encountered in the known C-terminal sequences, present an analytical challenge when conventional ion-exchange chromatography is used to analyze the results of the experiments. In summary, carboxypeptidase digestion of P-450's can yield useful data, but careful control of experimental conditions is necessary due to problems with the exopeptidases used and to the nature of the C-terminal region of these cytochromes.

3.3. Overall Sequences and Alignments

At the current time, nine complete and two nearly complete primary structures of P-450 cytochromes are known; these amino acid sequences

```
o-o ooooo oo   oooooo+ + + + +o     o oo   oo  o-++
MEFSLLLLLAFLAGLLLLLFRGHPKAHGRLPPGPSPLPVLGNLLQMDRKG      50

oo+  oo+o+-+o  -oo ooo    + oooo    - o+-  oo-   - o  + +
LLRSFLRLREKYGDVFTVYLGSRPVVVLCGTDAIREALVDQAEAFSGRGK      100

 oo   oo   o ooo    -+o+ o++o o   o+-o o ++ o--+o  --
IAVVDPIFQGYGVIFANGERWRALRRFSLATMRDFGMGKRSVEERIQEEA      150

+ oo--o++ +    oo-   ooo+ o    oo    ooo ++o-o+-  ooo+oo-
RCLVEELRKSKGALLDNTLLFHSITSNIICSIVFGKRFDYKDPVFLRLLD      200

ooo   o oo   o    oo-oo   oo++o    ++ oo+ o -o   oo    o-
LFFQSFSLISSFSSQVFELFPGFLKHFPGTHRQIYRNLQEINTFIGQSVE      250

+++    o-       +-oo-oooo+o-+-+ -   -o++  ooo oo ooo
KHRATLDPSNPRDFIDVYLLRMEKDKSDPSSEFHHQNLILTVLSLFFAGT      300

-     o+o ooooo+o +o -+o +-o- oo   ++   o--+ +o o -
ETTSTTLRYGFLLMLKYPHVTERVQKEIEQVIGSHRPPALDDRAKMPYTD      350

oo+-o  +o -oo  o  o +  o +-    o+ ooo  +  -oo oo    o+- +
AVIHEIQRLGDLIPFGVPHTVTKDTQFRGYVIPKNTEVFPVLSSALHDPR      400

oo-    o    +oo-    o++ -  oo  o  o ++o o - o  + -ooooo
YFETPNTFNPGHFLDANGALKRNEGFMPFSLGKRICLGEGIARTELFLFF      450

oo   o o    o   --o-o  +-   o  o    o o+oo +
TTILQNFSIASPVPPEDIDLTPRESGVGNVPPSYQIRFLAR      491
```

FIGURE 6. Primary structure of rabbit P-450 isozyme 2. Cationic, anionic, and hydrophobic residues are indicated in this case and succeeding figures by the symbols +, −, and 0, respectively. From Ref. 89 with permission.

are shown in Figs. 6 through 16. Bacterial P-450$_{cam}$, the first such structure to be completed, was done by Edman sequencing methodology by Yasunobu, Gunsalus, and colleagues[101,113,128] Later, Fujii-Kuriyama et al.[96] provided the first mammalian structure, that of rat P-450b, as deduced by analysis of cloned cDNA; the sequence of this cytochrome and that of rat P-450e have also been established through protein chemistry methods by Yuan et al.[84] The first complete mammalian amino acid sequence established through methods of protein chemistry was that of rabbit isozyme 2.[27,89,95,129] Since then, molecular biological techniques have yielded the sequences of rabbit isozyme 3b (clone 3),[121] rat isozymes c,[97,122,130] d,[123] and e,[131–134] mouse isozymes 1 and 3,[24,127,135] and bovine P-450$_{scc}$[124]; the lattermost cytochrome has also been studied extensively by Akhrem and colleagues.[136,137] Three-quarters of the amino acid se-

```
o-ooooo  o  o    oooo  oo+
MDLLIILGICLSCVVLLSLWK------------------------------    50


                                           oo-+  --o    +  o
------------------------------------------LIDRGEEFSGRGI   100

o  oo-+o  +  o  ooo     -+o+-  ++o  o  oo+  o  o  ++  o--+o  --
FPLFDRVTKGLGIVFSSGEKWKETRRFSLTVLRNLGMGKKTIEDRIQEEA         150

o  oo    o++             oooo  o      oo       ooo    +o-o----+o+  oo+
LCLIQALRKTNGSPCNPTFLLFCVPCNVICSVIFQNRFDYDDEKFKTLIK         200

oo+-  o-oo    oo  oo  oo  ooo+oo       +    oo+    -    o+ooo-+o
YLHENFELLGTPWIQLYNIFPILIHYLPGSHNQLFRNNDGQIKFVLEKVQ         250

-+  -  o-    +-oo-+ooo+o-+-++++   -o  o-  oo    oo-oo
EHQETLDSNNPRDFVDHFLIKMEKEKHKKQSEFTMDNLITTIWDVFSAGT         300

 -        o+o  ooooo++  -o    +o  --o-+oo  +++     o  -+  +o  o  -
DTTSNTLKFALLLLLKHPEITAKVQEEIEHVIGRHRSPCMQDRTRMPYTD         350

oo+-o  +oo-oo      o  +  o    -o-o    ooo    +      oo    o    ooo  -+
AVMHEIQRYVDLVPTSLPRAVTQDIEFNGYLIPKGTAIIPSLTSMLYNDK         400

-o    -+o-    +oo--    +o++  -ooo  o      ++    o  -  oo+o-ooooo
EFPNPEKFDPGHFLDESGKFKKSDYFMPFSAGKRACVGEGLVRMELFLLL         450

oo  +o  o+  oo-  +-o-     o-     o    o     o-o  oo
TTILQHFTLKPLVDPKDIDPTPVENGFGSVPPSYELCFVPA    491
```

FIGURE 7. Partial primary structure of rabbit P-450 isozyme 3b. From Refs. 105 and 121 with permission.

quence of rabbit isozyme 4,[106] and a partial cDNA sequence of bovine P-450$_{C-21}$[138] have been reported recently. In addition, the cDNA sequences of rabbit P-450's for which the proteins have yet to be purified have been determined by Kemper and co-workers[121] (clones 1 and 2) and by Zaphiropoulos, Folk, and Coon (unpublished data).[139] No microheterogeneity has been confirmed for any position of any of these primary structures, including rabbit isozyme 4, which at one time was thought perhaps to contain several distinct N-terminal chains[104] before the occurrence of n, n-1, and n-2 forms was established,[106] as already stated. The above findings argue strongly against the idea that the P-450's might by synthesized in a fashion similar to that of the immunoglobulins, with constant and variable regions.[27,140]

An alignment of rabbit isozymes 2 and 3b with rat isozymes b and e and bacterial P-450$_{cam}$ is shown in Fig. 17. As can readily be seen, all

```
      o     o o -oooo  oo oooo o+  + +o + o++o      o
    AMS PAA PL SV TELLLV SAV FCLV FW AV RAS RPKV PKG LKRL PG PSGV P--      50


    ------------------------------------------------------      100


          + -oo    oo              o  +++o  -  o+  o o
    -----GRPDL YSSSFIT--------••!  -WAARRRL AQDSL KSFSIAS      150


         oo--+o  - - oo +o -oo  o +o- o  ooo       oo
    NPASSSSCYL EEHV SQ EAENL ISRFQ EL MAAV GRFDPYSQ LVV SAANV IG      200


     o o        o-oo+    +oo-      o-oo oo+oo
    AMCFG----------LDVV RNSSKFV ETASSSSPV DFFPIV RYL P-----      250


      -o   +oo+oo    o+-+o--                  oo
    ---DFNQ RFL RFL--TV REHYED-------------------ANGG LI      300


     -+oo  oo -oo    o-  o   o o ooooo    ++ +
    PQ EKIV NLV NDIFG AG FDT ITTAL SW SL MYLV TN PRRQ R-----------      350


           +   o oo- ooo-oo    o             o  o+o
    ---------RPQ LPYL EAFIL EL FXXTSF------------TTL NG FH IP      400


    +-   ooo   o o +-   oo -  --o+ -+oo  -   o + o -+o oo
    KE CCIF IN QWQ IN HDPQ LWG DPEE FR PERFL TADG AA IN KPL AE KV TL FG      450


    o +++  o - o +o-ooooo ooo  o-o o               ++
    LG KRR CIG ETL ARW EV FL FL A ILL QXL EFSV PPG------------KH P      500


    + -+o   +   o +
    RCEHVQ ARPSF-K       513
```

FIGURE 8. Partial primary structure of rabbit P-450 isozyme 4. From Ref. 106 with permission.

five proteins show considerable relatedness with 37 of the 425 comparable positions conserved (8.7% identity). It should be mentioned in this connection that random alignments of five P-450 polypeptides would yield less than one identity per 500 compared positions.* The structural homology is not scattered but is localized in three regions, one around Cys residue 152 in the mammalian cytochromes, the second in the vicinity of position 390, and the third around Cys residue 436 in the mammalian cytochromes. The clustering of conserved positions suggests that these three regions are likely to function in some important structural or cat-

---

* In this review, *position* numbers are used for alignments of sequences, whereas *residue* numbers are used for individual sequences. These numbers need not be identical.

```
    o-   ooooo ooo oooooo+ + + +   o    + o oo   oo o-+
    MEPT ILLLLALLVGFLLLLVRGHPKSRGNFPPGPRPLPLLGNLLQLDRGG           50

    oo    oo  o+-+o  -oo  o+o    +  oooo     -  o+-  oo    --o   +
    LLNSFMQLREKYGDVFTVHLGPRPVVMLCGTDTIKEALVGQAEDFSGRGT           100

    o  oo-  oo+-o  ooo     -+o+  o++o  o    o+-o  o  ++  o--+o  --
    IAVIEPIFKEYGVIFANGERWKALRRFSLATMRDFGMGKRSVEERIQEEA           150

        oo--o++      o-   ooo     o     oo   ooo  -+o-o  -+ oo+oo-
    QCLVEELRKSQGAPLDPTFLFQCITANIICSIVFGERFDYTDRQFLRLLE           200

      ooo+  o  oo   o      oo-oo    oo+oo    ++ o + o  -oo-oo  +oo-
    LFYRTFSLLSSFSSQVFEFFSGFLKYFPGAHRQISKNLQEILDYIGHIVE           250

      +++    o-     +-oo- ooo+o-+-+   ++ -o++- ooo  oo  ooo
    KHRATLDPSAPRDFIDTYLLRMEKEKSNHHTEFHHENLMISLLSLFFAGT           300

     -       o+o  ooooo+o  +o  -+o  +-o-  oo    ++o   o--+ +o  o -
    ETSSTTLRYGFLLMLKYPHVAEKVQKEIDQVIGSHRLPTLDDRSKMPYTD           350

      oo+-o  +o  -oo o o  ++o  +-  oo+  ooo   +    -oo  oo     o+-
    AVIHEIQRFSDLVPIGVPHRVTKDTMFRGYLLPKNTEVYPILSSALHDPQ           400

     oo-+ -  o  -+oo-    o++ -  oo  o    ++o  o - o + -ooooo
    YFDHPDSFNPEHFLDANGALKKSEAFMPFSTGKRICLGEGIARNELFLFF           450

       oo   o o   +o   +-o-o  +-    o +o     ooo   +
    TTILQNFSVSSHLAPKDIDLTPKESGIGKIPPTYQICFSAR         491
```

FIGURE 9. Primary structure of rat P-450 isozyme b. From Refs. 84 and 96 with permission.

alytic roles in the native cytochromes. Of the residues represented among the conserved positions, Cys, Gly, Pro, Phe, Glu, Asp, and Arg are present in numbers significantly greater than in an average P-450 composition, and Asn, His, Val, and Trp are not represented. The absence of a conserved His casts doubt on this residue as providing either the proximal (fifth) or distal (sixth) ligand to the heme. Finally, when P-450$_{cam}$ is omitted from this alignment, 43% identity is observed for the remaining cytochromes; an alignment of only rabbit isozyme 2 and rat isozymes b and e shows 76% identity.

The optimal alignment of the partial structure of rabbit isozyme 4 with the complete sequences of rat isozymes c and d and with mouse isozymes 1 and 3 is shown in Fig. 18. All of these aryl hydrocarbon-inducible cytochromes are highly related, with 194 of the 355 comparable positions conserved (55% identity). Although structural homology is apparent throughout the aligned sequences, three regions of high conser-

```
 o  oo o   o      -ooo o   o o oooo+o + oo + o+      o o
MPSV YGFPAFTSATELLLAVTTFCLGFWVVRVTRTWVPKGLKSPPGPWGL       50

  oo +oo  o  +   +o  o +o    o -oo o+o     oooo  o  o+  o
PFIGHVLTLGKNPHLSLTKLSQQYGDVLQIRIGSTPVVVLSGLNTIKQAL      100

   o+  --o+ + -oo o oo      u u      uu  IIIo      o+ o
VKQGDDFKGRPDLYSFTLIANGQSMTFNPDSGPLWAARRRLAQNALKSFS      150

  o  -  o     oo--+o +- -ooo +o +oo -o +o- o+oooo o
IASDPTLASSCYLEEHVSKEAEYLISKFQKLMAEVGHFDPFKYLVVSVAN      200

  oo   o o ++o-+-- -oo oo o   -o -o    o -oo oo+oo
VICAICFGRRYDHDDQELLSIVNLSNEFGEVTGSGYPADFIPILRYLPNS      250

    o- o+-o ++oo oo++oo+-+o+ o-+ +o+-o - oo-+ -++o--
SLDAFKDLNKKFYSFMKKLIKEHYRTFEKGHIRDITDSLIEHCQDRRLDE      300

   o o --+oo ooo-oo  o- o  o o ooooo   +o ++o --
NANVQLSDDKVITIVFDLFGAGFDTITTAISWSLMYLVTNPRIQRKIQEE      350

  o- oo +-+  +o -+  o oo- ooo- o++  oo o o +  o+-  o
LDTVIGRDRQPRLSDRPQLPYLEAFILETFRHSSFVPFTIPHSTIRDTSL      400

    ooo + + ooo   o o +- -oo -  -o+ -+oo    o ++o -+
NGFYIPKGHCVFVNQWQVNHDQELWGDPNEFRPERFLTSSGTLNKHLSEK      450

  oooo o +++ o - o +o-ooooo ooo  o-o o   -+o-o   o o
VILFGLGKRKCIGETIGRLEVFLFLAILLQQMEFNVSPGEKVDMTPAYGL      500

  o++ + -+o o o+     +o
TLKHARCEHFQVQMRSSGPQHLQA      524
```

FIGURE 10. Primary structure of rat P-450 isozyme c. From Refs. 97 and 122 with permission.

vation are found, equivalent to those identified in Fig. 17. Of interest, the most divergent regions are found in the first 14 positions and in the vicinity of position 300; in the latter region, gaps also occur. The N-terminal divergence, as noted in Section 3.1, has proven to be a useful tool in the discrimination of related forms. Alignment of just the two rat proteins with the two mouse proteins shows 66% identity, and alignment of all of these with P-450$_{cam}$ indicates 9.5% identity, a value similar to that obtained from the results found in Fig. 17.

Additional perspective on the relatedness of the known primary structures of P-450's can be gained from systematic pairwise comparison of the various polypeptides at optimal alignment. A summary matrix of such comparisons is found in Table VI; relative to the random pairwise align-

```
o  o   oo o   -ooo     oo oooooo+   +   o + o+      o o oo
MAFSQYISLAPELLLATAIFCLVFWVLRGTRTQVPKGLKSPPGPWGLPFI         50

 +oo o +   +o o +o    o -oo o+o       oooo   o   o+   oo+
GHMLTLGKNPHLSLTKLSQQYGDVLQIRIGSTPVVVLSGLNTIKQALVKQ        100

 --o+ +  -oo o oo    + oo  -     oo  +++o   - o+ o o
GDDFKGRPDLYSFTLITNGKSMTFNPDSGPVWAARRRLAQDALKSFSIAS        150

 -    o    oo--+o +-    +oo +o +oo -o +o- o   oo- o   oo
DPTSVSSCYLEEHVSKEANHLISKFQKLMAEVGHFEPVNQVVESVANVIG        200

 o o + o ++  --oo oo+   +-oo- o      o-oo oo+oo      o+
AMCFGKNFPRKSEEMLNLVKSSKDFVENVTSGNAVDFFPVLRYLPNPALK        250

 +o+ o -  ooo o + o -+o -o +   o -o   oo++  -  o+-     oo
RFKNFNDNFVLSLQKTVQEHYQDFNKNSIQDITGALFKHSENYKDNGGLI        300

  -+oo  oo -oo     o- o    ooo ooooo  - +o ++o+--o- oo
PQEKIVNIVNDIFGAGFETVTTAIFWSILLLVTEPKVQRKIHEELDTVIG        350

 +-+   +o -+   o oo- ooo-oo+o   oo o o +   +-   o   o+o
RDRQPRLSDRPQLPYLEAFILEIYRYTSFVPFTIPHSTTRDTSLNGFHIP        400

 +-   ooo   o o +--+ o+- ooo+ -+oo   -    o-+ o -+oooo
KECCIFINQWQVNHDEKQWKDPFVFRPERFLTNDNTAIDKTLSEKVMLFG        450

 o +++ o -o   +o-ooooo ooo+ o-o o     o+o-o    o o o+ +
LGKRRCIGEIPAKWEVFLFLAILLHQLEFTVPPGVKVDLTPSYGLTMKPR        500

 -+o    o +o +
TCEHVQAWPRFSK         513
```

FIGURE 11. Primary structure of rat P-450 isozyme d. From Ref. 123 with permission.

ment value of 5.6 ± 0.7%, each sequence is significantly related to the rest, albeit to lesser and greater degrees. The most highly related pairs are rat isozymes b and e; rat isozyme d and mouse isozyme 3; and rat isozyme c and mouse isozyme 1 with 97, 93, and 92% identity, respectively. P-450$_{cam}$ and P-450$_{scc}$ are the most distantly related to the other proteins as well as to one another; the former protein appears to be more similar to phenobarbital-inducible rat isozymes b and e than to the others, while P-450$_{scc}$ is in general more related to the mammalian proteins than is P-450$_{cam}$. The values in Table VI also permit the subdivision of the microsomal cytochromes into classes; alignments indicate a close correspondence of rabbit isozyme 2 to rat isozymes b and e, of rabbit isozyme 4 to rat isozyme d and mouse isozyme 3, and of rat isozyme c to mouse isozyme 1. These findings are also consistent with the patterns of induc-

```
     o-    ooooo ooo oooooo+ + + +   o      + o oo   oo o-+
     MEPT ILLLLLALLV GFLLLLV RG HPKS RG NFPPG PR PL PLLG NLLQ LDRGG        50

     oo    oo o+-+o -oo o+o   + oooo    - o+-  oo    --o    +
     LLNSFMQLREKYG DV FTV HLG PR PVV ML CG T DT IKEALV GQ AE DFSG RGT     100

     o  oo-  oo+-o  ooo      o   o    o   o+-o  o  ++ o--+o  --
     IAV IEPIFKEYGV IFANG ERW KAL RRFSL AT MRD FG MG KRSV EER IQ E E A   100

        oo--o++      o-   ooo    o    oo   ooo -+o-o -+ oo+oo-
     QCLVEEL RKSQ GAPL DPT FL FQ CIT AN I ICS IV FG ERFDYT DRQ FL RLL E   200

     ooo+ o  oo    o    oo-oo    oo+oo    ++ o + o -oo-oo  +oo-
     LFYR TFSLL SSFS SQ VFEF FSG FL KY FPG AH RQ ISKNLQ EIL DYIG HIV E    250

     +++    o-     +-oo-  ooo+o-+-+   ++ -o++- ooo  oo  ooo
     KHRATL DPS APRD FI DT YLL RME KE KS NHH TEFHH ENL MISLL SL FFAG T    300

     -       o+o oooo o+o  +o o+o +-o-  oo    ++    o--+ +o o -
     ETG S T TL RYG FLL ML KY PHV TV KV Q KE IDQ VIG SH RPPSL DDR TKMP YT D   350

       oo+-o  +o  -o   o o  ++o +-  oo+ ooo   +   -oo oo    o+-
     AV IHE IQ RFA DL APIG LPH RV TKDT MFRG YLL PKN TEV YPIL SSAL HDPQ    400

     oo-+ -  o  -+oo-  -   o++  -  oo o    ++o o  - o +  -ooooo
     YFDH PDT FNPEH FL DA DG TL KKS EAFMPFS TG KR I CL G EG IARN EL FL FF  450

        oo   o o  +o   +-o-o  +-   o +o    ooo  +
     TT ILQ NFSV S SHL APKDI DL TPKESG IAKI PPT YQ ICFS AR        491
```

FIGURE 12. Primary structure of rat P-450 isozyme e. From Refs. 84 and 96 with permission.

tion of the cytochromes by xenobiotics. Generally, more structural homology is observed between the corresponding forms of different species than among the different forms found in a given species. Nonetheless, although these similarities are presently judged to represent homology (divergence from a common ancestral protein), the possibility of analogy (convergence) cannot be ruled out.[141] At present, however, no examples of primary structural convergence of proteins are known, but tertiary structural convergence of active site geometries has been described.[142]

3.4. Prominent Structural Features

3.4.1. Conserved Cysteine-Containing Regions in Relation to the Thiolate Ligand

A comparison of conserved Cys-containing regions in the known P-450 cytochrome structures is shown in Fig. 19. Such alignments could

```
oo o   oo      -ooo o oo o oooo+    + oo + o+        o o oo
MYGLPAFVSATELLLAVTVFCLGFWVVRATRTWVPKGLKTPPGPWGLPFI         50

+oo o +    +o o +o     o -oo o+o       oooo    o   o+   oo+
GHMLTVGKNPHLSLTRLSQQYGDVLQIRIGSTPVVVLSGLNTIKQALVRQ         100

--o+ + -oo o oo    + oo  -    oo  +++o    o+ o o
GDDFKGRPDLYSFTLITNGKSMTFNPDSGPVWAARRRLAQNALKSFSIAS         150

-          oo--+o +-    ooo +o +oo -o +o- o+oooo o   oo
DPTSASSCYLEEHVSKEANYLVSKLQKVMAEVGHFDPYKYLVVSVANVIC         200

o o   +o-+-- -oo oo o   -o -o     o  -oo oo+oo     o-
AICFGQRYDHDDQELLSIVNLSNEFGEVTGSGYPADFIPVLRYLPNSSLD         250

o+-o -+oo oo++oo+-+o+ o-+ +o+-o  - oo-+  -++o--
AFKDLNDKFYSFMKKLIKEHYRTFEKGHIRDITDSLIEHCQDRKLDENAN         300

o o --+oo ooo-oo   o-  o    o o ooooo    +o ++o --o-
VQLSDDKVITIVLDLFGAGFDTVTTAISWSLMYLVTNPRVQRKIQEELDT         350

oo +-+   +o -+   o oo-  ooo-  o++   oo o o +    +-   o  o
VIGRDRQPRLSDRPQLPYLEAFILETFRHSSFVPFTIPHSTTRDTSLNGF         400

oo +    ooo   o o +-+-oo -   -o+ -+oo      o-++o -+o o
YIPKGCCVFVNQWQVNHDRELWGDPNEFRPERFLTPSGTLDKRLSEKVTL         450

o o +++ o - o + -ooooo ooo   o-o+o    -+o-o    o o o+
FGLGKRKCIGETIGRSEVFLFLAILLQQIEFKVSPGEKVDMTPTYGLTLK         500

+ + -+o o o+       +o
HARCEHFQVQMRSSGPQHLQA      521
```

FIGURE 13. Primary structure of mouse P-450 isozyme 1. From Ref. 127 with permission.

prove to be of use in predicting the location of the Cys residue providing the proximal thiolate ligand to the heme, now accepted to be a general structural feature of all P-450's.[143-145] According to the sequence numbering convention of rabbit isozyme 2, three conserved regions containing Cys_{152}, Cys_{180}, and Cys_{436} are observed. The first and third of these Cys residues lie in the most highly conserved regions of the P-450 cytochromes[27,89,106,146] (see also Figs. 17 and 18 and discussion above), while a lesser degree of structural homology is seen in the region of Cys_{180}. Furthermore, the Cys_{180} region is not conserved in P-450$_{cam}$ or P-450$_{scc}$, nor the Cys_{152} region in P-450$_{scc}$. Thus, only Cys_{436} and the region bounding it are conserved in all currently known primary structures, a finding which suggests that Cys_{436} and its equivalent in the other structures may

```
o o   oo o  -ooo    oo oooooo+   +   o + o+      o o oo
MAFSQYISLAPELLLATAIFCLVFWMVRASRTQVPKGLKNPPGPWGLPFI      50

 +oo  o +    +o  o +o    o -oo o+o        oooo    o   o+    oo+
GHMLTVGKNPHLSLTRLSQQYGDVLQIRIGSTPVVVLSGLNTIKQALVRQ     100

 --o+  +  =oo  o  oo    I  o o   =     oo  +++o   -  o+  o  o
GDDFKGRPDLYSFTLITNGKSMTFNPDSGPVWAARRRLAQDALKSFSIAS     150

 -        oo--+o  +-    +oo +o +  o -o +o-  o   oo-  o   oo
DPTSASSCYLEEHVSKEANHLVSKLQKAMAEVGHFEPVSQVVESVANVIG     200

 o  o + o ++  --oo oo    +-oo-  o      o-oo oo+oo       o+
AMCFGKNFPRKSEEMLNIVNNSKDFVENVTSGNAVDFFPVLRYLPNPALK     250

 +o+ o  - ooooo + o -+o -o +    o -o     oo++  -  o+-    oo
RFKTFNDNFVLFLQKTVQEHYQDFNKNSIQDITSALFKHSENYKDNGGLI     300

 --+oo oo -oo    o-  o    o o ooooo o  o ++o+--o- oo
PEEKIVNIVNDIFGAGFDTVTTAITWSILLLVTWPNVQRKIHEELDTVVG     350

 +-+   +o -+   o oo- ooo-oo+o   oo o o +    +-  o  o+o
RDRQPRLSDRPQLPYLEAFILEIYRYTSFVPFTIPHSTTRDTSLNGFHIP     400

 +-+ ooo   o o +--+ o+- ooo+ -+oo      o-+   -+oooo
KERCIYINQWQVNHDEKQWKDPFVFRPERFLTNNNSAIDKTQSEKVMLFG     450

 o +++  o -o   +o-ooooo ooo +o-o o   o+o-o    o o o+
LGKRRCIGEIPAKWEVFLFLAILLQHLEFSVPPGVKVDLTPNYGLTMKPG     500

 -+o   o +o +
TCEHVQAWPRFSK      513
```

FIGURE 14. Primary structure of mouse P-450 isozyme 3. From Refs. 24 and 127 with permission.

provide the proximal thiolate ligand to the heme. No homology arguments have so far been proposed for the identity of the distal ligand in the low-spin cytochromes.

More direct evidence regarding the heme ligands is also available. In P-450$_{cam}$[63] and rabbit isozymes 2[89] and 4,[106] all of the Cys residues are found in the reduced state rather than as the disulfides; none of these, therefore, can be eliminated as candidates to provide the thiolate ligand. Chemical modification studies with P-450$_{cam}$[147] have indicated that the Cys$_{134}$ region (equivalent to that of Cys$_{152}$ in rabbit isozyme 2) provides the heme ligand, but thiol modification experiments with isozyme 2[148] show that Cys$_{152}$ is not the proximal ligand and suggest instead that Cys$_{436}$ functions in this role. However, X-ray crystallographic work with P-

```
   -  o         o   o   +o -+ooo-o-oo      o    o - o oo -    o
TTETIQSNANLAPLPPHVPEHLVFDFDMYNPSNLSAGVQEAWAVLQESNV          50

 -oo+       +oo  +    oo+- o--o++o    -   oo +-  - o-oo
PDLVRCNGGHWIATRGQLIREAYEDYRHFSSECPFIPREAGEAYDFIPTS         100

o-   -  + o+ o      oo o oo-+o-  +o -o       oo-  o+       o -
MDPPEQRQFRALANQVVGMPVVDKLENRIQELACSLIESLRPQGQCNFTE         150

 -o -   o  o+ooooo   o  ---o +o+oo  -  o + -   o o  - +-  oo-
DYAEPFPIRIFMLLAGLPEEDIPHLKYLTDQMTRPDGSMTFAEAKEALYD         200

ooo oo-  ++ +    - o oo      o + o   -- ++o   oooo      o-
YLIPIIEQRRQKPGTDAISIVANGQVNGRPITSDEAKRMCGLLLVGGLDT         250

oo oo o o-oo  +    -++ -oo  + -+o       --oo++o oo  -  +oo
VVNFLSFSMEFLAKSPEHRQELIQRPERIPAACEELLRRFSLVADGRILT         300

 -o-o+ o o++ -   ooo   oo  o--+-     o+o-o + +o +    o +
SDYEFHGVQLKKGDQILLPQMLSGLDERENACPMHVDFSRQKVSHTTFGH         350

  +o o    o ++-ooo o+-oo +o -o o       o ++   oo   o   o
GSHLCLGQSLARREIIVTLKEWLTRIPDFSIAPGAQIQHKSGIVSGVQAL         400

 ooo      + o
PLVWNPATTKAV        412
```

FIGURE 15. Primary structure of P-450$_{cam}$. From Refs. 101, 113, and 128 with permission.

450$_{cam}$ (see Chapter 13) supports the conclusion that Cys$_{355}$ (equivalent to Cys$_{436}$ in rabbit isozyme 2) provides the fifth ligand to the heme iron atom. The sixth ligand in both P-450$_{cam}$[149] and rabbit isozyme 2[150] is believed to come from an oxygen atom in water or an amino acid residue; Jänig et al.[151] have provided evidence from chemical modification for a protein Tyr residue having such a function in the case of the latter protein.

3.4.2. Amino Acid Compositions and Molecular Weights

A summary of properties of the nine P-450's for which complete primary structures are currently available is found in Table VII. The eight mammalian structures have between 481 and 524 residues with native molecular weights between 56,373 and 60,053 and calculated heme contents in the range from 16.6 to 17.7 nmole/mg protein. P-450$_{cam}$, which has about 80 amino acid residues less than the mammalian cytochromes,

```
o  +  + o -o      -   oo oo+oo+-+    +o+o++o- o +o  oo
ISTKTPRPYSEIPSPGDNGWLNLYHFWREKGSQRIHFRHIENFQKYGPIY          50

+-+o   o- oooo+   --o +oo+o-   o -+o-o   oo o++oo  + o o
REKLGNLESVYIIHPEDVAHLFKFEGSYPERYDIPPWLAYHRYYQKPIGV         100

óó++     o++- ḯooo    ◼◼  - o* oo nn  o  -oo oo+++o+
LFKKSGTWKKDRVVLNTEVMAPEAIKNFIPLLNPVSQDFVSLLHKRIKQQ         150

+oo  -o+--oo+o o-  o    ooo  -+o oo--  o   -   +oo-  oo+
GSGKFVGDIKEDLFHFAFESITNVMFGERLGMLEETVNPEAQKFIDAVYK         200

oo+   o  oo o   -oo+oo+  +  o+-+o   o- oo + -+o -ooo  -o+
MFHTSVPLLNVPPELYRLFRTKTWRDHVAAWDTIFNKAEKYTEIFYQDLR         250

++  -o+ o    ooo  oo+ -+ooo--o+   o -oo    o    o o o+oo
RKTEFRNYPGILYCLLKSEKMLLEDVKANITEMLAGGVNTTSMTLQWHLY         300

-o + o o   -oo+--oo    ++  -  -o +oo oo oo+   o+- o+o+ o
EMARSLNVQEMLREEVLNARRQAEGDISKMLQMVPLLKASIKETLRLHPI         350

 o o +o - -ooo -ooo  + oo o oo o +-   oo    -+o-   +o
SVTLQRYPESDLVLQDYLIPAKTLVQVAIYAMGRDPAFFSSPDKFDPTRW         400

o +-+-oo+o+  o o o o+   o ++o -o-o oooo+oo- o+o-o +o
LSKDKDLIHFRNLGFGWGVRQCVGRRIAELEMTLFLIHILENFKVEMQHI         450

 -o- oo ooo  -+ ooooo+ o -
GDVDTIFNLILTPDKPIFLVFRPFNQDPPQA        481
```

FIGURE 16. Primary structure of bovine adrenal P-450$_{scc}$. From Ref. 124 with permission.

has a calculated molecular weight of 46,857. Mean amino acid residue weights for all structures are about 113, except for P-450$_{scc}$, which shows an unusually high mean residue weight of 117. The former value is fairly typical for known protein sequences (e.g., 112 for cytochrome c[152] and 114 for detergent-solubilized cytochrome b_5[118]), although certain membranous polypeptides have somewhat lower values (e.g., 108 for bacteriorhodopsin[153] and 110 for thylakoid membrane protein[154]). All of the P-450's are relatively hydrophobic, with values between 32.3 and 38.5%; typical soluble proteins are usually < 30% hydrophobic, while extensively membrane-enmeshed polypeptides are > 40% hydrophobic. Finally, all of the P-450's contain similar numbers of acidic amino acid residues and also similar numbers of basic residues.

The molecular weights calculated from the complete sequences of the mammalian proteins are at variance with the corresponding values estimated by calibrated SDS–PAGE, as stated in Section 2.1 and shown

TABLE VI
Overall Homology among Primary Structures of P-450 Cytochromes[a]

Species	Isozyme	2	3b	4	b	e	c	d	1	3	cam	scc
Rabbit	2	—	47%	24%	77%	77%	25%	25%	26%	25%	13%	15%
	3b	200/427	—	18%	47%	47%	21%	21%	22%	20%	12%	15%
	4	86/355	62/353	—	24%	24%	64%	69%	65%	70%	13%	17%
Rat	b	379/491	200/425	86/355	—	97%	25%	25%	26%	25%	15%	14%
	e	376/491	201/425	86/355	478/491	—	25%	25%	26%	25%	15%	15%
	c	132/524	96/459	229/355	132/524	131/524	—	68%	92%	67%	12%	16%
	d	129/522	94/457	247/355	128/522	129/522	353/519	—	70%	93%	10%	15%
Mouse	1	137/521	99/456	229/355	134/521	133/521	483/524	359/516	—	71%	11%	16%
	3	131/521	93/457	248/355	130/521	128/521	347/519	478/513	365/516	—	11%	15%
P. putida	cam	63/476	51/428	53/409	73/476	71/476	61/505	51/502	57/504	54/502	—	12%
Bovine	scc	72/483	61/417	59/358	69/483	70/483	79/494	75/488	78/494	74/488	56/479	—

[a] Values at the upper right represent the percent of identical residues found in an optimal alignment of the respective cytochromes. Values at the lower left show the number of identical residues per total number of compared positions (including gaps). The sequences of rabbit isozymes 3b and 4 used are about 87 and 70% complete, respectively. Random pairwise alignments yielded an average value of $5.6 \pm 0.7\%$ identity.

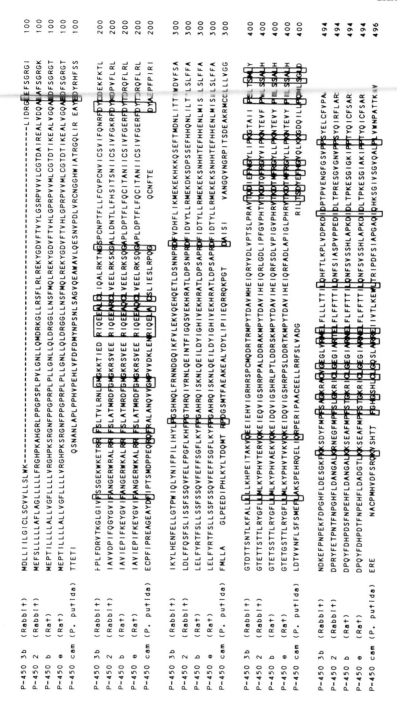

FIGURE 17. Optimal alignments of mammalian and bacterial cytochromes. See Figs. 6, 7, 9, 12, and 15 for references. Hyphens indicate unknown residues, and blank spaces indicate gaps inserted to obtain optimal alignment.

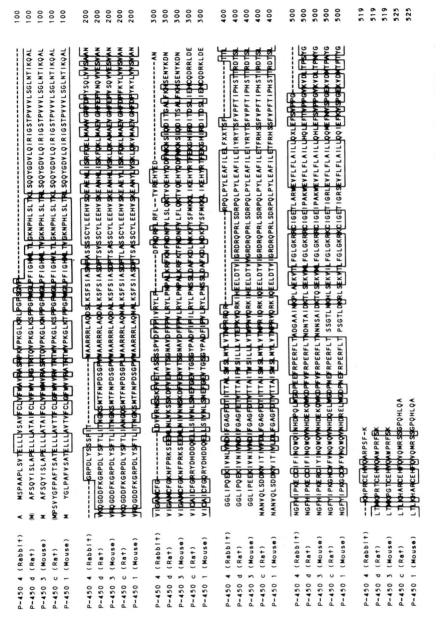

FIGURE 18. Optimal alignments of some closely related mammalian cytochromes. See Figs. 8, 10, 11, 13, and 14 for references. Hyphens indicate unknown residues, and blank spaces indicate gaps inserted to obtain optimal alignment.

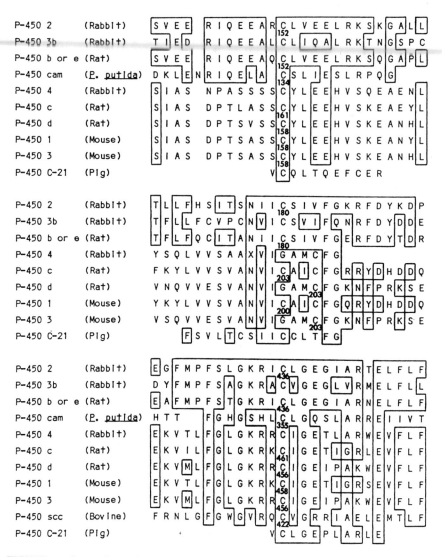

FIGURE 19. Comparison of the sequence of Cys peptides in various P-450 cytochromes. Homology with respect to rabbit isozyme 2 is shown by the enclosed areas. For references, see legends to Figs. 6–16 and Yuan et al.[54]

TABLE VII
Summary of Data Derived from Complete Sequences

Species	Isozyme	No. of residues	Molecular weight		Theoretical specific content	Mean residue weight	% hydrophobic[a]	% acidic	% basic
			Apoprotein	Holoenzyme					
Rabbit	2	491	55,721	56,373	17.7	113.5	37.3	10.6	14.7
Rat	b	491	55,941	56,593	17.6	113.9	36.9	11.4	14.7
	e	491	55,924	56,576	17.6	113.9	36.7	11.4	14.7
	c	524	59,401	60,053	16.6	113.4	36.1	10.3	13.7
	d	513	58,207	58,859	16.9	113.5	36.1	9.7	13.8
Mouse	1	521	58,923	59,575	16.7	113.1	35.5	10.6	13.8
	3	513	58,192	58,844	16.9	113.4	35.7	9.6	13.5
Bovine	scc	481	56,406	57,058	17.5	117.3	38.5	12.1	16.0
P. putida	cam	412	46,205	46,857	21.3	112.1	32.3	12.9	12.1

[a] This unweighted value is calculated as the sum of hydrophobic residues (F, I, L, M, V, W, Y) relative to total residues.

TABLE VIII
Comparison of Molecular Weights Calculated from Sequence
and Estimated by Electrophoresis

Species	Isozyme	Molecular weight		% deviance
		SDS–PAGE	Sequence	
Rabbit	2	49,500	55,721	−11.2%
Rat	b	51,200	55,941	−8.5%
	e	51,900	55,924	−7.2%
	c	54,300	59,401	−8.6%
	d	53,000	58,207	−8.9%
Mouse	1	55,000	58,923	−6.7%
	3	55,000	58,192	−5.5%
Bovine	scc	51,700	56,406	−8.3%
P. putida	cam	46,000	46,205	−0.4%

in Table VIII. For these proteins, the estimated molecular weights are low by approximately 10%. However, this is not unexpected for hydrophobic, aggregating proteins that generally bind more SDS per unit weight of polypeptide and thus migrate more rapidly during electrophoresis. P-450$_{cam}$, which is fairly hydrophobic but exists in aqueous solution as a monomer, shows little deviance between the actual and estimated molecular weights.

Amino acid compositions derived from the complete sequences of the cytochromes are shown in Table IX. The true compositions are all quite similar with low variation seen for all residues except Phe, Ile, Leu, Val, Asn, and Lys, which have standard deviations in the range of 6.1 to 7.5 residues. The number of Pro residues is remarkably similar for all the sequences; low deviance is also seen for the sum of Pro + Gly, or of all anionic, or of all charged residues. This observation suggests a possible role of helix-breaking and charged residues in the maintenance of the tertiary structure of each cytochrome. The total number of hydrophobic residues is well conserved for the mammalian proteins, even though such residues show large individual standard deviations. This is in accord with the finding by Tarr et al.[89] that 96% of the hydrophobic positions are conserved between rabbit isozyme 2 and rat isozyme b, though the proteins are only 77% identical overall. Again, this suggests the importance of hydrophobic side chains in maintenance of the tertiary structure.

3.4.3. Nonuniform Distribution of Certain Residues

As noted earlier (see Section 3.1), a highly hydrophobic region (attributed to retention of a signal peptide[104]) exists in all of the microsomal

TABLE IX
Amino Acid Compositions Determined from Complete Sequences

Amino acid residue	Rabbit 2	Rat b	Rat c	Rat d	Rat e	Mouse 1	Mouse 3	Bovine scc	P. putida cam	$\bar{X} \pm$ S.D.
A	24	23	25	24	24	24	27	22	31	24.9 ± 2.7
C	4	6	8	7	6	9	6	2	8	6.2 ± 2.2
D−	24	24	28	23	25	30	23	24	21	24.7 ± 2.7
E−	28	32	26	27	31	25	26	34	32	29.0 ± 3.3
FO	41	39	35	36	39	31	34	30	18	33.7 ± 6.9
G	35	33	34	29	33	34	28	23	25	30.4 ± 4.4
H+	14	17	16	13	17	14	13	15	12	14.6 ± 1.8
IO	27	30	31	30	30	26	29	32	26	27.8 ± 6.3
K+	20	24	28	35	24	28	31	33	12	26.1 ± 7.1
LO	64	63	59	53	63	57	49	51	42	55.7 ± 7.5
MO	8	11	8	8	11	8	9	14	10	9.7 ± 2.1
N	15	13	19	26	12	18	30	20	14	18.6 ± 6.1
P	32	30	29	33	31	29	33	30	30	30.8 ± 1.6
Q	16	17	24	23	17	23	23	19	24	20.7 ± 3.4
R+	38	31	28	23	31	30	25	29	26	29.0 ± 4.4
S	32	33	39	34	30	38	35	22	25	32.0 ± 5.6
T	26	27	31	31	30	34	30	23	19	27.9 ± 4.7
VO	30	23	34	39	22	40	41	29	24	31.3 ± 7.5
WO	1	1	7	8	1	7	9	9	4	5.2 ± 3.5
YO	12	14	15	11	14	16	12	20	9	13.7 ± 3.2
Total	491	491	524	513	491	521	513	481	412	493 ± 34.0
Sum of −	52	56	54	50	56	55	49	58	53	53.7 ± 3.0
Sum of +	72	72	72	71	72	72	69	77	50	69.7 ± 7.7
Sum of − & +	124	128	126	121	128	127	118	135	103	123.3 ± 9.0
Sum of O	183	181	189	185	180	185	183	185	133	178.2 ± 17.2
Sum of S & T	58	60	70	65	60	72	65	45	44	59.9 ± 9.9
Sum of G & P	67	63	63	62	64	63	61	53	55	61.2 ± 4.4

P-450's near the N-terminus; this is readily visualized by the annotation of Figs. 6 through 14. The signal sequences begin essentially at the N-terminus of rabbit isozymes 2 and 3b and rat isozymes b and e but are displaced some 15 residues in the case of rabbit isozyme 4, rat isozymes c and d, and mouse isozymes 1 and 3. The functional significance of this difference is not known, but the observation that at least rabbit isozyme 4 and rat isozymes c and d have never been isolated with an initiator Met, whereas the other rabbit and rat isozymes apparently always retain this residue, may be related to how these different signal peptides are bound to the microsomal membrane. The hydrophobic signal sequence is followed by a cationic halt-transfer signal which is presumably present to

TABLE X
Alignment of Proline Clusters

Species	Isozyme	Sequence
Rabbit	2	-R_{29}L P P G P S P L P V L-
	4	-K_{39}R L P G P S G V P-
Rat	b or e	-N_{29}F P P G P R P L P L L-
	c	-K_{42}S P P G P W G L P F I-
	d	-K_{39}S P P G P W G L P F I-
Mouse	1	-K_{39}T P P G P W G L P F I-
	3	-K_{39}N P P G P W G L P F I-
Bovine	scc	-K_4 T P R P Y S E I P S P-
P. putida	cam	-P_{13}L P P H V P E H L V F-
Collagen		-G P P^{a}G P X-
Bradykinin		R_1P P G F S P F R

^{a} This residue is Hyp or Ala in some cases.

limit the insertion of the N-terminal region into the phospholipid bilayer.[114] N-terminal signal sequences followed by halt-transfer signals have also been identified for other rabbit microsomal proteins, epoxide hydrolase,[116] and NADPH-cytochrome P-450 reductase.[119] Both P-450$_{scc}$ and P-450$_{cam}$ as isolated lack signal peptides, but the former cytochrome is known to be synthesized with a transient 39-residue signal sequence[124] that is removed during insertion into the mitochondrial membrane.[155] However, this transient signal sequence is quite different from those found in the microsomal cytochromes in that it is much less hydrophobic and contains many charged and generally cationic residues; this is also in contrast to the general structure of nascent signal sequences of the secretory proteins.[114] Thus, the mechanisms of membrane insertion for mitochondrial and microsomal P-450's may be fundamentally different. Beyond the N-terminal region, five to seven additional apolar regions of 15 to 20 residues occur in each protein.

Beyond the N-terminal signal sequence and halt-transfer signal in the microsomal cytochromes, a "proline cluster" occurs. This short sequence, which was first noted by Black et al.[27] for rabbit isozyme 2, is also apparently present near the N-terminus of P-450$_{cam}$ and P-450$_{scc}$. An alignment of these regions is found in Table X. While none of the Pro residues is invariant, four to five such residues are found in each region. Perhaps a more significant observation is that these regions contain up to seven helix-breaking residues each; a peptide with such a sequence may tend to form a compound β-turn, possibly resulting in a 3_{10}-helix.[89]

It is of interest to note that such a proline cluster also occurs in the peptide hormone bradykinin[156] and appears repeatedly in a prototype sequence of collagen.[157]

The binding site for N-linked oligosaccharides has been defined as Asn-X-(Ser or Thr).[158] Such sequences are found in all presently known P-450 primary structures, with Asn-X-Ser found most commonly. However, these are found randomly distributed within the sequences rather than being localized in a common region. Protein sequence analyses of rabbit[89,106] and rat[84] P-450's revealed no carbohydrate present at these sites, and only small amounts of sugar were found to be present by direct analysis of the microsomal cytochromes (see Section 2.5). Thus, while the binding sites exist, these do not appear to be utilized for complex carbohydrate attachment in the nonsteroidogenic P-450's; the sugars detected at low levels are likely to be of the simple O-linked type.

In each of the completely sequenced proteins, regions of about 50 residues occur that are essentially devoid of any of the following: Ala, Asp, Phe, His, Val, or Tyr. In these calculations, the rarely occurring residues Cys, Met, and Trp were not considered. Charged residues are distributed from N- to C-termini in a Gaussian fashion, while helix-breaking residues Pro plus Gly show an inverted Gaussian distribution in all of the sequences. Finally, 50% of Tyr residues in P-450$_{scc}$ occur within the first 100 residues.

3.4.4. Site of Rapid Proteolytic Cleavage

Treatment of rabbit isozyme 2 with endoproteinase Lys-C results in rapid scission of the Lys$_{274}$–Asp$_{275}$ peptide bond with concomitant loss of the native P-450 spectrum (Black and Coon, unpublished results).[159] Similar cleavage behavior was shown with rabbit isozyme 4, which is hydrolyzed at the amino side of Ala$_{295}$,[106] and with trypsin-treated bovine P-450$_{scc}$.[136,160] The alignments in Fig. 17 near position 275 and in Fig. 18 near position 300 show these sites relative to those of other cytochromes for which limited proteolytic studies have yet to be performed. With the exception of P-450$_{cam}$, Lys or Arg is found in each sequence at the corresponding respective position. It is possible, therefore, that all these P-450's will show similar rapid cleavage with trypsinlike proteases. Furthermore, the homology seen in this region is low, and gaps are needed for optimal alignment. As stated in the previous section, the central region of the cytochrome is also generally the most highly charged. Taken together, these data suggest that the site of rapid proteolysis exists as a surface loop in the tertiary structure of the cytochromes. Because of this accessibility and the fact that scission of a single peptide bond results in loss of native structure, the possibility may be considered that this site is utilized for protein degradation *in vivo*.

3.5. Variant Forms of P-450

Although evidence from the available primary structures indicates that the P-450's are unique polypeptides and contain no variable positions, structural variants of certain rabbit and rat isozymes isolated from different strains of the respective species have been identified. As noted in Section 2.3.2. HPLC analysis of tryptic digests revealed minor but detectable differences in the peptide maps of rat isozyme b or c, each isolated from both Long–Evans and Holtzman rats; these represent variant alleles at the same locus and are properly classified as pairs of allozymes.[86] In contrast, P-450b and e, which differ by only 13 residues,[84] are synthesized by separate genes and are thus classified as isozymes.[161] Dieter and Johnson[87] found functional and structural polymorphism in isozyme 3b related to strain differences in the New Zealand White rabbits used. The "6β$^-$" form of the cytochrome, prepared from an incipient inbred strain, has diminished progesterone 6β-hydroxylase activity and apparently lacks one tryptic peptide. Molecular biological investigations have revealed variant forms of rat isozyme e,[131–134] but the exact relationship among these remains uncertain. Other possibly variant P-450's purified from cholestyramine-treated rabbits and having the same electrophoretic behavior as isozyme 4 have been described by Boström and Wikvall[32] and by Chiang et al.[162]

4. Predictions Based on Sequence Data

4.1. Secondary Structure

Secondary structures have been predicted from the primary structures for rat isozyme b[146] and bacterial P-450$_{cam}$.[101,146] However, caution must be assumed with regard to the utility of these proposals, since the predictive methods used are usually correct to no better than 60%,[163] and the true structures will not be known until crystallographic data are available. Since the tertiary structure of P-450$_{cam}$ is now known (see Chapter 13), an evaluation of secondary structure predictions for this cytochrome is possible. Literature predictions were found to be 35[101] and 41%[146] correct, with α-helix estimates more accurate than those of β-sheet, β-turn, and random coil. The calculations from our laboratory yielded a value 49% correct, which is somewhat better but still far from satisfactory; α-helices were predicted with an accuracy of 71%. Thus, for the present, secondary structure predictions should be used with caution.

COMPARATIVE STRUCTURES OF P-450

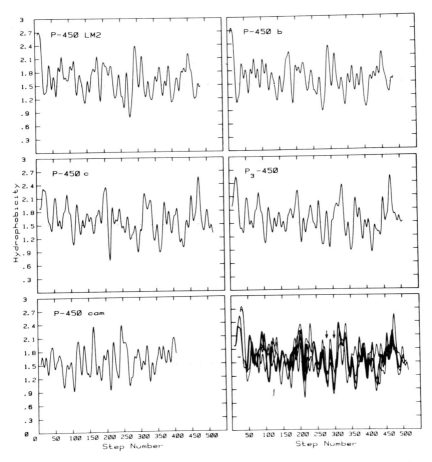

FIGURE 20. Zero-order hydrophobicity profiles for various P-450 cytochromes. Gaussian-smoothed results are shown for five P-450's, along with a composite at optimal alignment (lower right). Selected β-turn potentials are indicated by arrows.

4.2. Hydrophobicity Profiles

Zero-order hydrophobicity profiles have been calculated for a few P-450's[89,121,127,146] for which the primary structures are known. Because charged and hydrophobic residues appear to be highly conserved for P-450's, such plots of sequence polarity provide a useful way to visualize the distribution of hydrophobic free energy along the primary structure. This laboratory has utilized and scaled the composite parameters of Argos et al.[164] with Gaussian smoothing for such calculations; the profiles of five cytochromes are shown in Fig. 20, along with a composite overlay.

Although a few deviations are seen, all of the profiles are remarkably similar. Beyond the indication of sequence relatedness, this finding shows that the free energy potential that drives chain folding is similar for all of these polypeptides. This would imply[165] that the three-dimensional structures of the P-450's will all be similar. However, a clear difference between P-450$_{cam}$ and the other proteins is the lack of an N-terminal signal sequence in the former; how this difference may affect tertiary structure is not known.

The region seen near position 300 in the composite panel of Fig. 20 is of particular interest. All of the primary structures yield quite similar predictions in this region for both hydrophobicity and β-turns. To the N-terminal side of position 300 occurs the generally most polar section of the sequences and the location of the site of rapid proteolysis (see Section 3.4.4). To the C-terminal side occurs a strongly hydrophobic section. The crystal structure of P-450$_{cam}$ (see Chapter 13) shows that both β-turns indicated are predicted correctly and that the peaks near positions 275 and 300 correspond to the H and I helices of the bacterial protein. Furthermore, the latter helix runs along the distal surface of the heme, and its associated hydrophobicity likely provides the energetic driving force for substrate binding. The agreement of these predictions and the accuracy found in the case of P-450$_{cam}$ would argue that at least in this central region the tertiary structure of the bacterial protein may serve as a model for that of the mammalian cytochromes. Higher-order hydrophobicity calculations have been reported for only rabbit isozyme 2,[89] which was predicted to have an amphipathic helix near residue 320.[166] Such hydrophobic moment calculations[167] may prove useful in the refinement of structure prediction methods for hydrophobic proteins.

4.3. Membrane Topology

Membrane binding topology for rabbit isozyme 2 has been proposed based on amino acid sequence[89,95] and fluorescence energy transfer measurements.[168] All such models show the polypeptide chain to be extensively enmeshed in the membrane with polar segments protruding on one or both sides of the bilayer. In the case of the energy transfer study, the heme was suggested to be associated with a polypeptide segment within the membrane, and the N-terminus was determined to be located on the cytoplasmic side of the microsomal bilayer.[169] As support for the model, Tarr et al.[89] noted that rapid proteolysis was not observed for any polypeptide segment predicted to be membranous. However, further investigation is clearly required to shed light on the topological features of the P-450's; a simple topology such as was proposed for NADPH-cytochrome P-450 reductase[119] still remains a possibility.

5. Summary

The following conclusions concerning structural features of the P-450's can be drawn from the presently known amino acid sequences, either determined directly on the proteins or deduced from DNA.

The N-terminus of these cytochromes is free, and in the case of the microsomal P-450's the highly hydrophobic N-terminal region is apparently derived from an insertion signal peptide that has not undergone biological processing. In some instances, however, one or two of the terminal residues have been removed. The hydrophobic signal sequence is followed by a polar region that may possibly provide a halt-transfer signal and in turn by a proline cluster.

All complete P-450 sequences exhibit high hydrophobicity, which is distributed periodically throughout the primary structure. The alternating hydrophobic and hydrophilic stretches are capable of providing the respective membranous and nonmembranous segments of the mammalian cytochromes. Of the two cysteine-containing peptides located in two of the three regions that are highly conserved in the various forms of the enzyme, only that nearest the C-terminus contains an invariant cysteine residue. This residue may, therefore, provide the sulfur ligand to the heme iron atom. Comparisons of structural homology among the P-450's generally show less identity in the N-terminal region than in the remainder of the polypeptide chain and less in the cytochromes from a given source (e.g., rabbit, rat, or mouse hepatic microsomes) than in the corresponding cytochromes from different tissues (e.g., isozyme 2 from rabbit liver and lung) or from different species (e.g., rabbit isozyme 4, rat isozyme d, and mouse isozyme 3).

Since only a limited number of the cytochromes have been completely sequenced, some of the general conclusions reached may be useful primarily as hypotheses for further testing. However, the evidence presently available clearly indicates that all of the P-450's are structurally related. Furthermore, it should be emphasized that no evidence has been obtained for constant and variable regions comparable to those found in the immunoglobulins. Instead, the P-450 family appears to be comprised of a limited number of distinct but related gene products which in many instances are individually capable of binding and oxygenating a wide variety of physiological and xenobiotic substrates.

ACKNOWLEDGMENT. Part of the research described in this review was supported by Grant AM-10339 from the National Institutes of Health.

References

1. Haugen, D. A., van der Hoeven, T. A., and Coon, M. J., 1975, Purified liver microsomal cytochrome P-450: Separation and characterization of multiple forms, *J. Biol. Chem.* **250**:3567–3570.
2. Ryan, D. E., Thomas, P. E., Korzeniowski, D., and Levin, W., 1979, Separation and characterization of highly purified forms of liver microsomal cytochrome P-450 from rats treated with polychlorinated biphenyls, phenobarbital, and 3-methylcholanthrene, *J. Biol. Chem.* **254**:1365–1374.
3. Ohyama, T., Nebert, D. W., and Negishi, M., 1984, Isosafrole-induced cytochrome P_2-450 in DBA/2N mouse liver: Characterization and genetic control of induction, *J. Biol. Chem.* **259**:2675–2682.
4. Imai, Y., Hashimoto-Yutsudo, C., Satake, H., Girardin, A., and Sato, R., 1980, Multiple forms of cytochrome P-450 purified from liver microsomes of phenobarbital- and 3-methylcholanthrene-pretreated rabbits, *J. Biochem.* **88**:489–503.
5. Aoyama, T., Imai, Y., and Sato, R., 1982, Multiple forms of cytochrome P-450 from liver microsomes of drug-untreated rabbits: Purification and characterization, in *Microsomes, Drug Oxidations, and Drug Toxicity* (R. Sato and R. Kato, eds.), Wiley–Interscience, New York, pp. 83–84.
6. Miki, N., Sugiyama, T., Yamano, T., and Miyake, Y., 1981, Characterization of a highly purified form of cytochrome $P-450_{B1}$, *Biochem. Int.* **3**:217–223.
7. Slaughter, S. R., Wolf, C. R., Marciniszyn, J. P., and Philpot, R. M., 1981, The rabbit pulmonary monooxygenase system: Partial structural characterization of the cytochrome P-450 components and comparison to the hepatic cytochrome P-450, *J. Biol. Chem.* **256**:2499–2503.
8. Williams, D. E., Hale, S. E., Okita, R. T., and Masters, B. S. S., 1984, A prostaglandin ω-hydroxylase cytochrome P-450 ($P-450_{PG-\omega}$) purified from lungs of pregnant rabbits, *J. Biol. Chem.* **259**:14600–14608.
9. Ogita, K., Kusunose, E., Yamamoto, S., Ichihara, K., and Kusunose, M., 1983, Multiple forms of cytochrome P-450 from kidney cortex microsomes of rabbits treated with phenobarbital, *Biochem. Int.* **6**:191–198.
10. Ogita, K., Kusunose, E., Ichihara, K., and Kusunose, M., 1982, Multiple forms of cytochrome P-450 in kidney cortex microsomes of rabbits treated with 3-methylcholanthrene, *J. Biochem.* **92**:921–928.
11. Kusunose, E., Kuku, M., Ichihara, K., Yamamoto, S., Yano, I., and Kusunose, M., 1984, Hydroxylation of prostaglandin A_1 by the microsomes of rabbit intestinal mucosa, *J. Biochem.* **95**:1733–1739.
12. Guengerich, F. P., Dannan, G. A., Wright, S. T., Martin, M. V., and Kaminsky, L. S., 1982, Purification and characterization of liver microsomal cytochromes P-450: Electrophoretic, spectral, catalytic, and immunochemical properties and inducibility of eight isozymes isolated from rats treated with phenobarbital or β-naphthoflavone, *Biochemistry* **21**:6019–6030.
13. Lau, P. P., and Strobel, H. W., 1982, Multiple forms of cytochrome P-450 in liver microsomes from β-naphthoflavone-pretreated rats: Separation, purification, and characterization of five forms, *J. Biol. Chem.* **257**:5257–5262.
14. Waxman, D. J., Ko, A., and Walsh, C., 1983, Regioselectivity and stereoselectivity

of androgen hydroxylations catalyzed by cytochrome P-450 isozymes purified from phenobarbital-induced rat liver, *J. Biol. Chem.* **258**:11937–11947.
15. Kuwahara, S., Harada, N., Yoshioka, H., Miyata, T., and Omura, T., 1984, Purification and characterization of four forms of cytochrome P-450 from liver microsomes of phenobarbital-treated and 3-methylcholanthrene-treated rats, *J. Biochem.* **95**:703–714.
16. Waxman, D. J., and Walsh, C., 1982, Catalytic and structural properties of two new cytochrome P-450 isozymes from phenobarbital-induced rat liver: Comparison to the major induced isozymic form, in: *Cytochrome P-450: Biochemistry, Biophysics and Environmental Implications* (E. Hietanen, M. Laitinen, and O. Hänninen, eds.), Elsevier, Amsterdam, pp. 311–316.
17. Cheng, K.-C., and Schenkman, J. B., 1982, Purification and characterization of two constitutive forms of rat liver microsomal cytochrome P-450, *J. Biol. Chem.* **257**:2378–2385.
18. Kamataki, T., Maeda, K., Yamazoe, Y., Nagai, T., and Kato, R., 1983, Sex difference of cytochrome P-450 in the rat: Purification, characterization, and quantitation of constitutive forms of cytochrome P-450 from liver microsomes of male and female rats, *Arch. Biochem. Biophys.* **225**:758–770.
19. Elshourbagy, N. A., and Guzelian, P. S., 1980, Separation, purification, and characterization of a novel form of hepatic cytochrome P-450 from rats treated with pregnenolone-16α-carbonitrile, *J. Biol. Chem.* **255**:1279–1285.
20. Waxman, D. J., and Walsh, C., 1983, Cytochrome P-450 isozyme 1 from phenobarbital-induced rat liver: Purification, characterization, and interactions with metyrapone and cytochrome b_5, *Biochemistry* **22**:4846–4855.
21. Tamburini, P. P., Masson, H. A., Bains, S. K., Makowski, R. J., Morris, B., and Gibson, G. G., 1984, Multiple forms of hepatic cytochrome P-450: Purification, characterisation and comparison of a novel clofibrate-induced isozyme with other major forms of cytochrome P-450, *Eur. J. Biochem.* **139**:235–246.
22. Hayashi, S., Nashiro, M., and Okuda, K., 1984, Purification of cytochrome P-450 catalyzing 25-hydroxylation of vitamin D_3 from rat liver microsomes, *Biochem. Biophys. Res. Commun.* **121**:994–1000.
23. Negishi, M., and Nebert, D. W., 1979, Structural gene products of the *Ah* locus: Genetic and immunochemical evidence for two forms of mouse liver cytochrome P-450 induced by 3-methylcholanthrene, *J. Biol. Chem.* **254**:11015–11023.
24. Kimura, S., Gonzalez, F. J., and Nebert, D. W., 1984, Mouse cytochrome P_3-450: Complete cDNA and amino acid sequence, *Nucleic Acids Res.* **12**:2917–2928.
25. Dieter, H. H., Muller-Eberhard, U., and Johnson, E. F., 1982, Identification of rabbit microsomal cytochrome P-450 isozyme, form 1, as a hepatic progesterone 21-hydroxylase, *Biochem. Biophys. Res. Commun.* **105**:515–520.
26. Haugen, D. A., and Coon, M. J., 1976, Properties of electrophoretically homogeneous phenobarbital-inducible and β-naphthoflavone-inducible forms of liver microsomal cytochrome P-450, *J. Biol. Chem.* **251**:7929–7939.
27. Black, S. D., Tarr, G. E., and Coon, M. J., 1982, Structural features of isozyme 2 of liver microsomal cytochrome P-450: Identification of a highly conserved cysteine-containing peptide, *J. Biol. Chem.* **257**:14616–14619.
28. Koop, D. R., Morgan, E. T., Tarr, G. E., and Coon, M. J., 1982, Purification and characterization of a unique isozyme of cytochrome P-450 from liver microsomes of ethanol-treated rabbits, *J. Biol. Chem.* **257**:8472–8480.
29. Johnson, E. F., 1980, Isolation and characterization of a constitutive form of rabbit liver microsomal cytochrome P-450, *J. Biol. Chem.* **255**:304–309.
30. Koop, D. R., Persson, A. V., and Coon, M. J., 1981, Properties of electrophoretically

homogeneous constitutive forms of liver microsomal cytochrome P-450, *J. Biol. Chem.* **256**:10704–10711.
31. Bonfils, C., Dalet, C., Dalet-Beluche, I., and Maurel, P., 1983, Cytochrome P-450 isozyme LM$_{3b}$ from rabbit liver microsomes: Induction by triacetyloleandomycin, purification and characterization, *J. Biol. Chem.* **258**:5358–5362.
32. Boström, H., and Wikvall, K., 1982, Hydroxylations in biosynthesis of bile acids: Isolation of subfractions with different substrate specificity from cytochrome P-450$_{LM4}$, *J. Biol. Chem.* **257**:11755–11759.
33. Wikvall, K., 1984, Hydroxylations in biosynthesis of bile acids: Isolation of a cytochrome P-450 from rabbit liver mitochondria catalyzing the 26-hydroxylation of C$_{27}$-steroids, *J. Biol. Chem.* **259**:3800–3804.
34. Norman, R. L., Johnson, E. F., and Muller-Eberhard, U., 1978, Identification of the major cytochrome P-450 form transplacentally induced in neonatal rabbits by 2,3,7,8-tetrachlorodibenzo-*p*-dioxin, *J. Biol. Chem.* **253**:8640–8647.
35. Koop, D. R., and Coon, M. J., 1984, Purification of liver microsomal cytochrome P-450 isozymes 3a and 6 from imidazole-treated rabbits: Evidence for the identity of isozyme 3a with the form obtained by ethanol treatment, *Mol. Pharmacol.* **25**:494–501.
36. Guengerich, F. P., 1978, Separation and purification of multiple forms of microsomal cytochrome P-450, *J. Biol. Chem.* **253**:7931–7938.
37. Waxman, D. J., and Walsh, C., 1982, Phenobarbital-induced rat liver cytochrome P-450: Purification and characterization of two closely related isozymic forms, *J. Biol. Chem.* **257**:10446–10457.
38. Ryan, D. E., Thomas, P. E., and Levin, W., 1980, Hepatic microsomal cytochrome P-450 from rats treated with isosafrole: Purification and characterization of four enzymic forms, *J. Biol. Chem.* **255**:7941–7955.
39. Ryan, D. E., Thomas, P. E., and Levin, W., 1982, Purification and characterization of a minor form of hepatic microsomal cytochrome P-450 from rats treated with polychlorinated biphenyls, *Arch. Biochem. Biophys.* **216**:272–288.
40. Ryan, D. E., Iida, S., Wood, A. W., Thomas, P. E., Lieber, C. S., and Levin, W., 1984, Characterization of three highly purified cytochromes P-450 from hepatic microsomes of adult male rats, *J. Biol. Chem.* **259**:1239–1250.
41. Agosin, M., Morello, A., White, R., Repetto, Y., and Pedemonte, J., 1979, Multiple forms of noninduced rat liver cytochrome P-450: Metabolism of 1-(4′-ethylphenoxy)-3,7-dimethyl-6,7-epoxy-*trans*-2-octene by reconstituted preparations, *J. Biol. Chem.* **254**:9915–9920.
42. Larrey, D., Distlerath, L. M., Dannan, G. A., Wilkinson, G. R., and Guengerich, F. P., 1984, Purification and characterization of the rat liver cytochrome P-450 involved in the 4-hydroxylation of debrisoquine, a prototype for genetic variation in oxidative drug metabolism, *Biochemistry* **23**:2787–2795.
43. Huang, M.-T., West, S. B., and Lu, A. Y. H., 1976, Separation, purification, and properties of multiple forms of cytochrome P-450 from the liver microsomes of phenobarbital-treated mice, *J. Biol. Chem.* **251**:4659–4665.
44. Wang, H.-P., and Kimura, T., 1976, Purification and characterization of adrenal cortex mitochondrial cytochrome P-450 specific for cholesterol side chain cleavage activity, *J. Biol. Chem.* **251**:6068–6074.
45. Suhara, K., Gomi, T., Sato, H., Itagaki, E., Takamori, S., and Katagiri, M., 1978, Purification and immunochemical characterization of the two adrenal cortex mitochondrial cytochrome P-450-proteins, *Arch. Biochem. Biophys.* **190**:290–299.
46. Kashiwagi, K., Dafeldecker, W. P., and Salhanick, H. A., 1980, Purification and characterization of mitochondrial cytochrome P-450 associated with cholesterol side chain cleavage from bovine corpus luteum, *J. Biol. Chem.* **255**:2606–2611.

47. Ichikawa, Y., and Hiwatashi, A., 1982, The role of the sugar regions of components of the cytochrome P-450-linked mixed-function oxidase (monooxygenase) system of bovine adrenocortical mitochondria, *Biochim. Biophys. Acta* **705**:82–91.
48. Takemori, S., Suhara, K., Hashimoto, S., Hashimoto, M., Sato, H., Gomi, T., and Katagiri, M., 1975, Purification of cytochrome P-450 from bovine adrenocortical mitochondria by an "aniline-Sepharose" and the properties, *Biochem. Biophys. Res. Commun.* **63**:588–593.
49. Tilley, B. E., Watanuki, M., and Hall, P. F., 1977, Preparation and properties of side-chain cleavage cytochrome P-450 from bovine adrenal cortex by affinity chromatography with pregnenolone as ligand, *Biochim. Biophys. Acta* **493**:260–271.
50. Katagiri, M., Takemori, S., Itagaki, E., and Suhara, K., 1978, Purification of adrenal cytochrome P-450 (cholesterol desmolase and steroid 11β- and 18-hydroxylase), *Methods Enzymol.* **52**:124–132.
51. Kaminami, S., Ochi, H., Kobayashi, Y., and Takemori, S., 1980, Studies on the steroid hydroxylation system in adrenal cortex microsomes: Purification and characterization of cytochrome P-450 specific for steroid C-21 hydroxylation, *J. Biol. Chem.* **255**:3386–3394.
52. Hiwatashi, A., and Ichikawa, Y., 1981, Purification and reconstitution of the steroid 21-hydroxylase system (cytochrome P-450-linked mixed function oxidase system) of bovine adrenocortical microsomes, *Biochim. Biophys. Acta* **664**:33–48.
53. Bumpus, J. A., and Dus, K. M., 1982, Bovine adrenocortical microsomal hemeproteins P-450$_{17\alpha}$ and P-450$_{C-21}$: Isolation, partial characterization, and comparison to P-450$_{scc}$, *J. Biol. Chem.* **257**:12696–12704.
54. Yuan, P.-M., Nakajin, S., Haniu, M., Shinoda, M., Hall, P. F., and Shively, J. E., 1983, Steroid 21-hydroxylase (cytochrome P-450) from porcine adrenocortical microsomes: Microsequence analysis of cysteine-containing peptides, *Biochemistry* **22**:143–149.
55. Nakajin, S., Shively, J. E., Yuan, P.-M., and Hall, P. F., 1981, Microsomal cytochrome P-450 from neonatal pig testis: Two enzymic activities (17α-hydroxylase and C$_{17,20}$-lyase) associated with one protein, *Biochemistry* **20**:4037–4042.
56. Nakajin, S., Shinoda, M., and Hall, P. F., 1983, Purification and properties of 17α-hydroxylase from microsomes of pig adrenal: A second C$_{21}$ side-chain cleavage system, *Biochem. Biophys. Res. Commun.* **111**:512–517.
57. Nakajin, S., Shinoda, M., Haniu, M., Shively, J. E., and Hall, P. F., 1984, C$_{21}$ steroid side chain cleavage enzyme from porcine adrenal microsomes: Purification and characterization of the 17α-hydroxylase/C$_{17,20}$-lyase cytochrome P-450, *J. Biol. Chem.* **259**:3971–3976.
58. Graf, H., Ruf, H. H., and Ullrich, V., 1983, Prostacyclin synthetase, a P-450 enzyme, *Angew. Chem. Int. Ed. Engl.* **22**:487–488.
59. Kominami, S., Shinzawa, K., and Takamori, S., 1982, Purification and some properties of cytochrome P-450 specific for steroid 17α-hydroxylation and C$_{17}$–C$_{20}$ bond cleavage from guinea pig adrenal microsomes, *Biochem. Biophys. Res. Commun.* **109**:916–921.
60. Klotz, A. V., Stegeman, J. J., and Walsh, C., 1983, An arylhydrocarbon hydroxylating hepatic cytochrome P-450 from the marine fish *Stenotomus chrysops*, *Arch. Biochem. Biophys.* **226**:578–592.
61. Higashi, K., Ikeuchi, K., Karasaki, Y., and Obara, M., 1983, Isolation of immunochemically distinct form of cytochrome P-450 from microsomes of tulip bulbs, *Biochem. Biophys. Res. Commun.* **115**:46–52.
62. Yoshida, Y., and Aoyama, Y., 1984, Yeast cytochrome P-450 catalyzing lanosterol 14α-demethylation. I. Purification and spectral properties, *J. Biol. Chem.* **259**:1655–1660.

63. Dus, K., Katagiri, M., Yu, C.-A., Erbes, D. L., and Gunsalus, I. C., 1970, Chemical characterization of cytochrome P-450$_{cam}$, *Biochem. Biophys. Res. Commun.* **40**:1423–1430.
64. O'Keeffe, D. H., Ebel, R. E., and Peterson, J. A., 1978, Purification of bacterial cytochrome P-450, *Methods Enzymol.* **52**:151–157.
65. Berg, A., Ingelman-Sundberg, M., and Gustafsson, J.-A., 1979, Purification and characterization of cytochrome P-450$_{meg}$, *J. Biol. Chem.* **254**:5264–5271.
66. Dus, K., Goewert, R., Weaver, C. C., Carey, D., and Appleby, C. A., 1976, P-450 hemeproteins of *Rhizobium japonicum*: Purification by affinity chromatography and relationship to P-450$_{cam}$ and P-450$_{LM2}$, *Biochem. Biophys. Res. Commun.* **69**:437–445.
67. Appleby, C. A., 1978, Purification of *Rhizobium* cytochromes P-450, *Methods Enzymol.* **52**:157–166.
68. Wang, P., Mason, P. S., and Guengerich, F. P., 1980, Purification of human liver cytochrome P-450 and comparison to the enzyme isolated from rat liver, *Arch. Biochem. Biophys.* **199**:206–219.
69. Wang, P. P., Beaune, P., Kaminski, L. S., Dannan, G. A., Kadlubar, F. F., Larrey, D., and Guengerich, F. P., 1983, Purification and characterization of six cytochrome P-450 isozymes from human liver microsomes, *Biochemistry* **22**:5375–5383.
70. Lu, A. Y. H., and Miwa, G. T., 1980, Molecular properties and biological functions of microsomal epoxide hydrase, *Annu. Rev. Pharmacol. Toxicol.* **20**:513–531.
71. Poulson, L. L., and Ziegler, D. M., 1979, The liver microsomal FAD-containing monooxygenase: Spectral characterization and kinetic studies, *J. Biol. Chem.* **254**:6449–6455.
72. van der Hoeven, T. A., and Coon, M. J., 1974, Preparation and properties of partially purified cytochrome P-450 and reduced NADP-cytochrome P-450 reductase from rabbit liver microsomes, *J. Biol. Chem.* **249**:6302–6310.
73. French, J. S., Guengerich, F. P., and Coon, M. J., 1980, Interactions of cytochrome P-450, NADPH-cytochrome P-450 reductase, phospholipid, and substrate in the reconstituted liver microsomal enzyme system, *J. Biol. Chem.* **255**:4112–4119.
74. Dean, W. L., and Gray, R. D., 1982, Relationship between state of aggregation and catalytic activity for cytochrome P-450$_{LM2}$ and NADPH-cytochrome P-450 reductase, *J. Biol. Chem.* **257**:14679–14685.
75. Chiang, Y.-L., and Coon, M. J., 1979, Comparative study of two highly purified forms of liver microsomal cytochrome P-450: Circular dichroism and other properties, *Arch. Biochem. Biophys.* **195**:178–187.
76. Dus, K., Litchfield, W. J., Miguel, A. G., van der Hoeven, T. A., Haugen, D. A., Dean, W. L., and Coon, M. J., 1974, Structural resemblance of cytochrome P-450 isolated from *Pseudomonas putida* and from rabbit liver microsomes, *Biochem. Biophys. Res. Commun.* **60**:15–21.
77. Dus, K. M., 1982, Insights into the active site of the cytochrome P-450 haemoprotein family—A unifying concept based on structural considerations, *Xenobiotica* **12**:745–772.
78. Thomas, P. E., Lu, A. Y. H., Ryan, D., West, S. B., Kawalek, J., and Levin, W., 1976, Immunochemical evidence for six forms of rat liver cytochrome P-450 obtained using antibodies against purified rat liver cytochromes P-450 and P-448, *Mol. Pharmacol.* **12**:746–758.
79. Reik, L. M., Levin, W., Ryan, D. E., and Thomas, P. E., 1982, Immunochemical relatedness of rat hepatic microsomal cytochromes P-450c and P-450d, *J. Biol. Chem.* **257**:3950–3957.
80. Guengerich, F. P., Wang, P., Mason, P. S., and Mitchell, M. B., 1981, Immunological comparison of rat, rabbit, and human microsomal cytochromes P-450, *Biochemistry* **20**:2370–2378.

81. Chen, Y.-T., Lang, M. A., Jensen, N. M., Negishi, M., Tukey, R. H., Sidransky, E., Guenther, T. M., and Nebert, D. M., 1982, Similarities between mouse and rat liver microsomal cytochromes P-450 induced by 3-methylcholanthrene, *Eur. J. Biochem.* **122:**361–368.
82. Dean, W. L., and Coon, M. J., 1977, Immunochemical studies on two electrophoretically homogeneous forms of rabbit liver microsomal cytochrome P-450: P-450$_{LM2}$ and P-450$_{LM4}$, *J. Biol. Chem.* **252:**3255–3261.
83. Cleveland, D. W., Fischer, S. G., Kirschner, M. W., and Laemmli, U. K., 1977, Peptide mapping by limited proteolysis in sodium dodecyl sulfate and analysis by gel electrophoresis, *J. Biol. Chem.* **252:**1102–1106.
84. Yuan, P.-M., Ryan, D. E., Levin, W., and Shively, J. E., 1983, Identification and localization of amino acid substitutions between two phenobarbital-inducible rat hepatic microsomal cytochromes P-450 by microsequence analysis, *Proc. Natl. Acad. Sci. USA* **80:**1169–1173.
85. Ozols, J., Heinemann, F. S., and Johnson, E. F., 1981, Amino acid sequence of an analogous peptide from two forms of cytochrome P-450, *J. Biol. Chem.* **256:**11405–11408.
86. Ryan, D. E., Wood, A. W., Thomas, P. E., Walz, F. E., Jr., Yuan, P.-M., Shively, J. E., and Levin, W., 1983, Comparisons of highly purified hepatic microsomal cytochromes P-450 from Holtzman and Long–Evans rats, *Biochim. Biophys. Acta* **709:**273–283.
87. Dieter, H. H., and Johnson, E. F., 1982, Functional and structural polymorphism of rabbit microsomal cytochrome P-450 form 3b, *J. Biol. Chem.* **257:**9315–9323.
88. Yoshioka, H., Miyata, T., and Omura, T., 1984, Induction of a phenobarbital-inducible form of cytochrome P-450 in rat liver microsomes by 1,1-di(*p*-chlorophenyl)-2,2-dichloroethylene, *J. Biochem.* **95:**937–947.
89. Tarr, G. E., Black, S. D., Fujita, V. S., and Coon, M. J., 1983, Complete amino acid sequence and predicted membrane topology of phenobarbital-induced cytochrome P-450 (isozyme 2) from rabbit liver microsomes, *Proc. Natl. Acad. Sci. USA* **80:**6552–6556.
90. Botelho, L. H., Ryan, D. E., and Levin, W., 1979, Amino acid compositions and partial amino acid sequences of three highly purified forms of liver microsomal cytochrome P-450 from rats treated with polychlorinated biphenyls, phenobarbital, or 3-methylcholanthrene, *J. Biol. Chem.* **254:**5635–5640.
91. Botelho, L. H., Ryan, D. E., Yuan, P.-M., Kutny, R., Shively, J. E., and Levin, W., 1982, Amino-terminal and carboxy-terminal sequence of hepatic microsomal cytochrome P-450d, a unique hemeprotein from rats treated with isosafrole, *Biochemistry* **21:**1152–1155.
92. Haniu, M., Ryan, D. E., Iida, S., Lieber, C. S., Levin, W., and Shively, J. E., 1984, NH$_2$-terminal sequence analyses of four rat hepatic microsomal cytochromes P-450, *Arch. Biochem. Biophys.* **235:**304–311.
93. Ogishima, T., Okada, Y., Kominami, S., Takemori, S., and Omura, T., 1983, Partial amino acid sequences of two mitochondrial and two microsomal cytochrome P-450's from adrenal cortex, *J. Biochem.* **94:**1711–1714.
94. Katagiri, M., Takemori, S., Itagaki, E., Suhara, K., Gomi, T., and Sato, H., 1976, Characterization of purified cytochrome P-450$_{scc}$ and P-450$_{11\beta}$ from bovine adrenocortical mitochondria, in: *Iron and Copper Proteins* (K. T. Yasunobu, H. F. Mower, and O. Hayaishi, eds.), Plenum Press, New York, pp. 281–289.
95. Heinemann, F. S., and Ozols, J., 1982, The covalent structure of rabbit phenobarbital-induced cytochrome P-450: Partial amino acid sequence and order of cyanogen bromide peptides, *J. Biol. Chem.* **257:**14988–14999.

96. Fujii-Kuriyama, Y., Mizukami, Y., Kawajiri, K., Sogawa, K., and Muramatsu, M., 1982, Primary structure of a cytochrome P-450: Coding nucleotide sequence of phenobarbital-inducible cytochrome P-450 cDNA from rat liver, *Proc. Natl. Acad. Sci. USA* **79:**2793–2797.
97. Sogawa, K., Gotoh, O., Kawajiri, K., and Fujii-Kuriyama, Y., 1984, Distinct organization of methylcholanthrene- and phenobarbital-inducible cytochrome P-450 genes in the rat, *Proc. Natl. Acad. Sci. USA* **81:**5066–5070.
98. Tarr, G. E., 1985, Manual Edman sequencing system, in: *Microcharacterization of Polypeptides: A Practical Manual* (J. E. Shively, ed.), Humana Press, Clifton, N. J., in press.
99. Armstrong, R. N., Pinto-Coelho, C., Ryan, D. E., Thomas, P. E., and Levin, W., 1983, On the glycosylation state of five rat hepatic microsomal cytochrome P-450 isozymes, *J. Biol. Chem.* **258:**2106–2108.
100. Negishi, M., Jensen, N. M., Garcia, G. S., and Nebert, D. W., 1981, Structural gene products of the murine *Ah* complex: Differences in ontogenesis and glucosamine incorporation between liver microsomal cytochromes P_1-450 and P-448 induced by polycyclic aromatic compounds, *Eur. J. Biochem.* **115:**585–594.
101. Haniu, M., Armes, L. G., Yasunobu, K. T., Shastry, B. A., and Gunsalus, I. C., 1982, Amino acid sequence of the *Pseudomonas putida* cytochrome P-450. II. Cyanogen bromide peptides, acid cleavage peptides, and the complete sequence, *J. Biol. Chem.* **257:**12664–12671.
102. Hiwatashi, A., and Ichikawa, Y., 1980, Purification and organ-specific properties of cholecalciferol 25-hydroxylase system: Cytochrome P-450$_{D25}$-linked mixed function oxidase system, *Biochem. Biophys. Res. Commun.* **97:**1443–1449.
103. Goodwin, C. D., Cooper, B. W., and Margolis, S., 1982, Rat liver cholesterol 7α-hydroxylase: Modulation of enzyme activity by changes in phosphorylation state, *J. Biol. Chem.* **257:**4469–4472.
104. Haugen, D. A., Armes, L. G., Yasunobu, K. T., and Coon, M. J., 1977, Amino-terminal sequence of phenobarbital-inducible cytochrome P-450 from rabbit liver microsomes: Similarity to hydrophobic amino-terminal segments of preproteins, *Biochem. Biophys. Res. Commun.* **77:**967–973.
105. Coon, M. J., Black, S. D., Fujita, V. S., Koop, D. R., and Tarr, G. E., 1985, Structural comparison of multiple forms of cytochrome P-450, in: *Microsomes and Drug Oxidations, Proceedings of the 6th International Symposium* (A. R. Boobis, J. Caldwell, F. DeMatteis, and C. R. Elcombe, eds.), Taylor and Francis Ltd., London, pp. 42–51.
106. Fujita, V. S., Black, S. D., Tarr, G. E., Koop, D. R., and Coon, M. J., 1985, On the amino acid sequence of cytochrome P-450 isozyme 4 from rabbit liver microsomes, *Proc. Natl. Acad. Sci. USA* **81:**4260–4264.
107. Bar-Nun, S., Kreibich, G., Adesnick, M., Alterman, L., Negishi, M., and Sabatini, D. D., 1980, Synthesis and insertion of cytochrome P-450 into endoplasmic reticulum membranes, *Proc. Natl. Acad. Sci. USA* **77:**965–969.
108. Wrighton, S. A., Maurel, P., Schuetz, E. G., Watkins, P. B., Young, B., and Guzelian, P. S., 1985, Identification of the cytochrome P-450 induced by macrolide antibiotics in rat liver as the glucocorticoid responsive cytochrome P-450$_p$, *Biochemistry* **24:**2171–2178.
109. Cheng, K.-C., and Schenkman, J. B., 1983, Testosterone metabolism by cytochrome P-450 isozymes RLM_3 and RLM_5 and by microsomes: Metabolite identification, *J. Biol. Chem.* **258:**11738–11744.
110. Hayashi, S., Noshiro, M., and Okuda, K., 1984, Purification of cytochrome P-450 catalyzing 25-hydroxylation of vitamin D_3 from rat liver microsomes, *Biochem. Biophys. Res. Commun.* **121:**994–1000.

111. Akhrem, A. A., Lapko, V. N., Lapko, A. G., Shkumotov, V. M., and Chaschin, V. L., 1979, Isolation, structural organization and mechanism of action of mitochondrial steroid hydroxylating systems, *Acta Biol. Med. Ger.* **38:**257–273.
112. Tanaka, M., Zeitlin, S., Yasunobu, K. T., and Gunsalus, I. C., 1976, Current status of the sequence studies on the *Pseudomonas putida* camphor hydroxylase system, in: *Iron and Copper Proteins* (K. T. Yasunobu, H. F. Mower, and O. Hayaishi, eds.), Plenum Press, New York, pp. 263–269.
113. Haniu, M., Tanaka, M., Yasunobu, K. T., and Gunsalus, I. C., 1982, Amino acid sequence of the *Pseudomonas putida* cytochrome P-450. I. Sequence of tryptic and clostripain peptides, *J. Biol. Chem.* **257:**12657–12663.
114. Sabatini, D. D., Kreibich, G., Morimoto, T., and Adesnik, M., 1982, Mechanisms for the incorporation of proteins in membranes and organelles, *J. Cell Biol.* **92:**1–22.
115. DuBois, G. C., Appella, E., Ryan, D. E., Jerina, D. M., and Levin, W., 1982, Human hepatic microsomal epoxide hydrolase: A structural comparison with the rat enzyme, *J. Biol. Chem.* **257:**2708–2712.
116. Heinemann, F. S., and Ozols, J., 1984, The covalent structure of hepatic microsomal epoxide hydrolase. II. The complete amino acid sequence, *J. Biol. Chem.* **259:**797–804.
117. Mihara, K., Sato, R., Sakakibara, R., and Wada, H., 1978, Reduced NAD-cytochrome b_5 reductase: Location of the hydrophobic, membrane-binding region at the carboxyl-terminal end and the masked amino terminus, *Biochemistry* **17:**2829–2834.
118. Ozols, J., and Heinemann, F. S., 1982, Chemical structure of rat liver cytochrome b_5: Isolation of peptides by high-pressure liquid chromatography, *Biochim. Biophys. Acta* **704:**163–173.
119. Black, S. D., and Coon, M. J., 1982, Structural features of liver microsomal NADPH-cytochrome P-450 reductase: Hydrophobic domain, hydrophilic domain, and connecting region, *J. Biol. Chem.* **257:**5929–5938.
120. Yasunobu, K. T., Armes, L. G., Crabb, J. W., and Haniu, M., 1980, Comparison of the known chemical structures of some cytochrome P-450 enzymes, in: *Biochemistry, Biophysics, and Regulation of Cytochrome P-450* (J.-Å. Gustafsson, J. Carlsted-Duke, A. Mode, and J. Rafter, eds.), Elsevier, Amsterdam, pp. 49–56.
121. Leighton, J. K., DeBrunner-Vossbrinck, B. A., and Kemper, B., 1984, Isolation and sequence analysis of three cloned cDNAs for rabbit liver proteins that are related to rabbit cytochrome P-450 (form 2), the major phenobarbital-inducible form, *Biochemistry* **23:**204–210.
122. Yabusaki, Y., Shimizu, M., Murakami, H., Nakamura, K., Oeda, K., and Ohkawa, H., 1984, Nucleotide sequence of a full-length cDNA coding for 3-methylcholanthrene-induced rat liver cytochrome P-450$_{MC}$, *Nucleic Acids Res.* **12:**2929–2938.
123. Kawajiri, K., Gotoh, O., Sogawa, K., Tagashira, Y., Muramatsu, M., and Fujii-Kuriyama, Y., 1984, Coding nucleotide sequence of 3-methylcholanthrene-inducible cytochrome P-450d cDNa from rat liver, *Proc. Natl. Acad. Sci. USA* **81:**1649–1653.
124. Morohashi, K., Fujii-Kuriyama, Y., Okada, Y., Sogawa, K., Hirose, T., Inayama, S., and Omura, T., 1984, Molecular cloning and nucleotide sequence of cDNA for mRNA of mitochondrial cytochrome P-450 (P-450$_{scc}$) from bovine adrenal cortex mitochondria, *Proc. Natl. Acad. Sci. USA* **81:**4647–4651.
125. Ambler, R. P., 1967, Enzymic hydrolysis with carboxypeptidases, *Methods Enzymol.* **11:**155–166.
126. Hayashi, R., 1976, Carboxypeptidase Y, *Methods Enzymol.* **45:**568–587.
127. Kimura, S., Gonzalez, F. J., and Nebert, D. W., 1984, The murine *Ah* locus: Comparison of the complete cytochrome P_1-450 and P_3-450 cDNA nucleotide and amino acid sequences, *J. Biol. Chem.* **259:**10705–10713.

128. Haniu, M., Armes, L. G., Tanaka, M., Yasunobu, K. T., Shastry, R. S., Wagner, G. C., and Gunsalus, I. C., 1982, The primary structure of the monoxygenase cytochrome P-450$_{cam}$, *Biochem. Biophys. Res. Commun.* **105**:889–894.
129. Heinemann, F. S., and Ozols, J., 1983, The complete amino acid sequence of rabbit phenobarbital-induced liver microsomal cytochrome P-450, *J. Biol. Chem.* **258**:4195–4201.
130. Hines, R. N., Levy, J. B., Conrad, R. D., Iverson, P. L., Shen, M.-L., and Bresnick, E., 1985, Gene structure and nucleotide sequence for rat cytochrome P-450c, *Arch. Biochem. Biophys.* **237**:465–476.
131. Kumar, A., Raphael, C., and Adesnik, M., 1983, Cloned cytochrome P-450 cDNA: Nucleotide sequence and homology to multiple phenobarbital-induced mRNA species, *J. Biol. Chem.* **258**:11280–11284.
132. Phillips, I. R., Shephard, E. A., Ashworth, A., and Rabin, B. R., 1983, Cloning and sequence analysis of a rat liver cDNA coding for a phenobarbital-inducible microheterogeneous cytochrome P-450 variant: Regulation of its message level by xenobiotics, *Gene* **24**:41–52.
133. Mizukami, Y., Sogawa, K., Suwa, Y., Muramatsu, M., and Fujii-Kuriyama, Y., 1983, Gene structure of a phenobarbital-inducible cytochrome P-450 in rat liver, *Proc. Natl. Acad. Sci. USA* **80**:3958–3962.
134. Affolter, M., and Anderson, A., 1984, Segmental homologies in the coding and 3' noncoding sequences of rat liver cytochrome P-450e and P-450b cDNAs and cytochrome P-450e-like genes, *Biochem. Biophys. Res. Commun.* **118**:655–662.
135. Gonzalez, F. J., Mackenzie, P. I., Kimura, S., and Nebert, D. W., 1984, Isolation and characterization of full-length cDNA and genomic clones of mouse 3-methylcholanthrene-inducible cytochrome P$_1$-450 and P$_3$-450, *Gene* **29**:281–292.
136. Akhrem, A. A., Vasilevsky, V. I., Adamovich, T. B., Lapko, A. G., Shkumatov, V. M., and Chashchin, V. L., 1980, Chemical characteristics of the cholesterol side chain cleavage cytochrome P-450 (P-450$_{scc}$) from bovine adrenal cortex mitochondria, in: *Biochemistry, Biophysics, and Regulation of Cytochrome P-450* (J.-Å. Gustafsson, J. Carlstedt-Duke, A. Mode, and J. Rafter, eds.), Elsevier, Amsterdam, pp. 57–64.
137. Chashchin, V. L., Lapko, V. N., Adamovich, T. B., Lapko, A. G., Kuprina, N. S., and Akhrem, A. A., 1982, Primary structure of 20 S, 22 R-cholesterol hydroxylating cytochrome P-450 from adrenal cortex mitochondria. I. Peptides of exhaustive chymotryptic hydrolysis, *Bioorg. Khim.* **8**:1307–1320.
138. White, P. C., New, M. I., and Dupont, B., 1984, Cloning and expression of cDNA encoding a bovine adrenal cytochrome P-450 specific for steroid 21-hydroxylation, *Proc. Natl. Acad. Sci. USA* **81**:1986–1990.
139. Zaphiropoulos, P. G., Folk, W. R., and Coon, M. J., 1984, Isolation and characterization of a P-450-like gene, in: *6th International Symposium on Microsomes and Drug Oxidations* (Abstracts), p. 60.
140. Nebert, D. W., Eisen, H. J., Negishi, M., Lang, M. A., and Hjelmeland, L. M., 1981, Genetic mechanisms controlling the induction of polysubstrate monooxygenase (P-450) activities, *Annu. Rev. Pharmacol. Toxicol.* **21**:431–462.
141. Wu, T. T., Fitch, W. M., and Margoliash, E., 1974, The information content of protein amino acid sequences, *Annu. Rev. Biochem.* **43**:539–565.
142. Garavito, R. M., Rossman, M. G., Argos, P., and Eventoff, W., 1977, Convergence of active center geometries, *Biochemistry* **16**:5065–5071.
143. Cramer, S. P., Dawson, J. H., Hodgson, K. O., and Hager, L. P., 1978, Studies on the ferric forms of cytochrome P-450 and chloroperoxidase by extended x-ray absorption fine structure: Characterization of the Fe–N and Fe–S distances, *J. Am. Chem. Soc.* **100**:7282–7290.

144. White, R. E., and Coon, M. J., 1980, Oxygen activation by cytochrome P-450, *Annu. Rev. Biochem.* **49**:315–356.
145. Champion, P. M., Stallard, B. R., Wagner, G. C., and Gunsalus, I. C., 1982, Resonance Raman detection of an Fe–S bond in cytochrome P-450$_{cam}$, *J. Am. Chem. Soc.* **104**:5469–5472.
146. Gotoh, O., Tagashira, Y., Iizuka, T., and Fujii-Kuriyama, Y., 1983, Structural characteristics of cytochrome P-450: Possible location of the heme-binding cysteine in determined amino acid sequences, *J. Biochem.* **93**:807–817.
147. Haniu, M., Yasunobu, K. T., and Gunsalus, I. C., 1983, Heme-binding and substrate protected cysteine residues in P-450$_{cam}$, *Biochem. Biophys. Res. Commun.* **116**:30–38.
148. Black, S. D., and Coon, M. J., 1985, Studies on the identity of the heme-binding cysteinyl residue in rabbit liver microsomal cytochrome P-450 isozyme 2, *Biochem. Biophys. Res. Commun.* **128**:82–89.
149. Dawson, J. H., Anderson, L. A., and Sono, M., 1982, Spectroscopic investigations of ferric cytochrome P-450$_{cam}$ ligand complexes: Identification of the ligand *trans* to cysteinate in the native enzyme, *J. Biol. Chem.* **257**:3606–3617.
150. White, R. E., and Coon, M. J., 1982, Heme ligand replacement reactions of cytochrome P-450: Characterization of the bonding atom of the axial ligand *trans* to thiolate as oxygen, *J. Biol. Chem.* **257**:3073–3083.
151. Jänig, G.-R., Dettmer, R., Usanov, S. A., and Ruckpaul, K., 1983, Identification of the ligand *trans* to thiolate in cytochrome P-450$_{LM2}$ by chemical modification, *FEBS Lett.* **159**:58–62.
152. Dickerson, R. E., Takano, T., Eisenberg, D., Kallai, O. B., Samson, L., Cooper, A., and Margoliash, E., 1971, Ferricytochrome *c*. I. General features of the horse and bonito proteins at 2.8 Å resolution, *J. Biol. Chem.* **246**:1511–1535.
153. Khorana, H. G., Gerber, G. E., Herlihy, W. C., Gray, C. P., Anderegg, R. J., Nihei, K., and Biemann, K., 1979, Amino acid sequence of bacteriorhodopsin, *Proc. Natl. Acad. Sci. USA* **76**:5046–5050.
154. Zurawski, G., Bohnert, H. J., Whitfield, P. R., and Bottomley, W., 1982, Nucleotide sequence of the gene for the M_r 32,000 thylakoid membrane protein from *Spinacia aleracea* and *Nicotiana debneyi* predicts a totally conserved primary translation product of M_r 38,950, *Proc. Natl. Acad. Sci. USA* **79**:7699–7703.
155. Matocha, M. F., and Waterman, M. R., 1984, Discriminatory processing of the precursor forms of cytochrome P-450$_{scc}$ and adrenodoxin by adrenocortical and heart mitochondria, *J. Biol. Chem.* **259**:8672–8678.
156. Elliott, D. F., Lewis, G. P., and Horton, E. W., 1960, The structure of bradykinin-A plasma kinin from ox blood, *Biochem. Biophys. Res. Commun.* **3**:87–91.
157. Harper, E., and Kang, A. H., 1970, Studies on the specificity of bacterial collagenases, *Biochem. Biophys. Res. Commun.* **41**:482–487.
158. Hubbard, S. C., and Ivatt, R. J., 1981, Synthesis and processing of asparagine-linked oligosaccharides, *Annu. Rev. Biochem.* **50**:555–583.
159. Black, S. D., Tarr, G. E., Fujita, V. S., Yasunobu, K. T., and Coon, M. J., 1983, Site-specific cleavage of rabbit hepatic microsomal P-450 isozyme 2 by endoproteinase Lys-C, *Fed. Proc.* **42**:1774.
160. Chashchin, V. L., Vasilevsky, V. I., Shkumutov, V. M., and Akhrem, A. A., 1983, Separation and localization to the polypeptide chain of domain in the cholesterol-specific cytochrome P-450, *Bioorg. Khim.* **9**:1690–1692.
161. Walz, F. G., Jr., Vlasuk, G. P., Omiecinski, C. J., Bresnick, E., Thomas, P. E., Ryan, D. E., and Levin, W., 1982, Multiple immunoidentical forms of phenobarbital-induced rat liver cytochromes P-450 are encoded by different mRNAs, *J. Biol. Chem.* **257**:4023–4026.

162. Chiang, J. Y. L., Malmer, M., and Hutterer, F., 1983, A form of rabbit liver cytochrome P-450 that catalyzes the 7α-hydroxylation of cholesterol, *Biochim. Biophys. Acta* **750**:291–299.
163. Kabsch, W., and Sander, C., 1983, How good are predictions of protein secondary structure?, *FEBS Lett.* **155**:179–182.
164. Argos, P., Rao, J. K. M., and Hargrave, P. A., 1982, Structural prediction of membrane-bound proteins, *Eur. J. Biochem.* **128**:565–575.
165. Sweet, R. M., and Eisenberg, D., 1983, Correlation of sequence hydrophobicities measures similarity in three-dimensional protein structure, *J. Mol. Biol.* **171**:479–488.
166. Kanellis, P., Romans, A. Y., Johnson, B. J., Kercret, H., Chiovetti, R., Jr., Allen, T. M., and Segretst, J. P., 1980, Studies of synthetic peptide analogs of the amphipathic helix: Effect of charged amino acid residue topography on lipid affinity, *J. Biol. Chem.* **255**:11464–11472.
167. Eisenberg, D., Weiss, R. M., and Terwilliger, T. C., 1982, The helical hydrophobic moment: A measure of the amphiphilicity of a helix, *Nature* **299**:371–374.
168. Schwarze, W., Bernhardt, R., Jänig, G.-R., and Ruckpaul, K., 1983, Fluorescent energy transfer measurements on fluorescein isothiocyanate modified cytochrome P-450_{LM2}, *Biochem. Biophys. Res. Commun.* **113**:353–360.
169. Bernhardt, R., Ngoc Dao, N. T., Stiel, H., Schwarze, W., Friedrich, J., Jänig, G.-R., and Ruckpaul, K., 1983, Modification of cytochrome P-450 with fluorescein isothiocyanate, *Biochim. Biophys. Acta* **745**:140–148.

CHAPTER 7

Oxygen Activation and Transfer

PAUL R. ORTIZ de MONTELLANO

1. Introduction

The cytochrome P-450-catalyzed insertion of an oxygen into a substrate culminates a process that reduces molecular oxygen to a species equivalent, in terms of formal electron count and reactivity, to an oxygen atom. The catalytic sequence for microsomal cytochrome P-450 enzymes, which has been reviewed previously,[1-4] involves the following steps (Fig. 1): (1) binding of a substrate, (2) reduction of the two flavin prosthetic groups of cytochrome P-450 reductase by NADPH, (3) transfer of one of the two electrons thus made available to cytochrome P-450, (4) binding of molecular oxygen to give a ferrous cytochrome P-450–dioxygen complex, (5) transfer of a second electron from cytochrome P-450 reductase, or of an electron from cytochrome b_5, to the complex, (6) cleavage of the oxygen–oxygen bond with concurrent incorporation of the distal oxygen atom into a molecule of water, (7) transfer of the second oxygen atom to the substrate, and (8) dissociation of the product. The catalytic cycle for mitochondrial cytochrome P-450 enzymes differs in that the transfer of electrons from cytochrome P-450 reductase to the hemeprotein is mediated by adrenodoxin, an iron–sulfur protein (see Chapter 12). The steps in the catalytic sequence in which bonds to oxygen are made or broken are addressed in this chapter. The substrate-binding step is discussed in Chapter 3, however, and the electron transfer steps in Chapters 4, 5, 11, and 12.

The ferric prosthetic heme group of P-450 is most readily reduced to the ferrous state when the iron is coordinated to one rather than two axial

---

PAUL R. ORTIZ de MONTELLANO • Department of Pharmaceutical Chemistry, School of Pharmacy, University of California, San Francisco, California 94143.

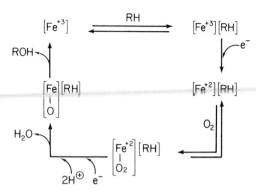

FIGURE 1. The basic catalytic cycle of cytochrome P-450. The prosthetic heme group of the enzyme is represented in this chapter by an iron in brackets or, occasionally, by an iron in a square of nitrogens.

ligands. The shift from the hexacoordinate to the pentacoordinate state that accompanies the binding of a substrate consequently favors reduction of the prosthetic heme group. Reduction of the enzyme to the ferrous state sets the stage for the binding and activation of molecular oxygen. The reduced enzyme, which forms a carbon monoxide complex with an absorption maximum at approximately 450 nm,[5,6] is stable in the absence of oxygen or alternative electron acceptors. Electron transfer to alkyl halocarbons or other reducible substrates can compete with the activation of molecular oxygen when the oxygen tension is low,[7] but reductive reactions constitute a minor sector of the catalytic spectrum of P-450. The binding of oxygen to ferrous P-450, the event that commits the enzyme to the production of some form of activated oxygen, is the starting point for the ensuing discussion.

2. Activation of Oxygen

2.1. Oxygen Binding

The binding of oxygen to ferrous $P-450_{LM_2}$,[8,9] bacterial $P-450_{cam}$,[10] or adrenal $P-450_{scc}$[11,12] yields a dioxygen complex $[Fe^{2+}][O_2]$ characterized by absolute maxima at approximately 420 and 558 nm and difference maxima at 440 and 590 nm. The binding of oxygen is reported, on the basis of stopped-flow studies with substrate-free $P-450_{LM_2}$, to occur in two spectroscopically distinguishable steps.[13] The first step, which occurs within the dead time of the instrument ($k > 60,000$ min^{-1}), yields an intermediate that is converted in a slower step ($k = 210$ min^{-1}) to the ferrous dioxygen species observed in steady-state microsomal studies. The rate constant for the second step is somewhat higher in the presence of a substrate.[13] The only complex observed when the $P-450_{LM_2}$ reaction is monitored at cryogenic temperatures rather than by stopped-flow techniques, however, is the final ferrous dioxygen complex, possibly because the transient intermediate decays to the ferrous dioxygen complex within the 5-sec dead time of the low-temperature instrument.[8] The binding of oxygen to two additional rabbit liver isozymes has been investigated by stopped-flow methods. The ferrous dioxygen complex of P-450$_{LM_4}$, an isozyme primarily in the high-spin state even in the absence of substrates, is formed in a monophasic process with a pseudo-first-order rate constant $k = 58$ sec^{-1}, while that of P-450$_{LM_{3b}}$, an isozyme that like P-450$_{LM_2}$ is primarily in the low-spin state, is completed in a kinetically less well-defined reaction within 100 msec.[14] The binding of oxygen to reduced rat liver microsomes at cryogenic temperatures reportedly involves two kinetic steps.[15] The first step in the activation of molecular oxygen thus occurs in a rapid if kinetically heterogeneous manner. The relevance of spectroscopically observed oxygen binding to normal catalytic turnover is underscored by the finding in single-turnover experiments that the ferrous dioxygen complex decays in parallel with the O-deethylation of 7-ethoxycoumarin.[16]

Limited structural information is available for the P-450$_{cam}$ ferrous dioxygen complex (see Chapter 12), but not for the more unstable dioxygen complexes of the hepatic enzymes. Mössbauer studies of the P-450$_{cam}$ dioxygen complex indicate the iron is diamagnetic and has properties generally similar to those of the iron in oxyhemoglobin.[17,18] The Mössbauer data, the analogy to oxyhemoglobin, and the accumulated information on model systems support a formal [Fe^{3+}][O$_2^-$] structure for the complex in which the iron is hexacoordinate (low-spin).

2.2. Decomposition of the Ferrous Dioxygen Complex

The P-450$_{LM_2}$ ferrous dioxygen complex, which has a half-life of approximately 11 hr at $-30°$C in water–glycerol,[8] autoxidizes cleanly to the ferric enzyme under stopped-flow conditions with a rate constant $k = 12$ min^{-1}.[13] The autoxidation rate reportedly increases from 12 to 90 min^{-1} in the presence of reduced cytochrome b_5 even though b_5 is not significantly oxidized under the conditions of the experiment.[13] Electron transfer from b_5 nevertheless does accelerate enzyme turnover and regeneration of the ferric enzyme.[15,19] In agreement with this, the steady-state microsomal concentration of the ferrous dioxygen complex is decreased by added b_5 but is increased by antibodies to b_5.[20] The ferric P-450 and reduced b_5 concentrations are furthermore inversely proportional to the concentration of the ferrous dioxygen complex.[20,21] The recent finding that the ferrous dioxygen P-450 complex can autoxidize by directly transferring an electron to *oxidized* b_5 provides a new dimension to the mechanisms by which b_5 may accelerate regeneration of ferric P-450.[260]

The steady-state balance between the ferric enzyme and the ferrous dioxygen complex shifts toward the latter when substrates are added due to accelerated reduction of the ferric enzyme caused by the associated change in the iron coordination state.[20] The steady-state concentration of the ferrous dioxygen complex decreases, on the other hand, as the pH[20,21] or ionic strength[20] decreases because the ability of b_5 to provide the second electron is inversely dependent on these parameters. The pH and ionic strength consequently can determine whether electron transfer from b_5 or from cytochrome P-450 reductase controls reduction of the ferrous dioxygen complex, although it remains to be determined whether the interaction of b_5 with all P-450 isozymes depends in a similar manner on pH and ionic strength. It is not known whether the changes in pH and ionic strength alter the reduction potential of the P-450 enzyme, interfere with the docking of P-450 reductase or b_5, or increase the resistance of the molecular conduit which carries the electron to the heme.

The ferrous dioxygen intermediate can autoxidize to the ferric enzyme by loss of superoxide or by acquisition of an electron followed by loss of hydrogen peroxide:

$$[Fe^{2+}][O_2] \leftrightarrow [Fe^{3+}][O_2^-] \rightarrow [Fe^{3+}] + O_2^-$$

$$[Fe^{3+}][O_2^-] + e^- \rightarrow [Fe^{3+}][O_2^{2-}] \xrightarrow{2H^+} [Fe^{3+}] + H_2O_2$$

The NADPH-dependent formation of hydrogen peroxide by liver microsomes[22-25] accounts quantitatively in a reconstituted P-450$_{LM_2}$ system for the discrepancy between the metabolites produced and the NADPH and oxygen consumed.[26] In the absence of a substrate, hydrogen peroxide is generated slowly with a 1:1:1 NADPH:O_2:H_2O_2 stoichiometry. The increase in NADPH and oxygen consumption caused by many substrates is quantitatively balanced by the metabolites that are formed. The reaction stoichiometry with substrates such as benzphetamine, however, is only maintained if allowance is made for the increased production of hydrogen peroxide. Hydrogen peroxide could result from autoxidation of P-450 reductase or b_5 rather than P-450,[27-29] but the fact that hydrogen peroxide formation is stimulated rather than inhibited by P-450 substrates,[24-26] is inhibited by carbon monoxide,[25,30,31] and is increased severalfold as the P-450$_{LM_2}$/P-450 reductase ratio is raised,[32] implicates P-450 as the primary source of the hydrogen peroxide. The formation of hydrogen peroxide does not distinguish between autoxidative loss of superoxide and hydrogen peroxide, however, because hydrogen peroxide is also the end product of superoxide elimination.

The direct quantitation of superoxide, which circumvents ambiguities associated with hydrogen peroxide measurements, indicates that the dissociation of superoxide from P-450 is an important autoxidative mechanism. The formation of superoxide, detected some time ago,[33-35] has been quantitated in hepatic microsomes from phenobarbital-pretreated rats[36,37] and in reconstituted P-450 systems.[32,38] Superoxide formation accounts in these systems for the carbon monoxide-inhibitable production of hydrogen peroxide.[37] The inhibition of superoxide and hydrogen peroxide formation by carbon monoxide is reversed in parallel, as expected for a single P-450-dependent process, when the microsomal incubation mixture is irradiated at 450 nm.[37] The baseline formation of superoxide observed in the absence of substrates, however, is resistant to inhibition by carbon monoxide, a result that suggests this fraction of the superoxide flux is produced by a mechanism independent of P-450 itself. Extrapolation of the superoxide data to zero protein concentration was required before the 1:1 stoichiometry of superoxide to hydrogen peroxide emerged because superoxide is quenched by protein. The failure to correct for superoxide quenching may explain the discrepancies reported by other laboratories between the formation of superoxide and hydrogen peroxide.[32,39] The oxidation of p-nitroanisole to p-nitrophenol and the formation of hydrogen peroxide are furthermore quantitatively and inversely related as the b_5/P-450$_{LM_2}$ ratio is varied in a reconstituted system.[29] The same inverse relationship is observed between superoxide formation and p-nitroanisole metabolism. The quantitative relationship between hydrogen peroxide and superoxide formation and the parallel suppression of both processes by b_5 clearly argue that the P-450$_{LM_2}$ ferrous dioxygen complex autoxidizes by dissociation of superoxide. P-450$_{cam}$ has also been shown to autoxidize by loss of superoxide (Chapter 12).[40]

The results on superoxide formation demonstrate that P-450$_{LM_2}$ autoxidizes by loss of this radical but do not require that other hepatic P-450 isozymes autoxidize by the same mechanism. Hydrogen peroxide formation and substrate turnover, for example, have been suggested to diverge after a second electron reduces the ferrous dioxygen complex to an [Fe^{3+}][O_2^{2-}] species because the binding of hexobarbital to the ferric enzyme correlates well with both hydrogen peroxide formation and hexobarbital hydroxylation.[41,42] This correlation, however, does not require a common intermediate beyond the ferrous dioxygen complex. The observation that the autoxidation of rabbit liver P-450$_{LM_4}$ yields hydrogen peroxide without the detectable formation of superoxide, on the other hand, argues strongly for an autoxidative mechanism that does not depend on elimination of superoxide.[14] The direct formation of hydrogen peroxide in the autoxidation of P-450$_{LM_4}$ requires the acquisition of a second electron from an undetermined source. The difference between the reaction trajectories that lead to cleavage of the dioxygen bond and autoxidative release of hydrogen peroxide, both of which require uptake of a second electron, is obscure. Recent work has shown that the ferrous dioxygen complex of P-450$_{LM_4}$ can autoxidize by electron transfer to oxidized b_5, but this autoxidative process ceases well before both reactants are exhausted. Pompon and Coon suggest that the ferrous dioxygen complex exists in two spectroscopically identical but structurally distinct forms, only one of which can autoxidize by electron transfer to b_5.[43] The existence of two ferrous dioxygen complexes, if in fact this occurs, provides a mechanism for rationalizing the two different consequences of acquiring a second electron. The source of the second electron in the autoxidative process is not known, but one possibility is that P-450$_{LM_4}$ has an oxidizable function in the active site that is oxidized in the process that generates hydrogen peroxide.

The two autoxidative mechanisms, one involving loss of superoxide and the other acquisition of an electron and loss of hydrogen peroxide, correspond in general terms to the two mechanisms proposed for the autoxidation of hemoglobin and myoglobin.[44-46] The dependence of superoxide release on oxygen concentration, and the acceleration caused by protons and nucleophilic anions, has led Caughey et al. to propose that superoxide does not simply dissociate from hemoglobin.[45] They suggest instead that the dissociation of oxygen makes the iron site available for an anionic ligand, and that binding of such a ligand facilitates electron transfer from the ferrous porphyrin to the departing oxygen. The enhanced autoxidation of microsomal[37] and bacterial[47] P-450 under acidic conditions, and of P-450$_{cam}$[48] and phenobarbital-inducible microsomal P-450[49] in the presence of azide and other nucleophilic anions, suggests that similar mechanisms may govern the loss of superoxide from P-450 and hemoglobin, although the detailed mechanism is still not firmly established even for hemoglobin.[50] The direct autoxidative loss of hydrogen peroxide from hemoglobin is particularly interesting in the context of P-450 because it is coupled to the oxidation of phenols and other oxidizable substrates.[44] This observation supports the proposal, if the hemoglobin parallel applies, that the autoxidation of P-450$_{LM_4}$ may involve electron transfer from an oxidizable group in the active site. The hemoglobin data furthermore imply that hydrogen peroxide formation from ferrous dioxygen P-450 complexes may be coupled, in the presence of highly oxidizable substrates, to one-electron oxidation of those substrates, although clear evidence for such a reaction in the P-450 system is not available.

2.3. Oxygen–Oxygen Bond Cleavage

The delivery of a second electron to the ferrous dioxygen complex by P-450 reductase or b_5 (Chapter 4) is required for normal catalytic turnover of P-450. The second electron initiates dioxygen bond cleavage and unmasks the catalytically active species. The details of this oxygen bond cleavage, the structure of the resulting oxygen complex, and the mechanism of the oxygen transfer are probably the least understood molecular aspects of the catalytic sequence because the steps that follow introduction of the second electron occur too rapidly for reaction intermediates to be observed. Our understanding of the final steps of the catalytic sequence therefore rests heavily on studies of the reactions of P-450 with artificial oxygen donors and on studies of less complicated hemeprotein and metalloporphyrin models. Answers to important questions, including how the atom of dioxygen distal to the iron is activated for elimination, whether the oxygen bond cleavage occurs by a homolytic or heterolytic mechanism, and where the oxidation equivalents are located in the catalytic complex once the oxygen–oxygen bond is broken, are only now beginning to emerge.

It appears most likely, but has not yet been clearly demonstrated, that a single oxygen atom is bound to the prosthetic heme iron in the catalytically active species. The electron inventory for such a complex indicates the oxygen bears six valence electrons if the iron is held in the ferric state. No metalloporphyrin or hemeprotein complex is known, however, that actually has this specific electron distribution. Formally equivalent complexes in which two electrons are transferred from the iron to the oxygen to give an $[Fe^{5+}][O^{2-}]$ species, or in which transfer of one electron yields an $[Fe^{4+}][O^{-}]$ complex with seven electrons on the oxygen, are also unknown. The two-electron oxidized hemeproteins that have been characterized are characterized by a tetravalent iron and a full electron octet on oxygen. The electron required to complete the oxygen octet is drawn from the porphyrin ring (Fig. 2A) or from an active-site amino acid residue (Fig. 2B). A porphyrin radical cation is therefore present in the former and a protein radical in the latter. The oxoiron complexes (Compounds I) generated when horseradish peroxidase and catalase react with hydrogen peroxide are examples of the former electron distribution, although the electron may be removed from a different porphyrin orbital in the two enzymes.[51–54] The porphyrin radical cation is also observed in model metalloporphyrin complexes (Chapter 1). Compound I of cytochrome c peroxidase, on the other hand, is an example of a structure with a protein-centered radical.[55] Compound I of lactoperoxidase appears to be a bridging example in which a porphyrin radical cation is in reversible equilibrium with a protein radical.[56,57] The hydrogen peroxide-mediated oxidation of hemoglobin or myoglobin, a final instructive example, yields an $[Fe^{4+}][O^{2-}]$ complex with a transient protein radical that is rapidly dissipated.[58,59] The electron distribution for a metalloporphyrin or hemeprotein complex with a thiolate ligand, however,

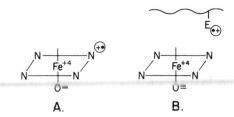

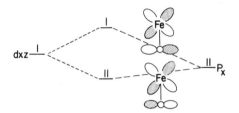

FIGURE 2. The known electron distributions for "Compound I" of the peroxidases and the d_{xz}–P_x orbital interaction that allows transfer of unpaired electron density from the iron to the oxygen.

has never been defined. The electron donor properties of a thiolate ligand differ markedly from those of an imidazole or phenoxide, a fact that may radically alter the electronic structure of the catalytic complex. Nevertheless, if the precedent offered by hemeprotein and metalloporphyrin models is valid, it is likely that significant unpaired electron density will be channeled into the p_x and p_y orbitals of the $[Fe^{4+}][O^{2-}]$ intermediate. As suggested by Groves and Nemo[60] (Chapter 1), low-spin Fe^{4+} has one electron in the d_{xz} and one in the d_{yz} orbitals. These orbitals have the proper symmetry to mix with the filled oxygen p_x and p_y orbitals (Fig. 2). The transfer of unpaired electron density to the oxygen, which determines its propensity to enter into radical reactions, is governed by the degree of orbital mixing and the extent to which unpaired electron density is localized on the oxygen in the resulting hybrid orbitals. The limiting structure for such a hybrid is one in which the iron is in the ferric state and the oxygen bears seven electrons.

The catalytic oxygen complex is normally generated by reductive cleavage of the dioxygen bond in the ferrous dioxygen complex, but the mechanism by which the enzyme facilitates this bond cleavage is not understood. Hamilton proposed in a theoretical discussion that the distal oxygen in the ferrous dioxygen complex may be acylated to make it a better leaving group prior to rupture of the dioxygen-bond.[61] The *chemical* validity of this proposal is confirmed by the demonstration in model studies (Chapter 1) that acylation of the dioxygen complex of manganese[62] and ferric[63] porphyrins facilitates formation of the corresponding oxomanganese and oxoiron species. The finding that an atom of labeled ox-

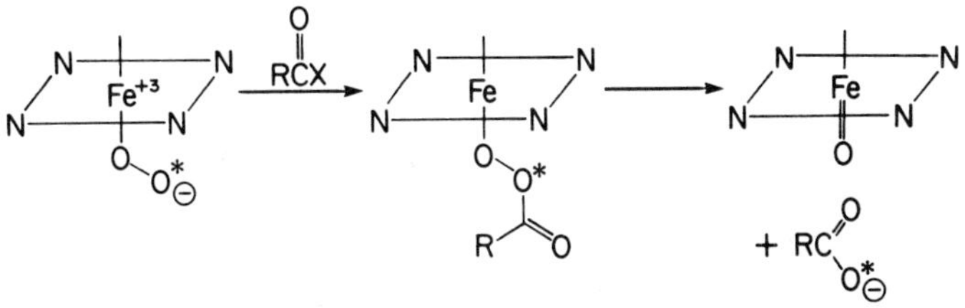

FIGURE 3. Mechanism proposed for the incorporation of labeled molecular oxygen into lipoic acid (RCOX) in lipoic acid-dependent turnover of P-450$_{cam}$.

ygen is stoichiometrically incorporated into the carboxyl group of lipoic acid when purified, photochemically reduced, P-450$_{cam}$ undergoes a single turnover in the presence of camphor, dihydrolipoic acid, and $^{18}O_2$ (Fig. 3), is consistent, in fact, with the operation of such a mechanism (Chapter 12).[64] The second electron required for oxygen activation in this model system is presumably provided by the sulfhydryl groups of dihydrolipoic acid. The catalytic turnover of P-450$_{cam}$ under these conditions is 1000 times slower than the reaction promoted by putidaredoxin, however, even if dihydrolipoic acid is present in high concentration.[47] The parallel incorporation of ^{18}O into lipoic acid and hydroxycamphor thus clearly demonstrates that acylation is coupled to oxygen activation in the model system, but does not establish that acylation occurs under normal turnover conditions. Indeed, it appears likely that the acylation elegantly demonstrated in the lipoic acid experiment is an artifact of the model system without relevance to the normal function of P-450. First of all, P-450$_{cam}$ does not require lipoic acid or any other carboxylic cofactor when the normal electron donor is present. Second, no experimental evidence exists for acylation of oxygen in any other P-450 system (or in any other hemeprotein), although the mechanistic ramifications of an acylation requirement have been discussed.[1] Third, and most important, no acylating function appears to be present in the active site of substrate-bound ferric P-450$_{cam}$ (Chapter 13). Finally, the fact that the dioxygen bond cleavage catalyzed by peroxidases and metalloporphyrins does not require oxygen acylation[65–68] makes recourse to such a cumbersome mechanism, in the absence of compelling evidence, unnecessary.

The finding that the reaction of superoxide with ferrous porphyrins and the electrochemical reduction of ferroporphyrin dioxygen complexes yield cyclic iron dioxygen complexes (Fig. 4) suggests the possible involvement of such a species in dioxygen bond cleavage.[69,70] Activation of oxygen through such a cyclic peroxide mechanism is unlikely, however, because structurally related model cycloperoxides fail to oxidize substrates as reactive as styrene.[71] Unless the thiolate ligand of P-450 radi-

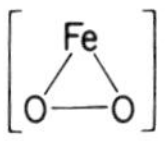

$$\begin{bmatrix} & \text{Fe} & \\ \text{O} & \!\!\!-\!\!\! & \text{O} \end{bmatrix}$$

FIGURE 4. The cyclic peroxoiron structure.

cally alters their chemistry, the chemical inertness of iron cycloperoxides precludes their involvement in P-450 catalysis.

The deployment of charged groups analogous to those that facilitate dioxygen bond cleavage in the peroxidases[65–68] could exert similar catalytic leverage in P-450, although their presence would require solution of logistic problems associated with preferential binding of lipophilic substrates in the active-site region also occupied by the oxygen activation machinery. The functionalities required for peroxidaselike activation of oxygen do not appear, in any case, to be present in the active site of ferric P-450$_{cam}$ (Chapter 13). In the absence of a gross structural rearrangement when P-450$_{cam}$ is reduced, dioxygen bond cleavage must occur in the absence of catalytic assistance by charged protein residues. The mechanism by which the dioxygen bond is cleaved thus remains unresolved. It is possible that a key role of the thiolate ligand in P-450 is to facilitate dioxygen bond cleavage in a lipophilic, electrostatically neutral, environment. The potential influence of the *trans*-axial ligand on dioxygen bond cleavage is evident in the shift from homolytic to heterolytic cleavage of alkyl hydroperoxides by metallotetraphenylporphyrin complexes when imidazole is made available as a *trans*-axial ligand.[72] The stronger electron donor properties of a thiolate ligand might similarly facilitate the more difficult heterolytic cleavage of dioxygen itself. Homolysis of the dioxygen bond to give a hydroxyl radical and a "Compound II" species cannot be completely ruled out but the peroxidase activity associated with homolytic cleavage of alkylperoxides by P-450 (see Section 2.4) argues against such a mechanism. A homolytic mechanism would require participation of both the hydroxyl radical and the iron-bound oxygen in the catalytic reaction.[1]

2.4. Peroxides as Oxygen Donors

The catalytic choreography of three membrane-bound proteins and two cosubstrates (NADPH and molecular oxygen) in normal P-450 function obscures the molecular details of the catalytic events. The discovery that hydroperoxides and other artificial oxygen donors support catalytic turnover of P-450 in the absence of electron transfer proteins or cosubstrates has therefore been exploited in mechanistic studies. It is now evident, however, that inferences drawn from work with peroxides must be carefully evaluated to determine their relevance to the normal function of the enzyme.

The decomposition of linoleic acid hydroperoxide by trypsin-treated hepatic microsomes (hemeproteins other than P-450 are removed by this treatment) is inhibited by cyanide or alternative substrates for the enzyme, and is elevated by phenobarbital pretreatment, to the same extent as the P-450 catalytic activity.[73,74] The decomposition of linoleic acid hydroperoxide by P-450 is coupled by a classic peroxidatic mechanism to the oxidation of N,N,N',N'-tetramethyl-p-phenylenediamine,[73] NADPH,[75] diaminobenzidine,[76] alcohols,[77] phenols,[76] and other peroxidase substrates. P-450 similarly catalyzes the peroxidatic decomposition of cumene hydroperoxide,[73] several regioisomers of cholesterol hydroperoxide,[74] pregnenolone 17α-hydroperoxide,[74] progesterone 17α-hydroperoxide,[74] and even, albeit to a smaller extent, hydrogen peroxide.[73] The peroxidase activity of P-450 exceeds that of other hemeproteins with most hydroperoxides,[74] although similar reactions are catalyzed by hematin and other hemeproteins.[73,76,78] A role for this activity in the hydroperoxide-dependent peroxidation of microsomal lipids[79] is suggested by the fact that reduced pyridine nucleotides attenuate lipid peroxidation, that the inhibition is enhanced by substrates that promote the reduction of P-450, and that carbon monoxide inhibits the peroxidative process.[80,81]

The mechanistic interest in reactions of P-450 with peroxides, however, stems from the occurrence of peroxygenative reactions in which an oxygen of the peroxide is incorporated into the substrate[82] rather than from the coupling of peroxide reduction to one-electron oxidation of electron donors. The cumene hydroperoxide-supported microsomal N-demethylations of dimethylaniline, aminopyrine, benzphetamine, propoxyphene, ethylmorphine, and methamphetamine are among the reactions of this type that have been reported.[83] The role of P-450 in these N-demethylation reactions is confirmed by the demonstration that N-methylaniline, formaldehyde, and cumyl alcohol are formed in stoichiometric amounts in the reaction of reconstituted P-450$_{LM2}$ with cumene hydroperoxide and N,N-dimethylaniline,[84] Hydrogen peroxide supports N-demethylation reactions much less effectively than alkyl hydroperoxides although aniline, ethanol, and benzene reportedly are oxidized by P-450$_{LM2}$ by mechanisms that depend, at least in part, on the *in situ* generation of hydrogen peroxide.[32,85,86] In retrospect, it is probable that the N-demethylation reactions stem, in part, from a peroxidative rather than peroxygenative process.[87,88] More definitive evidence for peroxygenative reactions is provided by the hydroperoxide-dependent O-dealkylation of alkyl ethers,[89,90] the hydroxylation of aromatic substrates,[90,91] and the hydroxylation of unactivated hydrocarbons.[92–96] Lauric acid,[93,94] androstenedione, testosterone, progesterone, 17β-estradiol, and 5β-cholestane-3α,7α,12α-triol[92,94–96] are among the hydrocarbons shown to be hydroxylated by a hydroperoxide-dependent mechanism. The hydroxylation of aromatic substrates, first demonstrated for biphenyl, benzpyrene, coumarin, and aniline,[90,91] occurs with the NIH shift characteristic of the NADPH-dependent reaction.[97] The oxygen incorporated into the epoxide of phenanthrene when the hydrocarbon is oxidized by hepatic microsomes and cumene hydroperoxide was shown to derive from a source other than the medium, although it was not specifically shown to derive from the peroxide rather than from molecular oxygen.[97] The cumene hydroperoxide-dependent hydroxylation of cyclohexane by reconstituted P-450$_{LM_2}$ similarly incorporates an oxygen from a source other than the medium.[84] Hydroperoxides do not support the turnover of all P-450 isozymes with equal facility. Differences in the regiochemistry of warfarin metabolism are observed when cumene hydroperoxide is substituted for NADPH and oxygen.[98] Cumene hydroperoxide furthermore supports ω-1- but not ω-hydroxylation of lauric acid,[93] and does not support the aromatization of sterols by placental P-450.[99]

The peroxygenase activities of rabbit liver microsomes[100] and reconstituted P-450$_{LM_2}$[101–103] have been analyzed by kinetic methods. An "ordered bi bi" kinetic sequence in which the substrate binds before the peroxide and the catalytic event occurs in a ternary enzyme/substrate/peroxide complex is suggested by (1) the observation that the O-demethylation of p-nitroanisole by rabbit liver microsomes and t-butylhydroperoxide gives rise to convergent double reciprocal plots when the substrate or peroxide concentration is varied and (2) the finding that cyanide inhibits the reaction competitively with respect to the peroxide but noncompetitively with respect to the substrate.[100] The same conclusion has been reached in a kinetic analysis of the hydroperoxide-dependent hydroxylation of toluene by reconstituted P-450$_{LM_2}$.[101–103] Two reversibly formed ternary complexes have, in fact, been observed with the purified enzyme. The first of these (C) is directly related to the hydroxylation event but has a low affinity for the peroxide and is difficult to detect, whereas the second (D) is readily detected but is a dead-end complex:

$$[\text{enzyme–substrate}] + \text{peroxide} \rightleftarrows C \rightleftarrows D$$

The difference spectrum of complex D, with a trough at 416 nm and a maximum at approximately 436 nm, is essentially identical to that observed when cumene hydroperoxide is added to hepatic microsomes from phenobarbital-pretreated rabbits.[104] An ill-defined but similar difference spectrum is seen when cumene hydroperoxide[95] or 20-hydroperoxycholesterol[105] is added to adrenal microsomes, but these spectra are rapidly replaced by reverse type I spectra due to the binding of metabolically formed cumenol or hydroxysterol to the ferric enzyme. Two intermediates can also be detected spectroscopically in the reaction of P-450$_{cam}$ with m-chloroperbenzoic acid[106] but the Soret band of the more

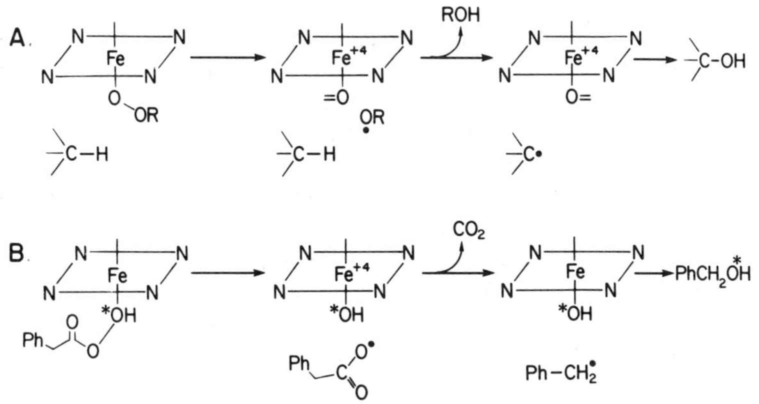

FIGURE 5. (A) Hypothetical mechanism for P-450-catalyzed hydrocarbon hydroxylations in which the first step is hydrogen abstraction by a homolytically generated oxygen radical and (B) the homolytic mechanism proposed to explain the decarboxylation of peroxyacids by P-450.

stable intermediate is at 405 nm rather than, as in complex D, at 425 nm. The Soret band in P-450$_{cam}$ is thus blue-shifted whereas that of P-450$_{LM_2}$ is red-shifted with respect to the 417-nm Soret band of the resting enzyme. The magnitude and direction of the P-450$_{LM_2}$ shift are comparable to those incurred in the conversion of ferric horseradish peroxidase to Compound II.[67] In contrast to the horseradish peroxidase system, however, the magnitude and position of the Soret band of complexes C and D depend on the peroxide used to generate the complex. The implication that the organic moiety of the peracid is retained in the complexes is supported by the observation that the rate of conversion of complex C to complex D depends directly on the electron-withdrawing ability of substituents on the peracid, that the rate of substrate hydroxylation is limited by the decay of complex C, and that the hydroxylation rate depends on the structure of both the substrate and the peracid.

The reversible formation of ternary complexes C and D is incompatible with a classical peroxidase mechanism in which irreversible formation of Compound I is followed by oxidation of the substrate in a reaction unaffected by the oxygen donor. Blake and Coon suggest that homolytic dioxygen bond cleavage produces an [Fe^{4+}][O^{2-}] species and an alkoxy radical and that these two species cooperate in the hydroxylation reaction (Fig. 5A). Support for such a homolytic mechanism is provided by the finding that P-450$_{LM_2}$ catalyzes the conversion of 2-phenylperacetic acid to benzyl alcohol, that the alcohol oxygen derives from the peracid, and that cyclohexane is concurrently hydroxylated if it is present.[107] Homolytic oxygen transfer to the iron, decarboxylation, and return of the iron-bound oxygen to the resulting benzyl radical, a reaction pathway with precedent in the hematin-catalyzed conversion of unsaturated lipid hydroperoxides to rearranged epoxy alcohols,[108,109] are suggested by the results (Fig. 9D). The data, however, do not require hydroxylation of cyclohexane by the species that is responsible for benzyl alcohol formation. Hydrogen abstraction from cyclohexane by the benzyl radical (or its carboxy radical precursor) is implied by the homolytic hydroxylation mechanism, but toluene, the product expected from such a reaction, was not detected. Perdeuteration of cyclohexane furthermore reduces cyclohexanol formation without altering benzyl alcohol formation. The implication that the decarboxylation and hydroxylation reactions occur by concurrent but independent mechanisms is strongly supported by differences in the kinetic parameters of the two reactions and by their differential susceptibility to inhibitors.[110] These differences have led to the conclusion that the decarboxylation and hydroxylation reactions involve "no common intermediate beyond the ferric resting state of the enzyme."[110]

The parallel operation of peroxidative and peroxygenative pathways readily rationalizes most of the differences between the metabolite profiles obtained with hydroperoxides and NADPH. The metabolism of benzo[a]pyrene to phenols with NADPH but quinones with cumene hydroperoxide,[111] and of chloroaniline to (4-chlorophenyl)-hydroxylamine with NADPH but 4-chloro-1-nitrobenzene with cumene hydroperoxide,[112] can be explained by secondary peroxidative oxidation of the initial phenol and hydroxylamine metabolites. The observation by EPR of nitrogen radicals in the hydroperoxide- but not NADPH-dependent N-demethylation of aminopyrine,[87,88,113] and of the nitroxide radical in the hydroperoxide- but not NADPH-supported oxidation of N-hydroxy-2-acetylaminofluorene,[114-116] is consistent with this explanation. Differences nevertheless exist that cannot be explained by the simple superposition of peroxidative and peroxygenative processes. The different ratio of primary, secondary, and tertiary methylcyclohexanols produced in the P-450$_{LM_2}$-catalyzed oxidation of methyl cyclohexane supported by cumene hydroperoxide and NADPH,[117] for example, probably results from distortion of the active site in the catalytically active substrate/peroxide/enzyme complex[101-103] because hydrocarbon hydroxylations are not characteristic of the peroxidative pathway.

Transient EPR signals are observed at approximately 2.01 g in incubations of hepatic microsomes with cumene hydroperoxide,[104] and of P-450$_{cam}$ with m-chloroperbenzoic acid,[118] that resemble those observed in the reactions of metmyoglobin with hydrogen peroxide or ethyl hydroperoxide.[119,120] The nature of the radicals in the P-450 systems remains unknown but their exclusive observation in peroxide-dependent turnover suggests they are probably a peroxidative artifact.

Hydrogen peroxide, which is closer in size and structure than the alkyl peroxides to the normal dioxygen species, supports the peroxygenase activity of P-450 less effectively than cumene hydroperoxide[83,84,92,113,117,121–124] and supports the peroxidase activity very poorly.[73] The product distribution for the hydrogen peroxide-supported oxidation of aniline[121] and benzo[a]pyrene[122] resembles that of the NADPH- rather than cumene hydroperoxide-supported process. The P-450$_{LM_2}$-catalyzed N-demethylation of aminopyrine supported by hydrogen peroxide furthermore proceeds,[113] as does the NADPH- but not cumene hydroperoxide-dependent reaction,[85,86] without the detectable formation of aminopyrine radicals. Hydrogen peroxide supports the hydroxylation of prostaglandins at different positions by liver microsomes from differentially induced rats.[124] contradicting the suggestion, based on work with purified rabbit P-450$_{LM_2}$, that carbon hydroxylations are not supported by this oxidant.[117] The discrepancies between the hydrogen peroxide- and NADPH-dependent N-demethylation rates for 15 secondary and tertiary amines,[123] and the differences in the products obtained with NADPH and H_2O_2 in other instances (e.g., prostaglandin hydroxylation),[124] indicate that differences exist in the abilities of different isozymes to turn over with H_2O_2. The low activity with H_2O_2 of the isozyme that ω-hydroxylates prostaglandins is particularly to be noted. It is not possible at this time to determine if the oxidizing species produced with NADPH differs significantly from that obtained with H_2O_2, but the fact that the prosthetic heme group is efficiently degraded with H_2O_2 but not NADPH (see Section 2.6) indicates that the two processes differ in some essential manner. The proposal that hydroxyl radicals are produced from autoxidatively generated superoxide and hydrogen peroxide[32,86] may be related to the differences in both the reaction regiochemistry and the stability of the enzyme.

2.5. Iodosobenzene and Other Single Oxygen Donors

The delivery of a single oxygen atom in the appropriate oxidation state simplifies the catalytic system even further and circumvents peroxidative processes associated with homolytic peroxide scission. The initial substitution of $NaIO_4$ and $NaClO_2$ for NADPH in the hydroxylation of steroids[92,94–96,125] and fatty acids[126] has given way to the use of iodosobenzene,[125–128] a lipophilic agent that more readily interacts with the membrane-bound enzymes. Iodosobenzene supports the catalytic turnover of P-450 in the absence of NADPH or molecular oxygen but the regioselectivities of the reactions supported by iodosobenzene and NADPH are not identical.[117,125,128] These differences could result from oxygen transfer in a ternary complex analogous to that implicated in the hydroperoxide-dependent reactions but could also result, in microsomal incubations, from preferential interaction of the oxygen donor with a subset of the P-450 isozymes.

The strong parallels between the iodosobenzene- and NADPH-dependent reactions of P-450, and the iodosobenzene-supported model reactions of metalloporphyrins (Chapter 1), provide the experimental basis for the widely accepted thesis that the catalytically active species is an iron-coordinated oxygen atom. The observation that labeled oxygen from $H_2^{18}O$ is incorporated in high yield into the products obtained from camphor,[129] cyclohexane,[130] and parathion[131] in the iodosobenzene-supported reaction is inconsistent, however, with the formation of a catalytic species identical to that obtained with NADPH if oxygen exchange occurs subsequent to formation of the catalytic oxoiron complex. The proposal that oxygen exchange occurs after the oxoiron complex is formed rests on the facts that (1) the oxygen of iodosobenzene is stable to uncatalyzed exchange[132] and only exchanges with the medium in the presence of intact microsomes,[130] (2) the oxygen in model oxomanganese and oxochromium porphyrin complexes exchanges with oxygen from the medium,[133,134] and (3) NMR proton relaxation studies suggest the active site of the *ferric* enzyme is accessible to water.[135] The exchange observed with iodosobenzene, however, contrasts sharply with the finding that oxygen from the medium is not incorporated into the products during (1) the NADPH-dependent microsomal hydroxylation of several sterols,[136,137] (2) the NADPH- or hydroperoxide-supported hydroxylation of camphor by P-450$_{cam}$,[129] (3) the NADPH- or hydroperoxide-dependent sulfoxidation of *p*-methylthioanisole by P-450$_{LM_2}$,[138,139] (4) the microsomal oxidation of a number of sulfur compounds with either NADPH or cumene hydroperoxide,[131] or, except for a trace, (5) the hydroxylation of cyclohexane catalyzed by reconstituted P-450$_{LM_2}$ and cumene hydroperoxide.[140] This mechanistic dichotomy can be rationalized in three ways: (1) the same oxoiron species is formed with iodosobenzene as with NADPH but iodosobenzene alters the enzyme structure in a manner that facilitates ox-

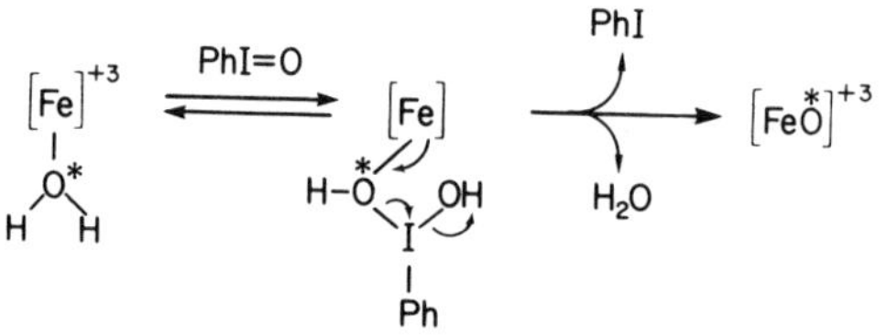

FIGURE 6. A mechanism for the introduction of an oxygen from the medium during iodosobenzene-supported P-450-catalyzed reactions.

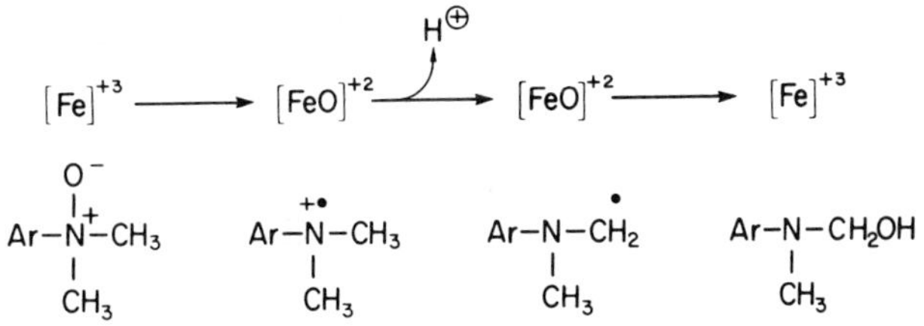

$$[Fe]^{+3} \longrightarrow [FeO]^{+2} \xrightarrow{H^{\oplus}} [FeO]^{+2} \longrightarrow [Fe]^{+3}$$

$$\underset{\underset{CH_3}{|}}{Ar-\overset{\overset{O^-}{|+}}{N}-CH_3} \qquad \underset{\underset{CH_3}{|}}{Ar-\overset{+\bullet}{N}-CH_3} \qquad \underset{\underset{CH_3}{|}}{Ar-\overset{\bullet}{N}-CH_2} \qquad \underset{\underset{CH_3}{|}}{Ar-N-CH_2OH}$$

FIGURE 7. Mechanism for the P-450-catalyzed N-dealkylation of dimethylaniline-N-oxide involving a one- but not a two-electron-oxidized state of the enzyme.

ygen exchange, (2) the oxidizing species obtained with iodosobenzene differs from that obtained with NADPH, or (3) the oxygen of *iodosobenzene* is exchanged in a reaction catalyzed by P-450 prior to oxygen transfer to the iron. The latter explanation is the simplest and most attractive. The catalytic involvement of P-450 and the high incorporation (> 98%) of oxygen from the medium into metabolites are readily explained if water coordinated to the iron adds to iodosobenzene to yield a trisubstituted iodine species that dissociates from the enzyme, resulting in exchange of the oxygen in iodosobenzene, or decomposes to the catalytically active species (Fig. 6).* The sixth ligand in P-450 is now believed to be an oxygen, possibly a water molecule, as required by such a mechanism. Similar mechanisms can be invoked to rationalize the finding that the tosylimine analogue of iodosobenzene ($MeC_6H_5SO_2N=IC_6H_5$) supports the P-450$_{LM_2}$-catalyzed hydroxylation rather than tosylamination of cyclohexane.[141]

Aryldimethylamine N-oxides also serve as single oxygen donors but their utility is compromised by the fact that, so far, they have only been found to support reactions in which the organic framework of the aryl N-oxide is itself the substrate. The N-oxide of N,N-dimethylaniline is thus converted to N-methylaniline by P-450$_{cam}$ but camphor is not simultaneously hydroxylated.[142] The N-oxide may serve as a donor of oxygen with seven rather than six electrons. Proton loss from the resulting nitrogen radical cation and collapse of the carbon radical with the Fe^{IV}–OH species could then yield N-dealkylated products without allowing alternative substrates to compete effectively for the catalytic species (Fig. 7).

2.6. Substrate-Independent Reactions

P-450 consumes one NADPH and one oxygen in each monooxygenative event, but this stoichiometry only holds empirically if allowance is

---

* I thank Professor John Groves for the suggestion that an iron-coordinated water molecule might be involved in the oxygen exchange observed with iodosobenzene.

made for oxygen and NADPH consumed in the autoxidative formation of hydrogen peroxide[26]:

$$O_2 + NADPH + H^+ + RH \longrightarrow H_2O + NADP + ROH$$
$$O_2 + NADPH + H^+ \longrightarrow H_2O_2 + NADP$$

Oxygen and NADPH are consumed in a 1:1 ratio in both of these reactions. In contrast, NADPH and oxygen are consumed in a 2:1 ratio in incubations of rabbit liver microsomes with perfluorohexane, a compound that binds to P-450 but is metabolically inert.[143] The reaction stoichiometry suggests the four-electron reduction of molecular oxygen to two molecules of water:

$$O_2 + 2\ NADPH + 2\ H^+ \rightarrow 2\ H_2O + 2NADP$$

The discrepancy between the O_2 and NADPH consumed and the H_2O_2 produced by rat liver microsomes in the absence of a substrate likewise is consistent with reduction of a fraction of the oxygen to two molecules of water.[39] The four-electron reduction of molecular oxygen is confirmed by a study with reconstituted P-450$_{LM_2}$ that also established that H_2O_2 is not a detectable intermediate in the reduction of oxygen to water.[144] It is probable, but unproven, that the oxoiron complex is reduced, in the absence of a hydroxylatable substrate, by electrons provided by P-450 reductase or b_5:

$$[Fe{=}O]^{3+} + 2H^+ \xrightarrow{2e^-} [Fe^{3+}] + H_2O$$

The prosthetic heme of P-450 is rapidly destroyed *in vitro* by H_2O_2,[145] alkyl hydroperoxides,[73,74,84,96,146] and iodosobenzene.[127,142] The heme is degraded primarily to hematinic acid and methylvinylmaleimide, although propentdyopents are formed as minor products.[147,148] The mechanism of this heme degradation is not known but apparently does not involve the formation of biliverdin.[148] The possibility that P-450 is oxidatively destroyed under physiological conditions is suggested by the finding that the 11β-hydroxylase of cultured adrenocortical cells, and the 17α-hydroxylase/C$_{17-20}$ lyase of cultured Leydig cells, are degraded by oxygen-dependent, antioxidant-inhibitable, processes.[149-151] The homolytic formation of radicals from peroxides may be responsible for the heme destruction caused by these agents, but the analogous destructive activity of iodosobenzene must be reconciled with evidence that it produces an oxoiron species similar or identical to that obtained with NADPH. The absence of significant enzyme self-degradation under normal turnover conditions suggests that the catalytic oxygen complex is not inherently destructive, although its reactivity could be masked under normal conditions if the enzyme is protected by the requirement that a substrate is bound prior to oxygen activation, or by NADPH-dependent reduction of activated oxygen to water in the absence of a receptive substrate. The observation of low-level chemiluminescence when P-450 reacts with iodosobenzene, alkylperoxides, or hydrogen peroxide provides one possible mechanism for heme degradation if the catalytic oxygen is not directly responsible for the reaction. The chemiluminescence is believed to result from the formation of singlet oxygen when the oxoiron complex reacts with the oxygen donor (Ph = phenyl)[152]:

$$[FeO]^{3+} + PhI{=}O \rightarrow [Fe{-}O{-}O{-}IPh] \rightarrow [Fe]^{3+} + O_2 + PhI$$

The formation of singlet oxygen in reactions that require relatively elevated concentrations of alternative oxygen donors is compatible with the stability of the enzyme under normal physiological conditions.

3. Mechanisms of Oxygen Transfer Reactions

3.1. Introduction

The oxygen activated by P-450 enzymes, as implied by their monooxygenative role is normally transferred to a substrate. The structurally related oxygen complex generated by the catalytic turnover of peroxidases is cleanly differentiated from that of P-450 because it functions as an electron sink rather than as an oxygen donor. This functional differentiation of P-450 from the peroxidases does not, however, preclude mechanisms for P-450 in which the two oxidation equivalents are committed individually rather than simultaneously, or in which electron transfer precedes oxygen transfer. The difference between a concerted mechanism with only one transition state and a nonconcerted mechanism with two or more transition states is illustrated schematically for the hydroxylation of a hydrocarbon in Fig. 8. A growing body of evidence suggests, in fact, that the transfer of oxygen from P-450 to a substrate is mediated, at least in some instances, by nonconcerted mechanisms. The evidence relevant to the concerted or nonconcerted nature of the principal reactions catalyzed by P-450 is addressed in this section.

3.2. Hydrocarbon Hydroxylation

The hydroxylation of a hydrocarbon generally occurs with retention of stereochemistry, as reported for the 7α- and 11α-hydroxylation of ste-

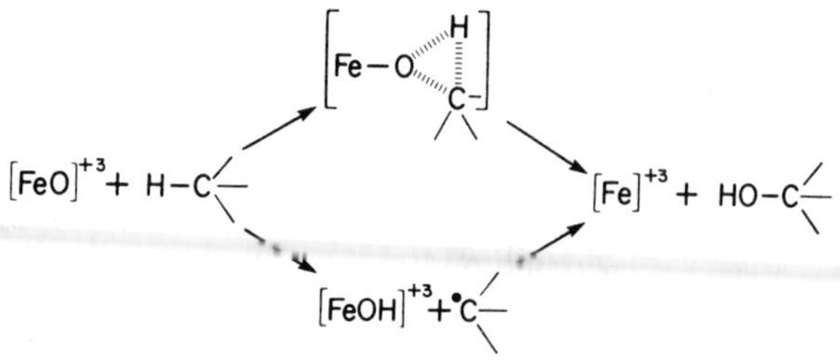

FIGURE 8. Schematic illustration of concerted versus nonconcerted hydrocarbon hydroxylation mechanisms.

roids,[153,154] the benzylic hydroxylation of ethyl benzene,[155] the ω- and ω-1-hydroxylation of lauric acid,[156] and the terminal hydroxylation of (1R) and (1S)-[1-^{3}H,^{2}H,^{1}H]octane.[157] This retention of stereochemistry, and the kinetic (rather than intrinsic) isotope effects obtained in early work (see later discussion), constitute the experimental support for the concerted insertion of oxygen into C–H bonds. Retention of stereochemistry is a valid measure of the concertedness of a reaction only if it can be assumed that stereochemical fidelity is not imposed by the active site. The *loss* of stereochemistry, in contrast, absolutely requires the intervention of an intermediate and therefore reliably implicates a nonconcerted mechanism. The scrambling of stereochemistry and regiochemistry observed in a number of hydroxylation reactions thus unambiguously points to a nonconcerted mechanism for at least those reactions. Stereochemical scrambling was first observed in the hydroxylation of *exo*-tetradeuterated norbornane by rabbit liver microsomes, which yields, among other products, the *endo*-alcohol with three rather than four deuterium atoms and the *exo*-alcohol with four deuterium atoms (Fig. 9).[158]

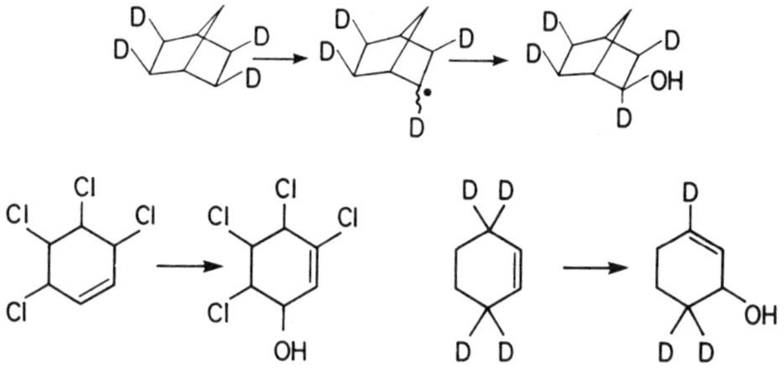

FIGURE 9. Hydrocarbon hydroxylations that require an intermediate and therefore a nonconcerted mechanism.

A second example of stereochemical scrambling is provided by the observation that P-450$_{cam}$ removes the *endo* or *exo* hydrogen from camphor but only delivers the hydroxyl to the *exo* position.[159] It appears from a recent abstract that even the hydroxylation of ethylbenzene, a reaction long considered to proceed with retention of configuration, is not stereospecific.[160] The allylic rearrangement of double bonds, the only type of regiochemical scrambling so far known, has been demonstrated to occur during the hydroxylation of 3,4,5,6-tetrachlorocyclohexene and related compounds by rat or housefly microsomes,[161] and of 3,3,6,6,-tetradeuterated cyclohexene, methylenecyclohexane, and β-pinene by P-450$_{LM_2}$ (Fig. 9).[162] This transposition of functionality requires the formation of an allylically delocalized (presumably free radical) intermediate but does not reveal whether the intermediate is generated by hydrogen abstraction (a hydroxylation reaction) or sequential loss of a π-bond electron and a proton (an aborted epoxidation). Epoxide metabolites accompany the allylically hydroxylated products except in the case of the tetrachlorinated cyclohexene.[161,162] The substantial isotope effects (k_H/k_D = 4-6) found for hydroxylation of the allylically deuterated substrates, however, are more consistent with rearrangement of the double bond during a hydroxylation rather than epoxidation process. Efforts to confirm the formation of a carbon radical intermediate by attaching a cyclopropyl ring to the hydroxylated carbon, an approach based on the rapid rearrangement of cyclopropylmethyl to 3-butenyl radicals (k = 1.3 × 10^8 sec^{-1}, 25°C),[163] have been unsuccessful.[164,165] The utility of the cyclopropyl ring as a mechanistic probe in the substrates that have been examined is compromised by its incorporation into a larger ring structure, however, because it is known that cyclopropylmethyl radicals in such structures open more slowly and sometimes are even more stable in the unopened form.[166]

The kinetic isotope effects for P-450-catalyzed hydrocarbon hydroxylations calculated from the rates of product formation from deuterated and undeuterated substrates are generally small. These small isotope effects provided early support for a concerted hydroxylation mechanism because the bent transition state required by such a process, as illustrated by the isotope effects for carbene insertions into C–H bonds (k_H/k_D = 0.9-2.5),[167,168] are theoretically predicted to involve small isotope effects.[169] The isotope effects calculated from the rates of metabolite formation, however, are inaccurate measures of the isotope effects intrinsic to the hydroxylation step if the catalytic rate is significantly determined by enzymatic steps other than the hydroxylation itself. The intrinsic isotope effects determined from the ratio of oxygen insertion into C–H and C–D bonds in otherwise identical sites in a single molecule, which avoids ambiguities due to multiple rate-limiting steps, are in fact quite large: (1) k_H/k_D = 11.5 for the hydroxylation of tetradeuterated norbornane (Fig.

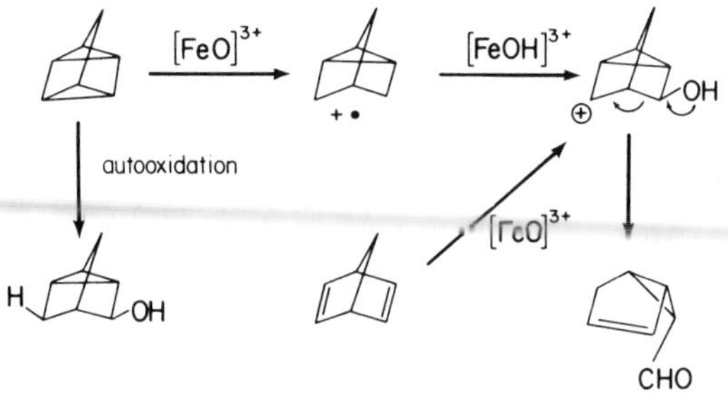

FIGURE 10. Electron transfer in the P-450-catalyzed oxidation of quadricyclane.

9),[158] (2) $k_H/k_D = 11$ for the benzylic hydroxylation of [1,1-^{2}H]-1,3-diphenylpropane,[170] and (3) $k_H/k_D = 10$ for the O-demethylation of p-trideuteromethoxy-anisole.[171] The intrinsic isotope effect for the O-dealkylation of 7-ethoxycoumarin, calculated from measurements of the deuterium and tritium isotope effects on V_{max}, yields an equally high value of $k_H/k_D = 13.5$.[172] It is consequently clear that the isotope effects for hydroxylation reactions are large but are usually masked in kinetic terms by other rate-determining steps. Isotope effects greater than 5 on V_{max}, indicative of a major C–H bond-breaking contribution to the overall enzymatic rate, have nevertheless been observed.[173] The large intrinsic isotope effects associated with carbon hydroxylation reactions are consistent with the correlation, in the absence of overriding steric or substrate positioning factors, between the strength of C–H bonds and their susceptibility to hydroxylation.[174,175] The isotope effects and the bond strength correlation indicate that the C–H bond is essentially broken in the transition state and thus provide strong support for a nonconcerted mechanism.

Most nonconcerted hydroxylation mechanisms require removal of the hydrogen by the activated oxygen followed by rapid collapse of the carbon intermediate with the activated hydroxyl (Fig. 8). The oxidation of a hydrocarbon by initial removal of an electron rather than a hydrogen has recently been demonstrated in the special case of a highly oxidizable hydrocarbon.[176] The result of this reaction is insertion into a *carbon–carbon* rather than carbon–hydrogen bond. Quadricyclane, a strained hydrocarbon with a very low oxidation potential ($E_{1/2} = 0.92$ V),[177] slowly autoxidizes in the presence of unreduced microsomes or trace metals to nortricyclanol. The NADPH-dependent microsomal oxidation of quadricyclane, however, yields the rearranged aldehyde that is also obtained as the principal product in the enzymatic oxidation of norbornadiene (Fig. 10). Control experiments have shown that norbornadiene is not an intermediate in the quadricyclane oxidation but rather that a common intermediate is probably generated from both substrates. Oxidation of quadricyclane to the radical cation, followed by capture of the carbon radical by the enzymatically activated oxygen, rationalizes these results because the resulting cationic intermediate is known to rearrange to the observed aldehyde. The radical cation is also generated in the autoxidative reaction but it is trapped by water rather than by the activated oxygen and thus gives a radical rather than cationic intermediate that does not rearrange.

In sum, the scrambling of stereochemistry in the hydroxylation of norbornane and camphor, the allylic rearrangements observed in the hydroxylation of unsaturated hydrocarbons, the correlation of reactivity with bond strength, the very large intrinsic isotope effects, and the dissociation of electron from oxygen transfer in the oxidation of quadricyclane provide firm evidence for nonconcerted hydroxylation mechanisms. The only available support for a concerted mechanism is provided by the retention of stereochemistry observed in a number of hydroxylations, but the demonstration that this is not universal suggests that stereochemical fidelity is imposed by the structural constraints of the active site rather than by the nature of the hydroxylation mechanism. The identity of the reaction intermediate remains ambiguous, but a radical is favored over a cation by the chemistry observed in model systems (see Chapter 1) and by the absence of the skeletal rearrangement products expected from cations.[158] The mechanism of P-450-catalyzed hydroxylation reactions therefore is best represented by the nonconcerted sequence in Fig. 8 in which the carbon radical obtained by hydrogen atom transfer to the activated oxygen complex collapses, in a second step, with the resulting equivalent of an iron-bound hydroxyl radical.

3.3. Olefin Epoxidation

The transfer of oxygen to carbon–carbon π-bonds proceeds, in the three diagnostic examples that have been examined, with retention of the olefin stereochemistry. The relevant reactions are the epoxidations of *cis*-stilbene,[178] oleic acid,[179] and *trans*-[1-^{2}H]-1-octene.[180] Retention of stereochemistry, however, as already noted (Section 3.2), does not unambiguously differentiate concerted from nonconcerted mechanisms. Asymmetric addition of the oxygen to the π-bond of styrene during its enzymatic epoxidation is suggested by the observation of an inverse isotope effect ($k_H/k_D = 0.93$) for deuterium on the internal but not terminal carbon of the olefinic bond.[181] The converse, no isotope effect for an internal deuterium but a large inverse isotope effect for a terminal deuterium, is observed when the epoxidation is carried out with *m*-chloroperbenzoic acid.[182] Similar isotope effects are expected for deuterium substitution at both ends of the π-bond if the two carbon–oxygen bonds are formed simultaneously and at comparable rates. The asymmetry in the oxygen

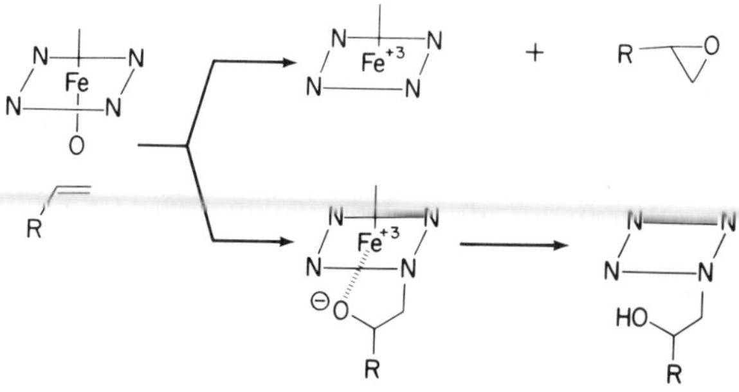

FIGURE 11. Heme alkylation during the oxidation of terminal olefins.

transfer, however, does not necessarily call for a nonconcerted mechanism because the two carbon–oxygen bonds could be formed simultaneously but at different rates. The differential isotope effects observed in the concerted chemical epoxidation emphasize the limited value of the secondary isotope effect data as a criterion for concertedness.

Evidence for an actual nonconcerted epoxidation mechanism is provided by the destruction of the enzyme associated with catalytic turnover of terminal olefins and by the rearrangements that accompany the metabolism of certain olefins. The first of these, the destruction of P-450 by terminal olefins (see Chapter 8), reflects N-alkylation of the prosthetic heme group of the enzyme. The epoxide metabolite is not responsible for heme alkylation even though the structure of the heme adduct indicates that oxygen transfer is required to activate the π-bond (Fig. 11). The results thus require epoxidation and heme alkylation to diverge at a point prior to epoxide formation.

The rearrangements observed during the oxidation of certain olefins constitute the second line of evidence for nonconcerted epoxidation mechanisms. The oxidative rearrangement of trichloroethylene to trichloroacetaldehyde, first thought to result from acid-catalyzed rearrangement of the epoxide metabolite, actually appears to occur during the oxygen transfer reaction because synthetic trichloroethylene oxide does not rearrange to trichloroacetaldehyde except under nonphysiological conditions (Fig. 12).[183,184] The metabolic formation of chloro- and dichloroacetic acids from 1,1-dichloroethylene, but not from the synthetically prepared epoxide, concurs with the view that chlorine migration occurs prior to epoxide formation.[185] Rearrangements have also been observed to occur in the metabolism of unhalogenated olefins. The microsomal oxidation of *trans*-1-phenylbutene yields 1-phenyl-1-butanone and 1-phenyl-2-butanone as minor products,[185] and that of styrene a trace of 2-phenylacetaldehyde,[186]

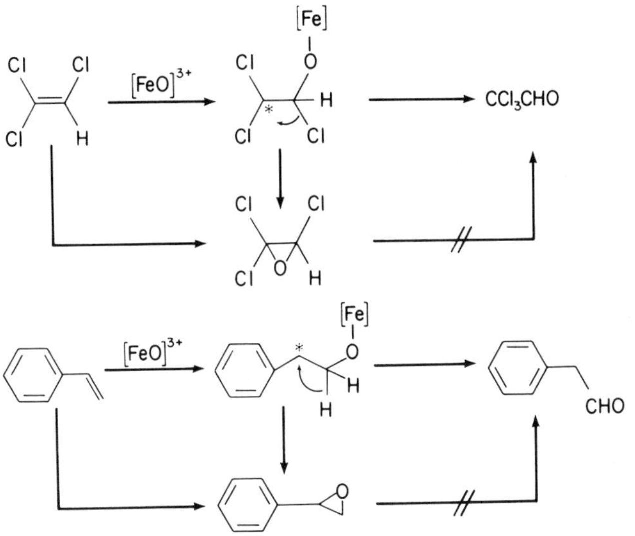

FIGURE 12. Rearrangements observed during the oxidation of olefins that appear to not involve the epoxide metabolites. The asterisk denotes a radical or cation center.

even though these rearranged products are not obtained from the corresponding epoxide metabolites under physiological conditions. The allylic scrambling of regiochemistry in the hydroxylation of cyclohexene and related hydrocarbons described in Section 3.2 could,[161,162] in principle, also be triggered by the removal of an electron from the π-bond in an epoxidation reaction. The rearrangements argue for the formation of a cationic intermediate that does not derive from the epoxide. This cationic species may be an intermediate in the epoxidation pathway, in which case the shift competes with epoxide formation, but may also stem from leakage of such an intermediate into a secondary reaction pathway. One attractive possibility for a divergent mechanism is for the epoxidation to proceed through an acyclic radical intermediate that closes normally to the epoxide but occasionally is oxidized to the corresponding cation by electron transfer to the heme center (Fig. 13).

The P-450-catalyzed oxidation of terminal acetylenes is accompanied by *quantitative* rearrangement of the acetylenic hydrogen to the vicinal carbon.[187–189] The hydrogen shift in the case of acetylenes is not prima facie evidence for a nonconcerted oxidation mechanism, however, because the expected unsaturated epoxide (oxirene) metabolite would itself rapidly rearrange to the observed products. The hydrogen migration must occur in the rate-determining step of the catalytic process because the

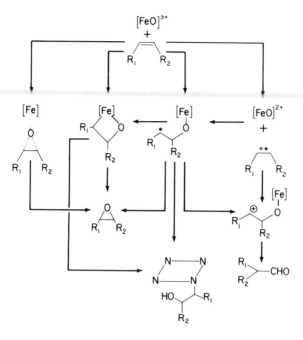

FIGURE 13. Alternative mechanisms for the oxidation of a carbon–carbon π-bond.

oxidation of aryl acetylenes to arylacetic acid metabolites is subject to a large isotope effect when the acetylenic hydrogen is replaced by deuterium.[189,190] The oxidation of aryl acetylenes by m-chloroperbenzoic acid yields the same products[187,190] and is subject to the same isotope effects as the enzymatic reaction.[190] This suggests that the hydrogen migration and the oxygen transfer actually occur simultaneously in both the chemical and enzymatic reactions. The enzymatic oxidation of terminal acetylenes, as of terminal olefins, results in alkylation of the prosthetic heme group of the enzyme (see Chapter 8). It is clear from the structures of the heme adducts that the heme is alkylated by a species produced by delivery of the activated oxygen to the internal carbon of the π-bond, whereas the metabolites follow from delivery of the oxygen to the external carbon. In agreement with this regiochemical difference, the destruction of P-450 by aryl acetylenes is not subject to a detectable isotope effect when deuterium is substituted for the acetylenic hydrogen. These results, which require the oxidative sequence leading to metabolites to diverge from that resulting in heme alkylation at a point prior to transfer of oxygen to the π-bond, are consistent with a stepwise reaction mechanism.

The evidence for nonconcerted π-bond oxidation, albeit more indirect than that marshalled in support of nonconcerted hydroxylation mechanisms, at the very least argues convincingly that P-450 enzymes under appropriate conditions catalyze nonconcerted epoxidation reactions. The precise nature of the nonconcerted mechanism remains to be defined, but the available data are most consistent with a pivotal radical species that partitions between epoxide formation, prosthetic heme alkylation, and oxidation to a cationic intermediate in which migration of a substituent competes with epoxide ring closure (Fig. 13). The radical intermediate may be preceded, as shown, by electron transfer from the olefin to the activated oxygen complex. The feasibility of electron transfer from a hydrocarbon to the activated oxygen is demonstrated by the oxidation of quadricyclane[176] discussed in the previous section. The data can also be explained by the formation of isomeric metallacyclobutane intermediates, either directly or by closure of the radical intermediate, that subsequently undergo an internal ligand transfer to give either the epoxide or the heme adduct, although the observed rearrangements still require the formulation of a cationic reaction branch. No actual evidence exists for metallacyclobutane intermediates in the enzymatic reaction but their intervention in the metalloporphyrin-catalyzed oxidation of π-bonds is suggested by indirect kinetic evidence (see Chapter 1).

3.4. Aromatic Hydroxylation

The introduction of a hydroxyl group into an aromatic ring by P-450 reflects, with few exceptions, epoxidation of the aromatic ring followed by chemical opening of the epoxide ring, migration of a hydride to the vicinal carbon, and keto–enol tautomerization (the "NIH shift") (Fig. 14a).[191] A fraction of the hydrogen that shifts is retained in the metabolite because it becomes equivalent to the proton at the receiving carbon and only one of the two hydrogens is lost in the tautomerization step. Deuterium substitution alters the proportion of the shifted atom that is retained but does not significantly alter the hydroxylation rate because the deuterium-sensitive tautomerization occurs after the rate-limiting enzymatic epoxidation. Quantitative loss of the hydrogen on the hydroxylated carbon and small kinetic isotope effects due to deuterium substitution are occasionally observed, however.[192,193] These unusual reactions, for the most part hydroxylations *meta* to a halide substituent, are possibly true hydroxylations in which oxygen is inserted directly into the C–H bond. The experimental support for a nonconcerted olefin epoxidation mechanism naturally suggests, however, the alternative that aromatic rings are also oxidized by nonconcerted mechanisms. The NIH shift, if this is true, could precede as well as follow epoxide ring formation (Fig. 14c). The

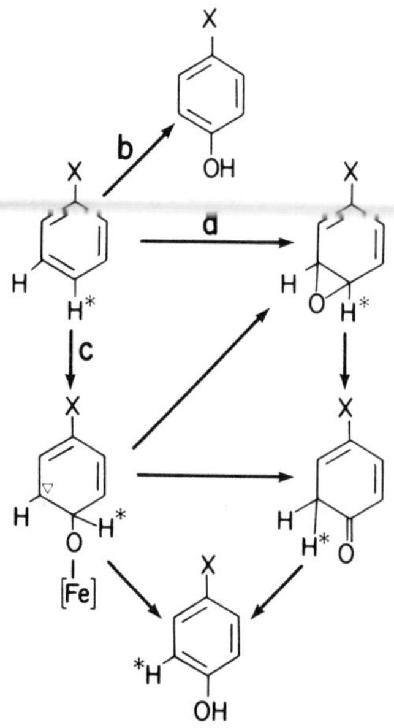

FIGURE 14. Alternative mechanisms for the hydroxylation of an aromatic ring. The asterisk denotes a labeled hydrogen and the triangle a radical or cation.

experimental evidence for a nonconcerted aromatic epoxidation mechanism, however, is at this time relatively weak.

The 7-hydroxylation of [7-^{2}H]warfarin proceeds with 77% retention of deuterium, a result compatible with a normal NIH shift mechanism but not with direct insertion of oxygen into the C–H bond.[194] The suggestion that this transformation is inconsistent with a conventional NIH shift mechanism involving an epoxide intermediate because the *meta* rather than *ortho* or *para* hydroxylated product is obtained is unwarranted. This regiochemical argument rests on the assumption that the substituent in warfarin relative to which the *ortho, meta,* and *para* positions are defined is the oxygen, whereas, in fact, the linear free energy substituent constants (σ_p) for an acetoxy (+0.31) and a methyl (−0.17)[195] suggest that the *alkyl* substituent is the dominant electron donor, and therefore that the hydroxyl group is introduced at the conventional *para* position. A related regiochemical argument for a nonconcerted hydroxylation, particularly as isotopic substitution rules out direct oxygen insertion, is provided by the formation of 3-hydroxybiphenyl as a minor metabolite of biphenyl. The 3-hydroxy metabolite reportedly is not formed from the synthetic epoxides of biphenyl when they are added to aqueous solution.[196,197] It was not demonstrated, however, that 3-hydroxybiphenyl is not formed from the epoxides when hepatic microsomes are present.

The net rates of hydroxylation of monosubstituted (H, F, Cl, Br, I) benzenes have been reported to correlate linearly with the σ^+ values for the substituents, although a similar correlation was not found for the individual *ortho* and *para* hydroxylation rates.[198] The data, however, are subject to ambiguities that preclude inferences concerning the concerted or nonconcerted nature of the reaction, although they do provide evidence for the development of positive charge in the transition state. The most comprehensive effort to "deconvolute" an aromatic hydroxylation process in order to obtain evidence for mechanisms that do not proceed through epoxide intermediates is provided by a study of the microsomal hydroxylation of a range of monosubstituted, deuterium-labeled, benzenes.[199] The results demonstrate the coexistence of an NIH shift mechanism and a mechanism in which the hydrogen at the reaction site is quantitatively eliminated, but do not actually discriminate between concerted and nonconcerted epoxidation mechanisms.

In sum, the experimental data support the parallel operation of two aromatic hydroxylation mechanisms, one in which the hydrogen at the site of reaction is partially retained (the NIH shift mechanism) and one in which it is completely eliminated (the direct hydroxylation mechanism). None of the data now available, however, *requires* a nonconcerted epoxidation mechanism, although such a mechanism is theoretically attractive and would be consistent with the evidence for such mechanisms in the oxidation of olefins and acetylenes.

3.5. Heteroatom Oxidation and Dealkylation

P-450 oxidizes nitrogen and sulfur to the corresponding oxides but has yet to be shown to similarly oxidize the more electronegative (and consequently less reactive) oxygen atom. P-450 also catalyzes the addition of a hydroxyl group to the carbon adjacent to heteroatoms in a reaction that ultimately eliminates the heteroatom, but the electronegativity of oxygen again sets it apart because this hydroxylation appears to stem from initial oxidation of the heteroatom in the case of sulfur or nitrogen but from direct carbon hydroxylation in the case of oxygen. The O-dealkylation of ethers and other oxygen derivatives therefore is most usefully viewed in the context of hydrocarbon hydroxylations rather than of heteroatom oxidations. The key step in oxidation or dealkylation of nitrogen or sulfur, on the other hand, appears to be the transfer of an electron from the heteroatom to the catalytic oxygen complex (Fig. 15). The resulting nitrogen or sulfur radical cation then collapses with the activated oxygen, yielding the oxide, or loses a proton from the adjacent carbon to give a heteroatom-substituted carbon radical that is trapped by the activated oxygen. The resulting α-hydroxylated product, identical to that

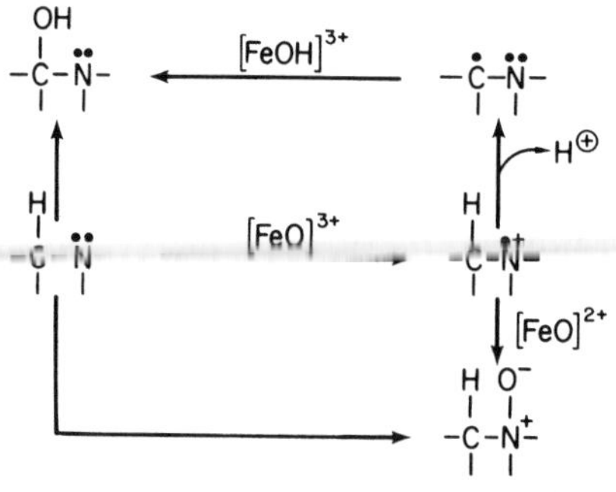

FIGURE 15. Mechanisms for the P-450-catalyzed oxidation and dealkylation of nitrogen. A similar scheme applies to sulfur.

expected from a conventional carbon hydroxylation reaction, in turn decomposes to the obseved dealkylation products.

Probably the most direct evidence for electron transfer to P-450 in the oxidation of a nitrogen-containing substrate is provided by the metabolism of 3,5-(bis)carbethoxy-2,6-dimethyl-4-ethyl-1,4-dihydropyridine (Fig. 16).[200] Catalytic turnover of this substrate by hepatic microsomes results both in transfer of the 4-ethyl moiety from the substrate to a nitrogen of the prosthetic heme group, an alkylation that inactivates the enzyme (see Chapter 8), and in extrusion of the 4-ethyl as a radical that can be spin trapped and isolated.[200] Catalytic generation of the ethyl radical can only be reasonably explained by one-electron oxidation of the dihydropyridine to a radical cation that aromatizes by ejecting the ethyl radical.

The P-450-catalyzed oxidation of a hydroxylamine (hydroxynorcocaine) yields an EPR-detectable nitroxide radical in a reaction reported not to depend on the peroxidative activity of the enzyme.[201] This observation contrasts with the finding that liver microsomes do not detectably

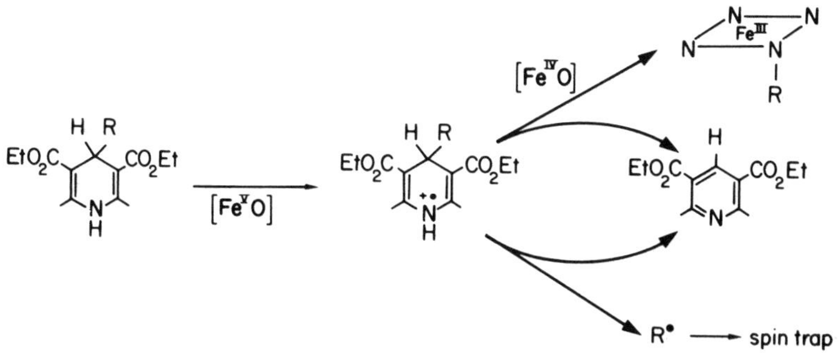

FIGURE 16. Proposed mechanism for the heme alkylation and radical formation associated with the oxidative metabolism of 4-alkyl-1,4-dihydropyridines.

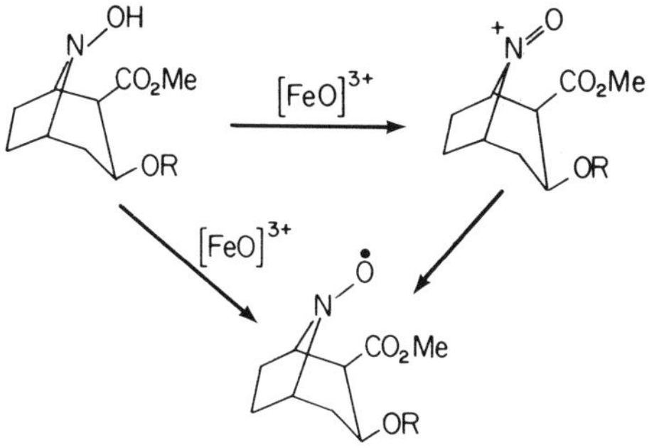

FIGURE 17. Alternatives for oxidation of the hydroxylamine of norcocaine to a nitroxide radical.

oxidize N-hydroxy-2-acetylaminofluorene to the nitroxide radical in the NADPH-dependent reaction but do so in the presence of lipid peroxides.[114,115] The radicals detected in a conflicting study of the NADPH-dependent microsomal metabolism of N-hydroxy-2-acetylaminofluorene presumably reflect peroxidative reactions occurring in the absence of added chelating agents.[116] The possibility cannot be excluded, however, that the hydroxylamine of norcocaine undergoes a two-electron oxidation to the nitroso product which disproportionates or is reduced by microsomal electron donors to the observed nitroxide radical (Fig. 17).

The isotope effects for the dealkylation of alkyl amines and thioethers provide useful information on the mechanisms by which these heteroatoms are oxidized. The O-dealkylation of 7-ethoxycoumarin, as expected for a carbon hydroxylation, is characterized by a large intrinsic isotope effect ($k_H/k_D = 13-14$).[172] In contrast, the intrinsic isotope effects for the N-dealkylation of amines, estimated from the ratio of deuterated products obtained from aryl amines bearing both deuterated and undeuterated N-methyl groups, fall in the range $k_H/k_D = 1.3-3.0$.[202-204] These small isotope effects are comparable to those measured for the dealkylation of amines by photochemical[205] or electrochemical[206] methods, reactions in which one-electron oxidation of the nitrogen is known to be the first step. The photochemical dealkylation of a mixture of dimethylaniline and di(trideuteromethyl)aniline thus proceeds without a measurable isotope effect, as does the metabolic reaction, whereas an isotope effect between 2 and 3 is observed for both the photochemical and enzymatic dealkylation of N-methyl-N-trideuteromethylaniline.[205] The product ratios for the electrochemical and enzymatic N-dealkylation of amines bearing two different N-alkyl groups are also very similar.[206] The intrinsic isotope effects for S-dealkylation reactions have not been measured and therefore cannot be discussed, but a correlation has been shown to exist between the acidity of the protons vicinal to the sulfur and the ratio of dealkylation to S-oxide formation.[139] The comparable intrinsic isotope effects and product ratios

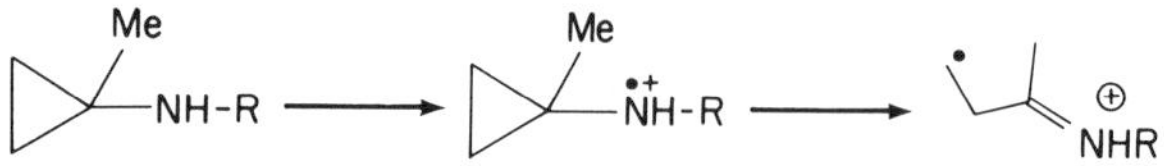

FIGURE 18. Ring opening associated with one-electron oxidation of a cyclopropyl amine.

for the chemical and enzymatic N-dealkylation reactions, the much lower intrinsic isotope effects for N-dealkylations than for carbon hydroxylations or O-dealkylations, and the limited information available on sulfur dealkylations, strongly support the proposal that N- and S-dealkylations are initiated by oxidation of the heteroatom to the corresponding radical cation (Fig. 18).

An interesting, but unexplained, difference exists in the isotope effects for the P-450-catalyzed, NADPH- or cumene hydroperoxide-dependent, N-demethylation of N-methyl-N-trideuteromethylaniline ($k_H/k_D <$ 3) and the analogous hydroperoxide-dependent reactions catalyzed by hemeproteins such as hemoglobin and horseradish peroxidase (k_H/k_D values as high as 10).[204] This difference is particularly surprising because N-dealkylation by horseradish peroxidase, for example, almost certainly involves oxidation of the nitrogen to the radical cation.

A linear free energy correlation ($\rho^+ = -0.16$) has been reported between the V_{max} values for the P-450-catalyzed sulfoxidation of four substituted thioanisoles and the one-electron potentials of the substrates, although a fifth value, that for p-chlorothioanisole, did not fit this correlation.[138] An analogous correlation ($\rho^+ = -0.2$) has been reported for oxidation of the four sulfoxides to the corresponding sulfones.[207] Correlations of reaction rates with oxidation potentials, however, particularly when based on a very limited number of compounds, are suggestive but do not require enzymatic oxidation of the sulfur to a radical cation because two-electron oxidation mechanisms also depend on the substrate oxidation potentials. A recent example of the coincidence of substituent effects for one- and two-electron oxidation mechanisms is provided by work on the chemical oxidation of NAD analogues.[208]

Cyclopropylamines inactivate P-450 even when there is no hydrogen on the cyclopropyl carbon adjacent to the nitrogen. The inactivation therefore cannot be explained simply by oxidation of the cyclopropyl amine to an iminium cation that alkylates the enzyme.[209–212] It has therefore been proposed that one-electron oxidation of the amine to the radical cation is followed by ring opening to the β-iminium radical and alkylation of the enzyme (Fig. 18). This theoretically attractive mechanism is supported by the recent demonstration that the oxidation potentials of a series of cyclopropyl amines and other cyclopropyl derivatives correlate reasonably well with the rate of enzyme inactivation.[212] The details of the inactivation mechanism, however, remain obscure except for inferences drawn from parallels with the inactivation of monoamine oxidase by the same cyclopropylamines.[211,213] Silverman and Yamasaki treated 2-phenylcyclopropylamine-inactivated monoamine oxidase with 2,4-dinitrophenylhydrazine and isolated cinnamaldehyde, the product expected from alkylation of the enzyme by the ring-opened imine invoked in the radical cation mechanism.[214]

4. Mechanisms of Biosynthetic P-450 Enzymes

4.1. Introduction

P-450 enzymes that are integrated into biosynthetic pathways are generally more substrate-specific than enzymes primarily devoted to xenobiotic metabolism and, in some instances, catalyze sequential oxidative reactions that terminate in carbon–carbon bond cleavage. The oxygen activation and substrate hydroxylation mechanisms, however, appear to be similar for all forms of P-450. Enzymes that catalyze straightforward reactions, such as the 7α-hydroxylation of cholesterol during bile acid synthesis, or consecutive hydroxylations without unusual consequences, such as the double hydroxylation of the 18-methyl group in aldosterone biosynthesis,[215] therefore require no further mechanistic discussion. The unique aspects of enzymes that result in carbon–carbon bond cleavage, however, and the mechanism of prostacyclin synthase, an enzyme that by some criteria is a member of the P-450 family, are reviewed in this section.

4.2. Cholesterol Side-Chain Cleavage

The C20–C22 bond of cholesterol is cleaved by a mitochondrial P-450 enzyme that catalyzes three sequential oxidative steps, each of which consumes one molecule of oxygen and one of NADPH. The three steps are 22(R)-hydroxylation, 20(S)-hydroxylation, and severance of the C20–C22 bond (Fig. 19) (see Chapters 10 and 11). The efficient conversion of cholesterol to pregnenolone is ensured by the fact that the hydroxylated intermediates bind to the enzyme up to 300 times more tightly than cholesterol[216] and by the increased stability of the ferrous dioxygen complex in each successive turnover.[12,217] The first two hydroxylations, which proceed with retention of configuration,[218,219] are unexceptional reactions but the final step, the oxidative carbon–carbon bond cleavage, differs from reactions reviewed in previous sections of this chapter. Mechanisms

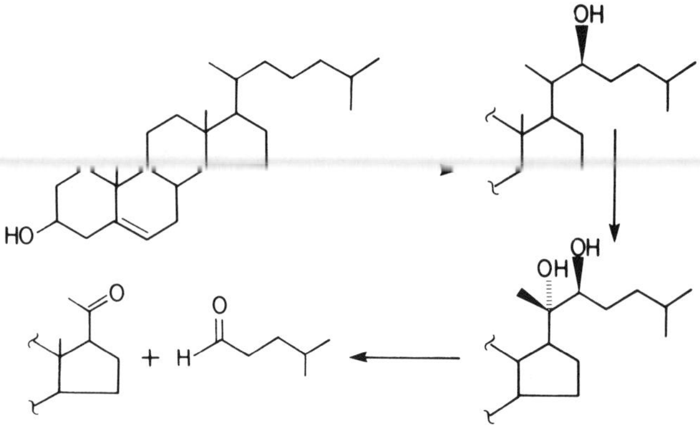

FIGURE 19. The intermediates in the triple turnover of P-450$_{scc}$ that cleaves the side chain of cholesterol.

that couple carbon–carbon bond cleavage to introduction of a third hydroxyl group at C22 are ruled out by the demonstration that the 22(S) hydrogen is retained in the 4-methylpentenal fragment that is excised in the reaction.[218] Two mechanisms for the carbon–carbon bond scission are plausible. In one, the activated oxygen complex generated in the third turnover is intercepted by one of the hydroxyl groups of 20(R),22(R)-dihydroxycholesterol. Proton removal from the hydroxyl adjacent to the resulting hydroperoxide then initiates carbon–carbon bond cleavage (Fig. 20). The unusual step in this mechanism is addition of the hydroxyl to the activated oxygen, the reverse of enzyme-catalyzed peroxide heterolysis. Addition of the hydroxyl to the iron rather than the oxygen of the

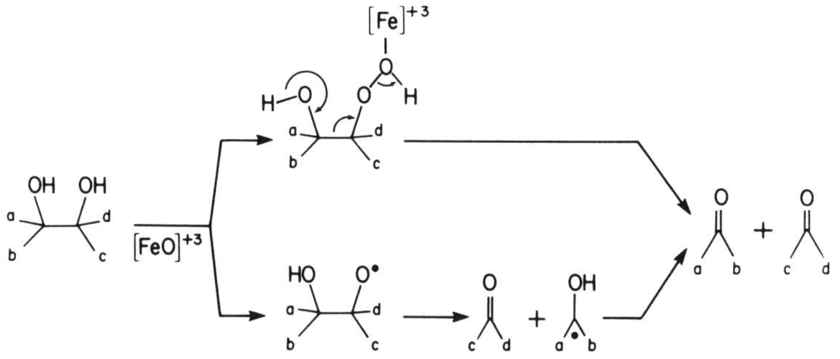

FIGURE 20. Alternative mechanisms for the final step in the cholesterol side-chain cleavage reaction. The letters represent the substituents on the carbons involved in the reaction.

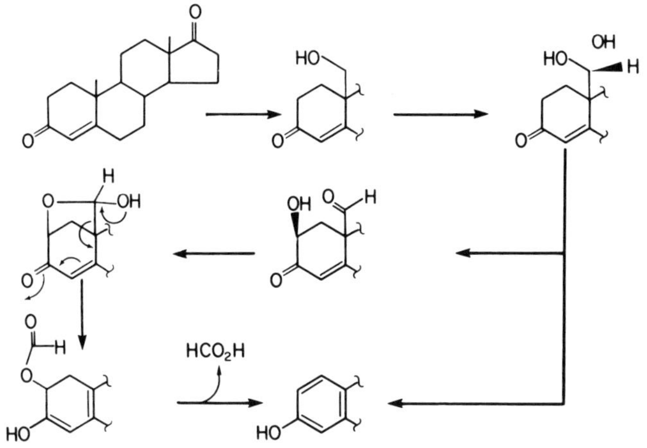

FIGURE 21. The intermediates in the catalytic turnover of aromatase.

catalytic complex, giving a transient species with the iron simultaneously bound to two oxygens, is conceivable as a variant of this mechanism. In the alternative mechanism, abstraction of the hydrogen from one of the hydroxyls by the activated oxygen complex produces an alkoxy radical. Homolytic scission of the carbon–carbon bond in this species and electron transfer from the resulting carbon radical to the protonated iron–oxygen complex completes the reaction process. The two reactions required in the second mechanism are well established for alkyl peroxides.

4.3. Aromatase

Aromatase, like P-450$_{scc}$, catalyzes a sequence of two carbon hydroxylations followed by an oxidative carbon–carbon bond cleavage (Fig. 21).[220] Three molecules of oxygen and three of NADPH are required for this catalytic sequence.[221] The 19-methyl group is eliminated in the final oxidative step, which also results in aromatization of the A-ring of androst-4-ene-3,17-dione and related substrates. The mechanism has primarily been studied with membrane-bound placental preparations, although progress in purification of the enzyme has been reported (see Chapter 10).[222] The aromatization reaction is not supported by cumene hydroperoxide or iodosobenzene,[99] in agreement with the common failure of substitute oxygen donors to replace the NADPH and oxygen requirements of biosynthetic P-450 enzymes.

The first 19-methyl hydroxylation catalyzed by human placental aromatase proceeds, as expected, with retention of configuration.[223] The second hydroxylation reaction, which removes the 19-pro-*R* hydrogen,[224,225] yields a gem diol that chemically eliminates water to give the 19-aldehyde. The small tritium isotope effect reported for the aromatization of [19-^{3}H]androst-4-ene-3,17-dione[226] is exclusively associated with the first hydroxylation step.[227] The second hydroxylation of the 19-methyl group proceeds without a detectable isotope effect. This difference in isotope effects is readily explained by the fact that the first hydroxylation can discriminate between the tritium and the hydrogen in a given methyl group, whereas the second hydroxylation stereospecifically removes the pro-*R* hydrogen and thus is constrained to react at that position whether it is occupied by a hydrogen or a tritium. The first isotope effect thus results from an internal competition, whereas an isotope effect in the second hydroxylation would have to result from the kind of *intermolecular* discrimination that is commonly suppressed in P-450 reactions. The differential isotope effects, contrary to the suggestion of the authors, do not require a difference in the binding sites or mechanisms of the reactions.[227] The final oxidative reaction, which stereospecifically removes the 1β- and 2β-hydrogens,[228–232] has been suggested to be a 2β-hydroxylation. The expected 2β-hydroxy-19-oxo sterol has been synthesized and shown to rapidly aromatize in the absence of the enzyme.[233] The 2β-hydroxy sterol has furthermore been detected in incubations carried out at low pH, conditions which slow down the aromatization step.[234] The importance of stereochemistry and the order of the hydroxylation reactions is demonstrated by the finding that the 2α-hydroxy-19-oxo sterol and analogues with a 19-hydroxyl rather than 19-oxo function are not aromatized chemically or enzymatically. Additional evidence for introdution of a 2β-hydroxyl group is provided by studies with antibodies that specifically recognize sterols with a 2β-hydroxyl function.[235] The antibodies inhibit the aromatization of androst-4-ene-3,17-dione but this inhibition is relieved when the antibody is presaturated with 2β,19-dihydroxyandrost-4-ene-3,17-dione. These results, and the finding that the formic acid eliminated in the aromatization step contains the first and third oxygen atoms consumed in the catalytic sequence,[236,237] have led to the postulate that intramolecular addition of the 2β-hydroxyl to the 19-aldehyde moiety to give a hemiacetal precedes rupture of the carbon–carbon bond (Fig. 21).[236] The recent report that the 2β-hydroxyl is not incorporated into the formic acid when the synthetic intermediate is incubated with placental microsomes,[237] however, conflicts with the finding that the third oxygen *is incorporated* into formic acid in the normal catalytic reaction. It is argued, on the basis of this finding, that 2β-hydroxylation is not an obligatory step in estradiol biosynthesis. Alternative mechanisms in which the third oxidation introduces a 4,5-epoxide,[238] a 1β-hydroxyl,[232] or a C_{19}-peroxide[236,239] have been excluded, as has the possibility that the substrate is covalently bound to the enzyme by Schiff base formation through the 3-keto function prior to catalytic turnover.[240] The exact nature of the third oxidative step therefore remains uncertain. The question of whether one or more active sites are involved in the aromatase sequence is under investigation[235] but the question may be semantic if, as suggested by the requirement that 19-hydroxylation precede 2β-hydroxylation, a single active site moves through different conformations as the substrate is processed.

4.4. Lanosterol 14-Demethylation

Removal of the 14-methyl group of lanosterol as formic acid in the biosynthesis of cholesterol[241,242] introduces a 14,15-double bond into the sterol framework (Fig. 22).[243–245] The opening 14-methyl hydroxylation is NADPH- and oxygen-dependent and is generally believed to be an unexceptional P-450-catalyzed hydroxylation.[246–249] The mechanisms of the second and third steps in the hepatic reaction, however, are subject to disagreement. The second step, oxidation of the 14-hydroxymethyl group to the aldehyde, and the third step, oxidative elimination of the aldehyde as formic acid with concomitant introduction of the 14,15-double bond,[246–248] have been reported to require NADPH and molecular oxygen.[246–248] The implication that the last two reactions are also catalyzed

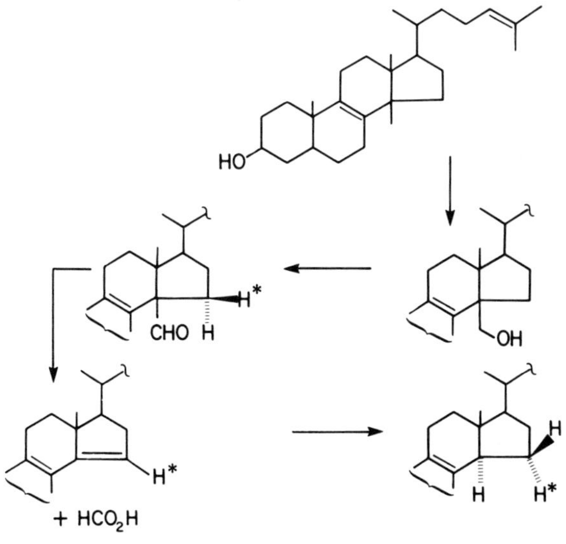

FIGURE 22. The probable intermediates in the 14-demethylation of lanosterol.

by P-450 is contradicted, however, by the report that NAD is the only cofactor required for the last two reactions.[249] The insensitivity of the last two steps, in contrast to the first hydroxylation, to inhibition by carbon monoxide has been invoked in support of a non-P-450 mechanism,[249,250] but the fact that P-450 enzymes are not uniformly susceptible to such inhibition renders this argument moot.[217,251,252] The sensitivity of aromatase[252,253] and P-450$_{scc}$[217] to carbon monoxide, for example, decreases drastically as the enzymes traverse the conformational and ligand states implicit in their multistep catalytic sequences. Oxidation of the 14-hydroxymethyl to an aldehyde is readily achieved by a double hydroxylation, if P-450 is involved, or by a dehydrogenation, if an NAD-dependent enzyme is required. The discrepancy in these reports remains to be resolved, but the purification from yeast of a P-450 enzyme that mediates the entire demethylation process demonstrates that a monooxygenase can catalyze the complete sequence.[254,255] Unfortunately, the NADPH and oxygen stoichiometry have not yet been successfully determined due to poor coupling of electron transport to oxygen consumption in the reconstituted yeast system.[255] Efforts to purify the liver enzyme(s) responsible for 14-demethylation are under way but are not yet advanced enough to resolve the discrepancy.[256]

The reaction that eliminates the 14-carbaldehyde group and introduces the 14,15-π-bond with stereospecific removal of the 15α-hydrogen remains undefined.[244,257,258] One possibility is that the enzyme introduces a 15α-hydroxyl that is eliminated in conjunction with formic acid, in analogy with the last step catalyzed by aromatase (Fig. 23). The observation that 14α-methyl-cholest-7-ene-3β,15β-diol is converted to cholesterol by rat liver homogenates despite the incorrect stereochemistry of the hydroxyl group,[259] however, illustrates the caution that must be exercised in interpreting studies of the fate of chemically synthesized hypothetical intermediates. Plausible alternatives to the introduction of a 15-hydroxyl include, for example, nucleophilic addition of the ferrous dioxygen species to the aldehyde to give an iron-coordinated alkyl peroxide that decomposes with concomitant abstraction of the 15α-hydrogen (Fig. 23). Further experimental work is required before a mechanism can be formulated with any confidence.

4.5. Prostacyclin Synthase

The enzyme responsible for the conversion of PGH$_2$ to prostacyclin, a physiologically important constituent of the arachidonic acid cascade, remains poorly characterized. The enzyme is not reduced by the normal microsomal electron transport chain but, when reduced chemically in the presence of carbon monoxide, exhibits the absorbance maximum at 450

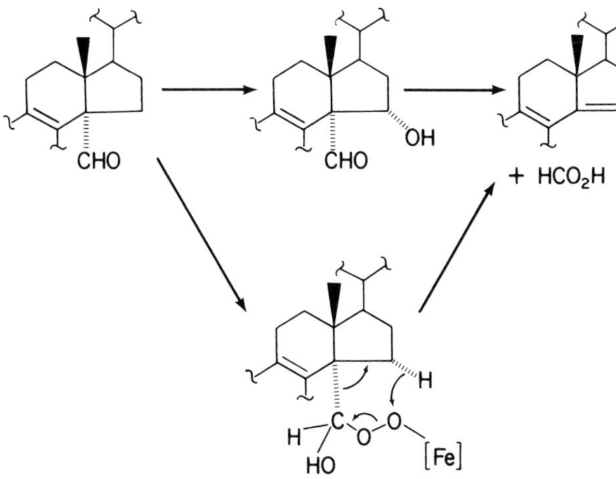

FIGURE 23. Alternative mechanisms for the final step in the 14-demethylation of lanosterol.

nm associated with a thiolate-ligated prosthetic heme group.[260,261] This has led to the suggestion that the enzyme, which catalyzes a peroxygenase reaction in which the peroxide donor and oxygen acceptor are in the same molecule, should be viewed as an unusual form of P-450 rather than as an unusual peroxidase. The transformation can be attractively rationalized by heterolytic scission of the *endo*-peroxide to give an iron–alkoxy complex that adds electrophilically to the vicinal double bond (Fig. 24). The resulting cation is then converted to prostacyclin by proton elimination. The initial step resembles the peroxygenase reactions catalyzed

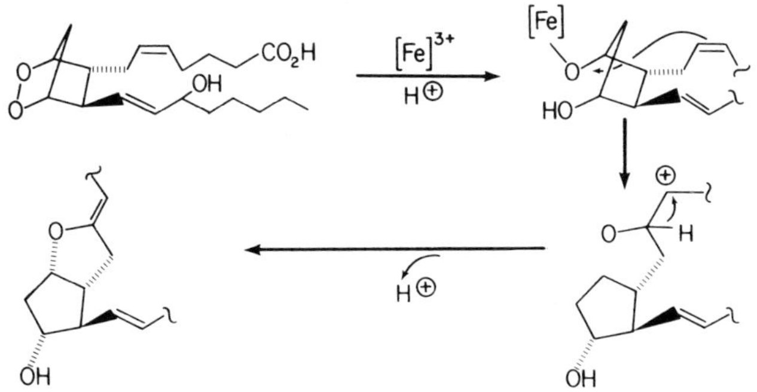

FIGURE 24. The mechanism proposed for conversion of the *endo*-cycloperoxide precursor to prostacyclin by an enzyme related to the P-450 class of enzymes.

by more conventional P-450 enzymes, with the exception that the oxygen transferred to the iron bears an alkyl substituent. The postulate (Section 2.3) that the thiolate ligand in P-450 enzymes facilitates cleavage of the dioxygen bond in the absence of assistance by charged species in the active site readily rationalizes the role of this ligand in the mechanism postulated for prostacyclin synthase. The validity of this mechanism, and the presence and role of the thiolate ligand, remain to be confirmed. The isolation of a related enzyme that catalyzes the synthesis of thromboxane from PGH_2 has been reported.[262]

5. Summary

The binding of oxygen to ferrous P-450 yields a dioxygen complex from which the ferric enzyme can be regenerated by autoxidative loss of superoxide or, in the specific case of P-450LM_4, by acquisition of a second electron coupled to loss of hydrogen peroxide. Electron transfer to the dioxygen complex, however, generally results in activation of oxygen to a metallooxygen species that oxidizes substrates, is reduced by additional electrons to (presumably) a molecule of water, destroys the prosthetic heme group, or reacts with oxygen donors to generate singlet oxygen. The key step in NADPH-dependent oxygen activation, cleavage of the dioxygen bond to yield a species equivalent to $[FeO]^{3+}$, is probably not assisted by acylation of the dioxygen terminus or, if the apparent absence of appropriate functionalities in the active site of P-450$_{cam}$ is general (Chapter 13), by ionic active-site residues. The unique electron donor properties of the thiolate ligand therefore are probably required to facilitate dioxygen bond cleavage in a neutral environment and to enhance and modulate the chemical reactivity of the resulting oxygen complex. A catalytically active complex similar to that obtained in normal turnover is obtained when peroxides are substituted for oxygen and NADPH, but experimental ambiguities inherent in the apparently competitive heterolytic and homolytic cleavage of the peroxide bond make it difficult to extrapolate arguments based on peroxide-supported reactions to the normal catalytic mechanism. The turnover of P-450 by iodosobenzene, on the other hand, appears to closely parallel the NADPH-dependent reaction although even here the data must be carefully evaluated to ensure their relevance.

The primary support for a concerted reaction mechanism is generally the absence of evidence for a nonconcerted process. The growing evidence for nonconcerted P-450-catalyzed oxygen transfer mechanisms therefore makes a concerted mechanism for any reaction promoted by these enzymes questionable, but does not rule out the possibility that the mechanism is substrate- and reaction-dependent. It is fair to say that no evidence exists that *requires* a concerted mechanism for any P-450-catalyzed reaction, whereas some of the available evidence cannot be explained except by nonconcerted mechanisms involving stepwise commitment of the two-oxidation equivalents.

ACKNOWLEDGMENT. The preparation of this chapter was aided by grant support from the National Institutes of Health.

References

1. White, R. E., and Coon, M. J., 1980, Oxygen activation by cytochrome P-450, *Annu. Rev. Biochem.* **49**:315-356.
2. Griffin, B. W., Peterson, J. A., and Estabrook, R. W., 1979, Cytochrome P-450: Biophysical properties and catalytic function, in: *The Porphyrins*, Volume 7 (D. Dolphin, ed.), Academic Press, New York, pp. 333-375.
3. Sato, R., and Omura, T. (eds.), 1978, *Cytochrome P-450*, Academic Press, New York.
4. Lambeth, J. D., Seybert, D. W., Lancaster, J. R., Salerno, J. C., and Kamin, H., 1982, Steroidogenic electron transport in adrenal cortex mitochondria, *Mol. Cell. Pharmacol.* **45**:13-31.
5. Omura, T., and Sato, R., 1984, The carbon monoxide-binding pigment of liver microsomes. I. Evidence for its hemoprotein nature, *J. Biol. Chem.* **239**:2370-2378.
6. Estabrook, R. W., Peterson, J., Baron, J., and Hildebrandt, A., 1972, The spectrophotometric measurement of turbid suspensions of cytochromes associated with drug metabolism, in: *Methods in Pharmacology*, Volume 2 (C. F. Chignell, ed.), Appleton–Century–Crofts, New York, pp. 303-350.
7. McLane, K. E., Fisher, J., and Ramakrishnan, K., 1983, Reductive drug metabolism, *Drug Metab. Rev.* **14**:741-799.
8. Bonfils, C., Debey, P., and Maurel, P., 1979, Highly purified microsomal cytochrome P-450: The oxyferro intermediate stabilized at low temperature, *Biochem. Biophys. Res. Commun.* **88**:1301-1307.
9. Estabrook, R. W., Hildebrandt, A. G., Baron, J., Netter, K. J., and Leibman, K. C., 1971, New spectral intermediate associated with cytochrome P-450 function in liver microsomes, *Biochem. Biophys. Res. Commun.* **42**:132-139.
10. Peterson, J. A., Ishimura, Y., and Griffin, B. W., 1972, *Pseudomonas putida* cytochrome P-450: Characterization of an oxygenated form of the hemoprotein, *Arch. Biochem. Biophys.* **149**:197-208.
11. Larroque, C., and Van Lier, J. E., 1980, The subzero temperature stabilized oxyferro complex of purified cytochrome P-450$_{scc}$, *FEBS Lett.* **115**:175-177.
12. Tuckey, R. C., and Kamin, H., 1982, The oxyferro complex of adrenal cytochrome P-450$_{scc}$: Effect of cholesterol and intermediates on its stability and optical characteristics, *J. Biol. Chem.* **257**:9309-9314.
13. Guengerich, F. P., Ballou, D. P., and Coon, M. J., 1976, Spectral intermediates in the reaction of oxygen with purified liver microsomal cytochrome P-450, *Biochem. Biophys. Res. Commun.* **70**:951-956.
14. Oprian, D. D., Gorsky, L. D., and Coon, M. J., 1983, Properties of the oxygenated form of liver microsomal cytochrome P-450, *J. Biol. Chem.* **258**:8684-8691.

15. Begard, E., Debey, P., and Douzou, P., 1977, Sub-zero temperature studies of microsomal cytochrome P-450: Interaction of Fe^{+2} with oxygen, *FEBS Lett.* **75**:52–54.
16. Anderson, K. K., Debey, P., and Balny, C., 1979, Subzero temperature studies of microsomal cytochrome P-450: O-dealkylation of 7-ethoxycoumarin coupled to single turnover, *FEBS Lett.* **102**:117–120.
17. Sharrock, M., Munck, E., Debrunner, P. G., Marshall, V., Lipscomb, J. D., and Gunsalus, I. C., 1973, Mössbauer studies of cytochrome $P-450_{cam}$, *Biochemistry* **12**:258–265.
18. Sharrock, M., Debrunner, P. G., Schulz, C., Lipscomb, J. D., Marshall, V., and Gunsalus, I. C., 1976, Cytochrome $P-450_{cam}$ and its complexes: Mossbauer parameters of the heme iron, *Biochim. Biophys. Acta* **420**:8–26.
19. Bonfils, C., Balny, C., and Maurel, P., 1981, Direct evidence for electron transfer from ferrous cytochrome b_5 to the oxyferrous intermediate of liver cytochrome P-450 LM_2, *J. Biol. Chem.* **256**:9457–9465.
20. Noshiro, M., Ullrich, V., and Omura, T., 1981, Cytochrome b_5 as electron donor for oxy-cytochrome P-450, *Eur. J. Biochem.* **116**:521–526.
21. Werringloer, J., and Kawano, S., 1980, The control of the cyclic function of liver microsomal cytochrome P-450: "Counterpoise"-regulation of the electron transfer reactions required for the activation of molecular oxygen, in: *Biochemistry, Biophysics and Regulation of Cytochrome P-450* (J. Å. Gustafsson, D. Carlstedt-Duke, A. Mode, and J. Rafter, eds.), Elsevier/North-Holland, Amsterdam, pp. 359–362.
22. Gillette, J. R., Brodie, B. B., and LaDu, B. N., 1957, The oxidation of drugs by liver microsomes: On the role of TPNH and oxygen, *J. Pharmacol. Exp. Ther.* **119**:532–540.
23. Thurman, R. G., Ley, H. G., and Scholz, R., 1972, Hepatic microsomal ethanol oxidation: Hydrogen peroxide formation and the role of catalase, *Eur. J. Biochem.* **25**:420–430.
24. Hildebrandt, A. G., and Roots, J., 1975, Reduced nicotinamide adenine dinucleotide phosphate (NADPH) dependent formation and breakdown of hydrogen peroxide during mixed function reactions in liver microsomes, *Arch. Biochem. Biophys.* **171**:385–397.
25. Werringloer, J., 1977, The formation of hydrogen peroxide during hepatic microsomal electron transport reactions, in: *Microsomes and Drug Oxidations* (V. Ullrich, J. Roots, A. G. Hildebrandt, R. W. Estabrook, and A. H. Conney, eds.), Pergamon Press, Elmsford, N.Y., pp. 261–268.
26. Nordblom, G. D., and Coon, M. J., 1977, Hydrogen peroxide formation and stoichiometry of hydroxylation reactions catalyzed by highly purified liver microsomal cytochrome P-450, *Arch. Biochem. Biophys.* **180**:343–347.
27. Grover, T. A., and Piette, L. H., 1981, Influence of flavin addition and removal on the formation of superoxide by NADPH-cytochrome P-450 reductase: A spin trap study, *Arch. Biochem. Biophys.* **212**:105–114.
28. Bosterling, B., and Trudell, J. R., 1981, Spin trap evidence of superoxide radical anions by purified NADPH-cytochrome P-450 reductase, *Biochem. Biophys. Res. Commun.* **98**:569–575.
29. Ingelman-Sundberg, M., and Johansson, I., 1980, Cytochrome b_5 as electron donor to rabbit liver cytochrome $P-450_{LM2}$ in reconstituted phospholipid vesicles, *Biochem. Biophys. Res. Commun.* **97**:582–589.
30. Bast, A., Savenije-Chapel, E. M., and Kroes, B. H., 1984, Inhibition of mono-oxygenase and oxidase activity of rat hepatic cytochrome P-450 by H_2-receptor blockers, *Xenobiotica* **14**:399–408.
31. Jeffery, E. H., and Mannering, G. J., 1983, Interaction of constitutive and phenobarbital-induced cytochrome P-450 isozymes during the sequential oxidation of benzph-

etamine: Explanation for the difference in benzphetamine-induced hydrogen peroxide production and 455-nm complex formation in microsomes from untreated and phenobarbital-treated rats, *Mol. Pharmacol.* **23**:748–757.
32. Ingelman-Sundberg, M., and Johansson, I., 1984, Mechanisms of hydroxyl radical formation and ethanol oxidation by ethanol-inducible and other forms of rabbit liver microsomal cytochromes P-450, *J. Biol. Chem.* **259**:6447–6458.
33. Debey, P., and Balny, C., 1973, Production of superoxide ions in rat liver microsomes, *Biochimie* **55**:329–332.
34. Bartoli, G. M., Galeotti, T., Palombini, G., Parisi, G., and Azzi, A., 1977, Different contributions of rat liver microsomal pigments in the formation of superoxide anions and hydrogen peroxide during development, *Arch. Biochem. Biophys.* **184**:276–281.
35. Auclair, C., De Prost, D., and Hakim, J., 1978, Superoxide anion production by liver microsomes from phenobarbital treated rat, *Biochem. Pharmacol.* **27**:355–358.
36. Kuthan, H., Ullrich, V., and Estabrook, R. W., 1982, A quantitative test for superoxide radicals produced in biological systems, *Biochem. J.* **203**:551–558.
37. Kuthan, H., and Ullrich, V., 1982, Oxidase and oxygenase function of the microsomal cytochrome P-450 monooxygenase system, *Eur. J. Biochem.* **126**:583–588.
38. Kuthan, H., Tsuji, H., Graf, H., Ullrich, V., Werringloer, J., and Estabrook, R. W., 1978, Generation of superoxide anion as a source of hydrogen peroxide in a reconstituted monooxygenase system, *FEBS Lett.* **91**:343–345.
39. Zhukov, A. A., and Archakov, A. I., 1982, Complete stoichiometry of free NADPH oxidation in liver microsomes, *Biochem. Biophys. Res. Commun.* **109**:813–818.
40. Sligar, S. G., Lipscomb, J. D., Debrunner, P. G., and Gunsalus, I. C., 1974, Superoxide anion production by the autoxidation of cytochrome P-450$_{cam}$, *Biochem. Biophys. Res. Commun.* **61**:290–296.
41. Heinemeyer, G., Nigam, S., and Hildebrandt, A. G., 1980, Hexobarbital-binding, hydroxylation and hexobarbital-dependent hydrogen peroxide production in hepatic microsomes of guinea pig, rat, and rabbit, *Naunyn-Schmiedebergs Arch. Pharmacol.* **314**:201–210.
42. Hildebrandt, A. G., Heinemeyer, G., and Roots, I., 1982, Stoichiometric cooperation of NADPH and hexobarbital in hepatic microsomes during the catalysis of hydrogen peroxide formation, *Arch. Biochem. Biophys.* **216**:455–465.
43. Pompon, D., and Coon, M. J., 1984, On the mechanism of action of cytochrome P-450: Oxidation and reduction of the ferrous dioxygen complex of liver microsomal cytochrome P-450 by cytochrome b$_5$, *J. Biol. Chem.* **259**:15377–15385.
44. Wallace, W. J., and Caughey, W. S., 1975, Mechanism for the autoxidation of hemoglobin by phenols, nitrite, and "oxidant" drugs: Peroxide formation by one electron donation of bound dioxygen, *Biochem. Biophys. Res. Commun.* **62**:561–567.
45. Wallace, W. J., Houtchens, R. A., Maxwell, J. C., and Caughey, W. S., 1982, Mechanism of autooxidation for hemoglobins and myoglobins: Promotion of superoxide production by protons and anions, *J. Biol. Chem.* **257**:4966–4977.
46. Satoh, Y., and Shikama, K., 1981, Autoxidation of myoglobin: A nucleophilic displacement mechanism, *J. Biol. Chem.* **256**:10272–10275.
47. Lipscomb, J. D., Sligar, S. G., Namtvedt, M. J., and Gunsalus, I. C., 1976, Autooxidation and hydroxylation reactions of oxygenated cytochrome P-450$_{cam}$, *J. Biol. Chem.* **251**:1116–1124.
48. Sligar, S. G., Shastry, B. S., and Gunsalus, I. C., 1976, Oxygen reactions of the P450 protein, in: *Microsomes and Drug Oxidations* (V. Ullrich, A. Hildebrandt, I. Roots, and R. W. Estabrook, eds.), Pergamon Press, Elmsford, N.Y., pp. 202–208.
49. Estabrook, R. W., Kawano, S., Werringloer, J., Kuthan, H., Tsuji, H., Graf, H., and Ullrich, V., 1979, Oxycytochrome P-450: Its breakdown to superoxide for the formation of hydrogen peroxide *Acta Biol. Med. Ger.* **38**:423–434.

50. Shikama, K., 1984, A controversy on the mechanism of autoxidation of oxymyoglobin and oxyhaemoglobin: Oxidation, dissociation, or displacement?, *Biochem. J.* **223**:279–280.
51. Dolphin, D., Forman, A., Borg, D. C., Fajer, J., and Felton, R. H., 1971, Compounds I of catalase and horse radish peroxidase: π-cation radicals, *Proc. Natl. Acad. Sci. USA* **68**:614–618.
52. Morishima, I., Takamuki, Y., and Shiro, Y., 1984, Nuclear magnetic resonance studies of metalloporphyrin π-cation radicals as models for compound I of peroxidases, *J. Am. Chem. Soc.* **106**:7666–7672.
53. Roberts, J. E., Hoffman, B. M., Rutter, R., and Hager, L. P., 1981, Electron-nuclear double resonance of horseradish peroxidase compound I, *J. Biol. Chem.* **256**:2118–2121.
54. La Mar, G. N., de Ropp, J. S., Latos-Grazynski, L., Balch, A. L., Johnson, R. B., Smith, K. M., Parish, D. W., and Cheng, R.-J., 1983, Proton NMR characterization of the ferryl group in model heme complexes and hemoproteins: Evidence for the $Fe^{IV}=O$ group in ferryl myoglobin and compound II of horseradish peroxidase, *J. Am. Chem. Soc.* **105**:782–787.
55. Hoffman, B. M., Roberts, J. E., Kang, C. H., and Margoliash, E., 1981, Electron paramagnetic and electron nuclear double resonance of the hydrogen peroxide compound of cytochrome c peroxidase, *J. Biol. Chem.* **256**:6556–6564.
56. Courtin, F., Michot, J.-L., Virion, A., Pommier, J., and Deme, D., 1984, Reduction of lactoperoxidase–H_2O_2 compounds by ferrocyanide: Indirect evidence of an apoprotein site for one of the two oxidizing equivalents, *Biochem. Biophys. Res. Commun.* **121**:463–470.
57. Courtin, F., Deme, D., Virion, A., Michot, J.-L., Pommier, J., and Nunez, J., 1982, The role of lactoperoxidase–H_2O_2 compounds in the catalysis of thyroglobulin iodination and thyroid hormone synthesis, *Eur. J. Biochem.* **124**:603–609.
58. Yonetani, T., and Schleyer, H., 1967, Studies on cytochrome c peroxidase: The reaction of ferrimyoglobin with hydroperoxides and a comparison of peroxide-induced compounds of ferrimyoglobin and cytochrome c peroxidase, *J. Biol. Chem.* **242**:1974–1979.
59. King, N. K., Looney, F. D., and Winfield, M. E., 1967, Amino acid free radicals in oxidized myoglobin, *Biochim. Biophys. Acta* **133**:65–82.
60. Groves, J. T., and Nemo, T. E., 1983, Aliphatic hydroxylation catalyzed by iron porphyrin complexes, *J. Am. Chem. Soc.* **105**:6243–6248.
61. Hamilton, G. A., 1974, Chemical models and mechanisms for oxygenases, in: *Molecular Mechanisms of Oxygen Activation* (O. Hayaishi, ed.), Academic Press, New York, pp. 405–448.
62. Groves, J. T., Watanabe, Y., and McMurry, T. J., 1983, Oxygen activation by metalloporphyrins: Formation and decomposition of an acylperoxymanganese(III) complex, *J. Am. Chem. Soc.* **105**:4489–4490.
63. Khenkin, A. M., and Shteinman, A. A., 1984, The mechanism of oxidation of alkanes by peroxo complexes of iron porphyrins in the presence of acylating agents: A model for activation of O_2 by cytochrome P-450, *J. Chem. Soc. Chem. Commun.* **1984**:1219–1220.
64. Sligar, S. G., Kennedy, K. A., and Pearson, D. C., 1980, Chemical mechanisms for cytochrome P-450 hydroxylation: Evidence for acylation of heme bound dioxygen, *Proc. Natl. Acad. Sci. USA* **77**:1240–1244.
65. Poulos, T. L., and Kraut, J., 1980, The stereochemistry of peroxidase catalysis, *J. Biol. Chem.* **255**:8199–8205.
66. Finzel, B. C., Poulos, T. L., and Kraut, J., 1984, Crystal structure of yeast cytochrome c peroxidase refined at 1.7-Å resolution, *J. Biol. Chem.* **259**:13027–13036.

67. Dunford, H. B., 1982, Peroxidases, in: *Advances in Inorganic Biochemistry* (G. Eichhorn and L. G. Marzilli, eds.), Elsevier, Amsterdam, pp. 41–68.
68. Dunford, H. B., and Araiso, T., 1979, Horseradish peroxidase. XXXVI. On the difference between peroxidase and metmyoglobin, *Biochem. Biophys. Res. Commun.* **89**:764–768.
69. McCandlish, E., Miksztal, A. R., Nappa, M., Sprenger, A. Q., Valentine, J. S., Stong, J. D., and Spiro, T. G., 1980, Reactions of superoxide with iron porphyrins in aprotic solvents: A high spin ferric porphyrin peroxo complex, *J. Am. Chem. Soc.* **102**:4268–4271.
70. Ataollah, S., and Goff, H. M., 1982, Characterization of superoxide–metalloporphyrin reaction products: Effective use of deuterium NMR spectroscopy, *J. Am. Chem. Soc.* **104**:6318–6322.
71. Welborn, C. H., Dolphin, D., and James, B. R., 1981, One-electron electrochemical reduction of a ferrous porphyrin dioxygen complex, *J. Am. Chem. Soc.* **103**:2869–2871.
72. Mansuy, D., Battioni, P., and Renaud, J.-P., 1984, In the presence of imidazole, iron- and manganese-porphyrins catalyze the epoxidation of alkenes by alkyl hydroperoxides, *J. Chem. Soc. Chem. Commun.* **1984**:1255–1257.
73. Hrycay, E. G., and O'Brien, P. J., 1971, Cytochrome P-450 as a microsomal peroxidase utilizing a lipid peroxide substrate, *Arch. Biochem. Biophys.* **147**:14–27.
74. Hrycay, E. G. and O'Brien, P. J., 1972, Cytochrome P-450 as a microsomal peroxidase in steroid hydroperoxide reduction, *Arch. Biochem. Biophys.* **153**:480–494.
75. Hrycay, E. G., and O'Brien, P. J., 1974, Microsomal electron transport. II. Reduced nicotinamide adenine dinucleotide-cytochrome b_5 reductase and cytochrome P-450 as electron carriers in microsomal NADH-peroxidase activity, *Arch. Biochem. Biophys.* **160**:230–245.
76. O'Brien, P. J., 1978, Hydroperoxides and superoxides in microsomal oxidations, *Pharmac. Ther. A* **2**:517–536.
77. Rahimtula, A. D., and O'Brien, P. J., 1977, The hydroperoxide dependent microsomal oxidation of alcohols, *Eur. J. Biochem.* **77**:210–211.
78. Hrycay, E. G., and O'Brien, P. J., 1971, The peroxidase nature of cytochrome P-420 utilizing a lipid peroxide substrate, *Arch. Biochem. Biophys.* **147**:28–35.
79. O'Brien, P. J., and Rahimtula, A., 1975, Involvement of cytochrome P-450 in the intracellular formation of lipid peroxides, *J. Agr. Food Chem.* **23**:154–158.
80. Lindstrom, T. D., and Aust, S. D., 1984, Studies on cytochrome P-450-dependent lipid hydroperoxide reduction, *Arch. Biochem. Biophys.* **233**:80–87.
81. Cavallini, L., Valente, M., and Bindoli, A., 1983, NADH and NADPH inhibit lipid peroxidation promoted by hydroperoxides in rat liver microsomes, *Biochim. Biophys. Acta* **752**:339–345.
82. Ishimaru, A., and Yamazaki, I., 1977, Hydroperoxide-dependent hydroxylation involving "H_2O_2-reducible hemoprotein" in microsomes of pea seeds: A new type enzyme acting on hydroperoxide and a physiological role of seed lipoxygenase, *J. Biol. Chem.* **252**:6118–6124.
83. Kadlubar, F. F., Morton, K. C., and Ziegler, D. M., 1973, Microsomal-catalyzed hydroperoxide-dependent C-oxidation of amines, *Biochem. Biophys. Res. Commun.* **54**:1255–1261.
84. Nordblom, G. D., White, R. E., and Coon, M. J., 1976, Studies on hydroperoxide-dependent substrate hydroxylation by purified liver microsomal cytochrome P-450, *Arch. Biochem. Biophys.* **175**:524–533.
85. Ingelman-Sundberg, M., and Ekstrom, G., 1982, Aniline is hydroxylated by the cytochrome P-450-dependent hydroxyl radical-mediated oxygenation mechanism, *Biochem. Biophys. Res. Commun.* **106**:625–631.

86. Johansson, I., and Ingelman-Sundberg, M., 1983, Hydroxyl radical-mediated cytochrome P-450-dependent metabolic activation of benzene in microsomes and reconstituted enzyme systems from rabbit liver, *J. Biol. Chem.* **258**:7311–7316.
87. Ashley, P. L., and Griffin, B. W., 1981, Involvement of radical species in the oxidation of aminopyrine and 4-aminoantipyrine by cumene hydroperoxide in rat liver microsomes, *Mol. Pharmacol.* **19**:146–152.
88. Griffin, B. W., 1982, Use of spin traps to elucidate radical mechanisms of oxidations by hydroperoxides catalyzed by hemoproteins, *Can. J. Chem.* **60**:1463–1473.
89. Rahimtula, A. D., and O'Brien, P. J., 1975, Hydroperoxide-dependent O-dealkylation reactions catalyzed by liver microsomal cytochrome P-450, *Biochem. Biophys. Res. Commun.* **62**:268–275.
90. Burke, D. M., and Mayer, R. T., 1975, Inherent specificities of purified cytochrome P-450 and P-448 toward biphenyl hydroxylation and ethoxyresorufin deethylation, *Drug Metab. Dispos.* **3**:245–253.
91. Rahimtula, A. D., and O'Brien, P. J., 1974, Hydroperoxide catalyzed liver microsomal aromatic hydroxylation reactions involving cytochrome P-450, *Biochem. Biophys. Res. Commun.* **60**:440–447.
92. Hrycay, E. G., Gustafsson, J.-Å., Ingelman-Sundberg, M., and Ernster, L., 1975, Sodium periodate, sodium chlorite, organic hydroperoxides, and H_2O_2 as hydroxylating agents in steroid hydroxylation reactions catalyzed by partially purified cytochrome P-450, *Biochem. Biophys. Res. Commun.* **66**:209–216.
93. Ellin, A., and Orrenius, S., 1975, Hydroperoxide-supported cytochrome P-450-linked fatty acid hydroxylation in liver microsomes, *FEBS Lett.* **50**:378–381.
94. Danielsson, H., and Wikvall, K., 1976, On the ability of cumene hydroperoxide and $NaIO_4$ to support microsomal hydroxylations in biosynthesis and metabolism of bile acids, *FEBS Lett.* **66**:299–302.
95. Gustafsson, J.-Å, Hrycay, E. G., and Ernster, L., 1976, Sodium periodate, sodium chlorite, and organic hydroperoxides as hydroxylating agents in steroid hydroxylation reactions catalyzed by adrenocortical microsomal and mitochondrial cytochrome P_{450}, *Arch. Biochem. Biophys.* **174**:440–453.
96. Hrycay, E. G., Gustafsson, J.-Å, Ingelman-Sundberg, M., and Ernster, L., 1976, The involvement of cytochrome P-450 in hepatic microsomal steroid hydroxylation reactions supported by sodium periodate, sodium chlorite, and organic hydroperoxides, *Eur. J. Biochem.* **61**:43–52.
97. Rahimtula, A. D., O'Brien, P. J., Seifried, H. E., and Jerina, D. M., 1978, The mechanism of action of cytochrome P-450: Occurrence of the "NIH shift" during hydroperoxide-dependent aromatic hydroxylations, *Eur. J. Biochem.* **89**:133–141.
98. Fasco, M. J., Piper, L. J., and Kaminsky, L. S., 1979, Cumene hydroperoxide-supported microsomal hydroxylations of warfarin—A probe of cytochrome P-450 multiplicity and specificity, *Biochem. Pharmacol.* **28**:97–103.
99. Kelly, W. G., and Stolee, A. H., 1978, Stabilization of placental aromatase by dithiothreitol in the presence of oxidizing agents, *Steroids* **31**:533–539.
100. Koop, D. R., and Hollenberg, P. F., 1980, Kinetics of the hydroperoxide-dependent dealkylation reactions catalyzed by rabbit liver microsomal cytochrome P-450, *J. Biol. Chem.* **255**:9685–9692.
101. Blake, R. C., and Coon, M. J., 1980, On the mechanism of action of cytochrome P-450: Spectral intermediates in the reactions of P-450$_{LM2}$ with peroxy compounds, *J. Biol. Chem.* **255**:4100–4111.
102. Blake, R. C., and Coon, M. J., 1981, On the mechanism of action of cytochrome P-450: Role of peroxy spectral intermediates in substrate hydroxylation, *J. Biol. Chem.* **256**:5755–5763.

103. Blake, R. C., and Coon, M. J., 1981, On the mechanism of action of cytochrome P-450: Evaluation of homolytic and heterolytic mechanisms of oxygen–oxygen bond cleavage during substrate hydroxylation by peroxides, *J. Biol. Chem.* **256**:12127–12133.
104. Rahimtula, A. D., O'Brien, P. J., Hrycay, E. G., Peterson, J. A., and Estabrook, R. W., 1974, Possible higher valence states of cytochrome P-450 during oxidative reactions, *Biochem. Biophys. Res. Commun.* **60**:695–702.
105. Larroque, C., and van Lier, J. E., 1983, Spectroscopic evidence for the formation of a transient species during cytochrome P-450$_{scc}$ induced hydroperoxysterol-glycol conversions, *Biochem. Biophys. Res. Commun.* **112**:655–662.
106. Wagner, G. C., Palcic, M. M., and Dunford, H. B., 1983, Absorption spectra of cytochrome P-450$_{cam}$ in the reaction with peroxy acids, *FEBS Lett.* **156**:244–248.
107. White, R. E., Sligar, S. G., and Coon, M. J., 1980, Evidence for a homolytic mechanism of peroxide oxygen–oxygen bond cleavage during substrate hydroxylation by cytochrome P-450, *J. Biol. Chem.* **255**:11108–11111.
108. Dix, T. A., and Marnett, L. J., 1983, Hematin-catalyzed rearrangement of hydroperoxylinoleic acid to epoxy alcohols via an oxygen rebound, *J. Am. Chem. Soc.* **105**:7001–7002.
109. Pace-Asciak, C. R., 1984, Arachidonic acid epoxides: Demonstration through [^{18}O]oxygen studies of an intramolecular transfer of the terminal hydroxyl group of (12s)-hydroperoxyeicosa-5,8,10,14-tetraenoic acid to form hydroperoxides, *J. Biol. Chem.* **259**:8332–8337.
110. McCarthy, M.-B., and White, R. E., 1983, Competing modes of peroxyacid flux through cytochrome P-450, *J. Biol. Chem.* **258**:11610–11616.
111. Capdevila, J., Estabrook, R. W., and Prough, R. A., 1980, Differences in the mechanism of NADPH- and cumene hydroperoxide-supported reactions of cytochrome P-450, *Arch. Biochem. Biophys.* **200**:186–195.
112. Hlavica, P., Golly, I., and Mietaschk, J., 1983, Comparative studies on the cumene hydroperoxide- and NADPH-supported N-oxidation of 4-chloroaniline by cytochrome P-450, *Biochem. J.* **212**:539–547.
113. Renneberg, R., Damerau, W., Jung, C., Ebert, B., and Scheller, F., 1983, Study of H_2O_2-supported N-demethylations catalyzed by cytochrome P-450 and horseradish peroxidase, *Biochem. Biophys. Res. Commun.* **113**:332–339.
114. Bartsch, H., and Hecker, E., 1971, On the metabolic activation of the carcinogen N-hydroxy-N-2-acetylaminofluorene. III. Oxidation with horseradish peroxidase to yield 2-nitrosofluorene and N-acetoxy-N-2acetylaminofluorene, *Biochim. Biophys. Acta* **237**:567–578.
115. Reigh, D. L., and Floyd, R. A., 1981, Evidence for a cytochrome P-420 catalyzed mechanism of activation of N-hydroxy-2-acetylaminofluorene, *Cancer Biochem. Biophys.* **5**:213–217.
116. Stier, A., Reitz, I., and Sackmann, E., 1972, Radical accumulation in liver microsomal membranes during biotransformation of aromatic amines and nitro compounds, *Naunyn-Schmiedebergs Arch. Pharmacol.* **274**:189–191.
117. Coon, M. J., White, R. E., and Blake, R. C., 1979, Mechanistic studies with purified liver microsomal cytochrome P-450: Comparison of O_2- and peroxide-supported hydroxylation reactions, in: *Oxidases and Related Redox Systems* (T. E. King, H. S. Mason, and M. Morrison, eds.), Pergamon Press, Elmsford, N.Y., pp. 857–885.
118. Wagner, G. C., and Gunsalus, I. C., 1982, Cytochrome P-450: Structure and states, in: *Biology and Chemistry of Iron, NATO Adv. Study Inst. Ser. C* **89**:405–412.
119. King, N. K., Looney, F. D., and Winfield, M. E., 1967, Amino acid free radicals in oxidized metmyoglobin, *Biochim. Biophys. Acta* **133**:65–82.
120. Yonetani, T., and Schleyer, H., 1967, Studies on cytochrome c peroxidase. The re-

121. Renneberg, R., Scheller, F., Ruckpaul, K., Pirrwitz, J., and Mohr, P., 1978, NADPH and H$_2$O$_2$-dependent reactions of cytochrome P-450$_{LM}$ compared with peroxidase catalysis, *FEBS Lett.* **96:**349–353.
122. Renneberg, R., Capdevila, J., Chacos, N., Estabrook, R. W., and Prough, R. A., 1981, Hydrogen peroxide-supported oxidation of benzo[a]pyrene by rat liver microsomal fractions, *Biochem. Pharmacol.* **30:**843–848.
123. Estabrook, R. W., Martin-Wixtrom, C., Saeki, Y., Renneberg, R., Hildebrandt, A., and Werringloer, J., 1984, The peroxidatic function of liver microsomal cytochrome P-450: Comparison of hydrogen peroxide and NADPH-catalyzed N-demethylation reactions, *Xenobiotica* **14:**87–104.
124. Holm, K. A., Engell, R. J., and Kupfer, D., 1985, Regioselectivity of hydroxylation of prostaglandins by liver microsomes supported by NADPH versus H$_2$O$_2$ in methylcholanthrene-treated and control rats: Formation of novel prostaglandin metabolites, *Arch. Biochem. Biophys.* **237:**477–489.
125. Gustafsson, J.-Å., Rondahl, L., and Bergman, J., 1979, Iodosylbenzene derivatives as oxygen donors in cytochrome P-450 catalyzed steroid hydroxylations, *Biochemistry* **18:**865–870.
126. Gustaffson, J.-Å, and Bergman, J., 1976, Iodine- and chlorine-containing oxidation agents as hydroxylating catalysts in cytochrome P-450-dependent fatty acid hydroxylation reactions in rat liver microsomes, *FEBS Lett.* **70:**276–280.
127. Lichtenberger, F., Nastainczyk, W., and Ullrich, V., 1976, Cytochrome P-450 as an oxene transferase, *Biochem. Biophys. Res. Commun.* **70:**939–946.
128. Berg, A., Ingelman-Sundberg, M., and Gustaffson, J.-Å, 1979, Purification and characterization of cytochrome P-450$_{meg}$, *J. Biol. Chem.* **254:**5264–5271.
129. Heimbrook, D. C., and Sligar, S. G., 1981, Multiple mechanisms of cytochrome P-450-catalyzed substrate hydroxylation, *Biochem. Biophys. Res. Commun.* **99:**530–535.
130. Macdonald, T. L., Burka, L. T., Wright, S. T., and Guengerich, F. P., 1982, Mechanisms of hydroxylation by cytochrome P-450: Exchange of iron–oxygen intermediates with water, *Biochem. Biophys. Res. Commun.* **104:**620–625.
131. Kexel, H., Schmelz, E., and Schmidt, H.-L., 1977, Oxygen transfer in microsomal oxidative desulfuration, in: *Microsomes and Drug Oxidations* (V. Ullrich, I. Roots, A. Hildebrandt, R. W. Estabrook, and A. H. Conney, eds.), Pergamon Press, Elmsford, N.Y., pp. 269–274.
132. Schardt, B. C., and Hill, C. L., 1983, Preparation of iodobenzene dimethoxide: A new synthesis of [^{18}O]iodosylbenzene and a reexamination of its infrared spectrum, *Inorg. Chem.* **22:**1563–1565.
133. Groves, J. T., and Kruper, W. J., 1979, Preparation and characterization of an oxoporphinatochromium(V) complex, *J. Am. Chem. Soc.* **101:**7613–7615.
134. Groves, J. T., Kruper, W. J., and Haushalter, R. C., 1980, Hydrocarbon oxidations with oxometalloporphinates: Isolation and reactions of a (porphinato)manganese(V) complex, *J. Am. Chem. Soc.* **102:**6375–6377.
135. Rein, H., Maricic, S., Janig, G. R., Vuk-Pavlovic, S., Benko, B., Ristau, O., and Ruckpaul, K., 1976, Haem accessibility in cytochrome P-450 from rabbit liver: A proton magnetic relaxation study by stereochemical probes, *Biochim. Biophys. Acta* **446:**325–330.
136. Hayano, M., Lindberg, M. C., Dorfman, R. I., Hancock, J. E. H., and von Doering, W. E., 1955, On the mechanism of the C-11β-hydroxylation of steroids: A study with H$_2$O^{18} and O$_2^{18}$, *Arch. Biochem. Biophys.* **59:**529–532.

137. Hayano, M., Saito, A., Stone, D., and Dorfman, R. I., 1956, Hydroxylation of steroids by microorganisms in the presence of $^{18}O_2$, *Biochim. Biophys. Acta* **21**:380–381.
138. Watanabe, Y., Iyanagi, T., and Oae, S., 1980, Kinetic study on enzymatic S-oxygenation promoted by a reconstituted system with purified cytochrome P-450, *Tetrahedron Lett.* **21**:3685–3688.
139. Watanabe, Y., Numata, T., Iyanagi, T., and Oae, S., 1981, Enzymatic oxidation of alkyl sulfides by cytochrome P-450 and hydroxyl radical, *Bull. Chem. Soc. Jpn.* **54**:1163–1170.
140. Nordblom, G. D., White, R. E., and Coon, M. J., 1976, Studies on hydroperoxide-dependent substrate hydroxylation by purified liver microsomal cytochrome P-450, *Arch. Biochem. Biophys.* **175**:524–533.
141. White, R. E., and McCarthy, M.-B., 1984, Aliphatic hydroxylation by cytochrome P-450: Evidence for rapid hydrolysis of an intermediate iron–nitrene complex, *J. Am. Chem. Soc.* **106**:4922–4926.
142. Heimbrook, D. C., Murray, R. E., Egeberg, K. D., Sligar, S. G., Nee, M. W., and Bruice, T. C., 1984, Demethylation of N,N-dimethylaniline and p-cyano-N,N-dimethylaniline and their N-oxides by cytochromes $P450_{LM2}$ and $P450_{cam}$, *J. Am. Chem. Soc.* **106**:1514–1515.
143. Staudt, H., Lichtenberger, F., and Ullrich, V., 1974, The role of NADH in uncoupled microsomal monooxygenations, *Eur. J. Biochem.* **46**:99–106.
144. Gorsky, L. D., Koop, D. R., and Coon, M. J., 1984, On the stoichiometry of the oxidase and monooxygenase reactions catalyzed by liver microsomal cytochrome P-450: Products of oxygen reduction, *J. Biol. Chem.* **259**:6812–6817.
145. Guengerich, F. P., 1978, Destruction of heme and hemoproteins mediated by liver microsomal reduced nicotinamide adenine dinucleotide phosphate-cytochrome P-450 reductase, *Biochemistry* **17**:3633–3639.
146. Jeffery, E., Kotake, A., Azhary, R. E., and Mannering, G. J., 1977, Effects of linoleic acid hydroperoxide on the hepatic monooxygenase systems of microsomes from untreated, phenobarbital-treated, and 3-methylcholanthrene-treated rats, *Mol. Pharmacol.* **13**:415–425.
147. Yoshinaga, T., Sassa, S., and Kappas, A., 1982, A comparative study of heme degradation by NADPH-cytochrome c reductase alone and by the complete heme oxygenase system: Distinctive aspects of heme degradation by NADPH-cytochrome c reductase, *J. Biol. Chem.* **257**:7794–7802.
148. Schaefer, W. H., Harris, T. M., and Guengerich, F. P., 1985, Characterization of the enzymatic and non-enzymatic peroxidative degradation of iron porphyrins and cytochrome P-450 heme, *Biochemistry* **24**:3254–3263.
149. Hornsby, P. J., 1980, Regulation of cytochrome P-450-supported 11β-hydroxylation of deoxycortisol by steroids, oxygen, and antioxidants in adrenocortical cell cultures, *J. Biol. Chem.* **255**:4020–4027.
150. Quinn, P. G., and Payne, A. H., 1985, Steroid product-induced, oxygen-mediated damage of microsomal cytochrome P-450 enzymes in Leydig cell cultures, *J. Biol. Chem.* **260**:2092–2099.
151. Klimek, J., Schaap, A. P., and Kimura, T., 1983, The relationship between NADPH-dependent lipid peroxidation and degradation of cytochrome P-450 in adrenal cortex mitochondria, *Biochem. Biophys. Res. Commun.* **110**:559–566.
152. Cadenas, E., Sies, H., Graf, H., and Ullrich, V., 1983, Oxene donors yield low-level chemiluminescence with microsomes and isolated cytochrome P-450, *Eur. J. Biochem.* **130**:117–121.
153. Bergstrom, S., Lindstredt, S., Samuelson, B., Corey, E. J., and Gregoriou, G. A., 1958, The stereochemistry of 7α-hydroxylation in the biosynthesis of cholic acid from cholesterol, *J. Am. Chem. Soc.* **80**:2337–2338.

154. Corey, E. J., Gregoriou, G. A., and Peterson, D. H., 1958, The stereochemistry of 11α-hydroxylation of steroids, *J. Am. Chem. Soc.* **80**:2338.
155. McMahon, R. E., Sullivan, H. R., Craig, J. C., and Pereira, W. E., 1969, The microsomal oxygenation of ethyl benzene: Isotopic, stereochemical, and induction studies, *Arch. Biochem. Biophys.* **132**:575–577.
156. Hamberg, M., and Bjorkhem, I., 1971, ω-Oxidation of fatty acids. I. Mechanism of microsomal ω1- and ω2-hydroxylation, *J. Biol. Chem.* **246**:7411–7416.
157. Shapiro, S., Piper, J. U., and Caspi, E., 1982, Steric course of hydroxylation at primary carbon atoms: Biosynthesis of 1-octanol from (1R)- and (1S)-[1-^{3}H,^{2}H,^{1}H;1-^{14}C]octane by rat liver microsomes, *J. Am. Chem. Soc.* **104**:2301–2305.
158. Groves, J. T., McClusky, G. A., White, R. E., and Coon, M. J., 1978, Aliphatic hydroxylation by highly purified liver microsomal cytochrome P-450: Evidence for a carbon radical intermediate, *Biochem. Biophys. Res. Commun.* **81**:154–160.
159. Gelb, M. H., Heimbrook, D. C., Malkonen, P., and Sligar, S. G., 1982, Stereochemistry and deuterium isotope effects in camphor hydroxylation by the cytochrome P-450$_{cam}$ monooxygenase system, *Biochemistry* **21**:370–377.
160. White, R. E., Bhattacharyya, A., Miller, J. P., and Favreau, L. V., 1985, Stereochemistry of aliphatic hydroxylation by cytochrome P-450 isozymes, *Fed. Proc.* **44**:474.
161. Tanaka, K., Kurihara, N., and Nakajima, M., 1979, Oxidative metabolism of tetrachlorocyclohexenes, pentachlorocyclohexenes, and hexachlorocyclohexenes with microsomes from rat liver and house fly abdomen, *Pestic. Biochem. Physiol.* **10**:79–95.
162. Groves, J. T., and Subramanian, D. V., 1984, Hydroxylation by cytochrome P-450 and metalloporphyrin models: Evidence for allylic rearrangement, *J. Am. Chem. Soc.* **106**:2177–2181.
163. Griller, D., and Ingold, K. U., 1980, Free-radical clocks, *Acc. Chem. Res.* **13**:317–323.
164. White, R. E., Groves, J. T., and McClusky, G. A., 1979, Electronic and steric factors in regioselective hydroxylation catalyzed by purified cytochrome P-450, *Acta Biol. Med. Ger.* **38**:475–482.
165. Sligar, S. G., Gelb, M. H., and Heimbrook, D. C., 1984, Bio-organic chemistry and cytochrome P-450-dependent catalysis, *Xenobiotica* **14**:63–86.
166. Wong, P. C., and Griller, D., 1981, A kinetic EPR study of the norbornenyl-nortricyclyl radical rearrangement, *J. Org. Chem.* **46**:2327–2329.
167. Simons, J. W., and Rabinovitch, B. S., 1963, Deuterium isotope effects in rates of methylene radical insertion into carbon–hydrogen bonds and across carbon carbon double bonds, *J. Am. Chem. Soc.* **85**:1023–1024.
168. Goldstein, M. J., and Dolbier, W. R., 1965, The intramolecular insertion mechanism of α-haloneopentyl lithium, *J. Am. Chem. Soc.* **87**:2293–2295.
169. O'Ferrall, R. A. M., 1970, Model calculations of hydrogen isotope effects for nonlinear transition states, *J. Chem. Soc. B* **1970**:785–790.
170. Hjelmeland, L. M., Aronow, L., and Trudell, J. R., 1977, Intramolecular determination of primary kinetic isotope effects in hydroxylations catalyzed by cytochrome P-450, *Biochem. Biophys. Res. Commun.* **76**:541–549.
171. Foster, A. B., Jarman, M., Stevens, J. D., Thomas, P., and Westwood, J. H., 1974, Isotope effects in O- and N-demethylations mediated by rat liver microsomes: An application of direct insertion electron impact mass spectrometry, *Chem. Biol. Interact* **9**:327–340.
172. Miwa, G. T., Walsh, J. S., and Lu, A. Y. H., 1984, Kinetic isotope effects on cytochrome P-450-catalyzed oxidation reactions: The oxidative O-dealkylation of 7-ethoxycoumarin, *J. Biol. Chem.* **259**:3000–3004.
173. Lu, A. Y. H., Harada, N., and Miwa, G. T., 1984, Rate-limiting steps in cytochrome

P-450-catalyzed reactions: Studies on isotope effects in the O-deethylation of 7-ethoxycoumarin, *Xenobiotica* **14**:19–26.
174. Frommer, U., Ullrich, V., and Staudinger, H., 1970, Hydroxylation of aliphatic compounds by liver microsomes. I. The distribution of isomeric alcohols, *Hoppe-Seylers Z. Physiol. Chem.* **351**:903–912.
175. White, R. E., McCarthy, M.-B., Egeberg, K. D., and Sligar, S. G., 1984, Regioselectivity in the cytochromes P-450: Control by protein constraints and by chemical reactivities, *Arch. Biochem. Biophys.* **228**:493–502.
176. Stearns, R. A., and Ortiz de Montellano, P. R., 1985, Cytochrome P-450-catalyzed oxidation of quadricyclane: Evidence for a radical cation intermediate, *J. Am. Chem. Soc.* **107**:4081–4082.
177. Gassman, P. G., and Yamaguchi, R., 1982, Electron transfer from highly strained polycyclic molecules, *Tetrahedron* **38**:1113–1122.
178. Watabe, T., and Akamatsu, K., 1974, Microsomal epoxidation of *cis*-stilbene: Decrease in epoxidase activity related to lipid peroxidation, *Biochem. Pharmacol.* **23**:1079–1085.
179. Watabe, T., Ueno, Y., and Imazumi, J., 1971, Conversion of oleic acid into *threo*-dihydroxystearic acid by rat liver microsomes, *Biochem. Pharmacol.* **20**:912–913.
180. Ortiz de Montellano, P. R., Mangold, B. L. K., Wheeler, C., Kunze, K. L., and Reich, N. O., 1983, Stereochemistry of cytochrome P-450-catalyzed epoxidation and prosthetic heme alkylation, *J. Biol. Chem.* **258**:4202–4207.
181. Hanzlik, R. P., and Shearer, G. O., 1978, Secondary deuterium isotope effects on olefin epoxidation by cytochrome P-450, *Biochem. Pharmacol.* **27**:1441–1444.
182. Hanzlik, R. P., and Shearer, G. O., 1975, Transition state structure for peracid epoxidation: Secondary deuterium isotope effects, *J. Am. Chem. Soc.* **97**:5231–5233.
183. Henschler, D., Hoos, W. R., Fetz, H., Dallmeier, E., and Metzler, M., 1979, Reactions of trichloroethylene epoxide in aqueous systems, *Biochem. Pharmacol.* **28**:543–548.
184. Miller, R. E., and Guengerich, F. P., 1982, Oxidation of trichloroethylene by liver microsomal cytochrome P-450: Evidence for chlorine migration in a transition state not involving trichloroethylene oxide, *Biochemistry* **21**:1090–1097.
185. Liebler, D. C., and Guengerich, F. P., 1983, Olefin oxidation by cytochrome P-450: Evidence for group migration in catalytic intermediates formed with vinylidene chloride and *trans*-1-phenyl-1-butene, *Biochemistry* **22**:5482–5489.
186. Mansuy, D., Leclaire, J., Fontecave, M., and Momenteau, M., 1984, Oxidation of monosubstituted olefins by cytochromes P-450 and heme models: Evidence for the formation of aldehydes in addition to epoxides and allylic alcohols, *Biochem. Biophys. Res. Commun.* **119**:319–325.
187. Ortiz de Montellano, P. R., and Kunze, K. L., 1981, Shift of the acetylenic hydrogen during chemical and enzymatic oxidation of the biphenylacetylene triple bond, *Arch. Biochem. Biophys.* **209**:710–712.
188. Ortiz de Montellano, P. R., 1985, Alkenes and alkynes, in: *Bioactivation of Foreign Compounds* (M. W. Anders, ed.), Academic Press, New York, pp. 121–155.
189. McMahon, R. E., Turner, J. C., Whitaker, G. W., and Sullivan, H. R., 1981, Deuterium isotope effect in the biotransformation of 4-ethynylbiphenyls to 4-biphenylacetic acids by rat hepatic microsomes, *Biochem. Biophys. Res. Commun.* **99**:662–667.
190. Ortiz de Montellano, P. R., and Komives, E. A., 1985, Branchpoint for heme alkylation and metabolite formation in the oxidation of aryl acetylenes by cytochrome P-450, *J. Biol. Chem.* **260**:3330–3336.
191. Jerina, D. M., and Daly, J. W., 1974, Arene oxides: A new aspect of drug metabolism, *Science* **185**:573–582.
192. Tomaszewski, J. E., Jerina, D. M., and Daly, J. W., 1975, Deuterium isotope effects during formation of phenols by hepatic monooxygenases: Evidence for an alternative to the arene oxide pathway *Biochemistry* **14**:2024–2030.

193. Preston, B. D. Miller, J. A., and Miller, E. C., 1983, Non-arene oxide aromatic ring hydroxylation of 2,2′,5,5′-tetrachlorobiphenyl as the major metabolic pathway catalyzed by phenobarbital-induced rat liver microsomes, *J. Biol. Chem.* **258**:8304–8311.
194. Bush, E. D., and Trager, W. F., 1982, Evidence against an abstraction or direct insertion mechanism for cytochrome P-450 catalyzed meta hydroxylations, *Biochem. Biophys. Res. Commun.* **104**:626–632.
195. Ritchie, C. D., and Sager, W. F., 1964, Structure–activity relationships, *Prog. Phys. Org. Chem.* **2**:323–400.
196. Billings, R. E., and McMahon, R. E., 1978, Microsomal biphenyl hydroxylation: The formation of 3-hydroxybiphenyl and biphenyl catechol, *Mol. Pharmacol.* **14**:145–154.
197. Swinney, D. C., Howald, W. N., and Trager, W. F., 1984, Intramolecular isotope effects associated with meta-hydroxylation of biphenyl catalyzed by cytochrome P-450, *Biochem. Biophys. Res. Commun.* **118**:867–872.
198. Burka, L. T., Plucinski, T. M., and MacDonald, T. L., 1983, Mechanisms of hydroxylation by cytochrome P-450: Metabolism of monohalobenzenes by phenobarbital-induced microsomes, *Proc. Natl. Acad. Sci. USA* **80**:6680–6684.
199. Hanzlik, R. P., Hogberg, K., and Judson, C. M., 1984, Microsomal hydroxylation of specifically deuterated monosubstituted benzenes: Evidence for direct aromatic hydroxylation, *Biochemistry* **23**:3048–3055.
200. Augusto, O., Beilan, H. S., and Ortiz de Montellano, P. R., 1982, The catalytic mechanism of cytochrome P-450: Spin-trapping evidence for one electron substrate oxidation, *J. Biol. Chem.* **257**:11288–11295.
201. Rauckman, E. J., Rosen, G. M., and Cavagnaro, J., 1982, Norcocaine nitroxide, a potential hepatotoxic metabolite of cocaine, *Mol. Pharmacol.* **21**:458–463.
202. Abdel-Monem, M. M., 1975, Isotope effects in enzymatic N-demethylation of tertiary amines, *J. Med. Chem.* **18**:427–430.
203. Miwa, G. T., Garland, W. A., Hodshon, B. J., Lu, A. Y. H., and Northrop, D. B., 1980, Kinetic isotope effects in cytochrome P-450-catalyzed oxidation reactions: Intermolecular and intramolecular deuterium isotope effects during the N-demethylation of N,N-dimethylphentermine, *J. Biol. Chem.* **255**:6049–6054.
204. Miwa, G. T., Walsh, J. S., Kedderis, G. L., and Hollenberg, P. F., 1983, The use of intramolecular isotope effects to distinguish between deprotonation and hydrogen atom abstraction mechanisms in cytochrome P-450- and peroxidase-catalyzed N-demethylation reactions, *J. Biol. Chem.* **258**:14445–14449.
205. Dopp, D., and Heufer, J., 1982, N-Demethylation of N,N-dimethylaniline by photoexcited 3-nitrochlorobenzene, *Tetrahedron Lett.* **23**:1553–1556.
206. Shono, T., Toda, T., and Oshino, N., 1982, Electron transfer from nitrogen in microsomal oxidation of amine and amide: Simulation of microsomal oxidation by anodic oxidation, *J. Am. Chem. Soc.* **104**:2639–2641.
207. Watanabe, Y., Iyanagi, T., and Oae, S., 1982, One electron transfer mechanism in the enzymatic oxygenation of sulfoxide to sulfone promoted by a reconstituted system with purified cytochrome P-450, *Tetrahedron Lett.* **23**:533–536.
208. Powell, M. F., Wu, J. C., and Bruice, T. C., 1984, Ferricyanide oxidation of dihydropyridines and analogues, *J. Am. Chem. Soc.* **106**:3850–3856.
209. Hanzlik, R. P., and Tullman, R. H., 1982, Suicidal inactivation of cytochrome P-450 by cyclopropylamines: Evidence for cation–radical intermediates, *J. Am. Chem. Soc.* **104**:2048–2050.
210. Macdonald, T. L., Zirvi, K., Burka, L. T., Peyman, P., and Guengerich, F. P., 1982, Mechanism of cytochrome P-450 inhibition by cyclopropylamines, *J. Am. Chem. Soc.* **104**:2050–2052.
211. Tullman, R. H., and Hanzlik, R. P., 1984, Inactivation of cytochrome P-450 and monoamine oxidase by cyclopropylamines, *Drug Metab. Dispos.* **15**:1163–1182.

212. Guengerich, F. P., Willard, R. J., Shea, J. P., Richards, L. E., and Macdonald, T. L., 1984, Mechanism-based inactivation of cytochrome P-450 by heteroatom-substituted cyclopropanes and formation of ring opened products, *J. Am. Chem. Soc.* **106**:6446–6447.
213. Silverman, R. B., and Yamasaki, R. B., 1984, Mechanism-based inactivation of mitochondrial monoamine oxidase by N-(1-methylcyclopropyl)benzylamine, *Biochemistry* **23**:1322–1332.
214. Silverman, R. B., 1983, Mechanism of inactivation of monoamine oxidase by *trans*-2-phenylcyclopropylamine and the structure of the enzyme–inactivator adduct, *J. Biol. Chem.* **258**:14766–14769.
215. Wada, A., Okamoto, M., Nonaka, Y., and Yamano, T., 1984, Aldosterone biosynthesis by a reconstituted cytochrome P-450$_{11\beta}$ system, *Biochem. Biophys. Res. Commun.* **119**:365–371.
216. Lambeth, J. D., Kitchen, S. E., Farooqui, A. A., Tuckey, R., and Kamin, H., 1982, Cytochrome P-450$_{scc}$–substrate interactions: Studies of binding and catalytic activity using hydroxycholesterols, *J. Biol. Chem.* **257**:1876–1884.
217. Tuckey, R. C., and Kamin, H., 1983, Kinetics of O_2 and CO binding to adrenal cytochrome P-450$_{scc}$: Effect of cholesterol, intermediates, and phosphatidylcholine vesicles, *J. Biol. Chem.* **258**:4232–4237.
218. Byon, C.-Y., and Gut, M., 1980, Steric considerations regarding the biodegradation of cholesterol to pregnenolone: Exclusion of (22S)-22-hydroxycholesterol and 22-ketocholesterol as intermediates, *Biochem. Biophys. Res. Commun.* **94**:549–552.
219. Burstein, S., Middleditch, B. S., and Gut, M., 1975, Mass spectrometric study of the enzymatic conversion of cholesterol to (22R)-22-hydroxycholesterol, (20R,22R)-20,22-dihydroxycholesterol, and pregnenolone, and of (22R)-22-hydroxycholesterol to the glycol and pregnenolone in bovine adrenocortical preparations, *J. Biol. Chem.* **250**:9028–9037.
220. Fishman, J., 1982, Biochemical mechanisms of aromatization, *Cancer Res.* **42**:3277s–3280s.
221. Thompson, E. A., and Siiteri, P. K., 1974, Utilization of oxygen and reduced nicotinamide adenine dinucleotide phosphate by human placental microsomes during aromatization of androstenedione, *J. Biol. Chem.* **249**:5364–5372.
222. Higashiyama, T., and Osawa, Y., 1984, Purification and partial characterization of two distinct human placental aromatase cytochromes P-450, *Fed. Proc.* **43**:2033.
223. Caspi, E., Arunachalam, T., and Nelson, P. A., 1983, Biosynthesis of estrogens: The steric mode of the initial C-19 hydroxylation of androgens by human placental aromatase, *J. Am. Chem. Soc.* **105**:6987–6989.
224. Osawa, Y., Shibata, K., Rohrer, D., Weeks, C., and Duax, W. L., 1975, Reassignment of the absolute configuration of 19-substituted 19-hydroxysteroids and stereomechanism of estrogen biosynthesis, *J. Am. Chem. Soc.* **97**:4400–4402.
225. Arigoni, D., Battaglia, R., Akhtar, M., and Smith, T., 1975, Stereospecificity of oxidation at C-19 in oestrogen biosynthesis, *J. Chem. Soc. Chem. Commun.* **1975**:185–186.
226. Miyairi, S., and Fishman, J., 1983, Novel method of evaluating biological 19-hydroxylation and aromatization of androgens, *Biochem. Biophys. Res. Commun.* **117**:392–398.
227. Miyairi, S., and Fishman, J., 1985, Radiometric analysis of oxidative reactions in aromatization by placental microsomes: Presence of differential isotope effects, *J. Biol. Chem.* **260**:320–325.
228. Brodie, H. J., Kripalani, K. J., and Possanza, G., 1969, Studies on the mechanisms of estrogen biosynthesis. VI. The stereochemistry of hydrogen elimination at C-2 during aromatization, *J. Am. Chem. Soc.* **91**:1241–1242.

229. Fishman, J., and Guzik, H., 1969, Stereochemistry of estrogen biosynthesis, *J. Am. Chem. Soc.* **91**:2805–2806.
230. Fishman, J., Guzik, H., and Dixon, D., 1969, Stereochemistry of estrogen biosynthesis, *Biochemistry* **8**:4304–4309.
231. Fishman, J., and Raju, M. S., 1981, Mechanism of estrogen biosynthesis: Stereochemistry of C-1 hydrogen elimination in the aromatization of 2β-hydroxy-19-oxoandrostenedione, *J. Biol. Chem.* **256**:4472–4477.
232. Townsley, J. D., and Brodie, H. J., 1968, Studies on the mechanism of estrogen biosynthesis. III. The stereochemistry of aromatization of C_{19} and C_{18} steroids, *Biochemistry* **7**:33–40.
233. Hosoda, H., and Fishman, J., 1974, Usually facile aromatization of 2β-hydroxy-19-oxo-4-androstene-3,17-dione to estrone: Implications in estrogen biosynthesis, *J. Am. Chem. Soc.* **96**:7325–7329.
234. Goto, J., and Fishman, J., 1977, Participation of a nonenzymatic transformation in the biosynthesis of estrogens from androgens, *Science* **195**:80–81.
235. Hahn, E. F., and Fishman, J., 1984, Immunological probe of estrogen biosynthesis: Evidence for the 2β-hydroxylative pathway in aromatization of androgens, *J. Biol. Chem.* **259**:1689–1694.
236. Akhtar, M., Calder, M. R., Corina, D. L., and Wright, J. N., 1982, Mechanistic studies on C-19 demethylation in oestrogen biosynthesis, *Biochem. J.* **201**:569–580.
237. Caspi, E., Wicha, J., Arunachalam, T., Nelson, P., and Spiteller, G., 1984, Estrogen biosynthesis: Concerning the obligatory intermediacy of 2β-hydroxy-10β-formylandrost-4-ene-3,17-dione, *J. Am. Chem. Soc.* **106**:7282–7283.
238. Morand, P., Williamson, D. G., Layne, D. S., Lompa-Krzymien, L., and Salvador, J., 1975, Conversion of an androgen epoxide into 17β-estradiol by human placental microsomes, *Biochemistry* **14**:635–638.
239. Covey, D. F., and Hood, W. F., 1982, A new hypothesis based on suicide substrate inhibitor studies for the mechanism of action of aromatase, *Cancer Res.* **42**:3327s–3333s.
240. Beusen, D. D., and Covey, D. F., 1984, Study of the role of Schiff base formation in the aromatization of androgen substrates by human placenta, *Fed. Proc.* **43**:330.
241. Alexander, K., Akhtar, M., Boar, R. B., McGhie, J. F., and Barton, D. H. R., 1972, The removal of the 32-carbon atom as formic acid in cholesterol biosynthesis, *J. Chem. Soc. Chem. Commun.* **1972**:383–385.
242. Mitropoulos, K. A., Gibbons, G. F., and Reeves, E. A., 1976, Lanosterol 14α-demethylase: Similarity of the enzyme system from yeast and rat liver, *Steroids* **27**:821–829.
243. Canonica, L., Fiecchi, A., Galli Kienle, M., Scala, A., Galli, G., Grossi Paoletti, E., and Paoletti, R., 1968, Evidence for the biological conversion of $\Delta^{8,14}$ sterol dienes into cholesterol, *J. Am. Chem. Soc.* **90**:6532–6534.
244. Gibbons, G. F., Goad, L. J., and Goodwin, T. W., 1968, The stereochemistry of hydrogen elimination from C-15 during cholesterol biosynthesis, *J. Chem. Soc. Chem. Commun.* **1968**:1458–1460.
245. Watkinson, I. A., Wilton, D. C., Munday, K. A., and Akhtar, M., 1971, The formation and reduction of the 14,15-double bond in cholesterol biosynthesis, *Biochem. J.* **121**:131–137.
246. Alexander, K. T. W., Akhtar, M., Boar, R. B., McGhie, J. F., and Barton, D. H. R., 1971, The pathway for the removal of C-32 in cholesterol biosynthesis, *J. Chem. Soc. Chem. Commun.* **1971**:1479–1481.
247. Akhtar, M., Freeman, C. W., Wilton, D. C., Boar, R. B., and Copsey, D. B., 1977, The pathway for the removal of the 14α-methyl group of lanosterol: The role of lanost-8-ene-3β,32-diol in cholesterol biosynthesis, *Bioorg. Chem.* **6**:473–481.

248. Akhtar, M., Alexander, K., Boar, R. B., McGhie, J. F., and Barton, D. H. R., 1978, Chemical and enzymic studies on the characterization of intermediates during the removal of the 14α-methyl group in cholesterol biosynthesis: The use of 32-functionalized lanostan derivatives, *Biochem. J.* **169**:449–463.
249. Pascal, R. A., Chang, P., and Schroepfer, G. J., 1980, Possible mechanisms of demethylation of 14α-methyl sterols in cholesterol biosynthesis, *J. Am. Chem. Soc.* **102**:6599–6601.
250. Gibbons, G. F., Pullinger, C. R., and Mitropoulos, K. A., 1979, Studies on the mechanism of lanosterol 14α-demethylation: A requirement for two distinct types of mixed-function-oxidase systems, *Biochem. J.* **183**:309–315.
251. Hansson, R., and Wikvall, K., 1982, Hydroxylations in biosynthesis of bile acids: Cytochrome P-450 LM$_4$ and 12α-hydroxylation of 5β-cholestane-3α,7α-diol, *Eur. J. Biochem.* **125**:423–429.
252. Meigs, R. A., and Ryan, K. J., 1971, Enzymatic aromatization of steroids. I. Effects of oxygen and carbon monoxide on the intermediate steps of estrogen biosynthesis, *J. Biol. Chem.* **246**:83–87.
253. Zachariah, P. K., and Juchau, M. R., 1975, Interactions of steroids with human placental cytochrome P-450 in the presence of carbon monoxide, *Life Sci.* **16**:1689–1692.
254. Yoshida, Y., and Aoyama, Y., 1984, Yeast cytochrome P-450 catalyzing lanosterol 14α-demethylation. I. Purification and spectral properties, *J. Biol. Chem.* **259**:1655–1660.
255. Aoyama, Y., Yoshida, Y., and Sato, R., 1984, Yeast cytochrome P-450 catalyzing lanosterol 14α-demethylation. II. Lanosterol metabolism by purified P-450$_{14DM}$ and by intact microsomes, *J. Biol. Chem.* **259**:1661–1666.
256. Trzaskos, J. M., Bowen, W. D., Shafiee, A., Fischer, R. T., and Gaylor, J. L., 1984, Cytochrome P-450-dependent oxidation of lanosterol in cholesterol biosynthesis: Microsomal electron transport and C-32 demethylation, *J. Biol. Chem.* **259**:13402–13412.
257. Ramm, P. J., and Caspi, E., 1969, The stereochemistry of tritium at carbon atoms 1, 7, and 15 in cholesterol derived from (3R,2R)-(2-^{3}H)-mevalonic acid, *J. Biol. Chem.* **244**:6064–6073.
258. Akhtar, M., Rahimtula, A. D., Watkinson, I. A., Wilton, D. C., and Munday, K. A., 1969, The status of C-6, C-7, C-15, and C-16 hydrogen atoms in cholesterol biosynthesis, *Eur. J. Biochem.* **9**:107–111.
259. Spike, T. E., Wang, A. H.-J., Paul, I. C., and Schroepfer, G. J., 1974, Structure of a potential intermediate in cholesterol biosynthesis, *J. Chem. Soc. Chem. Commun.* **1974**:477–478.
260. Ullrich, V., Castle, L., and Weber, P., 1981, Spectral evidence for the cytochrome P-450 nature of prostacyclin synthetase, *Biochem. Pharmacol.* **30**:2033–2036.
261. Graf, H., Ruf, H. H., and Ullrich, V., 1983, Prostacyclin synthase, a cytochrome P-450 enzyme, *Angew. Chem. Int. Ed. Engl.* **22**:487–488.
262. Haurand, M., and Ullrich, V., 1982, Isolation and characterization of thromboxane synthase as a cytochrome P-450 enzyme, *Hoppe-Seylers Z. Naturforsch.* **363**:972.

CHAPTER 8

Inhibition of Cytochrome P-450 Enzymes

PAUL R. ORTIZ de MONTELLANO and
NORBERT O. REICH

1. Introduction

The catalytic cycle of cytochrome P-450 (see Chapter 7) traverses three steps that are particularly vulnerable to inhibition: (1) the binding of substrates, (2) the binding of molecular oxygen subsequent to the first electron transfer, and (3) the catalytic step in which the substrate is actually oxidized. This chapter focuses on inhibitors that act at one of these three steps. Inhibitors that act at other steps in the catalytic cycle, such as quinones that interfere with the electron supply to the hemeprotein by accepting electrons directly from cytochrome P-450 reductase,[1-3] are not discussed here.

Cytochrome P-450 inhibitors can be divided into three mechanistically differentiable categories: (1) agents that bind reversibly, (2) agents that form quasi-irreversible complexes with the heme iron atom, and (3) agents that bind irreversibly to the protein or the prosthetic heme group, or that accelerate degradation of the prosthetic heme group without demonstrably binding to it. For the most part, inhibitors that interfere in the catalytic cycle prior to the actual oxidative event are reversible competitive or noncompetitive inhibitors. Agents that act during or subsequent to the oxygen transfer step, however, are generally irreversible or quasi-irreversible inhibitors and, in many instances, fall into the category of mechanism-based (or suicide) inhibitors. Extended lists of P-450 in-

---

PAUL R. ORTIZ de MONTELLANO and NORBERT O. REICH • Department of Pharmaceutical Chemistry, School of Pharmacy, University of California, San Francisco, California 94143.

hibitors, particularly of the more classical reversible agents, are available in recent reviews.[4,5] The emphasis in this chapter is on the mechanisms of inhibition; thus, most of the chapter is devoted to a discussion of agents that require catalytic turnover of the enzyme because the mechanisms of reversible competitive and noncompetitive inhibitors, despite their practical importance, are relatively straightforward.

2. Reversible Inhibitors

Inhibitors that compete reversibly with substrates for occupancy of the active site include substances that bind to its hydrophobic domain, that coordinate to the prosthetic heme iron atom, or that participate in specific hydrogen bonding or ionic interactions with specific active-site residues.[4,5] The first mechanism, simple competition for binding to the lipophilic domain of the active site, is evidenced by the competition that exists between the substrates of a given P-450 isozyme. This inhibition is optimal when the inhibitory substance is bound tightly but is a poor substrate. Inhibition by this mechanism is not particularly effective but, in appropriate situations, can cause physiologically relevant metabolic changes. A clear-cut example of such inhibition is provided by the mutual *in vitro* and *in vivo* inhibition of benzene and toluene metabolism.[6]

2.1. Coordination to Ferric Heme

The coordination of a strong ligand to the pentacoordinate iron, or the displacement of a weak ligand from the hexacoordinate heme by a strong ligand, shifts the enzyme from the high- to the low-spin form and gives rise to a "type II" binding spectrum with a Soret maximum at approximately 430 nm.[7-9] This spin state change is accompanied by a change in the redox potential of the enzyme that makes its reduction by cytochrome P-450 reductase more difficult.[10,11] This change in reduction potential, as much as physical occupation of the sixth coordination site, is responsible for the inhibition associated with the binding of strong ligands.

Ionic ligands such as cyanide bind preferentially to the ferric form of P-450.[12,13] The triple ferric positive charge is matched in the enzyme by the negative charges of its three ligands (the two porphyrin nitrogens and the thiolate), but the negative charges exceed the two positive charges of the ferrous iron. The cyanide binds more readily to the neutral (ferric) than the negative (ferrous) enzyme. In fact, cyanide binds more weakly to ferric P-450 than to ferric myoglobin because the thiolate ligand of P-450 places a higher electron density on the iron than does the imidazole ligand of myoglobin.[14] The chelation of ionic ligands is disfavored, in addition, by the lipophilic nature of the P-450 active site.[15]

2.2. Coordination to Ferrous Heme

Carbon monoxide, the simplest uncharged ligand, binds exclusively to the ferrous (reduced) form of P-450. Binding of carbon monoxide to the ferrous prosthetic heme group entails donation of electrons from the carbon to the iron through a σ-bond as well as back-donation of electrons from the occupied iron d-orbitals to the empty antibonding π-orbitals of carbon monoxide.[20] P-450 enzymes are so named because their carbon monoxide complexes have absorption maxima at approximately 450 nm.[16] The finding that the 450-nm absorption can be reproduced with model ferroporphyrins only with a thiolate ligand *trans* to the carbon monoxide provided key early evidence for a thiolate fifth ligand in P-450.[17] Inhibition by carbon monoxide is one of the hallmarks of processes catalyzed by P-450, although a number of the reactions catalyzed by biosynthetic P-450 isozymes are relatively resistant to inhibition by carbon monoxide.[18,19]

2.3. Heme Coordination and Lipophilic Binding

Inhibitors that bind to lipophilic regions of the protein, and simultaneously bind to the prosthetic heme iron, are inherently more effective than agents that depend on only one of these binding interactions. The activity of such agents as inhibitors of P-450 is governed both by their hydrophobic character and the strength of the bond between their heteroatomic lone pair and the prosthetic heme iron. The less effective agents, including alcohols, ethers, ketones, lactones and other structures in which the coordinating atom is an oxygen, only coordinate weakly to the prosthetic heme iron.[9,21–24] The Soret band of such complexes is found at approximately 415 nm.[7,9] The most effective reversible inhibitors, in contrast, interact strongly with both the protein and the prosthetic heme iron.[4,5] The binding of inhibitors that are strong iron ligands, as already noted for cyanide, gives rise to what is termed a type II difference spectrum with a Soret maximum at 430 nm.[7,9,25,26] For the most part, these powerful inhibitors are nitrogen-containing aliphatic and aromatic compounds.

Pyridine and imidazole derivatives have found particularly widespread utility as P-450 inhibitors.[4] Metyrapone (Fig. 1), one of the most frequently employed P-450 inhibitors, first gained prominence as an inhibitor of the 11β-hydroxylase that catalyzes the final step in cortisol biosynthesis.[27] This activity led to its use in the diagnosis and treatment

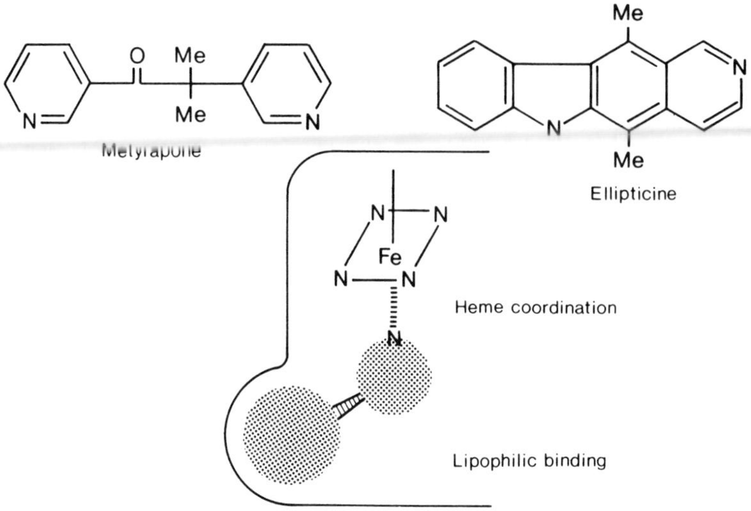

FIGURE 1. Schematic diagram of the two-point binding of agents with a lipophilic domain and a coordinating nitrogen function. The structures of two such agents are given.

of hypercortisolism (Cushing's syndrome) and other hormonal disorders.[28] The factors that determine the inhibitory potency of metyrapone and other nitrogen heterocycles are valid for most reversible inhibitors: (1) the intrinsic affinity of the ligand electron pair for the prosthetic heme iron, (2) the degree to which the intrinsic affinity of the ligand for the iron is moderated by steric interactions with substituents on the inhibitor,[29,30] (3) the lipophilicity of the nonligating portion of the inhibitor,[15,31] and, naturally, (4) the congruence between the geometry of the inhibitor and that of the active site (Fig. 1). The synergism between binding to the lipophilic domain and coordination with the heme iron is illustrated by the fact that imidazole and benzene individually are weak inhibitors, but when joined in phenylimidazole constitute a powerful inhibitor.[29] Optimization of the specificity of metyrapone, aminoglutethimide, and other classical inhibitors by structural modification continues to be of interest (see also Section 4).[31–33]

The ellipticines (Fig. 1), a relatively new but clinically interesting class of heterocyclic P-450 inhibitors, interact with both the ferrous and ferric forms of the enzyme.[34] The inhibition constants are in the 1–10 μM range but the inhibition is competitive in phenobarbital-treated rat liver microsomes and noncompetitive in microsomes from Arochlor-treated rats. This change in inhibitory mechanism may reflect interaction of the inhibitor with different forms of the enzyme, but may also be explained by alternative mechanisms (e.g., interaction of the inhibitor or its metabolites with P-450 reductase).

3. Catalysis-Dependent Inhibition

Several classes of inhibitors are known that are catalytically activated by the enzyme to transient species that irreversibly or quasi-irreversibly inhibit the enzyme. The inhibitory activity of the catalytically-generated species is superimposed on the reversible inhibition associated with binding of the parent structure to the ferric enzyme. Mechanism-based[35,36] (catalysis-dependent) inhibitors are potentially more enzyme-specific than reversible inhibitors because: (1) the inhibitor must first bind to the enzyme and therefore must satisfy the constraints imposed on classical inhibitors, (2) the inhibitor must then be catalytically activated and therefore must be acceptable as a substrate, and, finally, (3) reactive species produced by catalytic turnover irreversibly alter the enzyme and remove it permanently from the catalytic pool. Three classes of catalysis-dependent irreversible inhibitors of P-450 are known: (1) agents that bind covalently to the protein, (2) agents that quasi-irreversibly coordinate to the prosthetic heme iron, and (3) agents that alkylate or degrade the prosthetic heme group.

3.1. Covalent Binding to the Protein

Agents that inactivate P-450 by binding covalently to the protein after they are oxidatively activated by the same enzyme include a variety of sulfur compounds (e.g., carbon disulfide,[37-39] parathion,[40,41] diethyldithiocarbamate,[42] isothiocyanates,[43] thioureas,[44] and possibly mercaptosteroids[45]) and halogenated structures such as chloramphenicol.[46-49] The details of the mechanisms by which these compounds inactivate P-450 remain to be defined, although major progress has been made in elucidating the mechanisms of action of parathion and chloramphenicol. It has been shown for parathion that the protein is radiolabeled when [^{35}S]parathion is used but not when the radioactivity is present in the ethyl groups as ^{14}C.[40,41] Ninety percent of the ^{35}S label bound covalently to microsomal proteins is precipitated by antibodies to P-450 enzymes. Approximately 75% of the prosthetic heme of P-450 is lost in incubations of the enzyme with parathion, but no information is available on its fate. The bulk (50–75%) of the sulfur radiolabel is removed from the protein by cyanide or dithiothreitol, a fact consistent with the suggestion that it is present in the form of hydrodisulfides (RSSH), but the activity of the enzyme is not restored by these treatments. The covalent

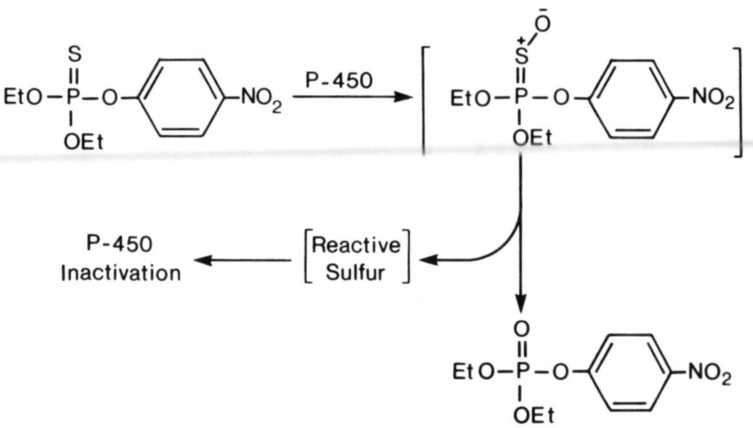

FIGURE 2. The oxidative activation of parathion. Direct experimental evidence for the structures of the intermediates in brackets is not available.

binding of sulfur, the well-established oxidation of sulfur compounds to S-oxides, and the formation of metabolites in which the sulfur is replaced by an oxygen, suggest the oxidative activation mechanism in Fig. 2.

In the case of chloramphenicol, not only has binding of [^{14}C]chloramphenicol to the apoprotein been correlated with loss of ethoxycoumarin deethylase activity, but proteolytic digestion of the protein has been shown to yield a single radiolabeled amino acid.[46-49] Hydrolysis of the modified amino acid yielded lysine and the chloramphenicol fragment shown in Fig. 3. Chloramphenicol is thus oxidized to an oxamyl chloride intermediate that either is hydrolyzed to the oxamic acid metabolite or acylates a critical lysine in the protein (Fig. 3). Acylation of the lysine apparently interferes with electron transfer from the reductase to the heme because the inactivated hemeprotein is still able to catalyze the deethylation of ethoxycoumarin supported by cumene hydroperoxide or iodosobenzene.[49]

The data on parathion and chloramphenicol clearly show that P-450 enzymes can generate reactive species that alkylate, acylate, or otherwise modify the protein skeleton. It is interesting, in view of the importance of protein alkylation in the mechanism-based inactivation of most enzymes,[35,36] that protein alkylation is much less important a mechanism for the inactivation of P-450 enzymes than is heme alkylation or degradation. This generalization appears to hold even if allowance is made for the fact that protein alkylation may have gone undetected in instances where enzyme inactivation has been attributed solely to heme alkylation. The data on parathion suggest, in fact, that heme destruction is required for enzyme inactivation even in some instances where protein modifi-

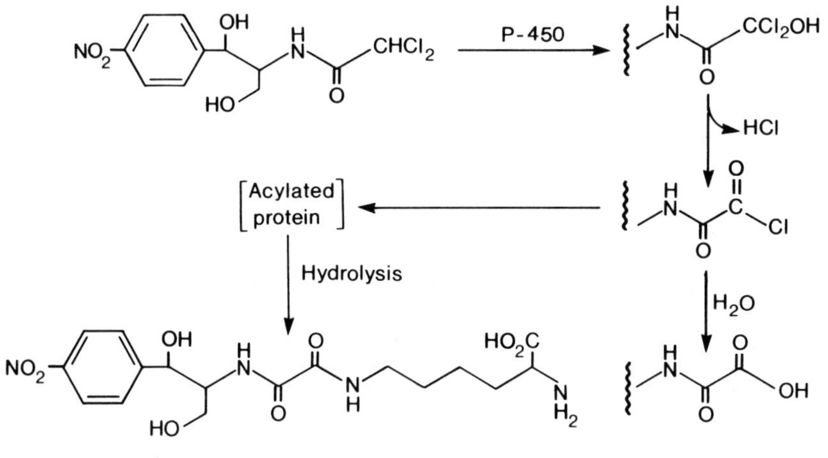

FIGURE 3. Mechanism proposed for the catalysis-dependent inactivation of P-450 by chloramphenicol. The acylated protein residue has been isolated and characterized.

cation clearly occurs. Approximately 4 nmole of radiolabeled sulfur binds covalently to the protein for each nanomole of heme chromophore that is lost. Catalytic turnover and sulfur activation apparently continue despite covalent attachment of sulfur to the protein until the heme itself is damaged.[40] Spironolactone (Fig. 4), an aldosterone antagonist employed as a diuretic and antihypertensive,[50,51] provides a related example. The inactivation of P-450 isozymes in the liver, adrenals, and testicles by this thiosteroid, which depends on the presence of the 7-thiol moiety and requires catalytic turnover of the enzyme, is closely linked to destruction of the prosthetic heme moiety.[45,52–54]

The greater vulnerability of P-450 to heme rather than protein modification may have several origins. First of all, the heme is perfectly positioned to intercept catalytically activated species. The active-site re-

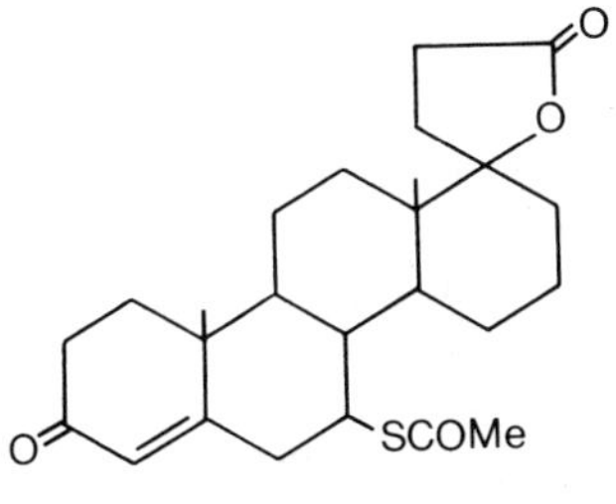

FIGURE 4. Structure of spironolactone.

gion, if the the X-ray crystal structure of P-450$_{cam}$ can be generalized (Chapter 13), may furthermore be relatively free of nucleophilic residues. Finally, and perhaps most importantly, the probable radical nature of the catalytically generated intermediates favors their capture by the heme. The absence of active-site nucleophiles is teleologically attractive, given that the active site of P-450 enzymes must be nonpolar (to facilitate the binding of lipophilic substrates) and resistant to the catalytically activated oxygen- and substrate-derived species generated by the enzyme (Chapters 3, 5, and 7). Preferential inactivation of P-450 enzymes by radicals that react with the heme rather than by protein-reactive electrophilic intermediates has practical consequences for the design of irreversible inhibitors.

3.2. Quasi-Irreversible Coordination to the Prosthetic Heme

Agents that are catalytically oxidized to intermediates that coordinate so tightly to the prosthetic heme of P-450 that they can only be displaced under special experimental conditions are considered in this section. The two major classes of such inhibitors are compounds with a dioxymethylene function and nitrogen compounds, usually amines, that are converted *in situ* to nitroso metabolites. 1,1-Disubstituted hydrazines and acyl hydrazines also appear to inhibit P-450, to some extent, by a related mechanism. Reductive coordination of halocarbons to the prosthetic heme under anaerobic conditions is covered in Section 3.4 because the reaction is closely associated with heme destruction.

3.2.1. Methylenedioxy Compounds

Aryl and alkyl methylenedioxy compounds, some of which are commercially employed as insecticide synergists,[55,56] are oxidized by P-450 to species that coordinate tightly to the prosthetic heme iron.[57] The catalytic role of the enzyme in unmasking the inhibitory species is confirmed by the time, NADPH, oxygen, and concentration dependence of the process, as well as by the finding that cumene hydroperoxide can substitute for the NADPH and oxygen requirement.[56-60] The ferrous complex is characterized by a difference absorption spectrum with maxima at 427 and 455 nm, whereas the ferric complex has a single absorption maximum at 437 nm.[57,59,60] The peaks at 427 and 455 nm are due to distinct complexes, although the structural relationship between the two complexes remains obscure.[57] The quasi-irreversible nature of the ferrous complex is evidenced by the fact that it can be isolated intact from animals treated with isosafrole. The catalytically active enzyme can be regenerated from the much less stable ferric complex, however, by incubation with lipo-

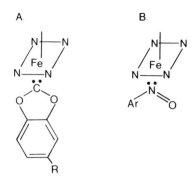

FIGURE 5. Structures proposed for the quasi-irreversible complexes formed during the catalytic turnover of (A) methylenedioxyphenyl compounds and (B) primary amines.

philic compounds that displace the inhibitor from the active site.[61,62] The ferrous complex is not disrupted by incubation with lipophilic compounds but can be broken by irradiation at 400–500 nm.[63,64] This indicates that the stability of the ferrous complex stems, as already described for reversible inhibitors (Section 2.3), from concurrent binding to the lipophilic active site[65] and to the ferrous prosthetic heme group. The weakening of the complex associated with the ferrous to ferric transition indicates that the activated species, like carbon monoxide, only binds to the ferrous heme. The evidence suggests the catalysis-dependent formation of a carbene–iron complex (Fig. 5A), although the actual structure of the hemeprotein complex remains to be established. Circumstantial evidence for a carbene complex is provided by the synthesis and characterization of model complexes.[66,67] The structural resemblance between carbenes and carbon monoxide readily rationalizes the 455-nm absorption maximum of the complexes. The absorbance maximum at 427 nm presumably is due to a different complex, perhaps to a carbene complex in which the *trans* ligand, as in P-420, is not a thiolate.[68] The carbene formulation rationalizes the incorporation of oxygen from the medium into the carbon monoxide derived from the dioxymethylene bridge (see below), and the observation that electron-withdrawing substituents increase the proportion of the carbon monoxide metabolite.[69] Addition of a hydroxyl to the iron-coordinated carbene would yield an iron-coordinated anion that could readily fragment into the observed catechol and carbon monoxide metabolites (Fig. 6). A different mechanism is required, however, to explain the fact that a fraction of the carbon monoxide incorporates an atom from molecular oxygen.

Association of the inhibitory activity with the dioxymethylene function, the importance of enzymatic turnover for inhibition, and enzymatic oxidation of the dioxymethylene group indicate that oxidation of that group is essential for quasi-irreversible inhibition. Free radical,[70]

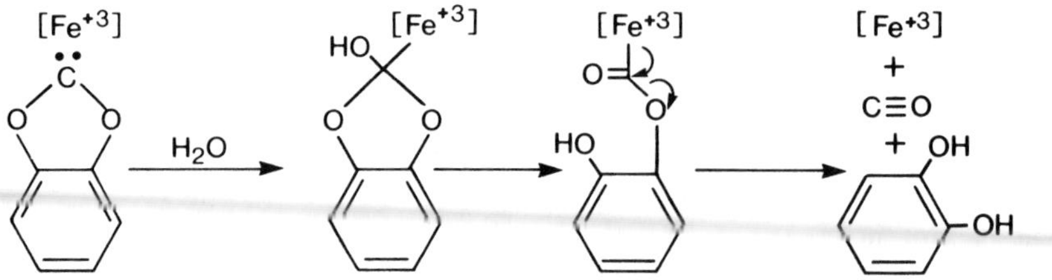

FIGURE 6. A possible mechanism for hydrolytic release of carbon monoxide from a carbene–iron complex. The prosthetic heme is shown as an iron in brackets.

carbocation[71] and carbanion[63] intermediates have been suggested, but the results are most consistent with formation of the carbene from the 2-hydroxylated metabolite or from a radical precursor of it (Fig. 7). Substituents on the dioxymethylene group, except for the ethoxy group, block complex formation.[56,64,72] The anomalous activity of the ethoxy-substituted compound is readily explained because o-dealkylation would yield the 2-hydroxylated metabolite obtained by oxidation of the unsubstituted compound.[64] The metabolism of aryldioxymethylene compounds to catechols, carbon monoxide, carbon dioxide, and formic acid is consistent with such a hydroxylation reaction,[56,69,73–75] particularly in view of the fact that carbon monoxide is evolved more slowly when the dioxymethylene hydrogens are replaced by deuteriums (k_H/k_D = 1.7–2.0). The observation of a similar isotope effect on the *in vivo* synergistic activity supports the hypothesis that carbon monoxide and complex formation are mechanistically linked.[76]

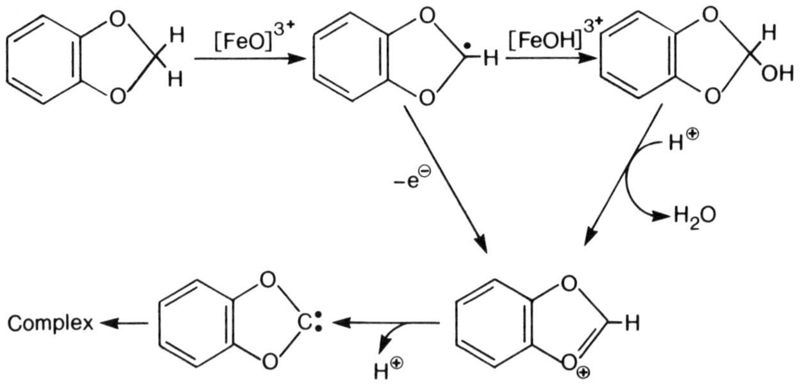

FIGURE 7. Alternative mechanisms for the oxidation of methylenedioxyphenyl compounds to carbenes via oxonium intermediates.

Oxidation of the dioxymethylene bridge to the carbene presumably involved in complex formation could occur by one of three mechanisms. Hydroxylation of the dioxymethylene bridge, followed by elimination of a molecule of water, could yield the oxonium ion (Fig. 7). Deprotonation of the acidic oxonium intermediate would provide the desired carbene. The oxonium intermediate could be obtained without actually passing through the 2-hydroxylated metabolite if the radical generated by removal of a hydrogen during the hydroxylation process transfers the unpaired electron to the oxidative species rather than collapsing with it (Fig. 7). Finally, the radical formed by hydrogen abstraction could collapse with the resulting $[FeOH]^{2+}$ species to give a carbon–iron complex in which the hydroxyl and the carbon are simultaneously bound to the iron.[66] Deprotonation and transfer of the oxygen from the iron to the carbon would then yield an intermediate that could give the carbene complex by eliminating a molecule of water or that could decompose to the carbon monoxide metabolite. It is not possible at this time to determine which of these mechanisms is operative.

3.2.2. Amines

The second, quite large, class of agents that form quasi-irreversible complexes [metabolic-intermediate (MI) complexes][77] with the heme of P-450 is composed of alkyl and aromatic amines, including a number of clinically useful amine antibiotics.[4,77–79] Oxidation of these amines yields intermediates that coordinate tightly to the ferrous heme and give rise to a spectrum with an absorbance maximum in the region of 445–455 nm.[77] Primary amines are required for complex formation, but secondary and tertiary amines are suitable precursors of the P-450 complexes if they are *N*-dealkylated *in situ* to the primary amines. The complexes formed with aromatic amines differ from those obtained with alkyl amines in that they are unstable to reduction by dithionite.[80] The normal competitive inhibition associated with the binding of amines does not, of course, require catalytic activation of the inhibitor, but activation is essential for formation of the tight, quasi-irreversible, complexes.[77,80,81] The primary amines appear to first be hydroxylated because the corresponding hydroxylamines also yield the complexes,[82] but the functionality that coordinates to the iron lies beyond the hydroxylamine in the oxidative scale because the hydroxylamines must also be oxidatively activated.[80,82] The chelated function appears, in fact, to be the nitroso group obtained by two-electron oxidation of the hydroxylamine (Fig. 5B).[81,82,84] The final oxidation may not actually require catalytic turnover of the enzyme because hydroxylamines autoxidize readily.[85] The conclusion that the nitroso function is involved in iron chelation is supported by the demonstration that apparently identical complexes are obtained from nitro compounds under reductive conditions.[86] The crystal structure of the complex between a nitroso compound and a model iron porphyrin indicates that, as expected, the iron binds to the nitrogen rather than the oxygen.[66]

3.2.3. 1,1-Disubstituted and Acyl Hydrazines

1,1-Disubstituted hydrazines, in contrast to monosubstituted hydrazines (see Section 3.3.4), are oxidized by P-450 to intermediates that chelate tightly to the prosthetic heme iron. The complexes, formed in a time-, NADPH-, and oxygen-dependent manner, are characterized by a ferric absorption maximum at approximately 438 nm and a ferrous maximum at 449 nm.[87] A similar transient complex with an absorbance maximum of 449 nm has been detected during the microsomal oxidation of isoniazid and other acyl hydrazines,[83,88] but the isoniazid complex falls apart on addition of ferricyanide and thus is only stable in the ferrous state.[89] Chemical model studies indicate that the oxidation of 1,1-dialkylhydrazines yields disubstituted nitrenes that form end-on complexes with the iron of metalloporphyrins. Specifically, nitrene complexes derived from 1-amino-2,2,6,6-tetramethyl-piperidine and several iron tetraarylporphyrins have been isolated and characterized by NMR, Mössbauer, and X-ray methods.[90,91] It therefore appears likely that the P-450 complexes formed during the metabolism of 1,1-disubstituted hydrazines, and possibly acyl hydrazines, are aminonitrene–iron complexes (Fig. 8). Ox-

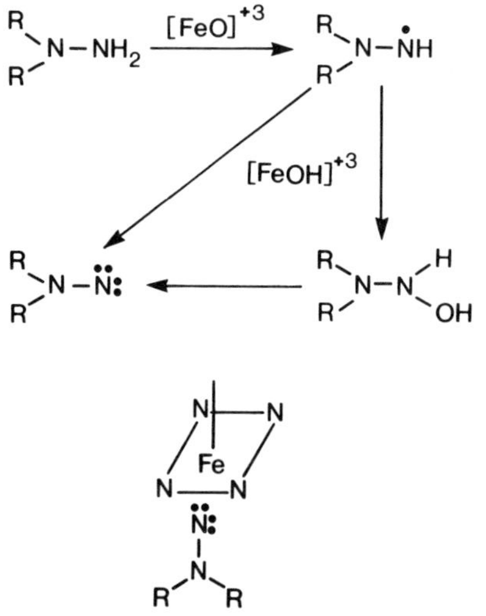

FIGURE 8. Oxidative routes to the nitrene–iron structures proposed for the complexes formed during the metabolism of 1,1-dialkylhydrazines.

idation of the dialkylhydrazines to the required aminonitrenes is readily rationalized by hydroxylation of the hydrazine or, more probably, by stepwise electron removal from the hydrazine (Fig. 8).

3.3. Covalent Binding to the Prosthetic Heme

It is now well documented that irreversible inactivation of P-450 frequently reflects covalent attachment of the inhibitor, or a fragment derived from it, to the prosthetic heme group. The evidence for a heme alkylation mechanism includes, in most instances, the demonstration of equimolar enzyme and heme loss and isolation and structural characterization of the resulting alkylated heme. It is important to note that a parallel loss of enzyme and heme is not sufficient, in the absence of explicit evidence for heme adduct formation, to conclude that heme alkylation is responsible for enzyme inactivation because alternative mechanisms exist for catalysis-dependent destruction of the prosthetic group (see Section 3.4). The possibility nevertheless must be kept in mind that heme adducts may be formed that are too unstable to be isolated. The technical difficulties inherent in quantitating heme adducts have prevented the quantitative correlation of heme adduct formation with enzyme inactivation. In the absence of such data, the possibility exists that mechanisms other than heme loss may operate even when heme alkylation is conclusively demonstrated.

3.3.1 Terminal Olefins

The P-450-catalyzed epoxidation of terminal olefins is paralleled, in many instances, by N-alkylation of the prosthetic heme group and inactivation of the enzyme.[92] The realization that self-catalyzed heme alkylation is a relatively common phenomenon evolved from early studies with 2-isopropyl-4-pentenamide (AIA) and 5-allyl-substituted barbiturates,[93,94] which established that the oxidative metabolism of homoallylic amides results in (1) equimolar loss of P-450 and microsomal heme, (2) accumulation of a green (red-fluorescent) porphyrin and, (3) derangement of the heme biosynthetic pathway.

A double bond with no more than one substituent is the only structural prerequisite for prosthetic heme alkylation by olefins. This is clearly demonstrated by the ability of ethylene, but not ethane, to destroy the prosthetic heme group, and by the inactivity of structures such as 3-hexene, cyclohexene, and 2-methyl-1-heptene.[95] Even monosubstituted olefins do not inactivate the enzyme if they are not accepted as substrates by the enzyme, if a group other than the double bond is oxidized, or if the double bond, as in styrene, is part of a conjugated system.[95] These observations suggest that heme alkylation by olefins is subject to strong steric constraints and is suppressed by substituents that delocalize charge or electron density from the double bond.

The structures of the N-alkylated porphyrins isolated from the livers of rats treated with a variety of olefins (ethylene, propene, octene, fluroxene, 2,2-diethyl-4-pentenamide, 2-isopropyl-4-pentenamide, and vinyl fluoride) have been unambiguously elucidated by spectroscopic methods. In all of the adducts, a porphyrin nitrogen is bound to the terminal carbon of the double bond and an oxygen to the internal carbon.[94,97] The oxygen in the ethylene and AIA adducts has been shown by ^{18}O studies to derive from molecular oxygen and therefore is presumed to be the catalytically activated oxygen.[96,97] The not unreasonable hypothesis that the heme adduct is formed by nucleophilic addition of the porphyrin nitrogen to the epoxide metabolite is ruled out because (1) the enzyme is not inactivated by the epoxides of olefins that inactivate the enzyme,[92,95] (2) the nitrogen and the oxygen add across the double bond in a *cis* fashion rather than in the *trans* fashion expected for addition of a nucleophile to an epoxide,[96] (3) the nitrogen reacts with the terminal rather than internal carbon of vinyl ethers even through the internal (oxygen-substituted) carbon is far more reactive in the corresponding epoxides[98] and (4) the pyrrole nitrogens of the heme are very poor nucleophiles and do not react with epoxides even under harsh chemical conditions. These results, in conjunction with the clear demonstration that enzyme turnover is required for enzyme inactivation, indicate that enzyme inactivation results from catalytic oxygen transfer to the double bond but is *not* mediated by the epoxide metabolite.

Linear olefins (ethylene, propene, octene) only detectably alkylate pyrrole ring D of the prosthetic group of the phenobarbital-inducible isozymes from rat liver, but heme alkylation by two "globular" olefins (2-isopropyl-4-pentenamide and 2,2-diethyl-4-pentenamide) is less regiospecific.[97,99] A detailed study of the regiochemistry and stereochemistry of heme alkylation by *trans*-[1-^2H]-1-octene shows (1) that the olefin stereochemistry is preserved during the alkylation reaction, and (2) that heme alkylation only occurs when the oxygen is delivered to the *re* face of the double bond even though stereochemical analysis of the epoxide metabolite indicates that the oxygen is delivered almost equally to *both* faces of the π-bond.[96] Heme alkylation is thus a highly regio- and stereospecific process.

A number of heme alkylation mechanisms are consistent with the available data, none of which involves single-step transfer of the oxygen to the π-bond (see Chapter 7, Fig. 13). The oxygen could add to the π-bond to give a transient carbon radical that alkylates the heme, closes to the epoxide, or transfers the unpaired electron to the heme before alkylating the heme or collapsing to the epoxide. Electrophilic oxygen addition could generate the cationic intermediate directly, but the observation that the enzyme is inactivated by fluroxene approximately every 100–200th turnover argues against such a mechanism. Electrophilic addition to the vinyl ether should result exclusively in oxygen addition to the terminal carbon because the charge would be stabilized on the internal carbon by the vicinal ether oxygen, but this orientation is contrary to that required by the isolated heme adduct.[98] The alkylation regiochemistry is readily reconciled, however, with mechanisms in which electron transfer to the activated oxygen complex precedes carbon–oxygen bond formation because the resulting radical cation could react with the oxygen at either end. Indeed, it is possible that the partition between adduct and metabolite formation could be determined by the regiochemistry (oxygen to inner or outer carbon) in the radical cation–oxygen recombination step.

The oxidation of olefins by P-450 can also be explained by initial 2 + 2 addition of the oxoiron complex to the π-bond to give one of the two metallacyclobutane intermediates generated from the two possible relative orientations of the olefin and the oxidative species. The metallacyclobutane could equally well result from collapse of the acyclic radical intermediate proposed in the radical mechanism discussed above. Indirect evidence for the formation of the metallacyclobutane intermediates in epoxidation reactions catalyzed by model metalloporphyrins has been reported (Chapter 1). The partitioning between epoxide formation and heme alkylation in such a mechanism would be determined by the relative importance of the two possible metallacyclobutane intermediates.

The information available on heme alkylation during P-450-catalyzed olefin epoxidation clearly demonstrates that double bond oxidation cannot be mediated exclusively by a mechanism in which both carbon–oxygen bonds are formed simultaneously, but is not yet sufficient to permit differentiation of the possible mechanisms. The details of the heme alkylation mechanism, the parameters that govern partitioning between epoxidation and heme alkylation, and the extent to which the heme alkylation mechanism mirrors that which results in epoxide formation, remain to be elucidated.

3.3.2. Terminal Acetylenes

The oxidation of terminal acetylenes to substituted acetic acids by P-450 (Chapter 7) is accompanied, as in the case of terminal olefins, by alkylation of the prosthetic heme group. The relationship between structure and activity resembles that for olefins, except that there are fewer exceptions to the rule that oxidation of triple bonds results in enzyme inactivation. This is evident in the fact that phenylacetylene, but not styrene, inactivates the enzyme,[100] and that internal acetylenes inactivate the enzyme, albeit without the formation of detectable heme adducts, whereas internal olefins do not.[95,101] Enzyme inactivation requires catalytic turnover of the enzyme and yields, in the case of terminal acetylenes, heme adducts similar to those obtained with terminal olefins.[98,99] The primary difference in the structure of the adducts is that addition of a hydroxyl group and a porphyrin nitrogen across the triple bond yields an *enol* structure that is isolated in the more stable keto form.[98,99] The other difference is that linear acetylenes react almost exclusively with the nitrogen of pyrrole ring A whereas, as noted before, linear olefins react with the nitrogen of pyrrole ring D.[99] This difference in the alkylation regiochemistry has been used to define the active site topography (Chapter 3).

The mechanisms written for the inactivation of P-450 by terminal olefins also apply to its inactivation by terminal acetylenes, with the proviso that an additional double bond is carried by all the reaction intermediates (see Chapter 7, Fig. 13) and that the oxidation of an acetylene is relatively difficult. The oxidation of a triple bond is furthermore distinguished by the fact that the carbon to which the oxygen is *initially* attached can be discerned, whereas this information is lost when an olefin is oxidized due to the symmetry of the final epoxide metabolite. The acetylene data demonstrate that the oxygen is added to the terminal carbon in the metabolite but the internal carbon in the heme adduct. This difference in the regiochemistry of oxygen addition, and the finding that replacement of the acetylenic hydrogen of phenylacetylene by deuterium decreases the rate of metabolite formation but leaves the rate of enzyme inactivation unaffected, indicates that commitment of a given catalytic turnover to metabolite formation or heme alkylation occurs prior to the oxygen transfer step (i.e., very early in the catalytic process).[100]

The acetylenic function is particularly useful for the construction of isozyme-selective or -specific irreversible inhibitors. It has been used (Section 4) as the enabling moiety in isozyme-specific inhibitors of P-450_{scc}, aromatase, the fatty acid and leukotriene B_4 hydroxylases, and P-450_c from rat liver.[102]

3.3.3. Dihydropyridines and Dihydroquinolines

The perturbation of heme biosynthesis and the decrease in hepatic P-450 concentrations that follow administration of 3,5-bis(carbethoxy)-2,4,6-trimethyl-1,4-dihydropyridine (DDC)[103–105] have been traced to its alkylation of the prosthetic heme of P-450.[94,106,107] Heme alkylation is detected if the substituent at position 4 of the dihydropyridine ring is a primary, unconjugated moiety (methyl, ethyl, propyl, *sec*-butyl), but not if it is an aryl (phenyl), secondary (isopropyl), or conjugated (benzyl) group.[108,109] The 4-aryl-substituted dihydropyridines do not inactivate the enzyme at all, but those with the secondary or conjugated substituents inactivate the enzyme but do not yield detectable heme adducts.

The mechanism(s) of destruction by analogues that do not give heme adducts remains obscure (see Section 3.4), but substantial progress has been made in clarifying the mechanism of analogues that alkylate the heme. The adducts consist of protoporphyrin IX with the 4-alkyl group of the parent substrate covalently attached to one of the nitrogen atoms.[108,110,111] Some evidence exists that different nitrogens are alkylated in different P-450 isozymes.[112,113] Oxidation of the 4-alkyl-1,4-dihydropyridines is therefore accompanied by transfer of the 4-alkyl group to the prosthetic heme. The nature of the activation step is clarified by two observations: (1) an N-ethyl moiety on the dihydropyridine does not interfere with enzyme inactivation, and (2) incubation of the 4-ethyl-dihydropyridine analogue with hepatic microsomes in the presence of a spin trap results in catalysis-dependent accumulation of the ethyl radical spin adduct.[108] These results require oxidation of the dihydropyridine to a radical cation that aromatizes by extruding the ethyl radical (see Chapter 7, Fig. 16). The spin trapping data indicate the ethyl group is extruded as the free radical, but do not rule out direct transfer of the ethyl group to the heme in the *alkylation* process. Enzyme inactivation is not attenuated by glutathione or by the radical trap, which suggests that once the free radical escapes from the active site into the medium, where it presumably reacts with the spin trap, it is no longer able to return and alkylate the heme. Regardless of whether the 4-alkyl group is transferred directly or reacts as the free radical, the initial site of reaction with the heme may be the nitrogen or the iron (see Section 3.3.4). The absence of heme adducts in the inactivation caused by the 4-isopropyl and 4-benzyl analogues can be explained by such a mechanism, not only because the iron–nitrogen shift is sensitive to steric effects,[112] but also because the more oxidizable secondary or benzylic moieties may be converted to the corresponding cations by electron loss to the iron in preference to undergoing the oxidative iron–nitrogen shift.

The oxidation of dihydropyridines to radical cations that aromatize by radical extrusion suggests that other structures may behave similarly. This possibility is confirmed by our recent demonstration that the P-450-catalyzed oxidation of 2,2-dialkyl-1,2-dihydroquinolines results in enzyme inactivation and heme adduct formation. One of the 2-alkyl substituents of the dihydroquinoline is found covalently attached in the heme adduct to a nitrogen of protoporphyrin IX.[114] The analogy with the action of 4-alkyl-1,4-dihydropyridines suggests the operation of a similar mechanism (Fig. 9).

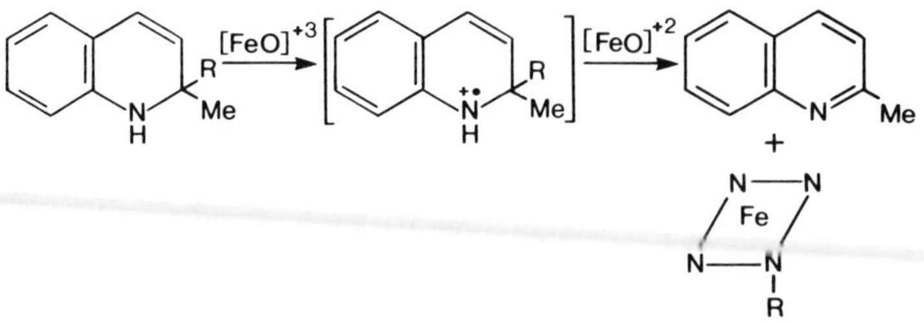

FIGURE 9. Mechanism proposed for oxidative activation of 2,2-dialkyl-1,2-dihydroquinolines to species that alkylate the prosthetic heme of P-450.

3.3.4. N-N Functions

The oxidative action of P-450 produces reactive intermediates from monosubstituted hydrazines, phenylhydrazones, 2,3-bis(carbethoxy)-2,3-diazabicyclo[2.2.0]hex-5-ene, sydnones, and 1-aminoaryltriazoles that efficiently destroy the enzyme by alkylating its prosthetic heme group.

Phenelzine, the alkylhydrazine for which the P-450-destructive mechanism is best understood, causes an approximately equimolar loss of enzyme and heme when incubated with hepatic microsomes.[115] This enzyme and heme loss is paralleled by the formation of a prosthetic heme adduct identified as N-(2-phenylethyl)protoporphyrin IX.[116] The 2-phenylethyl radical, as demonstrated by spin trapping experiments, is generated in the incubations. The formation of this radical suggests, but does not unambiguously establish, that the prosthetic heme is alkylated by the 2-phenylethyl radical. The radical, as already noted, may react initially with the nitrogen or the iron of the prosthetic heme group (Fig. 10).

The proposal that the radical binds to the iron and then shifts to the nitrogen rests primarily on the demonstration that the inactivation of myoglobin by arylhydrazines is mediated by such a mechanism, and on the analogy between the well-characterized myoglobin reaction and the inactivation of P-450 by phenylhydrazine. A transient complex with an absorbance maximum at 480 nm precedes irreversible destruction of the prosthetic heme group when P-450 is incubated with phenylhydrazine or with N-phenylhydrazones.[117,118] A complex with a long-wavelenth absorbance maximum also accompanies the inactivation of myoglobin, hemoglobin, or catalase by phenylhydrazine. Extraction of the prosthetic group from the myoglobin, hemoglobin, and catalase complexes under denaturing conditions yields N-phenylprotoporphyrin IX, although this N-phenylporphyrin is not obtained unless the proteins are denatured under aerobic (oxidative) conditions.[112,119-124] The P-450 complex is intrinsically less stable but otherwise is similar to the myoglobin complex,

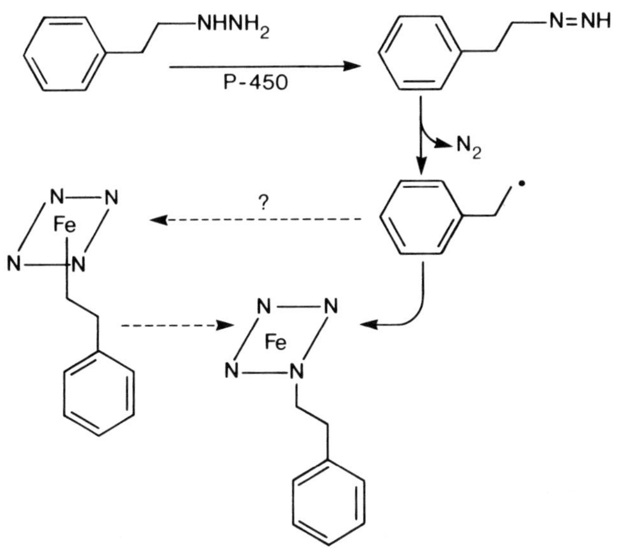

INHIBITION OF P-450 ENZYMES

FIGURE 10. Mechanism for alkylation of the prosthetic heme group during the oxidative metabolism of phenelzine. The possibility that an iron–alkyl intermediate precedes the N-alkylated product is indicated by the dashed lines.

although it has not been explicitly demonstrated that N-phenylprotoporphyrin IX is produced in the reaction. Spectroscopic and chemical studies, including an X-ray structural analysis of the myoglobin complex, unambiguously demonstrate that the phenyl group is bound to the prosthetic heme iron rather than to one of its nitrogens in the undenatured complexes.[120–124] The phenyl group therefore migrates from the iron to the nitrogen in an oxidative process set in motion by protein denaturation (Fig. 11). This migration is sensitive to steric effects because the aryl

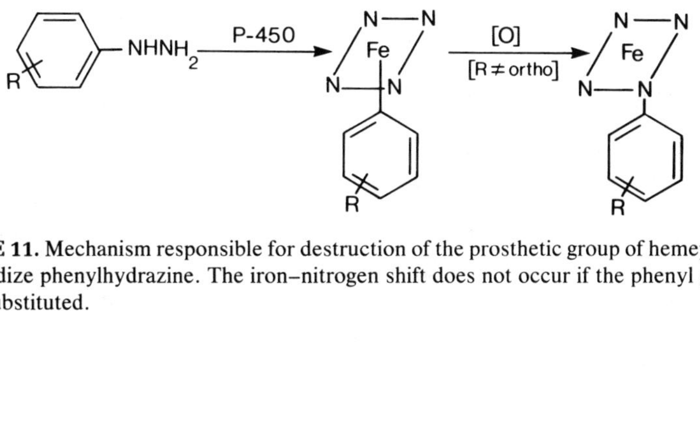

FIGURE 11. Mechanism responsible for destruction of the prosthetic group of hemeproteins that oxidize phenylhydrazine. The iron–nitrogen shift does not occur if the phenyl group is *ortho*-substituted.

moiety in aryl–iron complexes obtained from *ortho*-substituted phenylhydrazines does not undergo the oxidative shift.[112]

The possibility that alkyl radicals, like their aryl counterparts, bind to the iron before shifting to the nitrogen is supported by the observation that the type II complexes formed between alkyldiazenes and P-450 in the absence of oxygen are converted, when limited amounts of oxygen are introduced, to complexes with the absorption maximum in the vicinity of 480 nm characteristic of complexes with an iron–carbon σ-bond.[125] Model alkyl diazene–iron tetraphenylporphyrin complexes have been shown to exist under anaerobic conditions.[126] Alkyl–iron complexes, however, are much less stable than aryl–iron complexes.

1-Aminobenzotriazole (ABT) is oxidized by chemical reagents to benzyne, an exceedingly reactive species, and two molecules of nitrogen.[127] The P-450-catalyzed oxidation of ABT apparently follows a similar reaction course because benzyne, or its equivalent, has been shown to add across two of the nitrogens of the prosthetic heme group.[128,129] The benzyne may add directly to the two nitrogens, generating an *N,N*-bridged species that autoxidizes to the isolated bridged porphyrin, or could first bridge the iron and a nitrogen of the heme and subsequently rearrange to the *N,N*-bridged species (Fig. 12). A broad range of P-450 isozymes are inactivated by ABT without detectable toxic effects.[128–130] Destructive activity is retained if substituents are placed on the phenyl ring or on the exocyclic nitrogen, or if the phenyl framework is replaced by other aryl moieties.[129,131] It is not known if the oxidation of ABT to benzyne

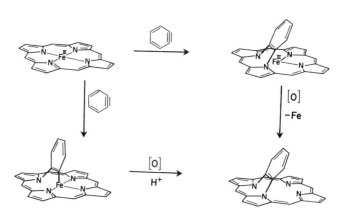

FIGURE 12. Alternative routes to the heme adduct obtained when benzyne is enzymatically generated from 1-aminobenzotriazole. The peripheral porphyrin substituents have been omitted for clarity.

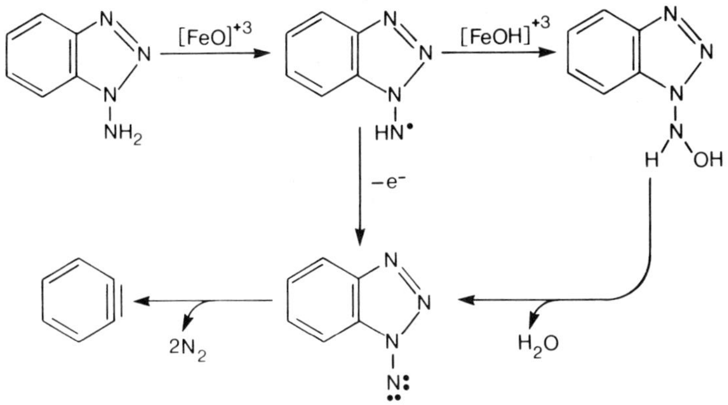

FIGURE 13. Alternative mechanisms for the oxidative generation of benzyne from 1-aminobenzotriazole.

involves hydroxylation of the exocyclic nitrogen or initial oxidation to a radical cation (Fig. 13). The similarity between the activation mechanism proposed for ABT and other 1,1-disubstituted hydrazines (Fig. 8) is to be noted.

Cyclobutadiene, which may exist as a rectangular structure with a singlet electronic state or as a square structure with a triplet electronic state, can be generated by chemical oxidation of 2,3-diazabicyclo[2.2.0]hex-5-ene.[132] Bis(carbethoxy)-2,3-diazabicyclo[2.2.0]hex-5-ene (DDBCH) is a mechanism-based irreversible inhibitor of P-450 that exploits the basic reactivity of the parent bicyclic system.[133] The bis(carbethoxy) derivative was selected for the enzymatic work because the parent bicyclic hydrazine autoxidizes too readily to be of biological utility. The prosthetic heme of the enzyme is converted in the inactivation reaction into the N-2-cyclobutenyl derivative. The secondary, allylic, N-alkyl moiety in this adduct makes it much less stable than other adducts, which bear primary, unactivated, N-alkyl groups. The failure of internal olefins and acetylenes to detectably alkylate the prosthetic heme group suggests, in fact, that secondary carbons are generally too sterically encumbered to react with the heme. The 2-cyclobutenyl adduct implicates cyclobutadiene, or a closely related species, in heme alkylation, although the precise nature of the reactive species has not been defined. The observed adduct is readily explained by addition of cyclobutadiene to a nitrogen of the heme to give a transient intermediate that abstracts a hydrogen from an active-site residue. The transient intermediate could be stabilized by bond formation between the iron of the heme and the carbon that eventually abstracts the active-site hydrogen (Fig. 14). As

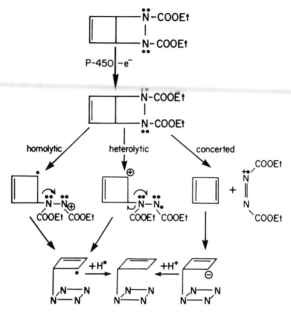

FIGURE 14. Alternative mechanisms for the oxidative generation of a cyclobutadienoid species that alkylates the prosthetic heme group of P-450.

shown, electron abstraction from DDBCH can lead to the observed adduct by several pathways that differ in whether the cyclobutadiene is neutral, cationic, or anionic.

The finding that a fluorescent hepatic pigment accumulates in the livers of dogs and rats administered a sydnone derivative[134] led to the recent demonstration that this sydnone is catalytically activated by P-450 to a species that alkylates the prosthetic heme group (Fig. 15).[135] The heme adduct isolated from rats treated with the sydnone has been identified as N-vinylprotoporphyrin IX. Activation of the sydnone by hydroxylation of the electron-rich zwitterionic carbon, followed by ring opening and elimination of the carboxylic fragment to give the diazo species (Fig. 15), is suggested by this finding. The chemical oxidation of sydnones has been shown to proceed by this mechanism.[135] The diazoalkane, in turn, reacts with the heme, possibly via an initial carbene complex, to give a nitrogen–iron bridged species. The formation of such bridged nitrogen–iron species has been documented in model porphyrin systems.[136,137] The negative charge on the carbon in the bridged intermediate finally eliminates the thiophenyl moiety and generates the N-vinyl adduct.[138]

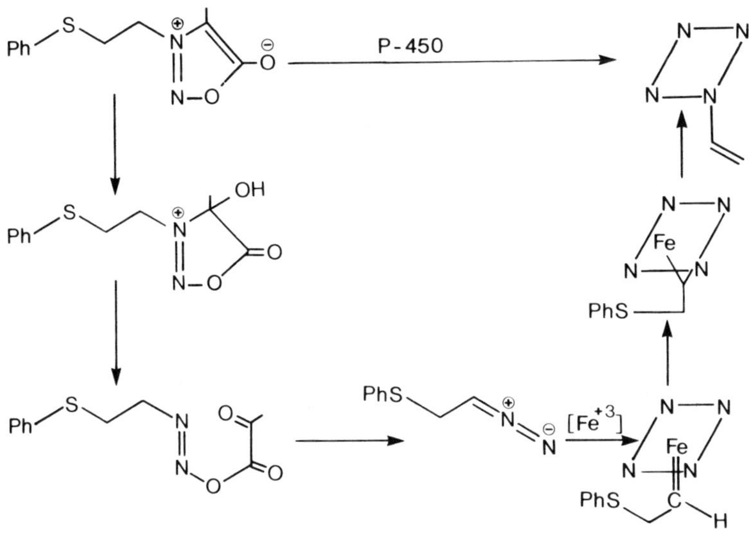

FIGURE 15. Mechanism proposed for the oxidative activation of a sydnone to a species that alkylates the prosthetic heme group of P-450.

3.4. Heme Degradation

P-450 is sometimes inactivated by mechanisms that involve destruction of the prosthetic heme group but that are not accompanied by the detectable formation of heme adducts. A clear example of such a P-450-destructive mechanism is that mediated by peroxides (Chapter 7), which degrade the prosthetic heme group to pyrrole and dipyrrole fragments.[139,140] The destructive mechanisms of halocarbons (CCl_4),[141] internal acetylenes (3-hexyne),[101] allenes (1,1-dimethylallene),[142] cyclopropylamines (N-methyl-N-benzylcyclopropylamine),[143,144] and benzothiadiazoles (5,6-dichloro-1,2,3-benzothiadiazole)[145] remain poorly characterized but are clearly members of this somewhat indeterminate class. The internal acetylenes are probably enzymatically oxidized to reactive species analogous to those obtained from the terminal acetylenes but, unlike the terminal analogues, do not detectably N-alkylate the prosthetic heme. The heme adducts expected from N-alkylation by disubstituted acetylenes have been sought but have not been found despite the fact that they should be at least as stable as the adduct obtained with cyclobutadiene (Section 3.3.4). Similar ambiguities cloud the destructive mechanisms of allenes, cyclopropyl amines (Chapter 7, Fig. 18), and benzothiadiazoles (Fig. 16). A hypothetical activation mechanism can be formulated for each of these functionalities but no experimental evidence is available to support the mechanisms.

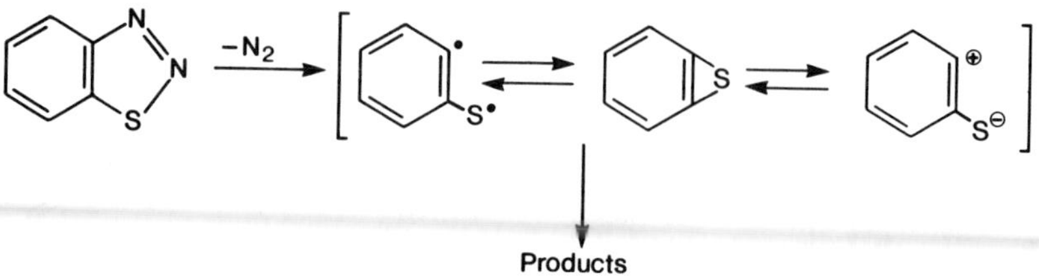

FIGURE 16. Possible mechanism for the oxidative activation of 5,6-dichlorobenzothiadiazole.

The destructive action of halocarbons, probably the most studied members of this class, was believed at one time to result from the secondary action of the concomitantly formed lipid peroxides. More recent evidence, however, suggests that substances like CCl_4 destroy the prosthetic heme group directly rather than by a process mediated by lipid peroxides.[146–149] The reductive metabolism of halocarbons, including CCl_4, gives rise to semistable complexes with Soret maxima between 450 and 500 nm.[150–153] Model studies, including the isolation and detailed characterization of a dichlorocarbene–metalloporphyrin complex,[154] suggest that the long-wavelength Soret bands are due to complexes of halocarbenes with the reduced heme. The finding that carbon monoxide is a product of the reduction of CCl_4 by P-450, a reaction that presumably occurs by a mechanism analogous to that proposed for the generation of carbon monoxide from methylenedioxyphenyl complexes (Section 3.2.1), supports this hypothesis.[155] Model porphyrin dichlorocarbene–iron complexes do, in fact, react with water to give carbon monoxide and with primary amines to give isonitriles.[155,156] Recent work with halothane, however, suggests that complexes in which the halocarbon is σ-bonded to the ferric prosthetic heme group are, in some instances, formed in preference to carbene complexes.[157,158] It has been argued that halocarbons *only* give unstable σ-bonded complexes, but the data invoked in support of this view are not convincing because they were obtained in an experimental system where the complexes with Soret maxima at long wavelengths were not observed.[159]

The causal steps that link iron–alkyl complex formation to irreversible destruction of the enzyme and its heme moiety are not known, although the results of model studies with diaryl- and carbethoxy-substituted carbene complexes suggest that the halogenated carbenes may migrate to the nitrogens of the porphyrin.[160–163] The *N*-haloalkyl adduct obtained by migration of a dichlorocarbene would probably be unstable toward reaction with water and would therefore not be detected by the methods that have been used to isolate other *N*-alkyl porphyrins. It is not possible at this time, however, to rule out release of the halogenated carbenes or alkyl moieties as reactive species that attack the protein or the heme in an undetermined manner.

4. Isozyme-Specific Inhibitors of Biosynthetic Enzymes

P-450 isozyme-specific inhibitors are of substantial potential importance not only as therapeutic, insecticidal, or herbicidal agents, but also as probes of the structures, mechanisms, and biological roles of specific isozymes. Efforts to develop isozyme-specific inhibitors have focused on biosynthetic enzymes because their inhibition is potentially of greater practical utility. In addition, their relatively high substrate specificity makes them more amenable to specific inhibition. The broad, overlapping, substrate specificities of isozymes that primarily metabolize xenobiotics make it difficult to obtain isozyme-specific rather than isozyme-selective inhibitors.[164] Isozyme-selective inhibitors of hepatic enzymes, examples of which are provided by safrole,[57] 2-isopropyl-4-pentenamide,[165] ellipticine,[34] the amphetamines,[77] 2,3-bis(carbethoxy)-2,3-diazabicyclo[2.2.0]hex-5-ene,[133] and 1-ethynylpyrene,[102] are fairly common. A note of caution is appropriate with respect to the inhibitor specificities reported in the literature. Inhibitors can only be claimed to be specific to the extent that they have been tested against the diversity of P-450 isozymes. The claim for specificity of inhibitors tested against only one or two isozymes necessarily is limited.

4.1. P-450$_{scc}$

The three oxidative steps required to cleave the side chain of cholesterol are catalyzed by a single P-450 enzyme (P-450$_{scc}$) (see Chapters 11 and 12). The long-standing efforts to develop specific, competitive, inhibitors of this enzyme are exemplified by the recent report that an analogue of glutethimide (Fig. 17A), unlike glutethimide itself, inhibits P-450$_{scc}$ without significantly inhibiting aromatase (P-450$_{arom}$).[31] The data reported on the inhibition are consistent with reversible binding of the glutethimide analogue to the prosthetic heme iron, but do not exclude the possibility, suggested by the structure of the inhibitor (see Section 3.2.3), that the agent is oxidized to a nitrene that is quasi-irreversibly coordinated. Potent reversible inhibitors (K_i = 40 to 700 nM) of P-450$_{scc}$ have also been constructed by placing an amino function on the cholesterol side chain at positions that favor chelation to the prosthetic heme iron (Fig. 17C).[166-168] The most potent of these inhibitors has a 22(R) amino group. The amine in this analogue occupies the position of the first hy-

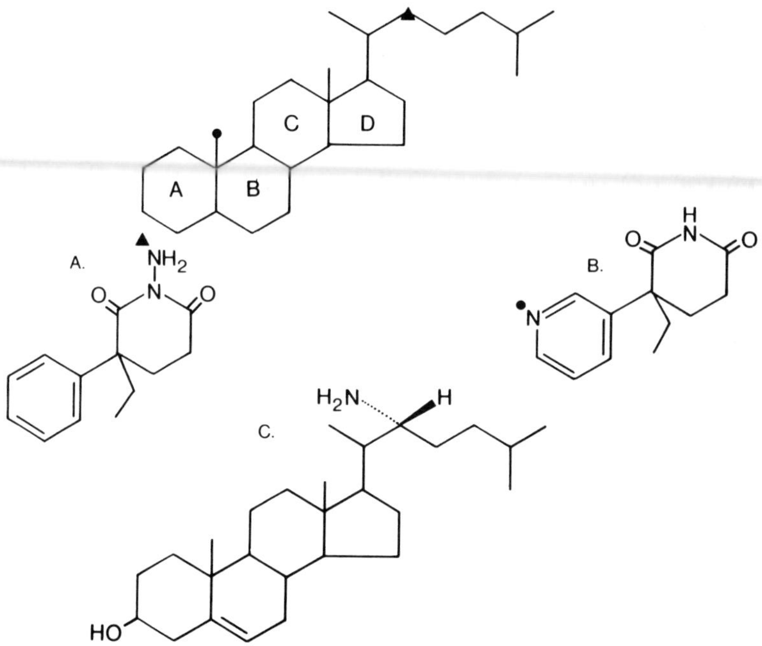

FIGURE 17. Inhibitors of P-450$_{scc}$ (A and C) or aromatase (B). The marked nitrogens in A and B presumably interact, respectively, with the prosthetic hemes of P-450$_{scc}$ and aromatase. The carbons of the sterol skeleton acted upon by these two different enzymes are marked with the same symbols.

droxyl group that would be inserted into the sterol side chain in the normal cleavage reaction.[166–169]

A number of mechanism-based, irreversible, inhibitors of P-450$_{scc}$ have recently been developed. Analogues of pregnendiol with an acetylenic group grafted into the side chain (Fig. 18A) inactivate the enzyme in a time-, concentration-, and NADPH-dependent manner.[170] The heme chromophore is destroyed in the inactivation reaction but an alkylated heme, in agreement with the data on the inactivation of hepatic isozymes by internal acetylenes (see Section 3.4), is not detected. A mechanism-based inhibitor of P-450$_{scc}$ is also obtained if the side-chain carbons beyond C-23 in 20-hydroxycholesterol are replaced by a trimethylsilyl group (Fig. 18B).[171] The one-electron chemical oxidation of 1-substituted 3-trimethylsilyl-1-propanols has been shown to yield ethylene, the trimethylsilyl radical, and an aldehyde (Fig. 18).[172] The inactivation of P-450$_{scc}$, if the chemical model is relevant, may therefore result from reaction of the enzyme with the trimethylsilyl radical, or from secondary oxidation of the ethylene liberated in the first catalytic turnover.

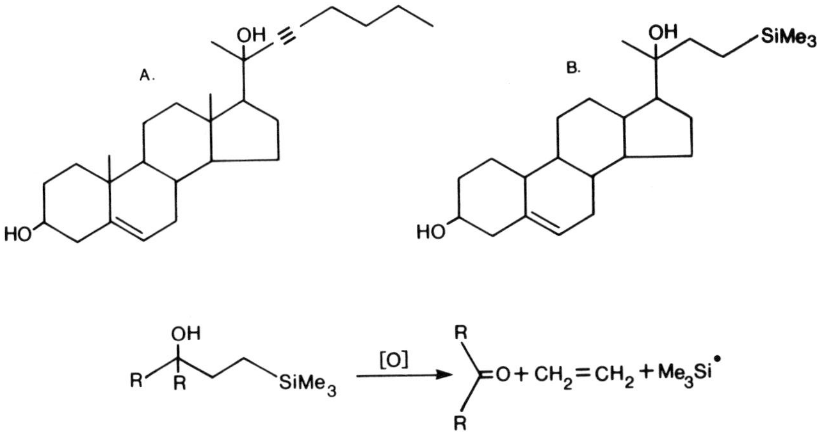

INHIBITION OF P-450 ENZYMES

[Structures A and B: steroid-based inhibitors with OH and SiMe3 substituents on side chains, both with HO group on A-ring]

$$R\underset{R}{\overset{OH}{\diagdown}}\diagup\diagdown SiMe_3 \xrightarrow{[O]} \underset{R}{\overset{R}{\diagdown}}C=O + CH_2=CH_2 + Me_3Si^\bullet$$

FIGURE 18. Two mechanism-based inhibitors (suicide substrates) of P-450$_{scc}$. The reaction that occurs when a 3-trimethylsilylpropanol undergoes a one-electron chemical oxidation is shown.

4.2. Aromatase

The three-step transformation catalyzed by aromatase, the enzyme that controls the conversion of androgens to estrogens, is discussed in Chapters 7, 10, and 11. Inhibitors of aromatase are potentially useful agents for the control of estrogen-dependent mammary tumors[173–175] and, possibly, for the control of coronary heart disease.[176] Aminoglutethimide, an inhibitor of aromatase, is currently used to treat hormone-dependent metastatic breast carcinoma, but its poor specificity, particularly its inhibition of P-450$_{scc}$, causes side effects that compromise its utility.[177] Replacement of the phenyl group in aminoglutethimide (Fig. 17B) by a pyridine moiety has recently been reported to yield an agent tht inhibits aromatase without inhibiting P-450$_{scc}$.[32] The basis for the enhanced specificity is not known, but the pyridine nitrogen is probably so positioned that it can coordinate with the heme of aromatase but not with that of P-450$_{scc}$. It is interesting, in this context, that an aminoglutethimide analogue with a nitrogen at the opposite end is more potent that glutethimide as an inhibitor of P-450$_{scc}$ but has little or no activity against aromatase (see Section 4.1).[31] The 19-methyl group and carbon 22 of the sterol side chain (Fig. 17) are separated by a distance roughly equal to the length of the aminoglutethimide structure. It is therefore tempting to speculate that aminoglutethimide is bound in roughly the same orientation relative to the sterol substrate in the active sites of P-450$_{scc}$ and P-450$_{arom}$, and that the differential inhibitory activity of the two analogues derives from positioning of the nitrogen so that it can coordinate with the prosthetic heme.

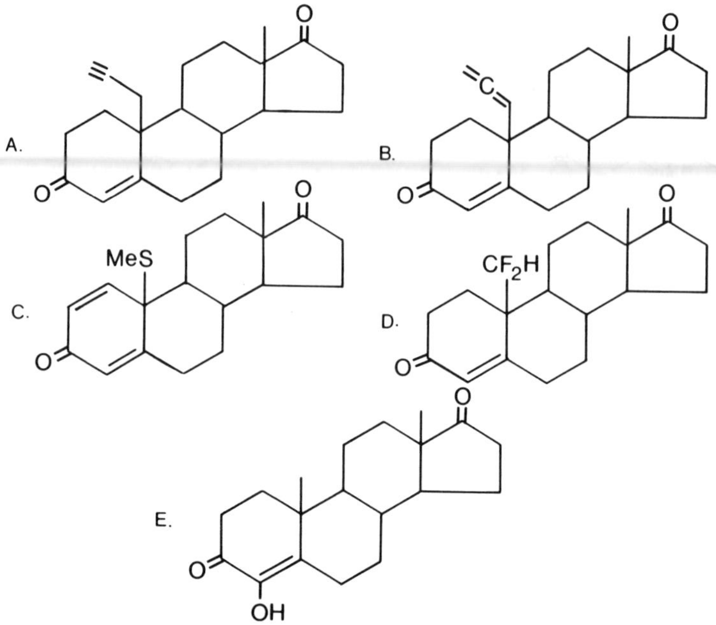

FIGURE 19. Mechanism-based inhibitors of aromatase.

The development of mechanism-based inhibitors for aromatase reflects, to some extent, the strategies developed for inactivation of the hepatic isozymes. Replacement of the normally hydroxylated methyl group (C19) by a propargylic or allenic moiety (Fig. 19A, B) converts the sterol into an irreversible inhibitor of aromatase.[178–180] The details of the mechanisms by which these acetylenic and allenic agents inactivate the enzyme remain obscure, but the heme alkylation mechanisms by which unsaturated functionalities destroy the hepatic enzymes probably also apply here (see Sections 3.3.2 and 3.4). A mechanism-based inhibitor is also obtained if the C19 methyl is replaced by a difluoromethyl group (Fig. 19D), presumably because the initial hydroxylation yields a difluoromethylalcohol that is converted to a reactive acyl fluoride by elimination of hydrogen fluoride.[181] Replacement of the C19 methyl by a methylthio moiety (Fig. 19C) also results in enzyme inactivation, but the inactivation is only partially dependent on NADPH.[182] Finally, 4-acetoxy- and 4-hydroxy-4-androstene-3,17-dione (Fig. 19E) irreversibly inactivate placental aromatase by a catalysis-dependent mechanism.[183,184] The 19-methyl group is required for the activity of these agents. A satisfactory mechanism for the inhibitory activity of the 4-substituted analogues remains elusive.

The time-dependent inactivation of aromatase by 10-hydroperoxy-4-estrene-3,3-dione has recently been described.[185] The inactivation, which is inhibited by NADPH or substrate, is partially reversed by dithiothreitol. Cumene hydroperoxide and H_2O_2 also inactivate the enzyme, but their destructive action is not prevented by the substrate or NADPH. The relationship of this inhibition to the relatively common destructive activity of hydroperoxides (Section 3.4) remains to be clarified.

4.3. Lanosterol 14-Demethylation

A key step in the biosynthesis of cholesterol is the P-450-catalyzed 14-demethylation of lanosterol (Chapters 7 and 11). The preferential inhibition of this enzyme by a number of substituted imidazoles, pyridines, pyrimidines, and other lipophilic heterocycles has been exploited successfully in the construction of clinically important antifungal agents.[186] Miconazole and its derivatives inhibit the 14-demethylase activity of fungi at extremely low (nM) concentrations,[187,188] but only inhibit the 14-demethylase activity of the mammalian host[189] or other P-450 activities (aminopyrine N-demethylase)[190] at higher (up to 100 μM) concentrations. Low doses of ketoconazole, however, appear to also inhibit the C_{17-20} lyase in man.[191] The substituted imidazoles and other nitrogen-based heterocyclic antifungal agents bind tightly to P-450 and give rise to type II binding spectra, as expected if the inhibition results from coordination of the inhibitors with the prosthetic heme group.[192] The 14-methyl sterols that accumulate in the membranes of susceptible fungi when the 14-demethylase is inhibited are believed to cause the changes in membrane permeability responsible for the antifungal action of the inhibitors.[193–195]

4.4. Fatty Acid and Leukotriene Hydroxylases

Fatty acids, including arachidonic acid and its derivatives, are hydroxylated at the ω- and ω-1 positions by P-450 enzymes in liver, kidney, intestine, adrenal, polymorphonuclear leukocyte, and lung microsomes.[196–197] The analogous enzymes of plants and bacteria introduce hydroxyl groups at the ω, ω-1, ω-2, ω-3, and ω-4 positions.[198,199] 11-Dodecynoic acid, the terminal acetylenic analogue of lauric acid, and 10-undecynoic acid inactivate the hepatic P-450 enzymes that hydroxylate lauric acid while minimally altering the spectroscopically measured concentration of P-450 or the benzphetamine or *N*-methyl-*p*-chloroaniline *N*-demethylase activities.[200] The acetylenic fatty acids similarly inactivate the plant P-450 enzyme responsible for in-chain hydroxylation of fatty acids.[201] The kinetics of the inactivation of lauric acid hydroxylases by 10-undecynoic acid imply that at least three rat liver microsomal isozymes catalyze the reaction. The *in vivo* utility of the acetylenic fatty acids, however, is unfortunately compromised by their toxicity. This difficulty has been circumvented by modifying the inhibitors so that they are not subject to β-oxidation. The results indicate that 10-undecynyl sulfate, the acetylenic analogue of 10-undecynoic acid in which the carboxyl group is replaced by a sulfate, is sufficiently less toxic to be used *in vivo* as a mechanism-based inactivator of the lauric acid hydroxylases.[202]

Leukotriene B_4, an important mediator of inflammation in humans, is primarily metabolized in polymorphonuclear leukocytes by ω-hydroxylation.[196] The ω-hydroxylated LTB_4 may be as important as the parent compound in mediating the inflammation response.[203] The LTB_4 ω-hydroxylase, which appears to be a P-450 enzyme, is inactivated by 15-hexadecynoic and 17-octadecynoic acids in whole cells or cell lysates.[204] The saturated analogues of these long-chain acetylenic acids are inactive, while 10-undecynoic acid, the short-chain acid, is much less effective. The long-chain acetylenic fatty acids may also inactivate the lung isozymes responsible for ω-hydroxylation of prostaglandins, but the effects of inhibitors on these enzymes are only now being examined.[205] Finally, preliminary data suggest that acetylenic terpenoids, which resemble acetylenic fatty acids, inactivate the P-450 enzyme responsible for the final epoxidation step in the biosynthesis of juvenile hormone by insects.[206]

5. Summary

The development of inhibitors of P-450 enzymes, and our understanding of the mechanisms of action of such inhibitors, have flowered over the past few years. Efforts to increase the potency and specificity of reversible inhibitors by modifying the lipophilic framework and the heme-coordinating nitrogen function have yielded a number of important clinical and agricultural agents. The more recent development of mechanism-based irreversible inhibitors (suicide substrates) for P-450 enzymes greatly enhances the potential specificity and utility of P-450 inhibitors. A variety of inactivator functionalities are now available for the construction of such mechanism-based inhibitors, each of which irreversibly inactivates the enzyme that catalyzes its oxidation. The differential, specific, inactivation of the fatty acid, leukotriene, and prostaglandin hydroxylases by terminal acetylenic acids of varying chain lengths illustrates the potential of this approach.

Although agents like chloramphenicol inactivate P-450 by reacting with its protein structure, by far the most vulnerable site for the action of mechanism-based inhibitors appears to be the prosthetic heme group. Identification of the prosthetic heme as the Achilles' heel of P-450 enzymes is of practical importance for the design of appropriate inhibitors. As borne out by experience, the functionalities that inactivate P-450 with highest specificity and efficiency are generally oxidized to radical or neutral reactive species rather than to cationic alkylating intermediates. The catalytic involvement of P-450 in its own inactivation, and the detailed reconstruction of the destructive event made possible by isolation and structural characterization of the prosthetic heme adducts, make the irreversible inhibitors important mechanistic and topological probes. The practical utility of mechanism-based P-450 inhibitors, which is only now being explored, is likely to surpass that of the more classical reversible agents.

ACKNOWLEDGMENTS. The preparation of this review was assisted by grants from the National Institutes of Health.

References

1. Rahimtula, A. D., and O'Brien, P. J., 1977, The peroxidase nature of cytochrome P-450, in: *Microsomes and Drug Oxidations* (V. Ullrich, I. Roots, A. Hildebrandt, R. W. Estabrook, and A. H. Conney, eds.), Pergamon Press, Elmsford, N.Y., pp. 210–217.
2. Yang, C. S., and Strickhart, F. S., 1974, Inhibition of mixed function oxidase activity by propylgallate, *Biochem. Pharmacol.* **23**:3129–3138.
3. Cummings, S. W., and Prough, R. A., 1983, Butylated hydroxyanisole-stimulated NADPH oxidase activity in rat liver microsomal fractions, *J. Biol. Chem.* **258**:12315–12319.
4. Testa, B., and Jenner, P., 1981, Inhibitors of cytochrome P-450s and their mechanism of action, *Drug. Metab. Rev.* **12**:1–117.
5. Netter, K. J., 1980, Inhibition of oxidative drug metabolism in microsomes, *Pharmacol. Ther. A* **10**:515–535.
6. Sato, A., and Nakajima, T., 1979, Dose-dependent metabolic interaction between benzene and toluene in vivo and in vitro, *Toxicol. Appl. Pharmacol.* **48**:249–256.
7. Jefcoate, C. R., 1978, Measurement of substrate and inhibitor binding to microsomal cytochrome P-450 by optical-difference spectroscopy, *Methods Enzymol.* **52**:258–279.
8. Kumaki, K., Sato, M., Kon, H., and Nebert, D. W., 1978, Correlation of type I, type II, and reverse type I difference spectra with absolute changes in spin state of hepatic microsomal cytochrome P-450 iron from five mammalian species, *J. Biol. Chem.* **253**:1048–1058.
9. Schenkman, J. B., Sligar, S. G., and Cinti, D. L., 1981, Substrate interactions with cytochrome P-450, *Pharmacol. Ther.* **12**:43–71.
10. Sligar, S. G., Cinti, D. L., Gibson, G. G., and Schenkman, J. B., 1979, Spin state control of the hepatic cytochrome P-450 redox potential, *Biochem. Biophys. Res. Commun.* **90**:925–932.
11. Guengerich, F. P., 1983, Oxidation–reduction properties of rat liver cytochromes P-450 and NADPH-cytochrome P-450 reductase related to catalysis in reconstituted systems, *Biochemistry* **22**:2811–2820.

12. Kitada, M., Chiba, K., Kamataki, T., and Kitagawa, H., 1977, Inhibition by cyanide of drug oxidations in rat liver microsomes, *Jpn. J. Pharmacol.* **27**:601–608.
13. Ho, B., and Castagnoli, N., 1980, Trapping of metabolically generated electrophilic species with cyanide ion: Metabolism of 1-benzylpyrrolidine, *J. Med. Chem.* **23**:133–139.
14. Sono, M., and Dawson, J. H., 1982, Formation of low spin complexes of ferric cytochrome P-450-CAM with anionic ligands. Spin state and ligand affinity comparison to myoglobin, *J. Biol. Chem.* **257**:5496–5502.
15. Backes, W. L., Hogaboom, M., and Canady, W. J., 1982, The true hydrophobicity of microsomal cytochrome P-450 in the rat: Size dependence of the free energy of binding of a series of hydrocarbon substrates from the aqueous phase to the enzyme and to the membrane as derived from spectral binding data, *J. Biol. Chem.* **257**:4063–4070.
16. Omura, T., and Sato, R., 1964, The carbon monoxide-binding pigment of liver microsomes. I. Evidence for its hemoprotein nature, *J. Biol. Chem.* **239**:2370–2378.
17. Collman, J. P., and Sorrell, T. N., 1975, A model for the carbonyl adduct of ferrous cytochrome P-450, *J. Am. Chem. Soc.* **97**:4133–4134.
18. Canick, J. A., and Ryan, K. J., 1976, Cytochrome P-450 and the aromatization of 16-alpha-hydroxytestosterone and androstenedione by human placental microsomes, *Mol. Cell. Endocrinol.* **6**:105–115.
19. Gibbons, G. F., Pullinger, C. R., and Mitropoulos, K. A., 1979, Studies on the mechanism of lanosterol 14-alpha-demethylation: A requirement for two distinct types of mixed-function-oxidase systems, *Biochem. J.* **183**:309–315.
20. Hanson, L. K., Eaton, W. A., Sligar, S. G., Gunsalus, I. C., Gouterman, M., and Connell, C. R., 1976, Origin of the anomalous Soret spectra of carboxycytochrome P-450, *J. Am. Chem. Soc.* **98**:2672–2674.
21. Cohen, G. M., and Mannering, G. J., 1972, Involvement of a hydrophobic site in the inhibition of the microsomal para-hydroxylation of aniline by alcohols, *Mol. Pharmacol.* **8**:383–397.
22. Testa, B., 1981, Structural and electronic factors influencing the inhibition of aniline hydroxylation by alcohols and their binding to cytochrome P-450, *Chem. Biol. Interact.* **34**:287–300.
23. Wattenberg, L. W., Lam, L. K. T., and Fladmoe, A. V., 1979, Inhibition of chemical carcinogen-induced neoplasia by coumarins and alpha-angelicalactone, *Cancer Res.* **39**:1651–1654.
24. Remmer, H., Schenkman, J., Estabrook, R. W., Sasame, H., Gillette, J., Narasimhulu, S., Cooper, D. Y., and Rosenthal, O., 1966, Drug interaction with hepatic microsomal cytochrome, *Mol. Pharmacol.* **2**:187–190.
25. Jefcoate, C. R., Gaylor, J. L., and Callabrese, R. L., 1969, Ligand interactions with cytochrome P-450. I. Binding of primary amines, *Biochemistry* **8**:3455–3463.
26. Schenkman, J. B., Remmer, H., and Estabrook, R. W., 1967, Spectral studies of drug interaction with hepatic mircrosomal cytochrome P-450, *Mol. Pharmacol.* **3**:113–123.
27. Dominguez, O. V., and Samuels, L. T., 1963, Mechanism of inhibition of adrenal steroid 11-beta-hydroxylase by methopyrapone (metopirone), *Endocrinology* **73**:304–309.
28. Temple, T. E., and Liddle, G. W., 1970, Inhibitors of adrenal steroid biosynthesis, *Annu. Rev. Pharmacol.* **10**:199–218.
29. Rogerson, T. D., Wilkinson, C. F., and Hetarski, K., 1977, Steric factors in the inhibitory interaction of imidazoles with microsomal enzymes, *Biochem. Pharmacol.* **26**:1039–1042.
30. Wilkinson, C. F., Hetarski, K., Cantwell, G. P., and DiCarlo, F. J., 1974, Structure–activity relationships in the effects of 1-alkylimidazoles on microsomal oxidation in vitro and in vivo, *Biochem. Pharmacol.* **23**:2377–2386.

31. Foster, A. B., Jarman, M., Leung, C.-S., Rowlands, M. G., and Taylor, G. N., 1983, Analogues of aminoglutethimide: Selective inhibition of cholesterol side-chain cleavage, *J. Med. Chem.* **26**:50–54.
32. Foster, A. B., Jarman, M., Leung, C.-S., Rowlands, M. G., Taylor, G. N., Plevey, R. G., and Sampson, P., 1985, Analogues of aminoglutethimide: Selective inhibition of aromatase, *J. Med. Chem.* **28**:200–204.
33. Hays, S. J., Tobes, M. C., Gildersleeve, D. L., Wieland, D. M., and Beierwaltes, W. H., 1984, Structure-activity relationship study of the inhibition of adrenal cortical 11-beta-hydroxylase by new metyrapone analogues, *J. Med. Chem.* **27**:15–19.
34. Lesca, P., Rafidinarino, E., Lecointe, P., and Mansuy, D., 1979, A class of strong inhibitors of microsomal monooxygenases: The ellipticines, *Chem. Biol. Interact.* **24**:189–198.
35. Walsh, C., 1982, Suicide substrates: Mechanism-based enzyme inactivators, *Tetrahedron* **38**:871–909.
36. Santi, D. V., and Kenyon, G. L., 1980, Approaches to the rational design of enzyme inhibitors, in: *Burger's Medicinal Chemistry*, 4th ed., Part 1, Wiley–Interscience, New York, pp. 349–391.
37. Dalvi, R. R., Poore, R. E., and Neal, R. A., 1974, Studies of the metabolism of carbon disulphide by rat liver microsomes, *Life Sci.* **14**:1785–1796.
38. De Matteis, F. A., and Seawright, A. A., 1973, Oxidative metabolism of carbon disulphide by the rat: Effect of treatments which modify the liver toxicity of carbon disulphide, *Chem. Biol. Interact.* **7**:375–388.
39. Bond, E. J., and De Matteis, F. A., 1969, Biochemical changes in rat liver after administration of carbon disulphide, with particular reference to microsomal changes, *Biochem. Pharmacol.* **18**:2531–2549.
40. Halpert, J., Hammond, D., and Neal, R. A., 1980, Inactivation of purified rat liver cytochrome P-450 during the metabolism of parathion (diethyl p-nitrophenyl phosphorothionate), *J. Biol. Chem.* **255**:1080–1089.
41. Neal, R. A., Kamataki, T., Lin, M., Ptashne, K. A., Dalvi, R., and Poore, R. Y., 1977, Studies of the formation of reactive intermediates of parathion, in: *Biological Reactive Intermediates* (D. J. Jollow, J. J. Koesis, R. Snyder, and H. Vaino, eds.), Plenum Press, New York, pp. 320–332.
42. Miller, G. E., Zemaitis, M. A., and Greene, F. E., 1983, Mechanisms of diethyldithiocarbamate-induced loss of cytochrome P-450 from rat liver, *Biochem. Pharmacol.* **32**:2433–2442.
43. El-hawari, A. M., and Plaa, G. L., 1979, Impairment of hepatic mixed-function oxidase activity by alpha- and beta-naphthylisothiocyanate: Relationship to hepatotoxicity, *Toxicol. Appl. Pharmacol.* **48**:445–458.
44. Lee, P. W., Arnau, T., and Neal, R. A., 1980, Metabolism of alpha-naphthylthiourea by rat liver and rat lung microsomes, *Toxicol. Appl. Pharmacol.* **53**:164–173.
45. Menard, R. H., Guenthner, T. M., Taburet, A. M., Kon, H., Pohl, L. R., Gillette, J. R., Gelboin, H. V., and Trager, W. F., 1979, Specificity of the in vitro destruction of adrenal and hepatic microsomal steroid hydroxylases by thiosterols, *Mol. Pharmacol.* **16**:997–1010.
46. Halpert, J., and Neal, R. A., 1980, Inactivation of purified rat liver cytochrome P-450 by chloramphenicol, *Mol. Pharmacol.* **17**:427–434.
47. Halpert, J., 1982, Further studies of the suicide inactivation of purified rat liver cytochrome P-450 by chloramphenicol, *Mol. Pharmacol.* **21**:166–172.
48. Halpert, J., 1981, Covalent modification of lysine during the suicide inactivation of rat liver cytochrome P-450 by chloramphenicol, *Biochem. Pharmacol.* **30**:875–881.
49. Halpert, J., Naslund, B., and Betner, I., 1983, Suicide inactivation of rat liver cytochrome P-450 by chloramphenicol in vivo and in vitro, *Mol. Pharmacol.* **23**:445–452.

50. Kagawa, C. M., 1960, Blocking the renal electrolyte effects of mineralcorticoids with an orally active steroidal spirolactone, *Endocrinology* **67:**125–132.
51. Saunders, F. J., and Alberti, R. L., 1978, *Aldactone: Spironolactone: A Comprehensive Review*, Searle, New York.
52. Menard, R. H., Guenthner, T. M., Kon, H., and Gillette, J. R., 1979, Studies on the destruction of adrenal and testicular cytochrome P-450 by spironolactone: Requirement for the 7-alpha-thio group and evidence for the loss of the heme and apoproteins of cytochrome P-450, *J. Biol. Chem.* **254:**1726–1733.
53. Stripp, B., Menard, R. H., Zampaglione, N. G., Hamrick, M. E., and Gillette, J. R., 1973, Effect of steroids on drug metabolism in male and female rats, *Drug. Metab. Dispos.* **1:**216–221.
54. Hamrick, M. E., Zampaglione, N. G., Stripp, B., and Gillette, J. R., 1973, Investigation of the effects of methyltestosterone, cortisone, and spironolactone on the hepatic microsomal mixed function oxidase system in male and female rats, *Biochem. Pharmacol.* **22:**293–310.
55. Casida, J. E., 1970, Mixed function oxidase involvement in the biochemistry of insecticide synergists, *J. Agr. Food Chem.* **18:**753–772.
56. Hodgson, E., and Philpot, R. M., 1974, Interaction of methylene dioxyphenol (1,3-benzodioxole) compounds with enzymes and their effects on mammals, *Drug Metab. Rev.* **3:**231–301.
57. Wilkinson, C. F., Murray, M., and Marcus, C. B., 1984, Interactions of methylenedioxyphenyl compounds with cytochrome P-450 and effects on microsomal oxidation, in: *Reviews in Biochemical Toxicology*, Volume 6 (E. Hodgson, J. R. Bend, and R. M. Philpot, eds.), Elsevier, Amsterdam, pp. 27–63.
58. Kulkarni, A. P., and Hodgson, E., 1978, Cumene hydroperoxide-generated spectral interactions of piperonyl butoxide and other synergists with microsomes from mammals and insects, *Pestic. Biochem. Physiol.* **9:**75–83.
59. Franklin, M. R., 1971, The enzymic formation of a methylene dioxyphenyl derivative exhibiting an isocyanide-like spectrum with reduced cytochrome P-450 in hepatic microsomes, *Xenobiotica* **1:**581–591.
60. Elcombe, C. R., Bridges, J. W., Nimmo-Smith, R. H., and Werringloer, J., 1975, Cumene hydroperoxide-mediated formation of inhibited complexes of methylenedioxyphenyl compounds with cytochrome P-450, *Biochem. Soc. Trans.* **3:**967–970.
61. Elcombe, C. R., Bridges, J. W., Gray, T. J. B., Nimmo-Smith, R. H., and Netter, K. J., 1975, Studies on the interaction of safrole with rat hepatic microsomes, *Biochem. Pharmacol.* **24:**1427–1433.
62. Dickins, M., Elcombe, C. R., Moloney, S. J., Netter, K. J., and Bridges, J. W., 1979, Further studies on the dissociation of the isosafrole metabolite–cytochrome P-450 complex, *Biochem. Pharmacol.* **28:**231–238.
63. Ullrich, V., and Schnabel, K. H., 1973, Formation and binding of carbanions by cytochrome P-450 of liver microsomes, *Drug Metab. Dispos.* **1:**176–183.
64. Ullrich, V., 1977, Mechanism of microsomal monooxygenases and drug toxicity, in: *Biological Reactive Intermediates* (D. J. Jollow, J. Kocsis, R. Snyder, and H. Vaino, eds.), Plenum Press, New York, pp. 65–82.
65. Murray, M., Wilkinson, C. F., Marcus, C., and Dube, C. E., 1983, Structure–activity relationships in the interactions of alkoxymethylenedioxybenzene derivatives with rat hepatic microsomal mixed-function oxidases in vivo, *Mol. Pharmacol.* **24:**129–136.
66. Mansuy, D., 1981, Use of model systems in biochemical toxicology: Heme models, in: *Reviews in Biochemical Toxicology*, Volume 3, (E. Hodgson, J. R. Bend, and R. M. Philpot, eds.), Elsevier, Amsterdam, pp. 283–320.
67. Mansuy, D., Battioni, J. P., Chottard, J. C., and Ullrich, V., 1979, Preparation of a

porphyrin-iron-carbene model for the cytochrome P-450 complexes obtained upon metabolic oxidation of the insecticide synergists of the 1,3-benzodioxole series, *J. Am. Chem. Soc.* **101**:3971–3973.
68. Dahl, A. R., and Hodgson, E., 1979, The interaction of aliphatic analogs of methylenedioxyphenyl compounds with cytochromes P-450 and P-420, *Chem. Biol. Interact.* **27**:163–175.
69. Anders, M. W., Sunram, J. M., and Wilkinson, C. F., 1984, Mechanism of the metabolism of 1,3-benzodioxoles to carbon monoxide, *Biochem. Pharmacol.* **33**:577–580.
70. Hansch, C., 1968, The use of homolytic, steric, and hydrophobic constants in a structure–activity study of 1,3-benzodioxole synergists, *J. Med. Chem.* **11**:920–924.
71. Hennessy, D. J., 1965, Hydride-transfering ability of methylene dioxybenzenes as a basis of synergistic activity, *J. Agr. Food Chem.* **13**:218–231.
72. Cook, J. C., and Hodgson, E., 1983, Induction of cytochrome P-450 by methylenedioxyphenyl compounds: Importance of the methylene carbon, *Toxicol. Appl. Pharmacol.* **68**:131–139.
73. Casida, J. E., Engel, J. L., Essac, E. G., Kamienski, F. X., and Kuwatsuka, S., 1966, Methylene-^{14}C-dioxyphenyl compounds: Metabolism in relation to their synergistic action, *Science* **153**:1130–1133.
74. Kamienski, F. X., and Casida, J. E., 1970, Importance of methylenation in the metabolism *in vivo* and *in vitro* of methylenedioxyphenyl synergists and related compounds in mammals, *Biochem. Pharmacol.* **19**:91–112.
75. Yu, L.-S., Wilkinson, C. F., and Anders, M. W., 1980, Generation of carbon monoxide during the microsomal metabolism of methylenedioxyphenyl compounds, *Biochem. Pharmacol.* **29**:1113–1122.
76. Metcalf, R. L., Fukuto, C. W., Fahmy, S.,,El-Azis, S., and Metcalf, E. R., 1966, Mode of action of carbamate synergists, *J. Agr. Food. Chem.* **14**:555–562.
77. Franklin, M. R., 1977, Inhibition of mixed-function oxidations by substrates forming reduced cytochrome P-450 metabolic-intermediate complexes, *Pharmacol. Ther. A* **2**:227–245.
78. Larrey, D., Tinel, M., and Pessayre, D., 1983, Formation of inactive cytochrome P-450 Fe(II)-metabolite complexes with several erythromycin derivatives but not with josamycin and midecamycin in rats, *Biochem. Pharmacol.* **32**:1487–1493.
79. Delaforge, M., Jaquen, M., and Mansuy, D., 1983, Dual effects of macrolide antibiotics on rat liver cytochrome P-450. Induction and formation of metabolite-complexes: A structure–activity relationship, *Biochem. Pharmacol.* **32**:2309–2318.
80. Mansuy, D., Beaune, P., Cresteil, T., Bacot, C., Chottard, J. C., and Gans, P., 1978, Formation of complexes between microsomal cytochrome P-450-Fe(II) and nitrosoarenes obtained by oxidation of arylhydroxylamines or reduction of nitroarenes in situ, *Eur. J. Biochem.* **86**:573–579.
81. Jonsson, J., and Lindeke, B., 1976, On the formation of cytochrome P-450 product complexes during the metabolism of phenylalkylamines, *Acta Pharm. Suec.* **13**:313–320.
82. Franklin, M. R., 1974, The formation of a 455 nm complex during cytochrome P-450-dependent N-hydroxylamphetamine metabolism, *Mol. Pharmacol.* **10**:975–985.
83. Muakkasah, S. F., Bidlack, W. R., and Yang, W. C. T., 1981, Mechanism of the inhibitory action of isoniazid on microsomal drug metabolism, *Biochem. Pharmacol.* **30**:1651–1658.
84. Mansuy, D., 1978, Coordination chemistry of cytochromes P-450 and iron-porphyrins: Relevance to pharmacology and toxicology, *Biochimie* **60**:969–977.
85. Lindeke, B., Anderson, E., Lundkvist, G., Jonsson, H., and Eriksson, S.-O., 1975, Autoxidation of N-hydroxyamphetamine and N-hydroxyphentermine: The formation

of 2-nitroso-1-phenyl-propanes and 1-phenyl-2-propanone oxime, *Acta Pharm. Suec.* **12**:183-198.
86. Mansuy, D., Gans, P., Chottard, J.-C., and Bartoli, J.-F., 1977, Nitrosoalkanes as Fe(II) ligands in the 455-nm-absorbing cytochrome P-450 complexes formed from nitroalkanes in reducing conditions, *Eur. J. Biochem.* **76**:607-615.
87. Hines, R. N., and Prough, R. A., 1980, The characterization of an inhibitory complex formed with cytochrome P-450 and a metabolite of 1,1-disubstituted hydrazines, *J. Pharmacol. Ther.* **214**:80-86.
88. Moloney, S. J., Snider, B. J., and Prough, R. A., 1984, The interactions of hydrazine derivatives with rat-hepatic cytochrome P-450, *Xenobiotica* **14**:803-814.
89. Muakkassah, S. F., Bidlack, W. R., and Yang, W. C. T., 1982, Reversal of the effects of isoniazid on hepatic cytochrome P-450 by potassium ferricyanide, *Biochem. Pharmacol.* **31**:249-251.
90. Mahy, J.-P., Battioni, P., Mansuy, D., Fisher, J., Weiss, R., Mispelter, J., Morgenstern-Badarau, I., and Gans, P., 1984, Iron porphyrin-nitrene complexes: Preparation from 1,1-dialkylhydrazines: Electronic structure from NMR, Mossbauer, and the magnetic susceptibility studies and crystal structure of the [tetrakis(p-chlorophenyl) porphyrinato] [(2,2,6,6-tetramethyl-1-piperidyl) nitrene]iron complex, *J. Am. Chem. Soc.* **106**:1699-1706.
91. Mansuy, D., Battioni, P., and Mahy, J. P., 1982, Isolation of an iron-nitrene complex from the dioxygen and iron porphyrin dependent oxidation of a hydrazine, *J. Am. Chem. Soc.* **104**:4487-4489.
92. Ortiz de Montellano, P. R., 1985, Alkenes and alkynes, in: *Bioactivation of Foreign Compounds* (M. Anders, ed.), Academic Press, New York, pp. 121-155.
93. De Matteis, F., 1978, Loss of liver cytochrome P-450 caused by chemicals, in: *Heme and Hemoproteins, Handbook of Experimental Pharmacology*, Volume 44 (F. De Matteis and W. N. Aldridge, eds.), Springer-Verlag, Berlin, pp. 95-127.
94. Ortiz de Montellano, P. R., and Correia, M. A., 1983, Suicidal destruction of cytochrome P-450 during oxidative drug metabolism, *Annu. Rev. Pharmacol. Toxicol.* **23**:481-503.
95. Ortiz de Montellano, P. R., and Mico, B. A., 1980, Destruction of cytochrome P-450 by ethylene and other olefins, *Mol. Pharmacol.* **18**:128-135.
96. Ortiz de Montellano, P. R., Mangold, B. L. K., Wheeler, C., Kunze, K. L., and Reich, N. O., 1983, Stereochemistry of cytochrome P-450-catalyzed epoxidation and prosthetic heme alkylation, *J. Biol. Chem.* **258**:4208-4213.
97. Ortiz de Montellano, P. R., Stearns, R. A., and Langry, K. C., 1984, The allylisopropylacetamide and novonal prosthetic heme adducts, *Mol. Pharmacol.* **25**:310-317.
98. Ortiz de Montellano, P. R., Kunze, K. L., Beilan, H. S., and Wheeler, C., 1982, Destruction of cytochrome P-450 by vinyl fluoride, fluroxene, and acetylene: Evidence for a radical cation intermediate in olefin oxidation, *Biochemistry* **21**:1331-1339.
99. Kunze, K. L., Mangold, B. L. K. Wheeler, C., Beilan, H. S., and Ortiz de Montellano, P. R., 1983, The cytochrome P-450 active site, *J. Biol. Chem.* **258**:4202-4207.
100. Ortiz de Montellano, P. R., and Komives, E. A., 1985, Branchpoint for heme alkylation and metabolite formation in the oxidation of aryl acetylenes, *J. Biol. Chem.* **260**:3330-3336.
101. Ortiz de Montellano, P. R., and Kunze, K. L., 1980, Self-catalyzed inactivation of hepatic cytochrome P-450 by ethynyl substrates, *J. Biol. Chem.* **255**:5578-5585.
102. Gan, L.-S., Acebo, A. L., and Alworth, W. L., 1984, 1-Ethynylpyrene, a suicide inhibitor of cytochrome P-450 dependent benzo(a)pyrene hydroxylase activity in liver microsomes, *Biochemistry* **23**:3827-3836.
103. De Matteis, F., Abbritti, G., and Gibbs, A. H., 1973, Decreased liver activity of porphyrin-metal chelatase in hepatic porphyria caused by 3,5-diethoxycarbonyl-1,4-dihydrocollidine: Studies in rats and mice, *Biochem J.* **134**:717-727.

104. De Matteis, F., and Gibbs, A., 1972, Stimulation of liver 5-aminolaevulinate synthetase by drugs and its relevance to drug-induced accumulation of cytochrome P-450, *Biochem. J.* **126:**1149–1160.
105. Gayarthri, A. K., and Padmanaban, G., 1974, Biochemical effects of 3,5-diethoxycarbonyl-1,4-dihydrocollidine in mouse liver, *Biochem. Pharmacol.* **23:**2713–2725.
106. Tephly, T. R., Gibbs, A. H., Ingall, G., and De Matteis, F., 1980, Studies on the mechanism of experimental porphyria and ferrochelatase inhibition produced by 3,5-diethoxycarbonyl-1,4-dihydrocollidine, *Int. J. Biochem.* **12:**993–998.
107. Cole, S. P. C. C., and Marks, G. S., 1984, Ferrochelatase and N-alkylated porphyrins, *Mol. Cell. Biochem.* **64:**127–137.
108. Augusto, O., Beilan, H. S., and Ortiz de Montellano, P. R., 1982, The catalytic mechanism of cytochrome P-450: Spin-trapping evidence for one-electron substrate oxidation, *J. Biol. Chem.* **257:**11288–11295.
109. De Matteis, F., Hollands, C., Gibbs, A. H., de Sa, N., and Rizzardini, M., 1982, Inactivation of cytochrome P-450 and production of N-alkylated porphyrins caused in isolated hepatocytes by substituted dihydropyridines: Structural requirements for loss of haem and alkylation of the pyrrole nitrogen atom, *FEBS Lett* **145:**87–92.
110. Tephly, T. R., Coffman, B. L., Ingall, G., Abou Zeit-Har, M. S., Goff, H. M., Tabba, H. D., and Smith, K. M., 1981, Identification of N-methylprotoporphyrin IX in livers of untreated mice and mice treated with 3,5-diethoxycarbonyl-1,4-dihydrocollidine: Source of the methyl group, *Arch. Biochem. Biophys.* **212:**120–126.
111. De Matteis, F., Gibbs, A. H., Farmer, P. B., and Lamb, J. H., 1981, Liver production of N-alkylated porphyrins caused by treatment with substituted dihydropyridines, *FEBS Lett.* **129:**328–331.
112. Ortiz de Montellano, P. R., and Kerr, D. E., 1985, Inactivation of myoglobin by ortho-substituted aryl hydrazines: Formation of prosthetic heme aryl-iron but not N-aryl adducts, *Biochemistry* **24:**1147–1152.
113. De Matteis, F., Gibbs, A. H., and Hollands, C, 1983, N-Alkylation of the haem moiety of cytochrome P-450 caused by substituted dihydropyridines. Preferential attack of different pyrrole nitrogen atoms after induction of various cytochrome P-450 isoenzymes, *Biochem. J.* **211:**455–461.
114. Lukton, D., and Ortiz de Montellano, P. R., 1985, Oxidative inactivation of cytochrome P-450 by a 2,2-dialkyl-1,2-dihydroquinoline, *Fed. Proc.* **44:**1399.
115. Muakkassah, W. F., and Yang, W. C. T., 1981, Mechanism of the inhibitory action of phenelzine on microsomal drug metabolism, *J. Pharmacol. Exp. Ther.* **219:**147–155.
116. Ortiz de Montellano, P. R., Augusto, O., Viola, F., and Kunze, K. L., 1983, Carbon radicals in the metabolism of alkyl hydrazines, *J. Biol. Chem.* **258:**8623–8629.
117. Jonen, H. G., Werringloer, J., Prough, R. A., and Estabrook, R. W., 1982, The reaction of phenylhydrazine with microsomal cytochrome P-450: Catalysis of heme modification, *J. Biol. Chem.* **257:**4404–4411.
118. Mansuy, D., Battioni, P., Bartoli, J.-F., and Mahy, J. -P., 1985, Suicidal inactivation of microsomal cytochrome P-450 by hydrazones, *Biochem. Pharmacol.* **34:**431–432.
119. Ortiz de Montellano, P. R., and Kunze, K. L., 1981, Formation of N-phenylheme in the hemolytic reaction of phenylhydrazine with hemoglobin, *J. Am. Chem. Soc.* **103:**581–586.
120. Saito, S., and Itano, H. A., 1981, Beta-meso-phenylbiliverdin IX-alpha and N-phenylprotoporphyrin IX, products of the reaction fo phenylhydrazine with oxyhemoproteins, *Proc. Natl. Acad. Sci. USA* **78:**5508–5512.
121. Augusto, O., Kunze, K. L., and Ortiz de Montellano, P. R., 1982, N-Phenylprotoporphyrin IX formation in the hemoglobin–phenylhydrazine reaction: Evidence for a protein-stabilized iron-phenyl intermediate, *J. Biol. Chem.* **257:**6231–6241.

122. Kunze, K. L., and Ortiz de Montellano, P. R., 1983, Formation of a sigma-bonded aryl-iron complex in the reaction of arylhydrazines with hemoglobin and myoglobin, *J. Am. Chem. Soc.* **105**:1380–1381.
123. Ortiz de Montellano, P. R., and Kerr, D. E., 1983, Inactivation of catalase by phenylhydrazine: Formation of a stable aryl-iron heme complex, *J. Biol. Chem.* **258**:10558–10563.
124. Ringe, D., Petsko, G. A., Kerr, D. E., and Ortiz de Montellano, P. R., 1984, Reaction of myoglobin with phenylhydrazine: A molecular doorstop, *Biochemistry* **23**:2–4.
125. Battioni, P., Mahy, J. -P., Delaforge, M., and Mansuy, D., 1983, Reaction of monosubstituted hydrazines and diazenes with rat-liver cytochrome P-450: Formation of ferrous-diazene and ferric sigma-alkyl complexes, *Eur. J. Biochem.* **134**:241–248.
126. Battioni, P., Mahy, J. -P., Gillet, G., and Mansuy, D., 1983, Iron porphyrin dependent oxidation of methyl- and phenylhydrazine: Isolation of iron(II)-diazene and sigma-alkyliron (III) (or aryliron(III)) complexes. Relevance to the reactions of hemoproteins with hydrazines, *J. Am. Chem. Soc.* **105**:1399–1401.
127. Campbell, C. D., and Rees, C. W., 1969, Reactive intermediates. Part III. Oxidation of 1-aminobenzotriazole with oxidants other than lead tetra-acetate, *J. Chem. Soc. C* **1969**:752–756.
128. Ortiz de Montellano, P. R., and Mathews, J. M., 1981, Autocatalytic alkylation of the cytochrome P-450 prosthetic haem group by 1-aminobenzotriazole: Isolation of an N,N-bridged benzyne–protoporphyrin IX adduct, *Biochem. J.* **195**:761–764.
129. Ortiz de Montellano, P. R., Mathews, J. M., and Langry, K. C., 1984, Autocatalytic inactivation of cytochrome P-450 and chloroperoxidase by 1-aminobenzotriazole and other aryne precursors, *Tetrahedron* **40**:511–519.
130. Costa, A. K., and Ortiz de Montellano, P. R., 1985, Dissociation of cytochrome P-450 inactivation and induction, *Fed. Proc.* **44**:652.
131. Mathews, J. M., and Bend, J. R., 1985, Analogs of 1-aminobenzotriazole (ABT) as isozyme selective suicide inhibitors of rabbit pulmonary cytochrome P-450, *Fed. Proc.* **44**:1466.
132. Whitman, D. W., and Carpenter, B. K., 1980, Experimental evidence for nonsquare cyclobutadiene as a chemically significant intermediate in solution, *J. Am. Chem. Soc.* **102**:4272–4274.
133. Stearns, R. A., and Ortiz de Montellano, P. R., 1985, Inactivation of cytochrome P-450 by a catalytically generated cyclobutadiene species, *J. Am. Chem. Soc.* **107**:234–240.
134. Stejskal, R., Itabashi, M., Stanek, J., and Hruban, Z., 1975, Experimental porphyria induced by 3-[2-(2,4,6-trimethylphenyl)-thioethyl]-4-methylsydnone, *Virchows Arch. B* **18**:83–100.
135. White, E. H., and Egger, N., 1984, Reaction of sydnones with ozone as a method of deamination: On the mechanism of inhibition of monoamine oxidase by sydnones, *J. Am. Chem. Soc.* **106**:3701–3703.
136. Chevrier, B., Weiss, R., Lange, M. C., Chottard, J. -C., and Mansuy, D., 1981, An iron(III)–porphyrin complex with a vinylidene group inserted into an iron–nitrogen bond: Relevance of the structure of the active oxygen complex of catalase, *J. Am. Chem. Soc.* **103**:2899–2901.
137. Latos-Grazynski, L., Cheng, R. -J., La Mar, G. N., and Balch, A. L., 1981, Reversible migration of an axial carbene ligand into an iron–nitrogen bond of a porphyrin: Implications for high oxidation states of heme enzymes and heme catabolism, *J. Am. Chem. Soc.* **103**:4271–4273.
138. Grab, L. A., Ortiz de Montellano, P. R., Sutherland, E. P., and Marks, G. S., 1985, Mechanism-based inactivation of cytochrome P-450 by sydnones and inhibition of ferrochelatase by the resulting heme adduct, *Fed. Proc.* **44**:1610.
139. Schaefer, W. H., Harris, T. M., and Guengerich, F. P., 1985, Characterization of the

enzymatic and non-enzymatic peroxidative degradation of iron porphyrins and cytochrome P-450 heme, *Biochemistry* **24**:3254–3263.
140. Nerland, D. E., Iba, M. M., and Mannering, G. J., 1981, Use of linoleic acid hydroperoxide in the determination of absolute spectra of membrane-bound cytochrome P-450, *Mol. Pharmacol.* **19**:162–167.
141. Guzelian, P. S., and Swisher, R. W., 1979, Degradation of cytochrome P-450 haem by carbon tetrachloride and 2-allyl-2-isopropylacetamide in rat liver in vivo and in vitro: Involvement of non-carbon monoxide-forming mechanisms, *Biochem. J.* **184**:481–489.
142. Ortiz de Montellano, P. R., and Kunze, K. L., 1980, Inactivation of hepatic cytochrome P-450 by allenic substrates, *Biochem. Biophys. Res. Commun.* **94**:443–449.
143. Hanzlik, R. P., Kishore, V., and Tullman, R., 1979, Cyclopropylamines as suicide substrates for cytochromes P-450, *J. Med. Chem.* **22**:759–761.
144. Macdonald, T. L., Zirvi, K., Burka, L. T., Peyman, P., and Guengerich, F. P., 1982, Mechanism of cytochrome P-450 inhibition by cyclopropylamines, *J. Am. Chem. Soc.* **104**:2050–2052.
145. Ortiz de Montellano, P. R., and Mathews, J. M., 1981, Inactivation of hepatic cytochrome P-450 by a 1,2,3-benzothiadiazole insecticide synergist, *Biochem. Pharmacol.* **30**:1138–1141.
146. De Groot, H., and Haas, W., 1981, Self-catalyzed O_2-independent inactivation of NADPH- or dithionite-reduced microsomal cytochrome P-450 by carbon tetrachloride, *Biochem. Pharmacol.* **30**:2343–2347.
147. Poli, G., Cheeseman, K., Slater, T. F., and Danzani, M. U., 1981, The role of lipid peroxidation in CCl_4-induced damage to liver microsomal enzymes: Comparative studies in vitro using microsomes and isolated liver cells, *Chem. Biol. Interact.* **37**:13–24.
148. Fernandez, G., Villaruel, M. C., de Toranzo, E. G. D., and Castro, J. A., 1982, Covalent binding of carbon to the heme moiety of cytochrome P-450 and its degradation products, *Res. Commun. Chem. Pathol. Pharmacol.* **35**:283–290.
149. De Groot, H., Harnisch, U., and Noll, T., 1982, Suicidal inactivation of microsomal cytochrome P-450 by halothane under hypoxic conditions, *Biochem. Biophys. Res. Commun.* **107**:885–891.
150. Reiner, O., and Uehleke, H., 1971, Bindung von Tetrachlorkohlenstoff an reduziertes mikrosomales Cytochrome P-450 und an Ham, *Hoppe-Seylers Z. Physiol. Chem.* **352**:1048–1052.
151. Cox, P. J., King, L. J., and Parke, D. V., 1976, The binding of trichlorofluoromethane and other haloalkanes to cytochrome P-450 under aerobic and anaerobic conditions, *Xenobiotica* **6**:363–375.
152. Roland, W. C., Mansuy, D., Nastainczyk, W., Deutschmann, G., and Ullrich, V., 1977, The reduction of polyhalogenated methanes by liver microsomal cytochrome P-450, *Mol. Pharmac.* **13**:698–705.
153. Mansuy, D., and Fontecave, M., 1983, Reduction of benzyl halides by liver microsomes: Formation of 478 nm-absorbing sigma-alkyl-ferric cytochrome P-450 complexes, *Biochem. Pharmacol.* **32**:1871–1879.
154. Mansuy, D., Lange, M., Chottard, J. C., Bartoli, J. F., Chevrier, B., and Weiss, R., 1978, Dichlorocarbene complexes of iron(II)-porphyrins—Crystal and molecular structure of $FE(TPP)(CCl_2)(H_2O)$, *Angew. Chem. Int. Ed. Engl.* **17**:781–782.
155. Ahr, H. J., King, L. J., Nastainczyk, W., and Ullrich, V., 1980, The mechanism of chloroform and carbon monoxide formation from carbon tetrachloride by microsomal cytochrome P-450, *Biochem. Pharmacol.* **29**:2855–2861.
156. Mansuy, D., Lange, M., Chottard, J. C., and Bartoli, J. F., 1978, Reaction du complexe carbenique Fe(II)(tetraphenylporphyrine)(CCl_2) avec les amines primaires: Formation d'isonitriles, *Tetrahedron Lett.* **33**:3027–3030.
157. Mansuy, D., and Battioni, J. -P., 1982, Isolation of sigma-alkyl-iron(III) or carbene-

iron(II) complexes from reduction or polyhalogenated compounds by iron(II)-porphyrins: The particular case of halothane $CF_3CHClBr$, *J. Chem. Soc. Chem. Commun.* **1982**:638–639.
158. Ruf, H. H., Ahr, H., Nastainczyk, W., Ullrich, V., Mansuy, D., Battioni, J.-P., Montiel-Montoya, R., and Trautwein, A., 1984, Formation of a ferric carbanion complex from halothane and cytochrome P-450: Electron spin resonance, ulvutronic spectra and model complexes, *Biochemistry* **23**:5300–5306.
159. Castro, C. E., Wade, R. S., and Belser, N. O., 1985, Biodehalogenation: Reactions of cytochrome P-450 with polyhalomethanes, *Biochemistry* **24**:204–210.
160. Callot, H. J., and Scheffer, E., 1980, Model for the in vitro transformation of cytochrome P-450 into "green pigments," *Tetrahedron Lett.* **21**:1335–1338.
161. Lange, M., and Mansuy, D., 1981, N-Substituted porphyrins formation from carbene iron-porphyrin complexes: A possible pathway for cytochrome P-450 heme destruction, *Tetrahedron Lett.* **22**:2561–2564.
162. Chevrier, B., Weiss, R., Lange, M., Chotard, J. C., and Mansuy, D., 1981, An iron(III)–porphyrin complex with a vinylidene group inserted into an iron–nitrogen bond: Relevance to the structure of the active oxygen complex of catalase, *J. Am. Chem. Soc.* **103**:2899–2901.
163. Olmstead, M. M., Cheng, R.-J., and Balch, A. L., 1982, X-Ray crystallographic characterization of an iron porphyrin with a vinylidene carbene inserted into an iron–nitrogen bond, *Inorg. Chem.* **21**:4143–4148.
164. Guengerich, F. P., Dannan, G. A., Wright, T. S., Martin, M. V., and Kaminsky, L. S., 1982, Purification and characterization of liver microsomal cytochromes P-450: Electrophoretic, spectral, catalytic, and immunochemical properties and inducibility of eight isozymes isolated from rats treated with phenobarbital or beta-naphthoflavone, *Biochemistry* **21**:6019–6030.
165. Ortiz de Montellano, P. R., Mico, B. A., Mathews, J. M., Kunze, K. L., Miwa, G. T., and Lu, A. Y. H., 1981, Selective inactivation of cytochrome P-450 isozymes by suicide substrates, *Arch. Biochem. Biophys.* **210**:717–728.
166. Kellis, J. T., Sheets, J. J., and Vickery, L. E., 1984, Amino-steroids as inhibitors and probes of the active site of cytochrome P-450$_{scc}$. Effects on the enzyme from different sources, *J. Steroid Biochem.* **20**:671–676.
167. Sheets, J. J., and Vickery, L. E., 1983, Active site-directed inhibitors of cytochrome P-450$_{scc}$: Structural and mechanistic implication of a side chain-substituted series of amino-steroids, *J. Biol. Chem.* **258**:11446–11452.
168. Sheets, J. J., and Vickery, L. E., 1982, Proximity of the substrate binding site and the heme-iron catalytic site in cytochrome P-450$_{scc}$, *Proc. Natl. Acad. Sci. USA* **79**:5773–5777.
169. Nagahisa, A., Foo, T., Gut, M., and Orme-Johnson, W. H., 1985, Competitive inhibition of cytochrome P-450$_{scc}$ by (22R)- and (22S)-22-aminocholesterol: Side chain stereochemical requirements for C-22 amine coordination to the active-site heme, *J. Biol. Chem.* **260**:846–851.
170. Nagahisa, A., Spencer, R. W., and Orme-Johnson, W. H., 1983, Acetylenic mechanism-based inhibitors of cholesterol side chain cleavage by cytochrome P-450$_{scc}$, *J. Biol. Chem.* **258**:6721–6723.
171. Nagahisa, A., Orme-Johnson, W. H., and Wilson, S. R., 1984, Silicon mediated suicide inhibition: An efficient mechanism-based inhibitor of cytochrome P-450$_{scc}$ oxidation of cholesterol, *J. Am. Chem. Soc.* **106**:1166–1167.
172. Trahanovsky, W. S., and Himstedt, A. L., 1974, Oxidation of organic compounds with cerium(IV). XX. Abnormally rapid rate of oxidative cleavage of (beta-trimethylsilylethyl)phenylmethanol, *J. Am. Chem. Soc.* **96**:7974–7976.

173. Brodie, A. M. H., Marsh, D., and Brodie, H. J., 1979, Aromatase inhibitors. IV. Regression of hormone-dependent, mammary tumors in the rat with 4-acetoxy-4-androstene-3,17-dione, *J. Steroid Biochem.* **10**:423–429.
174. Henderson, I. C., and Canellos, G. P., 1980, Cancer of the breast (The past decade), *N. Engl. J. Med.* **302**:78–90.
175. Santen, R. J., Worgul, T. J., Samojlik, E., Interrante, A., Boucher, A. E., Lipton, A., Harvey, H. A., White, D. S., Smart, E., Cox, C., and Wells, S. A., 1981, A randomized trial comparing surgical adrenalectomy with aminoglutethimide plus hydrocortisone in women with advanced breast cancer, *N. Engl. J. Med.* **305**:545–551.
176. Phillips, G. B., Castelli, W. P., Abbott, R. D., and McNamara, P. M., 1983, Association of hyperestrogenemia and coronary heart disease in men in the Framingham cohort, *Am. J. Med.* **74**:863–869.
177. Harris, A. L., Powles, T. J., Smith, I. E., Coombes, R. C., Ford, H. T., Gazet, J. C., Harmer, C. L., Morgan, M., White, H., Parsons, C. A., and McKinna, J. A., 1983, Aminoglutethimide for the treatment of advanced postmenopausal breast cancer, *Eur. J. Cancer Clin. Oncol.* **19**:11–17.
178. Metcalf, B. W., Wright, C. L., Burkhart, J. P., and Johnston, J. O., 1981, Substrate-induced inactivation of aromatase by allenic and acetylenic steroids, *J. Am. Chem. Soc.* **103**:3221–3222.
179. Covey, D. G., Hood, W. F., and Parikh, V. D., 1981, 10-Beta-propynyl-substituted steroids, *J. Biol. Chem.* **256**:1076–1079.
180. Marcotte, P. A., and Robinson, C. H., 1982, Synthesis and evaluation of 10-beta-substituted 4-estrene-3,17-diones as inhibitors of human placental microsomal aromatase, *Steroids* **39**:325–344.
181. Marcotte, P. A., and Robinson, C. H., 1982, Design of mechanism-based inactivators of human placental aromatase, *Cancer Res.* **42**:3322–3325.
182. Flynn, G. A., Johnston, J. O., Wright, C. L., and Metcalf, B. W., 1981, The time-dependent inactivation of aromatase by 17-beta-hydroxy-10-methylthioestra-1,4-dien-3-one, *Biochem. Biophys. Res. Acta* **103**:913–918.
183. Covey, D. F., and Hood, W. F., 1982, Aromatase enzyme catalysis is involved in the potent inhibition of estrogen biosynthesis caused by 4-acetoxy- and 4-hydroxy-4-androstene-3,17-dione, *Mol. Pharmacol.* **21**:173–180.
184. Brodie, A. M. H., Garrett, W. M., Hendrickson, J. R., Tsai-Morris, C.-H., Marcotte, P. A., and Robinson, C. H., 1981, Inactivation of aromatase in vitro by 4-hydroxy-4-androstene-3,17-dione and 4-acetoxy-4-androstene-3,17-dione and sustained effects in vivo, *Steroids* **38**:693–702.
185. Covey, D. E., Hood, W. F., Bensen, D. D., and Carrell, H. L., 1984, Hydroperoxides as inactivators of aromatase: 10-Beta-hydroperoxy-4-estrene-3,17-dione, crystal structure and inactivation characteristics, *Biochemistry* **23**:5398–5406.
186. Vanden Bossche, H., Willemsens, G., Cools, W., Marichal, P., and Lauwers, W., 1983, Hypothesis on the molecular basis of the antifungal activity of N-substituted imidazoles and triazoles, *Biochem. Soc. Trans.* **11**:665–667.
187. Vanden Bossche, H., Lauwers, W., Willemsens, G., Marichal, P., Cornelissen, F., and Cools, W., 1984, Molecular basis for the antimycotic and antibacterial activity of N-substituted imidazoles and triazoles: The inhibition of isoprenoid biosynthesis, *Pestic. Sci.* **15**:188–198.
188. Heeres, J., De Brabander, M., and Vanden Bossche, H., 1982, Ketoconazole: Chemistry and basis for selectivity, in: *Current Chemotherapy and Immunotherapy*, Volume 2 (P. Periti and G. G. Grossi, eds.), American Society of Microbiology, Washington, D.C., pp. 1007–1009.
189. Willemsens, G., Cools, W., and Vanden Bossche, H., 1980, Effects of miconazole and

ketoconazole on sterol synthesis in a subcellular fraction of yeast and mammalian cells, in: *The Host Invader Interplay* (H. Van den Bossche, ed.), Elsevier/North Holland, Amsterdam, pp. 691–694.
190. Murray, M., Ryan, A. J., and Little, P. J., 1982, Inhibition of rat hepatic microsomal aminopyrine N-demethylase activity by benzimidazole derivatives: Quantitative structure–activity relationships, *J. Med. Chem.* **25**:887–892.
191. Santen, R. J., Vanden Bossche, H., Symoens, J., Brugmans, J., and DeCoster, R., 1983, Site of action of low dose ketoconazole or androgen biosynthesis in men, *J. Clin. Endocrinol. Metab.* **57**:732–736.
192. Gahder, P., Mercer, E. I., Baldwin, B. C., and Wiggins, T. E., 1983, A comparison of the potency of some fungicides as inhibitors of sterol 14-demethylation, *Pestic. Biochem. Physiol.* **19**:1–10.
193. Nes, W. R., 1974, Role of sterols in membranes, *Lipids* **9**:596–612.
194. Yeagle, P. L., Martin, R. B., Lala, A. K., Lin, H.-K., and Block, K., 1977, Differential effects of cholesterol and lanosterol on artificial membranes, *Proc. Natl. Acad. Sci. USA* **74**:4924–4926.
195. Freter, C. E., Laderson, R. C., and Sibert, D. F., 1979, Membrane phospholipid alterations in response to sterol depletion of LM cells, *J. Biol. Chem.* **254**:6909–6916.
196. Shak, S., and Goldstein, I., 1984, Omega-oxidation is the major pathway for the catabolism of leukotriene B_4 in human polymorphonuclear leukocytes, *J. Biol. Chem.* **259**:10181–10187.
197. Kupfer, D., 1982, Endogenous substrates of monooxygenases: Fatty acids and prostaglandins, in: *Hepatic Cytochrome P-450 Monooxygenase System* (J. B. Schenkman and D. Kupfer, eds.), Pergamon Press, Elmsford, N.Y., pp. 157–190.
198. Matson, R. S., Stein, R. A., and Fulco, A. J., 1980, Hydroxylation of 9-hydroxystearate by a soluble cytochrome P-450-dependent fatty acid hydroxylase from *Bacillus megaterium*, *Biochem. Biophys. Res. Commun.* **97**:955–961.
199. Kupfer, D., 1980, Endogenous substrates of monooxygenases: Fatty acids and prostaglandins, *Pharmacol. Ther. A* **11**:469–496.
200. Ortiz de Montellano, P. R., and Reich, N. O., 1984, Specific inactivation of hepatic fatty acid hydroxylases by acetylenic fatty acids, *J. Biol. Chem.* **259**:4136–4141.
201. Salaun, J. P., Reichhart, D., Simon, A., Durst, F., Reich, N. O., and Ortiz de Montellano, P. R., 1984, Autocatalytic inactivation of plant cytochrome P-450 enzymes: Selective inactivation of the lauric acid in-chain hydroxylase from *Helianthus tuberosus* L. by unsaturated substrate analogs, *Arch. Biochem. Biophys.* **232**:1–7.
202. CaJacob, C. A., and Ortiz de Montellano, P. R., 1985, Sodium 10-undecynyl sulfate: A specific in vivo irreversible inhibitor of fatty acid hydroxylases, *Fed. Proc.* **44**:1611.
203. Clancy, R. M., Dahinden, C. A., and Hugli, T. E., 1984, Oxidation of leukotrienes at the ω end: Demonstration of a receptor for the 20-hydroxy derivative of leukotriene B_4 on human neutrophils and implications for the analysis of leukotriene receptors, *Proc. Natl. Acad. Sci. USA* **81**:5729–5733.
204. Shak, S., Reich, N. O., Goldstein, I. M., and Ortiz de Montellano, P. R., 1985, Leukotriene B_4 ω-hydroxylase in human polymorphonuclear leukocytes: Suicidal inactivation by acetylenic fatty acids, *J. Biol. Chem.* **260**:13023–13028.
205. Masters, B. S. S., Okita, R. T., Ortiz de Montellano, P. R., Reich, N. O., and Williams, D. E., 1985, Pulmonary cytochrome P-450-mediated fatty acid and prostaglandin (PG) hydroxylation—Inhibition by antibodies and suicide substrates, *Fed. Proc.* **44**:651.
206. Feyereisen, R., Farnsworth, D. E., Prickett, K. S., and Ortiz de Montellano, P. R., 1985, Suicidal destruction of cytochrome P-450 in the design of inhibitors of insect juvenile hormone biosynthesis, in: *Bioregulators for Pest Control*. American Chemical Society Symposium Series No. 276, American Chemical Society, Washington, D.C., pp. 255–266.

CHAPTER 9

Induction of Hepatic P-450 Isozymes
Evidence for Specific Receptors

HOWARD J. EISEN

1. Introduction

The mammalian liver is a major site of cytochrome P-450-mediated metabolism of endogenous and exogenous compounds. Multiple forms of P-450 have been purified from rodent liver.[1,2] Some of these P-450 proteins are expressed constitutively, while others are expressed only in response to specific hormonal or chemical stimuli. Examples of such chemical stimuli are polycyclic aromatic compounds, barbiturates, and steroids such as pregnenolone 16α-carbonitrile. A major goal of research has been to identify the cellular processes that "connect" a chemical stimulus to the expression of a specific P-450 protein. Cellular macromolecules have been identified that bind polycyclic aromatic compounds; among these macromolecules is an intracellular protein termed the Ah receptor.[3,4] This protein may be involved directly in the regulation of expression of P-450 genes. Hepatic receptors for barbiturates and pregnenolone 16α-carbonitrile have not been identified. In this chapter, recent data are discussed that deal with the problem of identifying cellular "receptors" for these compounds.

An alternative approach to the identifications of regulatory factors is based on direct analysis of the gene affected by a chemical stimulus. For example, transcription of the chicken liver vitellogenin gene is regulated by estrogens. The chromatin structure of the vitellogenin gene in

---

HOWARD J. EISEN • Laboratory of Developmental Pharmacology, National Institute of Child Health and Human Development, National Institutes of Health, Bethesda, Maryland 20205.

liver nuclei can be "probed" with the aid of cloned DNA of the vitellogenin gene. Estrogen treatment results in distinct changes in chromatin structure which appear to be initiated by the binding of estrogen receptor complexes to regulatory regions of the vitellogenin gene.[5] Recently, P-450 genes have been cloned.[6-10] In addition to analysis of chromatin structure, other experimental approaches are suggested in this chapter which may identify regulatory factors associated with the actions of barbiturates and pregnenolone 16α-carbonitrile on P-450.

2. Induction of P-450 Isozymes by Polycyclic Aromatic Compounds

2.1. Measurement of mRNA Species Induced by Polycyclic Aromatic Compounds

In rats, planar polycyclic aromatic compounds such as 3-methylcholanthrene (MC) and 2,3,7,8-tetrachlorodibenzo-p-dioxin (TCDD) induce two P-450 isozymes: P-450c and P-450d. The related mouse proteins are P_1-450 and P_3-450. Initially, estimates of mRNA concentration for these isozymes were obtained by translation *in vitro* of mRNA preparations and immunoprecipitation of the translation products. Because of extensive homology between P-450c and P-450d, polyclonal antibodies prepared to homogeneous P-450c often cross-react with P-450d.

Morville *et al.*[11] used antibodies to P-450c and P-450d to characterize the *in vitro* translation products of mRNA from MC-treated rats. The antisera were adsorbed using covalently immobilized purified P-450c or P-450d to remove cross-reacting antibodies. The *in vitro* translation products of P-450c and P-450d mRNA differed in molecular weight and were distinguished readily on SDS–polyacrylamide gels. Total poly(A^+) mRNA from MC-induced rat liver was fractionated on methylmercury-containing agarose gels; the mRNA for P-450c was estimated to be 4000 nucleotides in length, whereas the mRNA for P-450d was considerably shorter ($\sim$ 2000 nucleotides in length). Similar estimates of the sizes of MC-inducible P-450 mRNA were obtained by workers in other laboratories.[12,13]

Morville *et al.*[11] estimated that MC-inducible P-450 mRNA represented less than 1% of the total poly(A^+) mRNA. Because of the low relative abundance of MC-induced P-450 mRNA, it was recognized that cloning of DNA complementary to these mRNAs would require substantial enrichment of P-450 mRNA prior to cDNA synthesis, and the use of highly selective procedures for identification of cloned DNA containing

P-450 sequences. In order to clone DNA complementary to mouse liver P_1-450 mRNA, Negishi et al.[14] used an antibody to P_1-450 to identify P_1-450 mRNA by translation in vitro of poly(A^+) mRNA from MC-treated livers of C57BL/6 mice. Sucrose density gradient centrifugation was used to fractionate poly(A^+) mRNA; fractions ($\sim$ 23 S) enriched in P_1-450 mRNA were then used for synthesis of cDNA; cDNA was cloned by (dG–dC) tailing procedures into the PstI site of the plasmid pBR322. Differential hybridization was used to identify a cDNA clone (clone 46) that hybridized to [^{32}P]-DNA complementary to mRNA from MC-induced mouse liver but not to [^{32}P]-DNA complementary to mRNA from control preparations. Finally, hybridization-arrest experiments demonstrated that clone 46 hybridized to mRNA that programmed the synthesis in vitro of immunoprecipitable P_1-450.

With the use of similar methods, Kawajiri and co-workers isolated two groups of cDNA clones which represented DNA complementary to rat liver P-450c and P-450d mRNA.[15] Fagan et al. also isolated a DNA clone complementary to P-450d mRNA.[13] These studies utilized size fractionation of poly(A^+) mRNA to enrich for MC-induced P-450 mRNA. An alternative approach to the enrichment of MC-induced mRNA was used by Kimura et al.[16] These investigators used antibodies to precipitate polysomes containing nascent P_1-450 and P_3-450 peptides; poly(A^+) mRNA was cloned using the Okayama–Berg vector.[17] The nucleotide sequences have been determined of full-length cDNA clones for P_1-450 and P_3-450.[18]

From direct analysis of the nucleotide sequences of the MC-induced P-450 isozymes, it was evident that DNA "probes" prepared from regions of significant homology were likely to hybridize with mRNA for both P-450 isozymes. Such cross-reactions were observed by Northern hybridization.[15,19] Measurement of the individual mRNA species clearly required the use of DNA probes that did not cross-hybridize, or the separation of mRNA species prior to quantitation. Kawajiri et al.[15] used a probe prepared from a homologous region of P-450c and P-450d mRNA to quantitate these mRNA species. The individual mRNA species were separated by fractionation on formaldehyde-containing agarose gels prior to immobilization on nitrocellulose and hybridization to [^{32}P]cDNA. If the cDNA probes used for mRNA quantitation are specific for either isozyme, then other methods are suitable for quantitation of mRNA. These methods include hybridization reactions performed in solution, and the method described recently by Spradling et al.[20] The latter procedure involves partial alkaline hydrolysis of poly(A^+) mRNA, ^{32}P-labeling of 5' ends of the mRNA fragments, and hybridization to cDNA immobilized on nitrocellulose filters.

2.2. The Ah Receptor

The data obtained from recombinant-DNA studies in rodents indicated that only two P-450 isozymes were induced by MC. In addition to polycyclic aromatic hydrocarbons such as MC, other compounds were found to induce these P-450 isozymes. The halogenated biphenyl compounds and related compounds (such as chlorinated dibenzo-p-dioxins) have been studied extensively; TCDD is 10,000-fold more potent than MC as an inducer of P-450-mediated monooxygenase activities.[21,22] TCDD and certain other potent halogenated compounds produce a wide range of effects in animals and tissue culture cell lines. These effects include epidermal hyperplasia (manifested as chloracne in man), thymic atrophy, tumor-promotion activity, a "wasting syndrome," and death.[23] Therefore, the induction of P-450 isozymes represents only one aspect of the action of these compounds. Detailed structure–activity studies demonstrated a consistent rank-order and relative potency of inducers for each of these effects.[23] Among the chlorinated biphenyl derivatives, those compounds constrained to a planar configuration by steric factors were potent inducers; the planar dibenzo-p-dioxins and dibenzofurans required halogenation at certain positions to function as potent inducers. These data based on structure–activity relationships led to the hypothesis that a specific "receptor" was involved in the diverse biological effects of these compounds.[21]

Further evidence for a "receptor" site was obtained from studies with inbred mouse strains.[24] Inbred mouse strains differ in their sensitivity to polycyclic aromatic compounds. For example, MC induces P_1-450 and P_3-450 in C57BL/6 (B6) but not DBA/2N (D2) mice. TCDD induces P_1-450 and P_3-450 in both strains; however, the $EC_{50} \cong 20$ nmole/kg for D2, whereas the $EC_{50} \cong 1$ nmole/kg in B6 mice[22] (Fig. 1). This difference in sensitivity is regulated by a genetic locus termed the *Ah* locus (for *a*ryl *h*ydrocarbon responsiveness).[25] In genetic crosses among B6, D2, and hybrid progeny, the B6 (Ah^b/Ah^b) pattern of response to inducers was inherited as a simple Mendelian dominant trait. Of interest, the genetic difference in EC_{50} for TCDD was observed not only for P-450 induction but also for other diverse effects of TCDD. From these data, it seemed possible that a specific "receptor" controlled the multiple effects of TCDD. D2 (Ah^d/Ah^d) mice appeared to express a "defective" receptor with decreased affinity for inducers.

Poland *et al.*[26] provided the first direct evidence for such a receptor. Using [^{3}H]-TCDD as a radioligand, these workers identified saturable [^{3}H]-TCDD binding to macromolecules in B6 liver cytosolic fractions. Both the high affinity ($K_d < 1$ nM) for TCDD and the molecular specificity for binding were consistant with the structure–activity relationships ob-

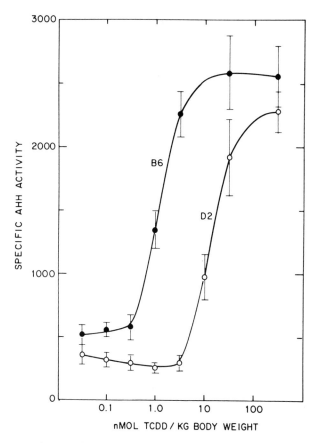

FIGURE 1. Hepatic AHH induction as a function of TCDD dosage in B6 (Ah^b/Ah^b) and D2 (Ah^d/Ah^d) inbred mice. A single intraperitoneal injection of TCDD in p-dioxane (0.5 ml/kg) was given 72 hr before mice were killed. Values plotted are means ± S.D. for groups of six mice. From Ref. 22 with permission.

tained from *in vivo* studies with inducers. Furthermore, no specific [^{3}H]-TCDD binding was detected in D2 liver cytosolic fractions. Subsequent studies have confirmed these findings. Various methods have been used to detect and characterize the cytosolic TCDD binding moieties. Okey et al.[27] used sucrose density gradient centrifugation and demonstrated that [^{3}H]-TCDD was bound specifically to a protein (~ 9 S) in B6 liver cytosolic fractions (Fig. 2). No specific binding was detected in D2 liver cytosolic fractions (Fig. 3). Even with other analytical techniques such as gel permeation chromatography,[28] no specific [^{3}H]TCDD binding was detected in D2 cytosolic fractions. These procedures all involved physical

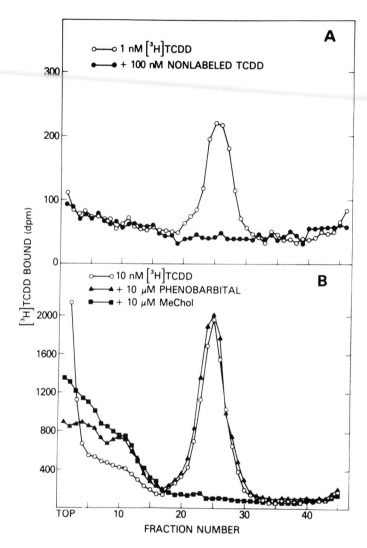

FIGURE 2. Detection of specific [^{3}H]-TCDD binding of a component from B6 hepatic cytosol. (A) Cytosol (1 mg protein/ml) from an 8-week-old B6 male was incubated with 1 nM [^{3}H]-TCDD in the absence of competitor (○) or in the presence of 100 nM nonlabeled TCDD (●) for 1 hr at 40°C. After dextran–charcoal treatment, gradients were centrifuged and fractionated. (B) Elimination of specific binding peak by MC but not by phenobarbital. Cytosol (5 mg protein/ml) was incubated with 10 nM [^{3}H]-TCDD in the absence of competitor (○), and in the presence of 10 μM MC (■) or 10 μM phenobarbital (▲). From Ref. 27 with permission.

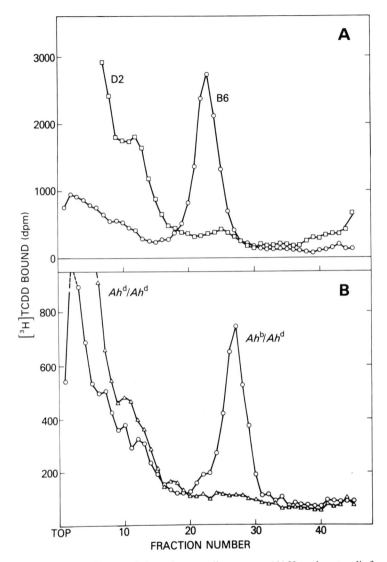

FIGURE 3. Genetic differences in hepatic cytosolic receptor. (A) Hepatic cytosolic fractions (5 mg protein/ml) from responsive B6 or nonresponsive D2 inbred mice were incubated at 4°C with 10 nM [^{3}H]-TCDD and analyzed by sucrose density gradients after dextran–charcoal treatment. (B) Hepatic cytosolic fractions from phenotypically responsive Ah^b/Ah^d heterozygote (3.1 mg protein/ml) and nonresponsive Ah^d/Ah^d homozygote (3.7 mg protein/ml) were incubated at 4°C with 10 nM [^{3}H]-TCDD and analyzed. From Ref. 27 with permission.

separation of free (unbound) radioligand from macromolecular-bound [^{3}H]-TCDD and required considerable time for analysis. The "rapid" dextran–charcoal assay[26] required multiple washes to decrease backgrounds of nonspecifically bound [^{3}H]-TCDD. Even such "rapid" assays might not be suitable for detection of a TCDD receptor complex that disassociated rapidly or was degraded rapidly in cytosolic fractions.

In the initial studies on TCDD binding, Poland et al.[26] observed that treatment of B6 mice *in vivo* with [^{3}H]-TCDD resulted in accumulation of specifically bound [^{3}H]-TCDD in liver nuclear fractions. Small quantities of specifically bound [^{3}H]-TCDD were also detected in D2 liver nuclear fractions. Okey et al.[27] observed the same phenomenon and demonstrated that the TCDD-binding moieties from B6 and D2 liver nuclear fractions had similar sedimentation coefficients ($\sim$ 6 S). These data indicated that a nuclear inducer–receptor complex was formed during treatment of mice *in vivo* with [^{3}H]-TCDD. The nuclear inducer–receptor complex remained strongly associated with the nucleus during extensive centrifugation (through 2.2 M sucrose) and washing by buffers that contained low ionic strength; high ionic strength (NaCl > 0.3 M) was required to solubilize these TCDD-binding moieties from the nuclear fractions. The experimental evidence indicated that the TCDD-binding protein from cytosolic fractions was the source of the nuclear TCDD-binding protein, although the mechanism of its formation was undetermined. What was the origin of the TCDD-binding moiety in D2 liver nuclear fractions, and could this moiety account for the induction of P_1-450 isozymes in D2 mice?

One approach to the analysis of the nuclear TCDD-binding proteins was to assess quantitative and temporal relationships between these moieties and the induction of P-450 isozymes. With the use of clone 46, P_1-450 mRNA was quantitated by R_0t analysis; these measurements were correlated to the levels of nuclear [(^{3}H)-TCDD–receptor complexes formed in mice following their treatment *in vivo* with [^{3}H]-TCDD[29] (Figs. 4–6). These studies confirmed previous observations that [^{3}H]-TCDD–receptor complexes accumulated in D2 nuclear fractions. Although the concentrations of nuclear [^{3}H]-TCDD–receptor complexes were lower in D2 (Ah^d/Ah^d) mice than in B6 (Ah^b/Ah^b) mice or (B6 × D2) F_1 hybrids (Ah^b/Ah^d), a good correlation was demonstrated between nuclear [^{3}H]-TCDD–receptor complexes and P_1-450 mRNA in mice with various Ah-locus genotypes. These data demonstrated also that the specific TCDD-binding moieties in D2 and B6 nuclei had similar physicochemical properties (sedimentation coefficients and Stokes radii). We have postulated that D2 mice contain a receptor moiety which is stabilized during formation *in vivo* of the nuclear ligand–receptor complex. Further analysis

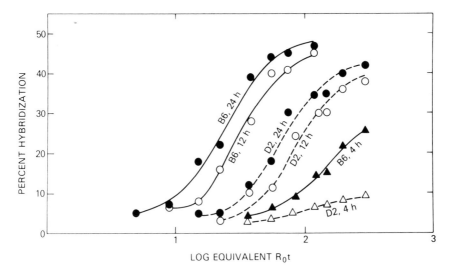

FIGURE 4. Hybridization of excess poly(A^+)-enriched RNA with the clone 46 probe, as a function of time following a single dose of TCDD (25 nmole/kg) given to B6 and D2 mice. From the livers of 10 mice per group, RNA was prepared at varying times after treatment with TCDD. R_0t analysis was performed. Since the clone 46 cDNA is double-stranded, 50% hybridization represents complete hybridization.[29]

of this phenomenon will require specific probes such as antireceptor antibodies and detailed knowledge of the mechanism(s) responsible for formation of the nuclear TCDD–Ah receptor complex.

2.3. Effect of Inducers on Transcription of P_1-450 and P_3-450 Genes

Although the studies outlined above demonstrated a good correlation between nuclear TCDD–Ah receptor complexes and the induction of P_1-450 mRNA, these experiments did not identify the step(s) in mRNA metabolism that was affected by TCDD. Gonzalez et al.[30] studied the effects of TCDD and MC on nuclear transcription of both P_1-450 and P_3-450 pre-mRNA. Nuclei were incubated with [^{32}P]-UTP under conditions that permitted elongation of nascent pre-mRNA transcripts. P_1-450 and P_3-450 pre-mRNA sequences were assayed by hybridization to specific cDNA probes for P_1-450 and P_3-450 mRNA. In B6 mice, transcription

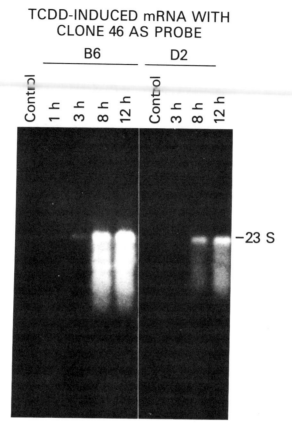

FIGURE 5. Northern blot of total poly(A$^+$)-enriched RNA hybridization to the clone 46 probe, as a function of time after B6 or D2 mice were treated with TCDD (25 nmole/kg). From the same samples used in Fig. 4, RNA (5 μg/ml) was electrophoresed and Northern hybridization performed using [^{32}P]cDNA (clone 46) as probe.[29]

rates with P_1-450 and P_3-450 were increased rapidly following treatment with MC or TCDD (Figs. 7, 8). MC did not affect transcription of P_1-450 pre-mRNA in D2 mice, whereas TCDD increased P_1-450 transcription in D2 mice. Although the relative transcription rates of P_1-450 and P_3-450 pre-mRNA were similar in MC-treated B6 mice, considerably more P_3-450 mRNA than P_1-450 mRNA accumulated in the liver. These data indicate that posttranscriptional processes (e.g., mRNA processing by splicing, or mRNA stabilization) may affect the levels of specific P-450 mRNA. Kawajiri et al.[15] have measured P-450c and P-450d mRNA levels in rats following treatment with various inducers. Following treatment of

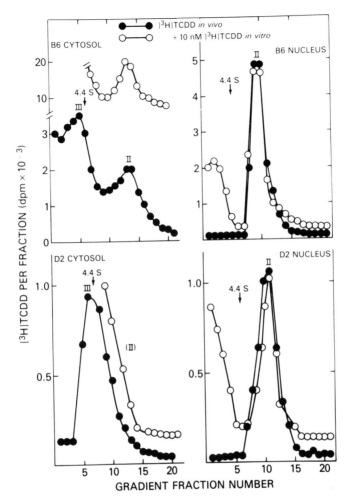

FIGURE 6. Velocity sedimentation analysis of B6 and D2 cytosolic fractions and nuclear extracts 4 hr after treatment with 50 nmole/kg [^{3}H]-TCDD. Mice (five per group) were treated with 50 nmole/kg [^{3}H]-TCDD. After 4 hr, livers were removed and cytosolic and nuclear fractions were prepared.[29] Portions of each sample were exposed *in vitro* to 10 mM [^{3}H]-TCDD for 1 hr at 4°C. All samples were then treated with dextran–charcoal to remove nonspecifically bound [^{3}H]-TCDD, following which sucrose density gradient centrifugation was performed. Peak III is the principal binding protein for MC in rodent liver; peak II is the Ah receptor.[28] Note the different values on the ordinate for samples from B6 and D2 mice. Note also that B6 cytosols contained additional peak II (Ah receptor) sites which were not occupied by this treatment *in vivo* with [^{3}H]-TCDD; D2 mice did not contain cytosolic peak II sites. The only peak II (Ah receptor) binding detected in nuclear fractions represented these sites occupied *in vivo*; no additional binding was detected in samples incubated *in vitro* with 10 mM [^{3}H]-TCDD.

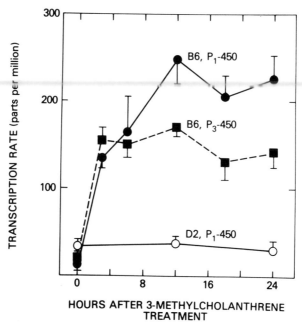

FIGURE 7. Transcription rates of the mouse P_1-450 and P_3-450 genes following MC treatment of B6 and D2 mice. Nuclei were isolated from B6 and D2 liver at various times following a single dose of MC. *In vitro* nuclear transcription was carried out[30]; approximately 4×10^6 to 6×10^6 dpm from each transcription reaction was hybridized in duplicate with each cDNA-containing filter and a pBR322 filter. Transcription rates were expressed as parts per million of the total disintegrations per minute hybridized, minus the adsorbed to filters containing pBR322. The radioactivity adsorbing to pBR322-bound filters was between 7 and 14 ppm of the total disintegrations per minute hybridized. Symbols and bars denote means ± S.D. for three duplicate determinations. pBR322-corrected control values used were 8 ± 4 and 16 ± 7 ppm for B6 and D2 P_1-450 gene transcription, respectively, and 10 ± 4 ppm for B6 P_3-450 gene transcription. The control values, expressed in parts per million per kilobase pair of cDNA insert, were 4.7 and 9.4 for B6 and D2 P_1-450, respectively, and 5.9 for B6 P_3-450 gene transcription.

rats with MC (25 mg/kg), P-450d mRNA reached considerably higher levels than P-450c mRNA.

2.4. Genetic Analysis of P-450 Induction in Tissue Culture Cell Lines

The genetic difference among mouse strains in ED_{50} for TCDD may be the result of a polymorphism affecting the TCDD-binding protein. The TCDD-binding protein has been termed the "gene product" of the *Ah*

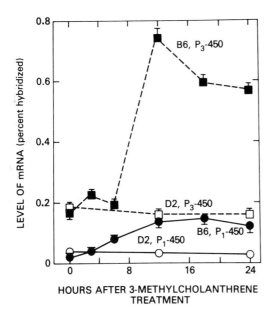

FIGURE 8. Hepatic P_1-450 and P_3-450 mRNA levels following MC treatment of B6 and D2 mice. Total poly(A^+)-enriched RNA was isolated from B6 and D2 liver at various times following MC treatment.[30] RNA was quantitated and the results are expressed as the percentage of total disintegrations per minute hybridized. Symbols and bars denote means ± S.D. for three duplicate determinations.

locus; however, this identification must be regarded as a hypothesis requiring further rigorous proof. For example, the *Ah* locus may actually encode a protein which modifies the TCDD-binding protein to a form which has a high affinity for TCDD. Because the TCDD-binding protein represents a minor component of the cell, it will be difficult to resolve many questions regarding its structure and function by conventional biochemical methods. Tissue culture cell lines provide alternative models for studying the TCDD- binding protein and its relationship to P-450 induction. Aryl hydrocarbon hydroxylase (AHH) activity is induced by polycyclic aromatic compounds in certain tissue culture cell lines such as the mouse hepatoma Hepa-1 and the rat hepatoma H4II-E. In other cell lines (rat hepatoma HTC and monkey kidney Vero), AHH activity is not induced. Okey et al.[31] demonstrated that cytosolic fractions of Hepa-1 and H4II-E cells contained specific, high-affinity binding sites for [^{3}H]-TCDD. HTC cells but not Vero cells contained similar binding sites. The intracellular TCDD-binding protein from Hepa-1 cells and B6 mouse liver cytosolic fractions had similar physicochemical properties. Therefore, the TCDD-binding protein in Hepa-1 cells was believed to be identical to the B6 (Ah^b/Ah^b) form of Ah receptor.

Further information regarding the mechanism of nuclear accumulation of TCDD–Ah receptor complexes was obtained by Okey et al.[31] When untreated Hepa-1 cells were fractionated into cytosolic and nuclear

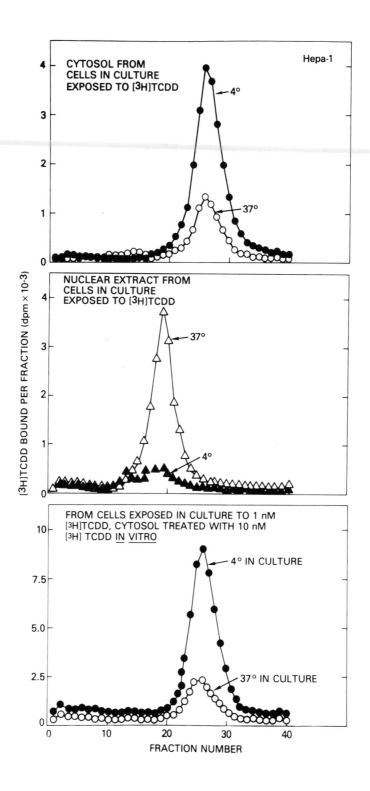

fractions, the ligand-free Ah receptor was predominantly associated with the cytosolic fractions. When intact Hepa-1 cells were treated *in vivo* with [^{3}H]-TCDD, accumulation of the TCDD–Ah receptor complex in nuclear fractions was a temperature-dependent process. At 4°C, the [^{3}H]-TCDD–Ah receptor complexes remained in cytosolic fractions, whereas at 37°C, the complexes accumulated in nuclear fractions. High ionic strength (NaCl > 0.3 M) was required to solubilize receptor complexes from nuclear fractions (Fig. 9).

Whitlock and Galeazzi[32] studied further the subcellular distribution of Ah receptors in Hepa-1 cells. In particular, these investigators considered the possibility that Ah receptors were located in the nucleus in untreated cells. The apparent localization of the Ah receptor in cytosolic fractions might represent an "artifact" related to dilution of the cell contents during homogenization. By homogenizing cells under conditions of minimal dilution, Whitlock and Galeazzi[32] showed that nuclear fractions contained most of the ligand-free Ah receptors; however, the association with nuclei was weak and simple dilution of the homogenate at low ionic strength was sufficient to redistribute the Ah receptors into cytosolic fractions. These data indicated that the nuclear "space" might be accessible to Ah receptors in intact cells. The Ah receptors appear to distribute between the nuclear and cytosolic fractions in proportion to the relative volume of these fractions. These data were consistent with previous observations regarding the subcellular distribution of estrogen receptors between nuclear and cytosolic fractions.[33] In cells treated with TCDD at 37°C, nuclear [^{3}H]-TCDD–Ah receptor complexes remained in nuclear fractions despite dilution and required treatment by high-ionic-strength buffers for solubilization. These studies raise interesting questions about the "true" subcellular distribution of the ligand-free Ah receptor, but confirm that the TCDD–Ah receptor complex formed *in vivo* at 37°C interacts strongly with the nucleus.

Treatment of Hepa-1 cells with TCDD led to a rapid increase in P_1-450 mRNA.[34] The rate of transcription of P_1-450 pre-mRNA was increased within minutes after addition of TCDD to Hepa-1 cultures. The TCDD–Ah receptor complex also was formed rapidly and accumulated rapidly in nuclear fractions (in the form requiring high ionic strength for solubilization). Although P_3-450 mRNA is the major TCDD-inducible P-450 mRNA in B6 mouse liver, P_3-450 mRNA was not detected in TCDD-

←

FIGURE 9. Temperature dependence of specific [^{3}H]-TCDD binding sites in intact Hepa-1 cells. (Top) Sucrose density gradient profiles of cytosolic fractions from cells incubated with 1 nM [^{3}H]-TCDD for 1 hr in culture at 4 or 37°C. (Middle) Nuclear extracts from these same cells. (Bottom) Following incubation of Hepa-1 cells with 1 nM [^{3}H]-TCDD for 1 hr in culture at 4 or 37°C, cytosolic cell-free fractions were then further treated *in vitro* with 10 nM [^{3}H]-TCDD for 1 hr at 4°C.

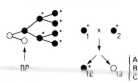

FIGURE 10. Scheme for selection of benzo[a] pyrene-resistant (BP^r) variants of the mouse hepatoma (Hepa-1) tissue culture cell line. Hepa-1 cells are treated with 20 μM BP. In wild-type cells (○), AHH activity is induced and the BP is converted to cytotoxic metabolites; these cells are killed. In variant cells (●), AHH activity is not induced; these cells survive. Genetic complementation groups are identified by fusion among BP^r clones. For example, fusion by polyethylene glycol of BP^r clone 1 and 2 can have two outcomes: (1) the fused clone ○$_{12}$ is inducible for AHH activity, indicating that the genetic defects complement one another; (2) the fused clone ●$_{12}$ is not inducible, indicating that the clones belong to two different complementation groups. Groups A–C are defined in Table I.

treated Hepa-1 cells (unpublished data). The reason for this lack of P$_3$-450 mRNA synthesis is not known. However, it was evident from these studies that tissue culture cell lines such as Hepa-1 provided an excellent model for studying P-450 induction.

AHH activity was induced in Hepa-1; neither the HTC nor the Vero cell lines responded to TCDD. HTC cells contained a specific TCDD-binding protein, whereas Vero cells did not. Thus, the HTC and Vero cell lines represented models of "resistance" to TCDD. However, it would be useful to develop methods for selection of large numbers of independently derived and cloned cell lines that exhibit such resistance to induction of AHH activity. Compounds such as benzo[a]pyrene (BP) are metabolized by P$_1$-450-mediated AHH activity. The metabolites are generally more polar than BP and are not retained by the cell. However, some of the BP metabolites are cytotoxic and react with cellular components. Because BP is fluorescent, Miller et al.[35] have used the fluorescence-activated cell sorter to separate cells on the basis of TCDD-induced AHH activity; these procedures have been used to isolate cells which are defective in metabolism of BP. Hankinson developed a selection procedure which was based on the cytotoxicity of BP metabolites[36] (Fig. 10).

The BP-resistant (BP^r) variants isolated by Hankinson comprised three complementation groups and thus represented defects involving three genes.[37] A fourth variant class was dominant in hybrids formed with wild-type Hepa-1 cells.[38] Miller et al.[35] isolated variants representing two complementation groups. In subsequent studies, the genetic defects were shown to be similar among variant cells derived independently in the two laboratories. It was pertinent to determine if these variant cells included those with defects affecting the Ah receptor.

Legraverend et al.[39] identified defects involving the Ah receptor among variants belonging to two of the complementation groups (Table I). Among variants from complementation group B, all had markedly

TABLE I
Ah Receptor Levels and Maximally Inducible AHH Activity in Hepa-1c1c7 Parent Line and Six Mutant Clones[a]

Cell culture line	Genetic characteristic	Ah receptor (fmole/mg protein)			Maximal AHH activity (units/ mg cellular protein)	
		Cytosol, in vitro treatment	Cytosol, exposure in culture	Nuclei, exposure in culture	BA[b] as inducer	TCDD as inducer
Hepa-1c1c7	wild-type	20	12	6	210	520
c1	group A		7.6	2.0	<0.4	<0.4
c2	group B	2.1	1.7	0.5	42	110
c3	dominant		18	3.4	22	16
c4	group C		15	0.3	<0.4	<0.4
c5	group A		6.0	3.4	<0.4	<0.4
c6	group B	1.0	0.5	0.4	4	3

[a] From Ref. 40.
[b] BA, benzo[a]anthracene.

decreased intracellular concentration of Ah receptors; however, the remaining Ah receptors appeared to undergo normal formation of nuclear [^{3}H]-TCDD–Ah receptor complexes. AHH activity was induced to a limited extent in these cells; the EC_{50} for induction of AHH activity by TCDD was similar in wild-type Hepa-1 and class B variants. Among class C variants, the intracellular concentration of Ah receptors was normal but no nuclear [^{3}H]-TCDD–Ah receptor complexes were formed *in vivo*. Among class A variants, intracellular Ah receptor content and formation of nuclear [^{3}H]-TCDD–Ah receptor complexes were apparently normal. The Hepa-1 variants isolated by Miller *et al.*[35] corresponded to the class B and class C variants. Of interest, one of the class B variant cells isolated by Miller *et al.*[35] contained a markedly decreased concentration of intracellular Ah receptors, and the EC_{50} for TCDD induction of AHH activity was increased in this variant.

The class A variants have been studied further by Hankinson *et al.*[40] These variants have apparently normal Ah receptors. Some did not express P_1-450 mRNA or AHH activity, whereas others showed partial responses to TCDD. One variant clone expressed high levels of mRNA that hybridized to [^{32}P]-DNA (clone 46) by Northern hybridization; however, AHH activity was not detectable in these cells. It was postulated that the mRNA was "defective" and could not be translated into enzymatically active P_1-450; however, further experiments are needed (e.g., cloning DNA complementary to this mRNA and determination of its nucleotide sequence). The class A variants may represent defects of the P_1-450 gene.

3. Induction of a P-450 Isozyme (P-450$_{PCN}$) by Steroids

3.1. Cloning of DNA Complementary to P-450$_{PCN}$ mRNA

Pregnenolone 16α-carbonitrile (PCN) and certain other steroids induce in liver a P-450 isozyme designated as P-450$_{PCN}$. Hardwick *et al.*[10] cloned cDNA complementary to P-450$_{PCN}$ mRNA by immunoprecipitating polysomes from PCN-treated rat liver. The mRNA coding for P-450$_{PCN}$ was approximately 2500 nucleotides in length. PCN treatment of rats resulted in a rapid increase in P-450$_{PCN}$ mRNA. Phenobarbital treatment also increased immunoprecipitable P-450$_{PCN}$ in rat liver; hence, the effects of PCN and phenobarbital were compared using cDNAs for P-450$_{PCN}$ and P-450b. The mRNAs for P-450b and P-450$_{PCN}$ differed in size and the cDNAs hybridized only to their respective mRNA. These data indicated that the cDNA probes did not share significant homology. Although the amino acid sequence and nucleotide sequence for P-450$_{PCN}$

have not been determined, it is likely that P-450$_{PCN}$ and P-450b share only limited homology. However, phenobarbital increased P-450$_{PCN}$ approximately twofold, and PCN increased P-450b mRNA approximately twofold. Analysis of genomic DNA following digestion by restriction enzymes demonstrated that several large restriction fragments were hybridized to the cDNA for P-450$_{PCN}$. These data indicated the possible presence of multiple genes containing sequences homologous to P-450$_{PCN}$.[10]

3.2. Glucocorticoids and the Induction of P-450$_{PCN}$

Synthetic glucocorticoids such as dexamethasone and methylprednisolone not only induce P-450$_{PCN}$ but also are more potent than PCN itself. Schuetz et al.[41] have examined in detail the effects of glucocorticoid hormones, PCN, and other steroids on the induction of P-450$_{PCN}$ in rat liver and in primary monolayer cultures of adult rat hepatocytes.

Primary monolayer cultures of rat hepatocytes were prepared by collagenase digestion of rat liver, and the hepatocytes were cultured on collagen-coated tissue culture dishes in serum-free medium. Because a principal question was the role of the glucocorticoid receptor in the induction of P-450$_{PCN}$, another glucocorticoid-induced enzyme activity [tyrosine aminotransferase (TAT) activity] was examined. In both rats and rat hepatocytes in monolayer culture, certain steroids differed with respect to their effect on TAT versus P-450$_{PCN}$. In rats, PCN induced P-450$_{PCN}$ but not TAT activity.[42] Dexamethasone induced both TAT and P-450$_{PCN}$ activity. TAT activity was induced at low doses of dexamethasone (ED$_{50}$ ~ 1 mg/kg), but P-450$_{PCN}$ was induced only at much higher doses (ED$_{50}$ ~ 50 mg/mg). This difference in ED$_{50}$ was also observed in the monolayer culture system.[42] In addition, the rank orders of relative potency of steroids were different for induction of TAT or P-450$_{PCN}$[41] (Table II).

PCN blocked effectively the induction by dexamethasone of TAT activity. PCN and dexamethasone both induced P-450$_{PCN}$. Furthermore, PCN and dexamethasone together had more than additive effects (e.g., synergistic effects) on the induction of P-450$_{PCN}$. Thus, PCN was a glucocorticoidlike agonist for induction of P-450$_{PCN}$ but was a glucocorticoid antagonist for induction of TAT activity. As noted by Shuetz and Guzelian,[42] these complex relationships cannot be explained by a mechanism involving a single receptor site. These data may indicate the presence of a distinct "receptor" for PCN and compounds such as spironolactone which induce P-450$_{PCN}$. Such a "receptor" might mediate the agonist effects of PCN on induction of P-450$_{PCN}$. The effects of PCN as a glucocorticoid antagonist are probably mediated by its interaction with the rat liver glucocorticoid receptor. Glucocorticoid antagonists such as pro-

TABLE II
Comparison of Steroid Concentration Giving Half-Maximal (ED$_{50}$) Induction of Tyrosine Aminotransferase (TAT) Activity or P-450$_{PCN}$ Synthesis in Cultured Hepatocytes[a]

Steroidal inducer	ED$_{50}$ TAT activity (nmole/liter)	Rank order	ED$_{50}$ P-450$_{PCN}$ synthesis (μmole/liter)	Rank order
Fluocinolone acetonide	0.33	1	6.25	7
Triamcinolone acetonide	0.79	2	9.50	9
Dexamethasone	3.30	3	0.70	1
Betamethasone	4.62	4	1.50	2
α-Methylprednisolone	26.40	5	9.00	8
Triamcinolone	72.6	6	3.30	3
Fluorocortisone	89.1	7	4.25	5
Corticosterone	363.0	8	47.50	10
Hydrocortisone	693.0	9	50.00	11
11β-Hydroxyprogesterone			50.00	11
PCN			5.50	6
Spironolactone			3.50	4

[a] Cultured hepatocytes were incubated in standard medium for 72 hr and then were exposed for 48 hr to medium containing one of the indicated steroids at concentrations ranging from 10^{-8} to 10^{-4} M (P-450$_{PCN}$) or from 10^{-11} to 10^{-5} M (TAT activity). At 120 hr in culture, cells were harvested for assay of TAT activity and were exposed to [^{3}H]leucine for 2 hr assay of P-450$_{PCN}$ synthesis. The results are given as the ED$_{50}$ calculated from plots of the data as the concentration of steroids that produced half-maximal increase in TAT or P-450$_{PCN}$ over control values[42].

gesterone and 11-deoxycorticosterone (cortexalone) bind with low affinity to the glucocorticoid receptor.[43] However, progesterone does not induce P-450$_{PCN}$. Studies of the interaction of PCN with the glucocorticoid receptor should be performed; such studies would probably clarify some of the complex actions of PCN.

4. Induction of P-450 Isozymes by Phenobarbital

4.1. Transcriptional Regulation of Phenobarbital-Inducible P-450 Isozymes

Two closely related P-450 isozymes (P-450b and P-450e) were induced by phenobarbital in rat liver. DNA complementary to mRNA for P-450b and P-450e has been cloned in several laboratories.[7,8,44] Because of the extensive homology in nucleotide sequence, quantitative measurement of mRNA using these cDNA probes included both P-450b and P-450e mRNA. The rat genome contained multiple genes that hybridized to P-

450b and P-450e cDNA.[45] Atchison and Adesnik[45] showed that treatment of rats with phenobarbital increased nuclear transcription of pre-mRNA which hybridized to P-450e cDNA. Under strigent conditions, hybridization was detected only to one of the genomic DNA clones homologous to P-450e. These data indicated that the other P-450e-related genes were not the source of P-450e mRNA transcripts and may be nontranscribed "pseudo"-genes.[45] Hardwick et al.[46] showed that phenobarbital increased rapidly (within 1 hr) nuclear transcription of the P-450b gene as well as transcription of the genes for NADPH-cytochrome P-450 oxidoreductase and epoxide hydrase. Transcription of the albumin gene was not increased by phenobarbital.

4.2. Bis[2-(3,5-dichloropyridyloxy)]benzene (TCPOBOP)—A Ligand for the Phenobarbital Receptor?

Poland et al.[47] showed that TCPOBOP was 650 times as potent as phenobarbital in mice as an inducer of P-450-mediated monooxygenase activities. TCPOBOP accumulated in liver and adipose tissue, was not metabolized significantly, and had a long-lasting effect in mice. Following a single dose of TCPOBOP, hepatic aminopyrine N-demethylase, epoxide hydrase, and liver weight remained elevated for over 20 weeks. In rats, TCPOBOP was considerably less potent ($ED_{50} = 1.2 \times 10^{-4}$ mole/kg) than in mice ($ED_{50} = 4.9 \times 10^{-7}$ mole/kg).[48]

Although [^{3}H]-TCPOBOP was prepared and was used to study tissue distribution and metabolism of the compound, no analysis of binding to cellular membrane fractions or intracellular molecules was reported. When [^{3}H]phenobarbital was used to investigate intracellular binding in rat liver cytosol, Tierney and Bresnick[49] were unable to detect macromolecular [^{3}H]phenobarbital complexes. [^{3}H]-TCPOBOP appeared to be more suitable as a radioligand for such studies. Although no evidence exists for a specific "phenobarbital" or "TCPOBOP receptor," it may be worthwhile to investigate structure–activity relationships among analogues of TCPOBOP to identify even more potent compounds. Rats may represent a species with genetically determined "resistance" to TCPOBOP; the species differences in ED_{50} may be useful for evaluation of ligand-binding studies.

4.3. Barbiturate Receptor Associated with the GABA Receptor

Barbiturates are anticonvulsants and hypnotic/sedatives. These effects on the central nervous system appear to involve a membrane-associated receptor that binds barbiturates and convulsants such as picrotoxin.[50] The synthetic "cage convulsant" t-butylbicyclophosphorothionate (TBPS) is the most potent ligand known for this receptor,[51] which has been termed the picrotoxin/barbiturate receptor. The latter is believed to form part of the GABA receptor "complex" which includes the GABA receptor, a "channel" for chloride ions, and the benzodiazepine receptor.

Barbiturates are relatively weak ligands for this receptor; hence, radioactive barbiturates are not used to study the picrotoxin/barbiturate receptor. Although the CNS picrotoxin/barbiturate receptor may have no direct relation to the induction by barbiturates of hepatic P-450, it is important to consider the possibility that a hepatic receptor for barbiturates may be associated with membrane fractions.

5. Isolation of Receptor Genes

5.1. Selection of AHH⁺ Revertants from Clones of AHH⁻ Hepa-1

The AHH⁻ mutants of Hepa-1 represent three genetic complementation groups. The defective gene function in a clone from group A can be corrected by fusion with clones from group B or group C[38] (Fig. 10). Gene transfer by DNA transfection might also be used to correct the defect in a clone from group A.[52] Various methods are currently used for "rescue" of such transfected genes; it should be possible to use these methods to clone DNA that complements the defects associated with groups A, B, C of the AHH⁻ Hepa-1 clones. However, it is necessary first to develop methods for selection of cells that have regained the capacity for AHH induction. Such revertant cells may be identified by the fluorescence-activated cell sorter.[35]

A novel method for selection of AHH⁺ revertants was proposed by Van Gurp and Hankinson.[53] Certain polycyclic aromatic compounds act as photosensitizing agents when cells containing the compound are irradiated with near-ultraviolet light. When excited by near-UV light, these compounds apparently excite molecular oxygen to a singlet state which reacts with cellular components such as membrane phospholipids. Cells containing high levels of induced AHH activity metabolize and eliminate these compounds and thus are protected from subsequent exposure to near-UV light. Because polycyclic aromatic compounds may be converted by AHH activity to cytotoxic metabolites, the ideal compound should be cytotoxic minimally in AHH⁺ cells. Van Gurp and Hankinson examined many polycyclic aromatic compounds; benzo(ghi)perylene was found to be cytotoxic minimally in AHH⁺ cells, but was an excellent sensitizing compound for near-UV phototoxicity. Under appropriate experimental conditions, AHH⁻ cells could be killed readily by near-UV light, whereas AHH⁺ cells survived.

If exogenous DNA (obtained from wild-type Hepa-1 cells) were introduced into AHH$^-$ cells, then any resulting AHH$^+$ cells can be selected among a large background of AHH$^-$ cells. The transfected DNA can be "tagged" by plasmid DNA and "rescued" by cloning into bacteriophage or cosmid vectors. Such methods have been used to clone several eukaryotic genes (such as oncogenes) for which suitable selection methods have been developed.[54]

5.2. Fusion Genes for Analysis of Induction of P-450 by Phenobarbital or Steroids

Somatic cell genetics and gene "rescue" techniques may also be useful for studying phenobarbital induction of P-450. However, there are no continuous tissue culture cell lines in which phenobarbital induces P-450b. Newman and Guzelian[55] have shown that primary monolayer cultures of adult hepatocytes are suitable for studying phenobarbital induction of P-450; however, these cells do not propagate in tissue culture. Furthermore, appropriate selection protocols have not been developed. Certain tissue culture cell lines may contain the cellular factors required for phenobarbital induction but may lack an inducible P-450b gene. Fusion genes consisting of the promoter region and 5' flanking sequences of the P-450b gene and coding regions of a gene such as herpes-simplex thymidine kinase may be suitable for "positive" and "negative" selection of phenobarbital-inducible thymidine kinase activity. Such hybrid gene constructions have been used extensively to study the regulation of eukaryotic gene expression.[56,57]

6. Interaction of Receptors and Regulatory Proteins with DNA and Chromatin

6.1. Evidence That Regulatory Proteins Bind Selectively to Specific DNA Sequences

If an inducer binds to a specific receptor (such as the Ah receptor), how might this molecular "signal" affect the transcription of a specific P-450 gene? The TCDD–Ah receptor complex has definite affinity for components in the cell nucleus. It is reasonable to postulate that the TCDD–Ah receptor complex interacts directly with regulated genes such as the P_1-450 gene. Unfortunately, the Ah receptor has not been purified and its interaction with nuclear components has not been studied rigorously. However, receptors for steroid hormones such as glucocorticoids and progesterone have been purified to homogeneity, and the interactions of these receptors with nuclear components have been studied in great detail. Experimental approaches that were used to study steroid hormone receptors may prove useful for analysis of the Ah receptor and other factors that regulate P-450 genes.

Steroid hormone–receptor complexes such as the glucocorticoid–receptor complex bind to DNA. Any sequence-specific interaction of high affinity would be difficult to detect in the large background of nonspecific interactions.[58] Using cloned DNA for the mouse mammary tumor virus (MMTV) genome, Payvar et al.[59,60] showed that certain regions of the MMTV genome were bound with high affinity to the glucocorticoid–receptor complex. One region that contained a specific-binding site was shown to be necessary for induction in vivo by glucocorticoids of MMTV-encoded mRNA. The LTR (long terminal repeat) of MMTV contained a glucocorticoid response element (GRE) which could make heterologous promoters respond in vivo to glucocorticoids. The GRE functioned in both orientations and at different locations when transposed into the promoter regions of other genes. These characteristics are similar to those of viral (enhancer) elements which are sequences that increase the transcriptional activity of viral promoters. It was postulated that binding of the glucocorticoid–receptor complex to the GRE was necessary to convert the region into an active enhancer of transcription. A "consensus" sequence has been identified in the regions 5' to the promoters of other glucocorticoid-responsive genes; and these regions have been shown to bind with high affinity to the glucocorticoid–receptor complex.[61] Although most of these studies have been done with highly purified receptor preparations, competitive binding assays using crude receptor preparations have given similar results. These assays measured the relative affinity of receptor preparations for cloned DNA sequences versus total DNA sequences (e.g., calf thymus DNA immobilized to cellulose).[62]

6.2. Alterations in Chromatin Structure Associated with Transcriptional Regulation

Nucleases such as DNase I have been used to examine the relative sensitivity to digestion of chromatin containing specific genes. For example, chicken hemoglobin genes in erythrocytes are more sensitive to DNase I digestion than hemoglobin genes in liver or brain.[63] These differences are believed to reflect differences in chromatin structure because isolated hemoglobin DNAs from these tissue were sensitive equally to digestion by DNase I. Additionally, defined regions of certain genes were "hypersensitive" to DNase I digestion. With the use of suitable [^{32}P]-DNA probes, "hypersensitive" regions have been identified among several genes including hemoglobin genes and the chicken vitellogenin gene.[5]

The latter gene was induced in liver by estrogens. Three groups of DNase I hypersensitive regions have been identified in the chicken liver vitellogenin gene. One region became hypersensitive in liver but not in other tissues during embryogenesis. Another region became hypersensitive after the first exposure to estrogens and remained hypersensitive even after estrogen withdrawal; a third region responded acutely to the presence of estrogens and disappeared when estrogens were withdrawn. Similar observations have been made for glucocorticoid-responsive genes such as the rat tryptophan oxygenase gene.[64] These DNase I hypersensitive regions are believed to represent localized regions of "perturbed" chromatin structure.

Exposure of *Drosophila* and other organisms to elevated temperature results in the rapid induction of heat-shock (HS) proteins. The "signal" that induces these genes is not known; however, heat-shock caused a defined pattern of DNase I hypersensitivity which can be detected with the use of cloned DNA for HS-protein genes. Wu[65] has shown that a specific protein from heat-shocked *Drosophila* becomes associated with one of the hypersensitivity regions. The "footprint" of this protein can be detected by exonuclease III digestion within the hypersensitive region. The presence of this protein in crude or partially purified extracts can be "assayed" using whole nuclei and exonuclease III digestion. Siebenlist et al.[66] have shown that transcription of the c-Myc oncogene in lymphoid cells was associated with the development of regions of DNase I hypersensitivity. One of these regions is the site at which a nuclear protein binds specifically to DNA. Using a nitrocellulose filterbinding assay to detect such high-affinity protein–DNA interactions, these investigators have partially purified this nuclear protein.

7. Conclusions

Polycyclic aromatic compounds, barbiturates, and PCN increase transcription of specific P-450 genes in rat liver. The nuclear transcription "runoff" assay currently in use measures the elongation of pre-mRNA transcripts in progress when nuclei were isolated. When these data are normalized to total RNA synthesis in isolated nuclei, the relative proportions of "runoff" synthesis for a specific transcript can be calculated. This measurement does not provide information about the initiation rate or the elongation rate for specific transcripts. Methods have been developed for measurement in intact cells of these parameters; glucocorticoid hormones, for example, increase the rate of initiation of transcripts of MMTV genes.[67] Recombinant DNA techniques, therefore, can provide considerable information about the early events that lead to the formation of an active transcription initiation complex. Recently, active transcription and elongation complexes have been partially purified.[68] Thus, it may be possible in the future to reconstitute *in vivo* the components necessary for activation of transcription of a gene such as P-450b. With currently available cloned P-450 DNA, it may also be possible to detect regions of DNase I hypersensitivity in chromatin following treatment of animals with inducers of P-450. Such approaches may be useful particularly for analysis of the action of barbiturates. It may be possible to retrace the "footprints" of barbiturate action from the P-450b gene *in situ* to the initial interaction of the compounds with the hepatocyte.

Considerable progress has been made toward the identification of a cellular receptor for polycyclic aromatic compounds. As outlined in this chapter, the Ah receptor can be cloned (in principle) with the use of appropriate selection procedures and DNA-mediated gene transfer. The identification of a "receptor" for inducers of $P-450_{PCN}$ may be difficult. Schuetz and Guzelian have demonstrated major differences in the actions of steroids on a glucocorticoid-inducible gene (TAT) and $P-450_{PCN}$. Even though glucocorticoids induce $P-450_{PCN}$, the rat liver glucocorticoid receptor may not be involved directly in the induction of $P-450_{PCN}$. PCN and its "receptor" might interact with a nucleotide sequence that differs from the consensus sequence for the glucocorticoid receptor. Even if we cannot easily identify this receptor with ligand-binding studies, its effects on chromatin structure may be apparent.

ACKNOWLEDGMENT. I am grateful to D. W. Nebert and to my colleagues in the Laboratory of Developmental Pharmacology, A. B. Okey, O. Hankinson, and P. Guzelian, for their help in preparing this chapter. I also wish to thank Ms. Linda A. Hamilton whose expert secretarial assistance was greatly appreciated.

References

1. Lu, A. Y. H., and West, S. B., 1980, Multiplicity of mammalian microsomal cytochromes P-450, *Pharmacol. Rev.* **31**:277–295.
2. Mannering, G. J., 1981, Hepatic cytochrome P-450 linked drug-metabolizing systems, in: *Concepts in Drug Metabolism*, Part B (P. Jenner and B. Testa, eds.), Dekker, New York.
3. Poland, A. P., Glover, E., and Kende, A. S., 1976, Stereospecific, high affinity binding of 2,3,7,8-tetrachlorodibenzo-*p*-dioxin by hepatic cytosol: Evidence that the binding species is the receptor for the induction of aryl hydrocarbon hydroxylase, *J. Biol. Chem.* **251**:4936–4946.
4. Okey, A. B., Bondy, G. P., Mason, M. E., Nebert, D. W., Forster-Gibson, C., Muncan, J., and Dufresne, M. J., 1980, Temperature-dependent cytosol-to-nucleus translocation

of the *Ah* receptor for 2,3,7,8-tetrachlorodibenzo-p-dioxin in continuous cell culture lines, *J. Biol. Chem.* **255**:11415–11422.
5. Burch, J. B. E., and Weintraub, H., 1983, Temporal order of chromatin structural changes associated with activation of the major chicken vitellogenin gene, *Cell* **33**:65–76.
6. Negishi, M., Swan, D. C., Enquist, L. W., and Nebert, D. W., 1981, Isolation and characterization of a cloned DNA sequence associated with the murine *Ah* locus and a 3-methylcholanthrene-induced form of cytochrome P450, *Proc. Natl. Acad. Sci. USA* **78**:800–804.
7. Mizukami, Y., Sogawa, K., Suwa, Y., Muramatsu, M., and Fujii-Kuriyama, Y., 1983, Gene structure of a phenobarbital-inducible cytochrome P-450 in rat liver, *Proc. Natl. Acad. Sci. USA* **80**:3958–3962.
8. Kumar, A., Raphael, C., and Adesnik, M., 1983, Cloned cytochrome P450 cDNA: Nucleotide sequence and homology to multiple phenobarbital-induced mRNA species, *J. Biol. Chem.* **258**:11280–11284.
9. Leighton, J. K., DeBrunner-Vossbrinck, B. A., and Kemper, B., 1984, Isolation and sequence analysis of three cloned cDNAs for rabbit liver proteins that are related to rabbit cytochrome P-450 (form 2), the major phenobarbital-inducible form, *Biochemistry* **23**:204–210.
10. Hardwick, J. P., Gonzalez, F. J., and Kasper, C. B., 1983, Cloning of DNA complementary to cytochrome P-450 induced by pregnenolone-16α-carbonitrile, *J. Biol. Chem.* **258**:10182–10186.
11. Morville, A. L., Thomas, P., Levin, W., Reik, L., Ryan, D. E., Raphael, C., and Adesnik, M., 1983, The accumulation of distinct mRNAs for the immunochemically related cytochromes P-450c and P-450d in rat liver following 3-methylcholanthrene treatment, *J. Biol. Chem.* **258**:3901–3906.
12. Negishi, M., and Nebert, D. W., 1981, Structural gene products of the *Ah* complex: Increases in large mRNA from mouse liver associated with cytochrome P_1-450 induction by 3-methylcholanthrene, *J. Biol. Chem.* **256**:3085–3091.
13. Fagan, J. B., Pastewka, J. V., Park, S. S., Guengerich, F. P., and Gelboin, H. V., 1982, Identification and quantitation of a 2.0-kilobase messenger ribonucleic acid coding for 3-methylcholanthrene-induced cytochrome P-450 using cloned cytochrome P-450 complementary deoxyribonucleic acid, *Biochemistry* **24**:6574–6580.
14. Negishi, M., Swan, D. C., Enquist, L. W., and Nebert, D. W., 1981, Isolation and characterization of cloned DNA sequence associated with the murine *Ah* locus and a 3-methylcholanthrene-induced form of cytochrome P-450, *Proc. Natl. Acad. Sci. USA* **78**:800–804.
15. Kawajiri, K., Gotch, O., and Tagashira, Y., 1984, Titration of mRNAs for cytochrome P-450c and P-450d under drug-inductive conditions in rat livers by their specific probes of cloned DNAs, *J. Biol. Chem.* **259**:10145–10149.
16. Kimura, S., Gonzalez, F. J., and Nebert, D. W., 1984, Mouse cytochrome P_3-450: Complete cDNA and amino acid sequence, *Nucleic Acids Res.* **12**:2917–2928.
17. Okayama, H., and Berg, P., 1983, A cDNA cloning vector that permits expression of cDNA inserts in mammalian cells, *Mol. Cell. Biol.* **3**:280–289.
18. Kimura, S., Gonzalez, F. J., and Nebert, D. W., 1984, The murine *Ah* locus, *J. Biol. Chem.* **259**:10705–10713.
19. Tukey, R. H., and Nebert, D. W., 1984, Structural gene products controlled by the *Ah* locus: Cloning, isolation and characterization of the mouse cytochrome P_3-450 complementary DNA, *Biochemistry* **23**:6003–6008.
20. Spradling, A. C., Digan, M. E., Mahowald, A. P., Scott, M., and Craig, E. A., 1980, Two clusters of genes for major chorion proteins of *Drosophila melanogaster*, *Cell* **19**:905–910.

21. Poland, A. P., and Glover, E., 1975, Genetic expression of aryl hydrocarbon hydroxylase by 2,3,7,8-tetrachlorodibenzo-p-dioxin: Evidence for a receptor mutation in genetically non-responsive mice, *Mol. Pharmacol.* **11**:389–398.
22. Poland, A. P., Glover, E., Robinson, J. R., and Nebert, D. W., 1974, Genetic expression of aryl hydrocarbon hydroxylase activity: Induction of monooxygenase activities and cytochrome P_1-450 formation by 2,3,7,8-tetrachlorodibenzo-p-dioxin in mice genetically "nonresponsive" to other aromatic hydrocarbons, *J. Biol. Chem.* **249**:5599–5606.
23. Poland, A. P., and Knutson, J. C., 1982, 2,3,7,8-Tetrachlorodibenzo-p-dioxin and related halogenated aromatic hydrocarbons: Examination of the mechanism of toxicity, *Annu. Rev. Pharmacol. Toxicol.* **22**:517–554.
24. Nebert, D. W., and Gelboin, H. V., 1969, The *in vivo* and *in vitro* induction of aryl hydrocarbon hydroxylase in mammalian cells of different species, tissues, strains, and developmental and hormonal states, *Arch. Biochem. Biophys.* **134**:76–89.
25. Eisen, H. J., Hannah, R. R., Legraverend, C., Okey, A. B., and Nebert, D. W., 1983, The Ah receptor: Controlling factor in the induction of drug-metabolizing enzymes by certain chemical carcinogens and other environmental pollutants, in: *Biochemical Actions of Hormones*, Volume X (G. Litwack, ed.), Academic Press, New York, p. 227.
26. Poland, A. P., Glover, E., and Kende, A. S., 1976, Stereospecific, high affinity binding of 2,3,7,8-tetrachlorodibenzo-p-dioxin by hepatic cytosol: Evidence that the binding species is the receptor for the induction of aryl hydrocarbon hydroxylase, *J. Biol. Chem.* **251**:4936–4946.
27. Okey, A. B., Bondy, G. P., Mason, M. E., Kahl, G. F., Eisen, H. J., Guenthner, T. M., and Nebert, D. W., 1979, Regulatory gene product of the Ah locus: Characterization of the cytosolic inducer–receptor complex and evidence for its nuclear translocation, *J. Biol. Chem.* **254**:11636–11648.
28. Hannah, R. R., Nebert, D. W., and Eisen, H. J., 1981, Regulatory gene product of the Ah complex: Comparison of 2,3,7,8-tetrachlorodibenzo-p-dioxin and 3-methylcholanthrene binding to several moieties in mouse liver cytosol, *J. Biol. Chem.* **256**:4584–4590.
29. Tukey, R. H., Hannah, R. R., Negishi, M., Nebert, D. W., and Eisen, H. J., 1982, The Ah locus: Correlation of intranuclear appearance of inducer–receptor complex with induction of cytochrome P_1-450 mRNA, *Cell* **31**:275–284.
30. Gonzalez, F. J., Tukey, R. H., and Nebert, D. W., 1984, Structural gene products of the Ah locus: Transcriptional regulation of cytochrome P_1-450 and P_3-450 mRNA levels by 3-methylcholanthrene, *Mol. Pharmacol.* **26**:117–121.
31. Okey, A. B., Bondy, G. P., Mason, M. E., Nebert, D. W., Forster-Gibson, C., Muncan, J., and Dufresne, M. J., 1980, Temperature-dependent cytosol-to-nucleus translocation of the Ah receptor for 2,3,7,8-tetrachloro-p-dioxin in continuous cell culture lines, *J. Biol. Chem.* **255**:11415–11422.
32. Whitlock, J. P., and Galaezzi, D. R., 1984, 2,3,7,8-Tetrachlorodibenzo-p-dioxin receptors in wild type and variant mouse hepatoma cells, *J. Biol. Chem.* **259**:980–985.
33. Welshons, W. V., Lieberman, M. E., and Gorski, J., 1984, Nuclear localization of unoccupied oestrogen receptors, *Nature* **307**:747–749.
34. Israel, D., and Whitlock, J. P., 1983, Induction of mRNA specific for cytochrome P_1-450 in wild type and variant mouse heptoma cells, *J. Biol. Chem.* **258**:10390–10394.
35. Miller, A. G., Israel, D., and Whitlock, J. P., 1983, Biochemical and genetic analysis of variant mouse hepatoma cells defective in the induction of benzo(a)pyrene-metabolizing enzyme activity, *J. Biol. Chem.* **258**:3523–3527.
36. Hankinson, O., 1981, Single-step selection of clones of a mouse hepatoma line deficient in aryl hydrocarbon hydroxylase, *Proc. Natl. Acad. Sci. USA* **76**:373–376.
37. Hankinson, O., 1981, Evidence that benzo(a)pyrene-resistant aryl hydrocarbon hydroxylase-deficient variants of mouse hepatoma line Hepa-1 are mutational in origin, *Somatic Cell Genet.* **7**:373–388.

38. Hankinson, O., 1983, Dominant and recessive aryl hydrocarbon hydroxylase-deficient mutants of mouse hepatoma line Hepa-1, and assignment of the recessive mutants to three complementation groups, *Somatic Cell Genet.* **9**:497–514.
39. Legraverend, C., Hannah, R. R., Eisen, H. J., Owens, I. S., Nebert, D. W., and Hankinson, O., 1982, Regulatory gene product of the *Ah* locus, *J. Biol. Chem.* **257**:6402–6407.
40. Hankinson, O., Anderson, R. D., Birren, B., Sander, F., Negishi, M., and Nebert, D. W., 1985, Mutations affecting the regulation of transcription of the cytochrome P_1-450 gene in mouse Hepa-1 cell cultures, *J. Biol. Chem.* **260**:1790–1795.
40. Legraverend, C., Hannah, R. R., Eisen, H. J., Owens, I. S., Nebert, D. W., and Hankinson, O., 1982, Regulatory gene product of the *Ah* locus, *J. Biol. Chem.* **257**:6402–6407.
41. Schuetz, E. G., Wrighton, S. A., Barwick, J. L., and Guzelian, P. S., 1984, Induction of cytochrome P-450 by glucocorticoids in rat liver. I. Evidence that glucocorticoids and pregnenolone 16α-carbonitrile regulate *de novo* synthesis of a common form of cytochrome P-450 in cultures of adult rat hepatocytes and in the liver *in vivo*. *J. Biol. Chem.* **259**:1999–2006.
42. Schuetz, E. G., and Guzelian, P. S., 1984, Induction of cytochrome P-450 by glucocorticoids in rat liver. II. Evidence that glucocorticoids regulate induction of cytochrome P-450 by a nonclassical receptor mechanism, *J. Biol. Chem.* **259**:2007–2012.
43. Munck, A., and Holbrook, N., 1984, Glucocorticoid receptors in rat thymic lymphocytes: A cyclic model, *J. Biol. Chem.* **259**:820–831.
44. Gonzalez, F. J., and Kasper, C. B., 1982, Cloning of DNA complementary to rat liver NADPH-cytochrome *c* (P-450). Oxidoreductase and cytochrome P-450b mRNAs: Evidence that phenobarbital augments transcription of specific genes, *J. Biol. Chem.* **257**:5962–5968.
45. Atchison, M., and Adesnik, M., 1983, A cytochrome P-450 multigene family: Characterization of a gene activated by phenobarbital administration, *J. Biol. Chem.* **258**:11285–11295.
46. Hardwick, J., Gonzalez, F. J., and Kasper, C. B., 1983, Transcriptional regulation of rat liver epoxide hydratase, NADPH-cytochrome P-450 oxidoreductase, and cytochrome P-450b genes by phenobarbital, *J. Biol. Chem.* **258**:8081–8085.
47. Poland, A., Mak, I., Glover, E., Boatman, R. J., Ebetino, F. H., and Kende, A. S., 1980, Bis[2-(3,5-dichloropyridyloxy)]benzene, a potent phenobarbital-like inducer of microsomal monooxygenase activity, *Mol. Pharmacol.* **18**:571–580.
48. Poland, A., Mak, I., and Glover, E., 1981, Species differences in responsiveness to 1,4-bis[2(3,5-dichloropyridyloxy)]-benzene, a potent phenobarbital-like inducer of microsomal monooxygenase activity, *Mol. Pharmacol.* **20**:442–450.
49. Tierney, B., and Bresnick, E., 1981, Differences in the binding of 3-methylcholanthrene and phenobarbital to rat liver cytosolic and nuclear protein fractions, *Arch. Biochem. Biophys.* **210**:729–739.
50. Olsen, R. W., Wong, E. H., Stauber, G. B., and King, R. G., 1984, Biochemical pharmacology of the γ-aminobutyric acid receptor/ionophore protein, *Fed. Proc.* **43**:2773–2778.
51. King, R. G., and Olsen, R. W., 1984, Solubilization of convulsant/barbiturate binding activity on the γ-aminobutyric acid/benzodiazepine receptor complex, *Biochem. Biophys. Res. Commun.* **119**:530–536.
52. Nebert, D. W., Hankinson, O., and Eisen, H. J., 1984, The Ah receptor: Binding specificity only for foreign chemicals, *Biochem. Pharmacol.* **33**:917–924.
53. Van Gurp, J. R., and Hankinson, O., 1983, Single-step phototoxic selection procedure for isolating cells that possess aryl hydrocarbon hydroxylase, *Cancer Res.* **43**:6031–6038.

54. Cooper, C. S., Park, M., Blair, D. G., Tainsky, M. A., Huebner, K., Croce, C. M., and Vande Woude, G. F., 1984, Molecular cloning of a new transforming gene from a chemically transformed human cell line, *Nature* **311**:29–33.
55. Newman, S., and Guzelian, P. S., 1982, Stimulation of *de novo* synthesis of cytochrome P-450 by phenobarbital in primary nonproliferating cultures of adult rat hepatocytes, *Proc. Natl. Acad. Sci. USA* **79**:2922–2926.
56. Karin, M., Haslinger, A., Holtgreve, H., Richards, R. I., Krauter, P., Westphal, H. M., and Beato, M., 1984, Characterization of DNA sequences through which cadmium and glucocorticoid hormones induce human metallothionein-II$_A$ gene, *Nature* **308**:513–518.
57. Mayo, K. E., Warren, R., and Palmiter, R. D., 1982, The mouse metallothionein-I gene is transcriptionally regulated by cadmium following transfection into human or mouse cells, *Cell* **29**:99–108.
58. Yamamoto, K. R., and Alberts, B. M., 1976, Steroid receptors: elements for modulation of eukaryotic transcription, *Annu. Rev. Biochem.* **45**:721–746.
59. Payvar, F., Wrange, Ö., Carlstedt-Duke, J., Okret, S., Gustafsson, J. Å., and Yamamoto, K. R., 1981, Purified glucocorticoid receptors bind selectively *in vitro* to a cloned DNA fragment whose transcription is regulated by glucocorticoids *in vivo*, *Proc. Natl. Acad. Sci. USA* **78**:6628–6632.
60. Payvar, F., DeFranco, D., Firestone, G. L., Edgar, B., Wrange, Ö., Okret, S., Gustafsson, J. Å., and Yamamoto, K. R., 1983, Sequence-specific binding of glucocorticoid receptor to MTV DNA at sites within and upstream of the transcribed region, *Cell* **35**:381–392.
61. Scheidereit, C., Geisse, S., Westphal, H. M., and Beato, M., 1983, The glucocorticoid receptor binds to defined nucleotide sequences near the promoter of mouse mammary tumor virus, *Nature* **304**:749–752.
62. Pfahl, M., 1982, Specific binding of the glucocorticoid–receptor complex to the mouse mammary tumor proviral promoter region, *Cell* **31**:475–482.
63. Weintraub, H., 1983, A dominant role for DNA secondary structure in forming hypersensitive structures in chromatin, *Cell* **32**:1191–1203.
64. Becker, P., Renkawitz, R., and Schütz, G., 1984, Tissue-specific DNaseI hypersensitive sites in the 5′-flanking sequences of the tryptophan oxygenase and the tyrosine aminotransferase genes, *EMBO J.* **3**:2015–2020.
65. Wu, C., 1984, Two protein-binding sites in chromatin implicated in the activation of heat-shock genes, *Nature* **309**:229–234.
66. Siebenlist, U., Hennighausen, L, Battey, J., and Leder, P., 1984, Chromatin structure and protein binding in the putative regulatory region of the c-myc gene in Burkitt lymphoma, *Cell* **37**:381–391.
67. Ucker, D. S., and Yamamoto, K. R., 1984, Early events in the stimulation of mammary tumor RNA synthesis by glucocorticoids: Novel assays of transcription rates, *J. Biol. Chem.* **259**:7416–7420.
68. Tolunay, H. E., Yang, L., Anderson, W. F., and Safer, B., 1984, Isolation of an active transcription initiation complex from HeLa cell-free extract, *Proc. Natl. Acad. Sci. USA* **81**:5916–5920.

CHAPTER 10

Regulation of Synthesis and Activity of Cytochrome P-450 Enzymes in Physiological Pathways

MICHAEL R. WATERMAN, MALIYAKAL E. JOHN, and
EVAN R. SIMPSON

1. Introduction

Regulation of the levels and activities of various forms of cytochrome P-450 has important implications in human biology. For example, it is well known that different individuals have different capacities to metabolize various drugs.[1] Such variation may result, in part, from an allelic distribution of P-450 isozymes;[2] or it may also result from individual variations in the level of a specific isozyme.[3,4] Also, deficiencies of different steroid hydroxylases lead to various disease states. For example, adrenal hyperplasia occurs in approximately one in 5000 births, making it one of the more common inborn errors of metabolism.[5] In at least 95% of these cases, the deficiency is detected as a decrease in steroid 21-hydroxylation in the adrenal cortex,[6] a reaction catalyzed by a specific form of P-450 ($P\text{-}450_{C21}$), leading to deficient cortisol biosynthesis. These are but a few examples which illustrate the reasons why investigation of the regulation of P-450 activities has become such an active area of research. In each example, understanding of the molecular basis of variations in P-450 gene expression will be necessary for the complete elucidation of the different phenotypes.

MICHAEL R. WATERMAN and MALIYAKAL E. JOHN • Departments of Biochemistry and Obstetrics and Gynecology, University of Texas Health Science Center, Dallas, Texas 75235. EVAN R. SIMPSON • Departments of Biochemistry and Obstetrics and Gynecology, and Cecil H. and Ida Green Center for Reproductive Biology Sciences, University of Texas Health Science Center, Dallas, Texas 75235.

TABLE I
Distribution of Some Cytochromes P-450 in Physiological Pathways

Pathway	Tissue	Cytochrome P-450	Subcellular localization
Steroid hormone biosynthesis	Adrenal cortex	$P\text{-}450_{scc}$	Mitochondria
		$P\text{-}450_{11\beta}$	Mitochondria
		$P\text{-}450_{17\alpha}$	Endoplasmic reticulum
		$P\text{-}450_{C21}$	Endoplasmic reticulum
	Testis	$P\text{-}450_{scc}$	Mitochrondria
		$P\text{-}450_{17\alpha}$	Endoplasmic reticulum
		$P\text{-}450_{aromatase}$	Endoplasmic reticulum
	Ovary	$P\text{-}450_{scc}$	Mitochondria
		$P\text{-}450_{17\alpha}$	Endoplasmic reticulum
		$P\text{-}450_{aromatase}$	Endoplasmic reticulum
	Placenta	$P\text{-}450_{scc}$	Mitochondria
		$P\text{-}450_{17\alpha}$	Endoplasmic reticulum
		$P\text{-}450_{aromatase}$	Endoplasmic reticulum
	Brain	$P\text{-}450_{aromatase}$	Endoplasmic reticulum
	Adipose tissue	$P\text{-}450_{aromatase}$	Endoplasmic reticulum
Vitamin D hydroxylation	Liver	$P\text{-}450_{\text{Vit. D 25OH}}$	Endoplasmic reticulum
		$P\text{-}450_{\text{Vit. D 25OH}}$	Mitochondria
	Kidney	$P\text{-}450_{\text{Vit. D 25OH}}$	Endoplasmic reticulum
		$P\text{-}450_{\text{25OH Vit. D 1}\alpha\text{OH}}$	Mitochondria
		$P\text{-}450_{\text{25OH Vit. D 24OH}}$	Mitochondria
	Intestine	$P\text{-}450_{\text{Vit. D 25OH}}$	Endoplasmic reticulum
		$P\text{-}450_{\text{25OH Vit. D 24OH}}$	Mitochondria
	Placenta	$P\text{-}450_{\text{25OH Vit. H 1}\alpha\text{OH}}$	Mitochondria

In the broadest sense, the various P-450 isozymes can be classified as: (1) those forms which metabolize physiological or endogenous substrates and (2) those forms which metabolize xenobiotic or exogenous substrates. In the preceding chapter, Dr. H. J. Eisen presented a discussion on the mechanisms of regulation of hepatic cytochromes P-450 by xenobiotic compounds. In this chapter, we will describe studies carried out on the regulation of various forms of P-450 by physiological compounds.

At the outset, it is important to note a fundamental difference between these two classes of P-450. Those P-450 isozymes which metabolize exogenous substrates are inducible *in vivo* and exposure to xenobiotics can lead to increased levels of specific forms of P-450 and/or their respective enzymatic activities. As has been shown in the case of several different xenobiotic metabolizing forms of P-450, changes in their levels result largely from changes in the amount of mRNA encoding these proteins.[7-9] In contrast, while those P-450 isozymes which metabolize endogenous substrates have the capacity to be induced, the constitutive levels of these enzymes are generally optimal for the particular function they serve. However, many of these P-450 isozymes carry out important physiological functions (i.e., glucocorticoid and sex hormone production, production of active forms of vitamin D, cholesterol and bile acid synthesis) and thus the mechanisms by which optimal levels of these forms are maintained are of considerable importance.

The number of forms of P-450 which metabolize endogenous substrates is very large. In this chapter, we will concentrate on the P-450 isozymes associated with (1) steroid hormone biosynthesis, (2) vitamin D hydroxylation, and (3) fatty acid hydroxylation. A number of the isozymes discussed in this chapter are listed in Table I. In addition, there are a number of steroid hydroxylases in the liver which hydroxylate substrates such as testosterone and progesterone. The role of these enzymes is unclear, as is their regulation. It is known, however, that several of these enzymes are sex specific and future studies on developmental aspects of their regulation will prove most interesting.

2. Steroidogenic Forms of P-450

A unique set of P-450 isozymes is localized in steroidogenic tissues, i.e., adrenal cortex, testis, and ovary. As will be seen, the activities of these forms of P-450 are regulated by peptide hormones which are targeted to these various tissues. The most detailed study of these enzymes has been carried out in the adrenal cortex where regulation of steroidogenesis is under the control of the peptide hormone, adrenocorticotropin

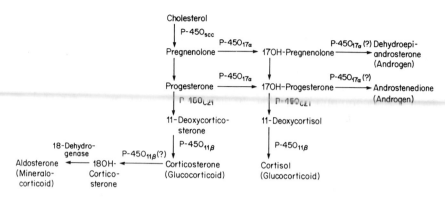

FIGURE 1. Steroidogenic pathways in the bovine adrenal cortex leading to the production of glucocorticoids, mineralocorticoids, and adrenal androgens. The involvement of various forms of cytochrome P-450 in these reactions is noted.

(ACTH). Therefore, emphasis will be placed on adrenocortical P-450 steroid hydroxylases in this section followed by brief descriptions of the regulation of the enzymes localized in the testis and ovary.

2.1. Adrenal Cortex

As can be seen from the pathway shown in Fig. 1, the adrenal cortex contains four distinct forms of P-450 which are involved in steroidogenesis, leading from cholesterol to cortisol. This pathway is particularly interesting in that it contains both microsomal and mitochondrial forms of P-450. The initial step which occurs in the mitochondrion is the conversion of cholesterol to pregnenolone. This step involves the cleavage of an isocaproic group from the cholesterol side chain and is known as the cholesterol side-chain cleavage reaction. This reaction utilizes 3 molecules of oxygen and 3 molecules of NADPH and is catalyzed by cholesterol side-chain cleavage cytochrome P-450 (P-450$_{scc}$).[10] Studies in several laboratories have led to the conclusion that this reaction proceeds via an initial hydroxylation at the 22-position of cholesterol to yield (22R)-hydroxycholesterol, which is subsequently converted to a dihydroxy intermediate (20R,22R) and then to pregnenolone.[11-13] The stoichiometry noted above suggests that the cleavage of the carbon–carbon bond (subsequent to the second hydroxylation) is an oxidative step. However, cleavage of the 20R,22R bond when carried out in the presence of ^{18}O does not result in the introduction of this isotope into either pregnenolone or isocapradehyde.[14] How oxygen is utilized in the cleavage of this bond remains to be elucidated (see Chapter 7).

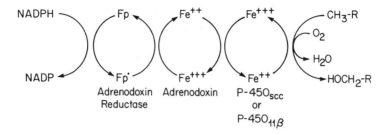

FIGURE 2. Electron transport system localized in steroidogenic mitochondria. This system is required for mitochondrial steroid hydroxylase activity.

The identification of the cholesterol side-chain cleavage reaction as one dependent on P-450 was based on carbon monoxide inhibition,[15,16] photochemical action spectrum having a maximum at 450 nm,[16] and finally partial purification and reconstitution of P-450$_{scc}$.[17,18] Several laboratories subsequently purified bovine or porcine P-450$_{scc}$ to homogeneity and characterized this enzyme.[19–29] P-450$_{scc}$ is an integral protein of the inner mitochondrial membrane as originally suggested by fractionation of the mitochondrion into its various compartments,[30] and recently established cytochemically.[26] The best available evidence indicates that the substrate binding site of cytochrome P-450$_{scc}$ is localized on the inner aspect of the inner mitochondrial membrane.[26] The molecular weight of bovine P-450$_{scc}$ is 49,000 by SDS–PAGE analysis.[21,31]

As can be seen in Fig. 2, reducing equivalents are provided to P-450$_{scc}$ via an electron transfer system localized in the mitochondrial matrix.[32] Electrons from NADPH are transferred to an FAD-containing flavoprotein known as NADPH-adrenodoxin reductase.[33] This enzyme, in turn, transfers electrons to an iron–sulfur protein known as adrenodoxin which then transfers electrons to P-450$_{scc}$.[32,34] Thus, adrenodoxin is capable of forming complexes, sequentially, with adrenodoxin reductase[35,36] and P-450$_{scc}$[37] (see Chapter 11). Although the transfer of reducing equivalents to P-450$_{scc}$ appears to be a very complex process, it seems to play no role in the regulation of P-450$_{scc}$ activity. This mitochondrial electron transport system resembles the soluble system found in the bacterium *Pseudomonas putida*, which transfers electrons from NADH to P-450$_{cam}$ to support camphor hydroxylation.[38]

Purified preparations of bovine adrenodoxin reductase consist of a monomeric protein of molecular weight 51,000 on SDS–PAGE[31] and molecular weight 54,000 based on amino acid composition.[35] This protein contains one mole of FAD per mole enzyme and shows a typical flavoprotein absorption spectrum.[39] Recently, evidence has been obtained indicating the presence of two forms of adrenodoxin reductase in adrenocortical mitochondria.[40] However, the role of two forms is unknown. The complete amino acid sequence of bovine adrenodoxin has been determined, consisting of 114 amino acids and a molecular weight of 12,500.[41] Each mole of adrenodoxin contains 2 moles of iron and 2 moles of labile sulfur, and the optical and electron spin resonance (ESR) properties of this iron–sulfur protein have been well characterized.[42–44]

A second form of P-450 is also known to reside in the inner membrane of adrenocortical mitochondria. This is the steroid 11β-hydroxylase or P-$450_{11\beta}$.[26,45] Reducing equivalents are transferred to this steroid hydroxylase by precisely the same system as described above for P-450_{scc}, namely adrenodoxin reductase and adrenodoxin. P-$450_{11\beta}$ catalyzes the conversion of 11-deoxycorticosterone and 11-deoxycortisol to corticosterone and cortisol, respectively, and thus catalyzes the final step in the adrenocortical steroidogenic pathway leading from cholesterol to glucocorticoids.[46,47] P-$450_{11\beta}$ is distinct from P-450_{scc} based on a variety of properties including optical and ESR spectra and amino acid composition.[39] The molecular weight of bovine P-$450_{11\beta}$ as estimated by SDS–PAGE is 48,000, somewhat smaller than that determined for P-450_{scc}.[31] As opposed to P-450_{scc}, P-$450_{11\beta}$ can be isolated in the low-spin ferric form which is easily converted to the high-spin ferric form upon addition of deoxycorticosterone. It is also found by ESR spectroscopy that the high-spin signal of P-$450_{11\beta}$ has a g-value of 7.8 while that of P-450_{scc} has a g-value of 8.1.[48,49]

An interesting aspect of P-$450_{11\beta}$ is that it has been found in several laboratories to catalyze 18-hydroxylation.[50–53] 18-Hydroxylation of corticosterone is on the pathway of the conversion of cholesterol to aldosterone, the key mineralocorticoid required for salt retention. Aldosterone biosynthesis occurs uniquely in the zona glomerulosa[54] (outer layer of cells) of the adrenal cortex while glucocorticoid biosynthesis occurs in the zona fasciculata–reticularis. Purified P-$450_{11\beta}$ from the zona fasciculata–reticularis can catalyze 18-hydroxylation reactions in reconstituted systems.[55] However, it still remains to be proven that the 18-hydroxylase from the zona glomerulosa which is required for salt retention is identical to the P-$450_{11\beta}$ required for glucocorticoid synthesis.

The microsomal components of the adrenocortical steroidogenic pathway include two forms of P-450 (P-450_{C21} and P-$450_{17\alpha}$) and the ubiquitous flavoprotein, NADPH-cytochrome P-450 reductase. The P-450 reductase supplies reducing equivalents from NADPH to each of the microsomal forms of P-450 and is presumably identical with the FAD/FMN-containing P-450 reductase which has been purified from liver and characterized in several laboratories.[56–58] P-450_{C21} played a pivotal role in the history of investigation into the nature of P-450. It is this enzyme which Estabrook, Cooper, and their colleagues used to first demonstrate a P-450 photochemical action spectrum with a maximum at 450 nm.[59] Recently, this protein has been purified[60,61] and determination of its primary sequence is in progress.[62] A form of P-450$_{C21}$ capable of catalyzing the steroid 21-hydroxylation of progesterone is found in most, if not all, tissues, including liver. However, the current hypothesis is that this enzyme is distinct from the adrenocortical P-450$_{C21}$. In fact, the physiological function of this extraadrenal P-450$_{C21}$ is unknown.

The other microsomal form of P-450, P-450$_{17\alpha}$, has also recently been purified from adrenal cortex.[63] As will be noted later, cytochrome P-450$_{17\alpha}$ is found in all steroidogenic tissues and the enzyme purified from immature pig testis has been characterized in greatest detail.[64,65] This enzyme resides at a branch point between the production of glucocorticoids and adrenal androgens (Fig. 1). Both pathways involve 17α-hydroxylation of either pregnenolone or progesterone. In the glucocorticoid pathway, the product of this reaction becomes substrate for P-450$_{C21}$. In the androgen pathway, the product of the 17α-hydroxylation step becomes the substrate for a side-chain cleavage reaction called the 17,20-lyase reaction. Hall and co-workers have convincingly established that in the testis the same protein (P-450$_{17\alpha}$) catalyzes both the 17α-hydroxylation step and the 17,20-lyase step.[64,65] It is reasonable to assume that the same is true in the adrenal cortex even though glucocorticoid production is the predominant pathway in this tissue while androgen production is the only pathway in other steroidogenic tissues. Elucidation of the nature of the active site(s) on P-450$_{17\alpha}$ and its regulation in different steroidogenic tissues will prove very interesting.

2.2. Ovary

P-450$_{scc}$ has been purified from bovine corpus luteum[66-68] and the properties of this enzyme appear to closely parallel those reported for P-450$_{scc}$ from bovine adrenal cortex. The identity of these two enzymes can only be confirmed by a comparison of primary amino acid sequence; however, it seems reasonable to assume in the absence of conflicting data that they are the same. Also, adrenodoxin reductase and adrenodoxin have been purified from corpus luteum and their properties are very similar to those of the two comparable enzymes from adrenal cortex.[69] To the best of our knowledge, P-450$_{17\alpha}$ has not been purified from ovarian tissue. As noted in Table IV, the ovary does not contain P-450$_{C21}$ and P-450$_{11\beta}$ as this steroidogenic tissue does not produce glucocorticoids or mineralocorticoids.

2.3. Testis

The testis is known to contain both the cholesterol side-chain cleavage system and P-450$_{17\alpha}$. Purification of the components of the cholesterol side-chain cleavage system from the testis has not been reported. Purification of P-450$_{17\alpha}$ from the testis has been noted previously in this chapter.[64,65]

3. ACTH Action on the Adrenal Cortex

ACTH has two actions on the adrenal cortex (an acute effect and a chronic effect) which can be separated temporally; each of which is found to be important in regulating activities of the various steroid hydroxylases. ACTH action is initiated by binding of the peptide hormone to specific receptors on the cell surface.[70] The binding of ACTH to its receptor results in activation of adenylate cyclase (also localized in the plasma membrane) leading to an increase in intracellular levels of cyclic AMP. While the precise details of events following elevation of cyclic AMP levels are not known, it has been observed that the amount of cyclic AMP bound to the regulatory subunit of cyclic AMP-dependent protein kinase is increased at all concentrations of ACTH which result in stimulation of steroidogenesis.[71] Therefore, it is likely that the actions of ACTH are mediated by activation of cyclic AMP-dependent protein kinases.

3.1. The Acute Effect

The rate-limiting step in the steroidogenic pathway is the cholesterol side-chain cleavage reaction.[72] Therefore, it can be imagined that increased concentrations of cholesterol in the vicinity of P-450$_{scc}$ will lead to increased activity of this enzyme as well as increased activity of the subsequent enzymes of the pathway. It appears that this is the mechanism by which steroidogenesis is regulated acutely. In other words, an increase in the flux of cholesterol into the side-chain cleavage reaction acutely increases the rate of production of all intermediates in the pathway leading to increased steroid output. The details of this acute action of ACTH are described in the following chapter. It should be noted that this action of ACTH occurs within seconds or minutes while the chronic action of ACTH described in the next section of this chapter takes hours to become manifest.

Thus, the acute aspect of ACTH regulation of adrenocortical steroid hydroxylases appears to have its locus primarily if not exclusively at the level of the availability of substrate (cholesterol) for the first step in the steroidogenic pathway, cholesterol side-chain cleavage. In response to ACTH binding to its receptor on the surface of the adrenocortical cell, adenylate cyclase activity is enhanced leading to an increased intracellular concentration of cyclic AMP. This leads to mobilization of cholesterol from its storage site in lipid droplets in the cytoplasm to the mitochondria. Under the influence of a cycloheximide-sensitive factor, cholesterol is transferred from the outer mitochondrial membrane to the inner mitochondrial membrane where it becomes available to P-450$_{scc}$ for conversion to pregnenolone and thereby leads to increased flux through the other steps in the adrenocortical steroidogenic pathway.

3.2. The Chronic Effect

While regulation of the acute action of ACTH occurs primarily or perhaps even exclusively at the level of the cholesterol side-chain cleavage reaction, the chronic effect of ACTH on the steroidogenic pathway appears to be associated with most, if not all, of the steps. As will be seen, the chronic effect of ACTH manifests itself in regulation of the synthesis of steroidogenic enzymes. The foundation for the elucidation of the chronic effect of ACTH and its mechanism resides in studies carried out in several laboratories which demonstrated that following hypophysectomy, a dramatic decrease in levels of both mitochondrial and microsomal cytochromes P-450 was observed in the adrenal cortex.[73-76] This was associated with losses in steroid hydroxylase activity and also both mitochondrial and microsomal systems necessary for reduction of these cytochromes P-450. In all cases, administration of ACTH to hypophysectomized animals resulted in partial restoration of P-450 levels and enzyme activities. More recently, decreases have been observed in levels of P-450$_{scc}$ and P-450$_{11\beta}$ upon hypophysectomy using immunofluorescence techniques, followed by an increase in these levels upon administration of ACTH.[39] These studies have led to the conclusion that by removing the source of ACTH (hypophysectomy), the factor necessary for regulation of optimal steroidogenic capacity was removed and that ACTH likely had a long-term role in maintenance of adrenocortical steroidogenic capacity.[77] However, the nature of this regulatory role of ACTH has only recently been elucidated. In particular, the development of a primary bovine adrenocortical cell culture system[78] and the purification of the bovine steroidogenic enzymes leading to production of antibodies specific for these proteins have made such studies possible.

3.2.1. Regulation of Synthesis of Mitochondrial Steroidogenic Enzymes

Initial studies on certain of these enzymes were carried out using the mouse Y-1 adrenocortical tumor cell line. Kowal noted that ACTH caused

TABLE II
ACTH-Mediated Induction of Steroid Hydroxylase
Biosynthesis in Bovine Adrenocortical Cells

	RNA translation		Cell labeling	
Enzyme	Time (hr)	Fold increase	Time (hr)	Fold increase
P-450$_{scc}$	36	6	36	7
P-450$_{11\beta}$	36	5	36	7
Adrenodoxin	36	3	36	3
P-450$_{C21}$	24	3	24	10
P-450$_{17\alpha}$	24	>20	24	>20

an increase in the adrenodoxin content of such cells as measured by ESR spectroscopy as well as observing an increase in 11β-hydroxylase activity.[79,80] Asano and Harding demonstrated that ACTH caused an increase in synthesis of adrenodoxin by determining the rate of incorporation of [^{3}H]leucine into this protein.[81] More recently, a systematic study of the synthesis of all the mitchondrial steroidogenic enzymes has been carried out using bovine adrenocortical cells in primary, monolayer culture. When treated with a pharmacological dose of ACTH (1 μM), increased rates of synthesis of P-450$_{scc}$, P-450$_{11\beta}$, adrenodoxin, and adrenodoxin reductase are observed.[82] In such studies, cells are treated with ACTH for varying periods of time and then pulse radiolabeled with [^{35}S]methionine. The cells are then lysed and the newly synthesized enzyme separated from all the other radiolabeled cell proteins by immunoisolation, and its synthesis quantitated by fluorography of the immunoisolates following SDS–PAGE. In the case of each of the mitochondrial components of this pathway, the temporal pattern of synthesis is found to be the same with an optimum being observed 36 hr following initiation of ACTH treatment (Table II). A similar result is observed for each enzyme following extraction of RNA from treated cells, *in vitro* translation of this RNA, and immunoisolation of newly synthesized proteins. Thus, as a first approximation, it seems that ACTH treatment of cultured cells leads to increased levels of mRNA specific for steroid hydroxylases which in turn leads to an increased rate of synthesis of these enzymes.[83–86] Furthermore, regulation of synthesis of the mitochondrial components of the steroidogenic pathway appears to occur in a coordinate fashion. As a result of the increased rate of synthesis, increased levels of mitochondrial P-450 and adrenodoxin, and increased activities of P-450$_{scc}$ and P-450$_{11\beta}$ were observed, indicating that increased synthesis results in increased enzyme mass.

While the ACTH receptor on the surface of adrenocortical cells is not well described, it has been clearly shown in several laboratories that upon binding of ACTH to its receptor, there is an increase in intracellular cAMP levels as a result of activation of adenylate cyclase. Thus, as has been documented for the acute response, it was postulated that the chronic effect of ACTH is mediated through cAMP. Using the bovine adrenocortical culture system, it has been shown that synthesis of P-450_{scc}, P-$450_{11\beta}$, and adrenodoxin is stimulated by analogues of cAMP.[87] In particular, dibutyryl cAMP and 8-bromo cAMP both stimulate the synthesis of these enzymes with a temporal pattern similar to that outlined above for ACTH. The increase in concentration of P-450_{scc} resulting from such treatment can be readily observed by immunofluorescence. Thus, cAMP analogues are able to mimic the chronic effect of ACTH. Furthermore, it has been shown that other effectors which activate adenylate cyclase and lead to an increased intracellular content of cAMP also lead to increased synthesis of the mitochondrial components. In particular, cholera toxin and prostaglandins E_2 and $F_{2\alpha}$ are effective.[88] It is interesting to note that prostaglandin $F_{2\alpha}$ is the least effective inducer and is also least effective in elevating cAMP levels. In fact, adrenodoxin synthesis is not induced by either prostaglandin. Thus, it appears that each mitochondrial component has a different sensitivity to cAMP.

These results substantiate the conclusion reached from studies involving hypophysectomized rats, that the chronic effect of ACTH involves regulation of synthesis of the mitochondrial steroidogenic enzymes and that this action of ACTH is mediated through cAMP. Furthermore, the regulation of synthesis appears to be at the transcriptional level based on the result of *in vitro* translation studies. However, it was considered possible that ACTH could also regulate the turnover of these enzymes and thus optimum steroidogenic capacity. It was found that ACTH treatment of cultured cells had no effect on the turnover rate of newly synthesized P-450_{scc} or adrenodoxin.[89] However, the half-life of P-$450_{11\beta}$ was slower in ACTH-treated cells (24 hr versus 16 hr), suggesting that turnover might be a factor in regulation of optimal P-$450_{11\beta}$ levels. It has also been noted that P-$450_{11\beta}$ is subject to inactivation through interaction with 11β-hydroxylated steroids (products of the 11β-hydroxylase reaction) which appear to act as pseudosubstrates.[90]

An interesting facet of the chronic effect of ACTH is that continued ACTH stimulation does not lead to sustained high levels of synthesis in all cases. Rather, beyond about 36 hr, the rate of synthesis of all the mitochondrial components of the steroidogenic pathway declines.[83,84,86,87] Desensitization of steroidogenesis by continued treatment of cells with ACTH has been known for a long time.[91] Apparently, this process can also come into play with respect to steroid hydroxylase

TABLE III
Molecular Weights of Precursor and Mature Forms of Mitochondrial Components of the Steroid Hydroxylase Pathway as Measured by SDS-PAGE

Protein	Precursor molecular weight	Mature molecular weight
P-450$_{scc}$	54,500	49,000
P-450$_{11\beta}$	53,500	48,000
Adrenodoxin	19,000	12,000
Adrenodoxin reductase	53,000	51,000

synthesis and surprisingly this desensitization is observed even in experiments using cAMP analogues in place of ACTH.[87] The mechanism of desensitization in either the acute or chronic response to ACTH is unknown.

An important characteristic of the mitochondrial steroidogenic components is that they are synthesized as higher molecular weight precursors in the cytoplasm and proteolytically processed upon insertion into the mitochondria[31,92-94] (Table III). Thus, P-450$_{scc}$, P-450$_{11\beta}$, adrenodoxin, and adrenodoxin reductase are encoded on nuclear genes. When the processing of the precursor form of P-450$_{scc}$ is studied *in vitro* (using an *in vitro* assay as described by Neupert[95]), it is found that mitochondria from bovine heart, liver, or kidney are incapable of importing and processing this precursor.[96] The same is found to be true for the P-450$_{11\beta}$ precursor molecule. On the other hand, these precursors are readily imported and processed in adrenocortical mitochondria. Adrenodoxin shows quite a different pattern in such studies. All the above-mentioned mitochondria are capable of efficiently importing and processing the adrenodoxin precursor molecule.[96] It has been found in other studies that the protease(s) required for processing mitochondrial precursors is found in the matrix compartment of mitochondria.[97] Matrix fractions from nonsteroidogenic mitochondria are incapable of processing P-450$_{scc}$ or P-450$_{11\beta}$ precursors although these precursors can be processed by the matrix fraction from steroidogenic mitochondria. However, the adrenodoxin precursor could be processed by the matrix fraction from all mitochondria tested.[96] Processing of the adrenodoxin precursor by adrenocortical mitochondria has also been demonstrated in another laboratory.[98] These studies indicate, for the first time, that specificity exists in the processing of mitochondrial precursor proteins. Furthermore, this specificity resides at the level of both recognition and proteolysis. The resolution of sites on adrenocortical mitochondria which are specific for the precursor forms of P-450$_{scc}$ and P-450$_{11\beta}$ will be important in furthering our understanding of mitochondrial biogenesis. Recently, it has been shown using cell cultures that the precursor forms of these mitochondrial components are formed and converted to mature enzymes *in situ*, illustrating the precursor–product relationship between the precursor and mature forms of these enzymes.

3.2.2. Regulation of Synthesis of Microsomal Steroidogenic Enzymes

In contrast to the mitochondrial enzymes, the synthesis of P-450$_{C21}$ in response to ACTH optimizes at 24 hr after initiation of ACTH treatment.[99] The temporal pattern is the same whether determined by labeling of cellular protein or by *in vitro* translation of cellular RNA. In addition to the difference in temporal pattern between this microsomal enzyme and the mitochondrial steroid hydroxylases, P-450$_{C21}$ also is synthesized as a mature enzyme rather than a higher molecular weight precursor. It has been observed that several other microsomal forms of P-450 are synthesized as mature forms.[100-102] Another interesting difference between P-450$_{C21}$ and the other enzymes examined, is that no increase in steroid 21-hydroxylase activity is observed in bovine adrenocortical cells in response to ACTH treatment even though an increase in synthesis is observed by two different criteria. The explanation for this dichotomy is unknown. In other studies, ACTH treatment was found to lead to no change[103] or a reduction[104] in steroid 21-hydroxylase activity. The result in this latter case was attributed to C_{19}-steroids serving as pseudosubstrates for P-450$_{C21}$. As observed for the mitochondrial components, the synthesis of P-450$_{C21}$ is also stimulated by cAMP analogues and by cholera toxin and prostaglandins E_2 and $F_{2\alpha}$.[87] Thus, the synthesis of this microsomal steroid hydroxylase appears to be regulated by a mechanism similar to that described above for the mitochondrial components of the steroidogenic pathway.

17α-Hydroxylase (P-450$_{17\alpha}$), the other microsomal steroid hydroxylase, is different in one very interesting respect from the other enzymes discussed. In the absence of ACTH, cultured bovine adrenocortical cells produce corticosterone as their major glucocorticoid product of the steroidogenic pathway.[105,106] Only when they are stimulated with ACTH or analogues of cAMP do they begin to produce significant quantities of cortisol as their final glucocorticoid product. The difference between corticosterone and cortisol is that cortisol is hydroxylated at the 17α-position while corticosterone is not. Thus, ACTH treatment leads to a dramatic increase in 17α-hydroxylase activity.[107] Subsequent studies have shown that this large increase in activity is the result of increased synthesis of P-450$_{17\alpha}$.[108] This increase could be observed by cell labeling, *in vitro* translation, or immunoblot analysis. In addition, no change was found in the K_m of 17α-hydroxylase in response to ACTH, only a change in V_{max} as expected from increased synthesis.[108] However, in all cases, no synthesis of P-450$_{17\alpha}$ could be detected in untreated cells. Thus, unlike the other steroid hydroxylases, P-450$_{17\alpha}$ diminishes to low, undetectable levels during cell culture. As a result of such a low level, the fold increase in synthesis of P-450$_{17\alpha}$ is much greater than that for any of the enzymes in this pathway. As described previously for P-450$_{C21}$, P-450$_{17\alpha}$ is also synthesized as a mature protein.[108] In addition, the temporal pattern of synthesis shows an optimum at 24 hr. Stimulation of 17α-hydroxylase activity has also been observed in studies where ACTH was administered *in vivo* to rabbits.[109]

As noted earlier, studies by Hall and co-workers have indicated that the 17α-hydroxylase activity and 17,20-lyase activity reside on the same polypeptide chain (P-450$_{17\alpha}$) in the testis.[64,65] Studies on the effect of ACTH on 17,20-lyase activity in adrenocortical cells have yet to be carried out, although measurement of steroid output of such cells does indicate that ACTH increases this activity as well as 17α-hydroxylase activity.[105,106] This is particularly interesting because in humans, only 17α-hydroxylase activity can be detected in the adrenal until just prior to puberty. At this time, adrenarche occurs and production of adrenal androgens (which require 17,20-lyase activity) begins.[110] Thus, these two activities may reside on the same polypeptide chain, but they have different developmental patterns in humans.

Preliminary results indicate that the microsomal flavoprotein, cytochrome P-450 reductase, is also increased in repsonse to ACTH. This enzyme is thought to be involved in both 17α-hydroxylase and steroid C21-hydroxylase activities.[58] Temporal studies indicate optimum synthesis of this enzyme at 24 hr and it is found that cAMP analogues also increase the synthesis of this reductase. However, treatment of cultured adrenal cells with phenobarbital does not lead to increased NADPH-cytochrome P-450 reductase. This is of interest because phenobarbital treatment does increase the synthesis of this enzyme in liver. It also should be mentioned that adrenal cytochrome b_5 has been implicated in 17,20-lyase activity[111,112] although no studies have been carried out on the synthesis of b_5 in steroidogenic tissues.

Thus, the synthesis of the microsomal components of the adrenocortical steroidogenic pathway appears to be regulated by ACTH via cAMP and this regulation appears to be at the level of transcription. However, the temporal pattern of increased synthesis of the microsomal enzymes is different from that of the mitochondrial enzymes. This may indicate that the synthesis of forms of P-450 in these two different subcellular compartments is controlled by different regulatory factors.

3.2.3. Site of Synthesis of Adrenocortical Steroid Hydroxylases

Following separation and isolation of cytoplasmic and membrane-bound ribosomes from bovine adrenal cortex, the site of synthesis of various components of the steroidogenic pathway has been determined.[113,114] Nascent adrenodoxin peptides were found to be associated with free, loosely bound and tightly bound ribosomes, while nascent adrenodoxin reductase peptides were found to be associated with free and loosely bound ribosomes.[113] Nascent P-450$_{scc}$ and P-450$_{11\beta}$ peptides were found associated with both free and bound ribosomes, while P-450$_{C21}$ nascent peptides were associated predominantly with bound ribosomes.[114] In several studies in liver, proteins destined for mitochondria have been found to be associated with free ribosomes. Why such proteins in the adrenal cortex are found to be associated with both classes of ribosomes remains to be determined. However, it seems clear that these proteins are synthesized as their complete precursor forms in the cytoplasm prior to their incorporation into mitochondria, rather than being inserted into mitochondria by a cotranslational process.

3.3. The Use of Recombinant DNA Technology in the Study of P-450 Gene Expression in the Adrenal Cortex

While the results of the experiments cited above clearly indicate that ACTH regulates the synthesis of adrenocortical steroid hydroxylases and that this chronic action is mediated by cAMP, the question remains as to the mechanism by which this regulation proceeds. *In vitro* translation studies suggest that elevated levels of cAMP lead to elevated levels of mRNA specific for the steroid hydroxylases. However, it also could be that preexisting mRNA becomes more active translationally or becomes more stable upon ACTH treatment. Thus, in order to elucidate the details of the ACTH-mediated regulation of steroid hydroxylase synthesis, it has been necessary to construct cDNA probes specific for these enzymes. In addition to providing primary sequence data on these enzymes, such probes are necessary for the quantitation of mRNA levels and the identification and characterization of genes encoding these enzymes. Once these studies have been completed, it will be possible to identify regulatory elements of these genes and thus investigate in detail their expression.

Preparation and identification of clones containing recombinant DNA molecules specific for these steroidogenic forms of P-450 have been undertaken in several laboratories. Three different laboratories have reported identification of clones specific for bovine adrenocortical P-450_{scc}.[115-117] In one case, the complete primary sequence of the enzyme (including the precursor segment) has been reported.[116] In another study, it has been established the ACTH treatment of cultured bovine adrenocortical cells leads to an increased accumulation of mRNA encoding P-450_{scc}.[117] A cDNA clone has also been constructed for bovine adrenocortical P-450_{C21} and has been used to investigate human steroid 21-hydroxylase deficiencies.[118,119] While it is too early in such studies to demonstrate the molecular basis of this disease, it appears that deficiency of enzymatic activity is related to gene deletion. Recently, cDNA clones specific for bovine P-$450_{11\beta}$ have been identified and used to establish that ACTH treatment also leads to accumulation of mRNA specific for P-$450_{11\beta}$.[120] Unpublished results indicate the identification of cDNA clones specific for bovine adrenodoxin and P-$450_{17\alpha}$. In both cases, ACTH treatment leads to accumulation of specific mRNA. In the case of adrenodoxin, sequence studies indicate that the precursor segment of this iron–sulfur protein is very different from that of P-450_{scc}. Perhaps this lack of homology provides the explanation for the difference in tissue specificity of processing of these two precursor proteins.

Through use of these cDNA probes in Northern hybridization analysis, the mRNA sizes for these proteins have been determined. Bovine P-450_{scc} contains 520 amino acids, and thus requires a nucleotide coding sequence of 1560 bases.[116] The P-450_{scc} mRNA is about 1900 nucleotides in length[117] (see Fig. 3). Bovine adrenocortical P-450_{C21} is estimated to contain about 450 amino acids and its mRNA is approximately 2000 nucleotides in length.[118] Porcine P-$450_{17\alpha}$ is about 54,000 daltons and the bovine enzyme is encoded by an mRNA approximately 1900 nucleotides in length. The bovine adrenodoxin precursor is 19,000 daltons or approximately 180 amino acids long. A protein of this size requires a nucleotide sequence of 540 bases. The adrenodoxin mRNA is approximately 1400 bases in length indicating substantial 5' and/or 3' untranslated regions. In this respect, bovine P-$450_{11\beta}$ mRNA is particularly interesting. The precursor protein is estimated to be 53,500 daltons and thus would be expected to be encoded by an mRNA of approximately 2 kilobases in length. Surprisingly, however, the major P-$450_{11\beta}$ mRNA is found to be about 4.3 kilobases long.[120] In addition, two larger P-$450_{11\beta}$ mRNA species of 7.2 and 6.2 kilobases have also been identified which can program P-$450_{11\beta}$ synthesis in an *in vitro* translation system. The reason for such a large mRNA encoding P-$450_{11\beta}$ and for multiple mRNA species is unknown. It should be noted, however, that in a few instances, unusually

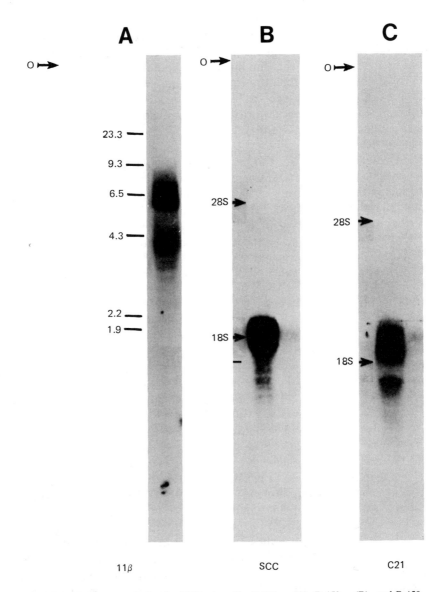

FIGURE 3. Northern analysis of mRNA sizes for P-450$_{11\beta}$ (A), P-450$_{scc}$ (B), and P-450$_{C21}$ (C) in bovine adrenocortical poly(A)$^+$ RNA. In A, the sizes are given in kilobases. In B and C, the size 18 S is approximately 2.0 kilobase.

large mRNA molecules have been identified for hepatic, xenobiotic-inducible forms of P-450.[121,122]

While cDNA clones specific for adrenocortical NADPH-cytochrome P-450 reductase have not been identified, such recombinant molecules have been characterized for rat liver NADPH-cytochrome P-450 reductase.[123] It seems likely that the hepatic enzyme is identical to the adrenocortical enzyme and thus the rat liver clone could be used to identify adrenocortical cDNA clones. It will be interesting to elucidate the gene structure for this enzyme in order to understand how the liver enzyme can be induced by phenobarbital while the adrenocortical enzyme can be induced by cAMP.

As mentioned above, through the use of cDNA probes it has been shown that ACTH treatment leads to accumulation of mRNA for each of the steroidogenic forms of P-450. In these same studies, it has been found that this action of ACTH can be mimicked by analogues of cAMP. Furthermore, in the case of bovine $P450_{scc}$, nuclear transcript runoff experiments indicate that the accumulation of mRNA results, at least in part, from increased transcription (M. E. John, 1985, unpublished results). It is necessary to qualify this statement for the moment as the effect of ACTH on mRNA stability has yet to be measured. Perhaps the ACTH action will manifest itself both at the level of increased transcription and increased mRNA stability, thus leading to the observed increase in accumulation of $P-450_{scc}$ mRNA. It seems likely that a similar mechanism will prove to be the case for all the components of the adrenocortical steroidogenic pathway. While transcription studies have yet to be carried out for all these enzymes, actinomycin D (a transcription inhibitor) is found to inhibit the ACTH-mediated accumulation of mRNA for all adrenocortical forms of P-450 as well as adrenodoxin.

Finally, the question remains as to what factors regulate the transcription of steroid hydroxylase genes. The connection between cAMP and the regulatory elements of these genes remains to be made. As a beginning in this direction, it has been observed that the ACTH-mediated accumulation of $P-450_{scc}$, $P-450_{11\beta}$, and $P-450_{17\alpha}$ mRNA is cycloheximide sensitive (M. E. John, 1985, unpublished results and Refs. 120 and 125). Cycloheximide is a translation inhibitor and these results lead to the hypothesis that protein synthesis is required for the cAMP-mediated accumulation of steroidogenic P-450 mRNAs. Such protein synthesis could result from transcription of a specific gene(s) or activation of a preexisting ribonuclear protein particle. Nevertheless, it appears that cAMP mediates the synthesis of this protein factor. The name steroid hydroxylase-inducing protein (SHIP) has been suggested for this factor.[126] It is unknown whether a specific protein is required for activation of transcription of each steroid hydroxylase gene, whether the genes encoding the mito-

TABLE IV
Tissue Distribution of Steroidogenic Forms of Cytochrome P-450 and the Peptide Hormones Which Regulate Their Levels and Activities

Tissue	P-450	Hormone	Steroid products
Adrenal cortex	P-450$_{scc}$ P-450$_{11\beta}$ P-450$_{17\alpha}$ P-450$_{C21}$	ACTH	Glucocorticoids Mineralocorticoids Androgens
Ovary	P-450$_{scc}$ P-450$_{17\alpha}$	FSH LH	Estrogen Testosterone
Testis	P-450$_{scc}$ P-450$_{17\alpha}$	LH	Testosterone Estrogen

chondrial and microsomal subsets of this pathway are regulated by different proteins, or whether one protein factor will regulate the synthesis of all these enzymes. Finally, the nature of the cAMP-dependent factors which regulate the synthesis of SHIP remains to be elucidated. Nevertheless, it is interesting to note that while the synthesis of steroidogenic forms of P-450 in adrenocortical cells in response to ACTH is sensitive to cycloheximide, synthesis of mouse P$_1$-450 in hepatocytes in response to the xenobiotic 2,3,7,8-tetrachlorodibenzo-*p*-dioxin (TCDD) is not inhibited by cycloheximide.[127]

3.4. Peptide Hormone Regulation of Steroid Hydroxylase Synthesis in Testis and Ovary

As noted in Table IV, other peptide hormones regulate the synthesis of steroid hydroxylases in other steroidogenic tissues. A limited number of studies have been carried out in these tissues. For example, it has been shown using rat or porcine Leydig cells in culture that either human chorionic gonadotropin (hCG) or luteinizing hormone (LH) treatment leads to increased synthesis of P-450$_{scc}$ and adrenodoxin.[128,129] This increase is also observed by *in vitro* translation of RNA isolated from such cells and furthermore these effects can be mimicked by analogues of cAMP. Using hypophysectomized rats, treatment with hCG was also shown to result in increases in both 17α-hydroxylase activity and 17,20-lyase activity,[130] implicating a role for hCG in the regulation of P-450$_{17\alpha}$ synthesis. Recently, both the above microsomal activities have been shown to increase in mouse Leydig cell cultures upon treatment with LH or 8-bromo-cAMP, further documenting the regulation of this enzyme in the testis by peptide hormones.[131]

In bovine ovarian granulosa cells, it has also been shown that the synthesis of P-450$_{scc}$ and adrenodoxin can be regulated by follicle-stimulating hormone (FSH).[132] This effect appears to be at the level of transcription as estimated from *in vitro* translation studies and is also mediated by analogues of cAMP.

Thus, it seems that the synthesis of steroid hydroxylases and related enzymes is regulated by the same mechanism in all steroidogenic tissues. The difference between tissues is the result of specificity for peptide hormones presumably at the level of their receptors. Once the peptide hormone binds to its receptor, it activates adenylate cyclase leading to increased levels of intracellular cAMP. It can be imagined that the subsequent steps in the ovary and testis will be the same as determined in the adrenal cortex. This mechanism is obviously of great physiological significance in that it provides optimal levels of glucocorticoids, mineralocorticoids, and sex hormones. It also has important implications in the ovarian cycle. Following ovulation, the granulosa cells remaining in the follicle undergo a process of luteinization leading to the development of the corpus luteum. The corpus luteum is an extremely active steroidogenic tissue producing progesterone which is important in implantation of a fertilized ovum in the uterine wall. The amount of P-450$_{scc}$ increases approximately 50-fold during the transition from granulosa to luteal cells.[133] Thus, in this particular case, there is a physiological induction of P-450$_{scc}$ which occurs on a regular monthly basis. The mechanism of this process will prove to be very interesting.

4. Role of P-450 in the Conversion of Androgens to Estrogens

The aromatization of androgens to estrogens is one of the most important but least understood of the reactions involved in steroid hormone biosynthesis. The pathway whereby aromatization of the A ring of androgens is believed to occur is indicated in Fig. 4. The sequence of reactions in the biosynthesis involves three enzymatic hydroxylations.[134] The first two take place on the C-19 methyl group resulting in its conversion to an aldehyde. The final and rate-determining hydroxylation step occurs at the 2-position subsequent to which the product rapidly and apparently nonenzymatically collapses to estrogen. Aromatization requires oxygen and NADPH. Thus, the enzyme system is of the mixed-function oxidase type. For every mole of estrogen formed, 3 moles of NADPH and 3 moles of oxygen are used, findings which are suggestive that three hydroxylation steps are involved in this reaction sequence.[135]

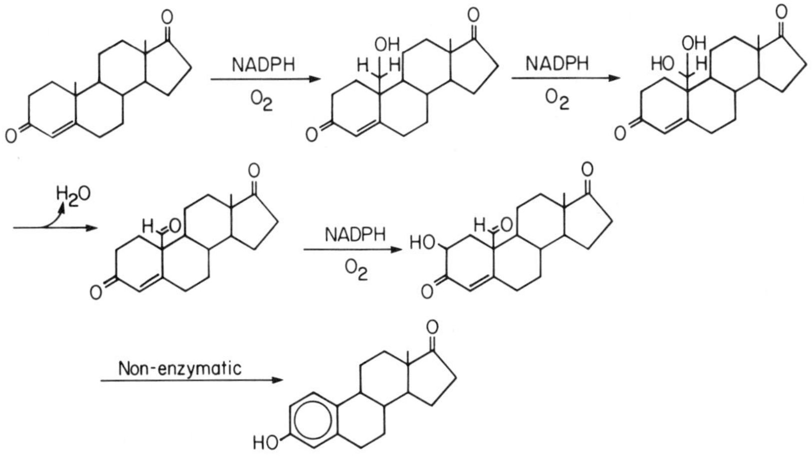

FIGURE 4. The androgen aromatization reactions leading to estrogen formation.

In addition to the granulosa cells of the ovaries, the aromatase enzyme complex occurs in a variety of tissues, both male and female; for example, it occurs in adipose tissue of both men and women[136]; in the Sertoli cells and Leydig cells of the testis[137,138]; in several regions of the brain[139]; and also in the placenta.[140] The aromatase enzyme complex is localized in the endoplasmic reticulum since conversion of C_{19} steroids to estrogens in placenta and other tissues where it occurs can be demonstrated in this subcellular compartment.[141] In addition, aromatase activity is also found in mitochondrial preparations of placenta and of other tissues.[142] Aromatase activity in mitochondria that have been disrupted by treatment with a high concentration of calcium is similar to aromatase activity of intact mitochondria when supported by an NADPH-generating system.[143] This finding is suggestive that aromatase activity of mitochondria is associated with the outer mitchondrial membrane. The properties of the aromatase system in the mitochondrial fraction of human term placenta are similar to those in the microsomal fraction, in terms of relative activities toward androstenedione, 19-nortestosterone, 16α-hydroxytestosterone, as well as sensitivity to carbon monoxide and to antibodies raised against hepatic microsomal NADPH-cytochrome c reductase.[144] When inner and outer mitochondrial membrane subfractions were assayed, aromatase activity was associated predominantly with outer mitochondrial membrane preparations.[144] This subcellular localization differs from that of cholesterol side-chain cleavage which is in the inner mitochondrial membrane. When the total mitochondrial fraction was separated into heavy, light, and very light mitochondria,[143] most of the aromatase activity was found in the latter fraction. Little activity was found in heavy mitochondria. Thus, aromatase activity in mitochondrial fractions may be due to contamination by endoplasmic reticulum vesicles. Nevertheless, when inosine diphosphatase activity was used as a marker for endoplasmic reticulum, aromatase activity in mitochondrial membranes could not be accounted for by microsomal contamination.[144]

4.1. Role of P-450 in Estrogen Biosynthesis

The question arose as to the involvement of P-450 and NADPH-cytochrome P-450 reductase, enzymes which are present in the endoplasmic reticulum, in the aromatase reaction. The aromatization of 19-norandrostenedione and 19-nortestosterone is inhibited by carbon monoxide. On the other hand, aromatization of androstenedione is insensitive to carbon monoxide[145,146] and is not inhibited by the P-450 inhibitor metyrapone. It is, however, inhibited by the P-450 inhibitors SKF 525-A and aminoglutethimide and antibodies raised against hepatic NADPH-cytochrome P-450 reductase. Furthermore, androstenedione and 19-nortestosterone interact with placental microsomes to form a type I optical difference spectrum, a finding that is indicative of substrate binding to low-spin ferric P-450.[147] The magnitude of the type I difference spectrum induced by androstenedione is almost equal to that of the reduced carbon monoxide difference spectrum of P-450 in placental microsomes.[147] Therefore, it seems that substrates for aromatase are also substrates for a P-450 species of placental microsomes, and that a substantial proportion of the P-450 in placental microsomes is involved in reactions of these steroids. Other substrates for the aromatase enzyme complex, namely 19-hydroxyandrostenedione and 19-oxyandrostenedione, also form type I difference spectra with P-450.[147] Competition for binding to P-450 exists among these substrates and androstenedione. These results suggest that all aromatization reactions may be carried out at the same active site and that a single enzyme complex is responsible for aromatization of all steroids.

The question as to how aromatization of one substrate can be carbon monoxide sensitive, whereas that of another is carbon monoxide insensitive, can be answered by pointing out that CO sensitivity is an imperfect criterion of P-450 involvement in any particular reaction.[148] For instance, CO insensitivity can occur when the flow of reducing equivalents from NADPH to P-450 is limited, i.e., when formation of the ferrous P-450 substrate complex is rate-limiting for the overall reaction.[148] The rate of aromatization of C_{19}-steroids, a CO-insensitive reaction, is 10 to 20 times the rate of aromatization of C_{18} steroids, a CO-sensitive reaction.[147] In the case of aromatization of C_{19} substrates, the flow of reducing equivalents to P-450 may indeed be rate-limiting, a property that would render the overall reaction CO insensitive. Interestingly, when aromatization of androstenedione is inhibited by 19-norandrostenedione, CO sensitivity of aromatization of androstenedione develops[149]; moreover, this CO sensitivity is reversed by light of 450-nm wavelength.[149] Final proof for the involvement of P-450 in aromatization and indeed for the existence of only one species of P-450 responsible for catalyzing the total reaction has been obtained as a result of the purification of the aromatase enzyme complex from human placental microsomes. In early studies utilizing detergent solubilization followed by chromatography on Bio-Gel P-10 and DEAE-cellulose, the enzyme system was resolved into two components, both of which were required for activity.[150] One component contained P-450 and the other, NADPH-cytochrome P-450 reductase. These of course are expected components of a P-450 system present in the endoplasmic reticulum. More recently, the aromatase P-450 of human placental microsomes has been purified utilizing immunoabsorption chromatography by means of monoclonal antibodies prepared against this enzyme. This immunopurified P-450, when reconstituted with pig liver NADPH-cytochrome P-450 reductase and dilauroylphosphatidylcholine, exhibited aromatase activity.[151]

4.2. Inhibitors of Aromatase Activity

Because a high proportion of breast tumors are estrogen-dependent, numerous attempts have been made to find inhibitors of aromatase which are useful clinically. Removal of the ovaries has only been partially successful as therapy since estrogens are also produced in adipose tissue from adrenal androgens. Thus, for example, aminoglutethimide has found some usefulness in the treatment of breast cancer. Attempts have been made to synthesize irreversible or suicide inhibitors of aromatase.[152,153] One such compound is MDL 18962.[154] Such compounds may offer greater specificity and more prolonged pharmacological action than agents currently available for clinical use. Recently, it has been found that certain naturally occurring and synthetic flavones are potent aromatase inhibitors.[155]

4.3. Regulation of Aromatase Activity

In most cells where it occurs, aromatase activity appears to be stimulated by cAMP and its analogues. This pertains to ovarian granulosa cells,[156] adipose stromal cells,[157] and human choriocarcinoma cells.[158] In cells which have hormone receptors coupled to adenylate cyclase, it

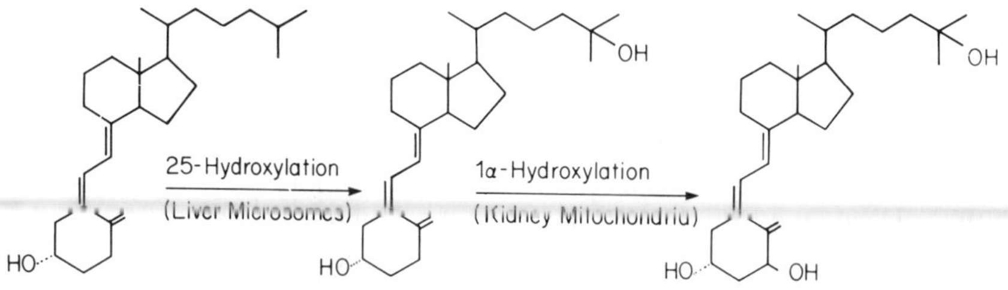

FIGURE 5. Cytochrome P-450-catalyzed hydroxylation reactions leading to production of active forms of vitamin D.

follows that such hormones can increase aromatase activity. This for example is true of FSH in ovarian granulosa cells[159] and testicular Sertoli cells,[137] of LH in Leydig cells,[138] and of ACTH in adipose stromal cells.[160] Aromatase activity in ovarian granulosa cells is regulated by a number of other factors which do not involve cAMP; for instance, it is stimulated by insulin and IGF I and inhibited by EGF.[161] While it is generally believed that aromatase activity is regulated primarily by changes in the synthesis of the enzyme, the mechanisms whereby this occurs are not understood at this time. However, with the recently reported availability of both polyclonal and monoclonal antibodies to this enzyme,[151] studies directed toward understanding the regulation of its synthesis should become feasible.

5. The Role of P-450 in Vitamin D Metabolism

The discovery that rickets and osteomalacia could be cured or prevented by ultraviolet light marked the beginning of the study of the metabolism of vitamin D and its biologically active metabolites. It is now recognized that the natural form of vitamin D_3, i.e., cholecalciferol, is produced in the skin by the action of ultraviolet light on the precursor, 7-dehydrocholesterol.[162] Cholecalciferol itself has little biological activity, since the active metabolite of cholecalciferol is 1,25-dihydroxycholecalciferol (Fig. 5) which is formed from cholecalciferol as a result of two hydroxylation reactions. As it is presently understood, all known oxidative metabolism of cholecalciferol must proceed through the 25-hydroxylated intermediate. This reaction occurs primarily in the liver although some hydroxylation can also be demonstrated in the intestine and kidney.[163] 25-Hydroxycholecalciferol is further hydroxylated at the 1α-position to produce 1α25-dihydroxycholecalciferol.[164,165] This reaction occurs in the kidney. In the nonpregnant animal, the kidney appears to be by far the most important site of 1α-hydroxylation. 25-Hydroxycholecalciferol-1α-hydroxylase may also be present in placental tissue[166] and consequently in addition to the kidney, the placenta is a potential site of 1α-hydroxylation.

5.1. Role of P-450 in the Formation of 1,25-Dihydroxycholecalciferol

Although not firmly established, is generally assumed that the site of 25-hydroxylation of cholecalciferol is the parenchymal cells of the liver. A specific cholecalciferol 25-hydroxylase is located in the endoplasmic reticulum of the liver.[167] This enzyme has been purified and shown to be a typical microsomal P-450 hydroxylation system. The reaction requires a specific form of microsomal P-450, NADPH-cytochrome P-450 reductase flavoprotein, as well as dilauroylphosphatidylcholine.[168] Upon reconstitution under these conditions, a specific activity of 2.3 nmole/min per mg protein has been achieved. Liver mitochondria also contain a 25-hydroxylase which will hydroxylate cholesterol as well as cholecalciferol. This system has been solubilized into three components: a species of P-450, an iron–sulfur protein, and a flavoprotein, which reduces the iron–sulfur protein utilizing NADPH.[169] The latter two components can be replaced with bovine adrenal adrenodoxin and adrenodoxin reductase. Since the K_m of this latter system for cholecalciferol is much greater than that of the microsomal system, the mitochondrial system may be responsible for converting higher concentrations of cholecalciferol to 25-hydroxycholecalciferol. This would explain the observation that the concentration of circulating 25-hydroxycholecalciferol is proportional to the concentration of vitamin D_3 administered, over a large range of concentration.[170] A function in the biosynthesis of bile acids has also been proposed for this mitochondrial enzyme.

In contrast to the microsomal 25-hydroxylase of the liver, the renal 1α-hydroxylase occurs in kidney mitochondria.[171,172] This system has been purified and shown to be a three-component system similar to the cholesterol side-chain cleavage and the 11β-hydroxylase systems of adrenal mitochondria. Thus, the 1α-hydroxylase has been shown to comprise a form of P-450, an iron–sulfur protein called renal ferridoxin or renodoxin, and a flavoprotein, NADPH-renodoxin reductase. The latter two enzymes can be replaced with adrenodoxin and adrenodoxin reductase in the reconstitued system. The specific activity of the purified preparation is 0.1 nmole/min per mg protein. Experiments designed to examine the sources of reducing equivalents for the mitochondrial 1α-hydroxylase have indicated that the NADPH required for the reaction is generated within the mitochondria via the oxidation of Krebs cycle intermediates, perhaps utilizing an energy-linked transhydrogenase.[173]

5.2. Regulation of Hydroxylases Involved in Cholecalciferol Metabolism

Most current evidence points to the fact that 25-hydroxylation is not a major site of regulation of cholecalciferol metabolism in the physiological sense, and the major site of regulation is at the level of the renal 1α-hydroxylase.[174] Lowering of blood calcium causes an increase in 1α-hydroxylation and, conversely, increasing blood calcium causes a suppression of hydroxylation. The responses are eliminated by parathyroidectomy[175] and it is now clear that the effects of blood calcium levels on 1α-hydroxylation are mediated primarily via parathyroid hormone (PTH), although a direct role of calcium to alter 1α-hydroxylation in renal cells cannot be ruled out. Specific binding sites for PTH in kidney have been demonstrated[176] and binding of the hormone to these sites causes an increase in cAMP formation. It may be, therefore, that the regulation of 1α-hydroxylase in kidney mitochondria by PTH is analogous to the regulation of cholesterol side-chain cleavage in the adrenal by ACTH.[177] Other hormones which have been claimed to have a role in the regulation of 1α-hydroxylation include prolactin, estrogens, and progesterone. The role for prolactin to stimulate 1α-hydroxylase activity has been reported[178]; however, this claim has not been substantiated[179] and therefore the role of this hormone remains unknown. *In vivo*, sex steroids, namely estrogen and progesterone, cause a dramatic increase in 1α-hydroxylase activity in chickens.[180] However, this cannot be demonstrated in cultures of renal cells[181] so it may be that this action of the sex steroid hormones is indirect. A major regulator of the 25-hydroxycholecalciferol 1α-hydroxylase appears to be 1,25-dihydroxycholecalciferol itself.[182] This compound suppresses 1α-hydroxylase both *in vivo*[182] and in kidney cells *in vitro*.[183] The mechanism appears to involve the translocation of a cytosolic receptor to the nucleus.[184]

5.3. Other Metabolites of Cholecalciferol

The pathway of metabolism of 25-hydroxycholecalciferol which has received greatest attention apart from 1α-hydroxylation is 24*R*-hydroxylation. 24*R*,25-dihydroxycholecalciferol is a major metabolite of vitamin D found in human and animal blood.[185] Although this metabolite is less active than 25-hydroxycholecalciferol in several bioassay systems, its presence in relatively high concentrations is suggestive of an as yet unsuspected role for this compound. 24-Hydroxylation occurs in a variety of tissues, but particularly in the kidney and intestine.[186] In the kidney the reaction occurs in the mitochondria and is catalyzed by a P-450-dependent system similar to that of the 1α-hydroxylase, although the P-450 responsible for 24-hydroxylation appears to differ from that responsible for 1α-hydroxylation.[187] Regulation of 24R-hydroxylation appears to occur in a reciprocal fashion to that of 1α-hydroxylation. Thus, under conditions of hypocalcemia where 1α-hydroxylation is increased, 24R-hydroxylation is suppressed. 1,25-Dihydroxycholecalciferol, which suppresses 1α-hydroxylase activity, causes an increase in 24R-hydroxylation.[188] Thus, the relative activities of 1α-hydroxylase and 24R-hydroxylase may be important determinants of the pattern of metabolism of 25-hydroxycholecalciferol and hence of the biological activities of the metabolites so formed. Other metabolites include 25-hydroxycholecalciferol-26,23-lactone[189] and 24-keto-1,25-dihydroxycholecalciferol.[190] The latter is found in intestinal mucosa of rats and has high affinity for the vitamin D cytosolic receptor. At present, the biological significance of the many pathways of metabolism of vitamin D is not clearly understood but this is an active area of investigation. Certainly, we can anticipate that in the near future the various forms of P-450 which carry out vitamin D hydroxylation will be further purified, antibodies raised against these enzymes, and regulation of their activities and synthesis studied in a fashion similar to that described for the steroid hydroxylases.

6. Role of P-450 in Lipid Hydroxylation Reactions

Lipids other than steroids serve as substrates for several different forms of P-450 which catalyze three distinct types of oxygenation reactions: ω- and ω-1-hydroxylation, epoxidation, and allylic oxidation.[191] ω-Oxidation of fatty acids contributes about 5% to the total oxidation of fatty acids in normal rat liver.[192] Rat liver cytochromes P-450 are able to catalyze the hydroxylation of fatty acids such as lauric acid (dodecanoic) in both ω- and ω-1-positions.[193–199] In addition, studies carried out using porcine and rat kidney cortex microsomes[200–202] suggest that cytochromes P-450 are involved in the ω-hydroxylation of lauric acid in these tissues as well. The same conclusion is reached by studies utilizing purified kidney enzymes in reconstituted assay systems. Such reactions have found renewed interest due to the pharmacological and toxicological responses associated with hypolipidemic drugs such as clofibrate [ethyl 2-(4-chlorophenoxy)-2-methyl propanoate], which is found to selectively increase rat liver microsomal 12- and 11-hydroxylations of lauric acid with no concomitant increase in the metabolism of pentobarbitone or benzo[a]pyrene[203-205] suggesting induction of specific forms of cytochrome P-450. This hypothesis is strengthened by the observation that no immunological cross-reactivity, nor similarity based on molecular weights or limited proteolytic analysis exists between clofibrate-, phenobarbital-, or β-naphthofluvone-inducible forms of P-450.[205]

Several studies have demonstrated that prostaglandins are metabolized by monooxygenases to ω-hydroxylated derivatives,[206-210] and thus the possibility exists that prostaglandins are the natural substrates for these forms of P-450 which may play a role in their elimination. An interesting example is prostaglandin ω-hydroxylase cytochrome P-450 (P-450$_{PG-\omega}$) which has been purified from lungs of pregnant rabbits.[211] The activity of this enzyme has been shown to increase 100-fold during late gestation[212] and it has been concluded that this enzyme is either absent or at low, undetectable levels in nonpregnant lung.[211] The mechanism by which the synthesis of P-450$_{PG-\omega}$ is regulated will prove very interesting as this appears to represent another example of physiological induction of P-450.

The metabolism of polyunsaturated fatty acids leading to olefin epoxide formation is also known to proceed in microsomes via P-450-catalyzed reactions.[213,214] While the biological implications of epoxyeicosatrienoic derivatives of arachidonic acid are only beginning to be explored, it can be anticipated that the regulation of both the levels and activities of the P-450 isozymes which catalyze these reactions in various tissues will prove to be of considerable importance.

Finally, P-450 has been implicated in allylic oxidation reactions of arachidonic acid in microsomes, leading to formation of hydroxyeicosatrienoic acids[215,216] via a lipoxygenase-like mechanism.[217-219] Such products have been shown to exhibit chemotactic activity and to be involved in inflammatory responses.[220] Thus, it appears that products of arachidonic acid generated via P-450-mediated reactions will prove to have potent and diverse biological functions. It is interesting to note that the major phenobarbital-inducible form of P-450 in liver has been found to possess this activity. This is an interesting example of a specific form of P-450 containing unrelated activities, one involved in xenobiotic metabolism and the other in metabolism of endogenous substrates.

These studies have demonstrated that cytochromes P-450 have the potential to take part in a variety of physiological functions as well as pharmacological responses and in this respect the regulation of their biosynthesis is an important issue. Characterization of distinct forms of P-450 induced by different xenobiotics[221] as well as constitutive isozymes[124,222] becomes critical in developing the tools necessary to study regulation.

7. Future Directions

As illustrated throughout this book, a large number of interesting forms of P-450 which metabolize endogenous substrates are known to exist in a variety of tissues. The mechanisms which regulate the levels and activities of these enzymes will prove important in our understanding of the regulation of a variety of different biological processes. In some instances, such as the steroidogenic systems described in this chapter, P-450 enzymes and their genes will prove to be important markers to be used in elucidation of the mechanism of peptide hormone action. In addition, as more of the forms of P-450 which metabolize endogenous compounds are purified, we can expect that the differences in mechanism of regulation between xenobiotic-inducible and physiologically inducible forms of P-450 will become apparent. Elucidation of gene structure encoding these enzymes will lead to a more detailed understanding of the evolutionary history of P-450, a field which is in its infancy. In some instances, the study of forms metabolizing endogenous substrates will also lead to a description of the molecular basis of human disease states. The construction of recombinant DNA molecules specific for these different forms of P-450 will lead to utilization of the techniques of site-directed mutagenesis in the examination of the mechanisms of their various hydroxylation reactions.

From our perspective, a particularly exciting prospect for the future involves the utilization of various forms of P-450 as tools to study a large and diverse group of physiological responses including hormone action and development. Thus, the fruits of the labors of those many investigators who have studied the varied aspects of the isozymes of P-450 and the reactions which they catalyze, will include the elucidation of the solutions to many fundamental questions in modern biology.

ACKNOWLEDGMENTS. We express our appreciation for the many contributions, both experimentally and intellectually, made by our colleagues over the past several years. In particular, we wish to acknowledge Drs. Ray DuBois, Robert Kramer, Bruria Funkenstein, Vijay Boggaram, Takashi Okamura, John McCarthy, Ian Mason, and Ronald Estabrook; as well as Mauricio Zuber, Bill Rainey, Marty Matocha, Albert Dee, Manorama John, and Leticia Cortez. Research support from the NIH (AM-28350 and HD-13234) and The Robert A. Welch Foundation (I-624) is also acknowledged.

References

1. Boobis, A. R., and Davies, D. S., 1984, Human cytochromes P-450, *Xenobiotica* **14**:151–185.

2. Rampersaud, A., and Walz, F. G., Jr., 1983, At least six forms of extremely homologous cytochromes P-450 in rat liver are encoded at two closely linked genetic loci, *Proc. Natl. Acad. Sci. USA* **80:**6542–6546.
3. Larrey, D., Distlerath, L. M., Dannan, G. A., Wilkinson, G. R., and Guengerich, F. P., 1984, Purification and characterization of the rat liver microsomal cytochrome P-450 involved in the 4-hydroxylation of debrisoquine, a prototype for genetic variation in oxidative drug metabolism, *Biochemistry* **23:**2787–2795.
4. Dieter, H. H., Muller-Eberhardt, U., and Johnson, E. F., 1982, Rabbit hepatic progesterone 21-hydroxylase exhibits a bimodal distribution of activity, *Science* **217:**741–743.
5. New, M. I., DuPont, B., Grumbach, K., and Levin, L. S., 1982, Congenital adrenal hyperplasia and related conditions, in: *The Metabolic Basis of Inherited Disease* (J. B. Stanbury, J. B. Wyngaarden, D. S. Fredrickson, J. L. Goldstein, and M. S. Brown, eds.), McGraw-Hall, New York, pp. 973–1000.
6. Kuhule, U., Chow, D., Rapaport, R., Pang, S., Levine, L. S., and New, M. I., 1981, The 21-hydroxylase activity in the glomerulosa and fasciculata of the adrenal cortex in congenital adrenal hyperplasia, *J. Clin. Endocrinol. Metab.* **52:**534–544.
7. Adesnick, M., Bar-Nun, S., Maschio, F., Zunich, M., Lippman, A., and Bard, E., 1981, Mechanism of induction of cytochrome P-450 by phenobarbital, *J. Biol. Chem.* **256:**10340–10345.
8. Bresnick, E., Brosseau, M., Levin, W., Reick, L., Ryan, D. E., and Thomas, P. E., 1981, Administration of 3-methylcholanthrene to rats increases the specific hybridizable mRNA coding for cytochrome P-450c, *Proc. Natl. Acad. Sci. USA* **78:**4083–4087.
9. Tuckey, R. H., Nebert, D. W., and Negishi, M., 1981, Structural gene product of the [Ah]complex: P_1-450 induction by use of a cloned DNA sequence, *J. Biol. Chem.* **256:**6969–6974.
10. Shikita, M., and Hall, P. F., 1974, The stoichiometry of the conversion of cholesterol and hydroxycholesterols to pregnenolone (3α-hydroxypregn-5-en-20-one) catalyzed by adrenal cytochrome P-450, *Proc. Natl. Acad. Sci. USA* **71:**1441–1445.
11. Burstein, S., Dinh, T., Co, N., Gut, M., Schleyer, H., Cooper, D. Y., and Rosenthal, O., 1972, Kinetic studies on substrate–enzyme interaction in the adrenal cholesterol side-chain cleavage system, *Biochemistry* **11:**2883–2891.
12. Burstein, S., Middleditch, B. S., and Gut, M., 1974, Enzymatic formation of (20R, 22R)-20,22-dihydroxy-cholesterol from cholesterol and a mixture of $^{16}O_2$ and $^{18}O_2$: Random incorporation of oxygen atoms, *Biochem. Biophys. Res. Commun.* **61:**642–647.
13. Hume, R., and Boyd, G. S., 1978, Cholesterol metabolism and steroid-hormone production, *Biochem. Soc. Trans.* **6:**893–898.
14. Takemoto, C., Nakano, H., Sato, H., and Tamaoki, B., 1968, Fate of molecular oxygen required by endocrine enzymes for the side-chain cleavage of cholesterol, *Biochim. Biophys. Acta* **152:**749–757.
15. Simpson, E. R., and Boyd, G. S., 1967, The cholesterol side-chain cleavage system of bovine adrenal cortex, *Eur. J. Biochem.* **2:**275–285.
16. Wilson, L. O., and Harding, B. W., 1970, Studies on adrenal cortical cytochrome P-450. III. Effect of carbon monoxide and light on steroid 11β-hydroxylation, *Biochemistry* **9:**1615–1620.
17. Simpson, E. R., and Boyd, G. S., 1967, Partial resolution of the mixed-function oxidase involved in the cholesterol side-chain cleavage reaction of bovine adrenal mitochondria, *Biochem. Biophys. Res. Commun.* **28:**945–950.
18. Bryson, M. J., and Sweat, M. L., 1968, Cleavage of cholesterol side chain associated with cytochrome P-450, flavoprotein, and nonheme iron-protein derived from the bovine adrenal cortex, *J. Biol. Chem.* **243:**2799–2804.

19. Shikita, M., and Hall, P. F., 1973, Cytochrome P-450 from bovine adrenocortical mitochondria: An enzyme for the side chain cleavage of cholesterol. I. Purification and properties, *J. Biol. Chem.* **248**:5598–5604.
20. Takemori, S., Suhara, K., Hashimoto, S., Hashimoto, M., Sato, H., Gomi, T., and Katagiri, M., 1975, Purification of cytochrome P-450 from bovine adrenocortical mitochondria by an "aniline-Sepharose" and the properties, *Biochem. Biophys. Res. Commun.* **63**:588–593.
21. Seybert, D. W., Lancaster, J. R., Jr., Lambeth, J. D., and Kamin, H., 1979, Participation of the membrane in the side chain cleavage of cholesterol, *J. Biol. Chem.* **254**:12088–12098.
22. Orme-Johnson, N. R., Light, D. R., White-Stevens, R. W., and Orme-Johnson, W. H., 1979, Steroid binding properties of beef adrenal cortical cytochrome P-450 which catalyzes the conversion of cholesterol into pregnenolone, *J. Biol. Chem.* **254**:2103–2111.
23. Hanukoglu, I., Spitsberg, V., Bumpus, J. A., Dus, K. M., and Jefcoate, C. R., 1981, Adrenal mitochondrial cytochrome P-450$_{scc}$, *J. Biol. Chem.* **256**:4321–4328.
24. Kido, T., and Kimura, T., 1981, Stimulation of cholesterol binding to steroid-free cytochrome P-450$_{scc}$ by poly (L-lysine), *J. Biol. Chem.* **256**:8561–8568.
25. Greenfield, N. J., Gerlimatos, B., Szwergold, B. S., Wolfson, A. J., Prasad, V. K., and Lieberman, S., 1981, Effects of phospholipid and detergent on the substrate specifity of adrenal cytochrome P-450$_{scc}$: Substrate binding and kinetics of cholesterol side chain oxidation, *J. Biol. Chem.* **256**:4407–4417.
26. Mitani, F., Shimizu, T., Urno, R., Ishimura, Y., Izumi, S., Komatsu, N., and Watanabe, K., 1982, Cytochrome P-450$_{11\beta}$ and P-450$_{scc}$ in adrenal cortex, *J. Histochem. Cytochem.* **30**:1066–1074.
27. Ichikawa, Y., and Hiwatashi, A., 1982, The role of the sugar regions of components of the cytochrome P-450-linked mixed-function oxidase (monooxygenase) system of bovine adrenocortical mitochondria, *Biochim. Biophys. Acta* **705**:82–91.
28. Chashchin, V. L., Vasilevsky, V. I., Shakumatov, V. M., and Akhrem, A. A., 1984, The domain structure of the cholesterol side chain cleavage cytochrome P-450 from bovine adrenocortical mitochondria, *Biochim. Biophys. Acta* **787**:27–38.
29. Lambeth, J. D., Kamin, H., and Seybert, D. W., 1980, Phosphatidylcholine vesicle reconstituted cytochrome P-450$_{scc}$: Role of the membrane in control of activity and spin state of the cytochrome, *J. Biol. Chem.* **255**:8282–8288.
30. Yago, N., Kobayashi, S., Sekiyama, S., Kurokawa, H., Iwai, V., Suzuki, I., and Ichii, S., 1970, Further studies on the submitochondrial localization of cholesterol side-chain cleaving enzyme system in hog adrenal cortex by sonic treatment, *J. Biochem.* **68**:775–783.
31. Kramer, R. E., DuBois, R. N., Simpson, E. R., Anderson, C. M., Kashiwagi, K., Lambeth, J. D., Jefcoate, C. R., and Waterman, M. R., 1982, Cell-free synthesis of precursor forms of mitochondrial steroid hydroxylase enzymes of the bovine adrenal cortex, *Arch. Biochem. Biophys.* **215**:478–485.
32. Omura, T., Sanders, E., Estabrook, R. W., Cooper, D. Y., and Rosenthal, O., 1966, Isolation from adrenal cortex of a nonheme iron protein and a flavoprotein functional as a triphosphopyridine nucleotide-cytochrome P-450 reductase, *Arch. Biochem. Biophys.* **117**:660–673.
33. Suhara, K., Ikeda, Y., Takemori, S., and Katagiri, M., 1972, The purification and properties of NADPH-adrenodoxin reductase from bovine adrenocortical mitochondria, *FEBS Lett.* **28**:45–47.
34. Suzuki, K., and Kimura, T., 1965, An iron protein as a component of steroid 11β-hydroxylase complex, *Biochem. Biophys. Res. Commun.* **19**:340–345.

35. Chu, J.-W., and Kimura, T., 1973, Studies on adrenal steroid hydroxylases: Complex formation of the hydroxylase components, *J. Biol. Chem.* **248**:5183–5187.
36. Lambeth, J. D., McCaslin, D. R., and Kamin, H., 1976, Adrenodoxin reductase: adrenodoxin complex: Catalytic and thermodynamic properties, *J. Biol. Chem.* **251**:7545–7550.
37. Lambeth, J. D., and Pember, S. O., 1983, Cytochrome P-450$_{scc}$–adrenodoxin complex reduction properties of the substrate-associated cytochrome and relation of the reduction states of heme and iron-sulfur centers to association of the proteins, *J. Biol. Chem.* **258**:5596–5602.
38. Tsai, R. L., Gunsalus, I. C., and Dus, K., 1971, Composition and structure of camphor hydroxylase components and homology between putidaredoxin and adrenodoxin, *Biochem. Biophys. Res. Commun.* **45**:1300–1306.
39. Mitani, F., 1979, Cytochrome P-450 in adrenocortical mitochondria, *Mol. Cell. Biochem.* **24**:21–43.
40. Suhara, K., Nakayama, K., Takikawa, O., and Katagiri, M., 1982, Two forms of adrenodoxin reductase from mitochondria of bovine adrenal cortex, *Eur. J. Biochem.* **125**:659–664.
41. Tanaka, M., Haniu, M., Yasunobu, K. T., and Kimura, T., 1973, The amino acid sequence of bovine adrenodoxin, *J. Biol. Chem.* **248**:1141–1157.
42. Palmer, G., Brintzinger, H., and Estabrook, R. W., 1967, Spectroscopic studies on spinach ferridoxin and adrenodoxin, *Biochemistry* **6**:1658–1664.
43. Kimura, T., Suzuki, K., Padmanabhan, R., Samejima, T., Terutani, O., and Ui, N., 1969, Studies on steroid hydroxylase: Molecular properties of adrenal iron-sulfur protein, *Biochemistry* **8**:4027–4031.
44. Padmanabhan, R., and Kimura, T., 1970, Studies on adrenal steroid hydroxylases: Extrinsic properties of the optical activity in adrenal iron-sulfur protein (adrenodoxin), *J. Biol. Chem.* **245**:2469–2475.
45. Sartre, M., Vignais, P. V., and Idelman, S., 1969, Distribution of the steroid 11β-hydroxylase and the cytochrome P-450 in membranes of beef adrenal cortex mitochondria, *FEBS Lett.* **5**:135–140.
46. Sweat, M. L., 1951, Enzymatic synthesis of 17-hydroxycorticosterone, *J. Am. Chem. Soc.* **73**:4056.
47. Wilson, L. D., Nelson, D. H., and Harding, B. W., 1965, A mitochondrial electron carrier involved in steroid hydroxylations, *Biochim. Biophys. Acta* **99**:391–393.
48. Alfano, J., Brownie, A. C., Orme-Johnson, W. H., and Beinert, H., 1973, Adrenal mitochondrial cytochrome P-450 and cholesterol side chain cleavage activity, *J. Biol. Chem.* **248**:7860–7864.
49. Simpson, E. R., and Williams-Smith, D. L., 1976, Electron paramagnetic resonance spectra of mitochondrial and microsomal cytochrome P-450 from the rat adrenal, *Biochim. Biophys. Acta* **449**:59–71.
50. Björkhem, I., and Karlmar, K.-E., 1975, 18-Hydroxylation of deoxycorticosterone by reconstituted systems from rat and bovine adrenals, *Eur. J. Biochem.* **51**:145–154.
51. Watanuki, M., Tilley, B. E., and Hall, P. R., 1977, Purification and properties of cytochrome P-450 (11β and 18-hydroxylase) from bovine adrenocortical mitochondria, *Biochim. Biophys. Acta* **483**:236–247.
52. Sato, H., Suhara, A. K., Itagaki, E., Takemori, S., and Katagiri, M., 1978, Properties of an adrenal cytochrome P-450 (P-450$_{11\beta}$) for the hydroxylations of corticosteroids, *Arch. Biochem. Biophys.* **190**:307–314.
53. Kim, C. Y., Sugiyama, T., Okamoto, M., and Yamano, T., 1982, Regulation of 18-hydroxycorticosterone formation in bovine adrenocortical mitochondria, *J. Steroid Biochem.* **18**:593–599.

54. Stachenko, J., and Giroud, C. J. P., 1959, Functional zonation of the adrenal cortex: Pathways of corticosteroid biogenesis, *Endocrinology* **64**:730–742.
55. Wada, A., Okamoto, M., Nonaka, Y., and Yamano, T., 1984, Aldosterone biosynthesis by a reconstituted cytochrome P-450$_{11\beta}$ system, *Biochem. Biophys. Res. Commun.* **119**:365–371.
56. Iyanagi, T., and Mason, H. S., 1973, Some properties of hepatic reduced nicotinamide adenine dinucleotide phosphate-cytochrome c reductase, *Biochemistry* **12**:2297–2308.
57. Yasukochi, Y., Peterson, J. A., and Masters, B.S.S., 1979, NADPH-cytochrome c (P-450) reductase, *J. Biol. Chem.* **254**:7097–7104.
58. Hiwatashi, A., and Ichikawa, Y., 1979, Physicochemical properties of reduced nicotinamide adenine dinucleotide phosphate-cytochrome P-450 reductase from bovine adrenocortical microsomes, *Biochim. Biophys. Acta.* **580**:44–63.
59. Estabrook, R. W., Cooper, D. Y., and Rosenthal, O., 1963, The light-reversible carbon monoxide inhibition of the steroid C-21 hydroxylase system of the adrenal cortex, *Biochem. Z.* **338**:741–755.
60. Kominami, S., Ochi, H., Kobayashi, Y., and Takemori, S., 1980, Studies on the steroid hydroxylation system in adrenal cortex microsomes, *J. Biol. Chem.* **255**:3386–3394.
61. Bumpus, J. A., and Dus, K. M., 1982, Bovine adrenocortical microsomal hemoproteins P-450$_{17\alpha}$ and P-450$_{C21}$, *J. Biol. Chem.* **257**:12696–12704.
62. Yuan, M., Nakajin, S., Haniu, M., Shinoda, M., Hall, P. F., and Shively, J. E., 1983, Steroid 21-hydroxylase (cytochrome P-450) from porcine adrenocortical microsomes: Microsequence analysis of cysteine containing peptides. *Biochemistry* **22**:143–149.
63. Nakajin, S., Shinoda, M., and Hall, P. F., 1983, Purification and properties of 17α-hydroxylase from microsomes of pig adrenal: A second C_{21} side-chain cleavage system, *Biochem. Biophys. Res. Commun.* **111**:512–517.
64. Nakajin, S., and Hall, P. F., 1981, Microsomal cytochrome P-450 from neonatal pig testis, *J. Biol. Chem.* **256**:3871–3876.
65. Nakajin, S., Hall, P. F., and Onoda, M., 1981, Testicular microsomal cytochrome P-450 for C_{21}-steroid side chain cleavage, *J. Biol. Chem.* **256**:6134–6139.
66. Kashiwagi, K., Dafeldecker, W. P., and Salhanick, H. A., 1980, Purification and characterization of mitochondrial cytochrome P-450 associated with cholesterol side chain cleavage from bovine corpus luteum, *J. Biol. Chem.* **255**:2606–2611.
67. Kashiwagi, K., Carraway, R. E., and Salhanick, H. A., 1982, Amino acid composition of cytochrome P-450$_{scc}$ from bovine corpus luteum, *Biochem. Biophys. Res. Commun.* **105**:110–116.
68. Tuckey, R. C., and Stevenson, P. M., 1984, Properties of bovine luteal cytochrome P-450$_{scc}$ incorporated into artificial phospholipid vesicles, *Int. J. Biochem.* **16**:497–503.
69. Tuckey, R. C., and Stevenson, P. M., 1984, Properties of ferredoxin reductase and ferrodoxin from the bovine corpus luteum, *Int. J. Biochem.* **16**:489–495.
70. Buckley, D. I., and Ramachandran, J., 1981, Characterization of corticotropin receptors on adrenocortical cells, *Proc. Natl. Acad. Sci. USA* **78**:7431–7435.
71. Sala, G. B., Hayashi, K., Catt, K. J., and Dufau, M. L., 1979, Adrenocorticotropin action in isolated adrenal cells, *J. Biol. Chem.* **254**:3861–3865.
72. Stone, D., and Hechter, O., 1954, Studies on ACTH action in perfused bovine adrenals: The site of action of ACTH in corticosteroidogenesis, *Arch. Biochem. Biophys.* **51**:457–469.
73. Doering, C. H., and Clayton, R. B., 1969, Cholesterol side-chain cleavage activity in the adrenal gland of the young rat: Development and responsiveness to adrenocorticotropic hormone, *Endocrinology* **85**:500–511.
74. Kimura, T., 1969, Effects of hypophysectomy and ACTH administration on the level of adrenal cholesterol side-chain desmolase, *Endocrinology* **85**:492–499.

75. Pfeiffer, D. R., Chu, J. W., Kuo, T. H., Chan, S. W., Kimura, T., and Tchen, T. T., 1972, Changes in some biochemical parameters including cytochrome P-450 after hypophysectomy and their restoration by ACTH administration in rats four months post hypophysectomy. *Biochem. Biophys. Res. Commun.* **48**:486–490.
76. Purvis, J. L., Canick, J. A., Mason, J. I., Estabrook, R. W., and McCarthy, J. L., 1973, Lifetime of adrenal cytochrome P-450 as influenced by ACTH, *Ann. N.Y. Acad. Sci.* **212**:319–342.
77. Simpson, E. R., and Waterman, M. R., 1983, Regulation by ACTH of steroid hormone biosynthesis in the adrenal cortex, *Can. J. Biochem. Cell Biol.* **61**:692–707.
78. Gospodarowicz, D., Ill, C. R., Hornsby, P. J., and Gill, G. N., 1977, Control of bovine adrenal cortical cell proliferation by fibroblast growth factor: Lack of effect of epidermal growth factor, *Endocrinology* **100**:1080–1089.
79. Kowal, J., 1969, Adrenal cells in tissue culture. III. Effect of adrenocorticotropin and 3′,5′-cyclic adenosine monophosphate on 11β-hydroxylase and other steroidogenic enzymes, *Biochemistry* **8**:1821–1831.
80. Kowal, J., Simpson, E. R., and Estabrook, R. W., 1970, Adrenal cells in tissue culture: On the specificity of the stimulation of 11β-hydroxylation by adrenocorticotropin, *J. Biol. Chem.* **245**:2438–2443.
81. Asano, K., and Harding, B. W., 1976, Biosynthesis of adrenodoxin in mouse adrenal tumor cells, *Endocrinology* **99**:977–987.
82. Waterman, M. R., 1982, ACTH-mediated induction of synthesis and activity of cytochrome P-450s and related enzymes in cultured bovine adrenocortical cells, *Xenobiotica* **12**:773–786.
83. DuBois, R. N., Simpson, E. R., Kramer, R. E., and Waterman, M. R., 1981, Induction of synthesis of cholesterol side chain cleavage cytochrome P-450 by ACTH in cultured bovine adrenocortical cells, *J. Biol. Chem.* **256**:7000–7005.
84. Kramer, R. E., Anderson, C. M., Peterson, J. A., Simpson, E. R., and Waterman, M. R., 1982, Adrenodoxin biosynthesis by bovine adrenal cells in monolayer culture: Induction by ACTH, *J. Biol. Chem.* **257**:14921–14925.
85. Kramer, R. E., Anderson, C. M., McCarthy, J. L., Simpson, E. R., and Waterman, M. R., 1982, Coordinate induction of synthesis of adrenal mitochondrial steroid hydroxylases by ACTH, *Fed. Proc.* **41**:1298.
86. Kramer, R. E., Simpson, E. R., and Waterman, M. R., 1983, Induction of 11β-hydroxylase by corticotropin in primary cultures of bovine adrenocortical cells, *J. Biol. Chem.* **258**:3000–3005.
87. Kramer, R. E., Rainey, W. E., Funkenstein, B., Dee, A., Simpson, E. R., and Waterman, M. R., 1984, Induction of synthesis of mitochondrial steroidogenic enzymes of bovine adrenocortical cells by analogs of cyclic AMP, *J. Biol. Chem.* **259**:707–713.
88. Boggaram, V., Simpson, E. R., and Waterman, M. R., 1984, Induction of synthesis of bovine adrenocortical cytochromes P-450$_{scc}$, P-450$_{11\beta}$, P-450$_{C21}$ and adrenodoxin by prostaglandins E$_2$ and F$_{2\alpha}$ and cholera toxin, *Arch. Biochem. Biophys.* **231**:271–279.
89. Boggaram, V., Zuber, M. X., and Waterman, M. R., 1984, Turnover of newly synthesized cytochromes P-450$_{scc}$ and P-450$_{11\beta}$ and adrenodoxin in bovine adrenocortical cells in monolayer culture: Effect of adrenocorticotropin, *Arch. Biochem. Biophys.* **231**:518–523.
90. Hornsby, P. J., 1980, Regulation of cytochrome P-450 supported 11β-hydroxylation of deoxycortisol by steroids, oxygen and antioxidants in adrenocortical cell cultures, *J. Biol. Chem.* **255**:4020–4027.
91. Rani, S., Keri, G., and Ramachandran, J., 1983, Studies on corticotropin-induced desensitization of normal rate adrenocortical cells, *Endocrinology* **112**:315–320.
92. DuBois, R. N., Simpson, E. R., Tuckey, J., Lambeth, J. D., and Waterman, M. R.,

1981, Evidence for a higher molecular weight precursor of cholesterol side chain cleavage cytochrome P-450 and induction of mitochondrial and cytosolic proteins by ACTH in adult bovine adrenal cells, *Proc. Natl. Acad. Sci. USA* **78:**1028–1032.
93. Nabi, H., and Omura, T., 1980, In vitro synthesis of adrenodoxin and adrenodoxin reductase: Existence of a putative large precursor form of adrenodoxin, *Biochem. Biophys. Res. Commun.* **97:**680–686.
94. Nabi, N., Kominami, S., Takemori, S., and Omura, T., 1980, In vitro synthesis of mitochondrial cytochrome P-450 (scc) and P-450 (11β) and microsomal cytochrome P-450 (C-21) by both free and bound polysomes isolated from bovine adrenal cortex, *Biochem. Biophys, Res. Commun.* **97:**687–693.
95. Harmey, M. A., Hallermayer, G., Korb, H., and Neupert, W., 1977, Transport of cytoplasmically synthesized proteins into the mitochondria in a cell free system from *Neurospora crassa*, *Eur. J. Biochem* **81:**533–544.
96. Matocha, M. F., and Waterman, M. R., 1984, Discriminatory processing of the precursor forms of cytochrome P-450$_{scc}$ and adrenodoxin by adrenocortical and heart mitochondria, *J. Biol. Chem.* **259:**8672–8678.
97. McAda, P. C., and Douglas, M. G., 1982, A neutral endopeptidase involved in the processing of an ATPase subunit precursor in mitochondria, *J. Biol. Chem.* **257:**3177–3182.
98. Omura, T., Ito, A., Okada, Y., Sagara, Y., and Ono, H., 1983, Proteolytic processing of enzyme precursors by liver and adrenal cortex mitochondria, in: *Protease Inhibitors: Medical and Biological Aspects* (N. Katunuma, H. Umezawa, and H. Holzer, eds.), Springer-Verlag, Berlin, pp. 307–312.
99. Funkenstein, B., McCarthy, J. L., Dus, K. M., Simpson, E. R., and Waterman, M. R., 1983, Effect of adrenocorticotropin on steroid 21-hydroxylase synthesis and activity in cultured bovine adrenocortical cells, *J. Biol. Chem.* **258:**9398–9405.
100. DuBois, R. N., and Waterman, M. R., 1979, Effect of phenobarbital administration to rats on the level of the in vitro synthesis of cytochrome P-450 directed by total rat liver RNA, *Biochem. Biophys. Res. Commun.* **90:**150–157.
101. Colbert, R. A., Bresnick, E., Levin, W., Ryan, D. E., and Thomas, P. E., 1979, Synthesis of liver cytochrome P-450b in a cell free protein synthesizing system, *Biochem. Biophys. Res. Commun.* **91:**886–891.
102. Bar-Nun, S., Kreibach, G., Adesnick, M., Alterman, L., Negishi, M., and Sabatini, D. D., 1980, Synthesis and insertion of cytochrome P-450 into endoplasmic reticulum membranes. *Proc. Natl. Acad, Sci USA* **77:**965–969.
103. Fevold, H. R., and Brown, R. L., 1978, The apparent lack of stimulation of rabbit adrenal 21-hydroxylase activity by ACTH, *J. Steroid Biochem.* **9:**583–584.
104. Hornsby, P. J., 1982, Regulation of 21-hydroxylase activity by steroids in cultured bovine adrenocortical cells: Possible significance for adrenocortical androgen synthesis, *Endocrinology* **111:**1092–1101.
105. Kramer, R. E., McCarthy, J. L., Simpson, E. R., and Waterman, M. R., 1983, Effects of ACTH on steroidogenesis in bovine adrenocortical cells in primary culture: Increased secretion of 17α-hydroxylated steroids associated with a refractoriness in total steroid output, *J. Steroid Biochem.* **18:**715–723.
106. DiBartolomeis, M. J., and Jefcoate, C. R., 1984, Characterization of the acute stimulation of steroidogenesis in primary bovine adrenal cortical cell cultures, *J. Biol. Chem.* **259:**10159–10167.
107. McCarthy, J. L., Kramer, R. E., Funkenstein, B., Simpson, E. R., and Waterman, M. R., 1983, Induction of 17α-hydroxylase (cytochrome P-450$_{17\alpha}$) activity by adrenocorticotropin in bovine adrenocortical cells maintained in monolayer culture, *Arch Biochem. Biophys.* **222:**590–598.

108. Zuber, M. X., Simpson, E. R., Hall, P. F., and Waterman, M. R., 1985, Effects of adrenocorticotropin on 17α-hydroxylase activity and cytochrome p450$_{17\alpha}$ synthesis in bovine adrenocortical cells, *J. Biol. Chem.* **260**:1842–1848.
109. Fevold, H. R., Wilson, P. L., and Slalina, S. M., 1978, Stimulated rabbit adrenal 17α-hydroxylase: Kinetic properties and a comparison with those of 3β-hydroxysteroid dehydrogenase, *J. Steroid Biochem.* **9**:1033–1041.
110. Schiebinger, R. J., Albertson, B. D., Cassorria, F. G., Bowyer, D. W., Gellhoed, G. W., Cutler, G. B., Jr., and Loriaux, D. L., 1981, The developmental changes in plasma adrenal androgens during infancy and adrenarche are associated with changing activities of adrenal microsomal 17-hydroxylase and 17,20-desmolase, *J. Clin. Invest.* **67**:1177–1182.
111. Katagiri, M., Suhara, K., Shiroo, M., and Fujimura, Y., 1982, Role of cytochrome b$_5$ in the cytochrome P-450-mediated C$_{21}$-steroid 17,20-lyase reaction, *Biochem. Biophys. Res. Commun.* **108**:379–384.
112. Onoda, M., and Hall, P. F., 1982, Cytochrome b$_5$ stimulates purified testicular microsomal P-450 (C$_{21}$ side-chain cleavage), *Biochem. Biophys. Res. Commun.* **108**:454–460.
113. Nabi, N., Ishikawa, T., Ohashi, M., and Omura, T., 1983, Contributions of cytoplasmic free and membrane-bound ribosomes to the synthesis of NADPH-adrenodoxin reductase and adrenodoxin of bovine adrenal cortex mitochondria, *J. Biochem.* **94**:505–515.
114. Nabi, N., Kominami, S., Takemori, S., and Omura, T., 1983, Contributions of cytoplasmic free and membrane-bound ribosomes to the synthesis of mitochondria cytochrome P-450(scc) and P-450(11β) and microsomal cytochrome P-450(C-21) in bovine adrenal cortex, *J. Biochem.* **94**:517–527.
115. Matteson, K. J., Chung, B., and Miller, W. L., 1984, Molecular cloning of DNA complementary to bovine adrenal P-450$_{scc}$ mRNA, *Biochem. Biophys. Res. Commun.* **120**:264–270.
116. Morahashi, K., Fujii-Kuriyama, Y., Okada, Y., Sogawa, Y., Hirose, T., Inayama, S., and Omura, T., 1984, Molecular cloning and nucleotide sequence of cDNA for mRNA of mitochondrial cytochrome P-450 (scc) of bovine adrenal cortex, *Proc. Natl. Acad. Sci. USA* **81**:4647–4651.
117. John, M. E., John, M. C., Ashley, P., MacDonald, R. J., Simpson, E. R., and Waterman, M. R., 1984, Identification and characterization of cDNA clones specific for cholesterol side-chain cleavage cytochrome P-450, *Proc. Natl. Acad. Sci. USA* **81**:5628–5632.
118. White, P. C., New, M. I., and DuPont, B., 1984, Cloning and expression of cDNA encoding a bovine adrenal cytochrome P-450 specific for steroid 21-hydroxylation, *Proc. Natl. Acad. Sci. USA* **81**:1986–1990.
119. White, P. C., New, M. I., and DuPont, B., 1984, HLA-linked congenital adrenal hyperplasia results from a defective gene encoding a cytochrome P-450 specific for steroid 21-hydroxylation, *Proc. Natl. Acad. Sci. USA* **81**:7505–7509.
120. John, M. E., John, M. C., Simpson, E. R., and Waterman, M. R., 1985, Cytochrome P-450$_{11\beta}$ gene expression: Transcriptional control by an ACTH regulated protein factor, *J. Biol. Chem.* **260**:5760–5767.
121. Negishi, M., and Nebert, D. W., 1981, Structural gene products of the *Ah* complex: Increases in large mRNA from mouse liver associated with cytochrome P$_1$-450 induction by 3-methylcholanthrene, *J. Biol. Chem.* **256**:3085–3091.
122. Omiencinski, C. J., Hines, R. N., Foldes, R. L., Levy, J. B., and Bresnick, E., 1983, Molecular induction by phenobarbital of a rat hepatic form of cytochrome P-450: Expression of a 4-kilobase messenger RNA, *Arch. Biochem. Biophys.* **227**:478–493.
123. Gonzales, F. J., and Kasper, C. B., 1982, Cloning of DNA complementary to rat liver NADPH cytochrome c (P-450) oxidoreductase and cytochrome P-450b mRNAs, *J. Biol. Chem.* **257**:5962–5968.

124. Schenkman, J. B., Jansson, I., Backes, W. L., Chen, K.-C., and Smith, C., 1982, Dissection of cytochrome P-450 isozymes (RLM): Four fractions of untreated rat liver microsomal proteins, *Biochem. Biophys. Res. Commun.* **107**:1517–1523.
125. Zuber, M. X., Simpson, E. R., and Waterman, M. R., 1985, Regulation of cytochrome P-450$_{17\alpha}$ activity and synthesis in bovine adrenocortical cells, *Ann. N.Y. Acad. Sci.* in press.
126. Waterman, M. R., John, M. E., Boggaram, V., Zuber, M. X., and Simpson, E. R., 1984, Mechanisms of induction of endogenous substrate oxidation, in: *Microsomes and Drug Oxidations* (A. R. Boobis, J. Caldwell, F. deMatteis, and C. R. Elcombe, eds.), Taylor & Francis, London, pp. 136–144.
127. Israel, D. I., and Whitlock, J. P., Jr., 1983, The Induction of mRNA specific for cytochrome P$_1$-450 in wild type and variant mouse hepatoma cells, *J. Biol. Chem.* **258**:10390–10394.
128. Anderson, C. M., and Mendelson, C. R., 1985, Regulation of steroidogenesis in rat Leydig cells in culture: Effect of human chorionic gonadtropin and dibutyrylcyclic AMP on the synthesis of cholesterol side chain cleavage cytochrome P-450 and adrenodoxin, *Arch. Biochem. Biophys.* **238**:378–387.
129. Mason, J. I., MacDonald, A. A., and Laptook, A., 1984, The activity and biosynthesis of cholesterol side-chain cleavage enzyme in cultured immature pig testis cells, *Biochim. Biophys. Acta* **795**:504–512.
130. Purvis, J. L., Canick, J. A., Latie, S. A., Rosenbaum, J. H., Hologgitas, J., and Menard, R. H., 1973, Life time of microsomal cytochrome P-450 and steroidogenic enzymes in rat testis influenced by human chorionic gonadotrophin, *Arch. Biochem. Biophys.* **159**:39–49.
131. Malaska, T., and Payne, A. H., 1984, Luteinizing hormone and cyclic AMP-mediated induction of microsomal cytochrome P-450 enzymes in cultured mouse Leydig cells, *J. Biol. Chem.* **259**:11654–11657.
132. Funkenstein, B., Waterman, M. R., and Simpson, E. R., 1984, Induction of synthesis of cholesterol side chain cleavage cytochrome P-450 and adrenodoxin by follicle-stimulating hormone, 8-bromo-cyclic AMP and low density lipoprotein in cultured bovine granulosa cells, *J. Biol. Chem.* **259**:8572–8577.
133. Funkenstein, B., Waterman, M. R., Masters, B.S.S., and Simpson, E. R., 1983, Evidence for the presence of cholesterol side chain cleavage cytochrome P-450 and adrenodoxin in granulosa cells, *J. Biol. Chem.* **258**:10187–10191.
134. Goto, J., and Fishman, J., 1977, Participation of a nonenzymatic transformation in the biosynthesis of estrogen from androgens, *Science* **195**:80–81.
135. Thompson, E. A., Jr., and Siiteri, P. K., 1974, Utilization of oxygen and reduced nicotinamide adenine dinucleotide phosphate by human placental microsomes during aromatization of androstenedione, *J. Biol. Chem.* **249**:5364–5372.
136. Schindler, A. E., Ebert, A., and Friedrich, E., 1972, Conversion of androstenedione to estrone by human fat tissue, *J. Clin. Endocrinol. Metab.* **35**:627–630.
137. Dorrington, J. H., and Armstrong, D. T., 1975, Follicle-stimulating hormone stimulates estradiol-17β synthesis in cultured Sertoli cells, *Proc. Natl. Acad. Sci. USA* **72**:2677–2681.
138. Valladares, L. E., and Payne, A. H., 1979, Induction of testicular aromatization by luteinizing hormone in mature rat, *Endocrinology* **105**:431–436.
139. Naftolin, F., Ryan, K. J., Davis, I. J., Reddy, V. V., Flores, F., Petro, Z., and Kuhn, M., 1975, The formation of estrogens by central neuroendocrine tissues, *Recent Prog. Horm. Res.* **31**:295–315.

140. Ryan, K. J., 1959, Metabolism of C-16-oxygenated steroids by human placenta: The formation of estriol, *J. Biol. Chem.* **234**:2006–2008.
141. Ryan, K. J., 1959, Biological aromatization of steroids, *J. Biol. Chem.* **234**:268–272.
142. Allen, F. A., Valdivia, E., and Colas, A. E., 1970, The aromatizing activity of placental mitochondrial and microsomal fractions, *Gynecol. Invest.* **1**:277–287.
143. Moorthy, K. B., and Meigs, R. A., 1978, Aromatization of steroids by mitochondrial preparations from human term placenta, *Biochim. Biophys. Acta* **528**:222–229.
144. Canick, J. A., and Ryan, K. J., 1978, Properties of the aromatase enzyme system associated with the mitochondrial fractions of human placenta, *Steroids* **32**:499–509.
145. Canick, J. A., and Ryan, K. J., 1976, Cytochrome P-450 and the aromatization of 16α-hydroxytestosterone and androstenedione by human placental microsomes, *Mol. Cell. Endocrinol.* **6**:105–115.
146. Meigs, R. A., and Ryan, K. J., 1968, Cytochrome P-450 and steroid biosynthesis in the human placenta, *Biochim. Biophys. Acta* **165**:476–482.
147. Thompson, E. A., Jr., and Siiteri, P. K., 1974, The involvement of human placental microsomal cytochrome P-450 in aromatization, *J. Biol. Chem.* **249**:5373–5378.
148. Estabrook, R. W., Franklin, M. R., and Hildebrandt, A. G., 1970, Factors influencing the inhibitory effect of carbon monoxide on cytochrome P-450-catalyzed mixed function oxidase reactions, *Ann. N.Y. Acad. Sci.* **174**:218–232.
149. Zachariah, P. K., and Judan, M. R., 1977, Inhibition of human placental mixed function oxidation with carbon monoxide: Reversal with monochromatic light, *J. Steroid Biochem.* **8**:221–228.
150. Thompson, E. A., Jr., and Siiteri, P. K., 1976, Partial resolution of the placental microsomal aromatase complex, *J. Steroid Biochem.* **7**:635–639.
151. Simpson, E. R., and Mendelson, C. R., 1984, Preparation and characterization of polyclonal antibodies against aromatase cytochrome P-450 in *Abstracts, 7th International Congress of Endocrinology*, Quebec City, p. 1450.
152. Covey, D. F., Hood, W. F., and Parikh, V. D., 1981, 10β-Propyryl-substituted steroids: Mechanism-based enzyme-activated irreversible inhibitors of estrogen biosynthesis, *J. Biol. Chem.* **256**:1076–1080.
153. Marette, P. A., and Robinson, C. H., 1982, Inhibition and inactivation of estrogen synthetase (aromatase) by fluorinated substrate analogues, *Biochemistry* **21**:2733–2738.
154. Johnston, J. O., Wright, C. L., and Metcalf, B. W., 1984, Biochemical and endocrine properties of a mechanism-based inhibitor of aromatase, *Endocrinology* **115**:776–785.
155. Kellis, J. T., Jr., and Vickery, L. E., 1984, Inhibition of human estrogen synthetase (aromatase) by flavones, *Science* **255**:1032–1034.
156. Fritz, I. B., Griswold, M. D., Louis, B. G., and Dorrington, J. H., 1976, Similarity of responses of cultured Sertoli cells to cholera toxin and FSH, *Mol. Cell. Endocrinol.* **5**:289–294.
157. Mendelson, C. R., Cleland, W. H., Smith, M. E., and Simpson, E. R., 1982, Regulation of aromatase activity of stromal cells from human adipose tissue, *Endocrinology* **111**:1077–1085.
158. Bellino, F. L., and Hussa, R. O., 1982, Estrogen synthetase stimulation by dibutyryl cyclic AMP in human choriocarcinoma cells in culture, *Endocrinology* **111**:1038–1044.
159. Dorrington, J. H., Moon, Y. S., and Armstrong, D. R., 1975, Estradiol-17β-biosynthesis in cultured granulosa cells from hypophysectomized immature rats: Stimulation by follicle-stimulating hormone, *Endocrinology* **97**:1328–1331.
160. Mendelson, C. R., Smith, M. E., Cleland, W. H. and Simpson, E. R., 1984, Regulation of aromatase activity of adipose stromal cells by catecholamines and adrenocorticotropin, *Mol. Cell. Endocrinol.* **37**:61–72.
161. Hsueh, A. J. W., Jones, P. B. C., Adashi, E. Y., Wang, C., Zhuang, L. Z., and Welsh,

T. H., Jr., 1983, Intraovarian mechanisms in the hormonal control of granulosa cell differentiation in rats, *J. Reprod. Fertil.* **69**:325–342.
162. Windaus, A., Schenk, F., and vonWelder, F., 1936, Über das antirachitisch wirksame Bestrahlungs-Produkt ans 7-Dehydrocholesterin, *Hoppe-Seylers Z. Physiol. Chem.* **241**:100–103.
163. Olson, E. B., Jr., Knutson, J. C., Bhattacharyya, M. H., and DeLuca, H. F., 1976, The effect of hepatectomy on the synthesis of 25-hydroxyvitamin D_3, *J. Clin. Invest.* **57**:1213–1220.
164. Frazer, D. R., and Kodicek, E., 1970, Unique biosynthesis by kidney of a biologically active metabolite, *Nature* **228**:764–766.
165. Gray, R., Boyle, I., and DeLuca, H. F., 1971, Vitamin D metabolism: The role of kidney tissue, *Science* **172**:1232–1234.
166. Tanaka, Y., Halloran, B., Schnoes, H. K., and DeLuca, H. F., 1979, *In vitro* production of 1,25-dihydroxyvitamin D_3 by rat placental tissue, *Proc. Natl. Acad. Sci. USA* **76**:5033–5035.
167. Madhok, T. C., and DeLuca, H. F., 1979, Characteristics of the rat liver microsomal enzyme system converting cholecalciferol into 25-hydroxycholecalciferol: Evidence for the participation of cytochrome P-450, *Biochem. J.* **184**:491–499.
168. Hayashi, S.-I., Nashiro, M., and Okuda, K., 1984, Purification of cytochrome P-450 catalyzing 25-hydroxylation of vitamin D_3 from rat liver microsomes, *Biochem. Biophys. Res. Commun.* **121**:994–1000.
169. Björkhem, J., Holmberg, I., Oftebro, H., and Pederson, J. I., 1980, Properties of a reconstituted vitamin D_3 25-hydroxylase from rat liver mitochondria, *J. Biol. Chem.* **255**:5244–5249.
170. Holick, M. F., and Clark, M. B., 1978, The photobiogenesis and metabolism of vitamin D, *Fed. Proc.* **37**:2567–2574.
171. Ghazarian, J. G., Jefcoate, C. R. Knutson, J. C., Orme-Johnson, W. H., and DeLuca, H. F., 1974, Mitochondrial cytochrome P-450: A component of chick kidney 25-hydroxycholecalciferol-1α-hydroxylase, *J. Biol. Chem.* **249**:3028–3033.
172. Hiwatashi, A., Nishii, Y., and Ichikawa, Y., 1982, Purification of cytochrome P-450$_{D1\alpha}$ (25-hydroxyvitamin D_3-1α-hydroxylase) of bovine kidney mitochondria, *Biochem. Biophys. Res. Commun.* **105**:320–327.
173. Ghazarian, J. G., and DeLuca, H. F., 1974, 25-Hydroxylase: A specific requirement for NADPH and a hemoprotein component in kidney mitochondria, *Arch. Biochem. Biophys.* **160**:63–72.
174. Omdahl, J. L., Gray, R. W., Boyle, I. T., Knutson, J., and DeLuca, H. F., 1972, Regulation and metabolism of 25-hydroxycholecalciferol by kidney tissue, *Nature New Biol.* **237**:63–64.
175. Garabedian, M., Holick, M. F., DeLuca, H. F., and Boyle, I. T., 1972, Control of 25-hydroxycholecalciferol metabolism by the parathyroid gland, *Proc. Natl. Acad. Sci. USA* **69**:1673–1675.
176. Zull, J. E., and Repke, D. W., 1972, The tissue localization of tritiated parathyroid hormone in thyroparathyroidectomized rats, *J. Biol. Chem.* **247**:2195–2199.
177. Ghazarian, J. G., 1979, Kidney cyclic nucleotide-mediated hydroxylations of 25-hydroxyvitamin D_3, in: *Molecular Endocrinology* (1. MacIntyre and M. Szelke, eds.), Elsevier/North-Holland, Amsterdam, pp. 333–340.
178. Spanos, E., Brown, D. J., Stevenson, J. C., and MacIntyre, I., 1981, Stimulation of 1,25-dihydroxycholecalciferol production by prolactin and related peptides in intact renal cell preparations *in vitro*, *Biochim. Biophys. Acta.* **672**:7–15.
179. Matsumoto, T., Horiuchi, N., Suda, T., Takahashi, H., Shunazawa, E., and Ogata, E., 1979, Failure to demonstrate stimulatory effect of prolactin on vitamin D metabolism in vitamin D-deficient rats, *Metabolism* **28**:925–927.

180. Castillo, L., Tanaka, Y., Wineland, M. J., Jowsey, J. O., and DeLuca, H. F., 1979, Production of 1,25-dihydroxyvitamin D_3 and formation of medullary bone in the egg-laying hen, *Endocrinology* **104**:1598–1601.
181. Trechsel, U., Bonjous, J. P., and Fleisch, H., 1979, Regulation of the metabolism of 25-hydroxyvitamin D_3 in primary cultures of chick kidney cells, *J. Clin. Invest.* **64**:206–217.
182. Tanaka, Y., Lorenc, R. S., and DeLuca, H. F., 1975, The role of 1,25-dihydroxyvitamin D_3 and parathyroid hormone in the regulation of chick 25-hydroxyvitamin D_3-24-hydroxylase, *Arch. Biochem. Biophys.* **171**:521–526.
183. Henry, H. L., 1979, Regulation of the hydroxylation of 25-hydroxyvitamin D_3 *in vivo* and in primary cultures of chick kidney cells, *J. Biol. Chem.* **254**:2722–2729.
184. Simpson, R. N., Franceschi, R. T., and DeLuca, H. F., 1980, Characterization of a specific, high affinity binding macromolecule for 1α,25-dihydroxyvitamin D_3 in cultured chick kidney cells, *J. Biol. Chem.* **255**:10160–10166.
185. Gray, R. W., Weber, H. P., Dominguez, J. H., and Lemann, J., Jr., 1974, The metabolism of vitamin D_3 and 25-hydroxyvitamin D_3 in normal and anephric humans, *J. Clin. Endocrinol. Metab.* **39**:1045–1056.
186. Knutson, J. C., and DeLuca, H. F., 1974, 25-Hydroxyvitamin D_3-24-hydroxylase: Cellular localization and properties, *Biochemistry* **13**:1543–1547.
187. Kulkowski, J. A., Chan, T., Martinez, J., and Ghazarian, J. G., 1979, Modulation of 25-hydroxyvitamin D_3-24-hydroxylase by aminophylline: A cytochrome P-450 monooxygenase system, *Biochem. Biophys. Res. Commun.* **90**:50–57.
188. Tanaka, Y., Lorenc, R. S., and DeLuca, H. F., 1975, The role of 1,25-dihydroxyvitamin D_3 and parathyroid hormone in the regulation of chick renal 25-hydroxyvitamin D_3-24-hydroxylase, *Arch. Biochem. Biophys.* **171**:521–526.
189. Wichmann, J. K., DeLuca, H. F., Schnoes, H. F., Horst, R. L., Shepard, R. M., and Jorgensen, N. A., 1979, 25-Hydroxyvitamin D_3 26,23-lactone: A new *in vivo* metabolite of vitamin D, *Biochemistry* **18**:4775–4780.
190. Napoli, J. L., Pramanik, B. C., Royal, P. M., Reinhardt, T. A., and Horst, R. L., 1983, Intestinal synthesis of 24-keto-1,25-dihydroxyvitamin D_3: A metabolite formed *in vivo* with high affinity for the vitamin D cytosolic receptor, *J. Biol. Chem.* **258**:9100–9107.
191. Capdevila, J., Saeki, Y., and Falck, J. R., 1984, The mechanistic plurality of cytochrome P-450 and its biological ramifications, *Xenobiotica* **14**:105–118.
192. Antony, G. J., and Landau, B. R., 1968, Relative contributions of α-, β- and ω-oxidative pathways to *in vitro* fatty acid oxidation in rat liver, *J. Lipid Res.* **9**:267–269.
193. Preiss, B., and Bloch, K., 1964, ω-Oxidation of long chain acids in rat liver, *J. Biol. Chem.* **239**:85–88.
194. Bjorkhem, D., and Danielsson, H., 1970, ω-Oxidation of branched chain fatty acids in rat liver homogenates, *Eur. J. Biochem.* **14**:473–477.
195. Das, M. L., Orrenius, S., and Ernster, L., 1968, On the fatty acid and hydrocarbon hydroxylation in rat liver microsomes, *Eur. J. Biochem.* **4**:519–523.
196. Lu, A. Y. H., and Coon, M. J., 1968, Role of hemoprotein P-450 in fatty acid ω-hydroxylation in a soluble enzyme system for liver microsomes. *J. Biol. Chem.* **243**:1331–1332.
197. Wada, F., Shibata, H., Groto, N., and Sakamoto, Y., 1968, Participation of the microsomal electron transport system involving cytochrome P-450 in ω-oxidation of fatty acids, *Biochim. Biophys. Acta* **162**:518–524.
198. Schenkman, J. B., Remmer, H., and Estabrook, R. W., 1967, Spectral studies of drug interaction with hepatic microsomal cytochrome, *Mol. Pharmacol.* **3**:113–123.
199. Orrenius, S., and Thor, H., 1969, Fatty acid interaction with the hydroxylative enzyme system of rat liver microsomes, *Eur. J. Biochem.* **9**:415–418.

200. Ichihara, K., Kusunose, E., and Kusunose, M., 1971, A fatty acid ω-hydroxylation system solubilized from porcine kidney cortex microsomes, *Biochim. Biophys. Acta* **239:**178–189.
201. Ellin, A., Jakobsson, S. V., Schenkman, J. B., and Orrenius, S., 1972, Cytochrome P-450 of rat kidney cortex microsomes: Its involvement in fatty acid ω- and (ω-1)-hydroxylation, *Arch. Biochem. Biophys.* **150:**64–71.
202. Jakobsson, S., Thor, H., and Orrenius, S., 1970, Fatty acid inducible cytochrome P-450 of rat kidney cortex microsomes, *Biochem. Biophys. Res. Commun.* **39:**1073–1080.
203. Gibson, G. G., Orton, T. C., and Tamburini, P. P., 1982, Cytochrome P-450 induction by clofibrate, *Biochem. J.* **203:**161–168.
204. Parker, G. L., and Orton, T. C., 1980, Induction by oxyisobutyrates of hepatic microsomal cytochrome P-450 with specificity towards hydroxylation of fatty acids, in: *Biochemistry, Biophysics and Regulation of Cytochrome P-450* (J. Å. Gustafsson, J. Carlstedt-Duke, A. Mode, and J. Rafter, eds.), Elsevier/North-Holland, Amsterdam, pp. 373–377.
205. Tamburini, P. P., Masson, H. A., Bains, S. K., Makowski, R. J., Morris, B., and Gibson, G. G., 1984, Multiple forms of hepatic cytochrome P-450: Purification, characterization and comparison of a novel clofibrate indued isozyme with other minor forms of cytochrome P-450, *Eur. J. Biochem.* **139:**235–246.
206. Samuelsson, B., Granstrom, E., Green, K., and Hamburg, M., 1971, Metabolism of prostaglandins, *Ann. N.Y. Acad. Sci.* **180:**138–163.
207. Iraelsson, U., Hamburg, M., and Samuelsson, B., 1969, Biosynthesis of 19-hydroxyprostaglandin A_1, *Eur. J. Biochem.* **11:**390–394.
208. Kupfer, D., Navarro, J., and Piccolo, D. E., 1978, Hydroxylation of prostaglandins A_1 and E_1 by liver microsomal monoxygenase: Characteristics of the enzyme system in the guinea pig, *J. Biol. Chem.* **253:**2804–2811.
209. Vatsis, K. P., Theoharides, A. D., Kupfer, D., and Coon, M. J., 1982, Hydroxylation of prostaglandins by inducible isozymes of rabbit liver microsomal cytochrome P-450: Participation of cytochrome b_5, *J. Biol. Chem.* **257:**11221–11229.
210. Okita, R. T., Parkhill, L. K., Yasukochi, Y., Masters, B. S. S., Theoharides, A. D., and Kupfer, D., 1981, The ω- and (ω-1)-hydroxylase activities of prostaglandins A_1 and E_1 and lauric acid by pig kidney microsomes and a purified kidney cytochrome P-450, *J. Biol. Chem.* **256:**5962–5964.
211. Williams, D. E., Hale, S. E., Okita, R. T., and Masters, B. S. S., 1984, A prostaglandin ω-hydroxylase cytochrome P-450 (P-450$_{PG\omega}$) purified from lungs of pregnant rabbits, *J. Biol. Chem.* **259:**4600–4608.
212. Powell, W. S., 1978, ω-Oxidation of prostaglandins by lung and liver microsomes: Changes in enzyme activity induced by pregnancy, pseudopregnancy, and progesterone treatment, *J. Biol. Chem.* **253:**6711–6716.
213. Oliw, E. H., Guengerich, F. P., and Oates, J. A., 1982, Oxygenation of arachidonic acid by hepatic monooxygenases: Isolation and metabolism of four epoxide intermediates, *J. Biol. Chem.* **257:**3771–3781.
214. Chacos, N., Falck, J. R., Wixtrom, C., and Capdevila, J., 1982, Novel epoxides formed during the liver cytochrome P-450 oxidation of arachidonic acid, *Biochem. Biophys. Res. Commun.* **104:**916–922.
215. Ford-Hutchinson, A. W., Bray, M. A., Doig, M. V., Shipley, M. E., and Smith, M. J. H., 1980, Leukotriene B, a potent chemokinetic and aggregating substance released from polymorphonuclear leukocytes, *Nature* **286:**264–265.
216. Borgeat, P., and Samuelsson, B., 1979, Arachidonic acid metabolism in polymorphonuclear leukocytes: Effects of ionophore A23187, *Proc. Natl. Acad. Sci. USA* **76:**2148–2152.

217. Capdevila, J., Chacos, N., Werringloer, J., Prough, R. A., and Estabrook, R. W., 1981, Liver microsomal cytochrome P-450 and the oxidative metabolism of arachidonic acid, *Proc. Natl. Acad. Sci. USA* **78:**5362–5366.
218. Capdevila, J., Parkhill, L., Chacos, N., Okita, R., Masters, B. S. S., and Estabrook, R. W., 1981, The oxidative metabolism of arachidonic acid by purified cytochrome P-450, *Biochem. Biophys. Res. Commun.* **101:**1357–1363.
219. Capdevila, J., Marnett, L. J., Chacos, N., Prough, R. A., and Estabrook, R. W., 1982, Cytochrome P-450-dependent oxygenation of arachidonic acid to hydroxyeicosatetraenoic acids. *Proc. Natl. Acad. Sci. USA* **79:**767–770.
220. Samuelsson, B., 1980, The leukotrienes: A new group of biologically active compounds including SRS-A, *Trends Pharmacol. Sci.* **1:**227–230.
221. White, R. E., and Coon, M. J., 1980, Oxygen activation by cytochrome P-450, *Annu. Rev. Biochem.* **49:**315–357.
222. Koop, D. R., Persson, A. V., and Coon, M. J., 1981, Properties of electrophoretically homogenous constitutive forms of liver microsomal cytochrome P-450, *J. Biol. Chem.* **256:**10704–10711.

CHAPTER 11

Cytochrome P-450 Enzymes in Sterol Biosynthesis and Metabolism

COLIN R. JEFCOATE

1. Introduction

The biosynthesis of sterols involves a series of both simple and complex cytochrome P-450-catalyzed monooxygenase reactions. These processes are generally distinguished from the multiplicity of reactions catalyzed by hepatic microsomal forms of P-450 by involving P-450 isozymes that exhibit a high degree of regio- and stereospecificity. The cell specificity of these monooxygenases has been noted in the previous chapter. In this chapter, we will examine the biochemical control of these steroidogenic cytochromes P-450. This is far more stringent than for the hepatic cytochromes P-450 because of the physiological importance of steroid hormones and bile acids. While the biosynthesis of these cytochromes is under hormonal control, this response takes a minimum of 12 hr for steroidogenic cytochromes P-450 (see Chapter 10). More common fluctuations in activity are required within minutes or hours. Here we will discuss the mechanisms of steroid hydroxylation and also some of the proposed mechanisms for rapid physiological control of steroidogenic enzymes.

Several of the steroidogenic cytochromes P-450 catalyze a complex sequence of oxygenase steps; for example, cholesterol side-chain cleavage and aromatase require three cycles of oxygenation to complete the full conversion, while 17,20-lyase requires two cycles. We will also look at whether these multiple processes derive from an ordered, coupled sequence of oxygenation steps. A characteristic of many steroidogenic

---

COLIN R. JEFCOATE • Department of Pharmacology, University of Wisconsin Medical School, Madison, Wisconsin 53706.

TABLE I
Distribution of Steroid Hydroxylases between Mitochondria and Microsomes

Reaction	Substrate	Tissues
Mitochondria		
20,22 Cleavage[a]	Cholesterol	Adrenal, testis, ovary, placenta
11β-, 18-HO[b]	Deoxycorticosterone	Adrenal
18-HO[a]	Corticosterone 18-Hydroxycorticosterone	Adrenal glomerulosa
1α-HO 24-HO	25-Hydroxyvitamin D_3	Kidney
26-HO	Cholesterol	Liver
Microsomes		
21-HO	17α-Hydoxyprogesterone[c]	Adrenal
17α-HO,[a] 17,20-lyase	Progesterone[d]	Adrenal, testis
Aromatase[a]	Androgens	Ovary, placenta
25-HO	Vitamin D	Liver
7α-HO	Cholesterol	Liver
12α-HO	3α,7α-dihydroxycholestane	Liver
25-HO	3α,7α,12α-Trihydroxycholestane	Liver
14α-HO	Dihydrolanosterol	Liver

[a] Multiple hydroxylases.
[b] HO, hydroxylation.
[c] Progesterone is also a substrate.
[d] Lyase activity measured with 17α-hydroxyprogesterone.

pathways is that reactions occur both in the endoplasmic reticulum and in the mitochondria. As noted in Chapter 10, these processes require very different electron transfer chains. In this chapter, we will examine the mechanism of the mitochondrial P-450 electron transfer chain and will look at the different possibilities for control offered by this process. We will also compare these mitochondrial P-450-dependent processes to the bacterial P-450-dependent process of camphor hydroxylation which uses a very similar electron transport process. The distribution of steroidogenic cytochromes P-450 between mitochondria and microsomes is summarized in Table I.

Four different mechanisms can potentially be used to rapidly affect the activity of P-450: control of electron transfer, control of substrate supply, protein modification, or allosteric activation. These mechanisms will be discussed with respect to steroidogenic cytochromes P-450 in subsequent sections. The distinct requirements of steroid synthesis in specific cell types may also involve discrete P-450 isozymes adapted to a particular function as recently reviewed by Lieberman et al.[1] An additional key feature of all cytochromes P-450 is that they are membrane-bound enzymes and that all of the potential control elements listed above may be changed by the membrane environment. The effects of phospholipids on P-450 have been discussed in Chapter 5, and here we will discuss this subject only with respect to cellular regulation.

2. Regulation of Mitochondrial Electron Transport to Steroid Hydroxylases

2.1. Flavoprotein Reduction by NADPH

Two mitochondrial enzymes catalyze the transfer of reducing equivalents from NADPH to P-450, a low-molecular-weight iron–sulfur protein or ferredoxin[2] and an NADPH-specific flavoprotein containing a single FAD.[3] These proteins have been most extensively studied following purification from bovine adrenal mitochondria. The ferredoxin (adrenodoxin) and flavoprotein (adrenodoxin reductase) from this source are both soluble single-subunit proteins (M^r 12,000 and 52,000, respectively). Adrenodoxin contains an active center consisting of two iron atoms, each coordinated by two cysteinyl sulfurs and bridged by two sulfur atoms. Although in the oxidized state both irons are in the $+3$ state, only a single electron can be introduced into a delocalized molecular orbital covering the whole Fe_2S_2 cluster. The FAD flavoprotein exhibits a 1000-fold preference for NADPH over NADH ($K_m = 6$ μM, 6 mM) and, by means of the intervention of the flavin semiquinone state, can divide the two reducing equivalents provided in one step by NADPH into two one-electron donations necessary for adrenodoxin redox activities and P-450 turnover. The thermodynamic and kinetic control of electron transfer between these proteins has been extensively studied by Kamin and co-workers and has recently been reviewed in detail.[4]

The first two steps in the reaction of NADPH with flavoprotein reductase can be observed under anaerobic conditions. Rapid complex formation by NADPH is followed by a slower hydride transfer (18 sec^{-1}) to generate the fully reduced flavin.[5] This process is characterized by a large isotope effect using $(4S)$-[^{2}H]-NADPH (7.5)[6] and generation of a characteristically long-wavelength charge transfer absorption (505–750 nM).[7] When $K_3Fe(CN)_6$ is included, reduction of this oxidant occurs at the same rate and with the same isotope effect as formation of reduced flavin. Clearly, these two one-electron steps involving the intermediate semiquinone state are limited by the same initial hydride transfer step. $NADP^+$ binds 1000 times more strongly to the fully reduced flavoprotein than the oxidized enzyme as evidenced by a 100-mV increase in the flavin oxidation–reduction potential in the presence of $NADP^+$. As a conse-

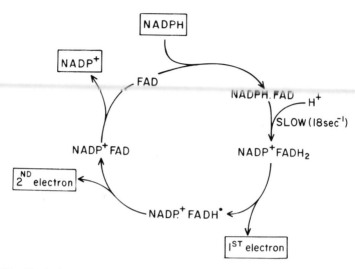

FIGURE 1. Catalytic cycles of NADPH-adrenodoxin reductase. The division of two reducing equivalents from NADPH into two one-electron transfer steps is emphasized. Reduction of FAD is rate-limiting.

quence, $NADP^+$ only dissociates *after* the reduced flavin has been reoxidized by two consecutive electron transfers to ferricyanide molecules (Fig. 1).

2.2. Reduction of Adrenodoxin by Adrenodoxin Reductase

Reduction of adrenodoxin by adrenodoxin reductase differs from reduction of $K_3Fe(CN)_6$ in two important respects: the two proteins form a strong complex,[8] and the electron transfer is thermodynamically unfavorable. These two phenomena are interlinked. Thus, the oxidation–reduction potential of adrenodoxin shifts by -50 to -100 mV on complex formation, indicating that the reduced form of adrenodoxin binds more weakly.[9] Since the isolated potentials of the reductase and adrenodoxin are similar (-290 mV), this shift causes reduction of adrenodoxin to be thermodynamically relatively unfavorable.

The affinity for the oxidized form is very strong at low ionic strength ($K_d = 10^{-8}$ M) but is decreased by specific cations and by increased pH.[10] In addition, at pH's above 7, electron transfer to adrenodoxin is aided by protonation of a specific histidine that may interact by H-bonding to the Fe_2S_2 cluster.[11] Indeed, this interaction probably accounts for the far more favorable oxidation–reduction potential of adrenodoxin than other ferredoxins which have pH-independent potentials of about −400 mV.[12]

Reduction of cytochrome c by NADPH requires both adrenodoxin and adrenodoxin reductase and has provided a valuable model to study electron transfer from this complex. Reduced adrenodoxin but not reduced flavoprotein can transfer electrons to cytochrome c. Reduction of cytochrome c by reduced adrenodoxin occurs at a far greater rate than NADPH-supported turnover mediated by the two proteins ($k > 400$ sec^{-1}).[13] Thus, although the NADPH-cytochrome c reduction results from electron transfer from reduced adrenodoxin, the rate is limited by other electron transfer processes within the adrenodoxin–adrenodoxin reductase complex. More detailed analysis of this reaction by stopped-flow spectrophotometry indicates that cytochrome c receives electrons from only two states of the complex.[13] In the first donor state, reduced adrenodoxin is associated with NADP$^+$-flavosemiquinone (contains two reducing equivalents), while in the second donor state, reduced adrenodoxin is associated with NADPH-oxidized flavin (contains three reducing equivalents). In the intermediate, one-reducing-equivalent state, the single electron is preferentially associated with the flavin rather than adrenodoxin and, therefore, is only slowly transferred to cytochrome c. The cycle, as postulated by Lambeth et al.,[4] is shown in Fig. 2. Although the electron transfer to oxidized adrenodoxin is slower than the reverse transfer (unfavorable potential difference), the reduction of cytochrome c pro-

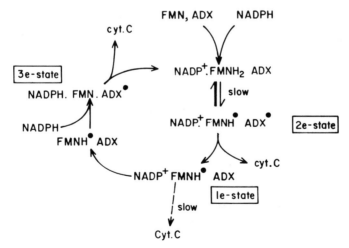

FIGURE 2. NADPH-cytochrome c reduction catalyzed by NADPH-adrenodoxin reductase and adrenodoxin. Note that rapid electron transfer to cytochrome c occurs in states with, respectively, two and three reducing equivalents but not with one reducing equivalent.

ceeds because the rate of the adrenodoxin–cytochrome c electron transfer greatly exceeds this rate of reverse transfer.

The rate of reduction of cytochrome c through this cycle (3.3 sec^{-1}) is slower than any step in forming reduced flavoprotein, is five-fold slower than ferricyanide reduction, and shows only a marginal isotope effect with (4S)-[^{2}H]-NADPH.[6] The slowness of the reaction and absence of an isotope effect must therefore be attributed to the only remaining unmeasured step, electron transfer between the reduced flavin and the Fe_2S_2 cluster of adrenodoxin. This is supported by the observation that the log of the rate of cytochrome c reduction under a variety of conditions (pH variation[10]; reconstitution with flavin analogues[6]) correlates closely with the potential difference between these centers. This relationship is typically found for such electron transfer processes. However, the two electrons in the catalytic cycle are transferred at similar rates, even though thermodynamically different in the proposed cycle, implying other controlling factors in the electron transfer reactions.

2.3. Reduction of Cytochromes by Reduced Adrenodoxin

A critical question to be resolved in these electron transfer reactions is whether the flavoprotein and the protein acceptor can bind simultaneously to reduced adrenodoxin. If the binding domains on adrenodoxin for the flavoprotein and the protein acceptor overlap, then dissociation of the flavoprotein–reduced adrenodoxin complex must precede electron transfer. Interestingly, work carried out by Kamin and co-workers indicates that the favored mechanism depends on the protein acceptor.[11,14] For cytochrome c reduction, the rate increases linearly as the ratio of adrenodoxin to flavoprotein is increased until a limit is reached at 1:1 ratio of the two proteins. Reduction occurs rapidly at low ionic strength and is inhibited at high ionic strength due to dissociation of the flavoprotein–adrenodoxin complex. These data are fully consistent with electron transfer to cytochrome c that is associated with the flavoprotein–adrenodoxin complex. A very different set of characteristics is observed for acetylated cytochrome c[14] and for both $P-450_{11\beta}$ and $P-450_{scc}$.[11] These reactions are characterized by slower turnover rates (5–60 min^{-1}). Little activity is seen at a 1:1 ratio of adrenodoxin to reductase, but activity increases rapidly as this ratio is exceeded. Finally, there is very little activity at *both* low and high ionic strength, resulting in a sharp dependence of activity on ionic strength. The unusual reaction characteristics of electron transfer to cytochromes P-450 and to acetylated cytochrome c can be explained by the obligatory dissociation of reduced adrenodoxin from the flavoprotein prior to complex formation with the acceptor protein and electron transfer.[11,14]

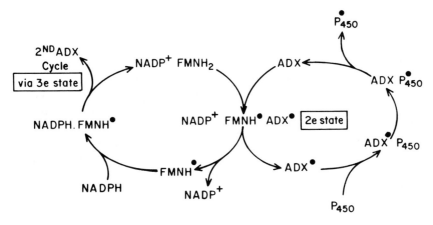

FIGURE 3. Reduction of cytochrome P-450 by NADPH-adrenodoxin reductase and adrenodoxin. Note that reduced adrenodoxin must dissociate from the reductase prior to reduction of P-450 (cf. Fig. 2).

This "shuttle mechanism" for reduction of cytochromes by adrenodoxin is shown in Fig. 3. At adrenodoxin concentrations lower than the flavoprotein concentration, the high-affinity binding of reduced adrenodoxin to the flavoprotein competes with the binding of reduced adrenodoxin to the acceptor protein. This is accentuated at low ionic strength which favors adrenodoxin–flavoprotein complex formation, and there is little complex formation between reduced adrenodoxin and oxidized cytochrome. For cytochrome c, direct transfer from the complex is independent of such competition. Nevertheless, detailed analysis of the effect of ionic strength and excess adrenodoxin on cytochrome c reduction rates suggests that an additional adjustment of the reduced adrenodoxin–flavoprotein complex must precede even electron transfer to cytochrome c.[4]

The effects of ionic strength on steroid hydroxylation with purified cytochromes P-450 are broadly consistent with an enzymatic cycle supported by a shuttling of adrenodoxin between flavoprotein and P-450. This is most evident when the stoichiometry of the three components approximates the proportions present in adrenal mitochondria (0.1 flavoprotein:1 adrenodoxin:1 P-450). Under these conditions, divalent ions (Ca^{2+}, Mg^{2+}) produce a sharp peak in the activities of both steroid monooxygenases (cholesterol side-chain cleavage and 11β-hydroxylation) at 10–30 times lower concentrations than stimulation by NaCl.[11] These responses to metal ions broadly parallel effects on the reduction of acetyl cytochrome c and also on the reduction of excess adrenodoxin, confirming their common origin in the shuttle mechanism of reduction. However, the peak rates of cholesterol side-chain cleavage and steroid 11β-hydroxylation differ substantially with the various ions, suggesting additional selective ion interactions.

Major support for the shuttle mechanism has been obtained from kinetic analysis of steroid hydroxylation at varying concentrations of adrenodoxin. The rates of both cholesterol side-chain cleavage and 11β-hydroxylation depend on the concentration of *free* reduced adrenodoxin in a manner consistent with the shuttle mechanism.[15] These experiments will be discussed later, together with studies of adrenodoxin complex formation. Variation of the adrenodoxin concentration also reveals effects of divalent ions that are not explained by the shuttle mechanism and which may result from effects of ions on complex formation between adrenodoxin and cytochromes P-450.

The differential binding of oxidized and reduced adrenodoxin to, respectively, the flavoprotein and P-450 is a key feature of the shuttle mechanism. This has been elegantly demonstrated by the effects of complex formation on oxidation–reduction potentials. Complex formation between the flavoprotein and adrenodoxin stabilizes the oxidized state of adrenodoxin relative to the reduced state (-90 mV shift in oxidation–reduction potential). In other words, electron transfer to adrenodoxin favors dissociation of adrenodoxin from the flavoprotein (K_d increases from 25 to 500 nM), thus facilitating the shuttle process. Conversely, upon complex formation of adrenodoxin with P-450$_{scc}$, the midpoint potential of the iron–sulfur protein decreases (-273 to -291 mV), indicating preferential binding of reduced adrenodoxin to oxidized P-450$_{scc}$ (K_d = 60 nM).[16] The K_d for oxidized adrenodoxin with reduced P-450 (120 nM) is 5 times higher than the K_d for the interaction of oxidized adrenodoxin with flavoprotein (K_d = 25 nM). Consequently, following electron transfer, a reverse shuttling of adrenodoxin back to the flavoprotein to start a further cycle is favored.

2.4. Control of Electron Transfer from Adrenodoxin during Steroid Hydroxylation

These electron transfer reactions are further controlled by the substrates for the cytochromes: cholesterol (P-450$_{scc}$) and deoxycorticosterone (P-450$_{11β}$). Both substrates, upon combining with their respective cytochromes, change the spin state from fully low spin to near fully high spin. Adrenodoxin binds only weakly to the substrate-free cytochrome (P-450$_{scc}$, K_d = 0.5 μM) and with a small spin state change (15% high spin). However, complex formation with cholesterol enhances adrenodoxin binding by six- to eight-fold (K_d = 7 × 10^{-8} M) and vice versa.[17,18]

Comparable measurements have not been carried out with P-450$_{11\beta}$ due to the instability of the enzyme in the absence of substrate. The comparable complex of bacterial proteins, putidaredoxin with P-450$_{cam}$, is far weaker even in the presence of substrate (K_d = 2.9 μM).[19] Adrenodoxin binding to P-450$_{scc}$ is substantially decreased by increases in ionic strength and this clearly contributes to the effects of ionic strength on cholesterol side-chain cleavage.[17] An increase in ionic strength also weakens the adrenodoxin–P-450$_{11\beta}$ interaction as can be seen from the effect of ionic strength on 11β-hydroxylation rates.[17]

Cholesterol binding to P-450$_{scc}$ shifts the midpoint potential from −412 mV to −305 mV, which is then comparable to the midpoint potential of adrenodoxin.[20] Cholesterol thus binds far more readily to the reduced cytochrome and, at the same time, facilitates reduction of the cytochrome by reduced adrenodoxin. Additional complex formation by either oxidized or reduced adrenodoxin causes a small negative shift in this potential but, as noted above, the shuttle reduction is favored by the preferred binding of reduced adrenodoxin to oxidized P-450$_{scc}$. Clearly, substrate availability provides the signal to initiate the shuttling of adrenodoxin, since the oxidation–reduction potential of the free cytochrome is too negative to permit significant electron transfer.

Kinetic analysis of the rates of cholesterol side-chain cleavage at various cholesterol and adrenodoxin concentrations indicates a random order of binding for cholesterol and adrenodoxin.[21] At a low cholesterol concentration (≤K_d), the K_m for adrenodoxin is comparable to the K_d (~200 nM). However, as the cholesterol concentration rises, the K_d decreases while the K_m rises and eventually becomes six-fold higher than the K_d. This difference between K_m and K_d is more normally observed when the rate-limiting step is faster than the dissociation of the reactants. However, the dissociation rates of adrenodoxin and cholesterol occur several orders of magnitude faster than each electron transfer cycle (2 sec^{-1}). Thus, an explanation for this discrepancy, which is constant over a broad range of ionic strengths, remains to be found.

High concentrations of flavoprotein (50 μM), adrenodoxin (200 μM), and P-450$_{scc}$ (24 μM) elute on Bio-Gel P-200 as a ternary complex.[22] However, quantitation of complex formation at lower concentrations of the proteins, by optical difference spectroscopy, indicates that binary complexes form with far higher affinity than the ternary complex which could not be detected.[15] This also indicates preferential formation of a binary adrenodoxin–flavoprotein complex in competition with a binary adrenodoxin–P-450 complex.[15] The concentration of free adrenodoxin is decreased by complex formation with the flavoprotein, and the K_d for the flavoprotein–adrenodoxin complex (5 nM) accounts fully for the effect of the flavoprotein on complex formation with P-450$_{scc}$. Consistent with these measurements of complex formation, the rates of cholesterol side-chain cleavage and 11β-hydroxylation are dependent on a Michaelis function of the free adrenodoxin concentration. Interestingly, the best fit is obtained with K_d for the flavoprotein–adrenodoxin interaction equal to 500 nM (i.e., weaker than in the analogous binding experiment). This is consistent with the presence of fully reduced adrenodoxin under the conditions used for cholesterol side-chain cleavage and with the much weaker binding of reduced adrenodoxin to the flavoprotein.[15]

Ca^{2+} ions influence cholesterol side-chain cleavage and 11β-hydroxylation in significantly different ways.[17] When the rate of reaction is measured as a function of the concentration of adrenodoxin reductase, 11β-hydroxylation reaches the same maximum turnover (80 min^{-1}) with both $CaCl_2$ (5mM) and NaCl (40mM). However, for cholesterol side-chain cleavage, the maximum activity obtained with $CaCl_2$ is six-fold less than the maximum activity obtained with $MgCl_2$ or NaCl (18 min^{-1}). $CaCl_2$ also inhibits cholesterol side-chain cleavage that is maximally stimulated by NaCl (ED_{50}, 30–80 μM). This selective inhibition by $CaCl_2$ is associated with a decreased turnover of reduced adrenodoxin by P-450$_{scc}$, even though complex formation is not inhibited.

The effect of ionic strength on these reactions is highly dependent on the concentration of adrenodoxin. When adrenodoxin is both saturating and fully reduced by excess flavoprotein reductase, increased ionic strength continuously increases the V_{max} for cholesterol side-chain cleavage (reaching 30 min^{-1}). In contrast, ionic strength does not increase V_{max} for 11β-hydroxylation. At each level of adrenodoxin, increased ionic

TABLE II
Kinetic Control of Cholesterol Side-Chain Cleavage and Deoxycorticosterone 11β-Hydroxylation

	Cholesterol side-chain cleavage	11β-Hydroxylation
Similarities		
Both activated by substrate (spin state change)		
Both activities are a Michaelis function of [ADX$_{free}$] (shuttle mechanism)		
Both activities increase and then decrease as ionic strength is increased with profiles dependent on; [ADX]; monovalent or bivalent cations		
Differences		
Increased ionic strength	Increases V_{max}	Does not affect V_{max}
Oxidized ADX	Strongly inhibits	Weakly inhibits
Ca	Inhibits	Activates like Mg^{2+}
Phospholipids, detergent (major effect)	Controls substrate binding	Stabilizes cytochrome

strength eventually decreases the rates of both cholesterol side-chain cleavage and 11β-hydroxylation, in part because of the increased K_d's for adrenodoxin binding to the cytochromes. At low ionic strength (10 mM HEPES), cholesterol side-chain cleavage stops while 11β-hydroxylase remains at a third of the peak value. Under these conditions, adrenodoxin is about 50% reduced. The reason that this high concentration of reduced adrenodoxin does not support cholesterol side-chain cleavage may be that the reaction is potently inhibited by oxidized adrenodoxin, particularly at low ionic strength. The dependence of the reaction rate upon the concentration of oxidized adrenodoxin is consistent with competitive inhibition of this type. For 11β-hydroxylation, oxidized and reduced adrenodoxin compete more equally for the cytochrome. The major features of electron transfer in cholesterol side-chain cleavage and 11β-hydroxylation, catalyzed by purified enzymes, are summarized in Table II.

3. Mechanism of Cholesterol Side-Chain Cleavage and Steroid Hydroxylases

3.1. Reaction Sequence for Cholesterol Side-Chain Cleavage

The catalytic cycle for monooxygenation by mitochondrial cytochromes P-450 is strictly analogous to camphor hydroxylation by P-450$_{cam}$ from *Pseudomonas putida* (Chapter 12) and involves two separate electron transfer steps. The oxidized cytochrome substrate complex accepts the first electron from reduced adrenodoxin. The second electron is transferred from reduced adrenodoxin to the ternary complex of reduced cytochrome, substrate, and molecular oxygen. During camphor hydroxylation, the two electron transfer steps occur at approximately equal rates, even though the second transfer is thermodynamically favored.

Cholesterol side-chain cleavage involves three consecutive monooxygenation steps and a total of six electron transfer steps per pregnenolone formed. Consequently, when V_{max} is 20–30 min^{-1},[17] the supporting electron transfer rate is 2–3 sec^{-1}. This compares with far faster electron transfer from reduced adrenodoxin to cytochrome c (> 400 sec^{-1}).[4] The maximum turnover rates are about ten times faster for camphor hydroxylation (~20 sec^{-1}).

Detailed study of the more complex turnover of cholesterol side-chain cleavage has required identification of the intermediate products. Burstein and co-workers first identified (22R)-hydroxycholesterol and (20R,22R)-dihydroxycholesterol as minor products during cholesterol side-chain cleavage.[23] Addition of equimolar amounts of adrenodoxin to

P-450$_{scc}$–cholesterol complexes, under anaerobic conditions followed by stoichiometric reduction and subsequent oxygenation, limits turnover to the first monooxygenation product. This, and subsequent single-turnover experiments, has established a sequence of 22R-hydroxylation, followed by 20R-hydroxylation, and a final oxidative cleavage of the diol to form pregnenolone and isocaproic aldehyde.[41] During continuous turnover, three sequential monooxygenations occur, even in the presence of high membrane cholesterol concentrations, because the affinity of P-450$_{scc}$ for the intermediate hydroxycholesterols far exceeds the affinity for cholesterol.[25] This, in part, reflects the stronger binding of cholesterol by membrane phospholipids as much as a preference of the active site of the cytochrome for hydroxycholesterol derivatives. Nevertheless, (22R)-hydroxycholesterol and (20R,22R)-dihydroxycholesterol displace cholesterol from the active site of P-450$_{scc}$ with very high affinity (K_d = 40–80 nM) in the absence of vesicles or detergent.[25]

3.2. Active Site of P-450$_{scc}$

The active site of P-450$_{scc}$ has been probed by means of a variety of cholesterol analogues that act as substrates, suicide substrates, and inhibitors. The interactions have been examined by means of optical absorption spectroscopy, low-termperature EPR spectroscopy, equilibrium constants, binding kinetics, and reaction kinetics. Optical and EPR spectra primarily measure the proportions of high- and low-spin states. However, since this equilibrium is temperature sensitive, comparisons can only be made if EPR spectra are measured by using a freezing procedure that is fast, relative to equilibria determining the spin state. Spin states are determined from optical spectra by measuring relative intensities in the Soret region or from the charge-transfer band at 645 nm that is characteristic of the high-spin state. With EPR spectra, integration must be made of second-derivative absorptions across a broad high-spin spectrum (g = 8–1.8) and a far narrower low-spin spectrum (g = 2.45–1.9). The second-derivative peaks indicate subtle differences in structure that are not apparent in optical absorption spectra; for example, high-spin substrate complexes of rat adrenal P-450$_{scc}$ and P-450$_{11\beta}$ exhibit distinct EPR spectra (g = 8.0, 7.9)[26] which are also different from spectra of the corresponding beef enzymes (g = 8.2, 8.1).[27] EPR spectra are particularly well suited to characterizing the cytochromes P-450 in optically turbid suspensions of mitochondria but require high concentrations of cytochrome (10 μM). Optical absorption changes are more suited to low concentrations of P-450 and spin states are determined in turbid suspensions from the difference spectra induced by pregnenolone or amines that both induce fully low-spin states.[28]

The chemical nature of the two spin states of P-450 has been discussed elsewhere in this volume. The fully high-spin state consists of a pentacoordinate heme-iron bound only to a cysteine-S^- from the protein. In the low-spin state, a hexacoordinate state is formed with an additional oxygen ligand which may be H_2O or a tyrosine from the protein.[29] Cholesterol and its analogues can greatly change the spin state equilibrium, both by perturbing the conformation of the protein and, in the case of hydroxyl analogues, by directly providing the sixth oxygen-ligand. In the absence of substrate, P-450$_{scc}$ is essentially completely low spin. Cholesterol combines with P-450$_{scc}$ to produce a predominately high-spin state (80%).[24,25] The interaction of hydroxylated derivatives depends on the position of substitution. The effectiveness at inducing a high-spin state is particularly sensitive in enzymatically active positions: $(20R,22R)$-dihydroxy $>$ $(22R)$ $>$ 20α $>$ pregnenolone. Substitution at other positions around the cholesterol rings can either increase the extent of the high-spin state (7β OH) or favor the low-spin state (25 OH) relative to cholesterol.[24] EPR spectra of P-450$_{scc}$, complexed by cholesterol analogues substituted with oxygens at positions 20 or 22, exhibit substantially perturbed low-spin heme spectra. In the most extreme case of $(22R)$-hydroxycholesterol, difference spectroscopy indicates that even the Soret optical absorption is shifted.[30] This perturbation of the low-spin spectra by the $(22R)$-hydroxy substituent probably derives from a direct but distorted interaction of the hydroxyl with the heme iron.

Proximity of the 22-position to the heme iron has also been indicated by studies of cholesterol analogues containing amino substituents in the side chain.[31] Substitution at positions 22 or 23 produces a high-affinity binding, potent competitive inhibition of cholesterol side-chain cleavage, and a shift in the Soret absorption typical of amine complexes (λ_{max} 422 nm instead of 417 nm). When the amino substituent is shifted in either direction along the side chain, weaker inhibition and a shift of the Soret maximum back to 417 nm ensues. Modification of the steroid rings seems less crucial to binding. Notably, cholesterol 3-sulfate is a good substrate, although it has been suggested that this may possibly involve a distinct form of the cytochrome.[1,32] As noted above, 7-hydroxy substituents also do not impede binding.[24]

The shape of the active site is clearly modified by the binding of adrenodoxin to the cytochrome. Adrenodoxin not only increases the affinity of P-450$_{scc}$ for cholesterol and other ligands forming predominantly high-spin complexes, but induces a significant (15%) increase in the high-spin state in the absence of substrate. In addition, the binding of a steroid such as pregnenolone, which stabilizes the low-spin state, is competitively weakened by adrenodoxin binding.[33]

The interaction of cholesterol with the active site of P-450$_{scc}$ can be modulated in two other ways. Lambeth and co-workers have discovered that cardiolipin binds to the cytochrome approximately stoichiometrically and enhances cholesterol binding by about three-fold, both independent of the additional stimulatory effect of adrenodoxin.[34] The effect is also found with monolysocardiolipin and more weakly with dilysocardiolipin and phosphatidylglycerol. Interestingly, the head group analogue, α-glycerophosphate, competitively blocks the binding of cardiolipin (K_i = 3 mM) but does not activate cholesterol binding. The nature of this interaction remains unknown but may be analogous to the activating effect of binding of 2–3 moles of cardiolipin to 1 mole of cytochrome oxidase.[35] Spin-label EPR studies indicate immobilization of only one fatty acid chain of cardiolipin by P-450$_{scc}$. This may be involved in improved fit of cholesterol in the active site.

A decrease in pH from 8 to 6 also enhances cholesterol binding by 20-fold but has no effect on adrenodoxin binding.[33] This change can be quantitatively explained by two substrate-free low-spin forms of the cytochrome interconverted through protonation of a group with pK_a 6.5. The pK_a for protonation is shifted to approximately 7.9 by cholesterol binding. This protonation exerts no effect on the binding of the product, pregnenolone, which competitively inhibits cholesterol binding. This suggests that the protonation activates binding of the side chain rather than the A–D rings of cholesterol. Since a positively charged group on the protein is unlikely to directly improve the binding of a saturated side chain, it seems most likely that an allosteric change in the active site is induced in this way. The similarity of the effects of protonation and cardiolipin on cholesterol binding to P-450$_{scc}$ suggests that the two phenomena may be closely related. For example, cardiolipin may interact with the protonated group and so facilitate this mechanism for enhanced cholesterol binding. A model which is consistent with the pH effect and with the binding of cholesterol, amino and hydroxy analogues, and pregnenolone is shown in Fig. 4.

The highly specific stereochemistry in the active site of P-450$_{scc}$ has been shown by demonstration of a stereospecific internal transfer of oxygen from 20-hydroperoxy cholesterols. The position of oxygen insertion to form the product, dihydroxycholesterol, is either 21 or 22 depending on whether the configuration of the initial hydroperoxide was R or S, respectively.[36] The R configuration does not fit the active site as closely as the S configuration, resulting in a detectable peroxy-Fe intermediate and a slower overall reaction. Evidently, this intramolecular oxene transfer is mediated by oxygen transfer to heme iron and presents a similar picture of the side chain binding in the active site to that obtained from the binding of hydroxy- and aminocholesterol analogues.

A model to describe spin state equilibria of P-450 has been proposed

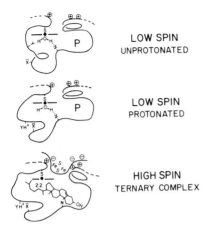

FIGURE 4. Proposed model for the active site of P-450$_{scc}$. —●— indicates the heme bound by cysteine-S. P represents the binding site for pregnenolone (low-spin complex). N is an amino acid group that facilitates the binding of water to heme (low-spin state). Y accepts a proton (pK 6.5) and induces a conformational change (YH$^+$ interacts with X$^-$) that facilitates cholesterol binding but does not directly affect the spin state (group N or cysteine-S).

by Sligar[37] based on work with P-450$_{cam}$ and has been applied to microsomal P-450 by Ruckpaul and co-workers.[38] In this model, the spin state of a P-450 complex is determined by the relative affinities of a ligand for conformations of the protein associated with, respectively, fully high- and low-spin states and the equilibrium constant between those conformations. The synergistic effects of adrenodoxin and cholesterol both on binding and on the spin state of P-450$_{scc}$ can be understood in terms of their selective and independent binding to a high-spin conformation of the cytochrome. In the absence of substrate, the low-spin conformation is favored (K_1 is estimated at 0.03). Cholesterol binds over 100 times more tightly to the high-spin conformation (K_3 increases to 3). Adrenodoxin activates cholesterol binding and also enchances the proportion of high spin in the ternary complex by independently favoring the high-spin conformation (K_1 is determined to be 0.15). Other ligands either preferentially bind to the low-spin conformation (pregnenolone) or bind to each conformation equally (H$^+$ and cardiolipin do not affect the spin state). 20α- and (22R)-hydroxycholesterol probably behave anomalously by preferentially binding to the high-spin conformation but then contribute a sixth hydroxyl ligand to form a partial low-spin state. These equilibria are summarized in Fig. 5. According to this model, the selective binding of adrenodoxin in the high-spin conformation of P-450$_{scc}$ accounts fully for the synergistic activation of cholesterol binding and the accompanying spin state changes.[33]

A fully high-spin substrate complex is seen for each of the highly selective steroidogenic cytochromes P-450 (SCC; aromatase; 11β-, 17α-, and 21-hydroxylases). This probably reflects a tightness of fit of the substrate to the active site that leaves insufficient space for the hydroxyl

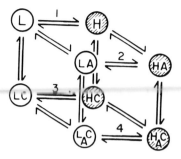

FIGURE 5. Spin state equilibria for P-450$_{scc}$ involving low-spin (L) and high-spin (H) conformations of the protein and the binding of cholesterol (C) and adrenodoxin (A). The observed spin state changes and the activation of cholesterol binding by adrenodoxin are fully consistent with the following approximate constants: $K_1 = 0.03$, $K_2 = 0.15$, $K_3 = 3$, $K_4 = 15$.

ligand of the low-spin state. These enzymes are different, in this respect, from the less specific liver microsomal enzymes in which only partially high-spin complexes are observed.[38]

3.3. Oxygen Activation by P-450$_{scc}$

As described in earlier chapters, the catalytic cycle of P-450 involves the combination with O_2 of a reduced cytochrome–substrate complex. This key intermediate then accepts the second electron prior to formation of the active perferryl oxygen which is transferred to the substrate. The oxygen–ferrous heme complex of P-450$_{scc}$ is only stable at subzero temperatures[24] and at 4°C decays by a first-order process (10^{-2} sec^{-1}). This decay releases O_2^- and, in the absence of adrenodoxin, no oxidation of cholesterol occurs. In a water/glycol mixture at $-30°C$, the oxygen–ferrous complex is stable for several hours, while at $-15°C$ there is a shift in the absorption maximum from 422 nm to 412 nm, suggesting two different states of the complex.[39]

The K_d value for oxygen binding to cholesterol–P-450$_{scc}$ (23 μM) is much higher than K_d values for oxygen binding to P-450$_{cam}$ and liver microsomal cytochromes P-450, suggesting a further means of controlling activity.[40] Interestingly, the reaction intermediates (22R)-hydroxycholesterol and (20R,22R)-dihydroxycholesterol stabilize this complex (2- and 4-fold). Isoelectronic CO complexes are destabilized by 150- and 300-fold, indicating a unique interaction between the 22-hydroxyl and O_2, while the same substituent interferes with CO complex formation. This stabilization of the oxygen complex derives from 150- to 1000-fold decreases in the dissociation rates of the respective complexes that more than offset slower association rates. The concerted monooxygenase sequence is further facilitated by 5- and 10-fold decreases by, respectively, (22R)-hydroxycholesterol and (20R,22R)-dihydroxycholesterol in the release of O_2^- (autoxidation). Again the dissociation rate for the CO complexes is insensitive to side chain substitution. Evidently, there are differences between the interactions of O_2 and CO with the heme which are distinguished by these substrates. Competition between CO and cholesterol and also the effect of cholesterol on the oxidation–reduction potential of P-450$_{scc}$ also confirm that the affinity for the substrate is much higher in the reduced state.[41]

3.4. Structure of P-450$_{scc}$

Sequencing of the gene for bovine P-450$_{scc}$ has provided the complete amino acid sequence but, as yet, little direct insight into the features of the active site.[42] P-450$_{scc}$ consists of 481 amino acids and contains only two cysteine residues (303 and 461). The second of these is located in a highly conserved region of other sequenced cytochromes P-450. Since the heme is coordinated by a cysteine-S$^-$, this has established that the heme is bound within 43 amino acids of the -terminus. Akhrem and co-workers[43] have carried out limited sequencing of the cytochrome that is fully consistent with this DNA sequence. Interestingly, these workers found that limited proteolysis of P-450$_{scc}$ split the protein into two polypeptide fragments (M_r 27,000 and 22,000) that remain associated and retain the full activity of the cytochrome. Apparently, P-450$_{scc}$ consists of two independently folded domains that remain rigidly associated following tryspin treatment. P-450$_{scc}$ is also a glycoprotein and neuraminidase treatment apparently inhibits reduction of the cytochrome by adrenodoxin, suggesting participation of the sugar in the binding of adrenodoxin or the subsequent electron transfer.[44]

3.5. Mechanism of 11β-Hydroxylation

P-450$_{11\beta}$ is readily separated from P-450$_{scc}$, due to a far greater lipid solubility that requires the use of high detergent concentrations during isolation.[45] Much less is known about the active site of P-450$_{11\beta}$. Maximum turnover numbers with the purified enzyme (~100 min^{-1}) require rates of electron transport that are comparable to those for side-chain cleavage and much slower than obtained in the bacterial camphor hydroxylase system. Substrates, deoxycorticosterone and deoxycortisol, convert the cytochrome to a fully high-spin form with an EPR spectrum that is clearly distinguishable from the spectrum of the high-spin cholesterol complex of P-450$_{scc}$. The pyridine derivative, metyrapone, binds extremely tightly to P-450$_{11\beta}$ ($K_d = 5 \times 10^{-8}$ M) in marked contrast to P-450$_{scc}$ and forms a low-spin complex typical of N-coordination to heme iron.[27]

P-450$_{11\beta}$, which catalyzes the hydroxylation of Δ^4-3-keto-C$_{21}$-steroids, is highly stereospecific for attack from the β side but is not fully selective for position 11. Both C-18 and C-19 angular methyl substituents provide C–H bonds in close proximity to the 11-position that also react with the activated oxygen complex. Thus, in the bovine enzyme, 10–15% of the total 11-deoxycorticosterone metabolism involves 18-hydroxylation, while in the rat enzyme, this increases to over 30% and significant 19-hydroxylation is observed.[45,46] Corticosterone is 1000 times less active than 11-deoxycorticosterone as a substrate for the purified bovine enzyme (0.02 min^{-1}). 18-Hydroxycorticosterone and aldosterone are formed from monooxygenation and subsequent oxygen-dependent dehydrogenation at position 18.[47] Aldosterone is also formed directly from 18-hydroxycorticosterone which is 3 times more reactive than corticosterone (0.06 min^{-1}). This reactivity at C-18 is also increased 4-fold by an adrenal mitochondrial lipid fraction, analogous to lipid effects on cholesterol side-chain cleavage.[47] The C-20,21 side chain is not critical, although 4-androstene-3,17-dione and testosterone are relatively poor substrates for 11β-hydroxylation.

These low alternative activities may be of physiological significance since aldosterone is produced about 1000 times slower than corticosterone and, under physiological conditions in adrenal mitochondria, P-450$_{11\beta}$ is largely depleted of competing substrates. One might question whether these minor activities derive from the same enzyme or a small contamination of such glomerulosa enzymes. However, during the purification of P-450$_{11\beta}$ from bovine adrenal mitochondria, 11β- and 18-hydroxylase activities copurify and are equally inhibited by metyrapone, CO, and antibodies raised against the purified enzyme.[48]

Aldosterone synthesis is exclusively located in the zona glomerulosa, while P-450$_{11\beta}$ is distributed in both glomerulosa and fasciculata tissue. This difference in steroidogenesis may be due to a distinct form of cytochrome in glomerulosa cells or, alternatively, to a unique activation of P-450$_{11\beta}$ (e.g., phospholipid). Control of the zona glomerulosa is different from that of the zona fasciculata, and steroidogenesis is elevated exclusively in the former by angiotensin II and potassium.[49] These responses are enhanced by *in vivo* sodium depletion. Kramer *et al.*[50] have demonstrated that, in mitochondria from rat adrenal glomerulosa, corticosterone also induces a type I spectral response. No such response is observed in adrenal fasciculata mitochondria. A sodium-depleted diet increases aldosterone synthesis from corticosterone in these mitochondria, while at the same time increasing this type I spectral response to corticosterone. By contrast, 11β-hydroxylation and the 11-deoxycorticosterone (DOC) type I binding to P-450$_{11\beta}$ are independent of the sodium level in the diet. Evidently, sodium depletion induces a change in glomerulosa mitochondria that results in an increased capacity of P-450 to form a substrate complex with corticosterone. The conversions of corticosterone to 18-hydroxycorticosterone and of the latter to aldosterone are each stimulated three-fold by sodium depletion, suggesting that a single form of P-450 is responsible for both steps. There is no increase in total P-450 after sodium depletion, even though the corticosterone-induced spectral change reaches 50% of the DOC-induced change. This suggests that a modification of a form of P-450$_{11\beta}$ activates corticosterone binding but not DOC binding.

Examination of the stoichiometry of the 11β-hydroxylase reaction with reconstituted purified enzymes indicates that a substantial amount of oxygen is reduced to superoxide radicals.[51] This is particularly favored by an excess of adrenodoxin and presumably involves an uncoupled dissociation of the oxygen–ferrous complex of P-450$_{11\beta}$. The reaction is, however, tightly coupled in mitochondria, such that one molecule of oxygen is consumed for each molecule of corticosterone formed.

4. The Role of the Mitochondrial Membrane in Steroid Hydroxylation

Reconstitution of P-450$_{scc}$ into a membrane is not necessary for catalytic turnover. However, the extreme insolubility of cholesterol in water means that either phospholipids or detergents (e.g., Tween 20) are necessary to provide access of cholesterol to the cytochrome. The binding of cholesterol to P-450$_{scc}$ is determined by the ratio of cholesterol to detergent of phospholipid, indicating exclusive equilibration of cholesterol with the cytochrome through the detergent/phospholipid membrane.[52] Detergents and phospholipids do not significantly affect the binding of adrenodoxin to the cytochrome, suggesting a clear separation of adrenodoxin- and lipid-binding domains on the cytochrome.[21] EPR studies using oriented membranes show that, unlike other integral membrane heme proteins, the heme plane of P-450$_{scc}$ lies parallel to the surface of the membrane.[4]

The activity of cholesterol side-chain cleavage is highly dependent on the nature of the phospholipid. For a series of phosphatidylcholines, parallel changes in K_m and K_d occur with little change in V_{max}, indicating that the affinity of the cytochrome for cholesterol is the source of these differences.[52] The order of these changes, diphytanoyl > dilinoleoyl > dioleoyl > dimyristoyl, does not correlate with effects of these lipids on physical properties of the membrane (fluidity, melting point, affinity for cholesterol). Presumably, these lipid effects are closely related to the stimulatory effect of cardiolipin on cholesterol binding, i.e., an allosteric change in the active site.

The mode of interaction of purified bovine adrenal P-450$_{scc}$ with

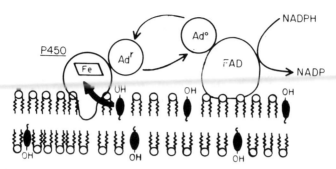

FIGURE 6. Cholesterol side-chain cleavage by P-450$_{scc}$ incorporated into a phospholipid bilayer. The SCC shuttling of ADX between peripherally bound flavoprotein (FAD) and P-450$_{scc}$ is shown together with the transfer of cholesterol from the bilayer to the incorporated cytochrome.

phospholipid vesicles depends on the enzyme preparation used. Tuckey and Kamin[53] report a slow association rate that depends on both the P-450$_{scc}$ and lipid concentration, while Kowluru et al.[54] observe over 100 times faster binding under comparable conditions. However, the former preparation of P-450$_{scc}$ appears to insert effectively into the membrane, while the latter P-450$_{scc}$ inserts weakly and exchanges rapidly between vesicles. Despite these differences between the P-450$_{scc}$ preparations in their interaction with the phospholipid vesicles, the rates of cholesterol side-chain cleavage reach comparable levels.

The side-chain cleavage rate depends only on effective transfer of cholesterol to the cytochrome and this can be achieved even with a relatively loose insertion. Nevertheless, Kowluru et al.[54] showed that phosphoinositides can substantially enhance the extent of incorporation of this preparation of P-450$_{scc}$ into phosphatidylcholine vesicles and also can increase the rates of cholesterol transfer. Interestingly, the rate of cholesterol transfer to P-450$_{scc}$, at high concentrations of cholesterol, is the same as the side-chain cleavage rate and evidently limits the rate of reaction. Since we know little of the state of P-450$_{scc}$ in the mitochondria, both strong and loose insertion may contribute to regulatory processes. A combined view of electron transfer and cholesterol transfer steps at membrane-bound P-450$_{scc}$ is shown in Fig. 6.

Much less is known about the incorporation of P-450$_{11\beta}$ into phospholipid vesicles. This enzyme is the most lipophilic of all purified cytochromes P-450 and requires high levels of detergent to remain soluble. The enzyme is greatly stabilized by transfer into phospholipid vesicles and also exhibits a high turnover in this environment (120 min^{-1}).[55] Interestingly, while phosphatidylcholine offers comparable stability to a nonionic detergent, a 1:1 mixture of phosphatidylcholine and phosphatidylethanolamine, which is close to the mitochondrial environment, offers much greater stabilization.

5. Control of Steroid Hydroxylation in Mitochondria

NADPH does not freely penetrate into intact mitochondria. Consequently, the activity of steroid hydroxylases depends on the generation of NADPH within the mitochondria. Three principal routes are available: (1) $NADP^+$-linked malic enzyme, (2) $NADP^+$-linked isocitrate dehydrogenase, and (3) energy-linked transhydrogenation of $NADP^+$ by NADH (from α-ketoglutarate dehydrogenase, malic dehydrogenase, or reverse electron flow from succinate). In bovine adrenal mitochondria, the respective rates for routes 1–3 are 360, 90, and 185 nmole/min per mg[56] but are very different in rat adrenal mitochondria where the malic enzyme has only low activity.[53] The relative activities for support of mitochondrial hydroxylases by tri- and dicarboxylic acid intermediates of the Krebs cycle are also dependent on the source of the mitochondria. In rat adrenal mitochondria, the order for supporting cholesterol side-chain cleavage is: isocitrate > malate > succinate.[57]

Interestingly, $P-450_{scc}$ and $P-450_{11\beta}$ may not derive reducing equivalents in the same way, at least in bovine mitochondria.[58] Thus, unlike 11β-hydroxylation, cholesterol side-chain cleavage is stimulated by cyanide when metabolism is supported by either succinate or malate. Since cyanide is effective through inhibition of cytochrome oxidase, this suggests different participation of the respiratory chain in supporting these hydroxylases, possibly through involvement of the energy-linked transhydrogenase. Involvement of the transhydrogenase is of major importance in tissues such as the human placental mitochondria. This involvement is indicated by an ADP stimulation of steroidogenesis through the provision of ATP to drive the transhydrogenation of NADH to NADPH.[59] The energy-linked transhydrogenation process also implies a close relationship between the respiratory chain and electron transfer to steroid hydroxylases.

A second point where the two mitochondrial electron transfer processes overlap is in competition of O_2. Recent work indicates that since $P-450_{scc}$ has a high K_m for O_2, activity is readily limited by oxygen depletion.[60] Reduction of O_2 by the respiratory chain may decrease the availability of O_2 to cytochromes P-450 for steroidogenesis.

6. Hormonal Regulation of Cholesterol Side-Chain Cleavage

The cellular regulation of steroidogenesis generally involves at least the hormonal activation of cholesterol side-chain cleavage. This activation process exhibits scarcely any delay before maximum stimulation is achieved. In adrenal cells, the conversion of pregnenolone and subsequent steroids to glucocorticoids is unaffected by hormonal stimulation. Similarly, hydroxycholesterol analogues undergo side-chain cleavage at the same rate, irrespective of hormonal stimulation.[61] This rate at least equals the maximum rate of cholesterol side-chain cleavage following ACTH stimulation. A unique characteristic of the ACTH stimulation of cholesterol side-chain cleavage is the susceptibility to protein synthesis inhibitors. Addition of inhibitors, such as cycloheximide or puromycin, decreases the ACTH-stimulated rate far more rapidly ($t_{1/2}$ = 3–4 min) than the degradation of known proteins. This rate of decline is also comparable to the loss of stimulation following removal of ACTH.[62] This sensitivity to protein synthesis inhibitors indicates the involvement of an exceptionally short-lived protein that must be continuously resynthesized to maintain ACTH stimulation of steroidogenesis. This again is specifically related to activation of cellular cholesterol metabolism, since protein synthesis inhibitors do not inhibit the side-chain cleavage of hydroxycholesterol analogues. These are general features of the hormonal regulation of steroidogenesis in other tissues, such as the testis and ovary, even though a distinct hormonal activation is involved.[63,64]

This regulation occurs at the level of cholesterol transport to P-450$_{scc}$. In rats *in vivo*[62] and in both rat and bovine adrenal cells in culture,[62,65] ACTH increases the movement of cholesterol to the mitochondria. When cholesterol side-chain cleavage is blocked by the inhibitor aminoglutethimide, ACTH increases the rate of mitochondrial cholesterol accumulation to a rate that is comparable to that of steroidogenesis in the absence of the inhibitor. However, although this movement of cholesterol is hormonally regulated, protein synthesis inhibitors do not inhibit mitochondrial cholesterol accumulation.[66]

The hormonal activation of cholesterol transfer to the mitochondria involves several discrete steps, and the limiting step may differ from tissue to tissue. Cholesterol is derived initially as cholesterol esters that are taken up by endocytosis of serum lipoproteins (LDL in most species, but HDL in the rat[67]). In the adrenal cell, there is normally a large store of cholesterol esters in the form of lipid droplets. ACTH activates cholesterol esterase via a cAMP-mediated phosphorylation resulting in an increase in cellular free cholesterol.[68] Little is known about the mechanism of transfer of cholesterol to the mitochondria, including whether it is sensitive to ACTH. However, in rat adrenals, this step is blocked by both microtubule and microfilament inhibitors.[62] The role of microfilaments has been elegantly demonstrated by introduction of both antiactin antibodies and the actin-binding DNA polymerase into adrenal cells.[69]

Isolated adrenal mitochondria retain at least part of the activation provided by ACTH. In addition, the inhibition of steroidogenesis by *in vivo* administration of protein synthesis inhibitors is retained by adrenal mitochondria after isolation.[70] Examination of P-450$_{scc}$ by either optical difference spectroscopy[61] or EPR spectroscopy indicates that, both in the absence of cellular activation and following inhibition of protein synthesis, the high-spin state is diminished.[71] This reflects a decrease in cholesterol–P-450$_{scc}$ complex formation.[28] A similar conclusion has been made from the inhibitory effect of cholesterol on the rate of combination of CO with P-450$_{scc}$ in mitochondria following ACTH activation.[41]

A careful analysis of the methods used for these experiments has shown that these correlations are only obtained when the adrenal glands are allowed to become anaerobic during isolation.[72] This has been clearly established by direct measurement of P-450 in whole glands. Within 5 sec after stopping blood flow to the adrenal gland, the EPR spectrum loses sensitivity to ACTH. However, at the same time, the P-450$_{scc}$ in ACTH-treated adrenals becomes substantially more reduced as oxygen is depleted.[73] These experiments are consistent with the hypothesis that, in the gland, *in vivo* cholesterol transfer to P-450$_{scc}$ is much slower than the turnover of cholesterol metabolism with the result that, irrespective of stimulation, the cytochrome is depleted of cholesterol. When oxygen is depleted, turnover ceases and cholesterol accumulation at P-450$_{scc}$ occurs in proportion to the ACTH activation of cholesterol transfer. The P-450$_{scc}$–cholesterol complex under near-anaerobic conditions is readily reduced, as observed in the EPR experiment.

The kinetics of cholesterol side-chain cleavage indicate a rapid phase of metabolism which is increased by ACTH action and a slow phase which is independent of ACTH action.[70] The rapid phase of metabolism is highly sensitive to O_2 concentration, confirming that this process is limited by turnover of P-450$_{scc}$, while the slow phase is insensitive to changes in O_2 concentration and probably reflects a cholesterol transport process.[60] Rapid phase metabolism is enhanced when mitochondrial cholesterol levels are enhanced by inhibition of P-450$_{scc}$ *in vivo* with aminoglutethimide. However, this rapid phase is decreased eight-fold when a protein synthesis inhibitor, such as cycloheximide, is added along with aminoglutethimide, even though mitochondrial cholesterol accumulation is unaffected. Cholesterol–P-450$_{scc}$ complex formation changes in parallel to these changes in the size of the reactive pool of cholesterol.[74]

The retention of diminished mitochondrial activity following cycloheximide treatment is dependent on mitochondrial integrity. The treatment of rats with ACTH in combination with aminoglutethimide causes accumulation of cholesterol in the inner mitochondrial membrane where P-450$_{scc}$ is located.[75] Cycloheximide treatment prevents this transfer of cholesterol to the inner membrane and, as a consequence, cholesterol accumulates in the outer membrane.[76] This location of cholesterol in the outer membrane and a barrier to transfer to the inner membrane then accounts for the inactivity of adrenal mitochondria following cycloheximide treatment.

Isolated inner membranes retain most of the differences in cholesterol metabolism that are observed in intact mitochondria. However, the various treatments affect the size of the reactive cholesterol pool rather than the initial rate. An unexpected characteristic of these inner membranes is that, in spite of four-fold differences in the level of reactive cholesterol, high cholesterol–$P-450_{scc}$ complex formation is uniformly observed. This suggests a substantial activation of binding affinity relative to intact mitochondria,[76] such that $P-450_{scc}$ is fully bound by even the low levels of cholesterol in inner membranes following low or inhibited activation.

An explanation of this inhibitory action of cycloheximide is provided by the recent isolation from ACTH-stimulated rat adrenals of a peptide (2200 daltons) that is removed by brief exposure of the animals to cycloheximide.[77] Addition of pure peptide to rat adrenal mitochondria, previously blocked by *in vivo* cycloheximide action, results in the activation of cholesterol side-chain cleavage.[77] Stimulation is also observed with sterol carrier protein-2,[78] a protein that is present in adrenal cytosol and readily binds cholesterol stoichiometrically. This protein also facilitates movement of cholesterol from lipid droplets to mitochondria, while the exact control of mitochondrial cholesterol movement remains to be elucidated. It seems that ACTH stimulates the formation of a 2200-dalton labile protein that activates intramitochondrial cholesterol movement, possibly by forming contacts between the outer and inner membranes. Some recent evidence suggests that the 2200-dalton peptide may derive from a larger precursor protein whose synthesis is rapidly stimulated by ACTH.[79]

Ca^{2+} may also participate in the cholesterol mobilization process.[80] Low concentrations of Ca^{2+} (0.1–1 mM) activate mitochondrial cholesterol side-chain cleavage but not 11β-hydroxylation and also increase the combination of cholesterol with $P-450_{scc}$. Ca^{2+} may act by facilitating contacts between the inner and outer membranes or by changing the availability of cholesterol by inducing phase transitions.

Previous work that has been discussed above indicates that cholesterol side-chain cleavage may be very sensitive to the phospholipid composition of the mitochondrial inner membrane. Farese[81] has reported that ACTH stimulates the synthesis of adrenal phospholipids (particularly phosphoinositides) within 10 min. Analogous changes have been observed in other steroidogenic tissues.[63] This work has recently been extended by Igarashi and Kimura[82] who have shown a rapid ACTH-induced increase in mitochondrial phospholipids, most notably with phosphatidyl-

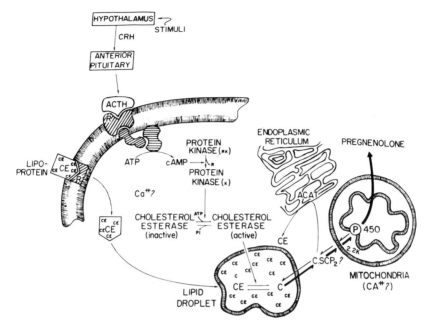

FIGURE 7. Schematic representation of the ACTH control of cholesterol metabolism in adrenal cells.

inositol (37–94%). This increase, like mitochondrial steroidogenesis, is inhibited by cycloheximide.[82] However, direct addition of these phospholipids to adrenal mitochondria fails to stimulate cholesterol metabolism under conditions where the 2200-dalton and sterol carrier proteins are very effective.[76] Nevertheless, a role for phospholipids in the regulation of cholesterol transfer or side-chain cleavage remains likely, possibly through regulation of intracellular calcium distribution which is now believed to be mediated by the diphosphoinositide hydrolysis product, inositol 1,4,5-triphosphate.[81] A schematic representation of the hormonal control of cholesterol side-chain cleavage is shown in Fig. 7.

7. Steroidogenesis in the Endoplasmic Reticulum

Four principal steroidogenic enzyme activities are located in the endoplasmic reticulum: steroid 21- and 17α-hydroxylases, 17,20-lyase, and aromatase activity that removes a C-19 methyl in converting androgens to estrogens. Each activity is supported by electron transfer from the microsomal flavoprotein, P-450 reductase. These enzymes have been purified from various microsomal sources by methods that are standard for the liver enzymes.

The kinetics of reduction of P-450$_{C21}$ have been extensively studied by Takemori and co-workers.[83] Like the liver microsomal cytochromes, reduction of P-450$_{C21}$ is substantially accelerated by the presence of substrate and exhibits two first-order rate constants (1 sec^{-1} and 0.05 sec^{-1}). The rate of fast-phase reduction is independent of detergent concentration, while the percentage decreases with increase in detergent concentration (70% to 20%). Detailed analysis suggests that the fast-phase reduction represents random collisions of the reductase and cytochrome. Direct 1:1 complex formation was detected during gel filtration of the combined proteins by shifts in their elution profiles.

The purified bovine P-450$_{C21}$ shows a five-fold preference in binding progesterone over 17α-hydroxyprogesterone, while both substrates induce nearly complete spin state changes. As noted previously, the extent of this spin state change is much greater than for substrate binding to liver microsomal cytochromes P-450. Like many other microsomal cytochromes P-450, the substrate-free enzyme exhibits a temperature-induced change from low to high spin.[84]

Steroid 17α-hydroxylases have been purified to homogeneity from both pig testis[85] and guinea pig adrenal microsomes.[86] When reconstituted with purified P-450 reductase, each purified enzyme has both 17α-hydroxylase and 17,20-lyase activities in a ratio of 2:1. With the pig testis enzyme, both activities show similar dependence on pH and temperature. A variety of inhibitors also exhibit identical K_i's for both activities.[85] These common characteristics of the two activities together indicate that they are catalyzed by the same protein. Like the 21-hydroxylase, both progesterone and 17α-hydroxyprogesterone induce a near-complete conversion of both adrenal and testis cytochromes to a high-spin state. Progesterone competitively inhibits the lyase activity on 17α-hydroxyprogesterone, confirming that there is a single site for both activities. Although the conversion of progesterone to androstenedione occurs via 17α-hydroxyprogesterone, this two-step reaction differs from cholesterol side-chain cleavage in that the intermediate, 17α-hydroxyprogesterone, binds comparably to progesterone. Formation of androstenedione from progesterone is, therefore, dependent on the competition between progesterone and 17α-hydroxyprogesterone.

As noted elsewhere in this volume, cytochrome b_5 may participate in stimulating microsomal oxygenase reactions. Electron transport from NADH to b_5 via b_5 reductase can enhance 17α-hydroxylase, lyase, and 21-hydroxylase activities on C-21 steroids.[87] Recent work by two groups indicates that the participation of b_5 can change the ratio of 17α-hydroxylation to 17,20-lyase activities for 17α-hydroxyprogesterone with either purified adrenal[88] or testis[89] cytochromes P-450. The reconstitution of b_5 with the guinea pig adrenal P-450 17α-lyase results in very different pH dependencies for stimulation of the two reactions. At pH 7.3, b_5 stimulates 17α-hydroxylation and 17,20-lyase activities by 1.7- and 7-fold, respectively. Comparable reconstitution of the pig testis P-450 and b_5 also indicates a greater sensitivity of the lyase process to b_5. Purified pig and rabbit liver b_5 produce very different effects from the pig testis b_5.

While 17α-hydroxylation and 17,20-lyase activity seem to be catalyzed by the same P-450 isozyme in bovine adrenal cells, ACTH induces progesterone 17α-hydroxylase activity without increasing 17,20-lyase activity on 17α-hydroxyprogesterone.[65] The more pronounced dependence of 17,20-lyase activity on the participation of b_5 may explain how these two activities can vary independently while still being catalyzed by a single form of P-450. Thus, an increase in the ratio of $P-450_{17\alpha}$ to b_5 during induction by ACTH could result in a disproportionate increase in 17α-hydroxylase activity.

The microsomal conversion of androgens to estrogens that occurs in the ovary and placenta requires 3 moles of O_2 and NADPH, suggesting that, like cholesterol side-chain cleavage, three consecutive monooxygenase steps are involved.[90] Two intermediates have been detected in the metabolism of androstenedione by placental microsomes: 19-hydroxyandrostenedione and 19-oxoandrostenedione. When metabolism was inhibited by KCN, the rates of formation of these two intermediates were both inhibited while the overall aromatization rate was only slightly affected. The rate of androstenedione 19-hydroxylation always exceeded the rate of estrogen formation, while the rate of formation of 19-oxoandrostenedione generally exceeded its rate of disappearance. These three observations suggest that the third oxidation step is rate-limiting.[90]

The establishment of the involvement of P-450 has proved troublesome. Thus, while aromatization of androstenedione and testosterone is insensitive to CO, even at low concentrations of O_2, aromatization of the 19-nor analogues is potently inhibited.[91] Substantial additional data now point to the involvement of P-450 in each stage of aromatization. Androstenedione, together with both identified intermediates from androstenedione and 19-norandrostenedione, all produce near-complete spin state changes in placental microsomal P-450.[92] In contrast to the insensitivity to CO, amines that bind to oxidized P-450 in competition with the substrate (aminoglutethimide, SKF 525A) inhibit androstenedione aromatization.[92] The involvement of P-450 is further supported by inhibition of aromatization of C_{19} and C_{18} steroids by antibodies against P-450 reductase.[92] As noted in Chapter 10, the rate of aromatization of substrates that are insensitive to CO (androstenedione) is at least 10 times faster than the CO-sensitive reactions. When androstenedione metabolism is slowed by competitive inhibition, CO sensitivity is restored.[93] These experimental results for androstenedione aromatization are all consistent with a rate-limiting first-electron transfer to the P-450–androgen complex. CO inhibition implies that second-electron transfer to the reduced P-450–O_2–androgen ternary complex is rate-limiting, at least when CO slows this step through competition with O_2. This may not readily occur if the first electron transfer step is, by far, the slower step.

Recent studies by Akhtar et al.[94] of the incorporation of $^{18}O_2$ into the products of aromatization have provided significant insight into the mechanism. This group studied the metabolism of a number of precursors, notably 19-hydroxy- and 19-oxo-4-androstene-3,17-dione in which the C-19 group was labeled with both ^2H and ^{18}O. Analysis of formic acid formed during aromatization of these intermediates indicated retention of ^{18}O. Use of the C-19 ^{16}O-intermediates, together with $^{18}O_2$, resulted in the incorporation of one atom of ^{18}O into formic acid. Aromatization of androstenedione in the presence of $^{18}O_2$ leads to complete incorporation of $^{18}O_2$ into *both* oxygens of formic acid. These findings imply that an atom of oxygen from the third step in aromatization is finally incorporated into the formic acid derived from C-19. This third step of aromatization may involve oxygenation at the 2β-position of 19-oxoandrostenedione since 2β-hydroxy-19-oxoandrostenedione has been trapped as an intermediate in the aromatization of both androstenedione and 19-oxoandrostenedione.[95] Based on this work, aromatization would require a cyclic intermediate such as that shown in Fig. 8 (step 3A). However, [2β-^{18}O,19-^3H]-2β-hydroxy-10β-formyl androst-4-ene-3,17-dione has recently been synthesized and the failure to obtain incorporation of ^{18}O into formic acid produced with placental aromatase strongly suggests that 2β-hydroxylation is not obligatory and that the third oxygen is incorporated at C-19 (Fig. 8, step 3B).[96] It has been suggested that distinct active sites are involved in these steps.[95] However, other work using placental microsomes indicates that androstenedione, 19-hydroxyandrostenedione, and 19-oxoandrostenedione are mutually competitive with K_i values equal to their respective K_m values.[97] This suggests that a single active site is involved and that the selective inhibition is derived from the different rates and limiting steps in the three steps of the overall reaction.

Recently, two fractions of aromatase activity have been purified from human placental microsomes, each fraction containing discrete P-450 isozymes (*a* and *b*) and associated reductases.[98] Aromatase activity was reconstituted only when the P-450 was combined with the reductase from the same fraction. These results indicate a multiplicity and unusual specificity in the P-450-dependent monooxygenations involved in estrogen biosynthesis.[98] Both aromatase fractions can use androstenedione or 16α-

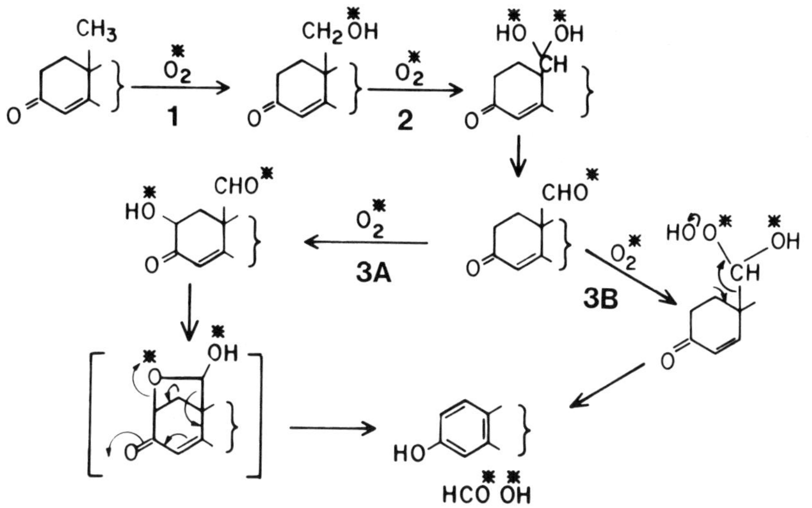

FIGURE 8. A mechanism of aromatization of androgens consistent with intermediate 2-hydroxylation and the release of doubly labeled formic acid from $^{18}O_2$.

hydroxytestosterone as substrate, but the ratio of activities differs by 30-fold between fractions, primarily through the increased metabolism of 16α-hydroxytestosterone by aromatase I, the minor form. Aromatase I and II can also be distinguished by bromo derivatives of androstenedione and testosterone which only inhibit aromatase I reactions.

8. Sterol Metabolism in Liver and Kidney

8.1. Activation of Vitamin D

Vitamin D is required for the regulation of calcium and phosphate homeostasis in avian and mammalian species.[99] When vitamin D_3 is deficient from the diet, the intestinal absorption of calcium and phosphate diminishes, resulting in impaired neuromuscular control, bone mineralization, and growth. This physiological activity has been shown to be due to a two-step activation of vitamin D_3 to 1,25-dihydroxyvitamin D_3. This process involves an initial 25-hydroxylation of vitamin D_3 in the liver, followed by 1-hydroxylation in kidney mitochondria (Chapter 10, Fig. 5). The actions of 1,25-dihydroxyvitamin D_3 on intestinal and bone cells are mediated by a receptor in the classical manner of steroid receptors. Parathyroid hormone, which is increased in deficiencies of calcium and vitamin D_3, most probably provides the stimulus to enhance 1α-hydroxylase activity in the kidney tubular cells through the mediation of cAMP. Interestingly, as in the case of cholesterol side-chain cleavage, this stimulation is rapidly lost in the presence of cycloheximide ($t_{1/2}$ 2.5–5 hr).[100] This decline seems far too fast for P-450 degradation and suggests a labile control factor analogous to that suggested for the adrenal cortex. In addition, Ca^{2+} stimulates 1α-hydroxylase activity in isolated kidney mitochondria in the concentration range 1–10 μM, while inhibiting this activity at higher concentrations (10–100 μM).[101]

The 100-fold higher concentration in plasma of 24,25-dihydroxyvitamin D_3 suggests that 24-hydroxylase is a more active enzyme, although again located in kidney mitochondria. The regulation of 24-hydroxylase activity is generally in the opposite direction to 1α-hydroxylation.[89] Although the physiological role of 24,25-dihydroxyvitamin D_3 remains elusive, there is some evidence of a complementary role to that of 1,25-dihydroxyvitamin D_3 in bone growth and in embryonic development. This reaction may also control the level of 1,25-dihydroxyvitamin D_3 through initiating a competing inactivation pathway.

8.2. Cholesterol Synthesis and Metabolism to Bile Acids

Cholesterol biosynthesis can be divided into two parts: the assembly of the C_{30} hydrocarbon, squalene, followed by the rearrangement and modification of this chain to form the C_{27} sterol. Squalene first undergoes the complex multiple cyclization to form lanosterol. Hydroxylases then play a key role in the removal of 3-methyl groups from the C_{30} sterol in order to form the C_{27} cholesterol. Three hydroxylation steps initiate the removal of methyl substituents at positions 4, 4', and 14α (Fig. 9). Reactions at positions 4 and 4' liberate CO_2, while the loss of the 14α-methyl

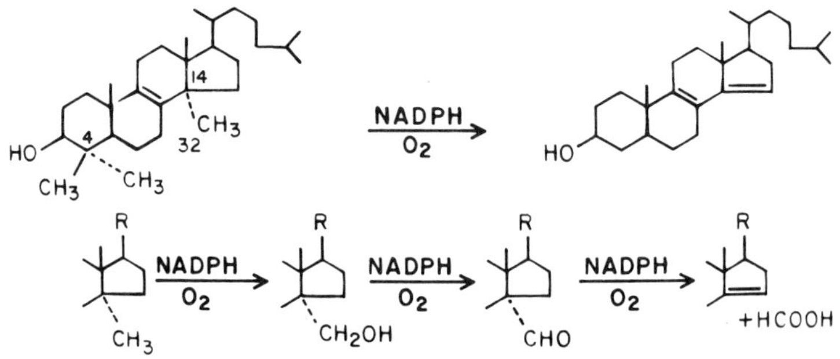

FIGURE 9. Oxidation of lanosterol at positions 4 and 14α during cholesterol biosynthesis.

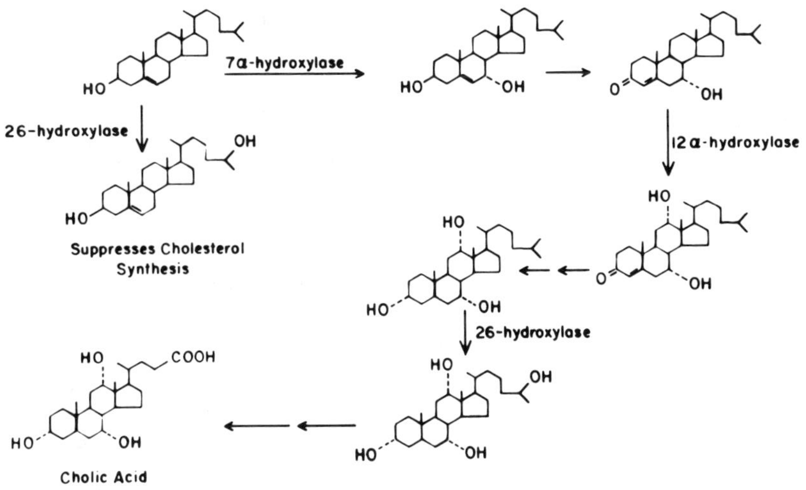

FIGURE 10. Hydroxylation steps in bile acid synthesis.

resembles aromatization in releasing formic acid. The former two reactions are mediated by b_5 and a microsomal iron protein, while 14α-hydroxylation involves a form of P-450.[102]

HMG-CoA reductase, the rate-limiting step in squalene and cholesterol biosynthesis, is also decreased through actions of low levels of hydroxy analogues of cholesterol such as 25-hydroxycholesterol.[103] Thus, the P-450-dependent processes present in liver microsomes and mitochondria may mediate the well-characterized inhibition of cholesterol synthesis by elevated levels of cholesterol. 25-Hydroxylanosterol is also a potent HMG-CoA reductase antagonist but may be generated from cyclization of diepoxysqualene which accumulates when squalene cyclization is blocked.[103]

Several modifications of both the rings and side chain of the cholesterol molecule are necessary for transformation to bile acids (Fig. 10). Ring metabolism involves 7α-hydroxylations, 3β-oxidation and isomerization, 12α-hydroxylation, and finally Δ^4 reduction to yield 5β-cholestane-3α,7α,12α-triol. The subsequent side-chain metabolism involves an initial 26-hydroxylation of this sterol. Further oxidation yields the 26-oic acid which is converted to a coenzyme A derivative. The removal of the three terminal carbon atoms to form cholic acid proceeds through a β-oxidation mechanism similar to that found in fatty acid catabolism.

Bile acids produced in the liver are secreted through the bile duct into the small intestine where they stimulate lipid absorption. Reabsorption from the intestine and passage back to the liver gives rise to enterohepatic circulation of bile acids. In man, bile acids circulate 6–10 times/day and only 10–15% is lost in feces. This loss is regenerated by bile acid synthesis which is then about 0.5 g/day in man.[104] If circulating bile acid is removed by a biliary fistula or decreased by binding to cholestyramine, a resin which binds bile acids, the synthesis of bile acids is increased several fold.[107] Feeding bile acids, on the other hand, decreases synthesis. The activity of the rate-limiting enzyme in bile acid synthesis, cholesterol 7α-hydroxylase, parallels these changes.

Cholesterol 7α-hydroxylase and 12α-hydroxylase are differentially regulated. Thus, the 7α-hydroxylase exhibits a diurnal rhythm following cholesterol synthesis, while 12α-hydroxylase remains constant.[105] However, the cholesterol level seems to be much less important in regulating 7α-hydroxylation than the bile acid level. The rapid enhancement of this activity by cholestyramine feeding and diurnal rise are suppressed by cycloheximide consistent with a half-time of the activity of 2–3 hr.[106] The rapidity of these effects suggests an additional labile regulatory protein.[107]

8.3. Liver and Kidney Mitochondrial Hydroxylases

Both kidney and liver mitochondria contain P-450.[108,109] Ferredoxins analogous to adrenodoxin have been purified from bovine liver mitochondria[110] and from chick kidney mitochondria.[111] In hepatic mitochondria, vitamin D_3 undergoes hydroxylation at the 25-position and cholesterol is also hydroxylated at C-26.[109] In addition, these same mitochondria catalyze 26-hydroxylation of 5β-cholestane-3α,7α,12α-triol (bile acid precursor) at a rate which is 500 times greater than hydroxylation of vitamin D_3.[112] This activity is only noncompetitively inhibited by vitamin D_3 and probably involves a distinct form of P-450. These activities are also found in liver microsomes which contain other specific enzymes that are important in vitamin D activation and bile acid synthesis (Section 8.4). Thus, again in the synthesis of these key physiological molecules, both mitochondrial and microsomal processes are involved.

P-450 levels in rat kidney mitochondria are about 100 times lower than in rat adrenal mitochondria (0.015 nmole/mg protein) and are unaffected by vitamin D deficiency which increases 1α-hydroxylase activity of 25-hydroxyvitamin D_3 and decreases 24-hydroxylase activity.[113] Unlike steroid hydroxylases, the substrate, 25-hydroxyvitamin D_3, does not induce a spectral change with kidney mitochondrial P-450. These activities are inhibited by CO in a light-reversible manner consistent with the involvement of P-450. However, 1α-hydroxylation is also inhibited by antioxidants such as menadione and p-phenylenediamine.[105] Recent experiments suggest that both monooxygenase and radical-mediated dioxygenase pathways may lead to 1α- and 24-hydroxylation of 25-hydroxyvitamin D_3.[114] Unlike mitochondrial steroid hydroxylations, 1α-hydroxylation is completely inhibited by uncouplers of oxidative phosphorylation, indicating that NADPH is derived from the energy-dependent transhydrogenase.[115]

P-450$_{D1α}$ has been purified to homogeneity from bovine kidney mitochondria.[116] The protein is immunologically similar but not identical to bovine adrenal P-450$_{scc}$. 1α-Hydroxylase activity can be reconstituted by addition of purified renodoxin and renodoxin reductase to the P-450 preparation.[111,116] The turnover, however, is 10,000 times lower than for cholesterol side-chain cleavage (0.008 min^{-1}). The kidney electron transfer proteins can be replaced by adrenodoxin and adrenodoxin reductase with little loss of activity. Evidently, differences between the mitochondrial ferredoxins from various species and tissues are minor. Presumably, similar shuttle mechanisms are involved as have been described for steroid hydroxylation. The apparently frequent duplication between microsomal and mitochondrial activities is poorly understood. There is also no indication of the difference between kidney mitochondria from vitamin D-deficient and replete animals that can explain the shifts between 1α- and 24-hydroxylation without changes in P-450 or renodoxin content. These results strongly suggest an additional control over the activity and selectivity of kidney mitochondrial P-450 that is determined by these physiological demands.

Recent work by Niranjan et al.[117] indicates that even xenobiotic metabolism is induced by phenobarbital and 3-methylcholanthrene in rat liver mitochondria. These reactions are supported in intact mitochondria by Krebs cycle intermediates but not by NADPH and, therefore, cannot represent microsomal contamination. Partially purified liver mitochondrial P-450 supports aflatoxin B_1 metabolism and benzo[a]pyrene metabolism when reconstituted with adrenodoxin and adrenodoxin reductase. Nevertheless, the mitochondrial cytochromes P-450 that are induced by phenobarbital and 3-methylcholanthrene are very similar to the microsomal cytochromes induced by these same agents, i.e., forms b and c, respectively. Xenobiotic metabolism, catalyzed by liver mitochondrial P-450, is also supported by microsomal P-450 reductase and is inhibited by antibodies against the corresponding microsomal cytochromes. Evidently, these mitochondrial isozymes retain the binding site for microsomal P-450 reductase but also gain the capacity to bind adrenodoxin.

8.4. Liver Microsomal Sterol Hydroxylases

Activities involved in cholesterol metabolism in the rabbit seem to be closely associated with a major methylcholanthrene-inducible form of P-450 in the rabbit (LM$_4$).[118] However, nearly pure P-450$_{LM_4}$ can only catalyze cholesterol 7α-hydroxylation when isolated after cholestyramine treatment, even though this treatment does not affect cholesterol binding to this cytochrome to induce a partially high-spin complex (K_s = 33 μM in 0.1% Triton X-100).[119] P-450$_{LM_4}$ from cholestyramine-fed rabbits has been further purified into two forms of the cytochrome. Only one form catalyzes 7α-hydroxylation, even though cholesterol binding is indistinguishable between the two forms.

12α-Hydroxylation of 7α-hydroxycholesterol also associates with P-450$_{LM_4}$, even though the activity is relatively insensitive to CO inhibition.[120] P-450$_{LM_4}$ from starved rabbits exhibits four-fold higher activity than from β-naphthoflavone-induced or untreated rabbits. The cytochrome from starved rabbits has a different amino acid composition and peptide map and, presumably, represents a distinct enzyme. This cytochrome is also isolated in a low-spin state as compared to the usual high-spin state of P-450$_{LM_4}$.

Interestingly, the 14-demethylation of lanosterol in rats is increased by isosafrole induction, suggesting that the active cytochrome is closely related to P-450$_d$, the equivalent in the rat of P-450$_{LM_4}$ in the rabbit.[102] Like cholesterol 7α-hydroxylation, C-14 demethylation is increased by cholestyramine feeding (sequesters bile acids in the gut). C-14 demethylation is also greatly suppressed by a cholesterol-rich diet. It is, therefore, very remarkable that three P-450-dependent reactions involved in cholesterol biosynthesis and metabolism and catabolism are all catalyzed by specific enzymes that are closely related to each other and to major xenobiotic-metabolizing forms. The relationship between the characteristics and control of constitutive cytochromes P-450 and xenobiotic-induced forms remains a key question for research on P-450.

Wikvall[121] has recently provided insight into the mechanism of physiological control of the cytochromes P-450 involved in bile acid synthesis. Reduced thiols (cysteine, dithiothreitol, and glutathione) and thioredoxin activate cholesterol 7α-hydroxylation, while GSSG causes inactivation. This suggests that the activity of this enzyme may depend on an equilibrium between disulfide and free sulfhydryl forms of the cytochrome. In addition, two modulatory proteins have been isolated from rat liver cytosol. One protein stimulates cholesterol 7α-hydroxylation in the presence of glutathione or thioredoxin, while the other protein inhibits 7α-hydroxylation. Proteins from rabbit liver microsomes and cytosol that modulate 12α-hydroxylase activity have also been isolated. Their action does not involve phosphorylation or dephosphorylation. However, in rats, 7α-hydroxylase activity is activated in microsomes by cAMP-dependent phosphorylation.[122] Clearly, differences between the two forms of rabbit liver P-450$_{LM_4}$ that bind cholesterol may derive from one or more such posttranslational modifications.

A constitutive form of P-450 from rat liver microsomes catalyzes 25-

hydroxylation of a variety of C_{27} sterols: 7α-hydroxycholesterol (0.36 min^{-1}), 7α,12α-dihydroxycholesterol (2.6 min^{-1}), vitamin D_3 (0.33 min^{-1}), and 1α-hydroxyvitamin D_3 (1.0 min^{-1}).[123] This same enzyme does not exhibit either 7α-, 12α-, or 26-hydroxylation of C_{27} sterols, even though these activities are much greater in untreated rat liver microsomes. However, the relatively high 25-hydroxylase activities are observed, in part, because solubilization and partial purification of this microsomal P-450 substantially activate (50 times) this reaction toward vitamin D_3 and 5β-cholestane derivatives. This form of P-450 also catalyzed hydroxylation of testosterone at two positions (2β, 1.0 min^{-1}; 16α, 0.9 min^{-1}) and demethylation of ethylmorphine (21 min^{-1}). These latter activities indicate that even for a highly specific sterol hydroxylase, smaller lipophilic molecules can occupy the active site and undergo hydroxylation. As might be expected, the smaller substrate is also free to adopt more than one configuration in the active site. For example, the reaction products for testosterone indicate hydroxylation at either end of the molecule. Other forms of liver microsomal P-450 hydroxylate testosterone at a variety of positions with selectivity that is dependent on the individual form.[124] Recent studies show that even closely related variants of rabbit liver P-450$_{LM3b}$ exhibit major changes in regioselectivity for progesterone.[125]

Out of 15 forms of P-450 that have been purified from uninduced rabbit liver microsomes by Aoyama *et al.*,[126] four forms catalyzed 25-hydroxylation of vitamin D_3, while two other forms produced either 26-hydroxyvitamin D_3 or a second unidentified product. This reflects part of a general trend for cytochromes P-450, as distinct from cytochromes of the "P-448" type, to catalyze ω- and ω-1-hydroxylation of long hydrocarbon chains. Other examples are provided by fatty acids (by nine forms of P-450) and prostaglandins. Presumably, the hydrocarbon chain fits into a long, narrow site with the terminal carbons adjacent to the heme. The specificity of these reactions between ω and ω-1 can be changed by participation of b_5, suggesting an additional perturbation on the active site.

It is clear that many forms of microsomal P-450 have important roles in the synthesis and metabolism of key natural products. However, many of these forms can also metabolize a variety of xenobiotics to provide a diversity of products (low regiospecificity). Mitochondrial forms of P-450 generally exhibit a high degree of specificity both for substrates and in product formation. The relationship between these many forms of P-450 must await more detailed sequence and structural characterization.

References

1. Lieberman, S., Greenfield, N. J., and Wolfson, A., 1984, A heuristic proposal for understanding steroidogenic processes, *Endocr. Rev.* **5**:128–148.

2. Suzuki, K., and Kimura, T., 1965, An iron protein as a component of steroid β-hydroxylase complex, *Biochem. Biophys. Res. Commun.* **19:**340–345.
3. Chu, J. W., and Kimura, T., 1973, Studies on adrenal steroid hydroxylases: Molecular and catalytic properties of adrenodoxin reductase (a flavoprotein), *J. Biol. Chem.* **248:**2089–2094.
4. Lambeth, J. D., Seybert, D. W., Lancaster, J. R., Salerno, J. C., and Kamin, H., 1982, Steroidogenic electron transport in adrenal cortex mitochondria, *Mol. Cell Biochem.* **45:**13–31.
5. Lambeth, J. D., and Kamin, H., 1976, Adrenodoxin reductase: Properties of the reduced enzyme with $NADP^+$ and NADPH, *J. Biol. Chem.* **251:**4299–4306.
6. Light, D. R., and Walsh, C., 1980, Flavin analogs as mechanistic probes of adrenodoxin reductase-dependent electron transfer to the cholesterol side chain cleavage cytochrome P-450 of the adrenal cortex, *J. Biol. Chem.* **255:**4264–4277.
7. Lambeth, J. D., and Kamin, H., 1977, Adrenodoxin reductase and adrenodoxin: Mechanisms of reduction of ferricyanide and cytochrome c, *J. Biol. Chem.* **252:**2908–2917.
8. Chu, J. W., and Kimura, T., 1973, Studies on adrenal steroid hydroxylases: Complex formation of the hydroxylase components, *J. Biol. Chem.* **248:**5183–5187.
9. Lambeth, J. D., McCaslin, D. R., and Kamin, H., 1976, Adrenodoxin reductase adrenodoxin complex: Catalytic and thermodynamic properties, *J. Biol. Chem.* **251:**7545–7550.
10. Lambeth, J. D., and Kamin, H., 1979, Adrenodoxin reductase adrenodoxin complex: Flavin to iron-sulfur electron transfer as the rate limiting step in the NADPH-cytochrome c reductase reaction, *J. Biol. Chem.* **254:**2766–2774.
11. Lambeth, J. D., Seybert, D. W., and Kamin, H., 1979, Ionic effects on adrenal steroidogenic electron transport: The role of adrenodoxin as an electron shuttle, *J. Biol. Chem.* **254:**7255–7264.
12. Sheridan, R. P., Allen, L. C., and Carter, C. W., 1981, Coupling between oxidation state and hydrogen bond conformation in high potential iron-sulfur protein, *J. Biol. Chem.* **256:**5052–5057.
13. Lambeth, J. D., Lancaster, J. R., and Kamin, H., 1981, Adrenodoxin reductase adrenodoxin complex: Rapid formation and breakdown of the complex and a slow conformational change in the flavoprotein, *J. Biol. Chem.* **255:**4667–4672.
14. Lambeth, J. D., Lancaster, J. R., and Kamin, H., 1981, Steroidogenic electron transport by adrenodoxin reductase and adrenodoxin: Use of acetylated cytochrome c as a mechanistic probe of electron transfer, *J. Biol. Chem.* **256:**3674–3678.
15. Hanukoglu, I., and Jefcoate, C. R., 1980, Mitochondrial cytochrome P-450$_{scc}$: Mechanism of electron transport by adrenodoxin, *J. Biol. Chem.* **255:**3057–3061.
16. Lambeth, J. D., and Pember, S. O., 1983, Cytochrome P-450$_{scc}$–adrenodoxin complex: Reduction properties of the substrate-associated cytochrome and relation of the reduction states of heme and iron-sulfur centers to association of proteins, *J. Biol. Chem.* **258:**5596–5602.
17. Hanukoglu, I., Privalle, C. T., and Jefcoate, C. R., 1981, Mechanisms of ionic activation of adrenal mitochondrial cytochromes P-450$_{scc}$ and P-450$_{11\beta}$, *J. Biol. Chem.* **256:**4329–4335.
18. Lambeth, J. D., Seybert, D. W., and Kamin, H., 1980, Phospholipid vesicle-reconstituted cytochrome P-450$_{scc}$: Mutually facilitated binding of cholesterol and adrenodoxin, *J. Biol. Chem.* **255:**138–143.
19. Sligar, S. G., and Gunsalus, I. C., 1976, A thermodynamic model of regulation: Modulation of redox equilibria in camphor monoxygenase, *Proc. Natl. Acad. Sci. USA* **73:**1078–1082.
20. Light, D. R., and Orme-Johnson, N. R., 1981, Beef adrenal cortical cytochrome P-

450 which catalyzes the conversion of cholesterol to pregnenolone: Oxidation-reduction potentials of the free, steroid-complexed, and adrenodoxin-complexed P-450, *J. Biol. Chem.* **256**:343–350.
21. Hanukoglu, I., Spitsberg, V., Bumpus, J. A., Dus, K. M., and Jefcoate, C. R., 1981, Adrenal mitochondrial cytochrome P-450$_{scc}$: Cholesterol and adrenodoxin interactions at equilibrium and during turnover, *J. Biol. Chem.* **256**:4321–4328.
22. Kido, T., and Kimura, T., 1979, The formation of binary and ternary complexes of cytochrome P-450$_{scc}$ with adrenodoxin and adrenodoxin reductase adrenodoxin complex: The implication in ACTH function, *J. Biol. Chem.* **254**:11806–11815.
23. Burstein, S., Middleditch, B. S., and Gut, M., 1975, Mass spectrometric study of the enzymatic conversion of cholesterol to (22R)-22-hydroxycholesterol, (20R,22R)-20,22-dihydroxycholesterol, and pregnenolone, and of (22R)-22-hydroxycholesterol to the glycol and pregnenolone in bovine adrenocortical preparations: Mode of oxygen incorporation, *J. Biol. Chem.* **250**:9028–9037.
24. Hume, R., Kelly, R. W., Taylor, P. L., and Boyd, G. S., 1984, The catalytic cycle of cytochrome P-450$_{scc}$ and intermediates in the conversion of cholesterol to pregnenolone, *Eur. J. Biochem.* **140**:583–591.
25. Orme-Johnson, N. R., Light, D. R., White-Stevens, R. W., and Orme-Johnson, W. H., 1979, Steroid binding properties of beef adrenal cortical cytochrome P-450 which catalyzes the conversion of cholesterol into pregnenolone, *J. Biol. Chem.* **254**:2103–2111.
26. Paul, D. P., Gallant, S., Orme-Johnson, N. R., Orme-Johnson, W. H., and Brownie, A. C., 1976, Temperature dependence of cholesterol binding to cytochrome P-450$_{scc}$ of the rat adrenal: Effect of adrenocorticotropic hormone and cycloheximide, *J. Biol. Chem.* **251**:7120–7126.
27. Jefcoate, C. R., Orme-Johnson, W. H., and Beinert, H., 1976, Cytochrome P-450 of bovine adrenal mitochondria: Ligand binding to two forms resolved by EPR spectroscopy, *J. Biol. Chem.* **251**:3706–3715.
28. Jefcoate, C. R., Simpson, E. R., Boyd, G. S., Brownie, A. C., and Orme-Johnson, W. H., 1973, The detection of different states of the P-450 cytochromes in adrenal mitochondria: Changes induced by ACTH, *Ann. N.Y. Acad. Sci.* **212**:243.
29. Jänig, G. R., Makower, A., Kraft, R., Rabe, H., and Ruckpaul, K., 1984, Identification of tyrosine as axial heme iron ligand in cytochrome P-450-LM2, *Xenobiotica* **14**(S1):49.
30. Jefcoate, C. R., 1977, Cytochrome P-450 of adrenal mitochondria: Steroid binding sites on two distinguishable forms of rat adrenal mitochondrial cytochrome P-450$_{scc}$, *J. Biol. Chem.* **252**:8788–8796.
31. Sheets, J. J., and Vickery, L. E., 1983, Active site-directed inhibitors of cytochrome P-450$_{scc}$: Structural and mechanistic implications of a side chain-substituted series of amino-steroids, *J. Biol. Chem.* **258**:11446–11452.
32. Greenfield, N. J., Gerolimatos, B., Szwergold, B. S., Wolfson, A. J., Prasad, V. V. K., and Lieberman S., 1981, Effects of phospholipid and detergent on the substrate specificity of adrenal cytochrome P-450$_{scc}$: Substrate binding and kinetics of cholesterol side chain oxidation, *J. Biol. Chem.* **256**:4407–4417.
33. Jefcoate, C. R., 1982, pH modulation of ligand binding to adrenal mitochondrial cytochrome P-450$_{scc}$, *J. Biol. Chem.* **257**:4731–4737.
34. Pember, S. O., Powell, G. L., and Lambeth, J. D., 1983, Cytochrome P-450$_{scc}$–phospholipid interactions: Evidence for a cardiolipin binding site and thermodynamics of enzyme interactions with cardiolipin, cholesterol, and adrenodoxin, *J. Biol. Chem.* **258**:3198–3206.
35. Robinson, N. C., Strey, F., and Talbert, L., 1980, Investigation of the essential boundary layer phospholipids of cytochrome c oxidase using Triton X-100 delipidation, *Biochemistry* **19**:3656–3661.

36. Larroque, C., and van Lier, J. E., 1983, Spectroscopic evidence for the formation of a transient species during cytochrome P-450$_{scc}$ induced hydroperoxysterol–glycol conversions, *Biochem. Biophys. Res. Commun.* **112**:655–662.
37. Sligar, S. G., Cinti, D. L., Gibson, G. G., and Schenkman, J. B., 1979, Spin state control of the hepatic cytochrome P450 redox potential, *Biochem. Biophys. Res. Commun.* **90**:925–932.
38. Ristau, O., Rein, H., Greschner, S., Jänig, G.-R., and Ruckpaul, K., 1979, Quantitative analysis of the spin equilibrium of cytochrome P 450 LM2 fraction from rabbit liver microsomes, *Acta Biol. Med. Ger.* **38**:177–185.
39. Larroque, C., and van Lier, J. E., 1980, The subzero temperature stabilized oxyferro complex of purified cytochrome P-450$_{scc}$, *FEBS Lett.* **115**:175–177.
40. Tuckey, R. C., and Kamin, H., 1983, Kinetics of O_2 and CO binding to adrenal cytochrome P-450$_{scc}$: Effect of cholesterol, intermediates, and phosphatidylcholine vesicles, *J. Biol. Chem.* **258**:4232–4237.
41. Mitani, F., Ilzuka, T., Ueno, R., Ishimura, Y., Kimura. T., Izumi, S., Komatsu, N., and Watanabe, K., 1982, Regulation of cytochrome P450 activities in adrenocortical mitochondria from normal rats and human neoplastic tissues, *Adv. Enzyme Regul.* **20**:213–231.
42. Morohashi, K., Fujii-Kuriyama, Y., Okada, Y., Sogawa, K., Hirose, T., Inayama, S., and Omura, T., 1984, Molecular cloning and nucleotide sequence of cDNA for mRNA of mitochondrial cytochrome P-450(SCC) of bovine adrenal cortex, *Proc. Natl. Acad. Sci. USA* **81**:4647–4651.
43. Chashchin, V. L., Vasilevsky, V. I., Shkumatov, V. M., and Akhrem, A. A., 1984, The domain structure of the cholesterol side-chain cleavage cytochrome P-450 from bovine adrenocortical mitochondria, *Biochim. Biophys. Acta* **787**:27–38.
44. Ichikawa, Y., and Hiwatashi, A., 1982, The role of the sugar regions of components of the cytochrome P-450-linked mixed-function oxidase (monooxygenase) system of bovine adrenocortical mitochondria, *Biochim. Biophys. Acta* **705**:82–91.
45. Sato, H., Ashida, N., Suhara, K., Itagaki, E., Takemori, S., and Katagiri, M., 1978, Properties of an adrenal cytochrome P-450 (P-450$_{11\beta}$) for the hydroxylations of corticosteroids, *Arch. Biochem. Biophys.* **190**:307–314.
46. Momoi, K., Okamoto, M., Fujii, S., Kim, C. Y., Miyake, Y., and Yamano T., 1983, 19-Hydroxylation of 18-hydroxy-11-deoxycorticosterone catalyzed by cytochrome P-450$_{11\beta}$ of bovine adrenocortex, *J. Biol. Chem.* **258**:8855–8860.
47. Okamoto, M., Wada, A., Onishi, T., Nonaka, Y., and Yamano, T., 1984, Cytochrome P450$_{11\beta}$ catalyzes production of aldosterone from corticosterone, *Sixth International Symposium on Microsomes and Drug Oxidations* (Abstracts), Brighton, England, p. 18.
48. Watanuki, M., Tilley, B. E., and Hall, P. F., 1978, Cytochrome P-450 for 11β- and 18-hydroxylase activities of bovine adrenocortical mitochondria: One enzyme of two?, *Biochemistry* **17**:127–130.
49. Haning, R., Tait, S. A. S., and Tait, J. F., 1970, *In vitro* effects of ACTH, angiotensins, serotonin and potassium on steroid output and conversion of corticosterone to aldosterone by isolated adrenal cells, *Endocrinology* **87**:1147–1167.
50. Kramer, R. E., Gallant, S., and Brownie, A. C., 1979, The role of cytochrome P-450 in the action of sodium depletion on aldosterone biosynthesis in rats, *J. Biol. Chem.* **254**:3953–3958.
51. Martsev, S. P., Bespalov, I. A., Chashchin, V. L., and Akhrem, A. A., 1982, Steroid 11β-hydroxylase system: Reconstitution and study of interactions among protein components, in: *Cytochrome P-450: Biochemistry, Biophysics and Environmental Implications* (E. Hietanen, M. Laitinen, and O. Hänninen, eds.), Elsevier, Amsterdam, pp. 413–420.

52. Lambeth, J. D., Kamin, H., and Seybert, D. W., 1980, Phosphatidylcholine vesicle reconstituted cytochrome P-450$_{scc}$: Role of the membrane in control of activity and spin state of the cytochrome, *J. Biol. Chem.* **255**:8282–8288.
53. Tuckey, R. C., and Kamin, H., 1982, Kinetics of the incorporation of adrenal cytochrome P-450$_{scc}$ into phosphatidylcholine vesicles, *J. Biol. Chem.* **257**:2887–2893.
54. Kowluru, R. A. George, R., and Jefcoate, C. R., 1983, Polyphosphoinositide activation of cholesterol side chain cleavage with purified cytochrome P-450$_{scc}$, *J. Biol. Chem.* **258**:8053–8059.
55. Lombardo, A., Defaye, G., Guidicelli, C., Monnier, N., and Chambaz, E. M., 1982, Integration of purified adrenocortical cytochrome P-450$_{11\beta}$ into phospholipid vesicles, *Biochem. Biophys. Res. Commun.* **104**:1638–1645.
56. Kimura, T., 1981, ACTH stimulation on cholesterol side chain cleavage activity of adrenocortical mitochondria, *Mol. Cell. Biochem.* **36**:105–122.
57. Peron, F. G., and Tsang, C. P. W., 1969, Further studies on corticosteroidogenesis. VI. Pyruvate and malate supported steroid 11β-hydroxylation in rat adrenal gland mitochondria, *Biochim. Biophys. Acta* **180**:445–458.
58. Hall, P. F., 1972, A possible role for transhydrogenation in side-chain cleavage of cholesterol, *Biochemistry* **11**:2891–2897.
59. Klimek, J., Boguslawski, W., and Zelewski, L., 1979, The relationship between energy generation and cholesterol side-chain cleavage reaction in the mitochondria from human term placenta, *Biochim. Biophys, Acta* **587**:362–372.
60. Stevens, V. L., Aw, T. Y., Jones, D. P., and Lambeth, J. D., 1984, Oxygen dependence of adrenal cortex cholesterol side chain cleavage: Implications in the rate-limiting steps in steroidogenesis, *J. Biol. Chem.* **259**:1174–1179.
61. Jefcoate, C. R., Simpson, E. R., and Boyd, G. S., 1974, Spectral properties of rat adrenal-mitochondrial cytochrome P-450, *Eur. J. Biochem.* **42**:539–551.
62. Crivello, J. F., and Jefcoate, C. R., 1980, Intracellular movement of cholesterol in rat adrenal cells: Kinetics and effects of inhibitors, *J. Biol. Chem.* **255**:8144–8151.
63. Strauss, J. F., III, Schuler, L. A., Rosenblum, M. F., and Tanaka, T., 1981, Cholesterol metabolism by ovarian tissue, *Adv. Lipid Res.* **18**:99–157.
64. Cooke, B. A., Dix, C. J., Magee-Brown, R., Janszen, F. H. A., and van der Molen, H. J., 1981, Hormonal control of Leydig cell function, *Adv. Cyclic Nucleotide Res.* **14**:593–609.
65. DiBartolomeis, M. J., and Jefcoate, C. R., 1984, Characterization of the acute stimulation of steroidogenesis in primary bovine adrenal cortical cell cultures, *J. Biol. Chem.* **259**:10159–10167.
66. Simpson, E. R., McCarthy, J. L. and Peterson, J. A., 1978, Evidence that the cycloheximide-sensitive site of adrenocorticotropic hormone action is in the mitochondrion, *J. Biol. Chem.* **253**:3135–3139.
67. Gwynne, J. T., and Hess, B., 1980, The role of high density lipoproteins in rat adrenal cholesterol metabolism and steroidogenesis, *J. Biol. Chem.* **255**:10875–10883.
68. Nishikawa, T., Mikami, K., Saito, Y., Tamura, Y., and Kumagai, A., 1981, Studies on cholesterol esterase in the rat adrenal, *Endocrinology* **108**:932–936.
69. Hall, P. F., 1984, The role of the cytoskeleton in hormone action, *Can. J. Biochem. Cell Biol.* **62**:653–665.
70. Simpson, E. R., Jefcoate, C. R., Brownie, A. C., and Boyd, G. S., 1972, The effect of ether anaesthesia stress on cholesterol-side-chain cleavage and cytochrome P450 in rat-adrenal mitochondria, *Eur. J. Biochem.* **28**:442–450.
71. Brownie, A. C., Simpson, E. R., Jefcoate, C. R., Boyd, G. S., Orme-Johnson, W. H., and Beinert, H., 1972, Effect of ACTH on cholesterol side-chain cleavage in rat adrenal mitochondria, *Biochem. Biophys. Res. Commun.* **46**:483–490.

72. Jefcoate, C. R., and Orme-Johnson, W. H., 1975, Cytochrome P-450 of adrenal mitochondria: *In vitro* and *in vivo* changes in spin states, *J. Biol. Chem.* **250:**4671–4677.
73. Williams-Smith, D. L., Simpson, E. R., Barlow, S. M., and Morrison, P. J., 1976, Electron paramagnetic resonance studies of cytochrome P-450 and adrenal ferredoxin in single whole rat adrenal glands: Effect of corticotropin, *Biochim. Biophys. Acta* **449:**72–83.
74. von Dippe, P. J., Tsao, K., and Harding, B. W., 1982, The effect of ether stress and cycloheximide treatment on cholesterol binding and enzyme turnover of adrenal cortical cytochrome P450$_{scc}$, *J. Steroid Biochem.* **16:**763–769.
75. Privalle, C. T., Crivello, J. F., and Jefcoate, C. R., 1983, Regulation of intramitochondrial cholesterol transfer to side-chain cleavage cytochrome P-450 in rat adrenal gland, *Proc. Natl. Acad. Sci USA* **80:**702–706.
76. Privalle, C. T., and Jefcoate, C. R., 1985, ACTH control of cholesterol side chain cleavage at adrenal mitochondrial cytochrome P-450$_{scc}$: Regulation of intramitochondrial cholesterol transfer, *J. Biol. Chem.* submitted for publication.
77. Pederson, R. C., and Brownie, A. C., 1983, Cholesterol side-chain cleavage in the rat adrenal cortex: Isolation of a cycloheximide-sensitive activator peptide, *Proc. Natl. Acad. Sci USA* **80:**1882–1886.
78. Vahouny, G. V., Chanderbhan, R., Noland, B. J., Irwin, D., Dennis, P., Lambeth, J. D., and Scallen, T. J., 1983, Sterol carrier protein$_2$: Identification of adrenal sterol carrier protein$_2$ and site of action for mitochondrial cholesterol utilization, *J. Biol. Chem.* **258:**11731–11737.
79. Krueger, R. J., and Orme-Johnson, N. R., 1983, Acute adrenocorticotropic hormone stimulation of adrenal corticosteroidogenesis: Discovery of a rapidly induced protein, *J. Biol. Chem.* **258:**10159–10167.
80. Simpson, E. R., 1979, Cholesterol side-chain cleavage, cytochrome P450, and the control of steroidogenesis, *Mol. Cell. Endocrinol.* **13:**213–227.
81. Farese, R. V., 1984, Phospholipids as intermediates in hormone action, *Mol. Cell. Endocrinol.* **35:**1–14.
82. Igarashi, Y., and Kimura, T., 1984, Adrenocorticotropic hormone-mediated changes in rat adrenal mitochondrial phospholipids, *J. Biol. Chem.* **259:**10745–10753.
83. Kominami, s., Hara, H., Ogishima, T., and Takemori, S., 1984, Interaction between cytochrome P-450 (P-450$_{C21}$) and NADPH-cytochrome P-450 reductase from adrenocortical microsomes in a reconstituted system, *J. Biol. Chem.* **259:**2991–2999.
84. Kominami, S., and Takemori, S., 1982, Effect of spin state on reduction of cytochrome P-450 (P-450$_{C21}$) from bovine adrenocortical microsomes, *Biochim. Biophys. Acta* **709:**147–153.
85. Nakajin, s., Hall, P. F., and Onoda, M., 1981, Testicular microsomal cytochrome P-450 for C$_{21}$ steroid side chain cleavage: Spectral and binding studies, *J. Biol. Chem.* **256:**6134–6139.
86. Kominami, S., Shinzawa, K., and Takemori, S., 1982, Purification and some properties of cytochrome P-450 specific for steroid 17α-hydroxylation and C$_{17}$–C$_{20}$ bond cleavage from guinea pig adrenal microsomes, *Biochem. Biophys. Res. Commun.* **109:**916–921.
87. Katagiri, M., Suhara, K., Shiroo, M., and Fujimura, Y., 1982, Role of cytochrome b$_5$ in the cytochrome P-450-mediated C$_{21}$-steroid 17,20-lyase reaction, *Biochem. Biophys. Res. Commun.* **108:**379–384.
88. Takemori, S., Kominami, S., and Shinzawa, K., 1984, Studies on cytochrome P450$_{17α,lyase}$ from guinea pig, in: *International Union of Biochemistry Symposium 134* (Abstracts), Indian Institute of Science, Bangalore.
89. Onoda, M., and Hall, P. F., 1982, Cytochrome b$_5$ stimulates purified testicular microsomal cytochrome P-450 (C$_{21}$ side-chain cleavage), *Biochem. Biophys. Res. Commun.* **108:**454–460.

90. Thompson, E. A., and Siiteri, P. K., 1974, Utilization of oxygen and reduced nicotinamide adenine dinucleotide phosphate by human placental microsomes during aromatization of androstenedione, *J. Biol. Chem.* **249**:5364–5372.
91. Meigs, R. A., and Ryan, K. J., 1968, Cytochrome P-450 and steroid biosynthesis in the human placenta, *Biochim. Biophys. Acta* **165**:476–482.
92. Thompson, E. A., Jr., and Siiteri, P. K., 1974, The involvement of human placental microsomal cytochrome P-450 in aromatization, *J. Biol. Chem.* **249**:5373–5378.
93. Zachariah, P. K., and Juchau, M. R., 1977, Inhibition of placental mixed function oxidation with carbon monoxide: Reversal with monochromatic light, *J. Steroid Biochem.* **8**:221–228.
94. Akhtar, M., Calder, M. R., Corina, D. L., and Wright, J. N., 1982, Mechanistic studies on C-19 demethylation in oestrogen biosynthesis, *Biochem. J.*, **201**:569–580.
95. Fishman, J., and Goto, H. L., 1982, Biochemical mechanism of aromatization, *Cancer Res.* (Suppl.) **42**:3277s–3280s.
96. Caspi, E., Wicha, J., Arunachalam, T., Nelson, P., and Spiteller, G., 1984, Estrogen biosynthesis: Concerning the obligatory intermediacy of 2β-hydroxy-10β-formylandrost-4-ene-3,17-dione, *J. Am. Chem. Soc.* **106**:7282–7283.
97. Reed, K. C., and Ohno, S., 1976, Kinetic properties of human placental aromatase: Application of an assay measuring ^3H_2O release from 1gb,2gb-^{3}H-androgens, *J. Biol. Chem.* **251**:1625–1631.
98. Osawa, Y., Tochigi, B., Higashiyama, T., Yarborough, C., Nakamura, T., and Yamamoto, T., 1982, Multiple forms of aromatase and response of breast cancer aromatase to antiplacental aromatase II antibodies, *Cancer Res.* (Suppl.) **42**:3299s–3306s.
99. Fraser, D. R., 1980, Regulation of the metabolism of vitamin D, *Physiol. Rev.* **60**:551–613.
100. Larkins, R. G., Macanley, S. J., and Macintyre, I., 1975, Inhibitors of protein and RNA synthesis and 1,25-dihydroxycholecalciferol formation *in vitro*, *Mol. Cell. Endocrinol.* **2**:193–202.
101. Horiuchi, N., Suda, T., Sasaki, S., Ogata, E., Ezawa, I., Sano, Y., and Shimazawa, E., 1975, The regulatory role of calcium in 25-hydroxycholecalciferol metabolism in chick kidney *in vitro*, *Arch. Biochem. Biophys.* **171**:540–548.
102. Trzaskos, J. M., Bowen, W. D., Shafiee, A., Fischer, R. T., and Gaylor, J. L., 1984, Cytochrome P-450-dependent oxidation of lanosterol in cholesterol biosynthesis: Microsomal electron transport and C-32 demethylation, *J. Biol. Chem.* **259**:13402–13412.
103. Panini, S. R., Sexton, R. C., and Rudney, H., 1984, Regulation of 3-hydroxy-3-methylglutaryl coenzyme A reduuctase by oxysterol by-products of cholesterol biosynthesis: Possible mediators of low density lipoprotein action, *J. Biol. Chem.* **259**:7767–7771.
104. Goad, L. J., 1984, Cholesterol biosynthesis and metabolism, in: *Biochemistry of Steroid Hormones*, 2nd ed. (H. L. J. Makin, ed.), Blackwell, Oxford, pp. 20–70.
105. Danielsson, H., and Sjövall, J., 1975, Bile acid metabolism, *Annu. Rev. Biochem.* **44**:233–253.
106. Mitropoulos, K. A., Balusubramanian, S., Gibbons, G. F., and Reeves, B. E. A., 1972, Diurnal variation in the activity of cholesterol 7α-hydroxylase in livers of fed and fasted rats, *FEBS Lett.* **27**:203–253.
107. Kwok, C. T., Burnett, W., and Hardie, I. R., 1981, Regulation of rat liver microsomal cholesterol 7α-hydroxylase: Presence of a cytosolic activator, *J. Lipid Res.* **22**:570–579.
108. Gray, R. W., Omdahl, J. L., Ghazarian, J. G., and DeLuca, H. F., 1972, 25-Hydroxycholecalciferol-1-hydroxylase: Subcellular location and properties, *J. Biol. Chem.* **247**:7528–7532.
109. Pedersen, J. I., Björkhem, I., and Gustafsson, J., 1979, 26-Hydroxylation of C_{27}-

steroids by soluble liver mitochondrial cytochrome P-450, *J. Biol. Chem.* **254:**6464–6469.
110. Pedersen, J. I., Oftebro, H., and Vänngard, T., 1977, Isolation from bovine liver mitochondria of a soluble ferredoxin active in a reconstituted steroid hydroxylation reaction, *Biochem. Biophys. Res. Commun.* **76:**666–673.
111. Pedersen, J. I., Ghazarian, J. G., Orme-Johnson, N. R., and DeLuca, H. F., 1976, Isolation of chick renal mitochondrial ferredoxin active in the 25-hydroxyvitamin D_3-1α-hydroxylase system, *J. Biol. Chem.* **251:**3933–3941.
112. Björkhem, I., Holmberg, I., Oftebro, H., and Pedersen, J. I., 1980, Properties of a reconstituted vitamin D_3 25-hydroxylase from rat liver mitochondria, *J. Biol. Chem.* **255:**5244–5249.
113. Ghazarian, J. G., Jefcoate, C. R., Knutson, J. C., Orme-Johnson, W. H., and DeLuca, H. F., 1974, Mitochondrial cytochrome P_{450}: A component of chick kidney 25-hydroxycholecalciferol-1α-hydroxylase, *J. Biol. Chem.* **249:**3026–3033.
114. Warner, M., 1983, 25-Hydroxyvitamin D hydroxylation: Evidence for a dioxygenase activity of solubilized renal mitochondrial cytochrome P-450, *J. Biol. Chem.* **258:**11590–11593.
115. Ghazarian, J. G., and DeLuca, H. F., 1974, 25-Hydroxylase: A specific requirement for NADPH and a hemoprotein component in kidney mitochondria, *Arch. Biochem. Biophys.* **160:**63–72.
116. Hiwatashi, A., Nishii, Y., and Ichikawa, Y., 1982, Purification of cytochrome P-450$_{D1\alpha}$ (25-hydroxyvitamin D_3-1α-hydroxylase) of bovine kidney mitochondria, *Biochem. Biophys. Res. Commun.* **105:**320–327.
117. Niranjan, B. G., Wilson, N. M., Jefcoate, C. R., and Avadhani, N. G., 1984, Hepatic mitochondrial cytochrome P-450 system: Distinctive features of cytochrome P-450 involved in the activation of aflatoxin B_1 and benzo(a)pyrene, *J. Biol. Chem.* **259:**12495–12501.
118. Hansson, R., and Wikvall, K., 1980, Hydroxylations in biosynthesis and metabolism of bile acids: Catalytic properties of different forms of cytochrome P-450, *J. Biol. Chem.* **255:**1643–1649.
119. Boström, H., 1983, Binding of cholesterol to cytochromes P-450 from rabbit liver microsomes, *J. Biol. Chem.* **258:**15091–15094.
120. Hansson, R., and Wikvall, K., 1982, Hydroxylations in biosynthesis of bile acids: Cytochrome P-450 LM_4 and 12α-hydroxylation of 5β-cholestane-3α,7α-diol, *Eur. J. Biochem.* **125:**423–429.
121. Wikvall, K., 1984, Purification and properties of the cytochrome P450 species involved in bile acid biosynthesis, in: *International Union of Biochemistry Symposium 134* (Abstracts), Bangalore, p. 134.
122. Goodwin, C. D., Cooper, B. W., and Margolis, S., 1982, Rat liver cholesterol 7α-hydroxylase: Modulation of enzyme activity by changes in phosphorylation state, *J. Biol. Chem.* **257:**4469–4472.
123. Andersson, S., Holmberg, I., and Wikvall, K., 1983, 25-Hydroxylation of C_{27}-steroids and vitamin D_3 by a constitutive cytochrome P-450 from rat liver microsomes, *J. Biol. Chem.* **258:**6777–6781.
124. Ryan, D. E., Iida, S., Wood, A. W., Thomas, P. E., Lieber, C. S., and Levin, W., 1984, Characterization of three highly purified cytochromes P450 from hepatic microsomes of adult male rats, *J. Biol. Chem.* **259:**1239–1250.
125. Johnson, E. F., and Schwab, G. E., 1984, Constitutive forms of rabbit-liver microsomal cytochrome P-450: Enzymatic diversity, polymorphism and allosteric regulation, *Xenobiotica* **14:**3–18.
126. Aoyama, T., Imai, Y., and Sato, R., 1984, Hydroxylation of vitamin D_3, prostaglandins and fatty acids by multiple forms of cytochrome P450 purified from rabbit liver microsomes, *Sixth International Symposium on Microsomes and Drug Oxidations* (Abstracts), Brighton, England, p. 44.

CHAPTER 12

Cytochrome P-450$_{cam}$ and Other Bacterial P-450 Enzymes

STEPHEN G. SLIGAR and RALPH I. MURRAY

1. Introduction

The purpose of this chapter is twofold. First, we will attempt to summarize the existing knowledge on the wide variety of bacterial cytochrome P-450 hemeproteins that have been discovered. As such, this represents the first comprehensive review of the bacterial cytochromes P-450. It should become evident to the reader that these cytochromes comprise a class just as diverse, if not more so, than their eukaryotic counterparts. Perhaps the best known and most extensively characterized of these bacterial enzymes is P-450$_{cam}$ isolated from *Pseudomonas putida*. From the initial isolation in Gunsalus's laboratory in 1967 and subsequent investigations in many laboratories, the numerous studies employing this cytochrome have shed much light on the mechanisms associated with the family of P-450-dependent monooxygenases as a whole. A second goal of this chapter is to review the detailed biophysical, biochemical, and molecular biological studies conducted with this system. It should be noted that "P-450$_{cam}$" and "bacterial P-450" are not synonymous. The various P-450 molecules described herein are in all probability only a very small fraction of those present in the myriad of bacterial cells in the biosphere. In view of the variety of substrate transformations that this enzyme family is able to catalyze, some of which are energetically very difficult, their distribution in nature is no doubt ubiquitous. One need only look through the American Type Culture Collection Catalogue[1] under such diverse genera as *Pseudomonas, Rhodococcus, Bacillus,* and *Nocardia* among others to

---

STEPHEN G. SLIGAR and RALPH I. MURRAY • Department of Biochemistry, University of Illinois, Urbana, Illinois 61801.

see the potential roles of yet undiscovered P-450 monooxygenases in the degradation and transformation of a plethora of compounds. Even if only a small portion of the enzymes involved in these various biotransformations are P-450 isozymes, there would correspond many hundreds of P-450 hemeproteins with definable specificities.

2. Bacterial Cytochromes P-450

In this section we will review the various P-450 systems documented to date (Table I). It has become accepted practice to affix a subscript to "P-450" to denote either the cell type of origin or the particular substrate metabolized. Insofar as this can be maintained unambiguously, we will adhere to this practice. Clearly, a new nomenclature would be less cumbersome in dealing with the myriad of P-450-catalyzed biotransformations. This is beyond the scope of this chapter.

2.1. P-450$_{meg}$

The ability of various strains of *Bacillus* to catalyze the hydroxylation of steroids in cell-free systems has been known for many years.[2-5] Gustafsson and co-workers investigated the steroid 15β-hydroxylase activity of *B. megaterium* (ATCC 13368) and found it to be catalyzed by

TABLE I
Bacterial P-450 Systems

Bacterium	Function	Reference
Acinetobacter calcoaceticus (EB104)	ω-Alkane hydroxylase	54–58
Bacillus megaterium (ATCC 13368)	15β-Hydroxylase of 3-oxo-Δ^4-steroids	6–9
B. megaterium (ATCC 14581)	ω-2-Fatty acid hydroxylase-epoxidase	26–34, 36, 48–51
Nocardia sp. (NH1)	*p*-Alkylphenyl ether dealkylase	126–128
Pseudomonas incognita	Linalool-10-methyl-hydroxylase Linalool-8-methyl-hydroxylase	106–110
P. putida (ATCC 17453)	Camphor-5-*exo*-hydroxylase	65–68
P. putida (JT810)	*p*-Cymene-7-hydroxylase	96, 97
P. putida (PL-W)	*p*-Cymene-7-hydroxylase	94–102
Rhizobium japonicum (CC 705)	Unknown, but see Ref.119	11, 116–120
Rhodococcus rhodochrous (ATCC 19067)	ω-Alkane hydroxylase	42, 43
Rhodococcus sp.	Camphor-6-*endo*-hydroxylase	89, 92, 93

a soluble P-450-dependent monooxygenase.[6-9] This "P-450$_{meg}$" was found to have a molecular weight of 52,000 and an isoelectric point at pH 4.9,[9] similar to the values reported for P-450$_{cam}$,[10] P-450$_c$[11] (*Rhizobium japonicum*), and P-450$_{npd}$.[12] P-450$_{meg}$ also resembles P-450$_{cam}$ and P-450c in its optical and EPR spectral characteristics.[9] The absolute optical spectrum of oxidized P-450$_{meg}$ has absorption maxima at 412 and 544 nm, while the reduced CO-liganded species shows absorption maxima at 420, 450, and 550 nm. The liquid-nitrogen-temperature EPR spectrum of reduced P-450$_{meg}$ is characteristic of that for low-spin P-450, with g-values at 1.92, 2.24, and 2.41. Berg et al.[6,7] determined that this 15β-steroid hydroxylase system consisted of three components: an NADPH-specific FMN-containing megaredoxin reductase; an iron–sulfur protein, megaredoxin; and P-450$_{meg}$. In both cell-free extracts and the reconstituted system, NADH could not substitute for NADPH. Thus, this is similar to the associated electron transfer system of the adrenal mitochondrial 11β-hydroxylase system, which consists of an NADPH-specific FAD-containing adrenodoxin reductase and adrenodoxin,[13] but different from both the P-450$_{cam}$ system, which consists of an NADH-specific FAD-containing putidaredoxin reductase and putidaredoxin, and the hepatic microsomal system, with an NADPH-specific FMN- and FAD-containing P-450 reductase.[14,15]

A variety of steroids are found to be processed by P-450$_{meg}$ in both cell-free and reconstituted systems, including progesterone and other 3-oxo-Δ^4-steroids,[6,7] but the K_m values of all the compounds tested were apparently too high to be of physiological importance. Berg and Rafter[16] investigated the inducibility of P-450$_{meg}$ and its ability to process nonsteroidal substances. They found that none of the inducer compounds tested were active in elevating levels of P-450$_{meg}$, and in some cases their presence actually lowered the production of the enzyme below its constitutive level. Of all the nonsteroidal compounds tested, only aniline was metabolized, forming *p*-aminophenol, but Berg and Rafter caution that this may be due to a general reactivity of hemeproteins[17,18] and not to specific substrate processing. With NADPH as the electron donor, all substrate hydroxylation was both stereo- and regiospecific with the 15β-alcohol being the only product.[6,7] However, in the exogenous-oxidant-supported hydroxylation of progesterone, the stereospecificity is relaxed and both 15α- and 15β-alcohols are formed, with the product ratio dependent on the oxidant used.[6,7]

It is well known that the membrane-bound microsomal cytochromes P-450 are dependent on phospholipid for activity,[19-22] with a twofold stimulation of hydroxylation activity following lipid activation in exogenous-oxidant-supported reactions.[23,24] Gustafsson and co-workers[9] showed that with purified P-450$_{meg}$, phosphatidylcholine stimulated both

NADPH- and NaIO$_4$-supported progesterone hydroxylation twofold. They postulated that since P-450$_{meg}$ is apparently a soluble protein, the observed stimulation might be due to lipid-facilitated interactions between the hydrophobic substrates and the enzyme.

Of interest with respect to the various pyridine nucleotide-supported P-450 systems is the ability of a reductase or an iron–sulfur protein from one system to complement its counterpart in another system. Work done in Gunsalus's laboratory[25] established that adrenodoxin, although similar in primary sequence, could not replace putidaredoxin in the P-450$_{cam}$ electron transport chain. Berg et al.[9] found a different situation in the P-450$_{meg}$ system in that both megaredoxin and adrenodoxin displayed cross-reactivities in the other system, generating the corresponding hydroxylated product. Furthermore, megaredoxin reductase could be replaced by rabbit liver microsomal P-450 reductase. Also, megaredoxin reductase, in conjunction with megaredoxin, was able to supply electrons to both P-450$_{11\beta}$ and P-450$_{LM_3}$, indicating that the electron transport chain of the B. megaterium steroid hydroxylase system is relatively nonspecific with respect to the terminal cytochrome component.

2.2. P-450$_{\omega-2}$

Fulco and co-workers[26–36] have partially purified and characterized a soluble P-450-dependent fatty acid hydroxylase-epoxidase from B. megaterium ATCC 14581. P-450$_{\omega-2}$ has been found to catalyze the ω-1-, ω-2-, and ω-3-hydroxylation of saturated long-chain fatty acids, alcohols, and amides.[26–34] Terminal hydroxylation was not observed with any substrate. This is in contrast to most other fatty acid hydroxylases of the mixed-function oxidase type, which act on the ω-methyl group of the alkyl chain and, usually to a lesser extent, at the ω-1-position.[35] The enzyme also catalyzed the epoxidation as well as the hydroxylation of various monounsaturated fatty acids[31,32,36] and the hydroxylation of 9-hydroxystearate to a mixture of dihydroxy derivatives.[33,34] In all cases, fatty acid methyl esters and alkanes were unreactive, in contrast to most other bacterial, yeast, and mammalian fatty acid hydroxylase systems which also act on alkyl hydrocarbons.[39–49]

Partially purified P-450$_{\omega-2}$ was isolated as a multiprotein complex with a molecular weight of 130,000[30,48] which could be fractionated into at least two protein components.[26] Neither fraction was active by itself, but the addition of bacterial ferredoxin to one partially restored enzymatic activity. The system showed an absolute requirement for NADPH.[26] The inability to separate the various redox components easily suggested to the authors that the P-450$_{\omega-2}$ system existed in vivo as a stable complex.

The dominant hydroxylated product of most substrates was the ω-2 alcohol,[26–30,33,34] even though the substrate carbon chains varied in length from 12 to 18 carbons, with pentadecanoic acid being the most active. Miura and Fulco[27] proposed a possible model for substrate interaction within the active site of the enzyme. Since neither hydrocarbons nor fatty acid methyl esters were processed, these authors suggested that hydrophobic interactions between the alkyl chain of the substrate and the enzyme were not sufficiently strong in themselves to effect efficient substrate binding. Because the presence of a relatively polar head group such as a carboxylate, alcohol, or amide was required for substrate processing, an electrostatic binding site for the substrate's polar functional group was postulated. To explain the type and distribution of oxygenated products obtained, Miura and Fulco suggested that the terminal methyl group is tightly bound and sequestered in a hydrophobic pocket where it is protected from oxidative attack and yet can still position the penultimate three methylene carbons in the active site for oxygenation. Data of Matson et al.[33] indicated that there was little hydrophobic interaction between the central region of the substrate molecule and the enzyme active site. These authors found 9-D-hydroxystearate to be a strong competitive inhibitor of palmitate hydroxylation and, as a substrate for P-450$_{\omega-2}$, the hydroxystearate molecule was more active than stearate in substrate-dependent NADPH oxidation. This suggested that the enzyme surface had a nonhydrophobic binding region that interacted with polar substituents near the center of the substrate chain.[33] If indeed such an enzyme active site with two strong substrate-binding domains existed, then variations in the total hydroxylation activity as a function of alkyl chain length would be related to the distance between the two regions. Presumably the most active substrate, pentadecanoic acid, had the optimal chain length to span these domains. To test this, Miura and Fulco[27] determined the rates of hydroxylation for a variety of *cis*-monounsaturated fatty acids. They reasoned that the presence of a *cis* double bond effectively shortened the fatty acid chain, so that a *cis*-monosaturated fatty acid longer than C_{15} would be more active than its saturated analogues, while the opposite would be true for *cis*-monounsaturated fatty acids shorter than 15 carbons. This is indeed what was observed.

Recent work by Fulco and co-workers[48–51] showed that P-450$_{\omega-2}$ activity could be induced as much as 100-fold by the well-known mammalian P-450 inducer, phenobarbital, as well as by other barbiturates. None of these compounds were substrates or apparently bound to the enzyme at the active site. No barbiturate activation of P-450$_{\omega-2}$ was observed in the cell-free system.[49] Of interest is that none of the fatty acids processed by this enzyme act as inducers of enzyme activity,[49] a situation similar to that found by Salaün et al.[52] With the P-450-dependent laurate hydroxylase from Jerusalem artichoke, in which phenobarbital but not laurate acts as an efficient inducer, preliminary evidence suggests that barbiturates, or their metabolites, cause an increase in *de novo* synthesis of P-450$_{\omega-2}$ rather than a stabilization of the enzyme *in vivo*.[51]

It appears that the P-450$_{\omega-2}$ system is not related to P-450$_{meg}$ described in the preceding section, which is isolated from another strain of *B. megaterium*. Progesterone, an active substrate for the latter system, was not hydroxylated in the P-450$_{\omega-2}$ system.[30] Also, the various electron transport components appear to be unrelated.[9,26] Berg *et al.*[7] have indicated that *B. megaterium* ATCC 13368, from which P-450$_{meg}$ has been purified, also contains a P-450-dependent fatty acid hydroxylating system. However, their system did not appear to require an iron–sulfur protein, suggesting that more work is needed to elucidate the relationship between the fatty acid hydroxylases from the two strains of *B. megaterium*.

2.3. P-450$_{oct}$

Cardini and Jurtshuk[42,43] partially purified a P-450-dependent alkane hydroxylase from *Corynebacterium* strain 7E1C (ATCC 19067, now classified as *Rhodococcus rhodochrous* ATCC 19067[1]). P-450$_{oct}$ was found to hydroxylate *n*-octane in cell-free extracts, with 1-octanol as the only product.[42,43] They resolved the enzyme system, by ammonium sulfate fractionation, into at least two protein components. One fraction contained P-450$_{oct}$ and the other a flavoprotein. The presence of a P-450 in the particulate fraction was established by its inhibition by low concentrations of CO, its insensitivity to cyanide, and the presence of a typical reduced, CO difference spectrum.[42,43] The enzyme had an absolute requirement for NADH as the electron donor and, by utilizing ^{18}O-labeled dioxygen,[43] the product oxygen atom was shown to originate from molecular oxygen. The authors found a sixfold induction of P-450$_{oct}$ levels when the cells were grown with *n*-octane as the sole carbon source.[43] Earlier studies[53] with this same *R. rhodochrous* strain demonstrated that the organism is capable of growth on a variety of hydrocarbons and other substrates. These included the *n*-alkanes from C_3 to C_{18}, terminally monounsaturated alkenes longer than ten carbons, and chloroalkanes such as 1-chlorooctane, 1-chlorododecane, 1-chlorohexadecane, and 1-chlorooctadecane. Although they did not investigate the metabolism of these compounds, it is more than likely that at least some of them were processed by P-450$_{oct}$ or a related P-450 system.

2.4. P-450$_{non}$

Recently, Asperger and co-workers[54–58] described the occurrence of a P-450 in *Acinetobacter calcoaceticus* strain EB104 grown on hexadecane. Asperger et al.[54] showed that this P-450 was present in hexadecane-grown cells but not detectable in the absence of the alkane. They later showed that alternative carbon sources such as succinate, glucose, or glycerol did not repress P-450$_{non}$ activity in cultures simultaneously supplemented with hexadecane.[58] This is a situation different from that found for P-450$_{cam}$, where catabolite repression of the camphor degradative enzymes was observed with succinate and glucose.[59-61] P-450$_{non}$ could be induced by the n-alkanes from hexane to hexadecane, but not by their alcohol or acid derivatives.[55,58] Nonane and decane were the most active inducers of enzyme levels. The cellular location of the cytochrome appeared to be cytoplasmic,[56] but the authors could not rule out a loose association with the plasma membrane. Purified P-450$_{non}$ was shown to have an absolute requirement for NADH as the electron source.[56] The participation of an iron–sulfur protein and an NADH-dependent ferrodoxin reductase was shown to be necessary for enzymatic turnover.[56] Kleber et al.[55] determined that P-450$_{non}$ functions in vivo as an ω-hydroxylase, catalyzing the initial hydroxylation of alkanes which are then converted to their corresponding acid derivatives by alcohol and aldehyde dehydrogenases and metabolized via β-oxidation. These workers investigated the ability of other Acinetobacter strains to grow at the expense of n-alkanes and found that of 15 strains tested, all but one gave good growth on long-chain alkanes.[54,55] Of interest is the fact that in only 7 of the 14 strains able to metabolize alkanes could the presence of a P-450 be established. The oxidases in the other strains are presently unknown, but many examples of non-P-450-dependent ω-hydroxylases have been reported in the literature.[35,37-39,62]

2.5. P-450$_{cam}$

P-450$_{cam}$ from P. putida (ATCC 17453) is by far the best-characterized bacterial P-450 system. To be discussed in subsequent sections of this review is the current level of understanding of the chemistry and physics of the P-450$_{cam}$ system. This section will provide a brief overview of genetic and molecular biological investigations that have been documented in this system. The enormous amount of knowledge pertaining to the genetic organization and regulation of this P-450 system is due in great part to the efforts of Gunsalus and co-workers over the past 15 years. Recent work in Gunsalus's and Sligar's laboratories has led to the cloning of the gene coding for P-450$_{cam}$, and the establishment of the DNA sequence for the coding and flanking regions. Development of broad-host-range shuttle vector systems in Schuler's laboratory at the University of Illinois has enabled the study of the expression of P-450$_{cam}$ in both E. coli and Pseudomonas.[299] Placing this gene behind the strong *tac* promoter has enabled Saelinger in Sligar's laboratory to overproduce P-450$_{cam}$ in *E. coli*. It is beyond the scope of this review to document all data concerning the regulation and control of peripheral metabolic processes. We will simply place the P-450$_{cam}$ discussion in perspective with the other bacterial P-450 systems described earlier. As will become obvious, however, the known genetics and bacterial physiology of this system far exceed those of any other bacterial P-450 system.

The nutritional diversity of *Pseudomonas* is well documented.[63,64] The ability of a microbe to metabolize the relatively inert monoterpene camphor as the sole source of carbon and energy provided investigators with P-450$_{cam}$. In 1959, Gunsalus and co-workers[65] isolated a saprophytic soil pseudomonad and began the study of the enzymes involved in camphor catabolism. Six years later, the pathway for camphor metabolism was elucidated by Hedegaard and Gunsalus.[59] The enzyme systems catalyzing the first three reactions of the pathway have since been purified in multigram quantities and extensively studied. The P-450$_{cam}$ system carries out the two-electron reduction of dioxygen with the concomitant stereospecific hydroxylation of the camphor molecule (I) at the 5-*exo* position.[59,66,68] Hydroxycamphor (II) is then oxidized to the diketone (III) by a dehydrogenase.[69] The third step is the lactonization of 5-ketocamphor (III) catalyzed by another mixed-function oxidase, ketolactonase I,[70–73] that yields an unstable molecule (IV) which rearranges to form the ring-opened cyclopentenone derivative of camphor (V) (Fig. 1).

The first three enzyme systems, the so-called early enzymes of the

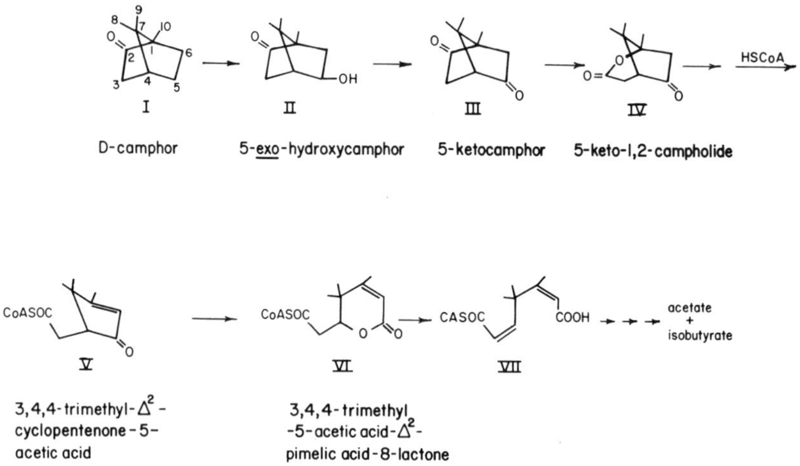

FIGURE 1. Metabolic degradation of camphor by *Pseudomonas putida*.

pathway, were all found to be induced by camphor and related compounds, such as bornane and its substituted derivatives.[74] Of the 21 compounds found to induce the early degradative enzymes, 9 were found to provide induction without supporting growth. Enzyme induction was also found with all the early metabolic intermediates of camphor dissimilation.[75] Catabolic repression of these enzymes was observed in the presence of succinate,[60,61] and to a lesser extent in the presence of glucose.[59-61] No detectable repression occurred with either citrate or glutamate as the carbon source.[60,61] The inability of the latter two compounds to cause repression of the camphor catabolic enzymes was invaluable for later genetic studies with strains carrying mutant loci within the pathway.[76,77] Because the enzymes of the early pathway appeared to be subject to coordinate biosynthetic control, Gunsalus et al.[78] suggested that they might be members of the same operon. The enzymes of the middle pathway, which convert the cyclopentenone derivate of camphor (V) to isobutyrate, were found to be induced by compound V.[74] Genetic studies[76,77] showed that the genes coding for the early and middle enzymes of camphor metabolism were clustered on a 240-kbp[79] transmissible plasmid, the "CAM plasmid." Cotransduction among point mutants in genes coding the first segment of camphor oxidation indicated two closely linked gene loci, specifying the reactions of early and middle pathway enzymes. This group of reactions before isobutyrate was determined to be plasmid encoded,[77] by both their ability to be cotransferred during conjugation[76,77] and their ease of mitomycin C curing.[77] Further degradation of isobutyrate is by way of chromosomally encoded late enzymes leading to chain debranching with the formation of propionate and succinate. These same chromosomal enzymes also mediate valine catabolism. Isobutyrate was found to be the principal inducer of these late enzymes.[80,81]

After the initial enzyme purification and genetic studies of the 1960s and early 1970s, much effort was focused on the various protein components of the P-450$_{cam}$ system, the NADH-specific FAD-containing putidaredoxin reductase, the iron–sulfur protein putidaredoxin, and P-450$_{cam}$. Not until the end of the 1970s, when the techniques necessary to study individual genes in greater detail came into general use, were the original inspired genetic studies of Rheinwald[76,77] revisited. The background genetic knowledge of the CAM plasmid was invaluable when recombinant DNA methodology was applied to this system. Koga,[82] working in Gunsalus's laboratory, cloned the gene coding for P-450$_{cam}$ into an RSF1010-derived broad-host-range vector, pKT240.[83] Selection was accomplished by mutant trans complementation of a P. putida strain with a defective P-450$_{cam}$ gene. The cloned P-450$_{cam}$ gene was expressed in both P. putida and E. coli at levels apparently above that of the CAM plasmid-encoded enzyme, although the instability and poor characterization of the pKT240 vector system prevented accurate quantitation. Schuler's development of a series of broad-host-range vectors with multiple cloning sites and easily selectable markers has allowed controlled expression of P-450 in many microorganisms.[299]

Sligar and co-workers[84] subcloned the P-450$_{cam}$ gene into the high copy-number pUC vectors,[85] thus placing the hemeprotein synthesis under control of the *lac* promoter in *E. coli*. In this system, the enzyme accounts for approximately 10% of the total soluble protein and is active in the *in vitro* hydroxylation of camphor. Gas-phase amino acid sequence analysis of the cloned protein indicated that a faithful P-450$_{cam}$ product is generated, with the first dozen N-terminal residues being identical to those reported for the enzyme coded for by the *P. putida* CAM plasmid. The total sequence of the P-450$_{cam}$ gene was determined by Unger in Sligar's laboratory (Fig. 2) with independent characterization by Lee in Gunsalus's laboratory. The reader should take note of the differences between the DNA and the published amino acid[86] sequences. The complete three-dimensional X-ray structure which has recently been solved by Poulos (Chapter 13) verifies the nucleotide-derived sequence. Efforts are currently under way in Sligar's, Gunsalus's, and Koga's laboratories to clone the genes coding for putidaredoxin reductase and putidaredoxin, as well as those coding for the other enzymes involved in camphor degradation.

Another bacterium, the orange-pigmented diphtheroid *Corynebacterium*, now classified as a *Rhodococcus*[87,88] isolated from soil by enrichment with terpin hydrate, was also found to grow on camphor as the sole source of carbon.[89] In contrast to the *P. putida* P-450$_{cam}$ enzyme that forms the 5-*exo*-alcohol of camphor, the diphtheroid P-450 forms the 6-*endo*-alcohol.[89–91] 6-hydroxycamphor is then oxidized to the 2,6-diketone[92,93] and metabolized in an analogous manner as the 2,5-diketone of the *P. putida* camphor pathway. The degradative enzyme of the *Rhodococcus* P-450$_{cam}$ pathway was induced in a similar manner by camphor[89] as was found for the *P. putida* pathway. Further development of this system has been minimal, though it would be of interest to compare these two camphor hydroxylases from two species with such a wide evolutionary separation.

2.6. P-450$_{cym}$

Bhattacharyya and co-workers[94] isolated a soil pseudomonad, PL-strain, by enrichment culture with the unsaturated monoterpene α-pinene as the carbon source. This strain, now known as *P. putida* strain PL-W[95] or strain JT101,[96,97] was capable of growth on α-pinene, β-pinene, limonene, 1-*p*-menthene, and *p*-cymene (4-isopropyltoluene) at rates com-

BACTERIAL P-450 ENZYMES

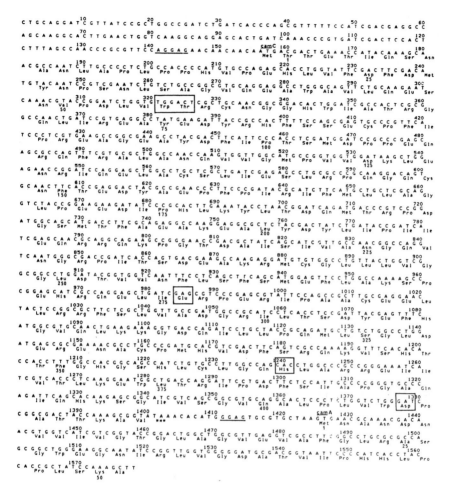

FIGURE 2. DNA sequence of the P-450$_{cam}$ gene from *P. putida*.

parable to that on glucose.[94] The metabolic pathway for the degradation of the aromatic hydrocarbon *p*-cymene has been studied by several investigators.[94,95,98–102] The methyl substituent on the aromatic ring was first oxidized by three enzymatic reactions to yield *p*-cumate[95,98–100] (4-isopropylbenzoate). Following this methyl oxidation, sequential ring hydroxylation gave 2,3-dihydroxy-*p*-cumate which was ring opened via the *meta* cleavage pathway.[95,99–102] The ring cleavage product was then degraded to common cellular metabolites.[95,98–100,102] The initial step in the catabolism of *p*-cymene, its hydroxylation to cumic alcohol (4-isopropylbenzyl alcohol), has been shown to be catalyzed by a P-450-dependent hydroxylase, P-450$_{cym}$. Initial attempts to purify P-450$_{cym}$ (p-cymene hydroxylase) were unsuccessful[100] due to the unstable nature of the enzyme. Recently, Gunsalus's laboratory has completed the purification and characterization of this multienzyme system. Madhyastha and Bhattacharyya[98,99] found that the enzymes involved in p-cymene degradation were inducible. Cells grown on glucose lacked the enzymes necessary for p-cymene catabolism in cell-free extracts.[95,98,99] Glucose did not repress the synthesis of the enzymes in the presence of p-cymene.[95,102] Ribbons and co-workers[97] have recently isolated another P. putida strain, JT810, capable of growth on p-cymene as well as related aromatic compounds. They found a p-cymene degradative pathway similar to that found for P. putida JT101. The first steps in the JT810 p-cymene pathway also involve oxidation of p-cymene to p-cumate, and in all probability the p-cymene hydroxylase is a P-450-dependent enzyme. It has been shown for both strains that the genes encoding the pathway enzymes reside on transmissible plasmids.[103,104]

2.7. P-450$_{lin}$

Bhattacharyya and co-workers[105] isolated a soil pseudomonad, a strain of *Pseudomonas incognita*, capable of growth on the monoterpene alcohol, linalool, as the sole source of carbon. The metabolism of linalool and the related compounds citronellol, limonene, geraniol, and nerol has been well characterized.[106–108] Linalool was found to be degraded by two main pathways[107]: stepwise oxidation of the 10-methyl group of linalool to linalool-10-carboxylic acid, and successive oxidation of the 8-methyl group to linalool-8-carboxylic acid via 8-hydroxylinalool and linalool-8-aldehyde. In both pathways, the initial hydroxylation is carried out by a P-450 monooxygenase. "P-450$_{10-lin}$" was found to be a soluble hemeprotein that carried out the 10-methyl hydroxylation of linalool.[108] "P-450$_{8-lin}$," on the other hand, was found associated with the particulate fraction and catalyzed the conversion of linalool to 8-hydroxylinalool. P-450$_{8-lin}$ showed a requirement for both NADH and dithiothreitol, although weak substrate turnover was seen with NADPH in place of NADH.[108] Rama Devi and Bhattacharyya[107] found that when both linalool and glucose were added to the culture medium, there was a significant reduction of linalool metabolites in cell extracts, similar to the glucose repression seen with the P-450$_{cam}$ system.[59–61]

Recent work in Gunsalus's laboratory has led to the purification of the components involved in the P-450$_{8-lin}$ system,[109] an NADH-dependent flavoprotein reductase (linredoxin reductase), an iron–sulfur protein (linredoxin), and P-450$_{8-lin}$. Heterologous recombination of the various proteins in P-450$_{8\text{-lin}}$ and P-450$_{\text{cam}}$ systems revealed a preference for their natural redox partners,[109] reminiscent of the heterologous recombination studies of adrenodoxin and putidaredoxin with P-450$_{11\beta}$, P-450$_{\text{scc}}$, and P-450$_{\text{cam}}$. Spectroscopic studies by Wagner et al.[110] with P-450$_{8\text{-lin}}$ and linredoxin showed these proteins to be similar to their counterparts in the system. The relationship between the redox components of the two linalool hydroxylase systems is not known.

Metabolism of acyclic monoterpenes has been demonstrated in other organisms. Seubert and co-workers[111-113] isolated a strain of *Pseudomonas citronellolis* capable of the degradation of citronellol, geraniol, and farnesol. Degradation is initiated by the oxidation of the alcohol functional group to a carboxylate followed by conversion to the corresponding acyl-CoA derivatives. Meehan and Coscia[114] isolated a microsomal P-450-dependent hydroxylase from *Vinca rosea* (periwinkle) capable of metabolizing geraniol and nerol to their corresponding 10-hydroxy derivatives, analogous to the P-450$_{10\text{-lin}}$ pathway in *P. incognita*. In the plant system, the monoterpenes have been implicated in the biosynthesis of cyclopentenoid monoterpene glucosides which are the nontryptamine segments of indole alkaloids.[115] Future efforts will hopefully ascertain the similarities between the cytochromes P-450 involved in the bacterial and higher plant 10-hydroxylation of linalool.

2.8. P-450 from *Rhizobium japonicum*

Appleby and co-workers[11,116-120] have identified at least six distinct cytochromes P-450 from *R. japonicum* strain CC 705 (Wisconsin 505). Of these hemeproteins, four have been purified, P-450a, P-450b, P-450b1, and P-450c,[11] named in the order of their chromatographic elution. These cytochromes were found to be soluble hemeproteins,[116] separable by anion-exchange chromatography,[11] both from symbiotic *R. japonicum*[53,55,117] cells grown anaerobically in a nitrate–organic carbon medium[118] and from aerobically grown bacteria.[11,118,119] Although their exact function remains unclear, Appleby et al.[119] postulated that the *R. japonicum* cytochromes P-450 might be involved in an oxidative phosphorylation pathway which supports dinitrogen-fixation. Dus et al.[120] reported that three of these cytochromes showed substantial cross-reactivity with anti-P-450$_{\text{cam}}$ antibodies: 54% for P-450c, 50% for P-450b, and 45% for P-450a. This can be compared to 60% cross-reactivity of hepatic P-450$_{\text{LM}_2}$ with anti-P-450$_{\text{cam}}$ antibodies.[121] The higher cross-reactivity of P-450c to anti-P-450$_{\text{cam}}$ antibodies, compared with the cross-reactivities of P-450a and P-450b, fits well with other similarities between the two cytochromes, including their absorption spectra,[11] their isoelectric points (4.5[10] and 4.9[11] for P-450$_{\text{cam}}$ and P-450c, respectively), and their amino acid compositions.[11,86] Preliminary data[11] also pointed toward strong cross-reactivity between the three *Rhizobium* cytochromes and anti-P-450$_{LM_2}$ antibodies, with P-450a being most closely related to P-450$_{LM_2}$. Of interest with regard to the identity of the heme axial ligand was the finding that P-450c (M_r 46,000)[11,120] contained only two cysteines.[120] Daniel and Appleby[118] were able to show that aerobically grown *R. japonicum* contained a small amount of soluble P-450 and that cells induced to grow anaerobically with nitrate as electron acceptor contained almost as much P-450 as did bacteroids, the form of *Rhizobium* when it colonizes the roots of leguminous plants and forms nodules. Fractionation of the above grown cells into their component cytochromes P-450 revealed that their distribution was dependent on the cells' mode of growth.[11,118] In the bacteroid, P-450c was the dominant species with lesser amounts of P-450a and P-450b. Extracts from anaerobic-nitrate cells contained a very large amount of a single new component, P-450b$_1$. In aerobically grown *R. japonicum*, both P-450a and P-450b were present. It appeared that in the transition of aerobically grown *R. japonicum* into nitrogen-fixing bacteroids, the constitutive cytochromes P-450 are increased in amount and a new species, P-450c, is induced, while a single new protein, P-450b$_1$, is induced[11] in anaerobic growth with nitrate. Peisach *et al.*[122] studied the optical and EPR properties of the different P-450 molecules. The absolute absorption spectrum of oxidized P-450a and the helium-temperature EPR were those of a pure low-spin ferric hemeprotein. P-450b, on the other hand, from both optical spectroscopy and low-temperature EPR, was shown to be a mixed-spin ferric P-450 (15% high spin at 1.4°K). P-450b$_1$ was also a mixture of high- and low-spin forms, with a high-spin ferric heme contribution of 90% by EPR. This result approaches the high-spin structure reported for the enzyme substrate complexes of P-450$_{cam}$[123] and P-450$_{LM_4}$.[124] Oxidized P-450c possessed optical and EPR properties similar to those of P-450b, but exhibited a smaller proportion of high-spin P-450 (5%). In addition to these spectroscopic distinctions, differences were found in the reaction of these cytochromes with exogenous ligands such as phenobarbital[11,122] and aliphatic amines.[11] By titration of the oxidized cytochromes with phenobarbital, P-450b and P-450c gave typical "type I" optical spectral shifts, indicating a shift to high-spin heme, although fivefold more phenobarbital was needed with P-450b to reproduce the P-450c spectrum. Neither P-450a nor P-450b$_1$ showed any spectral changes upon titration with phenobarbital. Kretovich et al.[125] detected a P-450 by difference spectroscopy in cell-free extracts of bacteriods isolated from lupine nodules. The P-450 was partially purified by chromatography. The relationship between this cytochrome and the various cytochromes P-450 characterized by Appleby and co-workers[11,116–120] is unknown.

2.9. P-450$_{npd}$

Cartwright and co-workers[12,126-128] isolated and purified a soluble P-450-dependent O-dealkylase from *Nocardia* strain NH1. They showed that this "P-450$_{npd}$" specifically dealkylated *p*-alkylphenyl ethers, such as anisic acid (4-methoxybenzoate), and isovanillate (4-methoxy-3-hydroxybenzoate); hence the subscript "npd," for *Nocardia para-dealkylase*.[12] P-450$_{npd}$ catalyzed the first enzymatic step in the catabolism of various *p*-alkylphenyl ethers for utilization as the sole source of carbon for energy.[126] Cartwright and Broadbent[128] fractionated the electron transfer system into at least two protein components, a 70,000-dalton NADH-specific flavin-containing reductase and P-450$_{npd}$ (45,000 daltons).[12] It was postulated that the reductase component may have been a complex between a flavoprotein and an iron–sulfur protein.[128] Although the reductase has only been partially purified due to its apparent instability,[128] P-450$_{npd}$ has been purified to homogeneity as judged by analytical ultracentrifugation and isoelectric focusing. These authors determined an isoelectric point of 4.94[12] for the hemeprotein, similar to the values of 4.5, 4.9, and 4.9 reported for P-450$_{cam}$,[10] P-450$_{meg}$,[9] and P-450c[120] (*R. japonicum*), respectively. A partial amino acid composition of P-450$_{npd}$ has also been determined.[12] Optical spectroscopy has shown that P-450$_{npd}$[12] exhibits spectroscopic properties similar to those of P-450$_{cam}$[129] in both the oxidized and reduced states and in the presence or absence of substrate. The EPR of P-450$_{npd}$ was that of a low-spin ferric hemeprotein, both in the substrate-free and bound forms at 15°K.

From the above discussion, it is clear that a wide variety of bacterial P-450 systems exist. These systems catalyze fascinating biotransformations of both scientific and commercial interest. Considering the wealth of scientific information provided by multifaceted investigations of the *Pseudomonas* P-450$_{cam}$ system, there is perhaps much undiscovered gold in these other bacterial monooxygenases. Hopefully, future studies of other systems which can be isolated in high yield and purity will uncover differences in protein–protein interaction and key rates such that new intermediates and structures can be discovered.

3. P-450$_{cam}$ System

The preceding section has documented the wide variety of bacterial P-450 monooxygenase systems that have been isolated and characterized to varying degrees. The best-known bacterial P-450 system is P-450$_{cam}$, isolated from the soil bacterium *P. putida*. In this organism, hydroxylation of the monoterpene camphor serves as the first step in the catabolism of this carbon- and energy-providing substrate. The unfolding of knowledge concerning this monooxygenase is rather unique as compared to classical enzymology. Isolated in the late 1950s in Gunsalus's laboratory at the University of Illinois, the system was microbiologically characterized, and the protein constituents resolved. Particularly interesting, however, was the early use of sophisticated resonance techniques to probe the active-site irons of the cytochrome and the iron–sulfur redoxin. This development, occurring prior to the completion of more classical substrate binding and steady-state kinetic experiments, was due in large part to the close interaction of the University of Illinois Department of Biochemistry with the research groups of Professors Debrunner and Frauenfelder in the Department of Physics. Thus, the bacterial P-450$_{cam}$ system provided much of the first spectroscopic characterizations of this unique class of hemeprotein systems.

Various chapters of this book have discussed the reaction cycle of the P-450 monooxygenases, and the present authors will also introduce yet another "wheel" depicting substrate binding, ferric–ferrous reduction of the P-450 heme iron, dioxygen binding to yield reduced and oxygenated protein, and second electron transfer followed by reductive scission of the dioxygen bond, hydrogen abstraction of the substrate, oxygen rebound, and final release of product and regeneration of the ferric hemeprotein (Fig. 3). In this soluble P-450$_{cam}$ system, as in other systems, there are two main areas of this cycle which provide key mechanistic questions of great relevance to modern biochemistry. The first relates to

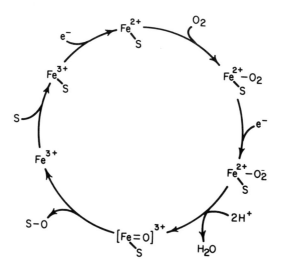

FIGURE 3. Proposed reaction cycle of P-450$_{cam}$.

the details surrounding the transfer of reducing equivalents between pyridine nucleotide, flavoprotein dehydrogenase, iron–sulfur redoxin, and the P-450 heme iron. With these characters in the play of biological electron transfer, one has represented many of the major classes of electron-transferring proteins and, as such, one has the potential of defining the mechanisms of intra- and interprotein electron and proton transfer events in a precise physical context. The second major area of scientific inquiry with regard to the reaction cycle of P-450 monooxygenases lies in the chemical details surrounding the reductive scission of the dioxygen bond from atmospheric dioxygen, activation of the substrate molecule, and oxygen atom transfer to yield an oxygenated product.

The remainder of this chapter is divided into two sections reflecting these important areas of research endeavor. Perhaps obviously missing in this discussion is a review of the spectroscopic characterization of the various stable states in the P-450 monooxygenase reaction cycle. This is an extensive omission, since by the end of 1984 nearly every conceivable form of spectroscopy available to the biophysicist and biochemist had been applied to the P-450 hemeprotein, including EPR, NMR, ENDOR, spin echo, magnetic susceptibility, Mössbauer, Raman, MCD, ORD, CD, fluorescence, ELDOR, and so on. However, the space limitations of the text preclude an extensive discussion of these spectroscopic investigations. Many of the techniques have, in addition, been the subject of recent review articles.[150,222,314] We are also omitting from our discussion of chemical mechanisms in P-450 oxygenation much of the general discussions of dioxygen bond scission, heme inactivation, and carbon chain attack that are excellently presented in this volume by Professor Ortiz de Montellano. Rather, we would like to focus on recent results in our laboratory that shed light on the interplay of the structure and function of the P-450 hemeprotein in the transfer of reducing equivalents to the heme active center and in the control of regiospecificity and stereochemistry in the oxygenation of the monoterpene substrate.

3.1. Structure–Function Relationship in Electron Transfer to P-450_{cam}

With the elucidation of the three-dimensional X-ray structure of P-450_{cam} by Poulos (Chapter 13), one is in a position to use this road map to define the controlling structural features in electron transfer and oxygenase chemistry. This translates into a description of the conformational changes and functionally important motions of the P-450_{cam} macromolecule. P-450_{cam} and its associated physiological redox transfer components provide an excellent system to study the linked dynamics of electron transfer and protein structure. Limiting in many cases is the ability to monitor internal equilibria in protein structure and, thus, deconvolute the free energy linkages operating to control physiological response. In the case of P-450$_{cam}$, and other d^5-iron systems, nature has provided an observable spin-state equilibrium.

The Ferric Spin Equilibrium and Its Relation to Electron Transfer

P-450$_{cam}$ is responsible for the hydroxylaation of camphor to form 5-*exo*-hydroxycamphor in the overall reaction cycle schematically illustrated in Fig. 3. The iron–sulfur protein putidaredoxin contains a ferredoxin-type 2Fe–2S center with antiferromagnetically coupled irons and is, thus, a transfer agent ultimately providing P-450$_{cam}$ with the two electrons needed to activate bound atmospheric oxygen.[67,130,131] As illustrated, these two reducing equivalents are provided in discrete one-electron transfer steps. The binding of the substrate camphor to the P-450$_{cam}$ hemeprotein, the formation of the putidaredoxin–P-450$_{cam}$ dienzyme complex, oxygen binding, and product release have been shown not to be rate-limiting in the overall metabolism.[132–135] Rather, in the normal reconstituted system, the rate-limiting step appears to be the transfer of the second reducing equivalent from putidaredoxin to oxygenated cytochrome.[133] The first electron transfer step, however, is rate-limiting in many cases in bacterial P-450$_{cam}$ when other substrates are metabolized, and in many of the hepatic P-450 systems, where it has been suggested that the electron transfer rates are determined by the percentage of high-spin ferric iron species initially present. We will discuss the first two coupled steps in the P-450$_{cam}$ reaction, substrate binding and ferric–ferrous reduction of the heme iron, in light of this spin equilibrium of the heme center.

The binding of camphor to P-450 elicits changes in the heme optical spectra, shifting the wavelength maximum of the Soret band from 417 nm to 390 nm with a concomitant modulation of the EPR spectra of the porphyrin-chelated iron.[67,123] The alterations upon binding of the substrate are due directly to a change in the iron spin state distribution from a primarily low-spin *d*-electron configuration ($S = \frac{1}{2}$) to a primarily high-spin configuration ($S = \frac{5}{2}$).[136,137] This spectrally observable change in the ligand field conformation provides a means to study the structure–function relationship of the ferric spin equilibrium. It was noted independently in 1975 by Sligar, studying the purified bacterial P-450$_{cam}$ system, and by Cinti *et al.*, using hepatic microsomal extracts, that this spectrally monitorable spin state equilibrium was strongly temperature dependent, allowing the determination of the enthalpies, entropies, and exact position of the thermal equilibrium for both the substrate-free and bound forms of the P-450 macromolecule. This development placed the classification of "type I" and "reverse type I" substrates on a less phenomenological level by showing that simple differential affinity of the substrate for the high- and low-spin conformations of the hemeprotein accounted for the observed optical difference spectra. Inasmuch as the thermal nature of this spin equilibrium and the relevant thermodynamic parameters linking it to the substrate affinities have been thoroughly documented and reviewed, we will not discuss in detail these aspects of the spin equilibrium in P-450 and other hemeproteins. Our concern in the next few paragraphs is to examine the structural basis for this observed spin state equilibrium, and to examine the electron transfer processes that are linked in rate and equilibria with the ferric spin state. The first question is probed by using high-pressure techniques to measure the volume change associated with the processes, while the latter, to be discussed first, brings us to a discussion of the rates and specificity of putidaredoxin–P-450 redox transfer events.

It was noted many years ago that the shift in the ferric spin equilibrium of P-450$_{cam}$ is accompanied by a redox potential change from roughly -300 mV to -170 mV when camphor binds to the cytochrome.[138] The physiological significance of this shift from the standpoint of the bacterial cell rests with the fact that the normal redox transfer partner, putidaredoxin, has a redox potential of -196 mV when bound to the cytochrome. Thus, reduction of P-450 is much more favorable in the presence of bound substrate.[138] Similar spin state changes in hepatic microsomal cytochromes P-450 were also shown to modulate the redox potential of this P-450 isozyme.[139–143] In light of this regulatory effect, it is relevant to examine the specificity of this first electron transfer event in the catalytic cycle. A wide variety of low-molecular-weight reductants (e.g., sodium dithionite) are chemically capable of producing singly reduced P-450$_{cam}$. In a series of elegant studies, Hintz and Peterson[144] investigated this reduction reaction extensively and found that the major active reducing species was the monomer form of dithionite, SO_2. Interestingly, in this case, the low-spin form of P-450 (substrate free) was reduced at a faster rate (6.4×10^5 M^{-1} sec^{-1}) than either low-spin, metyrapone-bound (1.4×10^5 M^{-1} sec^{-1}) or predominantly high-spin, camphor-bound (1.6×10^4 M^{-1} sec^{-1}) P-450$_{cam}$. However, when dithionite-reduced mediator dyes such as lumiflavin (-223 mV) and phenosafranine (-252 mV) were used, the camphor-bound species was more readily reduced than either the metyrapone-bound or camphor-free P-450$_{cam}$ species.[144] Debey et al.[145] also found that the reduction rate of P-450$_{cam}$ by a methyl viologen radical created *in situ* by pulse radiolysis was faster by an order of magnitude for the camphor-free cytochrome as compared to the rate observed for the camphor-bound species. However, these authors duly noted that hydrated electrons produced by pulse radiolysis are highly nonspecific and can react at many sites of the protein. Indeed, later experiments by Bazin et al.[146] revealed a linear correlation between the percentage of high-spin P-450$_{cam}$ species present (as produced by varying camphor and butanol concentrations) and the overall reduction rate when *in situ* reducing equivalents are produced by photoionizing internal tryptophan residues to produce a photoejected electron near the heme center. Thus, the determined ratio of reduction by nonphysiological reductants seems to be governed not only by the percentage of high-spin ferric heme of P-450$_{cam}$, but also by the mechanism and pathway of reduction. Such is the case even with physiological redox partners where multiple paths of electron/proton transfer exist. In the case of small molecules, for example, recent evidence[147] shows that the nonphysiological reductant, ferricyanide, binds to several sites of cytochrome c,[147] rendering a precise study of the reduction kinetics and mechanism for this system difficult.

A more self-consistent choice for reductant of P-450$_{cam}$ is its physiological partner, putidaredoxin. Electron transfer experiments performed between these two proteins under a number of different experimental conditions[133–135,148] revealed that electron transfer occurs between a dienzyme complex of the two protein species. The actual electron transfer rate exhibits saturation behavior as the putidaredoxin concentration is increased approximately fivefold over the concentration of P-450$_{cam}$.[133–135] Electron transfer rates between putidaredoxin and different ferric spin state populations of P-450$_{cam}$ have heretofore not been extensively documented. Pederson measured the first electron transfer between putidaredoxin and camphor-free P-450$_{cam}$ in the presence of CO and found that the forward rate for this first electron transfer event was two orders of magnitude slower (0.22 sec^{-1}) than the rate observed with camphor bound to the hemeprotein (41 sec^{-1}). However, for a complete quantitative study of the regulation of electron transfer rate through substrate-induced shift of the ferric spin equilibrium, it is necessary to be able to independently vary the redox, spin, and substrate binding equilibria. This can be accomplished in P-450$_{cam}$ through the use of various substrate analogues of camphor.[149–150] For example, the substrate analogue, 5-*exo*-bromocamphor, has been shown to induce a spin equilibrium which is 46% high spin and results in a redox potential of -246 mV.[149] This substrate was particularly interesting because the binding affinity of bromocamphor (2.9 μM) is very near that of camphor (2.4 μM) under identical conditions. This suggests that the overall binding free energy is independent of the ferric spin equilibrium free energy.[149] Other substrate analogues of camphor were found to perturb the ferric spin equilibria to varying degrees, with preliminary results indicating that the redox potential exhibits a linear correlation with the ferric spin equilibria.[338] In support of this hypothesis, experimental results obtained by Ruckpaul and co-workers, working with the P-450$_{LM_2}$ system, have shown that the first electron transfer rate between P-450 reductase and P-450$_{LM_2}$ is linearly correlated with the percentage of high-spin fraction present.[151] Recent experiments in Sligar's laboratory have measured the reversible first-order electron transfer for the reaction

$$\text{Pd}^r \cdot \text{P-450}^o \underset{k_{-1}}{\overset{k_1}{\rightleftharpoons}} \text{Pd}^o \cdot \text{P-450}^r$$

where Pd represents putidaredoxin and superscripts r and o refer to reduced and oxidized forms of the proteins. The observed rate constant is a combination of forward and back electron transfer rates $k_{obs} = k_1 + k_{-1}$. These experiments indicate that the reduction rate increases as the percentage of high-spin ferric species increases. Thus, for camphor-bound (95–100% high spin), fenchone-bound (60% high spin), and camphor-free (4% high spin) ferric protein, the observed reduction rate at saturating putidaredoxin concentrations is 110, 16, and 9 sec^{-1}, respectively.[338]

In preliminary investigations using P-450$_{cam}$, the spin state relaxation rates, as determined by temperature-jump spectroscopy, indicated that spin reequilibration was not rate-limiting in relation to the electron transfer event in both the substraate-free and bound forms.[152] The observed reduction kinetics are simple first-order in both cases. Hepatic P-450, on the other hand, shows biphasic kinetics when one monitors the increase in the ferrous CO-adduct of the heme upon reduction by its physiological electron donor in reconstituted systems. One possible explanation for the apparent "biphasic" kinetics seen in hepatic P-450 is that the rate-limiting step in the reduction sequence is the formation of the high-spin ferric species which then controls the reduction rate.[142] The "fast phase" observed in the electron transport scheme might thus represent the reduction of the fraction of high-spin P-450 present in a preequilibrium prior to the introduction of reducing equivalents. In this hypothesis, the rate of the slow phase is postulated to reflect the rate of formation of the high-spin species from the low-spin ferric form with only the high-spin form being reduced. Support for this hypothesis is provided by the observation of a direct correlation between the amount of initial high-spin species and the extent of the initial "burst" phase.[153] Computer simulations using an equilibrium model were consistent with this scheme. However, it is incorrect to ignore the occurrence of a reduced low-spin form of P-450 when examining reduction kinetics. As will be shown below, a significant transient population of the ferrous $S = 0$ can exist. In order to completely document all effects of substrate on reduction kinetics, we reexamined the bacterial P-450$_{cam}$ system.

The substrate analogues of camphor all induce type I spectral characteristics and from this spectral change two experimental quantities can be obtained. By standard optical titration experiments,[149] one quantity derived experimentally is the dissociation constant for a particular substrate. This dissociation constant, K_d, represents the total Gibbs free energy available for redox, spin, and other internal equilibrium processes. The maximum absorbance change (ΔA^{max}) observed in a standard difference spectrum, with the substrate saturating the cytochrome, is an independent quantity that measures the ability of the substrate to alter the thermal spin equilibria of the protein system. These two quantities, K_d and ΔA_{max}, can be expressed in terms of the microscopic equilibrium constants of a four-state model[132,136]:

$$\begin{array}{ccc} \text{P-450}_{HS} & \underset{K_4}{\rightleftarrows} & \text{P-450}_{HS}^S \\ K_1 \updownarrow & & \updownarrow K_2 \\ \text{P-450}_{LS} & \underset{K_3}{\rightleftarrows} & \text{P-450}_{LS}^S \end{array}$$

The various equilibrium constants, K_i, represent the binding and spin state equilibrium constants according to

$$K_1 = \frac{[\text{P-450}_{HS}]}{[\text{P-450}_{LS}]} \qquad K_2 = \frac{[\text{P-450}_{HS}^S]}{[\text{P-450}_{LS}^S]}$$

$$K_3 = \frac{[\text{P-450}_{LS}][S]}{[\text{P-450}_{LS}^S]} \qquad K_4 = \frac{[\text{P-450}_{HS}][S]}{[\text{P-450}_{HS}^S]}$$

where the subscripts HS and LS refer to the total high-spin ($S = \frac{5}{2}$) and low-spin ($S = \frac{1}{2}$) forms of the ferric cytochrome and the superscript S denotes the substrate-bound form. From free energy conservation and microreversibility, the equilibrium constants must satisfy the relation $K_1 K_3 = K_2 K_4$. The macroscopic quantities ΔA_{max} and K_d can be solved for in terms of these microscopic equilibrium constants:

$$K_d = \frac{1 + K_1}{1 + K_2} K_3$$

$$\Delta A_{max} = [\text{P-450}]_{total} \, \Delta\epsilon \, \frac{K_1 - K_2}{(1 + K_1)(1 + K_2)}$$

where

$$\Delta\epsilon = (\epsilon_{HS}^{391} + \epsilon_{LS}^{391}) - (\epsilon_{LS}^{417} + \epsilon_{HS}^{417})$$

In an optical titration of P-450$_{cam}$ by a variety of substrates, a clean isobestic point at 406 nm is observed, indicating that only two spectrally distinguishable states exist. The two observable parameters K_d and ΔA_{max}, together with a knowledge of the spin equilibrium of substrate-free protein, K_1, and the energy equality $K_1 K_3 = K_2 K_4$ can be used to uniquely determine K_2, K_3, and K_4.

Although many investigators using a variety of P-450 systems have reported dissociation constants for a variety of substrates, the extent of spectral change observed, ΔA_{max}, is also directly related to the microscopic spin state equilibrium constants. Further use of this parameter can separate the actual substrate binding free energies from the conformational free energies monitored by the spin state equilibria. As we have recently documented,[130] ΔA_{max} and K_d are two completely independent quantities and hence we stress again that lack of a large spectral change in any P-450 upon addition of substrate does not necessarily imply that there is no binding to the protein. Rather, it simply points to the near identity of the spin state equilibrium constants in the substrate-bound and free forms of the P-450.

The linked free energy model described above maintains conservation laws around the diagram for all thermodynamic state functions. The Gibbs free energy $\Delta G_1 + \Delta G_3 = \Delta G_2 + \Delta G_4$ was used before to deconvolute the microscopic dissociation constants. A complete thermodynamic description of the interaction of substrates with P-450$_{cam}$ must necessarily include documentation of the other relevant state functions such as entropy (ΔS), enthalpy (ΔH), and partial specific volume (ΔV). Enthalpies and entropies describing the regulation of spin state via substrate binding can be extracted by quantitating all equilibria as a function of temperature. Figure 4 presents a standard van't Hoff plot of the spin state equilibrium of P-450$_{cam}$ with a variety of substrate analogues saturating the cytochrome, with corresponding equilibrium constants and thermodynamic parameters summarized in Table II. Through these data, collected by M. Fisher and K. Carroway in our laboratory, one can see

TABLE II
Thermodynamic Parameters for the Ferric Spin Equilibrium in Substrate-Bound P-450$_{cam}$

Substrate	K	ΔG (kcal/mole)	ΔS (e.u.)	ΔH (kcal/mole)
Camphor	20.6	−1.79	25.4	5.78
Adamantanone	17.9	−1.71	28.8	6.86
Camphoroquinone	11.9	−1.47	24.6	5.85
Norcamphor	1.26	−0.136	19.1	5.01
TMCH	0.307	0.700	13.3	4.66

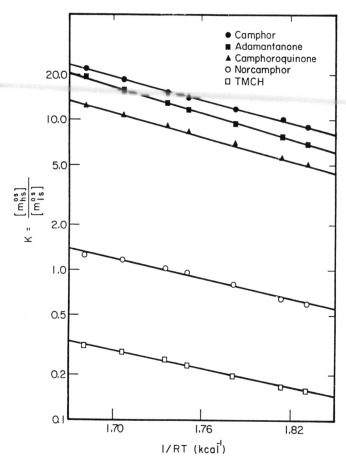

FIGURE 4. Temperature dependence of the spin equilibrium of substrate-bound forms of P-450$_{cam}$.

a profound feature of protein folding and regulatory equilibria. Strikingly obvious from Fig. 4 is the nearly identical enthalpies and widely varying entropies of the spin state equilibria. Thus, the spin equilibria of P-450 are poised by entropy factors and not enthalpy considerations. Since entropy reflects the accessible phase space of the system, the shift in spin equilibria via substrate association is by differential "tightening" of the structure. As will be presented shortly, this conformation equilibria can also be probed by high-pressure techniques. On a broader perspective, the use of entropy to control spin state equilibria is, in some sense, necessary in view of the desired regulation of the system's redox potential.

This follows from conservation of thermodynamic state functions and the dominant entropy factors in any adiabatic electron transfer process. In order to explore these implications further, we now examine the linkage of the P-450 oxidation–reduction potential.

The existence of a linkage between the redox properties of the P-450 heme center and the substrate and spin equilibria was noted over 10 years ago. First measurements of the association of substrate to oxidized and reduced P-450$_{cam}$ demonstrated a markedly higher affinity of the camphor substrate for the reduced form of the hemeprotein. From microreversibility, this must necessarily imply an effect of substrate on the oxidation–reduction free energy of the heme center. A shift of nearly 130 mV in the redox potential of P-450$_{cam}$ on binding of substrate was established. Since substrate binding also induced a large change in the spin state of the heme center, it was natural to ascertain the linkage relationships between these three equilibria. In 1976, a thermodynamic model was presented which showed that, at least in the case of P-450$_{cam}$, the majority of the free energy change in redox equilibria was contributed by the spin-state-linked conformational changes discussed above. Attempts to extend this model to the hepatic P-450 systems met with many experimental difficulties, including the varied reaction conditions used by different investigators and impure protein samples. Thus, though early reports on the lack of spectral and redox changes in P-450$_{LM_2}$ were shown to be incorrect, Guengerich[335] and others subsequently documented variations in hepatic P-450 redox potentials and spin states with a variety of different isozymes, but no obvious correlation between the spin state and measured potential was observed. Schenkman and co-workers[139] studied a partially purified P-450 preparation from uninduced rat liver, and found a linear correlation between redox potential and ferric spin state, though, as pointed out by Peterson (personal communication), no simple theory incorporating only the high- and low-spin ferric states and the ferrous $S = 2$ high-spin state could quantitatively account for the observed results. In addition, many attempts to measure redox potentials in P-450 systems have used organic dyes not only as mediators but also as system potential indicators. Since many of the dyes used are two-electron transfer agents, there is no clear way to relate the observed dye spectral changes with electron stoichiometry and one-electron potentials. For these reasons, we have endeavored to quantitate the redox potentials by cyclic voltammetry and electron transfer transient kinetics with highly purified, recrystallized samples of bacterial P-450$_{cam}$, using the ability of various substances to independently alter the ferric spin state equilibrium as described in preceding paragraphs. With the use of a thermodynamically complete equilibrium model, one can also explain the apparently conflicting data of others. As we will see, the free energies of redox potential and ferric spin equilibrium

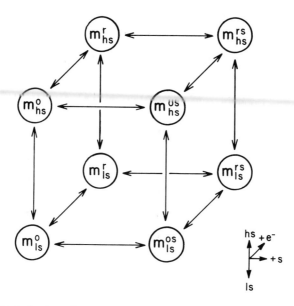

FIGURE 5. Linked spin, substrate binding, and redox equilibria in P-450$_{cam}$. o, oxidized; r, reduced; s, substrate-bound; ls, low spin; hs, high spin.

are indeed related, though partitioning of the total substrate binding energy into other conformer modes does occur. We will conclude this discussion by offering a model that also explains conflicting interpretations of the kinetics of electron transfer in P-450 systems. Emerging will be a demonstration that although simplified models may be appropriate for equilibrium diffusion, electron transfer in P-450 systems is kinetically controlled, and a thermodynamically complete model must be used.

Consideration of the three equilibria, substrate binding, spin state, and redox state of the P-450 hemeprotein, generates a "cube" model[136] of interacting equilibria, described by 12 "edges" or microscopic equilibrium constants (Fig. 5). Macroscopic constants, corresponding to the system observables such as fraction of ferric high spin, percentage of the system reduced, and fraction of hemeprotein saturated with substrate, are related to these 12 microscopic equilibrium constants as discussed previously. Free energy conservation and microreversibility considerations show that only 7 of these 12 constants are independent. Knowledge of these 7 equilibria then allows calculation of the fraction of each form present at any concentration of substrate or at any degree of reduction. The "kinetics" of transition of the system from any point in the "volume" of the cube in Fig. 5 to any other point are determined by the individual microscopic rate constants that make up the "edges." Considering, for simplicity, only the substrate-bound form of P-450, one has four states and equilibria:

$$P\text{-}450^S_{HS} (Fe^{3+}) \leftrightarrow P\text{-}450^S_{HS} (Fe^{2+})$$
$$\updownarrow \qquad\qquad \updownarrow$$
$$P\text{-}450^S_{LS} (Fe^{3+}) \leftrightarrow P\text{-}450^S_{LS} (Fe^{2+})$$

The observed redox potential reflects an equilibrium constant relating the total fraction of oxidized and reduced protein. It can be shown that

$$E_{obs} = \frac{RT}{F} \ln\left(\frac{K_2}{1 + K_2}\right) - \frac{RT}{F} \ln\left(\frac{K_1}{1 + K_1}\right) + \frac{RT}{F} \ln K_4$$

where K_1, K_2, and K_4 were defined earlier. Note that since $K_2/(1 + K_2)$ is the percentage of high-spin ferrous protein and $K_1/(1 + K_1)$ is the percentage of high-spin ferric form, the first two terms reflect the free energy differences in the redistribution of the spin states of ferrous and ferric states, whereas the last term is the microscopic redox potential for reducing a fully high-spin P-450 heme. This thermodynamic model is rigorously valid, although each of the states defined above may be further subdivided into other equilibria. Since Mössbauer studies[155] of P-450$_{cam}$ have shown that the ferrous form is almost completely $S = 2$ high spin, $K_2 \gg 1$, this approximation results in an effective "three-state" model which ignores the small population of ferrous low-spin species present at equilibrium. A major error, however, is to use this simplified three-state model to describe the reduction of P-450 hemeproteins. The system is kinetically controlled and a substantial population of the $S = 0$ low-spin ferrous state can occur during reduction. This can easily be seen in a schematic form as indicated below, where the individual rate constants can be used to calculate a family of "trajectories" of the system:

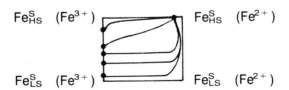

The various structural features of the P-450$_{cam}$ molecule can interact to provide additional redistributions of redox, substrate, and spin state free energies, through alterations in dielectric constant, ionic stabilizations, and dipole-induced interactions.[156] All of these factors may have a profound impact on substrate binding, spin state equilibrium, and redox potential.[136]

3.2. Conformational States of P-450$_{cam}$

With the elegant solution of the three-dimensional X-ray structure of P-450$_{cam}$ (Chapter 13), it is tempting to try to relate the internal equilibria of P-450$_{cam}$ described above to definable structural conformers. The existing X-ray structure results from the crystallization of a low-spin bisthiolate complex, with the thiol removed and substrate camphor added by dialysis. This results in a high-spin heme iron, as evidenced by the out-of-plane puckering of the porphyrin ring and movement of iron toward the axial liganding sulfur. This is not surprising: no significant barriers to iron movement would be expected and tunneling of chelated metals has been experimentally observed. It is not known, of course, whether the overall conformation of the protein is that of the low-spin thiol adduct, or the high-spin camphor-bound species. Constraints of crystal structure could easily prevent the global rearrangements that might be linked to substrate binding. In order to provide insight into the structural changes occurring in P-450$_{cam}$ that may be involved in catalytic function, we have employed several techniques to probe the effect of substrate binding on the macromolecule.

The rate-limiting problem in documenting protein conformational changes is the need for suitable spectroscopic or structural probes that reflect the conformer equilibria. Circular dichroism is often used to calculate bulk secondary structural features, in P-450$_{cam}$, significant differences between the camphor-bound and free forms of the cytochrome have been observed.[130] In order to ascertain whether large conformational changes accompanied the binding of substrate to P-450$_{cam}$, Lewis and Sligar[157] quantitated the radius of gyration of the protein in solution using small-angle X-ray scattering. Within experimental error, this parameter was identical in both the camphor-bound (23.9 ± 0.2 Å) and free (23.7 ± 0.2 Å) states, indicating that no large alteration in three-dimensional solution structure occurred. The existence of possible structural changes with the linked substrate binding and spin state equilibria in the P-450$_{cam}$ macromolecule can be probed using temperature-jump methods to alter the thermal spin state on a microsecond time scale. These experiments indicated that the spin state relaxation in both substrate-bound and free forms of the protein was quite slow by electronic standards, with relaxation times on a millisecond scale. Together with the large entropy values for these equilibria that were discussed in preceding paragraphs, the existence of "functionally important motions" of the P-450$_{cam}$ structure is perhaps indicated.

In order to further probe possible structural changes in P-450$_{cam}$, we have employed high-pressure techniques and sensitive second-derivative spectroscopy to examine the effects of substrate binding on structure. The final two topics of this section will focus on these results.

3.2.1. Structural Studies Using High-Pressure Techniques

The application of pressure to a system in equilibrium forces the system, by Le Chatelier's principle, to favor the equilibrium state which has the smaller volume. Thus, the volume change between equilibrium states for proteins obtained from the application of pressure to the system can measure a macroscopic conformational change, which depends not only on the differences between equilibrium states of the protein but also on differences in solvent interactions between the equilibrium states. The origins of a volume change in a protein are numerous but intuitively easy to grasp. The contributions to overall volume of a protein system include: (1) void volumes within the protein, (2) the intrinsic volume dictated by the atoms and bond lengths between them, and (3) the packing of solvent molecules around the protein. A negative volume change induced at higher pressures for protein systems can be the result of a combination of a number of phenomena. The breaking of an ionic salt linkage and the dissociation of a proton or ion can lead to a volume decrease due to increased electrostriction of the solvent around free charges. Stacking of aromatic residues also exhibits a negative volume change.[158] Pressure-induced substrate dissociation from proteins can result in an overall positive or negative volume change. Positive volume changes (for which the substrate-bound protein form is favored) can result from the interaction of aromatic groups and/or the existence of a so-called "soft" binding site.[159] The reduction in volume of the soft binding site under pressure is thought to arise from rotations about the backbone bonds. A negative volume change for substrate dissociation can result if the binding site is not compressible, since the substrate may not pack in the active site as well as solvent and therefore pressure-induced substrate dissociation would be favored.[159] The existence of pressure-sensitive sites which are removed from the active site may affect the binding of the substrate much the same as when allosteric effectors interact with proteins.

Hemeproteins have been shown to exhibit pressure-induced shifts in spin equilibria.[160] In general high-spin spectra have been transformed into low-spin spectra, prompting some investigators to suggest that the primary effect of pressure on the heme pocket is to shift from an open crevice (high spin) to a closed crevice (low spin) form of the globular protein. The pressure range for the transition from a high-spin form to a low-spin form depends on the protein structure. Recently, P-450$_{cam}$ was subjected to relatively low pressures (1–1600 bars) and changes in protein structure monitored by changes in the Soret absorption band. The characteristic absorption spectrum of the high-spin form (391-nm Soret max) was seen to shift to the low-spin spectrum (417-nm Soret max).[161,162] These results prompted the investigators to suggest that the spin equilibrium was pressure dependent and that the extent of the volume change depended on both potassium ion and proton concentrations.[161] In order to determine the extent of the volume change associated with the ferric spin equilibria, we employed a range of spin equilibria easily established through the use of various substrates and analogues.[168] This provided a consistent means of varying spin state equilibria without resorting to drastic changes in the solvent conditions as were used in the pressure experiments by other investigators.[161] Using pressure difference spectroscopy as outlined by Gibson and Carey,[163] we endeavored to keep solvent conditions constant (50 mM Tris·Cl, 100 mM KCl, pH 7.2 or 6.0) to avoid the introduction of extra variables into the system. Since small ligand associations with proteins can be altered at relatively low pressures (1–3000 bars)[159] substrate concentrations were selected to ensure that 99.9% of the P-450$_{cam}$ active site contained bound substrate. When pressure (1–800 bars) was applied to both the camphor-free and bound forms of P-450$_{cam}$ no discernible shifts of the spin equilibria to the low-spin form were seen. The typical high-spin to low-spin shift was seen only at substrate concentrations that did not saturate the P-450$_{cam}$ active site. This result strongly suggests that the observed spectral transition is due to substrate dissociation. Thus, treating the observed optical changes as substrate dissociation, the volume change between substrate-bound and free forms of the cytochrome is given by

$$\frac{\Delta G}{RT} = \ln K_d = P(\Delta V/RT) - \Delta S/R$$

The concentration of substrate-free protein and ultimately the dissociation constant for substrate binding is found by plotting $\ln (K_d)$ versus P/RT. Further evidence that pressure-induced substrate displacement is the cause of the observed spectral changes derives from the excellent agreement between a comparison of theoretically calculated differential absorbances of a titration curve at atmospheric pressure (K_d = 0.81 μM, pH 7.2) and a derived titration curve at 800 bars (K_d = 3.44 μM, pH 7.2).

The observed volume change for camphor dissociation at pH 7.2 is comparatively large (47 ml/mole) considering that the typical volume changes for most substrate–protein interactions are 5–15 ml/mole.[159] Another possible contribution to the large volume change could be the dissociation of exogenous or endogenous ions from the protein, permitting electrostriction of solvent around the resultant free charges. Indeed, the calculated volume change for camphor binding at higher proton concentrations (pH 6.0) is significantly decreased (30.8 ml/mole). Such decreases are not uncommon for simple proton dissociations. For example, the calculated volume change for the dissociation of a proton from the phosphate anion ($H_2PO_4^-$) is -24 ml/mole.[164] It is interesting to note that Sligar and Gunsalus[165] found that at 20°C, a proton bound preferentially to the high-spin form of the cytochrome regardless of the presence or absence of camphor. Nevertheless, one must not discount the possibility of changes in the hydration sphere of the protein at lower pH values resulting in a smaller volume change for substrate dissociation. Contrary to the above trend, increased potassium ion concentration increased the observed volume change for the substrate binding process. Thus, the large volume change seen for substrate dissociation may include the simultaneous binding of protons and/or potassium ions. Also of interest from these studies are the similar volume changes observed for camphor and related analogue binding (Table III) from which it is concluded that the initial spin equilibrium does not contribute to the large volume changes seen upon substrate dissociation. The change in ligand field structure which is linked to the chemical ferric spin equilibrium must therefore be localized near the heme moiety and represent a relatively small volume change. Indeed, inorganic complexes of globin show small volume changes (-4 ml/mole) accompanying a pressure-induced spin equilibrium shift.[166] This indicates that the large volume changes accompanying spin transitions[160] in proteins are due primarily to changes in protein structure.[164,167,168]

3.2.2. Local Motions of Tyrosine Residues

Since alterations in the ligand field structure around the porphyrin-chelated iron center do not involve "global" changes as indicated by the pressure studies[168] and small-angle X-ray scattering data[157] discussed

TABLE III
Substrate-Dependent Volume Changes

Substrate	K_D (μM)	% High spin	ΔV (pH 7.2) (ml/mole)	ΔV (pH 6.0) (ml/mole)
Camphor	0.81	93.75	47.0 ± 2.1	30.8 ± 2.1
Camphoroquinone	12.2	79.5	42.5 ± 5.9	ND[a]
Fenchone	17.5	61.5	40.0 ± 11.5	25.4 ± 4.7
Tetramethylcyclohexanone	34.0	18.0	46.5 ± 2.2	ND

[a] ND, not determined.

above, other spectroscopic probes must be developed to document localized protein structural changes which are intimately involved with changes in the ferric spin equilibrium. Recent studies by Ruckpaul and co-workers[169-171] have implicated a tyrosine residue as the sixth axial ligand in the low-spin form of P-450$_{LM_2}$. Although the spectroscopic data of Dawson et al.[172] and White[336] strongly indicate that the sixth axial ligand in the low-spin form is water, the possibility that this tyrosine residue is intimately involved in the large conformational change accompanying the change in the ferric spin equilibrium is still viable and consistent with available data. For example, chemical modification of one tyrosine residue of P-450$_{LM_2}$ is prevented by the presence of metyrapone, a compound known to provide a sixth axial ligand to ferric iron in the cytochromes P-450.[170] The three-dimensional X-ray structure of P-450$_{cam}$ places tyrosine-96 at the active site but, at least in this conformation of the protein, too far from the heme for axial coordination.

To determine whether tyrosine motions are coupled to the binding of substrate analogues to P-450$_{cam}$, second-derivative UV–visible spectroscopy as outlined by Ragone et al.[173] was used to determine any change in the degree to which tyrosine residues became exposed to solvent as the spectrally monitored ferric spin equilibrium was changed. This method defines an observable parameteric ratio between two peak-to-trough second-derivative absorbance changes to determine the degree of tyrosine exposure to solvent.[173,174] Although tryptophan residues largely mask the overall absorption of tyrosine residues, the second-derivative spectrum of tryptophan does not change significantly as its microenvironment changes polarity. Thus, for various proteins the percentage of tyrosine exposed to solvent agrees with that derived from the X-ray structure,[173] and hence this technique provides an excellent method to examine changes in the local microenvironments of tyrosine residues in relation to P-450$_{cam}$ conformational changes. Table IV presents the number of tyrosine residues exposed to solvent versus the spin state of the ferric protein as modulated by the binding of various substrates. The total degree of tyrosine solvent exposure is observed to decrease as the percentage of high-spin P-450$_{cam}$ is increased, and is consistent with the equivalent of one tyrosine becoming embedded in the interior of the protein matrix as the protein shifts to the high-spin form. Whether this observed change in tyrosine microenvironment is due to a single tyrosine motion or a combination of all nine tyrosine residues cannot be uniquely determined from this method. In light of the high-pressure and small angle X-ray scattering data, one would expect that the degree of exposure of the tyrosine residues would remain the same with different substrates. However, a linear correlation between ferric spin equilibrium of substrate-bound forms of the

TABLE IV
Degree of Tyrosine Exposure and High-Spin Fraction

Substrate	% high spin	No. of tyrosines exposed[a]
None	4	4.20
TMCH	14	4.06
Norcamphor	45	3.92
Fenchone	46	3.95
Camphenilone	55	3.92
Thiocamphor	61	3.83
5-Ketocamphor	64	3.83
3-endo-Bromocamphor	67	3.83
Adamantane	79	3.76
5-exo-Bromocamphor	81	3.78
Camphor	95	3.58
Adamantanone	99	3.66

[a] Average of 30 spectra.

cytochrome and the degree of exposure of tyrosine residues is observed. We suggest that a localized conformational change directly involved in controlling the spin equilibrium simultaneously changes the microenvironment of a single tyrosine residue. These experiments represent the first documentation of a functionally important conformational change of P-450 that is linked to the observed ferric spin equilibrium.

Previous workers have documented that potassium ions can increase the fraction of ferric high-spin species present at a constant camphor concentration.[137,175,339] Second-derivative absorbance measurements indicate that tyrosine exposure decreases with increasing high-spin and potassium ion concentration in the presence of a constant level of substrate. Here again, tyrosine exposure is found to be directly dependent on the percentage of high-spin species present and independent of the amount of substrate present. Thus, the change in tyrosine microenvironment is linked to the conformational change monitored by the spin state of the hemeprotein.

According to the recently published X-ray structure of camphor-bound P-450$_{cam}$[176] one tyrosine residue (Tyr-96) seems to be an integral part of the active site of P-450$_{cam}$ and is therefore an excellent candidate for the source of the observed changes in the second-derivative spectrum. From this X-ray structural analysis, the tyrosine hydroxyl proton and the carbonyl oxygen of the substrate camphor are sufficiently close to form a hydrogen bond. In order to test whether this hydrogen bond is central to the observed changes in tyrosine microenvironment, the degree of shift in tyrosine exposure to solvent upon binding of adamantane (which lacks a carbonyl function and yet is regioselectively hydroxylated)[154] was investigated. The data show that the degree of tyrosine exposure follows the same linear relationship with the percentage of high-spin P-450$_{cam}$ species present, suggesting that the proposed hydrogen bond between substrate carbonyl oxygen and Tyr-96 is not required for the manifestations of the spin-state-linked second-derivative absorbance changes. This eliminates any role for any perturbation of the Tyr-96 electronic structure due to alterations in hydrogen bond strength. It is important to note, however, that Tyr-96 is flanked by two phenylalanine residues (Phe-87 and Phe-98), and calculated distances from the X-ray analysis between the α-carbon on Tyr-96 and the α-carbon atoms on Phe-87 and Phe-98 are 7.30 and 6.05 Å, respectively.[176] Although these two Phe residues could provide the variability in the microenvironment of Tyr-96, the exact position and movement of the proposed essential tyrosine are not known.

Second-derivative UV–visible spectroscopy performed with the reduced and oxidized forms of the cytochrome reveals that these two macroscopic conformational states have very similar tyrosine microenvironments. Since the r_n value is actually a function of the absorbance of nine tyrosine residues, the fact that reduced and oxidized forms of the protein have similar r_n values indicates that the conformational changes induced by substrate binding in the oxidized form of P-450$_{cam}$ are retained when the cytochrome undergoes a single electron reduction. The effects of local and global structural variations in the P-450$_{cam}$ macromolecule on the chemistry and second electron transfer events in the detailed reaction cycle are currently under investigation in our laboratory.

3.3. Chemistry and Catalysis of P-450$_{cam}$

With the resolution of the camphor methylene hydroxylase into its component proteins (P-450$_{cam}$, putidaredoxin, and NADH-putidaredoxin reductase),[67] a model system was available for detailed study of the reaction chemistry of the cytochromes P-450. Although the net reaction stoichiometry for P-450 monooxygenases called for consumption of one molecule of molecular oxygen and two reducing equivalents from NADH during each hydroxylase cycle, the individual steps required to produce this net reaction had not yet been determined. Following the isolation of purified hydroxylase components, intensive studies of the structure, chemistry, and physics of P-450$_{cam}$ were initiated. Three intermediate states of the P-450 were quickly identified: ferric–substrate complex (m^{os}),[177,178] ferrous–substrate complex (m^{rs}) obtained by anaerobic reduction,[178,180] and the ternary oxy–ferrous–substrate complex ($m^{rs}_{O_2}$) formed by addition of oxygen in the absence of putidaredoxin.[178–181] This last complex is the predominant form of the protein observed spectrally in steady-state mixtures of the fully reconstituted hydroxylase.[25] The corresponding oxy–ferrous–substrate complexes have also been observed in the steady-state turnover of cytochromes P-450 in liver microsomes.[182] Thus, the initial portion of the reaction cycle was determined to be made up of two univalent reductions of the P-450 following the binding of substrate and oxygen, respectively.[148] Unfortunately, the $m_{O_2}^{rs}$ complex is also the last observable state prior to concomitant regeneration of ferric P-450 and release of product. The stopped-flow studies of Pederson et al.[133] showed that oxidation of Pd^r and regeneration of m^{os} occurred simultaneously and with no detectable intermediates as observed on a time scale of 0–1000 msec. Thus, the catalytic chemistry of oxygen activation and substrate oxidation is kinetically masked by the overall rate-limiting step of the reaction cycle, i.e., second electron transfer.[25,133,148,189]

In a strictly formal sense, there should be at least three additional intermediates corresponding to the two-electron reduced complex, the reactive oxygen complex, and the product complex. This formal assignment would be reduced by combination of these states in a concerted process, or increased by occurrence of multistep reactions between any of the formal states. The first portion of this section will address our current understanding of oxygen activation and cleavage by P-450$_{cam}$ and the second part will examine the mechanisms of substrate oxidation.

3.3.1. Putidaredoxin in Electron Transfer

To study the O_2 activation reaction, it is useful to begin with the stable complex immediately preceding oxygen cleavage and product formation: the oxy–ferrous–substrate complex. In the absence of putidaredoxin, $m_{O_2}^{rs}$ was readily isolated but was found to gradually regenerate m^{os} without formation of product. That this decay was first-order indicated it did not occur by disproportionation of the $m_{O_2}^{rs}$ complex.[25] Sligar et al.[183] showed that this reaction was an autoxidation in which superoxide radical anion dissociated from the P-450 in a rate-limiting first step, regenerating m^{os}. Subsequent disproportionation of superoxide produced singlet oxygen and peroxide. This autoxidation reaction is slow compared to the rate of turnover of the reconstituted camphor hydroxylase ($k = 0.01$ and 17 sec^{-1}, respectively, at 25°C).[25] While the autoxidation reaction of P-450$_{cam}$ is slow at ambient temperature and the complex can be further stabilized at low temperatures[184] ($t_{\frac{1}{2}} = 1.38$ min at 25°C, 48 hr at -30°C), the oxy–ferrous–substrate complexes of other P-450 isozymes are much less stable. They were first observed in liver microsomes by steady-state[182] and stopped-flow[185] measurements, and could only be isolated by use of low-temperature techniques.[186,187] Detailed autoxidation studies in Coon's laboratory indicated that the LM$_2$ isozyme behaved very similarly to bacterial P-450$_{cam}$, releasing superoxide in a first-order process, while LM$_4$ showed biphasic kinetics and no superoxide production.[337]

In the presence of either reduced or oxidized putidaredoxin, m$_{O_2}^{rs}$ undergoes rapid turnover with the reaction stoichiometry a function of the oxidation state of putidaredoxin.[25,179] With Pd^r in excess, the amount of product formed is equal to the amount of m$_{O_2}^{rs}$ available. The maximal amount of product formed with oxidized putidaredoxin (Pd^o) is nearly one-half the total m$_{O_2}^{rs}$ available,[189] suggesting that disproportionation of m$_{O_2}^{rs}$ provides the necessary reducing equivalents. In both cases, kinetic and fluorescence studies indicate that formation of a complex between Pd and P-450$_{cam}$ is required for production formation.[188,189] Although Pd is required for product formation from m$_{O_2}^{rs}$, it is not known whether it acts as an electron shuttle in the disproportionation of reducing equivalents or functions only in its role as an obligate effector for product formation (see Section 3.3.3).

3.3.2. Reactive Oxygen Intermediates Derived from Exogenous Oxidants

With the difficulties encountered in visualizing the chemistry of P-450-mediated O_2 activation and cleavage, metalloporphyrins and active oxygen donors were used to model the various hypothetical intermediates in the P-450 mechanism. The stability of molecular oxygen in an organic environment due to kinetic restriction of the spin-forbidden triplet/singlet couple in ionic reactions and the energetic restriction of radical reactions have been succinctly reviewed by Hamilton[190] with emphasis on the implication of these restrictions for biological oxidation reactions. Radical mechanisms are only feasible if the substrate or an intermediate (e.g., a coenzyme) can greatly stabilize the radical intermediate by delocalizing the unpaired electron in a highly conjugated system, while ionic mechanisms require a transition metal ion catalyst. Similar restrictions apply to radical reactions of reduced oxygen species including HO_2 and H_2O_2. H_2O_2 can react as a nucleophile or can be chemically altered to produce a species more reactive as an electrophile or radical reagent (e.g., a peracid or a peramide). Hydroxyl radicals, generated by reductive cleavage of H_2O_2 with a low-valence transition metal (Fenton chemistry), are highly reactive in solution and their utilization in enzyme-catalyzed reactions displaying the high degree of regiospecificity and stereoselectivity typical of P-450-catalyzed reactions would require careful substrate positioning and control of reaction coordinates by the protein structure.

The resemblance of oxygenase chemistry to the chemistry of carbenes and nitrenes first suggested that a six-electron oxygen species, analogous to a carbene or carbenoid species, could be the reactive intermediate in P-450 chemistry.[191,202] Addition of a second electron to the oxy–ferrous–substrate complex of P-450 would result in an activated O_2 intermediate with a formal structure of Fe(III)-peroxide.[190] Noting that electrons can be delocalized by resonance onto the metal, porphyrin, and proximal ligand, Hamilton felt that an Fe(III)-peroxide form of the oxenoid species was more likely than an Fe(V)=O complex (similar to peroxidase Compound I) because the oxygen atom insertion reaction by the latter species would require a transition state under unacceptable steric constraint.[190] The hydrogen abstraction/oxygen rebound mechanism proposed by Groves and McClusky[192] for hydroxylation by m-chloroperbenzoic acid and ferrous perchlorate provided an acceptable noninsertion mechanism for an Fe(V)=O reactant and was followed 2 years later by demonstration of a carbon radical intermediate in the liver microsomal P-450-catalyzed hydroxylation of norbornane.[193] Continued investigation with model porphyrin systems showed that aliphatic hydroxylation could be catalyzed by iron(III)– and manganese(III)–porphyrin complexes supported by exogenous oxidants such as iodosylbenzene,[194-197] indicating that an oxenoid mechanism is one acceptable explanation of P-450 chemistry (see Chapter 1). Turnover supported by oxygen atom donors had previously been demonstrated for P-450$_{cam}$[198] and for microsomal cytochromes P-450 from liver and adrenal cortex.[23,199-202] P-450-dependent oxygen transfer between iodosylbenzene and iodobenzene confirmed an overall "oxene transfer" function for P-450 and implied the occurrence of a transient $[FeO]^{3+}$ intermediate.[203] Further support for a P-450–oxene complex was provided by the total incorporation of solvent oxygen into product during iodosylbenzene-supported turnover of P-450$_{cam}$[204] and P-450b[205] (but see Chapter 7).

While the oxy–ferrous–substrate complex of P-450 is the last spectrally observable intermediate in reactions with NAD(P)H and O_2, a new spectral intermediate has been observed during exogenous-oxidant-supported reactions with P-450$_{cam}$,[133,198,206] liver microsomes,[202,207,208] and purified P-450$_{LM_2}$.[209] The characteristics of the spectrum suggested this new intermediate might be the P-450 equivalent of peroxidase Compound I. Turnover of Mn-P-450$_{cam}$ [prepared by reconstitution of apoprotein with aquo/acetato–manganese(III)–protoporphyrin IX] was examined in both reconstituted hydroxylase and exogenous-oxidant-supported reactions.[210] Addition of iodosylbenzene produced an intermediate with a spectrum characteristic of an $[MnO]^{3+}$ complex. The complex was able to epoxidize 5,6-dehydrocamphor[211] but could not support hydroxylation of camphor. Presumably, the oxo-thiolate complex of Mn–P-450 has a less active oxygen than the corresponding oxo(tetraphenylporphinato) manganese model systems which have been shown to support alkane hydroxylation.[196,197] Unfortunately, the instability of Mn(II)–P-450$_{cam}$ in the presence of molecular oxygen and the inability of Pd to reduce Mn–P-450 precluded observation of the activation and cleavage of O_2 in this system.[210] Thus, while model systems and exogenous oxidants have suggested some possible structures for the reactive oxygen species in the P-450 reaction cycle, the accuracy of these complexes as models for P-450 and the mechanism leading to them from the oxy–ferrous–substrate complex remain to be established. Further details on the chemistry of dioxygen activation and carbon chain functionalization are provided by Ortiz de Montellano in this volume.

3.3.3. Pd as an Effector in Dioxygen Activation

As noted earlier (Section 3.3.1), the role of Pd in the hydroxylation of camphor is to transfer electrons to P-450$_{cam}$ in two univalent reactions.[25,148] The reduction of m^{os} to m^{rs} can also be carried out by chemical and photochemical reductants including dithionite and proflavin/EDTA. In contrast to this, photochemical reduction of $m^{rs}_{O_2}$ in the absence of Pd results in regeneration of m^{rs} without product formation,[212] under conditions supporting a negligible autoxidation rate.[184] Although the $m^{rs}_{O_2}$ complex will autoxidize to m^{os} and superoxide in the absence of Pd,[183] addition of Pd° dramatically stimulates product formation in a reaction presumed to occur by disproportionation of $m^{rs}_{O_2}$. The highly specific requirement for Pd to mediate product formation, while other reducing agents were shown to be competent in reducing m^{os}, led to the conclusion that Pd has an effector role in the camphor hydroxylase in addition to its role in electron transfer.[189] Coincident to these studies, Sligar observed changes in the fluorescence spectrum of the single tryptophan in Pd that indicated this solvent-exposed, C-terminal residue was in a solvent-free, hydrophobic environment in the P-450$_{cam}$–Pd complex. These observations led to the hypothesis that the aromatic indole side chain of the tryptophan might be involved in electron transfer to P-450$_{cam}$. Following total removal of the penultimate glutamine and terminal tryptophan by treatment with carboxypeptidase A, the modified Pd (des-Trp-Pd) was found to have retained all the UV–visible and EPR spectral characteristics of the iron–sulfur center and the reduction potential of the native enzyme, indicating that the prosthetic group was intact. Kinetic analysis of the ability of des-Trp-Pd to support product formation in the reconstituted hydroxylase showed, however, that less than 2% of the activity of an equivalent amount of native Pd was retained.[213] The rate constant for Pd°-supported turnover of $m^{rs}_{O_2}$ was also found to have decreased 50-fold, suggesting that the missing tryptophan had a role in the effector function. More recently, examination of the rate of reduction of m^{os} by des-Trp-

Pd has revealed a change in both K_m and V_{max}, demonstrating that loss of the glutamine and tryptophan residues have affected electron tranfer as well as binding affinity.

The specificity of the requirement for Pd as effector in camphor hydroxylation was examined in kinetic studies of the turnover of $m_{O_2}^{rs}$ with several proteins, apoproteins, amino acids, cofactors, and small reductants. The results showed that a number of iron–sulfur proteins were able to support product formation at lower efficiency, while product yields with b_5 were even greater than with Pd. Of particular interest were the findings that the apoproteins of rubredoxin (*P. aerogenes*) and b_5, and a variety of small organic molecules, were also able to effect product formation. Common to the effector molecules identified was the presence of a free carboxyl group. Considering the evidence for a peracid-generated $[FeO]^{3+}$ complex and the various effector studies, Sligar et al.[198] proposed a model for oxygen activation by P-450$_{cam}$ in which Pd provides an amino acid carboxyl group for transient formation of a peracid function as the means for activation and cleavage of molecular oxygen. The requirements for such a model include that the two-electron reduced molecular oxygen–iron complex carry out a nucleophilic attack on the carbonyl carbon of the amino acid, forming a tetrahedral intermediate. With hydroxide ion (or H_2O) as the leaving group, a peracid complex with the iron is formed. This complex then undergoes dioxygen bond scission, producing the reactive oxygen $[FeO]^{3+}$ and regenerating the amino acid carboxyl. Two problems with this proposed mechanism need to be examined in detail: (1) the mechanism would require a one- or two-electron reduced iron–oxygen complex to have sufficient nucleophilicity to attack the amino acid carbonyl, and (2) the traditional view of monooxygenase chemistry, in which one atom of molecular oxygen is incorporated into substrate while the second is immediately reduced to water, would have to be modified to allow a net effect of reduction to water while at least one reaction cycle would be required before a label from O_2 would appear in the solvent. Sligar et al.[214–216] addressed the latter problem in single-turnover studies with dihydrolipoic acid as a small-molecule effector on the $m_{O_2}^{rs}$ complex. The results showed concomitant incorporation of $^{18}O_2$ into 5-*exo*-hydroxycamphor and the effector carboxyl group. It is of interest to note that the penultimate glutamine of Pd could potentially act as the effector by forming a transient peramide on its side chain. Initial arguments against this alternative pointed out that this could lead to loss of NH_3 from the tetrahedral intermediate in place of OH, converting the glutamine to glutamic acid. Experiments with dihydrolipoamide showed that it could function as the effector and that the amide structure was specifically retained during the reaction.[217] Since incorporation of $^{18}O_2$

into the dihydrolipoamide has not been examined, it cannot be concluded at this time that dihydrolipoic acid and amide act through the same mechanism. While these results are consistent with the model, a crucial experiment remains the demonstration of oxygen incorporation into the carboxyl group of tryptophan in Pd (or into glutamine). All attempts to date to find an unambiguous procedure for this experiment have met with difficulty and the question remains unanswered. The nucleophilicity of iron-bound O_2^{2-} has been addressed in a manganese–porphyrin model system. Groves et al.[218] demonstrated that addition of an acid chloride to an Mn(II)–O_2 species at $-20°C$ resulted in an intermediate, spectrally identical to the species formed by addition of the corresponding peracid to Mn(III)–porphyrin, that was competent in alkene epoxidation. The interpretation of these results postulates nucleophilic attack by oxygen, displacing Cl^-, and forming the iron–peracid intermediate. Thus, the feasibility of the peracid intermediate mechanism has been demonstrated, although proof awaits demonstration of the required intermediates in the native hydroxylase.

An additional possible complication arises from the recent demonstration[219] of O_2-dependent, multiple turnovers of P-450$_{cam}$ with phenazine methosulfate mediating electron transfer from NADH to P-450$_{cam}$ in the absence of both Pd and NADH-putidaredoxin reductase. Unlike the previously identified small organic effectors, phenazine methosulfate does not contain a free carboxylic acid or amide function and appears to define a new class of effector. However, the rate of product turnover is exceedingly small and could be due to a completely different reaction mechanism, perhaps involving bound peroxide.

3.3.4. Chemistry of Atmospheric Dioxygen Bond Cleavage

The evidence noted previously for the spectral similarity of the oxy–P-450 intermediate and peroxidase Compound I, together with the ability of oxene donors to support P-450-catalyzed oxidations of substrates, had led to a general acceptance of a Compound I-type intermediate as the reactive oxygen species in the P-450 reaction cycle (Fig. 6). This interpretation came into serious question when the results of several experiments could apparently not be explained by the heterolytic mechanism leading to "Compound I." Blake and Coon[220] observed that the "Compound I-type spectrum" occurring as an intermediate in the reaction of P-450$_{LM2}$[209] with peroxy compounds varied in the wavelength of maxima and minima and in amplitude as the organic portion of the peroxy compound was modified. In addition, peroxyphenylacetic acid was found to undergo decarboxylation when mixed with P-450$_{cam}$, P-450$_{LM2}$, or P-450$_{LM4}$, in a reaction characteristic of an intermediate

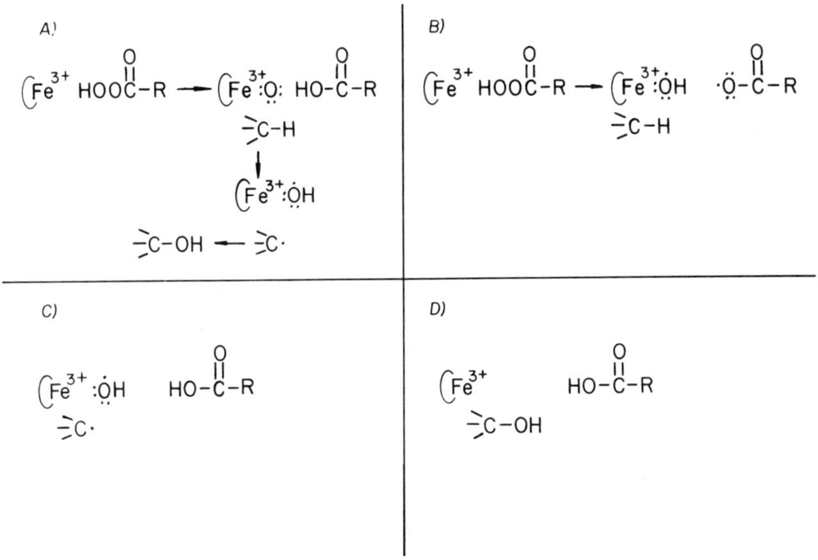

FIGURE 6. Possible chemistries of dioxygen activation and substrate hydroxylation. (A) Heterolytic mechanism: dioxygen cleavage, substrate activation (hydrogen abstraction), and rebound. (B) Homolytic mechanism: dioxygen cleavage. (C) Homolytic mechanism: substrate activation. (D) Homolytic mechanism: rebound.

carboxy-radical species, implying that the dioxygen bond of the peracid had undergone homolytic cleavage.[221] These observations, together with other inconsistencies found when comparing cytochromes P-450 and peroxidases, led to a new mechanistic proposal in which cleavage of the dioxygen bond of peroxide, peracid, and peramide intermediates occurs homolytically, producing an iron-bound hydroxyl radical and an organo-oxy radical (Fig. 6B).[222] A substrate radical produced by hydrogen atom abstraction by the organo-oxy radical (Fig. 6C) would recombine with the iron-bound hydroxyl radical to generate the product (Fig. 6D). Further experimentation, however, has contradicted the interpretation of the above data and raises questions about the viability of the homolytic mechanism as presented in Fig. 6B–D. Although the decarboxylation of peroxyphenylacetic acid by P-450$_{LM_2}$ suggests a radical (homolytic) mechanism, an elegant kinetic analysis by McCarthy and White[223] demonstrated that hydroxylation of substrate and decarboxylation of peroxyphenylacetic acid occur in independent pathways that diverge from the ferric–substrate P-450 complex. Thus, the decarboxylation may occur with homolytic cleavage while substrate oxidation occurs on a separate heterolytic pathway.

Blake and Coon[224,225] elegantly determined the kinetics of a model system in which cumene hydroperoxide was the oxidant and toluene the substrate. Examination of the linear free energy relationship for a series of *para-* and *meta-*substituted substrates and oxidants led them to conclude that a linear free energy relationship existed between the overall rate constant for product formation and both the oxidant and substrate. Such a result could be accounted for by a rate-limiting step in which both substrate and oxidant participated (abstraction of H· from RH by RO·, Fig. 6C) or by contributions to the overall rate of reaction by two separate reaction steps, one involving substrate and the other requiring oxidant. The authors argued against the latter interpretation because the reaction rate changes associated with a range of substrate and oxidant derivatives should cause significant changes in the relative contributions of each step to the overall reaction rate. Such changes would be clearly visible in the linear free energy relationships and were not detected, although only two derivatives of cumene hydroperoxide were used, spanning a range of σ values from only 0.17 to -0.06 out of a maximum range of 0.78 to -0.83. In recent work by Lee and Bruice,[226] the linear free energy correlation for oxygen transfer to ClFe(III)TPP was determined for a series of peracids and hydroperoxides (ROOH) as a function of the pK_a values of the corresponding hydroxides (ROH). Note the similarity of this reaction to heterolytic cleavage of the peroxy bond (ROOH $\rightarrow$ RO$^-$ + OH$^+$; ROH $\rightarrow$ RO$^-$ + H$^+$). The results showed a sharp break in the correlation at $pK_a \sim 11$. Although both phases of the correlation had negative slopes, the large difference in the magnitude of the slopes for each phase was interpreted to indicate a heterolytic dioxygen bond scission for the more reactive donors (peracids and very active hydroperoxides) and a homolytic bond scission for the less reactive hydroperoxides. These results underscore the potential importance of using a wide range of activities in the oxygen donor studies of Blake and Coon. The pK_a range for the hydroxides of the cumene hydroperoxides used by Blake and Coon is 15–15.6, placing these data in the middle of the homolytic phase of Lee and Bruice's results. A second observation is the correlation of the Hammett ρ values with mechanism. In a homolytic mechanism, the rate limiting step consumes cumene oxy-radical and generates the benzyl radical. Typical ρ values for free radical reactions are small. Hence, one would predict that ρ for both cumene hydroperoxide and toluene derivatives would be small. In contrast to this prediction, the Hammett value observed[225] yielded $\rho = -1.6$, toluenes; $\rho = 5.0$, cumene hydroperoxides. The substantial positive slope for the cumene hydroperoxides indicates substantial stabilization of a negative charge in the transition state by electron withdrawing substituents-a result easier to account for by heterolytic cleavage of the peroxy bond.

BACTERIAL P-450 ENZYMES

Thus, the mechanism of oxygen activation and the identity of the reactive oxygen intermediate remain hidden in a "black box," although the size of the box has been considerably reduced. We must hope that continued investigation will soon lead to a full understanding of the mechanisms of oxygen activation and cleavage.

3.3.5. Substrate Activation and Oxidation

In order to gain further insight into the chemical mechanisms of hydroxylation, one can focus on the carbon skeleton undergoing oxygenation. The bacterial P-450$_{cam}$, the liver microsomal isozymes, and the adrenal steroid-metabolizing cytochromes P-450 demonstrate significant differences in their regioselectivity and stereochemical control of substrate hydroxylation. The focus here will primarily be on the bacterial enzyme isolated from *Pseudomonas*,[129] with brief comments on the adrenal and liver isozymes for comparison. This emphasis is natural in light of the extensive and rigorous stereochemical analysis which has been developed for P-450$_{cam}$.

Perhaps the most intriguing observation concerning the stereochemistry of P-450$_{cam}$-catalyzed hydroxylation at the 5-*exo*-position of camphor is the apparent lack of complete stereochemical selectivity at the level of substrate activation, and rigorous stereochemical selection in the subsequent hydroxylation step. These separate steps in the reaction pathway are easily conceptualized as events prior to formation of a planar carbon radical intermediate and reactions subsequent to formation of this species. Before considering the stereochemistry of these distinct events, the evidence for such a planar species should be mentioned. This aspect of substrate activation has been reviewed previously,[227] and is further discussed in Chapter 7.

With the camphor analogue, pericyclocamphanone, it has been shown that the 6-*exo*- and 6-*endo*-alcohol are produced in NADH-supported turnovers.[233] No hydroxylation was detected at the 5-position. With the natural substrate camphor, 5-*exo*-hydroxycamphor is the sole product. This observation is consistent with the occurrence of an intermediate planar carbon center at the 5-position with the normal substrate camphor, since the severe ring strain (in pericyclocamphanone) would make it extremely difficult to generate a planarized carbon at the 5-position. Such a planar intermediate might be postulated to be either a radical species or a carbonium ion, whereas a carbanion produced by proton abstraction would result in a pyramidal rather than a planar carbon center. Such a geometry would be considerably less destabilized by the ring strain found in pericyclocamphanone. Therefore, the extreme sensitivity to ring strain at the 5-position precludes any carbanion sp^3 character. Rates of solvolysis of bridgehead carbons to form carbonium ions[228,229] and generation of carbon radicals[230] demonstrate a large range of rate effects depending on the system, so that comparison of these rates makes it difficult to predict which is actually more sensitive to ring strain. Even with 6-ketopericyclocamphanone in which the 6-position is inaccessible to hydroxylation, no 5-alcohol product was detected.[233] This emphasizes the extreme inertness of the tertiary carbon toward hydroxylation, presumably due to lack of planarizability. It is interesting to note that with 6-keto-pericyclocamphanone and a complete reconstituted enzyme system, NADH is rapidly oxidized, presumably leading to reduction of bound dioxygen to H_2O_2. Thus, 6-keto-pericyclocamphanone represents an uncoupler of the system, allowing efficient electron transfer to the high-spin form of the cytochrome with all reducing equivalents appearing as peroxide. With this substrate and the camphor hydroxylase, no rearranged products are detected. This is in contrast to what is observed in model systems. In iodosobenzene-supported hydroxylation of norcarane by Mn(III)TPP, the expected 2-*exo*- and 2-*endo*-alcohols are produced along with 3-hydroxymethyl cyclohexene and the 3-substituted alkyl chlorides.[197] These latter products represent: (1) rearrangement products resulting from formation of a carbon radical at the 2-position with subsequent ring opening to give the cyclohexenyl products and (2) solvent chlorine abstraction by the radical intermediate. These observations are consistent with a long-lived radical intermediate in the model systems and a fast radical recapture in the enzymatic system. This difference may reflect an enzyme-imposed juxtaposition of substrate radical and active oxygen species resulting in rate enhancement for radical recapture; alternatively, this observation may actually represent a chemical difference in the reactivity of the Mn-bound radical recapture species and the corresponding Fe-bound species. The latter interpretation is supported by the observation that with Fe(III)TPP, iodosobenzene, and norcarane, no alkyl chloride products were detected,[194] suggesting radical recapture and oxygen transfer before ligand transfer from solvent. With the hepatic P-450$_{LM_2}$, norcarane is hydroxylated to give 2-*endo*- and 2-*exo*-alcohol, with only trace amounts of rearrangement, suggesting that, like the camphor hydroxylase, radical recapture is fast.[231] As implied above, the initial species generated by hydrogen abstraction is presumed to be a radical as opposed to a carbonium ion. Studies on the hydroxylation of norbornane by P-450$_{LM_2}$[193,231] also favor a radical species. The authors observed 2-*endo*-norborneol as a product. Carbonium ions on the norbonyl skeleton have been shown to selectively produce the 2-*exo*-alcohol. Obviously, the difference in solution chemistry and events at the active site could conceivably be great enough to afford changes in stereochemistry, but usually an enzyme imposes greater stereochemical selectivity. The example cited above demonstrates a case in which the enzyme would exercise less stereochemical selectivity than if it were proceeding by the same mechanism as in free solution. Therefore, a carbonium ion intermediate seems unlikely. For the discussion which follows, a carbon-centered radical is assumed to best represent the nature of the activated substrate.

The stereochemistry of the step(s) resulting in oxygen insertion has been elucidated predominantly through the use of deuterium isotope effects.[232] Toward this end, Gelb et al. synthesized camphor analogues deuterated at either the 5-*exo*-position or the 5-*endo*-position as well as at the 6-*endo*-position. Both inter- and intramolecular isotope effects were analyzed extensively. These authors found that when looking at initial velocities of NADH oxidation in a complete reconstituted system, a small but significant intermolecular isotope effect was observed. It is relatively rare in enzymatic reaction pathways to observe a single rate-determining step, particularly when many steps are present as is the case with the classic P-450 reaction cycle. When NADH oxidation rates were compared with 5-*endo*-deutero-5-bromocamphor and 5-*endo*-protio-5-bromocamphor, there was also a small but significant difference,[232] consistent with a largely masked primary intermolecular isotope effect. The magnitude of these effects is not surprising in view of the results obtained by Gunsalus and co-workers[25,133] where stopped-flow techniques were used to convincingly demonstrate that the major contributing factor to the rate-determining step in camphor hydroxylation is transfer of the second electron from Pd. Furthermore, by looking directly at the breakdown of the ferrous–oxygenated P-450$_{cam}$ in the presence of deuterated or nondeuterated 5-*exo*-bromocamphor,[232] only very small differences in the decay rates were found. As this approach allows one to compare rates of chemical steps after formation of the one-electron reduced oxygen-bound heme, it might have been expected to maximize any observed isotope effects. The fact that the isotope effects were still small is consistent with a later step, the input of a second electron, being the primary rate-determining step.

Intramolecular isotope effects are more revealing in terms of the actual stereochemistry of hydrogen abstraction. Intra- and intermolecular isotope effects are masked by different factors so it is not immediately apparent that any intramolecular effects need be as small as the intermolecular effects. Intramolecular effects were determined by measuring the deuterium content of the 5-*exo*-hydroxycamphor produced from either 5-*endo*-deuterocamphor or 5-*exo*-deuterocamphor, using NADH, iodosobenzene, H_2O_2, and mCPBA-supported oxygenations. In all cases, a significant amount of 5-deuterium- and 5-hydrogen-containing 5-*exo*-hydroxycamphor was obtained (Fig. 7). It is important to notice that the hydrogen/deuterium alcohol ratios obtained from 5-*exo-d*-camphor were

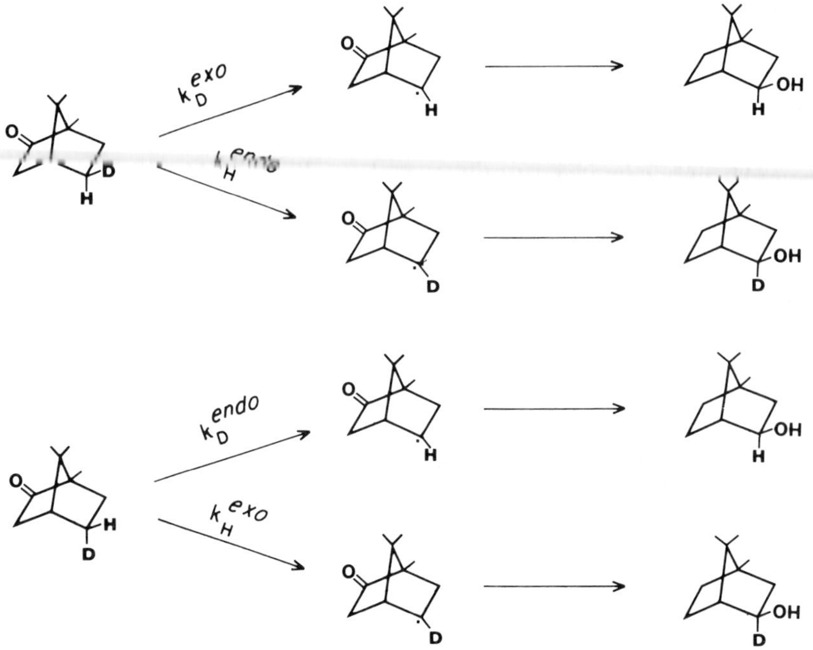

FIGURE 7. Metabolism of deuterated camphor analogues by P-450$_{cam}$.

significantly lower than the ratios obtained from the 5-*endo*-*d*-camphor. Clearly, there is some preference for *exo* abstraction, but this selectivity is not absolute. The following step, oxygen insertion, is completely selective as no *endo*-alcohol is obtained. Also, for a given substrate, the isotopic ratios are nearly identical whether turnover was achieved with the natural enzyme components and O_2, single oxygen oxidants, or dioxygen-containing oxidants. This result suggests similar hydrogen abstraction species for each of these oxidants. The distinction between the stereochemical options before and after generation of a planar radical species is best demonstrated pictorially (Fig. 7), with the kinetic parameters associated with reaction of 5-*exo*-*d*-camphor and 5-*endo*-*d*-camphor depicted. Several parameters must be clearly defined in order to analyze the possible reaction pathways and deconvolute the intramolecular isotope effects. k_H^{exo} is the rate of hydrogen abstraction from the *exo*-position and k_D^{exo} the corresponding rate of deuterium extraction. k_H^{endo} and k_D^{endo} represent the analogous rates of abstraction from the *endo*-position. The intrinsic isotope effect for the abstraction from the *exo*-position of camphor is $I^{exo} = k_H^{exo}/k_D^{exo}$, and similarly the intrinsic isotope effect for abstraction from the *endo*-position is $I^{endo} = k_H^{endo}/k_D^{endo}$. There will also be a partitioning effect due to geometrical or spatial orientation of the enzyme active site which will result in a preference for the enzyme to abstract either the *exo* or *endo* hydrogen. This selection, termed G, is k_H^{exo}/k_H^{endo}. Other parameters can be determined experimentally. By measuring the fraction of 5-*exo*-hydroxycamphor, which contains deuterium when 5-*exo*-d-camphor is the substrate, one obtains the ratio $A = k_H^{endo}/k_D^{exo}$. The same product ratio starting with 5-*endo*-d-camphor as substrate yields $B = k_H^{exo}/k_D^{endo}$. It should be kept in mind that if the hydrogen/deuterium abstraction step is reversible, then A and B may more accurately represent equilibrium isotope effects rather than true kinetic isotope effects. This may also explain the fact that the observed intermolecular effects are small. It should be apparent that if the enzyme were completely selective for 5-*exo* hydrogen abstraction, then $A = 0$, which is obviously not the case experimentally observed. The isotopic selection for hydrogen or deuterium at a single position then can be expressed in terms of A, B, and G:

$$I^{exo} = \frac{k_H^{exo}}{k_D^{exo}} = \frac{k_H^{endo}}{k_D^{exo}} \frac{k_H^{exo}}{k_H^{endo}} = AG$$

$$I^{endo} = \frac{k_H^{endo}}{k_D^{endo}} = \frac{k_H^{exo}}{k_D^{endo}} \frac{k_H^{endo}}{k_H^{exo}} = \frac{B}{G}$$

This analysis assumes complete isotopic purity. The relationship between these parameters is conveniently depicted in Fig. 8. I^{exo} affords a line of slope A, whereas I^{endo} defines a hyperbola when plotted as a function of G, the geometrical selection factor. If the isotope effect for abstraction from the *exo*-position was identical to the isotope effect for abstraction from the *endo*-position, then the intersection of the two curves would identify the geometrical partition effect for the system. This is not likely, assuming that the abstracting species is the same for abstraction from either position, in which case the geometry of the transition state for *exo* extraction would be different from the geometry of *endo* abstraction. Actual values for G, and therefore for I^{endo} and I^{exo}, cannot be obtained without the corresponding data from tritium-substituted camphor. Even without actual values for I^{endo}, I^{exo}, or G, these stereochemical experiments provide a good deal of information about the mechanism of camphor hydroxylation. Direct insertion of oxygen into the C–H bond would not be consistent with the partial loss of stereochemistry observed. A value of G much greater than unity would be consistent with a single species which abstracts hydrogen and inserts oxygen selectively at the *exo* face of the camphor skeleton. Still, there is the possibility that an

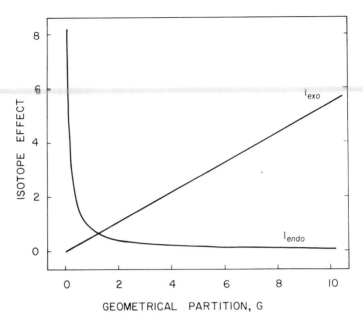

FIGURE 8. Graphical relation between isotope effects and geometrical selection. $G = k_H^{exo}/k_H^{endo}$, $I_{exo} = k_H^{exo}/k_D^{exo}$, $I_{endo} = k_H^{endo}/k_D^{endo}$.

abstracting species placed symmetrically between the *endo* and *exo* hydrogens could result in a G value of 1, with subsequent asymmetric placement of the oxygen atom selectively at the *exo* face.

The specificity of oxygen atom transfer from the *exo* face of the camphor skeleton is further demonstrated by the epoxidation of 5,6-dehydrocamphor.[211] High-resolution NMR and $^{18}O_2$-labeling studies indicate that the product is completely the *exo* isomer and one atom from molecular oxygen is inserted to form the epoxide. This further emphasizes the apparent geometry of the activated oxygen species as being more closely situated at the *exo* face of camphor. As for hydroxylation of the natural substrate, the stereochemistry of the epoxidation reaction is the same for the NADH-, iodosobenzene-, and H_2O_2-driven turnovers. Epoxidations in model systems have been well documented.[234] As with the hydroxylation reactions, differences in the stereochemistry of epoxidation by Mn(III)TPP versus Fe(III)TPP may reflect differences in the lifetime of the radical species generated.

The absolute stereoselectivity of oxygen insertion at the *exo* face of camphor is very reminiscent of the rigorous stereochemical integrity of product formation observed with the P-450$_{scc}$ enzyme. Here, cholesterol is hydroxylated at the 22- and 20-position to afford (20R, 22R)-dihydroxycholesterol as an intermediate in the production of pregnenolone.[235-237] No other stereoisomeric products are obtained with those steroid-metabolizing activities. Nevertheless, the stereochemical parallel to P-450$_{cam}$ cannot be extended to the hydrogen abstraction step since evidence for such a chemical step is tenuous at best in these systems. When 22R-[22-^{3}H]cholesterol was incubated with a side-chain cleavage preparation, the 22R-[22-^{3}H]-22-hydroxycholesterol intermediate had negligible loss of radioactivity. This suggests selectivity in the hydrogen abstraction if it occurs. On the other hand, the loss of even a small amount of tritium may in fact suggest possible abstraction of either enantiotopic hydrogen in the natural cholesterol, especially when one consdiers the large isotopic selectivity expected for hydrogen rather than tritium. So, whereas, the insertion of the oxygens into the cholesterol side chain demonstrates strict stereochemical selectivity, a systematic investigation of the stereochemistry of the putative hydrogen abstraction steps has not been performed.

As mentioned previously, when norbornane is hydroxylated by P-450$_{LM_2}$, both the 2-*exo*- and 2-*endo*-alcohols are produced.[193] Also, the four *exo*-hydrogens were replaced by deuterium, and the deuterium-containing product alcohols were examined. It was found that the *exo*-2-norborneol contained four deuteriums, whereas the *endo*-2-norborneol contained three deuteriums. This demonstrates an interesting contrast to the P-450$_{cam}$ stereochemistry, assuming both P-450$_{LM_2}$ and P-450$_{cam}$ proceed via a similar hydrogen-abstracted substrate species. These norborneol products demonstrate *endo* abstraction followed by *exo* rebound, and *exo* abstraction followed by *endo* rebound. Thus, both P-450$_{LM_2}$ and P-450$_{cam}$ are capable of *endo* or *exo* hydrogen abstraction, although P-450$_{cam}$ displays absolute stereochemical control in the oxygen transfer step.

This lack of complete stereochemical integrity of oxygen insertion is perhaps explained by a great degree of substrate motion at the active site of P-450$_{LM_2}$ as compared to P-450$_{cam}$. A comparison of substrate regioselection has been made directly between these two isozymes using camphor, adamantanone, and adamantane as substrates.[154] As expected, P-450$_{LM_2}$ gave both 5-*exo*- and 5-*endo*-hydroxycamphor, a result analogous to what was observed with norbornane. Also evident were 3-*exo*- and 3-*endo*-hydroxycamphor. The greater stereoselectivity of oxygen insertion by P-450$_{cam}$ was observed with the other substrates as well (Table V). With P-450$_{cam}$ as the catalyst, 5-hydroxyadamantanone and 1-adamantanol were produced from the nonhydroxylated substrates. As the 5- and 1-carbons of these substrates are tertiary, only one isomeric alcohol is possible for these products. With P-450$_{LM_2}$, adamantanone was processed to 4-*anti*- and 5-hydroxyadamantanone, while adamantane was hydroxylated at both the 1- and 2-positions. The authors predict which

TABLE V
Hydroxylation of Alicyclic Substrates by Various Cytochromes P-450[a]

Enzyme	Substrate	K_d	X_{max}^{HS}	K_{eq}	Rate	Regioselectivity (%)
P-450$_{cam}$	D-Camphor	2.9	0.94	16	60	5-*exo* (100)
	Adamantanone	3.5	0.98	49	52	5- (100)
	Adamantane	50	0.99	99	43	1- (100)
P-450$_{LM2}$	D-Camphor	44	0.37	0.59	1.3	3-*endo* (16)
						5-*exo* (14)
						5-*endo* (63)
						Other (7)
	Adamantanone	65	0.25	0.33	1.9	4-*anti* (57)
						5- (43)
	Adamantane	2.0	0.54	1.2	1.6	1- (91)
						2- (9)
P-450$_{LM4}$	D-Camphor	—[b]	—[b]	—[b]	0.006	5-*exo* (100)

[a] Rates are expressed as total mole product per mole enzyme per minute; regioselectivity is expressed as the positions of hydroxylation of the particular substrate, with the corresponding percentage of the total product in parentheses. Dissociation constants (K_d) and mole fractions of high-spin heme were determined from titrations of the ferric cytochromes with the respective substrates. Dissociation constants are expressed in micromolar values.[154]
[b] Quantity not measured.

positions of the adamantane skeleton would be most reactive on a strictly chemical basis, i.e., stability of generated radicals, steric hindrance, and so on, if no binding site restrictions were in effect, and point out that the observed product population corresponds to that predicted.

Results obtained with norcamphor and P-450$_{cam}$ may offer insight into the mechanism of reduced substrate motion at this enzyme active site. In contrast to the absolute stereospecificity of *exo* oxygen insertion observed with camphor, we have recently found that norcamphor affords two products. It is tempting to speculate that the 8,9-*gem*-dimethyl groups of the camphor molecule interact with the protein to restrict substrate motion. The X-ray crystal structure of the bacterial enzyme demonstrates the proximity of valine 295 and these methyl groups. Norcamphor does not possess these 8,9 methyl groups. In the absence of such a substrate anchor, increased motion at the active site might be expected to yield multiple products. Examination of the three-dimensional X-ray structure of P-450$_{cam}$ (Chapter 13) also shows that a hydrogen bond between Tyr-96 and the keto group of camphor plays a key role in substrate positioning. In order to ascertain the role of this hydrogen bond in the stereochemistry and regiospecificity of hydrogen atom abstraction and oxygen atom transfer, several camphor analogues were synthesized where the C=O group of the ketone was replaced by C=S. These thioketone analogues have virtually identical structures to the corresponding ketones and yet are incapable of hydrogen bond formation. When the thioketone analogue of camphor was metabolized by P-450$_{cam}$, multiple products were again observed.

3.4. Bleomycin as a P-450 Model System

The bleomycins constitute a family of glycopeptide antibiotics which differ only in their terminal amide functional groups,[238-242] and are employed clinically for the treatment of squamous cell carcinomas, Hodgkin's disease, and certain other cancers and lymphomas.[243,244] These antineoplastic antibiotics are isolated as one-to-one copper(II) complexes from cultures of *Streptomyces verticillus*.[245-247] The therapeutic efficacy of the bleomycins is believed to be related to their ability to degrade DNA,[248,249] a process that has been shown, *in vitro*, to proceed in the presence of appropriate metal ions and a source of oxygen.[250-258] The unique biological activity and interesting properties of bleomycin have resulted in intensive structural, synthetic, and mechanistic investigations. As a result of these studies, there is increasing evidence of strong similarities between the mechanism and properties of bleomycin and those of the P-450 family. Although bleomycin does not contain a macrocyclic polyaromatic structure[259,260] such as heme, it nevertheless is an effective metal chelator,[261,262] and readily forms an iron drug complex reminiscent of P-450. Both bleomycin and P-450 bind oxygen and carry out site-specific oxidations. Both systems activate dioxygen by parallel pathways and (Fig. 9) combine with O_2 to form an oxygenated complex.[222,263] Mössbauer and ^{1}H-NMR spectral studies have determined that, analogous to P-450, a rhombic high-spin ferrous complex is present.[155,264] The EPR-silent bleomycin–Fe(II)–O_2 is formed rapidly at a rate which is proportional to the O_2 concentration. This complex has, like its corresponding NO, CO, and C_2H_5NCO analogues, a much greater visible absorbance than bleomycin–FE(II).[263,265] These oxygen analogues, two of which are known to be Fe(II) ligands in ternary complexes with bleomycin, reversibly inhibit DNA degradation by Fe(II)–bleomycin and O_2.[265] On this basis, it is thought that the O_2 is bound as a metal ligand in the bleomycin–Fe(II)–O_2 complex in similar fashion to P-450. This belief is further substantiated by studies on bleomycin–Co(II)–O_2 complexes.[266,267] The oxygenated complex is a prerequisite for reactivity of both ferrous P-450 and ferrous bleomycin. Evaluation of the oxygenated P-450[123,155,268-270] and the bleomycin–Fe(II)–O_2 complex by Mössbauer spectroscopy[264,271] has demonstrated that both complexes are diamagnetic and have quadrupole splittings outside the range of values observed for low-spin ferrous complounds. These results, along with those from ^{57}Fe-EPR[272] and ^{1}H-NMR[273] evaluations, have led to the proposal that the electronic structure

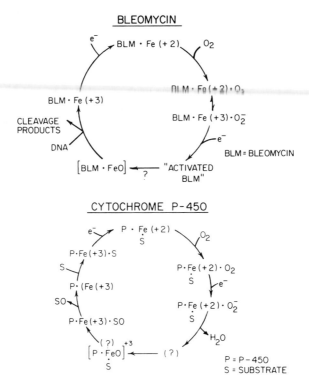

FIGURE 9. Comparison of oxygen activation by bleomycin and P-450.

of the bleomycin–Fe(II)–O_2 is a superoxide anion bound to a low-spin ferric ion. A single electron reduction then converts bleomycin–Fe(II)–O_2 to an unstable transient ferric species, "activated bleomycin,"[272,274] which is believed to be either the penultimate or ultimate agent for site-specific oxidation. The reductant, *in situ*, is proposed to be a second bleomycin–Fe(II) complex yielding bleomycin–Fe(III) in a one-to-one mixture with activated bleomycin.[272,274] Activation by this mechanism is in contrast to the flavoprotein or iron–sulfur protein-mediated one-electron reduction of the oxygenated P-450 complex in the mammalian microsomal or bacterial and mitochondrial systems, respectively.[25,275] Other reductants for this first one-electron reduction of the bleomycin–Fe(II)–O_2 complex have yet to be reported. Activated bleomycin and bleomycin–Fe(III) have identical optical spectra but can be resolved quantitatively by EPR.[272] The exact structure of activated bleomycin is not known with certainty, but ^{57}Fe-EPR spectra and EPR studies using $^{17}O_2$ for activation indicate it is a ferric complex containing at least one bound oxygen normally derived from O_2 as one of the metal ligands.[272] The electronic structure and oxidation state of activated bleomycin are similar to the one electron reduced P-450 Fe(II)–O$_2$ complex, a P-450 ferric peroxide complex.[222] Scission of the dioxygen bond in the P-450 ferric peroxide complex yields a species which is believed to be capable of substrate hydrogen abstraction resulting in substrate oxidation through a free radical intermediate. This type of mechanism has not been demonstrated in activated bleomycin. The most detailed model for bleomycin activity, however, proposes that DNA cleavage occurs via abstraction of a hydrogen from the C4' of DNA by some free radical.[276,277] EPR spin-trapping methods have demonstrated that oxygen radicals in the form of superoxide anion and hydroxyl radical are produced from bleomycin–Fe(II)–O$_2$ systems[278–280] although enzymatic assays for these species have failed to detect them in significant levels.[263] This result is not surprising if one assumes a close association of the radical with drug precluding availability to the enzyme. Experiments using various radical scavengers as inhibitors have provided mixed results.[250,252,281–286] The fact that certain radical scavengers have been demonstrated to protect against the oxidative activity of bleomycin,[250,252,282,283,285,286] however, lends credibility to the hypothesis of a mechanism of action involving, at least in part, one or more free radical species. It should also be noted that bleomycin oxidizes DNA by four electron equivalents whereas the P-450 system oxidizes its substrate by two electron equivalents.[222,276] The lack of the conjugated macrocyclic structure in bleomycin prohibits the removal of an electron from Fe(III) to yield Fe(IV), as proposed for hemeproteins, and thus the loss of an electron from the drug moiety. It has therefore been proposed that the oxidizing equivalents in activated bleomycin must reside in the iron-bound oxygen originating in O$_2$.[264]

Activated bleomycin, like activated P-450, can also be generated anaerobically from bleomycin–Fe(III) and hydrogen peroxide[272,274] and is indistinguishable from that obtained aerobically. Alkyl peroxides, peracids, and iodosobenzene have also been used to give an oxidatively active bleomycin although they have not been fully evaluated spectrally.[257,287] The ternary activated bleomycin rapidly breaks down to yield ferric bleomycin concurrently with the formation of oxidized product.[272,274] The ferric species can then be regenerated to ferrous bleomycin by a one-electron reduction. Numerous reductants have been used for this regeneration step. Predominant among these reductants are such sulfhydryl reagents as dithiothreitol, 2-mercaptoethanol, cysteine, and glutathione.[288,289] NADPH-cytochrome P-450 reductase has also been reported to activate at least two metallobleomycins.[290–293] EPR and Mössbauer properties are remarkably similar to those of low-spin ferric P-450 and are summarized in Table VI. Crystal field analyses of EPR spectra for ferric bleomycin are nearly the same as those of ferric P-450 in spite

TABLE VI
Selected Physical Parameters of Bleomycin versus P-450$_{cam}$

	Mössbauer		
	E_Q (mm/sec)	δ (mm/sec)	Reference
Activated bleomycin (iron)	3 ± 0.2	0.10 ± 0.07	264
Ferric bleomycin	3 ± 0.2	0.20 ± 0.07	264
Ferrous P-450 without substrate	2.85	0.30 ± 0.05	155

	EPR	
	g value	Reference
Activated bleomycin (iron)	1.94, 2.17, 2.26	39
Ferric bleomycin	1.89, 2.18, 2.45	39
Ferrous P-450	1.91, 2.26, 2.45	155
Cobalt(II)–bleomycin	2.272, 2.025	266
Cobalt(II)–bleomycin–O_2 complex	2.007, 2.098	266
Cobalt(II)–P-450 with substrate	2.31, 2.03	295
Cobalt(II)–P-450 without substrate	2.32, 2.03	295
Cobalt(II)–P-450 with substrate–O_2 complex	2.008, 2.079	295
Cobalt(II)–P-450 without substrate–O_2 complex	—, 2.067	295

	Redox potentials	
	$E°$ (mV)	Reference
Ferric bleomycin	+129	296
Ferric P-450 without substrate	−340	138
Ferric P-450 with substrate	−170	138

of the fact that bleomycin lacks the thiolate iron ligand of P-450.[264,268] Parallel EPR studies have also been carried out for cobalt-substituted bleomycin[294] and P-450,[295] the results of which are also included in Table VI. The nitrogen ligation and absence of the thiolate ligand in bleomycin are reflected in the differing redox potentials between ferric bleomycin ($E° = +129$ mV)[296] and ferric P-450 ($E°'_{-s} = -340$ mV, $E°'_{+s} = -170$ mV).[138] The bleomycin redox potential is, however, very similar to the nitrogen-ligated hemeproteins that bind and/or activate O_2, hemoglobin ($E°' = +144$ mV),[297] and tryptophan oxygenase ($E°' = +105$ mV).[298] The earlier discussion also noted that bleomycin can form complexes with various oxygen analogues and antagonists. Drug complexes with CO, NO, C_2H_5NCO, HO^-, N_3^-, CN^-, and CH_3NH_2, giving distinct visible absorption spectra, have been reported reminiscent of P-450 and other hemeproteins.[265,300] Addition of these oxygen analogues to ferrous bleomycin results in absorption intensification and shifting. This binding is reversible and bleomycin can be returned to the ferrous drug complex by purging with argon or converted to the bleomycin–Fe(II)–O_2 complex by purging with oxygen, as is the case with P-450.[265] As noted previously, CO and ethyl isocyanide inhibit the oxidative activity of the drug and give validity to the hypothesis that these compounds are acting as oxygen antagonists competing for the O_2 binding site, as they are believed to act in inactivating P-450.[265] Cyanide, however, a potent inhibitor of P-450 reactivity, enhances the drug's ability to degrade DNA. This enhancement is proposed to be a result of cyanide-enhanced dissociation of a transiently oxygenated Fe(II) drug intermediate.[265,301] EPR studies of the NO complex of Fe(II)–bleomycin show a superhyperfine coupling similar to that noted for the corresponding P-450 complex.[300,302] Extensive spectroscopic studies on bleomycin–iron complexes of these oxygen analogues have demonstrated, with the exception of CN complexes, that Mössbauer and crystal field parameters of these complexes are similar to those of the corresponding hemeprotein complexes.[264]

A wide range of metals, in addition to Fe(II) and Fe(III), can be complexed by bleomycin. These include Co(II), Co(III), Cu(I), Cu(II), Zn(II), Ni(II), Ni(III), Ga(III), Mn(II), Mn(III), Ca(II), and several trivalent lanthanide ions.[254–258,261,262,303–313] Metallobleomycin systems containing Fe(II), Fe(III), Cu(II), Co(III), Mn(II), and Mn(III) have been shown, upon activation, to degrade DNA *in vitro*.[250,258] The Mn(II)-substituted bleomycin, however, cannot be activated reductively[210] which parallels the behavior of Mn(II)–P-450.[82] Furthermore, systems utilizing Mn(III)-, Cu(II)-, and Fe(III)-substituted drug complexes have been shown to mediate oxygen transfer to definite substrates[257,287] in a similar manner as previous studies involving other P-450 model systems.[194,218,315–317]

These parallels noted above between bleomycin and P-450 prompted our laboratory to recently carry out a preliminary study evaluating the reactivity of ferric bleomycin toward a set of six oxidative chemistries characteristic of P-450 and chloroperoxidase, a heme-containing protein that also shares many spectral and magnetic properties with P-450.[318–327].

The results of this survey along with results from related investigations[221] are summarized in Table VII. Bleomycin was found to be unable to catalyze the decarboxylation of phenylperacetic acid or the peroxyacid/peroxide-supported hydroxylation of cumene, two typical P-450 chemistries.[328,329] Nevertheless, the ferric drug complex was able to mediate oxygen evolution from *m*-chloroperbenzoic acid and chlorination of monochlorodimedone with hydrogen peroxide and chloride ion, reactivities demonstrated by chloroperoxidase but not P-450$_{cam}$. Ferric bleomycin was also found to be chemically competent in promoting iodosobenzene-, peroxide-, or peroxyacid-dependent *N*-demethylation of *N,N*-

TABLE VII
Summary of the Chemical Reactivities of Ferric Bleomycin as Compared to Chloroperoxidase and Cytochrome P-450

	Bleomycin	P-450	Chloroperoxidase
Phenylperacetic acid decarboxylation	$-^{328}$	$+^{329}$	$-^{329}$
Cumene hydroxylation	$-^{328}$	$+^{329}$	$-^{329}$
Olefin expoxidaton	$+^{287}$	$+^{329}$	$+^{329}$
m-Chloroperbenzoic acid-supported O_2 evolution	$+^{328}$	$-^{328}$	$+^{332}$
Peroxide-supported monochlorodimedone chlorination	$+^{328}$	$-^{328}$	$+^{333}$
Peroxide/peracid-supported N-dimethylation of N,N-dimethylaniline	$+^{328}$	$+^{334}$	$+^{331}$
N-Oxide-supported N-demethylation of N,N-dimethylaniline	$-^{328}$	$+^{330}$	$-^{331}$
o-Dianisidine peroxidation	$+^{328}$	$+^{328}$	$+^{328}$

dimethylaniline as has been reported for both heme enzymes.[330,331] This was not achieved when N,N-dimethylaniline-N-oxide was used as the oxidant. P-450 has previously been shown to be able to utilize the N-oxide,[330] while chloroperoxidase does not have this reactivity.[331] The peroxidation of o-dianisidine, common to both proteins, was also shown to be effectively catalyzed by ferric bleomycin. Thus, of the set of oxidative chemistries evaluated in this survey, the reactivities demonstrated by bleomycin align more closely with the reactivities of chloroperoxidase than with those of P-450. Hence, in spite of the evidence demonstrating parallels between P-450 and bleomycin with respect to their dioxygen activation pathways, the electronic and magnetic parameters of certain intermediates in their respective iron systems, and the effects of various oxygen analogues and characteristics of their substituted systems, the utility of this metallodrug as a P-450 model system is still in question, in light of the results of the differences noted toward the various oxidative chemistries.

4. Conclusion

From the extensive reference list which follows it is obvious that an enormous amount of research effort has been directed at the bacterial P-450 systems. Many problems have been solved, but new vistas open as definition proceeds to molecular events. Perhaps in the next decade, attempts to describe electron transfer events and oxygenase chemistry in precise physical and chemical terms will be made. As new technologies such as molecular biology and various new spectroscopic techniques are brought to bear on P-450, much new and exciting insight will undoubtedly be obtained.

ACKNOWLEDGMENTS. The authors' research discussed in this review was supported in part by NIH Grants GM-31756 and GM-33775. S.G.S. gratefully acknowledges receipt of a Research Career Development grant from the National Institutes of Health.

We express our sincere thanks to the following graduate students for providing substantial input into the research effort and the writing of this review: Mr. Benjamin Unger, for the review of bacterial P-450 systems and for conducting the molecular biology studies of P-450$_{cam}$; Mr. William Atkins, who is responsible for the section on enzyme regio- and stereo-chemistry; Mr. Guy Padbury, for conducting the bleomycin research endeavors and the writing of that section of this review; Mr. Mark Fisher, for conducting the experiments on high-pressure effects, tyrosine microenvironment, redox potential, electron transfer rates, and spin state equilibrium. Mr. Kermit Carraway, though "only" an undergraduate, belongs in this category for documenting the entropy control of spin state equilibria. The graduate students in our laboratory provide much of the excitement of scientific inquiry and, to a large extent, are responsible for the very productive environment which exists.

We wish to acknowledge the editorial contributions of Ms. Jean Lewis and Ms. Tammy Houghland, who are responsible for typesetting, composition, and proofreading, and the numerous fruitful scientific discussions with Professors R. E. White, K. Suslick, J. Groves, and I. C. Gunsalus.

References

1. Daggett, P. M., Gherna, R. L., Pienta, P., Nierman, W., Hsu, H., Brandon, B., and Alexander, M. T. (eds.), 1982 *American Type Culture Collection Catalogue of Strains I*, American Type Culture Collection, Rockville, Md.
2. Fried, J., Thoma, R. W., Perlman, D., Herz, J. E., and Borman, A., 1955, The use of microorganisms in the synthesis of steroid hormones and hormone analogues, *Recent Prog. Horm. Res.* **11**:149–181.
3. McAleer, W. J., Jacob, T. A., Turnbull, L. B., Schoenewaldt, E. F., and Stoudt, T. H., 1958, Hydroxylation of progesterone by *Bacillus cereus* and *Bacillus megaterium*, *Arch. Biochem. Biophys.* **73**:127–130.
4. Shiraska, M., Ozaki, M., and Sugawara, S., 1961, Studies on microbiological transformation of steroid compounds, *J. Gen. Appl. Microbiol.* **7**:341–352.
5. Wilson, J. E., and Vestling, C. S., 1965, A cell-free steroid hydroxylating system from *Bacillus megaterium*, strain KM, *Arch. Biochem. Biophys.* **110**:401–404.

6. Berg, A., Carlström, K., Gustafsson, J.-Å., and Ingleman-Sunberg, M., 1975, Demonstration of a cytochrome P-450-dependent steroid 15β hydroxylase in *Bacillus megaterium*, *Biochem. Biophys. Res. Commun.* **66**:1414–1432.
7. Berg, A., Gustafsson, J.-Å., Ingelman-Sunberg, M., and Carlström, K., 1976, Characterization of a cytochrome P-450-dependent steroid hydroxylase system present in *Bacillus megaterium*, *J. Biol. Chem.* **251**:2831–2838.
8. Berg, A., Carlström, K., Ingleman-Sundberg, M., Rafter, J., and Gustafsson, J.-Å., 1977, Studies on a cytochrome P-450-dependent hydroxylase system active on steroids in *Bacillus megaterium*, in: *Microsomes and Drug Oxidations* (V. Ullrich, J. Roots, A. Hildebrandt, R. W. Estabrook, and A. H. Conney, eds.), Pergamon Press, Elmsford, N.Y., pp. 377–384.
9. Berg, A., Ingelman-Sundberg, M., and Gustafsson, J.-Å., 1979, Purification and characterization of cytochrome P-450$_{meg}$, *J. Biol. Chem.* **254**:5264–5271.
10. Dus, K., Katagiri, M., Yu, C.-A., Erbes, D. L., and Gunsalus, I. C., 1970, Chemical characterization of cytochrome P-450$_{cam}$, *Biochem. Biophys. Res. Commun.* **40**:1423–1430.
11. Appleby, C. A., and Daniel, R. M., 1965, *Rhizobium* cytochrome P-450: A family of soluble, separable hemoproteins, in: *Oxidases and Related Redox Systems*, Volume 2 (T. E. King, H. S. Mason, and M. Morrison, eds.), University Park Press, Baltimore, pp. 515–528.
12. Broadbent, D. A., and Cartwright, N. J., 1974, Bacterial attack on phenolic ethers: Electron acceptor-substrate binding proteins in bacterial O-dealkylases: Purification and characterization of cytochrome P-450$_{npd}$ of *Nocardia*, *Microbios* **9**:119–130.
13. Omura, T., Sanders, E., Estabrook, R. W., Cooper, D. Y., and Rosenthal, O., 1966, Isolation from adrenal cortex of a nonheme iron protein and flavoprotein functional as a reduced triphosphopyridine nucleotide-cytochrome P-450 reductase, *Arch. Biochem. Biophys.* **117**:660–673.
14. Lu, A. Y. H., and Coon, M. J., 1968, Role of hemoprotein P-450 in fatty acid ω-hydroxylation in a soluble enzyme system from liver microsomes, *J. Biol. Chem.* **243**:1331–1332.
15. Lu, A. Y. H., Kuntzman, R., West, S., Jacobson, M., and Conney, A. H., 1972, Reconstituted liver microsomal enzyme system that hydroxylates drugs, other foreign compounds, and endogenous substrates. II. Role of the cytochrome P-450 and P-448 fractions in drug and steroid hydroxylations, *J. Biol. Chem.* **247**:1727–1734.
16. Berg, A., and Rafter, J. J., 1981, Studies on the substrate specificity and inducibility of cytochrome P-450$_{meg}$, *Biochem. J.* **196**:781–786.
17. Mieyal, J. J., Ackerman, R. S., Blumer, J. L., and Freeman, L. W., 1976, Characterization of enzyme-like activity of human hemoglobin, *J. Biol. Chem.* **251**:3436–3441.
18. Mieyal, J. J., and Blumer, J. L., 1976, Acceleration of the autooxidation of human oxyhemoglobin by aniline and its relation to hemoglobin-catalyzed aniline hydroxylation, *J. Biol. Chem.* **251**:3442–3446.
19. Vore, M., Lu, A. Y. H., Kuntzman, R., and Conney, A. H., 1972, Organic solvent extraction of liver microsomal lipid. II. Effect on the metabolism of substrates and binding spectra of cytochrome P-450, *Mol. Pharmacol.* **10**:963–974.
20. Haugen, D. A., van der Hoeven, T. A., and Coon, M. J., 1975, Purified liver microsomal cytochrome P-450: Separation and characterization of multiple forms, *J. Biol. Chem.* **250**:3567–3570.
21. Ingleman-Sundberg, M., 1977, Protein–lipid interactions in the liver microsomal hydroxylase system, in: *Microsomes and Drug Oxidations* (V. Ullrich, J. Roots, A. Hildebrandt, and A. H. Conney, eds.), Pergamon Press, Elmsford, N.Y., pp. 67–75.
22. Narasimhulu, S., 1977, Relationship between the membrane lipids and substrate-cy-

tochrome P-450 binding reaction in bovine adreno-cortical microsomes, in: *Microsomes and Drug Oxidations* (V. Ullrich, J. Roots, A. Hildebrandt, R. W. Estabrook, and A. H. Conney, eds.), Pergamon Press, Elmsford, N.Y., pp. 119–126.
23. Nordblom, G. D., White, R. E., and Coon, M. J., 1976, Studies on hydroperoxide-dependent substrate hydroxylation by purified liver microsomal cytochrome P-450, *Arch. Biochem. Biophys.* **175**:524–533.
24. Ingelman-Sundberg, M., 1977, Phospholipids and detergents as effectors in the liver microsomal hydroxylase system, *Biochim. Biophys. Acta* **488**:225–234.
25. Tyson, C. A., Lipscomb, J. D., and Gunsalus, I. C., 1972, The roles of putidaredoxin and P-450$_{cam}$ in methylene hydroxylation, *J. Biol. Chem.* **247**:5777–5784.
26. Miura, Y., and Fulco, A. J., 1974, (ω-2) hydroxylation of fatty acids by a soluble system from *Bacillus megaterium*, *J. Biol. Chem.* **249**:1880–1888.
27. Miura, Y., and Fulco, A. J., 1975, ω-1, ω-2, and ω-3 hydroxylation of long-chain fatty acids, amides and alcohols by a soluble enzyme system from *Bacillus megaterium*, *Biochim. Biophys. Acta* **388**:305–317.
28. Hare, R. S., and Fulco, A. J., 1975, Carbon monoxide and hydroxymercuribenzoate sensitivity of a fatty acid (ω-2) hydroxylase from *Bacillus megaterium*, *Biochem. Biophys. Res. Commun.* **65**:665–672.
29. Ho. P. P., and Fulco, A. J., 1976, Involvement of a single hydroxylase species in the hydroxylation of palmitate at the ω-1, ω-2, and ω-3 positions by a preparation from *Bacillus megaterium*, *Biochim Biophys. Acta* **431**:249–256.
30. Matson, R. S., Hare, R. S., and Fulco, A. J., 1977, Characteristics of a cytochrome P-450-dependent fatty acid ω-2 hydroxylase from *Bacillus megaterium*, *Biochim. Biophys. Acta* **487**:487–494.
31. Buchanan, J. F., and Fulco, A. J., 1978, Formation of 9,10-epoxypalmitate and 9,10-dihydroxy-palmitate from palmitoleic acid by a soluble system from *Bacillus megaterium*, *Biochem. Biophys. Res. Commun.* **85**:1254–1260.
32. Michaels, B. C., Ruettinger, R. T., and Fulco, A. J., 1980, Hydration of 9,10 epoxypalmitic acid by a soluble enzyme from *Bacillus megaterium*, *Biochem. Biophys. Res. Commun.* **92**:1189–1195.
33. Matson, R. S., Stein, R. A., and Fulco, A. J., 1980, Hydroxylation of 9-hydroxystearate by a soluble cytochrome P-450-dependent fatty acid hydroxylase from *Bacillus megaterium*, *Biochem. Biophys. Res. Commun.* **97**:955–961.
34. Matson, R. S., and Fulco, A. J., 1981, Hydroxystearates as inhibitors of palmitate hydroxylation catalyzed by the cytochrome P-450 monoxygenase from *Bacillus megaterium*, *Biochem. Biophys. Res. Commun.* **103**:531–535.
35. Fulco, A. J., 1974, Metabolic alterations of fatty acids, *Annu. Rev. Biochem.* **43**:215–241.
36. Ruettinger, R. T., and Fulco, A. J., 1981, Epoxidation of unsaturated fatty acids by a soluble cytochrome P-450-dependent system from *Bacillus megaterium*, *J. Biol. Chem.* **256**:5728–5734.
37. Kusunose, M., Kusunose, E., and Coon, M. J., 1964, Enzymatic ω-oxidation of fatty acids. I. Products of octanoate, decanoate, and laurate oxidation, *J. Biol. Chem.* **239**:1374–1380.
38. Kusunose, M., Kusunose, E., and Coon, M. J., 1964, Enzymatic ω-oxidation of fatty acids. II. Substrate specificity and other properties of the enzyme system, *J. Biol. Chem.* **239**:2135–2139.
39. Peterson, J. A., Basu, D., and Coon, M. J., 1966, Enzymatic ω-oxidation. I. Electron carriers in fatty acid and hydrocarbon hydroxylation, *J. Biol. Chem.* **241**:5162–5164.
40. Peterson, J. A., Kusunose, M., Kusunose, E., and Coon, M. J., 1967, Enzymatic ω-oxidation. II. Function of rubredoxin as the electron carrier in ω-hydroxylation, *J. Biol. Chem.* **242**:4334–4340.

41. Lu, A. Y. H., and Coon, M. J., 1968, Role of hemoprotein P-450 in fatty acid ω-hydroxylation in a soluble enzyme system from liver microsomes, *J. Biol. Chem.* **243:**1331–1332.
42. Cardini, G., and Jurtshuk, P., 1968, Cytochrome P-450 involvement in the oxidation of n-octane by cell-free extracts of *Corynebacterium* sp. strain 7E1C, *J. Biol. Chem.* **243:**6070–6072.
43. Cardini, G., and Jurtshuk, P., 1970, The enzymatic hydroxylation of n-octane by *Corynebacterium* sp. strain 7E1C, *J. Biol. Chem.* **245:**2789–2796.
44. Lebeault, J. M., Lode, E. T., and Coon, M. J., 1971, Fatty acid and hydrocarbon hydroxylation in yeast: Role of cytochrome P-450 in *Candida tropicalis*, *Biochem. Biophys. Res. Commun.* **42:**413–419.
45. Frommer, U., Ullrich, V., Staudinger, H., and Orrenius, S., 1972, The monooxygenation of n-heptane by rat liver microsomes, *Biochim. Biophys. Acta* **280:**487–494.
46. Duppel, W., Lebeault, J. M., and Coon, M. J., 1973, Properties of a yeast cytochrome P-450-containing enzyme system which catalyzes the hydroxylation of fatty acids, alkanes, and drugs, *Eur. J. Biochem.* **36:**583–592.
47. Gallo, M., Bertrand, J. C., Roche, B., and Azoulay, E., 1973, Alkane oxidation in *Candida tropicalis*, *Biochim. Biophys. Acta* **296:**624–638.
48. Narhi, L. O., and Fulco, A. J., 1982, Phenobarbital induction of a soluble cytochrome P-450-dependent fatty acid monooxygenase in *Bacillus megaterium*, *J. Biol. Chem.* **257:**2147–2150.
49. Fulco, A. J., Kim, B. H., Matson, R. S., Narhi, L. O., and Ruettinger, R. T., 1983, Nonsubstrate induction of a soluble bacterial cytochrome P-450 monooxygenase by phenobarbital and its analogs, *Mol. Cell. Biochem.* **53/54:**155–161.
50. Kim, B. H., and Fulco, A. J., 1983, Induction by barbiturates of a cytochrome P-450-dependent fatty acid monooxygenase in *Bacillus megaterium*: Relationship between barbiturate structure and inducer activity, *Biochem. Biophys. Res. Commun.* **116:**843–850.
51. Nahri, L. O., Kim, B. H., Stevenson, P. M., and Fulco, A. J., 1983, Partial characterization of a barbiturate-induced cytochrome P-450-dependent fatty acid monooxygenase from *Bacillus megaterium*, *Biochem. Biophys. Res. Commun.* **116:**851–858.
52. Salaün, J. P., Benveniste, I., Reichhard, D., and Durst, F., 1981, Induction and specificity of a (cytochrome P-450)-dependent laurate in-chain-hydroxylase from higher plant microsomes, *Eur. J. Biochem.* **119:**651–655.
53. Kester, A. S., and Foster, J. W., 1963, Diterminal oxidation of long-chain alkanes by bacteria, *J. Bacteriol.* **85:**859–869.
54. Asperger, O., Naumann, A., and Kleber, H. P., 1981, Occurrence of cytochrome P-450 in *Acinetobacter* strains after growth on *n*-hexadecane, *FEMS Microbiol. Lett,* **11:**309–312.
55. Kleber, H. P., Claus R., and Asperger, O., 1983, Enzymologie der n-Alkanoxidation bei *Acinetobacter*, *Acta Biotechnol.* **3:**251–260.
56. Asperger, O., Muller, R., and Kleber, H. P., 1983, Isolierung von Cytochrom P-450 und des entsprechenden reduktase Systems aus *Acinetobacter calcoaceticus*, *Acta Biotechnol.* **3:**319–326.
57. Haferburg, D., Asperger, O., Lohs, U., and Kleber, H. P., 1983, Regulation der Alkanverwertung bei *Acinetobacter calcoaceticus*, *Acta Biotechnol.* **3:**371–374.
58. Asperger, O., Naumann, A., and Kleber, H. P., 1984, Inducibility of cytochrome P-450 in *Acinetobacter calcoaceticus* by *n*-alkanes, *Appl. Microbiol. Biotechnol.* **19:**398–403.
59. Hedegaard, J., and Gunsalus, I. C., 1965, Mixed function oxidation. IV. An induced methylene hydroxylase in camphor oxidation, *J. Biol. Chem.* **240:**4038–4043.

60. Jacobson, L. A. Bartholomaus, R. C., and Gunsalus, I. C., 1966, Repression of malic enzyme by acetate in *Pseudomonas, Biochem. Biophys. Res. Commun.* **24**:955-960.
61. Gunsalus, I. C., Bertland, A. U., and Jacobson, L. A., 1967, Enzyme induction and repression in anabolic and catabolic pathways, *Arch. Mikrobiol.* **59**:113-122.
62. Azoulay, E., Chouteau, J., and Davidovics, G., 1963, Isolement et caractèrisation des enzymes résponsables de l'oxydation des hydrocarbures, *Biochim. Biophys. Acta* **77**:554-567.
63. den Dooren de Jong, L. E., 1926, Bijdrage tot de Kennis van het Mineralisatieproces, Thesis, Thechnische Hogeschool, Delft, the Netherlands, Nijgh and Van Ditmar, Rotterdam, *Biol. Abstr.* **1**:12696.
64. Stainer, R. Y., Palleroni, N. J., and Doudoroff, M., 1966, The aerobic pseudomonads: A taxonomic study, *J. Gen. Microbiol.* **43**:159-271.
65. Bradshaw, W. H., Conrad, H. E., Corey, E. J., Gunsalus, I. C., and Lednicer, D., 1959, Degradation of (+)-camphor, *J. Am. Chem. Soc.* **81**:5007.
66. Cushman, D. W., Tsai, R. L., and Gunsalus, I. C., 1967, The ferroprotein component of a methylene hydroxylase, *Biochem. Biophys. Res. Commun.* **26**:577-583.
67. Katagiri, M., Ganguli, B. N., and Gunsalus, I. C., 1968, A soluble cytochrome P-450 functional in methylene hydroxylation, *J. Biol. Chem.* **243**:3543-3546.
68. Yu, C. A., Gunsalus, I. C., Katagiri, M., Suhara, K., and Takemori, S., 1974, Cytochrome P-450$_{cam}$. I. Crystallization and properties, *J. Biol. Chem.* **249**:94-101.
69. Paisley, N. S., 1961, Purification and characterization of two dehydrogenases acting on 5-ketocamphor, Ph.D. dissertation, University of Illinois, Urbana-Champaign.
70. Conrad, H. E., DuBus, R., and Gunsalus, I. C., 1961, An enzyme system for cyclic ketone lactonization, *Biochem. Biophys. Res. Commun.* **6**:293-297.
71. Conrad, H. E., and DuBus, R., Namtved, M. J., and Gunsalus, I. C., 1965, Mixed function oxidation. II. Separation and properties of the enzymes catalyzing camphor lactonization, *J. Biol. Chem.* **240**:495-503.
72. Conrad, H. E., Lieb, K., and Gunsalus, I. C., 1965, Mixed function oxidation. III. An electron transport complex in camphor ketolactonization, *J. Biol. Chem.* **240**:4029-4037.
73. Yu, C. A., 1969, Components and properties of the ketolactonase I enzymatic oxygenase system, Ph.D. dissertation, University of Illinois, Urbana-Champaign.
74. Hartline, R. L., and Gunsalus, I. C., 1971, Induction specificity and catabolite repression of the early enzymes in camphor degradation by *Pseudomonas putida, J. Bacteriol.* **106**:468-478.
75. Gunsalus, I. C., and Marshall, V. P., 1971, Monoterpene dissimilation: Chemical and genetic models, *CRC Crit. Rev. Microbiol.* **1**:291-310.
76. Rheinwald, J. G., 1970, The genetic organization of peripheral metabolism: A transmissible plasmid controlling camphor oxidation in *Pseudomonas putida*, M.S. thesis, University of Illinois, Urbana-Champaign.
77. Rheinwald, J. G., Chakrabarty, A. M., and Gunsalus, I. C., 1973, A transmissible plasmid controlling camphor oxidation in *Pseudomonas putida, Proc. Natl. Acad. Sci USA* **70**:885-889.
78. Gunsalus, I. C., Conrad, H. E., Trudgill, P. W., and Jacobson, L. A., 1965, Regulation of catabolic metabolism, *Isr. J. Med. Sci.* **1**:1099-1119.
79. Chakrabarty, A. M., 1976, Plasmids in *Pseudomonas, Annu. Rev. Genet.* **10**:7-30.
80. Jacobson, L. A., 1967, Enzyme induction and repression in the catabolism of (+)-camphor by *Pseudomonas putida*, Ph.D. dissertation, University of Illinois, Urbana-Champaign.
81. Marshall, V. P., and Sokatch, J. R., 1970, Synthesis of valine catabolic enzymes in *Pseudomonas putida, Fed. Proc.* **30**:1167.

82. Koga, H., and Gunsalus, I. C., 1982, Cloning and expression of CAM P-450 genes in *Pseudomonas putida* and *Escherichia coli*, *Proc. 55th Annu Meet. Jpn. Biochem. Soc.* **10**:82–84.
83. Bagdasarian, M. M., Amann, E., Lurz, R., Rückert, B., and Bagdasarian, M., 1983, Activity of the hybrid *trp-lac* (*tac*) promoter of *Escherichia coli* in *Pseudomonas putida*: Construction of broad-host-range, controlled-expression vectors, *Gene* **26**:273–282.
84. Unger, B. P., Gunsalus, I. C., and Sligar, S. G., 1986, Nucleotide sequence of the *Pseudomonas putida* cytochrome P-450$_{cam}$ gene and its expression in *Escherichia coli*, *J. Biol. Chem.* in press.
85. Vieria, J., and Messing, J., 1982, The pUC plasmids, an M13mp7-derived system for insertion mutagenesis and sequencing with synthetic universal primers, *Gene* **19**:259–268.
86. Haniu, M., Armes, L. G., Yasunobu, K. T., Shastry, B. A., and Gunsalus, I. C., 1982, Amino acid sequence of the *Pseudomonas putida* cytochrome P-450. II. Cyanogen bromide peptides, acid cleavage peptides, and the complete sequence, *J. Biol. Chem.* **257**:12664–12671.
87. Goodfellow, M., and Alderson, G., 1977, the actinomycete-genus *Rhodococcus*: A home for the '*rhodochrous*' complex, *J. Gen. Microbiol.* **100**:99–122.
88. Goodfellow, M., and Minnikin, D. E., 1981, Introduction to the coryneform bacteria, in: *The Prokaryotes: A Handbook on Habitats, Isolation, and Identification of Bacteria*, Volume 2 (M. P. Starr, H. Stolp, H. G., Trüper, A. Balous, and H. G. Schlegel, eds.), Springer-Verlag, Berlin, pp. 1812–1826.
89. Kay, J. W. D., Conrad, H. E., and Gunsalus, I. C., 1962, Camphor degradation: Hydroxy intermediate formed by a soil diphtheroid, *Bacteriol. Proc.* **30**:108.
90. Chapman, P. J., Cushman, D., Kuo, J. F., LeGall, J., and Gunsalus, I. C., 1964, Pathways in microbial metabolism of camphor enantiomers, *Abstracts of Papers*, National Meeting of the American Chemical Society, Chicago, American Chemical Society, Washington, D.C., p. 4Q.
91. Gunsalus, I. C., Trudgill, P. W., Cushman, D., and Conrad, H. E., 1965, Stereospecific biological oxygenation of terpenoids, *Symp. Recent Adv. Chem Terpenoids*, Poona, India, p. 3.
92. Chapman, P. J., Kuo, J. F., and Gunsalus, I.C., 1963, Camphor oxidation: 2,6-Diketocamphane pathway in diphtheroid, *Fed. Proc.* **22**:296.
93. Chapman, P. J., Meerman, G., Gunsalus, I. C., Srinivsan, R., and Rinehart, K. L., 1966, A new acyclic acid intermediate in camphor oxidation, *J. Am. Chem. Soc.* **88**:618–619.
94. Shukla, O. P., Moholay. M. N., and Bhattacharyya, P. K., 1968, Microbiological transformations of terpenes. Part X. Fermentation of α- and β-pinenes by a soil pseudomonad (PL-strain), *Indian J. Biochem.* **5**:79–91.
95. DeFrank, J. J., and Ribbons, D. W., 1977, *p*-Cymene pathway in *Pseudomonas putida*: Initial reactions, *J. Bacteriol.* **129**:1356–1364.
96. Wigmore, G. J., and Ribbons, D. W., 1980, *p*-Cymene pathway in *Pseudomonas putida*: Selective enrichment of defective mutants by using halogenated substrate analogs, *J Bacteriol.* **143**:816–824.
97. Wigmore, G. J., and Ribbons, D. W., 1981, Selective enrichment of *Pseudomonas* spp. defective in catabolism after exposure to halogenated substrates, *J. Bacteriol.* **146**:920–927.
98. Madhyastha, K. M., and Bhattacharyya, P. K., 1968, Microbiological transformations of terpenes. Part XII. Fermenation of *p*-cymene by a soil pseudomonad (PL-strain), *Indian J. Biochem.* **5**:102–111.
99. Madhyastha, K. M., and Bhattacharyya, P. K., 1968, Microbiological transformations

of terpenes. Part XIII. Pathways for degradation of *p*-cymene in a soil pseudomonad (PL-strain), *Indian J. Biochem.* **5:**161–167.
100. Madhyastha, K. M., Rangachari, P. N., Raghabendra, R. M., and Bhattacharyya, P. K., 1968, Microbiological transformations of terpenes. Part XV. Enzyme systems in the catabolism of *p*-cymene in PL-strains, *Indian J. Biochem.* **5:**167–173.
101. DeFrank, J. J., and Ribbons, D. W., 1976, The *p*-cymene pathway in *Pseudomonas putida* PL: Isolation of a dihydrodiol accumulated by a mutant, *Biochem. Biophys. Res. Commun.* **70:**1129–1135.
102. DeFrank, J. J., and Ribbons, D. W., 1977, *p*-Cymene pathway in *Pseudomonas putida*: Ring cleavage of 2,3-dihyroxy-*p*-cumate and subsequent reactions, *J. Bacteriol.* **129:**1365–1374.
103. Anderson, B. N., Wigmore, G. J., and Ribbons, D. W., 1979, Two degradative plasmids which encode the *meta*-cleavage of 4-substituted-2,3-dihydroxy-benzoates, *Abstracts of Papers*, Annual Meeting of the American Society of Microbiologists, American Society of Microbiologists, Washington, D.C., p. 199.
104. Gunsalus, I. C., and Yen, K. M., 1981, Metabolic plasmid organization and distribution, in: *Molecular Biology, Pathogenicity, and Ecology of Bacterial Plasmids* (S. B. Levy, R. C. Clowes, and E. L, Koenig, eds.), Plenum Press, New York, pp. 499–509.
105. Madhyastha, K. M., Bhattacharyya, P. K., and Vaidyanathan, C. S., 1977, Metabolism of a monoterpene alcohol, linalool, by a soil pseudomonad, *Can. J. Microbiol.* **23:**230–239.
106. Rama Devi, J., and Bhattacharyya, P. K., 1977, Microbiological transformations of terpenes. Part XXII. Fermentation of geraniol, nerol, and limonene by a soil pseudomonad, *Pseudomonas incognita* (linalool strain), *Indian J. Biochem. Biophys.* **14:**288–291.
107. Rama Devi, J., and Bhattacharyya, P. K., 1977, Microbiological transformations of terpenes. Part XXIV. Pathways of degradation of linalool, geraniol, nerol, and limonene by *Pseudomonas incognita* (linalool strain), *Indian J. Biochem. Biophys.* **14:**359–363.
108. Rama Devi, J., Bhat, S. G., and Bhattacharyya, P. K., 1978, Microbiological transformation of terpenes. Part XXV. Enzymes involved in the degradation of linalool in the *Pseudomonas incognita*, linalool strain, *Indian J. Biochem. Biophys.* **15:**323–327.
109. Ullah, A. H. J., Bhattacharyya, P. K., Bakthavachalam, J., Wagner, G. C., and Gunsalus, I. C., 1983, Linalool 8-hydroxylase: A heme thiolate monooxygenase, *Fed. Proc.* **42:**1897.
110. Wagner, G. C., Jung, C., Shyansunder, E., and Bowne, S., 1983, Spectral comparisons of the heme thiolate and iron sulfur proteins from linalool and camphor monooxygenases, *Fed. Proc.* **42:**1897.
111. Seubert, W., 1960, Degradation of isoprenoid compounds by microorganisms. I. Isolation and characterization of an isoprenoid-degrading bacterium *Pseudomonas citronellolis*, *J. Bacteriol.* **79:**426–434.
112. Seubert, W., Fass, E., and Remberger, U., 1963, Untersuchungen über den bakteriellen Abbau von Isoprenoiden. III. Reinigung und Eigenschafter der geranyl Carboxylase, *Biochem. Z.* **338:**245–264.
113. Seubert, W., and Fass, E., 1964, Untersuchungen uber den bakteriellen Abbau von Isoprenoiden. V. Der Mechanismus des Isoprenoidabbaues, *Biochem. Z.* **341:**35–41.
114. Meehan, T. D., and Coscia, C. J., 1973, Hydroxylation of geraniol and nerol by a monooxygenase from *Vinca rosea*, *Biochem. Biophys. Res. Commun.* **53:**1043–1048.
115. Escher, S., Loew, P., and Arigoni, D., 1970, The role of hydroxygeraniol and hydroxynerol in the biosynthesis of loganin and indole alkaloids, *Chem. Commun.* **13:**823–825.
116. Appleby, C. A., 1967, A soluble hemoprotein P-450 from nitrogen-fixing *Rhizobium* bacteroids, *Biochim. Biophys. Acta* **147:**399–402.

117. Appleby, C. A., 1969, Electron transport systems of *Rhizobium japonicum*. I. Haemoprotein P-450, other CO-reactive pigments, cytochromes and oxidases in bacteriods from N_2-fixing root nodules, *Biochim. Biophys. Acta* **172**:71–87.
118. Daniel, R. M., and Appleby, C. A., 1972, Anaerobic-nitrate, symbiotic and aerobic growth of *Rhizobium japonicum*: Effects on cytochrome P-450, other hemoproteins, nitrate and nitrite reductases, *Biochim. Biophys. Acta* **275**:347–354.
119. Appleby, C. A., Turner, G. I., and Macnicol, P. K., 1973, Involvement of oxyleghaemoglobin and cytochrome P-450 in an efficient oxidative phosphorylation pathway which supports nitrogen fixation in *Rhizobium*, *Biochim. Biophys. Acta* **387**:461–474.
120. Dus, K., Goewert, R., Weaver, C. C., Carey, D., and Appleby, C. A., 1976, P-450 hemoproteins of *Rhizobium japonicum*: Purification by affinity chromatography and relationship to P-450$_{cam}$ and P-450$_{LM2}$, *Biochem. Biophys. Res. Commun.* **69**:437–445.
121. Dus, K., Litchfield, W. J., Miguel, A. G., Van der Hoeven, T. A. Haugen, D. A., Dean, W. L., and Coon, M. J., 1974, Structural resemblance of cytochrome P-450 isolated from *Pseudomonas putida* and from rabbit liver microsomes, *Biochem. Biophys. Res. Commun.* **60**:15–21.
122. Peisach, J., Appleby, C. A., and Blumberg, W. E., 1972, Electron paramagnetic resonance and temperature dependent spin state studies of ferric cytochrome P-450 from *Rhizobium japonicum*, *Arch. Biochem. Biophys.* **150**:725–732.
123. Tsai, T., Yu, C. A. Gunsalus, I. C., Peisach, J., Blumberg, W., Orme-Johnson, W. H., and Beinart, H., 1970, Spin-state changes in cytochrome P-450$_{cam}$ on binding specific substrates, *Proc. Natl. Acad. Sci. USA* **66**:1157–1163.
124. Peisach, J., and Blumberg, W. E., 1970, Electron paramagnetic resonance study of the high- and low-spin forms of cytochrome P-450 in liver and in liver microsomes from a methylcholanthrene-treated rabbit, *Proc. Natl. Acad. Sci. USA* **67**:172–179.
125. Kretovich, V. L., Melik-Sarkisyan, S. S., and Matus, V. K., 1972, Cytochrome composition of yellow lupine nodules, *Biochemistry (USSR)* **37**:590–597.
126. Cartwright, N. J., Holdom, K. S., and Broadbent, D. A., 1971, Bacterial attack on phenolic ethers: Dealkylation of higher ethers and further observations on O-demethylases, *Microbios* **3**:113–130.
127. Broadbent, D. A., and Cartwright, N. J., 1971, Bacterial attack on phenolic ethers: Resolution of a *Nocardia* O-demethylase and purification of a cytochrome P-450 component, *Microbios* **4**:7–12.
128. Cartwright, N. J., and Broadbent, D. A., 1974, Bacterial attack on phenolic ethers: Preliminary studies on systems transporting electrons to the substrate binding components in bacterial O-dealkylases, *Microbios* **10**:87–96.
129. Gunsalus, I. C., and Wagner, G. C., 1978, Bacterial P-450$_{cam}$ methylene monooxygenase components: Cytochrome M, putidaredoxin, and putidaredoxin reductase, *Methods Enzymol.* **52**:166–188.
130. Peterson, J. A., 1971, Camphor binding by *Pseudomonas putida* cytochrome P-450, *Arch. Biochem, Biophys.* **144**:678–693.
131. Gunsalus, I. C., and Lipscomb, J. D., 1972, Component dynamics in oxygen reduction by cytochrome P-450$_{cam}$, in: *The Molecular Basis of Electron Transport* (J. Schultz and B. F. Cameron, eds.), Academic Press, New York, pp. 176–196.
132. Griffin, B. W., and Peterson, J. A., 1972, Camphor binding by *Pseudomonas putida* cytochrome P-450: Kinetics and thermodynamics of the reaction, *Biochemistry* **11**:4740–4746.
133. Pederson, T. C., Austin, R. H., and Gunsalus, I. C., 1977, Redox and ligand dynamics in P-450$_{cam}$–putidaredoxin complexes, in: *Microsomes and Drug Oxidations* (V. Ullrich, J. Roots, A. Hildebrandt, R. W. Estabrook, and A. H. Conney, eds.), Pergamon Press, Elmsford, N.Y., pp. 275–283.

134. Hintz, M. J., and Peterson, J. A., 1981, The kinetics of reduction of cytochrome P-450$_{cam}$ by reduced putidaredoxin, *J. Biol. Chem.* **256:**6721–6728.
135. Hintz, M. J., Mock, D. M., Peterson, L. L., Tuttle, K., and Peterson, J. A., 1982, Equilibrium and kinetic studies of the interaction of cytochrome P-450$_{cam}$ and putidaredoxin, *J. Biol. Chem.* **257:**14324–14332.
136. Sligar, S. G., 1976, Coupling of spin, substrate, and redox equilibria in cytochrome P-450, *Biochemistry* **15:**5399–5406.
137. Philson, S. B., 1976, Proton relaxation studies of the interaction of small molecules with cytochrome P-450, Ph.D. dissertation, University of Illinois, Urbana–Champaign.
138. Sligar, S. G., and Gunsalus, I. C., 1976, A thermodynamic model of regulation: Modulation of redox equilibria in camphor monooxygenase, *Proc. Natl. Acad. Sci. USA* **73:**1078–1082.
139. Sligar, S. G., Cinti, D. L., Gibson, G. G., and Schenkman, J. B., 1979, Spin state control of the hepatic cytochrome P-450 redox potential, *Biochem. Biophys. Res. Commun.* **90:**925.
140. Gibson, G. G., Cinti, D. L., Sligar, S. G., and Schenkman, J. B., 1980, The effect of microsomal lipids on the spin state of purified de-lipidated cytochrome P-450, *Biochem. Soc. Trans.* **8:**101.
141. Backes, W. L., Sligar, S. G., and Schenkman, J. B., 1980, Cytochrome P-450 reduction exhibits burst kinetics, *Biochem. Biophys. Res. Commun.* **97:**860–866.
142. Backes, W. L., Sligar, S. G., and Schenkman, J. B., 1982, Kinetics of hepatic cytochrome P-450 reduction: Correlation with spin state of ferric heme, *Biochemistry* **21:**1324–1330.
143. Rein, H., Ristau, O., Misselwitz, R., Buder, E., and Rickpaul, K., 1979, The importance of the spin equilibrium in cytochrome P-450 for the reduction rate of the heme iron, *Acta Biol. Med. Ger.* **38:**187–200.
144. Hintz, M. J., and Peterson, J. A., 1980, The kinetics of reduction of cytochrome P-450$_{cam}$ by the dithionite anion monomer, *J. Biol. Chem.* **255:**7317–7325.
145. Debey, P., Land, E. J., Santus, R., and Swallow, A. J., 1979, Electron transfer from pyridinyl radicals, hydrated electrons, $CO_2\cdot^-$ and $O_2\cdot^-$ to bacterial cytochrome P-450, *Biochem. Biophys. Res. Commun.* **86:**953–960.
146. Bazin, M., Pierre, J., Debey, P., and Santus, R., 1982, One electron photoreduction of bacterial cytochrome P-450 by ultraviolet light, *Eur. J. Biochem.* **124:**539–544.
147. Williams, R. J. P., Moore, G. R., Robinson, M. N., and Williams, G., 1984, The properties of cytochrome-c, *Abstracts of Papers*, 8th International Biophysics Congress, Bristol, U.K., p. 191.
148. Hui Bon Hoa, G., Begard, E., Debey, P., and Gunsalus, I. C., 1978, Two univalent electron transfers from putidaredoxin to bacterial cytochrome P-450$_{cam}$ at subzero temperature, *Biochemistry* **17:**2835–2839.
149. Gould, P., Gelb, M., and Sligar, S. G., 1981, Interaction of 5-bromocamphor with cytochrome P-450$_{cam}$, *J. Biol. Chem.* **256:**6686–6691.
150. Murray, R., Fisher, M., Debrunner, P., and Sligar, S., 1985, Structure and chemistry of cytochrome P-450, in: *Metalloproteins*, Volume I (P. Harrison, ed.), Macmillian & Co., London, pp. 157–206.
151. Schwarze, W., Blanck, J., Ristau, O., Jänig, G. R., Pommerening, K., Rein, H., and Ruckpaul, K., 1985, Spin state control of cytochrome P-450 reduction and catalytic activity in a reconstituted P-450$_{LM2}$ system as induced by a series of benzphetamine analogues, *Chemico-Biol. Interact.*, in press.
152. Cole, P., and Sligar, S. G., 1981, Temperature-jump measurement of the spin state relaxation rate of cytochrome P-450$_{cam}$, *FEBS Lett.* **133:**252–254.
153. Tamburini, P. P., Gibson, G. G., Backes, W. L., Sligar, S. G., and Schenkman, J.

B., 1984, Reduction kinetics of purified rat liver cytochrome P-450: Evidence for a sequential reaction mechanism dependent on the hemoprotein spin state, *Biochemistry* **23:**4526–4533.
154. White, R. E., McCarthy, M., Egeberg, K. D., and Sligar, S. G., 1984, Regioselectivity in the cytochromes P-450: Control by protein constraints and by chemical reactivities, *Arch. Biochem. Biophys.* **228:**493–502.
155. Sharrock, M., Debrunner, P. G., Schultz, C., Lipscomb, J. D., Marshall, V., and Gunsalus, I. C., 1976, Cytochrome P-450$_{cam}$ and its complexes: Mossbauer parameters of the heme iron, *Biochim. Biophys. Acta* **420:**8–26.
156. Schejter, A., Aviram, I., and Goldkorn, T., 1980, The contribution of electrostatic factors to the oxidation–reduction potentials of c-type cytochromes, in: *Electron Transport and Oxygen Utilization* (E. C. Ho, ed.), Elsevier, Amsterdam, pp. 95–99.
157. Lewis, B. A., and Sligar, S. G., 1983, Structural studies of cytochrome P-450 using small angle X-ray scattering, *J. Biol. Chem.* **258:**3599–3601.
158. Visser, A. J. W. G., Li, T. M., Drickamer, H. G., and Weber, G., 1977, Volume changes in the formation of internal complexes of flavinyltryptophan peptides, *Biochemistry* **16:**4883–4886.
159. Weber, G., and Drickamer, H. G., 1983, The effect of high pressure upon proteins and other biomolecules, *Q. Rev. Biophys.* **16:**89–112.
160. Ogunmula, G. B., Zipp, A., Chen, F., and Kauzmann, W., 1977, Effects of pressure on visible spectra of complexes of myoglobin, hemoglobin, cytochrome c, and horseradish peroxidase, *Proc. Natl. Acad. Sci. USA* **74:**1–4.
161. Hui Bon Hoa, G., and Marden, M. C., 1982, The pressure dependence of the spin equilibrium in camphor-bound ferric cytochrome P-450, *Eur. J. Biochem.* **124:**311–315.
162. Marden, M. C., and Hui Bon Hoa, G., 1982, Dynamics of the spin transition in camphor-bound ferric cytochrome P-450 versus temperature, pressure and viscosity, *Eur. J. Biochem.* **129:**111–117.
163. Gibson, Q., and Carey, F., 1975, Effect of pressure on the absorption spectrum of some heme compounds, *Biochem. Biophys. Res. Commun.* **67:**747–751.
164. Heremans, K., 1980, Biophysical chemistry at high pressure, *Rev. Phys. Chem Jpn.* **50:**259–273.
165. Sligar, S. G., and Gunsalus, I. C., 1979, Proton coupling in the cytochrome P-450 spin and redox equilibria, *Biochemistry* **18:**2290–2295.
166. Messana, C., Cerdonia, M., Shenkin, P., Noble, R. W., Fermi, G. Perutz, R. N., and Perutz, M. F., 1978, Influence of quaternary structure of the globin on thermal spin equilibria in different methemoglobin derivatives, *Biochemistry* **17:**3652–3661.
167. Morishima, I., Ogawa, S., and Yamada, H., 1980, High pressure proton nuclear magnetic resonance studies of hemoprotein: Pressure-induced structural change in heme environments of myoglobin, hemoglobin and horseradish peroxidase, *Biochemistry* **19:**1569–1575.
168. Fisher, M. T., Scarlata, S. F., and Sligar, S. G., 1985, High pressure investigation of cytochrome P-450$_{cam}$ spin and substrate binding equilibria, *Arch. Biochem. Biophys.* **240:**456–463.
169. Ruckpaul, K., Rein, H., Ballou, D. P., and Coon, M. J., 1980, Analysis of interactions among purified components of the liver microsomal cytochrome P-450 containing monoxygenase system by second derivative spectroscopy, *Biochim. Biophys. Acta* **626:**41–56.
170. Janig, G. R., Dettmer, R., Usanov, S. A., and Ruckpaul, K., 1983, Identification of the ligand trans to thiolate in cytochrome P-450$_{LM2}$ by chemical modification, *FEBS Lett.* **159:**58–62.

171. Janig, G. R., Makower, A., Rabe, H., Bernhardt, R., and Ruckpaul, K., 1984, Chemical characterization of cytochrome P-450$_{LM2}$: Characterization of tyrosine as axial heme iron ligand trans to thiolate, *Biochim. Biophys. Acta* **787**:8–18.
172. Dawson, J. H., Andersson, L. A., and Sono, M., 1982, Spectroscopic investigations of ferric cytochrome P-450$_{cam}$ ligand complexes, *J. Biol. Chem.* **257**:3606–3617.
173. Ragone, R., Colonna, G., Balestrieri, C., Servillo, L., and Irace, G., 1984, Determination of tyrosine exposure in proteins by second-derivative spectroscopy, *Biochemistry* **23**:1871–1875.
174. Servillo, L., Colonna, G., Balestrieri, C., Ragone, R., and Irace, G., 1982, Simultaneous determination of tyrosine and typtophan residues in proteins by second-derivative spectroscopy, *Anal. Biochem.* **126**:251–257.
175. Lang, R., Bonfils, C., and Debey, P., 1977, The low spin to high spin transition of camphor bound cytochrome P-450, *Eur. J. Biochem.* **79**:623–628.
176. Poulos, T. L., Finzel, B. C., Gunsalus, I. C., Wagner, G. C., and Kraut, J., 1985, The 2.6-Å crystal structure of *Pseudomonas putida* cytochrome P-450, *J. Biol. Chem.*, in press.
177. Gunsalus, I. C., 1968, A soluble methylene hydroxylase system: Structure and role of cytochrome P-450$_{cam}$ and iron–sulfur protein components, *Hoppe-Seylers Z. Physiol. Chem.* **349**:1610–1613.
178. Gunsalus, I. C., Tyson, C. A., and Lipscomb, J. D., 1973, Cytochrome P-450 reduction and oxygenation systems, in: *Oxidases and Related Redox Systems*, Volume 2 (T. E. King, H. S. Mason, and M. Morrison, eds.), University Park Press, Baltimore, pp. 583–603.
179. Ishimura, Y., Ullrich, V., and Peterson, J. A., 1971, Oxygenated cytochrome P-450 and its possible role in enzymic hydroxylation, *Biochem. Biophys. Res. Commun.* **42**:140–146.
180. Tyson, C. A.. Tsai, R., and Gunsalus, I. C., 1970, Fast reaction studies on the camphor P-450 hydroxylase system, *J. Am. Oil. Chem. Soc.* **47**:343A–344A.
181. Peterson, J. A., Ishimura, Y., and Griffin, B. W., 1972, *Pseudomonas putida* cytochrome P-450: Characterization of an oxygenated form of the hemoprotein, *Arch Biochem. Biophys.* **149**:197–208.
182. Estabrook, R. W., Hildebrandt, A. G., Baron, J., Netter, K. J., and Leibman, K., 1971, A new spectral intermediate associated with cytochrome P-450 function in liver microsomes, *Biochem. Biophys. Res. Commun.* **42**:132–139.
183. Sligar, S. G., Lipscomb, J. D., Debrunner, P. G., and Gunsalus, I. C., 1974, Superoxide anion production by the autoxidation of cytochrome P-450$_{cam}$, *Biochem. Biophys. Res. Commun.* **61**:290–296.
184. Eisenstein, L., Debey, P., and Douzou, P., 1977, P-450$_{cam}$: Oxygenated complexes stabilized at low temperature, *Biochem. Biophys. Res. Commun.* **77**:1377–1383.
185. Guengerich, F. P., Ballou, D. P., and Coon, M. J., 1976, Spectral intermediates in the reaction of oxygen with purified liver microsomal cytochrome P-450, *Biochem. Biophys. Res. Commun.* **70**:951–956.
186. Larroque, C., and VanLier, J. E., 1980, The subzero temperature stabilized oxyferro complex of purified cytochrome P-450$_{scc}$, *FEBS Lett.* **115**:175–177.
187. Bonfils, C., Andersson, K. K., Maurel, P., and Debey, P., 1980, Cytochrome P-450 oxygen intermediates and reactivity at subzero temperatures, *J. Mol. Cata.* **7**:299–308.
188. Gunsalus, I. C., Sligar, S. G., and Debrunner, P. G., 1975, Product formation from ferrous oxy-cytochrome P-450, *Biochem. Soc. Trans.* **3**:821–835.
189. Lipscomb, J. D., Sligar, S. G., Namtvedt, M. J., and Gunsalus, I. C., 1976, Autoxidation and hydroxylation reactions of oxygenated cytochrome P-450$_{cam}$, *J. Biol. Chem.* **251**:1116–1124.

190. Hamilton, G. A. 1974, Chemical models and mechanisms for oxygenases, in: *Molecular Mechanism of Oxygen Activation* (O. Hayaishi, ed.), Acadmenic Press, New York, pp. 405–451.
191. Hamilton, G. A., 1964, Oxidation by molecular oxygen. II. The oxygen atom transfer mechanism for mixed-function oxidases and the model for mixed-function oxidases, *J. Am. Chem. Soc.* **86**:3391–3392.
192. Groves, J. T., and McClusky, G. A., 1976, Aliphatic hydroxylation via oxygen rebound: Oxygen transfer catalyzed by iron, *J. Am. Chem. Soc.* **98**:859–861.
193. Groves, J. T., McClusky, G. A., White, R. E., and Coon, M. J., 1978, Aliphatic hydroxylation by highly purified liver microsomal cytochrome P-450: Evidence for a carbon radical intermediate, *Biochem. Biophys. Res. Commun.* **81**:154–160.
194. Groves, J. T., Nemo, T. E., and Myers, R. S., 1979, Hydroxylation and epoxidation catalyzed by iron–porphine complexes: Oxygen transfer from idosylbenzene, *J. Am. Chem. Soc.* **101**:1032–1033.
195. Chang, C. K., and Kuo, M.-S., 1979, Reaction of iron (III) porphyrins and iodosylxylene: The active oxene complex of cytochrome P-450, *J. Am. Chem. Soc.* **101**:3413–3415.
196. Hill, C. L., and Schardt, B. C., 1980, Alkane activation and functionalization under mild conditions by a homogeneous manganese (III) porphyrin iodosylbenzene oxidizing system, *J. Am. Chem. Soc.* **102**:6374–6375.
197. Groves, J. T., Kruper, W. J., Jr., and Haushalter, R. C., 1980, Hydrocarbon oxidations with oxometalloporphinates: Isolation and reactions of a (porphinato) manganese (V) complex, *J. Am. Chem. Soc,* **102**:6375–6377.
198. Sligar, S. G., Shastry, B. S., and Gunsalus, I. C., 1977, Oxygen reactions of the P-450 heme protein, in: *Microsomes and Drug Oxidations* (V. Ullrich, I. Roots, A. Holdebrandt, R. W., Estabrook, and A. H. Conney, eds.), Pergamon Press, Elmsford, N.Y., pp. 202–209.
199. Kadlubar, F. F., Morton, K. C., and Ziegler, D. M., 1973, Microsomal catalyzed hydroperoxide-dependent C-oxidation of amines, *Biochem. Biophys. Res. Commun.* **54**:1255–1261.
200. Rahimtula, A. D., and O'Brien, P. J., 1974, Hydroperoxide catalyzed liver microsomal aromatic hydroxylation reactions involving cytochrome P-450, *Biochem. Biophys. Res. Commun.* **60**:440–447.
201. Hrycay, E. G., Gustafsson, J.-A., Ingelman-Sundberg, M., and Ernster, L., 1975, Sodium periodate, sodium chlorite, organic hydroperoxides, and H_2O_2 as hydroxylating agents in steroid hydroxylation reactions catalyzed by partially purified cytochrome P-450, *Biochem. Biophys. Res. Commun.* **66**:209–216.
202. Lichtenberger, F., Nastainczyk, W., and Ullrich, V., 1976, Cytochrome P-450 as an oxene transferase, *Biochem. Biophys. Res. Commun.* **70**:939–946.
203. Burka, L. T., Thorsen, A., and Guengerich, P., 1980, Enzymatic monooxygenation of halogen atoms: Cytochrome P-450 catalyzed oxidation of iodobenzene by iodosobenzene, *J. Am. Chem. Soc.* **120**:7615–7616.
204. Heimbrook, D. C., and Sligar, S. G., 1981, Multiple mechanisms of cytochrome P-450-catalyzed substrate hydroxylations, *Biochem. Biophys. Res. Commun.* **99**:530–535.
205. MacDonald, T. L, Burka, L. T., Wright, S. T., and Guengerich, F. P., 1982, Mechanisms of hydroxylation by cytochrome P-450: Exchange of iron–oxygen intermediates with water, *Biochem. Biophys. Res. Commun.* **104**:620–625.
206. Wagner, G. C., Palcic, M. M., and Dunford, H. B., 1983, Absorption spectra of cytochrome P-450$_{cam}$ in the reaction with peroxy acids, *FEBS Lett.* **156**:244–248.
207. Rahimtula, A. D., O'Brien, P. J., Hrycay, E. G., Peterson, J. A., and Estabrook, R.

W., 1974, Possible higher valence states of cytochrome P-450 during oxidative reactions, *Biochem. Biophys. Res. Commun.* **60**:695–702.
208. Hrycay, E. G., Gustafsson, J.-A., Ingelman-Sundberg, M., and Ernster, L., 1976, The involvement of cytochrome P-450 in hepatic microsomal steroid hydroxylation reactions supported by sodium periodate, sodium chlorite, and organic hydroperoxides, *Eur. J. Biochem.* **61**:43–52.
209. Coon, M. J., Blake, R. C., II, Oprian, D. D., and Ballou, D. P., 1979, Mechanistic studies with purified components of the liver microsomal hydroxylation system: Spectral intermediates in reaction of cytochrome P-450 with peroxy compounds, *Acta Biol. Med Ger.* **38**:449–458.
210. Gelb, M. H., Toscano, W. A., Jr., and Sligar, S. G., 1982, Chemical mechanisms for cytochrome P-450 oxidation: Spectral and catalytic properties of a manganese-substituted protein, *Proc. Natl. Acad. Sci. USA* **79**:5758–5762.
211. Gelb, M. H., Malkonen, R., and Sligar, S. G., 1982, Cytochrome P-450$_{cam}$ catalyzed epoxidation of dehydrocamphor, *Biochem. Biophys. Res. Commun.* **104**:853–858.
212. Debey, P., and Balny, C., 1978, Light-activated reactions in bacterial cytochrome P-450 cycle: A study at subzero temperatures, *Biochem. Soc. Trans.* **6**:1289–1292.
213. Sligar, S. G., Debrunner, P. G., Lipscomb, J. D., Namtvedt, M. J., and Gunsalus, I. C., 1974, A role of the putidaredoxin COOH-terminus in P-450$_{cam}$ (cytochrome m) hydroxylations, *Proc. Natl. Acad. Sci. USA* **71**:3906–3910.
214. Sligar, S. G., Kennedy, K. A., and Pearson, D. C., 1980, Tetrahedral intermediates in cytochrome P-450 hydroxylations, in: *Microsomes, Drug Oxidations, and Chemical Carcinogenesis* (M. J. Coon, A. H. Conney, R. W. Estabrook, H. V. Gelboin, J. R. Gillette, and P. J. O'Brien, eds.), Academic Press, New York, pp. 391–393.
215. Sligar, S., Besman, M., Gelb, M., Gould, P., Heimbrook, D., and Person, D., 1980, Occurrence and role of an acyl-peroxide intermediate in oxygen dependent P-450 hydroxylations, in: *Biochemistry, Biophysics and Regulations of Cytochrome P-450* (J.-A. Gustafsson, J. Carlstedt-Duke, A. Mode, and J. Rafter, eds.), Elsevier/North-Holland, Amsterdam, 379–382.
216. Sligar, S. G., Kennedy, K. A., and Pearson, D. C., 1980, Chemical mechanisms for cytochrome P-450 hydroxylation: Evidence for acylation of heme-bound dioxygen, *Proc. Natl. Acad. Sci. USA* **77**:1240–1244.
217. Heimbrook, D. C., 1983, Active oxygen of cytochrome P-450, Ph.D. dissertation, Yale University, New Haven.
218. Groves, J. T., Watanabe, Y., and McMurry, T. J., 1983, Oxygen activation by metalloporphyrins: Formation and decomposition of an acylperoxymanganese (III) complex, *J. Am. Chem. Soc.* **105**:4489–4490.
219. Elbe, K. S., and Dawson, J. H., 1984, NADH and oxygen dependent multiple turnovers of cytochrome P-450$_{cam}$ without putidaredoxin and putidaredoxin reductase, *Biochemistry* **23**:2068–2073.
220. Blake, R. C., II, and Coon, M. J., 1980, On the mechanism of action of cytochrome P-450: Spectral intermediates in the reaction of P-450$_{LM2}$ with peroxy compounds, *J. Biol. Chem.* **255**:4100–4111.
221. White, R. E., Sligar, S. G., and Coon, M. J., 1980, Evidence for a homolytic mechanism of peroxide oxygen–oxygen bond cleavage during substrate hydroxylation by cytochrome P-450, *J. Biol. Chem.* **255**:11108–11111.
222. White, R. E., and Coon, M. J., 1980, Oxygen activation by cytochrome P-450, *Annu. Rev. Biochem.* **49**:315–356.
223. McCarthy, M.-B., and White, R. E., 1983, Competing modes of peroxyacid flux through cytochrome P-450, *J. Biol. Chem.* **258**:11610–11616.
224. Blake, R. C., II, and Coon, M. J., 1981, On the mechanism of action of cytochrome

P-450: role of peroxy spectral intermediates in substrate hydroxylation, *J. Biol. Chem.* **256**:5755–5763.
225. Blake, R. C., II, and Coon, M. J., 1981, On the mechanism of action of cytochrome P-450: Evaluation of homolytic and heterolytic mechanisms of oxygen–oxygen bond cleavage during substrate hydroxylation by peroxides, *J. Biol. Chem.* **256**:12127–12133.
226. Lee, W. A., and Bruice, T. C., 1985, Homolytic and heterolytic oxygen–oxygen bond scissions accompanying oxygen transfer to iron (III) porphyrins by percarboxylic acids and hydroperoxides: A mechanistic criterion for peroxidase and cytochrome P-450, *J. Am. Chem. Soc.* **107**:513–514.
227. Sligar, S. G., Gelb, M. H., and Heimbrook, D. C., 1984, Bio-organic chemistry and cytochrome P-450 dependent catalysis, *Xenobiotica* **14**:63–86.
228. Fort, R. C., and Schleyer, P. Von R., 1964, Adamantane: Consequences of the diamondoid structure, *Chem. Rev.* **64**:277–300.
229. Gleicher, G. J., and Schleyer, P. Von R., 1967, Conformational analysis of bridgehead carbonium ions, *J. Am. Chem. Soc.* **89**:582–593.
230. Humphrey, L. B., Hodgson, B., and Pincock, R. E., 1968, Bridgehead free radical stability: The decomposition of t-butoxyperoxy esters of the 1-adamantyl, 1-bicyclo [2.2.2]octyl, and 1-norbornyl systems, *Can. J. Chem.* **46**:3099–3103.
231. White, R. E., Groves, J. T., and McCluskey, G. A., 1979, Electronic and steric factors in regioselective hydroxylation catalyzed by purified cytochrome P-450, *Acta Biol. Med. Ger.* **38**:475–482.
232. Gelb, M. H., Heimbrook, D. C., Malkonen, P., and Sligar, S. G., 1982, Stereochemistry and deuterium isotope effects in camphor hydroxylation by the cytochrome P-450$_{cam}$ monoxygenase system, *Biochemistry* **21**:370–377.
233. Sligar, S. G., Gelb, M. H., and Heimbrook, D. C., 1982, Bioorganic studies of the P-450$_{cam}$ reaction mechanism, in: *Microsomes, Drug Oxidations and Drug Toxicity* (R. Sato and R. Kato, eds.), Wiley–Interscience, New York, pp. 155–161.
234. Powell, M. F., Pai, E. F., and Bruice, T. C., 1984, Study of (tetraphenylporphinato) manganese (III)-catalyzed epoxidation and demethylation using p-cyano-N,N-dimethylaniline N-oxide as oxygen donor in a homogeneous system: Kinetics, radiochemical ligation studies, and reaction mechanism for a model of cytochrome P-450, *J. Am. Chem. Soc.* **106**:3277–3285.
235. Burstein, S., and Gut, M., 1976, Intermediates in the conversion of cholesterol to pregnenolone: Kinetics and mechanism, *Steroids* **29**:115–131.
236. Burstein, S., Nickolsen, R. C., Byon, C. Y., Kimball, H. L., and Gut, M., 1977, Specificity of the cholesterol side-chain cleavage enzyme system of adrenal cortex, *Steroids* **30**:439–453.
237. Byon, C. Y., and Gut, M., 1980, Steric considerations regarding the biodegradation of cholesterol to pregnenolone: Exclusion of (22S)-22-hydroxycholesterol and 22-ketocholesterol as intermediates, *Biochem. Biophys. Res. Commun.* **94**:549–552.
238. Carter, S. K., Crooke, S. T., and Umezawa, H. (eds.), 1978, *Bleomycin: Current Status and New Developments*, Academic Press, New York.
239. Hecht, S. M. (ed.), 1979, *Bleomycin: Chemical, Biochemical and Biological Aspects*, Springer-Verlag, Berlin.
240. Burger, R. M., Peisach, J., and Horwitz, S. B., 1981, Mechanism of bleomycins: *In vitro* studies, *Life Sci.* **28**:715–727.
241. Povirk, L. F., 1983, Bleomycin, in: *Molecular Aspects of Anti-Cancer Drug Action* (S. Neidle and M. J. Waring, eds.), Macmillan & Co., London, pp. 157–182.
242. Fujii, A., Takita, T., Maeda, K., and Umezawa, H., 1973, Chemistry of bleomycin. XI. Structure of the terminal amines, *J. Antibiot.* **26**:398–399.
243. Crooke, S. T., and Bradner, W. T., 1977, Bleomycin: A review, *J. Med. (Westbury, N.Y.)* **1**:333–428.

244. Freedman, M. A., 1978, A review of the bleomycin experience in the United States, *Recent Results Cancer Res.* **63**:152–178.
245. Umezawa, H., Maeda, K., Takeuchi, T., and Okami, Y., 1966, New antibiotics, bleomycin A and B, *J. Antibiot. Ser. A* **19**:200–209.
246. Umezawa, H., Suhara, Y., Takeuchi, T., and Okami, Y., 1966, Purification of bleomycins, *J. Antibiot.* **19**:210–215.
247. Ishizuka, M., Takayama, H., Takeuchi, T., and Umezawa, H., 1967, Activity and toxicity of bleomycin, *J. Antibiot. Ser. A* **20**:15–24.
248. Suzuki, H., Nagai, K., Yamaki, H., Tanaka, N., and Umezawa, H., 1968, Mechanism of action of bleomycin: Studies with the growing culture of bacterial and tumor cells, *J. Antibiot.* **21**:379–386.
249. Umezawa, H., 1977, Recent studies on bleomycin, *Lloydia* **40**:67–81.
250. Onishi, T., Iwata, H., and Takagi, V., 1975, Effects of reducing and oxidizing agents on the action of bleomycin, *J. Biochem.* **77**:745–752.
251. Sausville, E. A., Peisach, J., and Horwitz, S. B., 1976, A role of ferrous ion and oxygen in the degradation of DNA by bleomycin, *Biochem. Biophys. Res. Commun.* **73**:814–822.
252. Sausville, E. A., Peisach, J., and Horwitz, S. B., 1978, Effect of chelating agents and metal ions on the degradation of DNA by bleomycin, *Biochemistry* **17**:2740–2746.
253. Sausville, E. A., Stein, R. W., Peisach, J., and Horwitz, S. B., 1978, Properties and products of the degradation of DNA by bleomycin and iron (II), *Biochemistry* **17**:2746–2754.
254. Chang, C.-H., and Meares, C. F., 1982, Light-induced nicking of deoxyribonucleic acid by cobalt (III) bleomycins, *Biochemistry* **21**:6332–6334.
255. Chang, C.-H., and Meares, C. F., 1984, Cobalt-bleomycins and deoxyribonucleic acid: Sequence-dependent interactions, action spectrum for nicking, and indifference to oxygen, *Biochemistry* **23**:2268–2274.
256. Barton, J. K., and Raphael, A. L., 1984, Photoactivated stereospecific cleavage of double-helical DNA by cobalt (III) complexes, *J. Am. Chem Soc.* **106**:2466–2468.
257. Ehrenfeld, G., Murugesan, N., and Hecht, S., 1984, Activation of oxygen and mediation of DNA degradation by manganese-bleomycin, *Inorg. Chem.* **23**:1496–1498.
258. Burger, R. M., Freedman, J. H., Horwitz, S. B., and Peisach, J., 1984, DNA degradation by manganese (II)-bleomycin plus peroxide, *Inorg. Chem.* **23**:2215–2217.
259. Takita, T., Muraoka, Y., Nakatani, T., Fujii, A., Umezawa, Y., Naganawa, H., and Umezawa, H., 1978, Chemistry of bleomycin, XIX. Revised structures of bleomycin and phleomycin, *J. Antibiot.* **31**:801–804.
260. Takita, T., 1978, Review of the structural studies on bleomycin, in: *Bleomycin: Chemical, Biochemical and Biological Aspects* (S. M. Hecht, ed.), Springer-Verlag, Berlin, pp. 37–47.
261. Nunn, A. D., 1976, Interactions between bleomycins and metals, *J. Antibiot.* **29**:1102–1108.
262. Dabrowiak, J. C., 1980, The coordination chemistry of bleomycin: A review, *J. Inorg. Biochem.* **13**:317–337.
263. Burger, R. M., Horwitz, S. B., Peisach, J., and Wittenberg, J. B., 1979, Oxygenated iron bleomycin: A short-lived intermediate in the reaction of ferrous bleomycin with O_2, *J. Biol. Chem.* **254**:12299–12302.
264. Burger, R. M., Kent, T. A., Horwitz, S. B., Münck, E., and Peisach, J., 1983, Mössbauer study of iron bleomycin and its activation intermediates, *J. Biol. Chem.* **258**:1559–1564.
265. Burger, R. M., Peisach, J., Blumberg, W. E., and Horwitz, S. B., 1979, Iron-bleomycin interactions with oxygen and oxygen analogues: Effects on spectra and drug activity, *J. Biol. Chem.* **254**:10906–10912.

266. Sugiura, Y., 1978, Oxygen binding to cobalt (III)-bleomycin, *J. Antibiot.* **31**:1206–1208.
267. Sugiura, Y., 1980, Monomeric cobalt (II)-oxygen adducts of bleomycin antibiotics in aqueous solution: A new ligand type for oxygen binding and effect of axial Lewis base, *J. Am. Chem. Soc.* **102**:5216–5221.
268. Chevion, M., Peisach, J., and Blumberg, W. E., 1977, Imidazole, the ligand trans to mercaptide in ferric cytochrome P-450: An EPR study of proteins and model compounds, *J. Biol. Chem.* **252**:3637–3645.
269. Sharrock, M. Münck, E., Debrunner. P. G., Marshall, V., Lipscomb, J. D., and Gunsalus, I. C., 1973, Mossbauer studies of cytochrome P-450$_{cam}$, *Biochemistry* **12**:258–265.
270. Champion, P. M., Lipscomb, J. D., Münck, E., Debrunner, P., and Gunsalus, I. C., 1975, Mossbauer investigations of high-spin ferrous heme proteins. I. Cytochrome P-450, *Biochemistry* **14**:4151–4158.
271. Dabrowiak, J. C., 1982, Bleomycin, *Adv. Inorg. Biochem.* **4**:69–113.
272. Burger, R. M., Peisach, J., and Horwitz, S. B., 1981, Activated bleomycin: A transient complex of drug, iron and oxygen that degrades DNA, *J. Biol. Chem.* **256**:11636–11644.
273. Sugiura, Y., Suzuki, T., Muraoka, Y., Umezawa, Y., Takita, T., and Umezawa, H., 1981, Deglyco-bleomycin-iron complexes: Implications for iron-binding site and role of the sugar portion in bleomycin antibiotics, *J. Antibiot.* **39**:1232–1235.
274. Kuramochi, H., Katsutoshi, T., Takita, T., and Umezawa, H., 1981, An active intermediate formed in the reaction of bleomycin-Fe (II) complex with oxygen, *J. Antibiot.* **34**:576–582.
275. Iyanagi, T., Anon, F. K., Imai, Y., and Mason, H. S., 1978, Studies on the microsomal mixed function oxidase system: Redox properties of detergent-solubilized NADPH-cytochrome P-450 reductase, *Biochemisty* **17**:2224–2230.
276. Giloni, L., Takeshita, M., Johnson, F., Iden, C., and Grollman, A. P., 1981, Bleomycin-induced strand scission of DNA: Mechanism of deoxyribose cleavage, *J. Biol. Chem.* **256**:8608–8615.
277. Wu, J. C., Kozarich, J. W., and Stubbe, J., 1983, The mechanism of free base formation from DNA by bleomycin, *J. Biol. Chem.* **258**:4694–4697.
278. Oberley, L. W., and Buettner, G. R., 1979, The production of hydroxyl radical by bleomycin and iron (II), *FEBS Lett.* **97**:47–49.
279. Sugiura, Y., and Kikuchi, T., 1978, Formation of superoxide and hydroxy radicals in iron (III)-bleomycin-oxygen system: Electron spin resonance detection by spin trapping, *J. Antiobiot.* **31**:1310–1312.
280. Sugiura, Y., 1979, The production of hydroxyl radical from copper (I) complex systems of bleomycin and tallysomycin: Comparsion with copper (II) and iron (II) systems, *Biochem. Biophys. Res. Commun.* **90**:375–383.
281. Muller, W. W. G., Yamazaki, Z. I., Breter, H. J., and Zahn, R. K., 1972, Action of bleomycin on DNA and RNA, *Eur. J. Biochem.* **31**:518–525.
282. Ishida, R., and Takahashi, T., 1975, Increased DNA chain breakage by combined action of bleomycin and superoxide radical, *Biochem. Biophys. Res. Commun.* **66**:1432–1438.
283. Lown, J. W., and Sim, S. K., 1977, The mechanism of the bleomycin-induced cleavage of DNA, *Biochem. Biophys. Res. Commun.* **77**:1150–1157.
284. Rodriguez, L. O., and Hecht, S. M., 1982, Iron (II)-bleomycin: Biochemical and spectral properties in the presence of radical scavengers, *Biochem. Biophys. Res. Commun.* **104**:1470–1476.
285. Cunningham, M. L., Ringrose, P. S., and Lokesh, B. R., 1983, Bleomycin cytotoxicity is prevented by superoxide dismutase *in vitro*, *Cancer Lett* **21**:149–153.

286. Cunningham, M. S., Ringrose, P. S., and Lokesh, B. R., 1984, Inhibition of the genotoxicity of bleomycin by superoxide dismutase, *Mutat. Res.* **135**:199–202.
287. Murugesan, N., Ehrenfeld, G. M., and Hecht, S. M., 1982, Oxygen transfer from bleomycin–metal complexes, *J. Biol. Chem.* **257**:8600–8603.
288. Antholine, W. E., and Petering, D. H., 1979, On the reaction of iron bleomycin with thiols and oxygen, *Biochem. Biophys. Res. Commun.* **90**:384–389.
289. Povirk, L. F., 1979, Catalytic release of deoxyribonucleic acid bases by oxidation and reduction of an iron–bleomycin complex, *Biochemistry* **18**:3989–3995.
290. Scheulen, M. E., Kappus, H., Thyssen, D., and Schmidt, C. G., 1982, DNA chain breakage by bleomycin in the presence of ferric ions and NADPH-cytochrome P-450 reductase, in: *Microsomes, Drug Oxidations and Drug Toxicity* (R. Sato and R. Kato, eds.), Wiley–Interscience, New York, pp. 553–554.
291. Scheulen, M. E., Kappus, H., Thyssen, D., and Schmidt, C. G., 1982, Redox cycling of iron (III)-bleomycin by NADPH-cytochrome P-450 reductase, *Biochem. Pharmacol.* **30**:3385–3388.
292. Scheulen, M. E., and Kappus, H., 1983, The activation of oxygen by bleomycin is catalyzed by NADPH-cytochrome P-450 reductase in the presence of iron ions and NADPH, in: *Oxygen Radicals in Chemistry and Biology, Proceedings 3rd International Conference* (W. Bors, M. Saran, and D. Tart, eds.), de Gruyter, Berlin, pp. 425–433.
293. Kilkuskie, R. E., MacDonald, T. L., and Hecht, S. M., 1984, Bleomycin may be activated for DNA cleavage by NADPH-cytochrome P-450 reductase, *Biochemistry* **23**:6165–6171.
294. Garneir-Suillerot, A., Albertini, J.-P., and Tosi, L., 1981, Mononuclear and binuclear Co (III)-dioxygen adducts of bleomycin: Circular dichroism and electron paramagnetic resonance studies, *Biochem. Biophys. Res. Commun.* **102**:499–506.
295. Wagner, G. C., Gunsalus, I. C., Wang, M.-Y. R., and Hoffman, B. M., 1981, Cobalt-substituted cytochrome P-450$_{cam}$, *J. Biol. Chem.* **256**:6266–6273.
296. Melnyk, D. L., Horwitz, S. B., and Peisach, J., 1981, Redox potential of iron-bleomycin, *Biochemistry* **20**:5327–5331.
297. Taylor, J. F., and Hastings, A. B., 1939, Oxidation–reduction potentials of the methemoglobin–hemoglobin system, *J. Biol. Chem.* **131**:649–662.
298. Feigelson, P., and Brady, F. O., 1974, Heme-containing dioxygenases, in: *Molecular Mechanisms of Oxygen Activation* (O. Hayaishi, ed.), Academic Press, New York, pp. 87–133.
299. Werneke, J. M., Sligar, S. G., and Schuler, M. A., 1985, Development of broad host range vectors for expression in *Pseudomonas*, *Gene* in press.
300. Sugiura, Y., Suzuki, T., Kawabe, H., Tanaka, H., and Watanabe, K., 1982, Spectroscopic studies on bleomycin-iron complexes with carbon monoxide, nitric oxide, isocyanide, azide, and cyanide and comparison with iron-porphyrin complexes. *Biochim. Biophys. Acta* **716**:38–44.
301. Sugiura, Y., Ogawa, S., and Morishima, I., 1980, Unusual interaction of iron-bleomycin with cyanide, *J. Am. Chem. Soc.* **102**:7945–7947.
302. Ebel, R. E., O'Keeffe, D. H., and Peterson, J. A., 1975, Nitric oxide complexes of cytochrome P-450, *FEBS Lett.* **55**:198–201.
303. Lenkinski, R. E., Peerce, B. E., Dallas, J. L., and Glickson, J. D., 1980, Interactions of gallium (III) with bleomycin antibiotics, *J. Am. Chem. Soc.* **102**:131–135.
304. Lenkinski, R. E., and Dallas, J. L., 1979, An NMR investigation of the kinetics of dissociation of the zinc (II) complex of bleomycin antibiotics, *J. Am. Chem. Soc.* **101**:5902–5906.
305. Lenkinski, R. E., Peerce, B. E., Pillai, R. P., and Glickson, J. D., 1980, Calcium (III) and the trivalent lanthanide ion complexes of the bleomycin antibiotics: Potentiometric, fluorescence and ^{1}H-NMR studies, *J. Am. Chem. Soc.* **102**:7088–7093.

306. Freedman, J. H., Horwitz, S. B., and Peisach, J., 1982, Reduction of copper (II)-bleomycin: A model for *in vivo* drug activity, *Biochemistry* **21**:2203–2210.
307. Chang, C.-H., Dallas, J. L., and Meares, C. F., 1983, Identification of a key structural feature of cobalt (III)-bleomycins: An exogenous ligand (i.e., hydroperoxide) bound to cobalt, *Biochem. Biophys. Res. Commun.* **110**:959–966.
308. Antholine, W., Hyde, J. S., Sealy, R. C., and Petering, D. H., 1984, Structure of cupric bleomycin: Nitrogen and proton couplings from EPR and electron nuclear double resonance spectroscopy, *J. Biol. Chem.* **259**:4437–4440.
309. Oppenheimer, N. J., Rodriguez, L. O., and Hecht, S. M., 1979, Proton nuclear magnetic resonance study of the structure of bleomycin and the zinc-bleomycin complex, *Biochemistry* **18**:3439–3445.
310. Oppenheimer, N. J., Rodriguez, L. O., and Hecht, S. M., 1980, Metal binding to modified bleomycins: Zinc and ferrous complexes with an acetylated bleomycin, *Biochemistry* **19**:4096–4103.
311. Oppenheimer, N. J., Chang, C., Rodriguez, L. O., and Hecht, S. M., 1981, Copper (I) bleomycin: A structurally unique oxidation–reduction active complex, *J. Biol. Chem.* **256**:1514–1517.
312. Sugiura, Y., Ishizu, K., and Miyoshi, K., 1979, Studies of metallobleomycins by electronic spectroscopy, electron spin resonance spectroscopy, and potentiometric titration, *J. Antibiot.* **32**:453–461.
313. Ishizu, K., Murata, S., Miyoshi, K., Sugiura, Y., Takita, H., and Umezawa, H., 1981, Electrochemical and ESR studies on Cu (II) complexes of bleomycin and its related compounds, *J. Antibiot.* **34**:994–1000.
314. Gunsalus, I. C., and Sligar, S. G., 1977, Oxygen reduction by the P-450 monoxygenase system, *Adv. Enzymol.* **47**:1–44.
315. Groves, J. T., and Kruper, W. J., Jr., 1979, Preparation and characterization of an oxoporphinatochromium (V) complex, *J. Am. Chem. Soc.* **101**:7613–7615.
316. Smegal, J. A., and Hill, C. L., 1983, Synthesis, characterization, and reaction chemistry to a bis(iodosylbenzene)-metalloporphyrin complex, $[PhI(OAc)O]_2Mn^{IV}TPP$: A complex possessing a five-electron oxidation capability, *J. Am. Chem. Soc.* **105**:2920–2922.
317. Collman, J. P., Brauman, J. I., Meunier, B., Raybuck, S. A., and Kodadek, T., 1984, Epoxidation of olefins by cytochrome P-450 model compounds: Mechanism of oxygen atom transfer, *Proc. Natl. Acad. Sci. USA* **81**:3245–3248.
318. Champion, P., Münck, E., Debrunner, P., Hollenberg, P., and Hager, L. P., 1973, Mossbauer investigations of chloroperoxidase and its halide complexes, *Biochemistry* **12**:426–435.
319. Hollenberg, P. F., and Hager, L. P., 1973, The cytochrome P-450 nature of the CO complex of ferrous chloroperoxidase, *J. Biol. Chem.* **248**:2630–2633.
320. Chiang, R., Makina, R., Spomer, W. E., and Hager, L. P., 1975, Chloroperoxidase: P-450 type absorption in the absence of sulfhydryl groups, *Biochemistry* **14**:4166–4171.
321. Champion, P. M., Chiang, R., Münck, E., Debrunner, P., and Hager, L. P., 1975, Mossbauer investigations of high spin ferrous heme proteins. II. Chloroperoxidase, horseradish peroxidase and hemoglobin, *Biochemistry* **14**:4159–4165.
322. Champion, P. M., Remba, R. D., Chiang, R., Fitcher, D. B., and Hager, L. P., 1976, Resonance Raman spectra of chloroperoxidase, *Biochim. Biophys. Acta* **446**:486–492.
323. Dawson, J. H., Trudell, J. R., Barth, G., Liner, R. E., Bunnenberg, E., Djerassi, C., Chiang, R. L., and Hager, L. P., 1976, Chloroperoxidase: Evidence for P-450 type heme environment from magnetic circular dichroism spectroscopy, *J. Am. Chem. Soc.* **98**:3709–3710.
324. Remba, R. D., Champion, P. M., Fitcher, D. B., Chiang, R., and Hager, L. P., 1979, Resonance Raman investigations of chloroperoxidase, horseradish peroxidase and cytochrome c using Soret band laser excitation, *Biochemistry* **18**:2280–2290.

325. Hollenberg, P. F., Hager, L. P., Blumberg, W. E., and Peisach, J., 1980, An electron paramagnetic resonance study of the high and low spin forms of chloroperoxidase, *J. Biol. Chem.* **255**:4801–4807.
326. Palcic, M. M., Rutter, R., Araiso, T., Hager, L. P., and Dunford, H. B., 1980, Spectrum of chloroperoxidase compound I, *Biochem. Biophys. Res. Commun.* **94**:1123–1127.
327. Sono, M., Dawson, J. H., and Hager, L. P., 1984, The generation of a hyperporphyrin spectrum upon thiol binding to ferric chloroperoxidase: Further evidence of endogenous thiolate ligation to the ferric enzyme, *J. Biol. Chem.* **259**:13209–13216.
328. Padbury, G., and Sligar, S. G., 1985, Chemical reactivities of bleomycin, *J. Biol. Chem.* **260**:7820–7823.
329. McCarthy, M. B., and White, R. E., 1983, Functional differences between peroxidase compound I and the cytochrome P-450 reactive intermediate, *J. Biol. Chem.* **258**:9153–9158.
330. Heimbrook, D. C., Murray, R. I., Egeberg, K. D., Sligar, S. G., Nee, M. W., and Bruice, T. C., 1984, Demethylation of N,N-dimethylaniline and p-cyano-N,N-dimethylaniline and their N-oxides by cytochromes P-450$_{LM2}$ and P-450$_{cam}$, *J. Am. Chem. Soc.* **106**:1514–1515.
331. Kedderis, G. L., Koop, D. R., and Hollenberg, P. F., 1980, N-Demethylation reactions catalyzed by chloroperoxidase, *J. Biol. Chem.* **255**:10174–10182.
332. Hager, L. P., Doubek, D. L., Silverstein, R. M., Lee, T. T., Thomas, J. A., Hargis, J. H., and Martin, J. C., 1973, The mechanism of oxygen evolution from peroxyacid by chloroperoxidase: A contribution to the structure of compound I, in: *Oxidases and Related Redox Systems* (T. E. King, H. S. Mason, and M. Martin, eds.), University Park Press, Baltimore, pp. 311–332.
333. Hager, L. P., Morris, D. R., Brown, F. S., and Eberwein, H., 1966, Chloroperoxidase. II. Utilization of halogen anions, *J. Biol. Chem.* **241**:1769.
334. Miwa, G. T., Walsh, J. S., Kedderis, G. L., and Hollenberg, P. F., 1983, The use of intramolecular isotope effects to distinguish between deprotonation and hydrogen atom abstraction mechanisms in cytochrome P-450 and peroxidase-catalyzed N-demethylation reactions, *J. Biol. Chem.* **258**:14445–14449.
335. Guengerich, F. P., 1983, Oxidation–reduction properties of rat liver cytochromes P-450 and NADPH-cytochrome P-450 reductase related to catalysis in reconstituted systems, *Biochemistry* **22**:2811–2820.
336. White, R. E., and Coon, M. J., 1982, Heme ligand replacement reactions of cytochrome P-450: Characterization of the bonding atom of the axial ligand trans to thiolate as oxygen, *J. Biol. Chem.* **257**:3073–3083.
337. Oprian, D. D., Gorsky, L. D., and Coon, M. J., 1983, Properties of the oxygenated form of liver microsomal cytochrome P-450, *J. Biol. Chem.* **258**:8684–8691.
338. Fisher, M. T., and Sligar, S. G., 1985, Control of heme protein redox potential. Linear free energy relationship between ferric spin state and potential, *J. Am. Chem. Soc.* **107**:5018.

CHAPTER 13

The Crystal Structure of Cytochrome P-450$_{cam}$

THOMAS L. POULOS

1. Introduction

From the preceding chapters it is clear that a high-resolution crystal structure of P-450 is very desirable and that the obvious candidate for crystallographic studies is P-450$_{cam}$. Indeed, some of the earliest studies on P-450$_{cam}$ resulted in its crystallization by Yu et al. in 1974.[1] However, these crystals, designated orthorhombic I,[2] were unsuitable for X-ray diffraction studies and it was not until 1982 that the first X-ray-quality crystals were reported.[2] Gunsalus's group and especially the efforts of G. C. Wagner were instrumental in reaching this milestone.

Initially, we succeeded in obtaining tetragonal crystals from polyethylene glycol solutions but these crystals did not diffract well beyond 3 Å and the long c axis (~ 255 Å) presented problems with respect to data collection. Therefore, we continued to search for suitable crystals and succeeded in obtaining another orthorhombic form (designated orthorhombic II) from ammonium sulfate solutions. In order to obtain orthorhombic II crystals, it was first necessary to strip the enzyme of any bound camphor, concentrate the enzyme to about 1.0 mM in the presence of high concentrations of dithiothreitol, and finally initiate crystallization by layering the P-450 solution over a 60% to 70% ammonium sulfate solution. Occasionally, one obtains the original orthorhombic I crystals. Our experience in crystallizing P-450$_{cam}$, which I relate here at the risk of contributing to the mystique and folklore surrounding protein crystallization, indicates that a delicate balance between protein concentration,

---

THOMAS L. POULOS • Genex Corporation, Gaithersburg, Maryland 20877.

ammonium sulfate concentration, and temperature must be achieved in order to obtain orthorhombic II crystals. To maximize the yield of good crystals requires the calibration of each freshly prepared ammonium sulfate solution with respect to temperature. Thiols are probably required for crystallization to prevent intermolecular disulfide bond formation.[3] Once in the crystalline state, however, P-450$_{cam}$ can be stored free of thiols and converted to the camphor-bound complex. All crystallographic work to date has been carried out with crystals in the ferric camphor-bound state.

2. Solution of the Structure and Accuracy of the Model

Data for parent and five heavy atom derivatives were collected using an automated diffractometer at the University of California, San Diego. From these data we obtained a 2.8-Å multiple isomorphous replacement (MIR) electron density map. The MIR map was sufficiently clear to obtain an accurate outline of a single P-450$_{cam}$ molecule, orient the heme, and correctly determine the direction of most helical segments and β pairs. However, an unambiguous N- to C-terminal tracing of the polypeptide chain was not possible, nor could we fit the entire sequence to the MIR map. In hindsight, these difficulties were due to a break in the polypeptide chain electron density between residues 216 and 226 and distortions in the electron density near heavy atom binding sites. To circumvent these problems, we adopted a "phase combination" procedure successfully employed in the refinement of cytochrome c peroxidase.[4] First, we derived a trial structure by emphasizing the fit of idealized secondary structure to the MIR map and wherever possible, side chains were included. The size, shape, and location of electron density envelopes were used as a guide for selecting side chains without strict adherence to the known sequence.[5] Only residues 242–332 could be unambiguously located, primarily because of the aforementioned problems as well as ambiguities in how reasonably well-defined helices and β pairs should be connected. The resulting model was at a considerably more primitive state than most MIR models.

Nevertheless, we subjected the crude trial structure to several rounds of restrained least-squares refinements.[6] The residual, R, to be minimized in the least-squares process is $\sum | F_o - F_c | / \sum F_o$, where F_o represents the observed structure factors derived from the intensity data and F_c, the computed structure factors obtained from the Fourier transform of the model. During refinement, atomic positional parameters are adjusted so as to minimize R. For proteins the data do not extend to atomic resolution so it is essential to constrain the model within specified limits of ideal bond angles, distances, and nonbonded contacts. If constraints are not applied, R may drop but the stereochemistry will rapidly diverge from anything that remotely resembles a protein.

In the majority of protein structure refinements, the computed (F_c) phases are usually more accurate than MIR phases so an electron density map computed with F_c phases is much clearer than an MIR map. With P-450$_{cam}$, however, the starting model was so crude that the F_c electron density map was not an improvement. However, it was possible to obtain a clearer electron density map by using a combination of F_c and MIR phases. Obtaining a higher-quality electron density map in this way is possible because for some reflections the F_c phase is more accurate while for others the MIR phase is a better estimate of the "true" phase so that by combining the two, one effectively selects the best phase from either source. Since phases are represented mathematically as probability distributions, one is able to assess the accuracy of a phase by determining the "figure of merit" which is related to the sharpness of the probability distribution and lies between 0 and 1.0, with 1.0 representing the best phase estimate. By monitoring the overall figure of merit for all reflections, one is able to quantitatively judge whether or not phase combination generates a better set of phases. For P-450$_{cam}$, the figure of merit increased from 0.54 to 0.65 after the first cycle of phase combination. More importantly, the resulting electron density allowed more of the sequence to be correctly located. After several such cycles of refinement, phase combination, and model building, the figure of merit increased to 0.78 at 2.6 Å, enabling a complete solution of the structure.

Our current refined model contains all but residues 1–9. The electron density for residues 1–9 is present but not sufficiently clear to warrant the inclusion of these side chains. During the latter stages of refinement, the DNA sequence of P-450$_{cam}$ became available (Sligar and Murray, this volume). The DNA sequence indicates that a dipeptide, Trp–Thr, should be inserted after Val-54 and that residue 361 is His and not Ser. Inclusion of these changes into our atomic model improved the fit to the electron density map, thereby confirming that the protein sequence deduced from the DNA sequence is correct. The current residual, R, is 0.234 and the model has been constrained to give a root mean square deviation from ideal bond distances of 0.037 Å. Temperature factors have not been refined nor have ordered solvent molecules been included. The upper estimate for errors in the atomic coordinates is $\sim \pm 0.3$ Å.

3. The Structure

3.1. Overall Topography

P-450$_{cam}$ is an asymmetrically shaped protein resembling a triangular prism only $\sim$ 30 Å thick with a maximum dimension of $\sim$ 60 Å (Fig. 1).

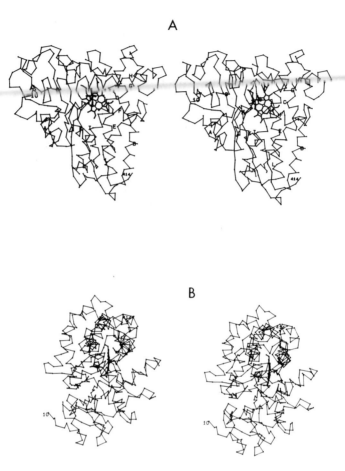

FIGURE 1. Two stereoscopic views of the P-450$_{cam}$ α-carbon backbone model showing the heme and camphor. The first residue in the model, Asn-10, and the C-terminus Val-414 are labeled. (A) View along the crystallographic Z axis. The 12 helical segments are labeled A–L. (B) View along the crystallographic Y axis.

Twelve helical segments, labeled A–L in Fig. 1A, dominate the structure and account for ~ 45% of the residues. Approximately 15% of the structure is divided between four antiparallel β pairs while there is no parallel β structure or extended sheet structure. This predominance of helical structure is a characteristic shared by most hemeproteins for which X-ray structures are available.

Examination of Fig. 1 reveals that the helices are arranged in three layers, one stacked on top of another with the heme embedded between layers. The plane of the heme lies in the crystallographic $X-Y$ plane so that the heme normal is almost perfectly aligned along the crystallographic Z axis, giving rise to the observed dichroism of orthorhombic II crystals.[7,8] The regular triangular shape of P-450$_{cam}$ also accounts for the efficient packing of four molecules in the unit cell which results in a relatively low water content in the crystals.[2]

P-450$_{cam}$ consists of three principal structural motifs: antiparallel helical bundle, parallel helical bundle, and a mix of helices and antiparallel β pairs. Helices E, F, G, and I form what Richardson[9] has termed an antiparallel Greek key helical bundle. This sort of helical topography is quite common and is found in such diverse proteins as the globins, cytochrome c peroxidase, thermolysin, lysozyme, and papain.[9] Helices I and L are nearly parallel to one another with an interhelical angle of $-32°$. Such an arrangement resembles the type II parallel alignment found in other proteins[9] though it is not as common as the antiparallel topography. Helices D and L are also in approximate parallel alignment. The remainder of the structure is a mix of helices or "random" coil interspersed by β pairs.

The number of residues in a helical conformation are about evenly divided between the N- and C-terminal halves of the molecule while a majority of the β structure is in the C-terminal half. Most of the helices, however, cluster together on the right side of the molecule (see Fig. 1A). As a result, P-450$_{cam}$ does not fold into an N- and C-terminal domain as do many other proteins. Instead, the polypeptide chain begins on the left half of the molecule, crosses over to the right, back to the left, and finally one more time to the right where the C-terminus Val-414 is situated near helix D.

3.2. Location of the Heme

The heme is sandwiched between the proximal helix (helix L) and the distal helix (helix I). The proximal helix runs along the molecular surface while the long distal helix runs through the center of the molecule where it provides hydrophobic groups that interact with the heme (Fig. 2). In addition to the heme thiolate bond, the heme is held in place by the aforementioned hydrophobic interactions and by hydrogen bonding interactions between the heme propionates and Arg-299 and His-355 (Fig. 2 and 3).

In both the globins and cytochrome c peroxidase (CCP), the heme is also embedded between two helices. In these proteins, however, the proximal and distal helices are antiparallel while in P-450$_{cam}$ they are parallel. Moreover, in the globins and CCP, the axial histidine ligand extends from the C-terminal end of the proximal helix, while in P-450$_{cam}$,

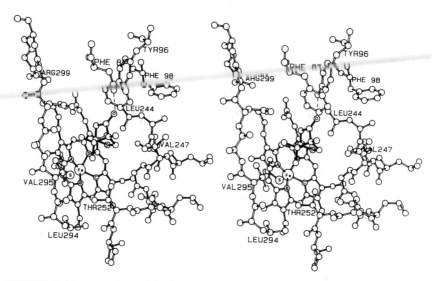

FIGURE 2. Stereoscopic model of the camphor binding pocket. The camphor carbonyl oxygen and C5 carbon atoms are labeled. Pyrrole ring nitrogens are also labeled. Several residues forming the top (toward the viewer) of the pocket have been removed for clarity.

the thiolate ligand extends from the N-terminal end (see Section 3.3) One additional difference is that the proximal helix in CCP is not exposed at the molecular surface while in P-450$_{cam}$ it is.

The asymmetric distribution of substituents (vinyls, methyls, propionates) of protoporphyrin IX makes the two faces of the heme ster-

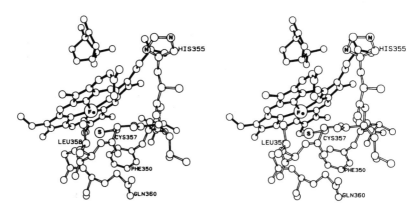

FIGURE 3. Stereoscopic model of P-450$_{cam}$ in the vicinity of Cys-357.

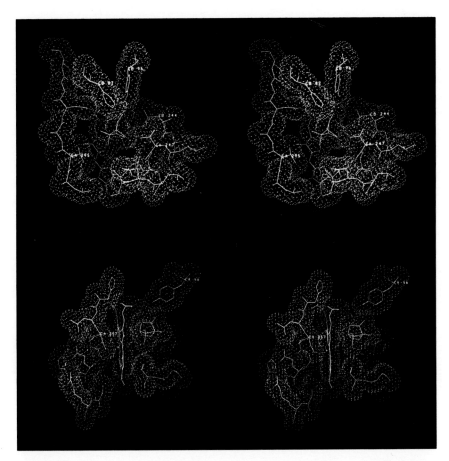

FIGURE 4. Dot surface representation showing the topography of the camphor binding site. The van der Waals surface of the protein is blue, the camphor green, and the heme red. (Upper) Viewed along the crystallographic Z axis. Residues shown include part of the distal helix (244–252) contacting both the heme and camphor. Also shown is one strand of an antiparallel β pair (294–299) containing Val-295, which contacts the camphor, and Arg-299, which forms a hydrogen bond or ion pair with the propionate of pyrrole ring A. (Lower) Viewed along the crystallographic Y axis. Part of the highly conserved region surrounding the thiolate ligand (355–364) is shown in addition to Tyr-96, which denotes a hydrogen bond to the camphor carbonyl oxygen atom.

eochemically distinct. Therefore, it is of interest to note that the stereochemically unique face available to O_2 and other ligands is exactly the same in P-450$_{cam}$, microsomal P450s,[10] the globins,[11] and catalase,[12] whereas CCP exhibits the opposite orientation.[13]

It should be evident from Fig. 1 that no edge or face of the heme is accessible at the molecular surface. This is in sharp contrast to what is observed in cytochromes where the heme edge is available at the molecular surface.[14] In comparing the crystal structures of several hemeproteins including the cytochromes, CCP, catalase, and now P-450$_{cam}$, one finds that in enzymes the heme is not accessible while in the cytochromes it is.[15] These differences in heme edge accessibility have significant implications regarding electron transfer mechanisms. Owing to the location of the heme in both CCP and P-450$_{cam}$, it is most unlikely that electron transfer occurs via direct contact between the donor group which in the case of P-450$_{cam}$ is the iron–sulfur center of putidaredoxin and the acceptor heme. Cytochromes, on the other hand, are able to form complexes where direct contract between electron transfer groups is possible. Precisely why there is this difference in heme accessiblity and, therefore, electron transfer distances is related to the chemistry of heme enzymes. The catalytic cycle of heme enzymes involves the production of highly reactive radical and iron–oxygen intermediates. If the heme were accessible at the molecular surface, then nonspecific, random discharge of these electrophilic intermediates would become more probable.[15] Therefore, it is necessary to sequester the heme within the protein's insulating environment.

Location of the heme well within the protein raises interesting questions as to whether or not protein groups participate in the electron transfer reaction. In the hypothetical model of the CCP–cytochrome c complex, a network of hydrogen bonds is predicted to form an electron transfer conduit from the peroxidase surface to the heme interior.[15] The feasibility of this proposal has received experimental support since two of the hydrogen bonding groups, a heme propionate[16] and a surface histidine,[17] have been implicated in the electron transfer process using chemical modification methods. We might look for parallels in the P-450–putidaredoxin system.

3.3. Axial Heme Ligands

Extending from the N-terminus of the proximal helix (helix L), Cys-357 provides the expected [18,19] axial thiolate ligand (Fig. 2–4). No electron density appears in the remaining axial coordination position, indicating that P-450$_{cam}$ in the ferric, camphor-bound state is probably pentacoordinate. This is not to say, however, that there are no water molecules near the iron atom. As discussed further in Section 5, a pocket just large enough for an O_2 molecule is situated between the camphor and distal helix above the iron atom. At present, this pocket looks "empty," but we suspect that it is actually filled with loosely bound or disordered water molecules which are not visible at 2.6 Å. Upon reduction and oxygenation, these solvent molecules are displaced by an incoming O_2 molecule.

While the proximal helix is on the surface and not buried as it is in CCP, Cys-357 is not fully accessible to the external milieu. Instead, the thiolate group is sequestered in a surface pocket formed by Phe-350, Leu-358, and Gln-360 (Fig. 3). There are no residues close enough to Cys-357 capable of a hydrogen bonding interaction as there is for the proximal ligand in the globins[20] and catalase.[12]

At present, we can only speculate and place structural limits as to the nature of the sixth ligand in the camphor-free, low-spin state. There is a preference for an oxygen ligand[21] though the structure shows that no oxygen-containing side chain is capable of interacting with the iron atom without a considerable disruption of secondary structure. A more reasonable hypothesis and one which is consistent with the available spectroscopic evidence is that in the camphor-free state, the substrate pocket is filled with water molecules, one of which serves as the sixth ligand.

3.4. Camphor Binding Site

Figures 2–4 depict the topography at the camphor binding site. As anticipated, the substrate pocket is lined with hydrophobic residues. Particularly noteworthy is Phe-87, Tyr-96, Leu-244, Val-247, and Val-295. The camphor molecule sits about 4 Å above pyrrole ring A directly adjacent to the O_2 binding site. Electron density corresponding to the camphor is extremely well defined and the orientation shown in Fig. 2–4 is based on the following criteria. First, a thin finger of electron density connects the camphor with the side chain hydroxyl group of Tyr-96, indicating a strong hydrogen bonding interaction between Tyr-96 and the camphor carbonyl oxygen atom. The geometry is also quite good for hydrogen bonding. By placing the camphor carbonyl oxygen atom at this position, the remainder of the camphor fits optimally in only one way, as shown in Fig. 2–4. Overall, the aromatic and aliphatic side chains surrounding the substrate provide an excellent "hand in glove" fit.

Figure 4 depicts the van der Waal's surface of the camphor site represented as color-coded dots. As shown in Figs. 3 and 4, the distal helix completely masks pyrrole ring C and parts of ring B and D, leaving ring A free for camphor binding. Such an arrangement is especially interesting when compared to the topography predicted for microsomal P-450. Ortiz de Montellano's group[22] employed a series of suicide substrate

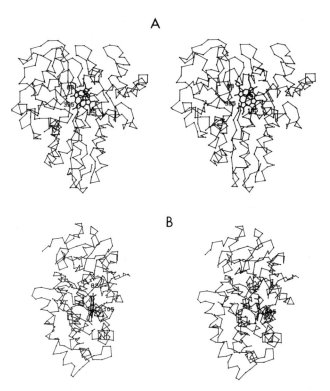

FIGURE 5. α-Carbon backbone models of P-450$_{cam}$ showing two possible routes to the substrate pocket.

alkenes and alkynes capable of alkylating the heme to establish which surface of the heme and which pyrroles are available for substrate interactions. The picture emerging from these elegant studies is essentially that shown in the upper diagram of Fig. 4 (see Fig. 6 and Ref. 22), suggesting that the substrate binding environment in the immediate vicinity of the heme is very much the same in P-450$_{cam}$ and microsomal P-450 cytochromes.

3.5. Substrate Access Channel

As shown in Fig. 1 and as concluded from examination of several space-filling diagrams, the camphor molecule is buried so one obvious question is how the camphor gains access to the heme. There appear to be two possibilities (Fig. 5). In Fig. 5A, three sharp reversals in the polypeptide chain centered at residues 88, 185, and 395 are juxtaposed close to one another where they form a distinct depression in the molecular surface directly above the camphor molecule. However, Phe-87, Phe-193, and Ile-395 form a hydrophobic cap which effectively seals off the camphor from the external milieu. Therefore, if this depression provides access to the active site, it must be capable of opening up to allow entry of a camphor molecule. Such a proposal is reasonable for several reasons. First, the surface turns which form the proposed access channel entry are often quite flexible structures. Second, in the absence of camphor, the substrate pocket is most likely occupied by solvent molecules. Therefore, it would be thermodynamically favored if, in the absence of substrate, the proposed channel remains open in order to allow a solvent continuum to be maintained much as found in the CCP access channel.[4,15] Third, in the absence of camphor, hydrophobic groups contacting the camphor, especially Phe-87, must be hydrated if they remain in their camphor-bound positions. However, movement of Phe-87 away from the camphor site could alleviate an entropically unfavored hydration as well as open up the access channel.

One other possible entrance to the active site is shown in Fig. 5B. Here, residues 88–105 define a loop which provides an opening but, again, movement of the protein would be required.

One additional reason for suspecting a fair degree of flexibility in the substrate pocket is the well-known ability of $P-450_{cam}$ to form iron-bound complexes with molecules considerably larger than camphor such as metyrapone and phenylimidazole.

4. Comparisons with Eukaryotic P-450

4.1. Sequences

Alignment of the known P-450 sequences (see Chapter 6) reveals two regions of good homology centered on $Cys-136_{cam}$ and $Cys-357_{cam}$. Both regions have been implicated as providing the heme thiolate ligand[5,23-26] The refined $P-450_{cam}$ structure provides unambiguous proof that Cys-357 is the ligand, thereby strongly implicating the homologous cysteine in microsomal P-450 ($Cys-436_{LM_2}$) as the ligand. Furthermore, the recently published nucleotide sequence of the mitochondrial cholesterol side-chain cleavage P-450[27] shows that this P-450 has only two cysteines, one of which (Cys-461) exhibits good homology with $Cys-357_{cam}$ and $Cys-436_{LM_2}$.

Other key residues contacting the proximal heme surface are also conserved. For example, Phe-350 and Leu-358 contact the proximal heme surface on either side of the Cys-357 thiolate ligand (Fig. 3). The hom-

TABLE I
Alignment of Distal Helices in Various P-450 Cytochromes[a]

				*			*				*						
$P-450_{cam}$		G	L	L	L	V	G	G	L	D	T_{252}	V	V	N	F	L	S	F
P-450d	(rat)	N	D	I	F	G	A	G	F	E	T_{319}	V	T	T	A	I	F	W
P3-450	(mouse)	N	D	I	F	G	A	G	F	D	T_{319}	V	T	T	A	I	T	W
P-450c	(rat)	F	D	L	F	G	A	G	F	D	T_{325}	I	T	T	A	I	S	W
P1-450	(mouse)	L	D	L	F	G	A	G	F	D	T_{322}	V	T	T	A	I	S	W
$P-450_{LM2}$	(rabbit)	L	S	L	F	F	A	G	T	E	T_{302}	T	T	S	T	T	L	R
P-450b	(rat)	L	S	L	F	F	A	G	T	E	T_{302}	T	S	S	T	T	L	R
P-450e	(rat)	L	S	L	F	F	A	G	T	E	T_{302}	T	G	S	T	T	L	R
$P-450_{scc}$	(bovine)	T	E	M	L	A	G	G	V	N	T_{329}	T	S	M	T	L	Q	W

[a] Hydrophobic stretch running over the heme distal surface is underlined. Residues with an asterisk contact the camphor molecule.

ologue to Phe-350$_{cam}$ in all P-450 cytochromes for which sequence data are available is also Phe while the Leu-358$_{cam}$ homologue is Leu, Ile, or Val. In addition, His-355$_{cam}$ is in a position to hydrogen bond with a heme propionate (Fig. 3) while the corresponding residue in eukaryotic P-450 is Arg, a functionally conservative change. Such striking similarities strongly suggest a common three-dimensional arrangement of residues near the thiolate ligand.

Structural similarities extend to the distal heme side as well. A hydrophobic stretch corresponding to the distal helix (helix I) centered on Thr-252$_{cam}$ aligns with a homologous hydrophobic stretch in microsomal P-450 (Chapter 6 and Table I). Immediately preceding Thr-252$_{cam}$ and the homologous Thr in eukaryotic P-450 is an Asp or Glu with the exception of P-450$_{scc}$ which has an Asn. Asp-251$_{cam}$ extends from the distal helix away from the heme to form an internal ion pair with Arg-186. Significantly, sequence alignments indicate that residue 186 is also an Arg (Arg-242) in the P-450d class of enzymes, suggesting a possible conservation of the Arg-186–Asp-251 ion pair.

Asp-251$_{cam}$ is preceded by a hydrophobic stretch common to all P-450 cytochromes which in P-450$_{cam}$ contacts both the substrate and heme. Of particular interest is the Gly-248–Gly-249 sequence in P-450$_{cam}$ which is either Gly–Gly or Ala–Gly in eukaryotic P-450. The distal helix is distorted in this region such that Gly-248 does not participate in a normal helical hydrogen bonding pattern but, instead, the Gly-248 carbonyl oxygen atom accepts a hydrogen bond from the side chain oxygen atom of Thr-252. Both a small residue, Gly or Ala, and a distortion of the helix are required here owing to the close approach of residue 248 to the heme distal surface. In addition, Thr at position 252 may be important even though a Ser could satisfy the same hydrogen bonding requirements be-

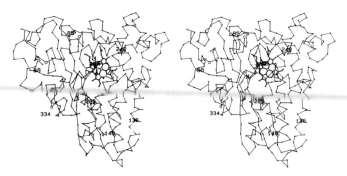

FIGURE 6. α-Carbon backbone model of P-450$_{cam}$ showing the location of all eight cysteine residues.

cause the Thr-252 side chain methyl group provides an additional point of hydrophobic contact with the heme and, in P-450$_{cam}$, with the substrate.

These sequence and probable structural homologies coupled with the suicide substrate inhibition work of Kunze et al.[22] described in Section 3.4, provide strong support for the view that the environment and topography in the vicinity of the O$_2$ binding site is very similar in all P-450 cytochromes.

Precisely why there is homology in the Cys-136$_{cam}$ peptide is not clear. As shown in Fig. 6, Cys-136 is well removed from the heme and cannot play a direct role in catalysis. That the mitochondrial P-450 has no sequence related to the Cys-136$_{cam}$ indicates that this region of P-450 does not play a critical role in P-450 function.

4.2. Three-Dimensional Structures

How similar are the structures of eukaryotic and prokaryotic P-450 cytochromes? At present the answer to this question is not know, but the available evidence suggests that all P-450 cytochromes share a common structure. Variations in surface loops, especially at the N-terminus, account for the membrane binding properties of eukaryotic P-450. Alterations in the size, shape, and accessibility to the substrate pocket account for the variety of substrates accepted by different P-450 cytochromes. Even interactions with specific electron transfer partners could be explained by small modulations in the steric and electrostatic properties on the enzyme's surface. Such variability, while seemingly large, could be accommodated in a large protein like P-450 without significantly altering the archetypal structure which we believe P-450$_{cam}$ represents.

Support for this view stems from the many functional and molecular properties shared by all P-450 cytochromes discussed in earlier chapters. Furthermore, as outlined in the preceding section, the available sequence data coupled with the X-ray structure strongly implicate a conserved three-dimensional arrangement of residues on the proximal heme surface and, possibly, on the distal surface. Sequence alignments, however, are not altogether straightforward since several gaps must be introduced into the P-450$_{cam}$ sequence for maximal alignments since eukaryotic P-450 cytochromes are larger. Nevertheless, all such gaps occur in regions where insertions into the P-450$_{cam}$ structure could be tolerated without grossly affecting the basic structure, namely, on the surface.

Finally, it would be most unusual for enzymes with such strikingly similar physical and functional properties to exhibit very different three-dimensional structures. Indeed, one of the most important principles which has emerged from protein crystallography is that there is a much stronger tendency to conserve tertiary rather than primary structure even over a wide phylogenetic range. A good example is dihydrofolate reductase where the bacterial and the larger eukaryotic structures are essentially the same.[28] Other examples include catalase,[12] glyceraldehyde phosphate dehydrogenase,[29,30] and lactate dehydrogenase.[31,32] We expect the same conservation to be maintained in P-450.

If one then accepts the postulate that a common three-dimensional structure is maintained by all P-450 cytochromes, an important challenge is to understand what changes are required in order to introduce substrate specificity or variability into the basic design. It is a difficult problem because unlike the immunoglobins, the substrate binding domain in P-450 is not confined to a "variable" region concentrated in one section of the polypeptide chain but is instead spread out throughout the polypeptide chain. Therefore, alterations in substrate specificity require many sequence changes throughout the molecule. Hopefully, the future will witness additional P-450 structures. Until then, a careful analysis of P-450 sequences with widely different substrate specificities may assist in answering this question.

5. Mechanism

In considering the P-450-catalyzed O_2 activation reaction, a good place to start is with peroxidases because (1) P-450 is often considered a peroxidase and (2) much more is known about how peroxidases work. Of interest is the question of whether the O–O bond cleaves heterolytically or homolytically. Peroxidases cleave the ROOH bond by employing a heterolytic mechanism exclusively, thereby forming species resembling RO^- and ^+OH in the transition state. The crystal structure of CCP has revealed how transition state stabilization is achieved. An invariant histidine and arginine are ideally positioned in the CCP active site for acid/base catalysis and charge-stabilized assisted heterolytic rupture of the RO–OH bond.[15,33–35] Once the RO–OH bond is broken, an oxygen atom with only six valence electrons is bound to the iron atom. This species rearranges to give Fe(IV)–OH and a protein-centered radical, while in the plant peroxidases the radical resides on the heme ring.[36–38] An obvious question is why other hemeproteins such as the globins exhibit poor peroxidase activity. Comparison of various crystal structures provides the answer. The globins have no residue analogous to the peroxidase arginine nor is the globin distal histidine suitably oriented to function as an acid/base catalyst as it is in peroxidase. As a result, the globins cannot stabilize the separation of charges required in a heterolytic mechanism. Instead, the globins probably cleave peroxides to generate oxygen radicals which destroy the protein.[39]

It is now generally accepted that molecular oxygen is first reduced to the peroxide level in P-450 and that the peroxide O–O bond is cleaved. Therefore, we might expect mechanistic parallels between peroxidase and P-450 and look for a P-450-catalyzed heterolytic cleavage of the peroxide bond. There are, however, certain problems with this view. First, the environment surrounding the O_2 binding site in P-450$_{cam}$ is more similar to the globins than it is to CCP or catalase in that the P-450$_{cam}$ pocket is more hydrophobic. This difference is not too unexpected since a nonpolar environment is more conducive to O_2 binding than the more polar peroxidase active site. Second, and most importantly, there are no residues in the P-450$_{cam}$ active site analogous to the peroxidase histidine or arginine capable of promoting a heterolytic process. In fact, with the exception of Thr-252, the O_2 binding site is surrounded by hydrophobic groups including the camphor. If we assume that heterolysis requires acid/base catalysis and/or charge stabilization especially when charge separation occurs in the protein's interior, and since P-450$_{cam}$ has no apparent means of doing so, it is possible that P-450$_{cam}$ operates by a homolytic process. Evidence in support of a homolytic process has been obtained primarily in the laboratories of Coon[40,41] and White.[42] These authors conclude that the catalytic "push" required to achieve homolysis is provided by the thiolate ligand from which an electron is transferred to the peroxide, giving a sulfur-centered radical:

$$S^- - Fe^{3+} - O1 - O2 \rightarrow \dot{S} - Fe^{3+} - O1H^- + \cdot O2H \qquad (1)$$

While the crystal structure cannot reveal if a sulfur-centered radical is reasonable, it is worthwhile to note that Cys-357 resides in a protected pocket (see Fig. 3).

The structure provides two additional limitations on proposed mechanisms of O_2 activation. Acylation of an active site carboxylate[43] or the C-terminus carboxylate of putidaredoxin by the iron-bound peroxide intermediate have been postulated.[44] Both possiblities appear unlikely since P-450$_{cam}$ has no active site glutamate or aspartate nor does it appear structurally feasible for the C-terminus end of putidaredoxin to penetrate deep into the hydrophobic substrate pocket. Alternatively, activation of an amide group to give an amidyl radical has been proposed.[41] This, too, is unlikely since P-450$_{cam}$ has no side chain amide group near the O_2 binding site which then would require the unlikely participation of a peptide amido nitrogen. On the other hand, a homolytic process would generate sufficiently active species to carry out the required hydrogen abstraction and hydroxylation reactions. In addition, free ·OH as the hydrogen-abstracting species offers an explanation for why camphor hydroxylation proceeds in two steps, one nonstereoselective and one stereoselective. In an elegant set of experiments, Gelb et al.[45] found that either the camphor C5 *exo* or *endo* hydrogen atoms are removed while hydroxylation occurs only at the C5 *exo* position. As these authors suggest, one possible explanation is that the hydrogen-abstracting and hydroxylating species are different. Homolysis releases ·O2H (reaction 1) which is now free to interact with either the C5 *endo* or *exo* hydrogen atoms while the iron-bound "activated" O1 atom contacts C5 only at the *exo* face. One serious objection in postulating the release of ·OH, however, is that the reactivity of such a radical is impossible to control and is perfectly capable of attacking protein groups as well as substrate molecules. Nevertheless, the reaction could proceed in a single step where O2 abstracts a hydrogen atom as the O1–O2 bond undergoes homolytic cleavage, thereby avoiding the release of ·OH. If such a concerted process is to function, then we must also invoke a rather rigid transition state where both O1 and O2 contact C5 of the camphor simultaneously. Simple model building experiments where an O_2 molecule is positioned into the active site indicate that the stereochemistry for such a transition state is quite good. In constructing a hypothetical P-450–camphor–O_2 complex, we used the oxymyoglobin[46] and model heme oxy complexes[47,48] as a guide. The Fe–O1 bond length was fixed at 2.0 Å and the Fe–O1–O2 bond angle at 120°. The only degree of freedom allowed is the position of O2 by rotation about the Fe–O1 bond. There are a limited number of possibilities for the location of O2 and the most sterically accessible is the pocket between Thr-252 and the C5 camphor carbon atom which is just large enough for an O_2 molecule (see Fig. 2). The upper diagram in Fig. 4 shows more clearly the open pocket just above the iron atom between the camphor molecule and distal helix. It is likely that O2 contacts both C5 *exo* and *endo* hydrogen atoms while O1 contacts only the *exo* face of C5.

While the preceding discussion supports a homolytic mechanism, the available data are also consistent with a heterolytic, oxene mechanism. For example, product analysis indicates that the hydrogen-abstracting reaction is relatively insensitive to the oxidant used whether it be the full NADH/O_2 system or iodosobenzene.[43] Since iodosobenzene presumably forms an oxene-type intermediate, one could argue that a single iron-bound oxygen atom carries out both the hydrogen abstraction and oxygen insertion steps. Therefore, we cannot choose at present between either a homolytic or a heterolytic mechanism.

ACKNOWLEDGMENTS. Work on the P-450$_{cam}$ X-ray structure was initiated while I was a member of Professor J. Kraut's group in the Chemistry Department at the University of California, San Diego, and was supported there by NSF Grant PCM79-14595 and NIH Grant RR-00757. Purified P-450$_{cam}$ was supplied by Professor I. C. Gunsalus's group at the University of Illinois. I also wish to thank members of the Protein Engineering Department at Genex and, in particular, Dr. Barry Finzel whose expertise in crystallographic refinement proved invaluable and Dr. Robert Ladner for implementation of graphics and plotting software.

References

1. Yu, C.-A., Gunsalus, I. C., Katagiri, M., Suhara, K., and Takemori, S., 1974, Cytochrome P-450$_{cam}$: Crystallization and properties, *J. Biol. Chem.* **249**:94–101.
2. Poulos, T. L., Perez, M., and Wagner, G. C., 1982, Preliminary crystallographic studies on cytochrome P-450$_{cam}$, *J. Biol. Chem.* **257**:10427–10429.
3. Lipscomb, J. D., Harrison, J. E., Dus, K. M., and Gunsalus, I. C., 1978, Cytochrome P-450$_{cam}$: Ss-dimer and -SH derivative reactivities, *Biochem. Biophys. Res. Commun.* **83**:771–778.
4. Finzel, B., 1983, Crystallographic refinement of cytochrome c peroxidase at 1.7 Å resolution, Ph.D. thesis, University of California, San Diego.
5. Haniu, M., Armes, L. G., Yasunobu, K. T., Shastry, B. A., and Gunsalus, I. C., 1982, Amino acid sequence of the *Pseudomonas putida* cytochrome P450, *J Biol. Chem.* **257**:12664–12671.
6. Hendrickson, W. A., and Konnert, J. H., 1980, Stereochemically restrained crystallographic least-squares refinement of macromolecular structures, in:*Computing in Crystallography* (R. Diamond, S. Ramseshan, and K. Venkatesan, eds.), Indian Institute of Science, Bangalore, pp. 1301–1323.
7. Devaney, P., Wagner, G. C., Debrunner, P. G., and Gunsalus, I. C., 1980, Single crystal ESR of cytochrone P-450$_{cam}$ from *Pseudomonas putida, Fed. Proc.* **39**:1139.
8. Gunsalus, I. C., Wagner, G. C., and Debrunner, P. G., 1980, Probes of cytochrome P450 structure, in:*Microsomes, Drug Oxidation and Chemical Carcinogenesis* (M. J. Coon, A. H. Conney, R. W. Estabrook, H. V. Gelboin, J. R. Gillette, and P. J. O'Brien, eds.), Academic Press, New York, pp. 233–242.

9. Richardson, J., 1981, The anatomy and taxonomy of protein structure, *Adv. Protein Chem.* **34**:167–339.
10. Ortiz de Montellano, P. R., Kunze, K. L., and Beilan, H. S., 1983, Chiral orientation of prosthetic heme in the cytochrome P450 active site, *J. Biol. Chem.* **258**:45–47.
11. Perutz, M. F., Muirhead, H., Cox, J. M., and Guaman, L. G., 1968, Three-dimensional Fourier synthesis of horse oxyhaemoglobin at 2.8 Å resolution: The atomic model, *Nature* **219**:131–139.
12. Murthy, M. R. N., Reid, T. J., Sicignano, A., Tanaka, N., and Rossmann, M. G., 1981, Structure of beef liver catalase, *J. Mol. Biol.* **152**:465–499.
13. Poulos, T. L., Freer, S. T., Alden, R. A., Xuong, N. H., Edwards, S. L., Hamlin, R. C., and Kraut, J., 1978, Crystallographic determination of the heme orientation and location of the cyanide binding site in cytochrome c peroxidase, *J. Biol. Chem.* **257**:10427–10429.
14. Salemme, F. R., 1977, Structure and function of cytochrome c, *Annu. Rev. Biochem.* **46**:299–329.
15. Poulos, T. L., and Finzel, B. C., 1984, Heme enzyme structure and function, in:*Peptide and Protein Reviews*, Volume 4 (M. T. W. Hearn, ed.), Dekker, New York, pp. 115–171.
16. Asakura, T., and Yonetani, T., 1969, Studies on cytochrome c peroxidase: Recombination of apoenzyme with protoheme dialkyl esters and etioheme, *J. Biol. Chem.* **244**:4573–4579.
17. Bosshard, H. R., Banziger, J., Hasler, T., and Poulos, T. L., 1984, The ctyochrome c peroxidase–cytochrome c electron transfer complex: The role of histidine residues, *J. Biol. Chem.* **259**:5683–5690.
18. Champion, P. M., Gunsalus, I. C., and Wagner, G. C., 1978, Resonance Raman investigations of cytochrome P-450$_{cam}$ from *Pseudomonas putida*, *J. Am. Chem. Soc.* **100**:3743–3751.
19. Champion, P. M., Stallard, B. R., Wagner, G. C., and Gunsalus, I. C., 1982, Resonance Raman detection of an Fe–S bond in cytochrome P-450$_{cam}$, *J. Am. Chem. Soc.* **104**:5469–5472.
20. Valentine, J. S., Sheridan, R. P., Allen, L. C., and Kahn, P. L., 1979, Coupling between oxidation state and hydrogen bond conformation in heme proteins, *Proc. Natl. Acad. Sci. USA* **76**:1009–1013.
21. Dawson, J. H., Andersson, L. A., and Sono, M., 1982, Spectroscopic investigations of cytochrome P-450$_{cam}$–ligand complexes: Identification of the ligand trans to cysteinate in the native enzyme, *J. Biol. Chem.* **257**:3606–3617.
22. Kunze, K. L., Mangold, B. L. K., Wheeler, C., Beilan, H. S., and Ortiz de Montellano, P. R., 1983, The cytochrome P450 active site: Regiospecificity of prosthetic heme alkylation by olefins and acetylenes, *J. Biol, Chem.* **258**:4202–4207.
23. Fujii-Kuriyama, Y., Mizumaki, Y., Kawajiri, K., Sugawa, K., and Muramatsu, M. C., 1982, Primary structure of a cytochrome P450: Coding nucleotide sequence of phenobarbital inducible cytochrome P450 cDNA from rat liver, *Proc. Natl. Acad. Sci. USA* **79**:2793–2797.
24. Heinemann, F. S., and Ozols, J., 1982, The covalent structure of rabbit phenobarbital-induced cytochrome P450: Partial amino acid sequence and order of cyanogen bromide peptides, *J. Biol. Chem.* **257**:14988–14999.
25. Black, S. D., Tarr, G. E., and Coon, M. J., 1982, Structural features of isozyme 2 of liver microsomal cytochrome P450, *J. Biol. Chem.* **257**:14616–14619.
26. Tarr, G. E., Black, S. D., Fujita, V. S., and Coon, M. J., 1983, Complete amino acid sequence and predicted membrane topology of phenobarbital induced cytochrome P450 (isozyme 2) from rabbit liver microsomes, *Proc. Natl. Acad. Sci. USA* **80**:6552–6556.

27. Morohashi, K., Fujii-Kuriyama, Y., Okada, Y., Sogawa, K., Hirose, T., and Inayama, S., 1984, Molecular cloning and nucleotide sequence of cDNA for mRNA of mitochondrial cytochrome P450 (SCC) of bovine adrenal cortex, *Proc. Natl. Acad. Sci. USA* **81**:4647–4651.
28. Volz, K. W., Matthews, D. A., Alden, L. A., Freer, S. T., Hansch, C., Kaufman, B. T., and Kraut, J., 1982, Crystal structure of avian dihydrofolate reductase containing phenyltriazine and NADPH, *J. Biol. Chem.* **257**:2528–2536.
29. Buehner, M., Ford, G. C., Olsen, K. W., and Rossmann, M. G., 1974, Three dimensional structure of D-glyceraldehyde-3-phosphate dehydrogenase, *J. Mol. Biol.* **90**:25–49.
30. Blesecker, G., Harris, J. I., Theiry, J. C., Walker, J. E., and Wonacott, A. J., 1977, Sequence and structure of D-glyceraldehyde-3-phosphate dehydrogenase from *Bacillus stearothermophilus*, *Nature* **266**:328–333.
31. Eventoff, W., Rossmann, M. G., Taylor, S. S., Torf, H. J., Meyer, H., Keil, W., and Kiltz, H. H., 1977, Structural adaptions of lactate dehydrogenase isozymes, *Proc. Natl. Acad. Sci. USA* **74**:2677–2681.
32. Rossman, M. G., 1983, Structure–function relationships of NAD-dependent dehydrogenases, in: *Biological Oxidations* (H. Sund and V. Ullrich, eds.) Springer-Verlag, Berlin, pp. 34–54.
33. Poulos, T. L., and Kraut, J., 1980, The stereochemistry of peroxidase catalysis, *J. Biol. Chem.* **255**:8199–8205.
34. Jones, P., and Suggett, A., 1968, The catalase–hydrogen peroxide system, *Biochem. J.* **110**:621–629.
35. Poulos, T. L., 1982, The peroxidase mechanism and the structure of cytochrome c peroxidase, in: *Molecular Structure and Biological Activity* (J. E. Griffin and W. L. Duax, eds.), Elsevier, Amsterdam, pp. 79–90.
36. Ariaso, T., Miyoshi, K., and Yamazaki, I., 1976, Mechanisms of electron transfer from sulfite to horseradish peroxidase-hydroperoxide compounds, *Biochemistry* **15**:3059–3063.
37. Dolphin, D., Forman, A., Borg, D. C., Fajer, J., and Felton, R. H., 1971, Compounds I of catalase and horse radish peroxidase: π-Cation radicals, *Proc. Natl. Acad. Sci. USA* **68**:614–618.
38. Aasa, R., Vanngard, T., and Dunford, H. B., 1975, EPR studies of compound I of horseradish peroxidase, *Biochim. Biophys. Acta* **391**:259–264.
39. King, N. K., and Winfield, M. E., 1963, The mechanisms of myoglobin oxidation, *J. Biol. Chem.* **238**:1520–1528.
40. White, R. E., Sligar, S. G., and Coon, M. J., 1980, Evidence for a homolytic mechanism of peroxide oxygen–oxygen bond cleavage during substrate hydroxylation by cytochrome P450, *J. Biol. Chem.* **255**:1108–1111.
41. White, R. E., and Coon, M. J., 1980, Oxygen activation by cytochrome P450, *Annu. Rev. Biochem.* **49**:315–356.
42. McCarthy, M. B., and White, R. E., 1983, Functional differences between compound I and the cytochrome P450 reactive oxygen intermediate, *J. Biol. Chem.* **258**:9153–9158.
43. Hamilton, G., 1974, Chemical models and mechanisms for oxygenases, in: *Molecular Mechanisms of Oxygen Activation* (O. Hayashi ed.), Academic Press, New York, pp. 405–451.
44. Sligar, S. G., Shastry, B. S., and Gunsalus, I. C., 1977, Oxygen reactions of the P450 heme protein, in: *Microsomes and Drug Oxidation* (V. Ullrich, I. Routs, A. Hildebrandt, R. W. Estabrook, and A. H. Cooney, eds.), Pergamon Press, Elmsford, N.Y., pp. 202–209.
45. Gleb, M. H., Heimbrook, D. C., Malkonen, P., and Sligar, S. G., 1982, Stereochemistry

and deuterium isotope effects in camphor hydroxylation by the cytochrome P-450$_{cam}$ monoxygenase system, *Biochemistry* **21**:370–377.
46. Phillips, S. E., 1980, Structure and refinement of oxymyoglobin at 1.6 Å resolution, *J. Mol. Biol.* **142**:531–554.
47. Jameson, G. B., Molinaro, F. S., Ibers, J. A., Collman, J. P., Brauman, J. I., Rose, E., and Suslick, K. S., 1978, Structural changes upon oxygenation of an iron (II) (porphyrinato) (imidazole) complex, *J. Am. Chem. Soc.* **100**:6769–6770.
48. Jameson, G. B., Rodley, G. A., Robinson, W. T., Gagne, R. R., Reed, C. A., and Collman, J. R., 1978, Structure of a dioxygen adduct of (1-methylimidazole)-*meso*-tetrakis (α,α,α,α-o-pivalamidophenyl) porphinatoiron (II): An iron dioxygen model for the heme component of oxymyoglobin, *Inorg. Chem.* **17**:850–857.

APPENDIX

Rat Hepatic Cytochrome P-450
Comparative Study of Multiple Isozymic Forms

DAVID J. WAXMAN

1. Multiplicity of P-450 Isozymes

The occurrence of multiple cytochrome P-450 isozymes in mammalian hepatic tissue and their contribution to the capacity of liver for oxidative metabolism of structurally diverse xenobiotics and endogenous substrates has been the subject of a number of excellent reviews.[e.g., 8,9,20,31] In the past few years, there has been a significant increase in the number of new P-450 forms or isozymes* described, with an even greater complexity predicted by recent studies on the molecular biology of P-450 (see Ref. 1 for review). To date, the most detailed biochemical studies of hepatic microsomal P-450's have been carried out in the rabbit and in the rat. The structural information presently available for this isozyme family has been reviewed by Black and Coon (this volume) and will not be reiterated here. Multiple rabbit hepatic P-450 isozymes, designated 1, 2, 3a, 3b, 3c, 4, 5, and 6 on the basis of their relative mobility on SDS gels† have been purified and characterized in detail by several laboratories and appear fairly well defined (Refs. 9, 14 and references therein). By contrast, many laboratories have actively participated in the isolation and biochemical

---

* The terms *P-450 form* and *P-450 isozyme* are used interchangeably. Rat hepatic P-450's are identified in this review by the designations adopted by this laboratory and, in the case of forms f, g, and UT-H, by the designations of Levin and Guengerich, respectively (see Table I).

† All known mammalian P-450's exhibit apparent molecular weights of 48,000–60,000 when analyzed on SDS–polyacrylamide gels.

DAVID J. WAXMAN • Department of Biological Chemistry and Dana–Farber Cancer Institute, Harvard Medical School, Boston, Massachusetts 02115.

characterization of rat hepatic P-450's, and it has become increasingly difficult to assess the relationships between rat P-450 preparations described by different laboratories. These difficulties stem in part from the large number of individual rat P-450's under study—at least 13, by our count (Table I); by the complication of strain differences, already established for two rat hepatic P-450's[41]; and by the absence of a generally accepted nomenclature for the rat P-450 isozymes. The approach taken in the designation of rabbit hepatic P-450's does not work well for the rat proteins since significant mobility differences are seen when corresponding preparations are analyzed in different laboratories or even within the same laboratory using reagents for electrophoresis obtained from different suppliers. Similarly, the λ_{max} of the Fe^{2+}-CO complex, although characteristic of each individual isozyme, does not identify it in a unique and unambiguous fashion. The aim of this review is not, however, to introduce a universally acceptable nomenclature for rat hepatic P-450's but is, rather, to summarize briefly and in a comparative manner the biochemical and immunochemical properties of rat hepatic P-450's in the hope of facilitating the comparison and identification of equivalent or corresponding forms described by different laboratories.

2. Comparison of P-450 Forms Isolated in Different Laboratories

In Table I, ten P-450 isozymes studied by this laboratory are compared to a similar number of forms isolated by Levin's and Guengerich's groups. Also shown are equivalent forms purified by investigators in several other laboratories as well as the designations given to analogous P-450 forms isolated from rabbit liver. The identification here of corresponding or equivalent rat P-450 forms is made by comparison of at least three or four of the following properties for each P-450 preparation: λ_{max} of the Fe^{2+}-CO complex; catalytic specificity in a reconstituted system; immunochemical properties; N-terminal sequence*; mobility on SDS gels relative to well-defined P-450 isozymes; age and sex specificity of expression; and response to characteristic monooxygenase inducers. As summarized in Table I, these analyses indicate that at least 13 distinct P-450's are expressed in and can be purified from rat hepatic tissue.

Additional rat hepatic P-450 preparations which are distinguishable from those listed in Table I include the following:

1. P-452.[50] This clofibrate-induced form exhibits high lauric acid hy-

---

* N-terminal sequences for various rat, rabbit, and other P-450's can be found in Black and Coon (this volume).

TABLE I
Rat Hepatic P-450's: Designations Used by Different Laboratories for Corresponding Forms

Waxman et al.[53,54,56,a]	Levin et al.[43,46,b]	Guengerich et al.[21,29,c]	Fe^{2+}-CO λ_{max}	Designations for equivalent rat P-450's	Analogous rabbit P-450's[d]
PB-1	—	PB-C	450	—	—[e]
PB-2a	—	PCN-E	449–450	P-450p[15,62]	3c
2c	h	UT-A	451	RLM5,[6,f] P-450-male[26,g]	—
2d	i	UT-I	449	P-450-female,[26] P-450 15β[32]	—
3	a	UT-F	452	—	2[j]
PB-4	b[h]	PB-B	450 }[i]	P-450,[34,50] fraction C,[59] PB-1,[28] P-450 PB[19]	—
PB-5	e	PB-D	450		—
6	—	—	450	—	—
BNF-B	c	BNF-B	447	P-448,[34] MC-1,[28] P-448 MC,[19] MC-I/MC-II,[48,k] MC-2,[60] P-447[50]	6
ISF-G	d	ISF-G	447	MC-2,[28] P-448 HCB,[19] MC-1,[60] ISF-P-450,[16] form 5[30]	4
—	f	—	447.5	—	—
—	g	—	447.5	—	—
—	—	UT-H	449	—	—

[a] P-450 forms 1–6 are numbered in their relative order of elution from Whatman DEAE–cellulose (DE52), with the prefix PB indicating significant induction of the paticular isozyme by phenobarbital. The two major polycyclic hydrocarbon-induced forms, BNF-B and ISF-G, are designated according to the nomenclature of Ref. 21.
[b] Individual P-450's are designated by nondescriptive letters (a, b, c, and so on) assigned sequentially as new isozymes are purified and characterized by these investigators.
[c] P-450's are assigned arbitrary letters (A, B, C, and so on) prefixed by UT, PB, PCN, BNF, or ISF to indicate whether the isozyme is more readily obtained from untreated rats or from rats pretreated with the indicated inducing agents. This does not imply, however, that the agent designated is necessarily the most efficacious inducer of the isozyme in question.
[d] Analogous forms are identified on the basis of primary structure comparisons (Black and Coon, this volume) and on the basis of their similar response to monooxygenase inducers. Rat hepatic P-450's corresponding to rabbit forms 1, 3a, 3b, and 5 cannot be identified at this time.
[e] Analogous rabbit forms have not been described.
[f] Note, however, that the N-terminal sequence reported by this group differs from that reported by Waxman[53] and by Haniu et al.[22] at residues 12 and 13.
[g] This identification should be viewed as tentative in the absence of information on the steroid hydroxylase activities of P-450-male.
[h] Strain difference characterize this major phenobarbital-induced form[41] such that P-450b, isolated from Long–Evans rats, can be distinguished from P-450 PB-4 (or P-450 PB-B), isolated from Sprague–Dawley rats.[44]
[i] It is unclear whether the forms studied by these investigators correspond to P-450 PB-4, to P-450 PB-5, or to a mixture.
[j] Although the primary structure of rabbit form 2 is 77% identical to rat forms PB-4 and PB-5, it exhibits the high activity characteristic of form PB-4 rather than the low activity associated with PB-5.
[k] These two forms can be distinguished by HPLC and yet exhibit indistinguishable catalytic properties and N-terminal sequences. Both appear equivalent to P-450 BNF-B.

droxylase activity (turnover = 43 min^{-1} P-450^{-1}). Although spectrally similar to forms 2c and 3, it appears distinct insofar as it does not catalyze formation of the major monohydroxy testosterone metabolites characteristic of these two P-450's (Table II).

2. P-450 PB-2.[28] Several properties reported for this form (including its two- to fourfold induction by phenobarbital) are similar to those of P-450 PB-1 (Table I). The N-terminal sequences of the two preparations do, however, differ at residues 5 and 6.

3. P-450 RLM3.[6,7] Although some of the properties of this isozyme appear similar to those of P-450g (Table I), the two preparations differ in N-terminal sequence at residue 10 and in the λ_{max} of the Fe^{2+}–CO complex.

4. Other forms. Insufficient information is available for evaluation of the relationship between the P-450 forms included in Table I and the P-450 preparations given the following designations: P-450 and P-451,[3] forms 1, 2, 3, and 4,[30] P-450 MC-III,[48] and PB-1 and PB-2.[60]

3. Influence of Induction Status and Sex on Isozyme Expression

Although at least 13 distinct P-450's are thus expressible in rat hepatic tissue, the levels of individual isozymes are strikingly dependent on the animal's age, sex, and history of exposure to foreign compounds which can serve as monooxygenase inducers. Immunoquantitation of P-450 isozyme levels in uninduced and in variously induced rats has been performed by a number of investigators[e.g., 18,21,25,40,51] and has been useful in identifying effective inducers for 7 of the 13 rat P-450 forms (Table II). These studies highlight the complexity of P-450 induction in rat liver as summarized below:

1. Although many monooxygenase inducers increase by 2- to 3-fold the total P-450/mg protein (P-450 specific content), there is often a differential response of individual isozymes to the same inducing agent. Thus, phenobarbital (PB) administration increases PB-4 ≥ 30-fold and PB-1 ~ 2- to 4-fold while, at the same time, it decreases P-450 2c levels an estimated 20–40%.

2. Structurally unrelated compounds can effectively induce the same P-450 isozyme. Thus, both PB and Aroclor 1254 are effective inducers of form PB-4.

3. The induction of each isozyme appears to be regulated independently. Thus, although PB induces both PB-1 and PB-2a severalfold, only PB-2a is induced upon 16α-cyanopregnenolone (PCN) administration.

TABLE II
Rat Hepatic P-450's: Monooxygenase Induction and Sex Specificity[a]

P-450 form	Typical inducers[b]	Fold induction	Sex specificity[c]
PB-1	PB	2–4	—[d]
PB-2a	PB, PCN, DEX	2–5	♂[e]
2c	—[f]	S[g]	♂
2d	—	—	♀
3	BNF, PB	~2	♀/♂ ~ 1.4
PB-4	PB, ACLR	≥30[h]	—
PB-5	PB, ACLR	≥20	—
6	—	—	—
BNF-B	BNF, MC, ACLR	≥30	—
ISF-G	ISF, BNF, ACLR	≥20	—
f	—	—	—
g	—	—	♂
UT-H	—	—	—

[a] Data obtained from the references indicated in Table I and in the text.
[b] Included are representative examples of agents commonly used to induce the indicated P-450's. PB, phenobarbital; PCN, 16α-cyanopregnenolone; DEX, dexamethasone; BNF, β-naphthoflavone; ACLR, Aroclor 1254, MC, 3-methylcholanthrene; ISF, isosafrole.
[c] Isozymes that are present at ≥10-fold higher levels in adult rats of one sex as compared to the other are termed sex specific.
[d] These P-450's are expressed at similar levels in adult rats of both sexes.
[e] The male-specific expression of this P-450 is abolished upon xenobiotic administration (see text and Ref. 57).
[f] No inducing agents have been described for these P-450's
[g] Expression of form 2c is suppressed by a variety of monooxygenase inducers. In the case of some polybrominated biphenyls, this results in a 90% decrease in immunoreactive 2c.[12]
[h] Minimum values due to the difficulty in obtaining accurate estimates of the low levels of these forms in uninduced rat liver.

4. The response of P-450 forms PB-1, 3, PB-4/PB-5, BNF-B, and ISF-G to various inducing agents is qualitatively and quantitatively equivalent in female and male rats. One notable exception is the ≥ 70-fold increase in form PB-2a upon administration of PCN to adult female rats. This striking induction contrasts with a 4- to 5-fold increase obtained in adult male rats and effectively abolishes the sex-specific expression of P-450 PB-2a.[57]

Immunochemical and catalytic studies on the sex specificity of P-450 isozyme expression (Table II) have established that P-450 2c is male-specific and undergoes a developmental induction at puberty.[33,53] P-450 2d is developmentally induced in female rats although it is also expressed at significant levels in immature males at 3–4 weeks of age.[33,57] The male specificity of P-450 PB-2a largely reflects its developmental suppression in female rats.[57] Each of these three P-450's mediates a characteristic microsomal steroid hydroxylase activity (Table III) and is regulated by neonatal androgen secretions. Purification studies suggest that P-450g is

TABLE III
Rat Hepatic P-450's: Isozyme-Specific Monooxygenase Activities

P-450 form	Characteristic activity	Turnover number[a] (min^{-1} P-450^{-1})	Substrate concentration (μM)	Catalytic activity exhibited by other P-450's[b] %	Catalytic activity exhibited by other P-450's[b] P-450 form	References
PB-1	S-Warfarin 7-hydroxylase	1.2	300	20	BNF-B	21, 55,[h]
PB-2a	Androstenedione 6β-hydroxylase	0.1–0.5[c]	25	(variable)[d]	several	57
2c	Testosterone 2α-hydroxylase	3.5	25	≤2	—[e]	7, 46, 56
2d	Steroid disulfate 15β-hydroxylase	2.0	25	≤2	—	32, 47
3	Testosterone 7α-hydroxylase	17.6	25	≤2	—	56, 61
PB-4	Androstenedione 16β-hydroxylase	12	25	≤2	—	56, 61
6	Testosterone hydroxylase[f]	0.5	25	≤10	—	[h]
BNF-B	7-Ethoxyresorufin O-deethylase	6.9	20	8	ISF-G	21, 27
UT-H	Debrisoquine 4-hydroxylase	3.5	1000	5	2c	29
PB-5	—[g]					
ISF-G	—					
f	—					
g	—					

[a] Typical values obtained at the substrate concentrations indicated. Decreased isozymic specificity may occur at higher substrate concentration.
[b] Ratio of catalytic activities exhibited by other P-450's (identified in this column) to those exhibited by the P-450 form identified in the first column, expressed as a percentage.
[c] Range of activities observed for various preparations. Purified isozyme appears generally inactive.
[d] Although several rat hepatic P-450's catalyze androstenedione 6β-hydroxylation in purified, reconstituted systems, antibody inhibition experiments have established that, irrespective of its low 6β-hydroxylase activity in purified systems (footnote c), P-450 PB-2a is the major isozymic contributor to androstenedione 6β-hydroxylase activity both in uninduced and in variously induced rat hepatic microsomes.[57]
[e] Catalytic activities are at or below the limits of detection for all other P-450's examined.
[f] Unique monohydroxy testosterone metabolite not identified.
[g] Although the catalytic specificities of these forms are distinct when examined using a series of monooxygenase substrates, isozyme-specific catalytic reactions have not been reported for any of these forms. P-450 PB-5 exhibits a substrate specificity profile which is quite similar to that of PB-4; its catalytic activity is, however, only 5–15% that of P-450 PB-4.[45,54]
[h] D. S. Waxman, unpublished results.

also a male-specific P-450 isozyme.[46] Although not sex-specific, P-450 3 and its corresponding microsomal testosterone 7α-hydroxylase activity are expressed at ~ 40% higher levels in adult females as compared to adult males.[57] By contrast, no significant sex differences characterize eight other P-450 forms (Table II).

Taken together, these findings demonstrate that the isozymic composition of rat hepatic tissue is markedly dependent on the animals' sex and induction status and, as such, these factors should be considered when undertaking purification of specific P-450 forms. Thus, PB-induced rat liver is a preferred tissue for purification of forms PB-1, PB-4, and PB-5, PCN-induced adult male liver is preferred as the source for PB-2a, uninduced adult male liver for forms 2c and g, uninduced adult female liver for form 2d, and so on (Table II).

4. Isozyme-Specific Monooxygenase Substrates

The broad and overlapping substrate specificities characteristic of mammalian liver microsomal P-450's contributes in a significant way to the difficulties encountered when attempting to identify corresponding P-450 preparations studied in different laboratories. Catalytic studies of P-450's using isozyme-specific* monooxygenase substrates would provide an effective approach to the unambiguous identification of corresponding P-450 forms. Such substrates have now been reported for nine rat hepatic P-450's (Table III). In several cases, it has been shown that the majority (≥ 80%) of the corresponding microsomal activity can be inhibited by antibodies monospecific for the appropriate P-450 form. Thus, 7α-hydroxytestosterone corresponds to a P-450 3-specific microsomal metabolite of testosterone and 16β-hydroxyandrostenedione to a P-450 PB-4-specific microsomal metabolite of androstenedione.[57] The majority of the isozyme-specific substrates identified thus far are steroid hormones, consistent with the hypothesis that these compounds serve as physiological substrates for mammalian liver P-450's.

5. Identification of Immunochemically Related P-450's

Immunochemical studies have proven useful for the identification of P-450 isozymes exhibiting close structural homologies. Thus, PB-4 and

---

* Although the monooxygenase reactions detailed in Table III appear specific for the P-450 forms studied thus far, the possibility that other forms (not yet isolated) are also active in these hydroxylations cannot be excluded.

TABLE IV
Immunochemically Cross-Reactive Rat P-450 Isozymes

A. Cross-reactivities identified using heterosera

Cross-reactive P-450 forms	References
PB-1, 2d	53
2c, 2d	26, 53
PB-4, PB-5	21, 45, 54
PB-4, 6	b
PB-4, f	46
BNF-B, ISF-G	19, 21, 42

B. Cross-reactivities identified using monoclonal antibodies

P-450 immunogen	Cross-reactive P-450 forms	Frequency[a]	Reference
PB-2a	PB-4, PB-5	3/11	39
PB-4	3	1/4	58
PB-4	PB-5	4/4	58
BNF-B	ISF-G	3/9	52
BNF-B	ISF-G	1/2	58

[a] Number of monoclonal antibodies which exhibit cross-reactivity to the P-450 forms listed in column 2 relative to the total number of monoclonal antibodies examined.
[b] D. J. Waxman, unpublished results.

PB-5 are immunochemically indistinguishable (Table IV) and exhibit 97% amino acid sequence identity.[2,17,63] P-450's BNF-B and ISF-G exhibit some shared immunochemical determinants as demonstrated using heterosera and, more recently, using monoclonal antibodies (Table IV). These two isozymes are now known to exhibit 68% overall amino acid sequence homology, with much greater homologies found along localized segments of the polypeptide chain.[49] High degrees of amino acid sequence homology probably also account for the cross-reactivities among other P-450 forms which can be detected using heterosera (Table IVA). By contrast, cross-reactivity detected in studies using limited numbers of monoclonal antibodies (e.g., the cross-reactivity of PB-4 and PB-5 with PB-2a and that of PB-4 with form 3; Table IVB) may reflect the presence of a limited number of shared epitopes rather than extensive sequence homology.

6. Biosynthetic P-450 Isozymes from Hepatic Tissue

In addition to the P-450's active in foreign compound metabolism described above, several distinct P-450's present in mammalian hepatic

TABLE V
Biosynthetic P-450 Isozymes Isolated from Hepatic Microsomes

	Species	Specific content (nmole P-450/mg protein)	Turnover number (min^{-1} P-450^{-1})	Fold purification/P-450	Ref.
Vitamin D$_3$ 25-hydroxylase	Rat	15.6	0.34	168 (17)[a]	4
	Rat	15.1	0.15	253 (25)[a]	24
Cholesterol 7α-hydroxylase	Rat	14.5	0.03[b]	0.7[c]	23
	Rabbit[d]	10.8	0.04	NR[e]	5
7α-Hydroxy-4-cholestene-3-one 12α-hydroxylase	Rabbit	9.0	11.9	59.5	36
Taurodeoxycholate 7α-hydroxylase	Rat	13.5	4.2	10.5	35

[a] Corrected values (in parentheses) take into account the ≥10-fold activation of microsomal D$_3$ 25-hydroxylase which occurs upon detergent solubilization, possibly the consequence of removal of an uncharacterized microsomal inhibitor.
[b] Values as high as 0.3 min^{-1} P-450^{-1} have recently been reported upon inclusion of cytosolic activators in the reconstituted system.[11]
[c] Turnover number actually decreases during purification.
[d] Preparation appears to correspond to rabbit P-450 form 4 containing cholesterol 7α-hydroxylase as a minor isozymic contaminant.
[e] NR, not reported.

microsomes have been identified on the basis of specific biosynthetic reactions which they catalyze. These include:

1. Vitamin D_3 25-hydroxylase, which generates 25-hydroxyvitamin D_3, the obligatory substrate for the kidney mitochondrial 1α-hydroxylase P-450 which, in turn, catalyzes formation of the physiologically active hormone, 1α,25-dihydroxyvitamin D_3[13]
2. Cholesterol 7α-hydroxylase, which catalyzes the first and rate-limited step in the conversion of cholesterol to bile acids[17,18]
3. 7α-Hydroxy-4-cholesten-3-one 12α-hydroxylase, which helps regulate the cholic acid/chenodeoxycholic acid ratio in bile
4. Taurodeoxycholate 7α-hydroxylase, involved in rehydroxylation of taurocholate during its enterohepatic circulation[10]

P-450 preparations active in each of these biosynthetic reactions have been obtained from either rat or rabbit liver (Table V). The 12α-hydroxylase and the taurodeoxycholate 7α-hydroxylase P-450's both exhibit respectable turnover numbers in reconstituted systems, suggesting that they are of a moderate to high degree of isozymic purity. By contrast, the low activities reported for the vitamin D_3 25-hydroxylase and cholesterol 7α-hydroxylase preparations suggest that these biosynthetic P-450's may be only minor isozymic constituents in the samples analyzed. Thus, although the physiologically active vitamin D_3 25-hydroxylase does not exhibit the age and sex dependence characteristic of P-450 2c (Table II), the vitamin D_3 25-hydroxylase preparations described (Table V) correspond to P-450 2c in N-terminal sequence and in their catalytic, spectral, and chromatographic properties.

Important, unresolved questions regarding these biosynthetic liver P-450's include the following:

1. Are the biosynthetic P-450's of liver endoplasmic reticulum of broad or of narrow substrate specificity? Do they metabolize xenobiotics?
2. Do xenobiotic-metabolizing P-450's contribute to these biosynthetic hydroxylations?
3. Can foreign compounds perturb the metabolic pathways for, e.g., vitamin D_3 activation or for bile acid biosynthesis by either induction or suppression of such biosynthetic P-450's?
4. Do any of the P-450 forms included in Table I correspond to these or to other biosynthetic P-450's?

7. Summary

Studies on the isolation and characterization of rat hepatic P-450's active in both foreign compound and endogenous substrate metabolism have been ongoing in many laboratories for the past several years. In this review, corresponding P-450 preparations reported by various investigators have been identified and their most characteristic properties summarized. It is proposed that all future characterizations of new hepatic P-450 preparations include, at a minimum, a determination of each of the following parameters to facilitate the unambiguous identification of corresponding P-450 forms: (1) λ_{max} of Fe^{2+}-CO complex, (2) monooxygenase activity with isozyme-specific substrates, (3) response to classic P-450 inducers, and (4) N-terminal sequence. Finally, it is suggested that these same properties be determined in studies of biosynthetic P-450's to help clarify their relationship to xenobiotic-metabolizing P-450's found in mammalian hepatic tissue.

ACKNOWLEDGMENTS. Studies carried out in the author's laboratory were supported in part by Grants AM-33765 from the National Institutes of Health and BC-462 from the American Cancer Society.

References

1. Adesnik, M., and Atchison, M., 1985, Genes for cytochrome P-450 and their regulation, *Crit. Rev. Biochem.* in press.
2. Atchison, M., and Adesnik, M., 1983, A cytochrome P-450 multigene family: Characterization of a gene activated by phenobarbital administration, *J. Biol. Chem.* **258**:11285–11295.
3. Agosin, M., Morello, A., White, R., Repetto, Y., and Pedemonte, J., 1979, Multiple forms of noninduced rat liver cytochrome P-450: Metabolism of 1-(4'-ethylphenoxyl)-3,17-dimethyl-6,7-epoxy-*trans*-2-octene by reconstituted preparations, *J. Biol. Chem.* **254**:9915–9920.
4. Andersson, S., Holmberg, I., and Wikvall, K., 1983, 25-Hydroxylation of C27-steroids and vitamin D_3 by a constitutive cytochrome P-450 from rat liver microsomes, *J. Biol. Chem.* **258**:6777–6781.
5. Bostrom, H., and Wikvall, K., 1982, Hydroxylations in biosynthesis of bile acids: Isolation of subfractions with different substrate specificity from cytochrome P-450 LM4, *J. Biol. Chem* **257**:11755–11759.
6. Cheng, K.-C., and Schenkman, J. B., 1982, Purification and characterization of two constitutive forms of rat liver microsomal cytochrome P-450, *J. Biol. Chem.* **257**:2378–2385.
7. Cheng, K.-C., and Schenkman, J. B., 1983, Testosterone metabolism by cytochrome P-450 isozymes RLM3 and RLM5 and by microsomes: Metabolite identification, *J. Biol. Chem.* **258**:11738–11744.
8. Conney, A. H., 1982, Induction of microsomal enzymes by foreign chemicals and carcinogenesis by polycyclic aromatic hydrocarbons: G. H. A. Clowes Memorial Lecture, *Cancer Res.* **42**:4875–4917.
9. Coon, M. J., and Koop, D. R., 1983, P-450 oxygenases in lipid transformation, *The Enzymes* **16**:645–677.
10. Danielsson, H., and Sjovall, J., 1975, Bile acid metabolism, *Annu. Rev. Biochem.* **44**:233–253.

11. Danielsson, H., and Kalles, I., and Wikvall, K., 1984, Regulation of hydroxylations in biosynthesis of bile acids: Isolation of a protein from rat liver cytosol stimulating reconstituted cholesterol 7α-hydroxylase activity, *J. Biol. Chem.* **259**:4258–4262.
12. Dannan, G. A., Guengerich, F. P., Kaminsky, L. S., and Aust, S. D., 1983, Regulation of cytochrome P-450: Immunochemical quantitation of eight isozymes in liver microsomes of rats treated with polybrominated biphenyl congeners, *J. Biol. Chem.* **258**:1282–1288.
13. DeLuca, H. F., 1981, The transformation of a vitamin into a hormone: The vitamin D story, *Harvey Lect.* **75**:333–379.
14. Dieter, H. H., and Johnson, E. F., 1982, Functional and structural polymorphism of rabbit microsomal cytochrome P-450 form 3b, *J. Biol. Chem.* **257**:9315–9323.
15. Elshourbagy, N. A., and Guzelian, P. S., 1980, Separation, purification, and characterization of a novel form of hepatic cytochrome P-450 from rats treated with pregnenolone-16α-carbonitrile, *J. Biol. Chem.* **255**:1279–1285.
16. Fisher, G. J., Fukushima, H., and Gaylor, J. L., 1981, Isolation, purification, and properties of a unique form of cytochrome P-450 in microsomes of isosafrole-treated rats, *J. Biol. Chem.* **256**:4388–4394.
17. Fujii-Kuriyama, Y., Mizukami, Y., Kawajiri, K., Sogawa, K., and Muramatsu, M., 1982, Primary structure of a cytochrome P-450: Coding nucleotide sequence of phenobarbital-inducible cytochrome P-450 cDNA from rat liver, *Proc. Natl. Acad. Sci. USA* **79**:2793–2797.
18. Goldstein, J. A., and Linko, P., 1984, Differential induction of two 2,3,7,8-tetrachlorodibenzo-p-dioxin inducible forms of cytochrome P-450 in extrahepatic versus hepatic tissues, *Mol. Pharmacol.* **25**:185–191.
19. Goldstein, J. A., Linko, P., Luster, M. I., and Sundheimer, D. W., 1982, Purification and characterization of a second form of hepatic cytochrome P-448 from rats treated with a pure polychlorinated biphenyl isomer, *J. Biol. Chem.* **257**:2702–2707.
20. Guengerich, F. P., 1979, Isolation and purification of cytochrome P-450 and the existence of multiple forms, *Pharmac. Ther.* **6**:99–121.
21. Guengerich, F. P., Dannan, G. A., Wright, S. T., Martin, M. V., and Kaminsky, L. S., 1982, Purification and characterization of rat liver microsomal cytochromes P-450: Electrophoretic, spectral, catalytic and immunochemical properties and inducibility of eight isozymes isolated from rats treated with phenobarbital or β-naphthoflavone, *Biochemistry* **21**:6019–6030.
22. Haniu, M., Ryan, D. E., Iida, S., Lieber, C. S., Levin, W., and Shively, J. E., 1984, NH_2-terminal sequence analysis of four rat hepatic microsomal cytochromes P-450, *Arch. Biochem. Biophys.* **235**:304–311.
23. Hansson, R., and Wikvall, K., 1980, Hydroxylations in biosynthesis and metabolism of bile acids: Catalytic properties of different forms of cytochrome P-450, *J. Biol. Chem.* **255**:1643–1649.
24. Hayashi, S., Noshiro, M., and Okuda, K., 1984, Purification of cytochrome P-450 catalyzing 25-hydroxylation of vitamin D_3 from rat liver microsomes, *Biochem. Biophys. Res. Commun.* **121**:994–1000.
25. Heuman, D. M., Gallagher, E. J., Barwick, J. L., Elshourbagy, N. A., and Guzelian, P. S., 1982, Immunochemical evidence for induction of a common form of hepatic cytochrome P-450 in rats treated with pregnenolone-16α-carbonitrile or other steroidal or non-steroidal agents, *Mol. Pharmacol.* **21**:753–760.
26. Kamataki, T., Maeda, K., Yamazoe, Y., Nagai, T., and Kato, R., 1983, Sex difference of cytochrome P-450 in the rat: Purification, characterization and quantitation of constitutive forms of cytochrome P-450 from liver microsomes of male and female rats, *Arch. Biochem. Biophys.* **225**:758–770.

27. Kaminsky, L. S., Guengerich, F. P., Dannan, G. A., and Aust, S. D., 1983, Comparison of warfarin metabolism by liver microsomes of rats treated with a series of polybrominated biphenyl congeners and by the component-purified cytochrome P-450 isozymes, *Arch. Biochem. Biophys.* **225**:398–404.
28. Kuwahara, S., Harada, N., Yoshioka, H., Miyata, T., and Omura, T., 1984, Purification and characterization of four forms of cytochrome P-450 from liver microsomes of phenobarbital-treated and 3-methylcholanthrene-treated rats, *J. Biochem.* **95**:703–714.
29. Larrey, D., Distlerath, L. M., Dannan, G. A., Wilkinson, G. R., and Guengerich, F. P., 1984, Purification and characterization of the rat liver microsomal cytochrome P-450 involved in the 4-hydroxylation of debrisoquine, a prototype for genetic variation in oxidative drug metabolism, *Biochemistry* **23**:2787–2795.
30. Lau, P. P., and Strobel, H. W., 1982, Multiple forms of cytochrome P-450 in liver microsomes from β-naphthoflavone-pretreated rats: Separation, purification, and characterization of five forms, *J. Biol. Chem.* **257**:5257–5262.
31. Lu, A. Y. H., and West, S. B., 1980, Multiplicity of mammalian microsomal cytochromes P-450, *Pharmacol. Rev.* **31**:277–295.
32. MacGeoch, C., Morgan, E. T., Halpert, J., and Gustafsson, J.-A., 1984, Purification, characterization, and pituitary regulation of the sex-specific cytochrome P-450 15β-hydroxylase from liver microsomes of untreated female rats, *J. Biol. Chem.* **259**:15433–15439.
33. Maeda, K., Kamataki, T., Nagai, T., and Kato, R., 1984, Postnatal development of constitutive forms of cytochrome P-450 in liver microsomes of male and female rats, *Biochem. Pharmacol.* **33**:509–512.
34. Masuda-Mikawa, R., Fujii-Kuriyama, Y., Negishi, M., and Tashiro, Y., 1979, Purification and partial characterization of hepatic microsomal P-450's from phenobarbital and 3-methylcholanthrene-treated rats, *J. Biochem.* **86**:1383–1394.
35. Murakami, K., and Okuda, K., 1981, Purification and characterization of taurodeoxycholate 7α-monooxygenase in rat liver, *J. Biol. Chem.* **256**:8658–8662.
36. Murakami, K., Okada, Y., and Okuda, K., 1982, Purification and characterization of 7α-hydroxy-4-cholesten-3-one 12α-monooxygenase, *J. Biol. Chem.* **257**:8030–8035.
37. Myant, N. B., and Mitropoulos, K. A., 1977, Cholesterol 7α-hydroxylase, *J. Lipid Res.* **18**:135–153.
38. Ozasa, S., and Boyd, G. S., 1981, Cholesterol 7α-hydroxylase of rat liver: Studies on the solubilisation, resolution and reconstitution of the enzyme complex, *Eur. J. Biochem.* **119**:263–272.
39. Park, S. S., Waxman, D. J., Miller, H., Guengerich, F. P., and Gelboin, H. V., 1985, Preparation and characterization of monoclonal antibodies to pregnenolone-16α-carbonitrile-induced rat liver cytochrome P-450, *Biochem. Pharmacol.* submitted.
40. Pickett, C. B., Jeter, R. L., Morin, J., and Lu, A. Y. H., 1981, Electroimmunochemical quantitation of cytochrome P-450, cytochrome P-448, and epoxide hydrase in rat liver microsomes, *J. Biol. Chem.* **256**:8815–8820.
41. Rampersaud, A., and Walz, F. G., Jr., 1983, At least six forms of extremely homologous cytochromes P-450 in rat liver are encoded at two closely linked genetic loci, *Proc. Natl. Acad. Sci. USA* **80**:6542–6546.
42. Reik, L. M., Levin, W., Ryan, D. E., and Thomas, P. E., 1982, Immunochemical relatedness of rat hepatic microsomal cytochromes P-450c and P-450d, *J. Biol. Chem.* **257**:3950–3957.
43. Ryan, D. E., Thomas P. E., Reik, L. M., and Levin, W., 1982, Purification, characterization and regulation of five rat hepatic microsomal cytochrome P-450 isoenzymes, *Xenobiotica* **12**:727–744.
44. Ryan, D. E., Wood, A. W., Thomas, P. E., Walz, F. G., Jr., Yuan, P.-M., Shively, J.

E., and Levin, W., 1982, Comparisons of highly purified hepatic microsomal cytochromes P-450 from Holtzman and Long-Evans rats, *Biochim. Biophys. Acta* **709**:273–283.
45. Ryan, D. E., Thomas, P. E., and Levin, W., 1982, Purification and characterization of a minor form of hepatic microsomal cytochrome P-450 from rats treated with polychlorinated biphenyls, *Arch Biochem. Biophys.* **216**:272–288.
46. Ryan, D. E., Iida, S., Wood, A. W., Thomas, P. E., Lieber, C. S., and Levin, W., 1984, Characterization of three highly purified cytochromes P-450 from hepatic microsomes of adult male rats, *J. Biol. Chem.* **259**:1239–1250.
47. Ryan, D. E., Dixon, R., Evans, R. H., Ramanathan, L., Thomas, P. E., Wood, A. W., and Levin, W., 1984, Rat hepatic cytochrome P-450 isozyme specificity for the metabolism of the steroid sulfate 5α-androstane-3α,17β-diol-3,17-disulfate, *Arch. Biochem. Biophys.* **233**:633–642.
48. Sakari, T., Soga, A., Yabusaki, Y., and Ohkawa, H., 1984, Characterization of three forms of cytochrome P-450 isolated from liver microsomes of rats treated with 3-methylcholanthrene, *J. Biochem.* **96**:117–126.
49. Sogawa, K., Gotoh, O., Kawajiri, K., and Fujii-Kuriyama, Y., 1984, Distinct organization of methylcholanthrene- and phenobarbital-inducible cytochrome P-450 genes in the rat, *Proc. Natl. Acad. Sci. USA* **81**:5066–5070.
50. Tamburini, P. P., Masson, H. A., Bains, S. K., Makowski, R. J., Morris, B., and Gibson, G. G., 1984, Multiple forms of hepatic cytochrome P-450: Purification, characterization and comparison of a novel clofibrate-induced isozyme with other major forms of cytochrome P-450, *Eur. J. Biochem.* **139**:235–246.
51. Thomas, P. E., Reik, L. M., Ryan, D. E., and Levin, W., 1981, Regulation of three forms of cytochrome P-450 and epoxide hydrolase in rat liver microsomes: Effects of age, sex, and induction, *J. Biol. Chem.* **256**:1044–1052.
52. Thomas, P. E., Reik, L. M., Ryan, D. E., and Levin, W., 1984, Characterization of nine monoclonal antibodies against rat hepatic cytochrome P-450c: Delineation of at least five spatially distinct epitopes, *J. Biol. Chem.* **259**:3890–3899.
53. Waxman, D. J., 1984, Rat hepatic cytochrome P-450 isoenzyme 2c: Identification as a male-specific, developmentally induced steroid 16α-hydroxylase and companion to a female-specific cytochrome P-450 isoenzyme, *J. Biol. Chem.* **259**:15481–15490.
54. Waxman, D. J., and Walsh, C., 1982, Phenobarbital-induced rat liver cytochrome P-450: Purification and characterization of two closely related isozymic forms, *J. Biol. Chem.* **257**:10446–10457.
55. Waxman, D. J., and Walsh, C., 1983, Cytochrome P-450 isozyme 1 from phenobarbital-induced rat liver: Purification, characterization, and interactions with metyrapone and cytochrome b_5, *Biochemistry* **22**:4846–4855.
56. Waxman, D. J., Ko, A., and Walsh, C., 1983, Regioselectivity of androgen hydroxylations catalyzed by cytochrome P-450 isozymes purified from phenobarbital-induced rat liver, *J. Biol. Chem.* **258**:11937–11947.
57. Waxman, D. J., Dannan, G. A., and Guengerich, F. P., 1985, Regulation of rat hepatic cytochrome P-450: Age-dependent expression, hormonal imprinting and xenobiotic inducibility of sex-specific isoenzymes, *Biochemistry* **24**:4409–4417.
58. Waxman, D. J., Park, S. S., and Gelboin, H. V., 1985, Isozymic specificities of monoclonal antibodies raised to rat hepatic cytochromes P-450, manuscript in preparation.
59. West, S. B., Huang, M.-T., Miwa, G. T., and Lu, A. Y. H., 1979, A simple and rapid procedure for the purification of phenobarbital-inducible cytochrome P-450 from rat liver microsomes, *Arch. Biochem. Biophys.* **193**:42–50.
60. Wolf, C. R., Moll, E., Friedberg, T., Oesch, F., Buchmann, A., Kuhlmann, W. D., and Kunz, H. W., 1984, Characterization, localization and regulation of a novel pheno-

barbital-inducible form of cytochrome P-450 reductase, glutathione transferases and microsomal epoxide hydrolase, *Carcinogenesis* **5**:993–1001.
61. Wood, A. W., Ryan, D. E., Thomas, P. E., and Levin, W., 1983, Regio- and stereoselective metabolism of two C-19 steroids by five highly purified and reconstituted rat hepatic cytochrome P-450 isozymes, *J. Biol. Chem.* **258**:8839–8847.
62. Wrighton, S. A., Maurel, P., Schultz, E. G., Watkins, P. B., Young, B., and Guzelian, P. S., 1985, Identification of the cytochrome P-450 induced by macrolide antibiotics in rat liver as the glucocorticoid-responsive, cytochrome P-450p, *Biochemistry* **24**:2171–2178.
63. Yuan, P.-M., Ryan, D. E., Levin, W., and Shively, J. E., 1983, Identification and localization of amino acid substitutions between two phenobarbital-inducible rat hepatic microsomal cytochromes P-450 by micro sequence analyses, *Proc. Natl. Acad. Sci. USA* **80**:1169–1173.

Index

Absorption anisotropy, decay of, 100, 126, 129
ABT: *see* 1-Aminobenzotriazole
Acetanilide, binding to P-450, 79
Acetylene, 288
Acetylenes
  destruction of P-450, 287, 288, 295, 298, 300–302
  isotope effects, 288
  oxidative metabolism, 287, 288, 295
Acinetobacter calcoaceticus, 434
ACTH, 137, 348, 352–362, 367, 370, 408–411, 413
ACTH receptor, 355
Active sites
  chirality, 78, 400
  flexibility, 230, 399, 477, 514
  lipophilicity, 78, 135, 275, 280, 512, 518
  location relative to membrane surface, 135, 136, 204
  P-450$_b$, 83, 84, 512–513
  P-450$_c$, 80–83
  P-450$_{cam}$, 459–462, 509–514
  P-450$_{scc}$, 399–401
  structure, 83, 222, 226, 280, 433, 462, 478, 505–520
  topology, 77–84, 276, 288, 399, 478, 505–520
Adamantane, 8, 461, 477, 478
Adamantanone, 451, 461, 477, 478
Adenylate cyclase, 352, 353, 355, 356
Adrenal androgens, 357; *see also* Androgens
Adrenal cortex, 347, 348–353, 365
Adrenodoxin, 92, 93, 349, 354, 356, 360, 363, 369, 389–397, 401, 403, 405, 419, 432
Adrenodoxin reductase, 92, 93, 389–397

Ah receptor, 315, 318–323, 325–332, 337, 340, 349
  defects in cultured cells, 330–332
  see also Receptors
  subcellular distribution, 327–332
AIA, *see* 2-Isopropyl-4-pentenamide
Aldosterone
  antagonists, 279
  biosynthesis, 249, 350, 404
N-Alkylprotoporphyrin IX, *see* Heme adducts
Allenes, P-450 destruction by, 295, 300
Amines
  MI complex formation, 283
  oxidative metabolism, 231, 283
Amino acid composition, 171–175, 196–199
  as criterion of structural distinctness, 172, 420
Amino terminal region
  absence of microheterogeneity, 184
  by Edman degradation, 176–180
  initiator methionine, 176, 198
  isozyme comparisons, 176–180
  proline cluster, 199–200
  sequences, 162
  signal peptides, 176, 198, 199, 202
  transfer sequences, 198, 199
p-Aminobenzoic acid, oxidation by HRP, 42, 44–46
1-Aminobenzotriazole, 292
  analogues of, 292
Aminoglutethimide, 276, 299, 366, 367, 408, 409, 413
Aminopyrine, 50, 103, 104, 227, 230, 231, 301
Aminosterols as topological probes, 399
1-Amino-2,2,6,6-tetramethylpiperidine, 284
2′-AMP as P-450 reductase inhibitor, 105

541

Amphetamine, 78, 297
Androgens, 364, 365, 388
Androst-4-ene-3,17-dione, 122, 137, 227, 251, 252, 366, 413, 431, 530
Angiotensin, stimulation of 11β-hydroxylase by, 404
Aniline, 51, 106, 227, 228, 231, 431
Anisic acid, 443
Anisole, 238
Antibodies, 252, 404, 531, 532
 cross reactivity, 167, 168, 316, 441, 531, 532
 to cytochrome b₅, 105–106
 monoclonal, 167–168, 367, 532
 to P-450, 167–168, 252, 367, 531–532
 to P-450 reductase, 105–106, 140, 413
Antifungal agents, 301
Arachidonic acid
 in cooxidation reactions, 53–55
 in membranes, 119
 oxidation of, 372
Arginine residues
 P-450, 515
 peroxidases, 65, 515
Aromatase, 228, 346, 364–367, 388
 inhibition of, 254, 297, 299–301, 367, 415
 mechanism, 251–253, 413–415
 purification, 251, 414
 spin state, 401
Aryl hydrocarbon hydroxylation in cultured cells, 327–332, 336
 activity-deficient mutants, 330, 336
Ascorbic acid, 18, 39, 46, 64
Autoxidation, *see appropriate hemeprotein*
Axial ligands
 catalase, 30, 36
 horseradish peroxidase, 6, 30, 34, 36
 metalloporphyrin models, 11, 12, 66, 226
 P-450, 3, 189–91, 399, 401–403, 460, 511, 514
 redox potentials, relation to, 11, 12, 274
 spin state, relation to, 401, 402
Azide, 16, 222
Azurin, 39

Bacillus megaterium, 430–434
Barbiturates, 285, 315, 335, 336
 receptor for neurological effects, 335
Benzene, 227, 245, 274, 276
Benzenes, halogenated, 244

Benzo(ghi)perylene, in selection of AHH revertant cells, 336
Benzo[*a*]pyrene
 cooxidation, 53
 isolation of cells resistant to, 330
 one-electron oxidation, 63
 oxidation by P-450, 80, 106, 128, 227, 230, 231, 330
Benzo[*a*]pyrene, 7,8-dihydroxy 7,8 dihydro-
 cooxidation, 53–55
 epoxidation stereochemistry, 53
 oxidation by P-450, 57–59, 61, 63
 oxidation during lipid peroxidation, 56, 59
Benzothiadiazoles, 295
Benzphetamine, 106, 107, 121, 134–137, 220, 227, 301
Benzyne, 292
Bile acids
 biosynthesis, 249, 417, 418
 regulation, 420
 role, 417, 418
Biliverdin, formation of, 234
Biphenyl, 227, 244
Bis[2-(3,5-dichloropyridyloxy)]benzene, 335
Bisulfite, reaction with HRP, 33, 44, 45
Bleomycin, 479–484
Brain, 346
5-Bromocamphor, 448, 473
Butylated hydroxyanisole, 60–62
tert-Butylbenzene, 16
tert-Butylbicyclophosphorothionate, 335
tert-Butylhydroperoxide, 20, 43

Calcium, effects on
 cholesterol side chain cleavage, 396, 410, 411
 1α-hydroxylase, 416
Camphor, 451
 binding, 512
 induction of P-450, 437
 metabolism, 225, 232, 436, 465
 stereochemistry of hydroxylation, 4, 237, 436, 471–479, 519
Camphoroquinone, 451
Carbenes, 237, 282, 464
 complexes with prosthetic heme iron, 281, 294, 296

INDEX

N,N-bis (Carbethoxy)-2,3,-diazabicylo [2.2.0]hex-5-ene, 290, 293, 297
3,5,-bis(Carbethoxy)-2,6-dimethyl-4-alkyl-1,4-dihydropyridines
  effect on heme biosynthesis, 288
  free radical products, 246, 289
  metabolism of, 246
  P-450 destruction by, 246, 288, 289
Carbon disulfide, 277
Carbon monoxide
  P-450 inhibition, 58, 220, 221, 227, 254, 275, 349, 366, 404, 418, 434
  inhibition of substrate binding, 402, 409
  insensitivity to inhibition by, 413, 414, 420
  as ligand, 79, 275, 281
  as metabolite, 281, 282, 296
  spectrum of reduced P-450 complex, 217, 275, 281, 526
Carbon tetrachloride, 296
Carboxy terminal sequences, 180–182
Carboxypeptidases, 180–182
Cardiolipin, 131, 134, 138, 400, 405
Catalase
  axial ligands, 30, 36, 65
  catalytic cycle, 33
  catalytic residues, 65
  Compound I, 30, 33, 37, 223
  Compound II, 33
  inactivation, 290
  molecular weight, 30
  reaction with peroxides, 14, 35, 48
  spectra, 36, 37
  substrate oxidation, 33, 47, 48
Catalytic cycle, P-450, 2, 3, 90, 223, 273
  bacterial, 397, 444, 462
  microsomal, 217
  mitochondrial, 217, 397
Catalytic rates, 225, 397, 403, 406, 419, 446, 463, 473, 530, 533
cDNA, 167, 168, 316, 317, 323, 332, 334, 335, 359, 360
Cell lines, 327, 330, 336, 337, 353, 367
Cetyltrimethylammonium bromide in P-450 denaturation, 166
CHAPS, 121
Chemiluminescence, 235
Chloramphenicol, 277, 278
Chlorin, 38
Chloroaniline, 230
Chlorobenzene, 106, 244
1-Chlorooctane, 434

1-Chlorodecane, 434
1-Chlorohexadecane, 434
1-Chlorooctadecane, 434
Chloroperoxidase, 35, 483
  axial ligands, 31, 36
  Compound I, 37
  EPR, 37
  molecular weight, 31
  peroxide bond cleavage by, 14
  reactions, 33, 47, 51, 484
  spectra, 36, 37
Cholate dialysis, 121–123
Cholecalciferol, see vitamin D
Cholera toxin, 355
5β-Cholestan-3α,7α,12α-triol, 418
Cholesterol
  biosynthesis, 253–254, 301, 416–418
  esterase, 408, 411
  20-hydroperoxide, 400
  7α-hydroxylase, 249, 388, 417, 420, 533, 534
  12α-hydroxylase, 417, 418, 533
  26-hydroxylase, 388, 417, 418
  in membranes, 129, 137, 138, 410
  see also P-450, biosynthetic forms
  in regulation of steroidogenesis
    acute, 136, 137, 352, 353, 408–410
    chronic, 353
    desensitization, 356
  side chain cleavage, 137, 249–251, 254, 352–357, 388–405
  3-sulphate, metabolism of, 399
  transport, 407–408
Cholestyramine, 202, 418, 420
Cholic acid, 417
Chromatin, 316, 337, 338
Chromyl chloride, olefin epoxidation by, 18, 19
Chrysene, 80
Chymotrypsin, see Peptide maps
Citronellol, bacterial metabolism, 441
Circular dichroism, 134, 139, 456
Clofibrate, 164, 371, 526
Colloidal platinum, 18
Complementation groups, 336
Conformational changes of P-450 enzymes, 130, 134, 399, 400, 401, 405, 453, 456–462
Cooxidation
  hematin-dependent, 54–56
  hemoglobin-dependent, 56, 57
  PGH synthase-dependent, 52–54

Corpus luteum, 164, 351, 364
Corticosterone, 334, 357, 404
Corticosterone-18-hydroxylase, 350, 388
Cortisol, 275, 357
"Counterpoise regulation" of P-450 turnover, 108
p-Cresol, oxidation by HRP, 32, 44, 46, 50
Cross linking of P-450, 100, 129, 130, 139, 140
Cryogenic studies, 219, 463
Crystal structure
  cytochrome C peroxidase, 65, 509, 511
  P-450$_{cam}$, 9, 65, 77, 83, 438, 456, 505–520
Cumene hydroperoxide, 41, 49, 227–230, 248, 251, 278, 280, 301, 470
  and lipid peroxidation, 58–64, 227
  metalloporphyrin reactions, 20, 48
  transfer of oxygen to substrate, 46, 47, 227, 228
Cupric phenanthroline, 140
Cyanide
  inhibition by, 227, 228, 274, 413, 434
  as ligand, 43, 274
Cyclic AMP, 352, 353, 355–357, 362–364, 367, 370, 406, 418, 420
Cyclobutadiene, 293
1,5,9-Cyclododecatriene epoxidation, 7
Cycloheptane, 8
Cyclohexane, 228, 230, 232, 233
Cyclohexene, 4, 17, 19, 20, 225, 241
Cycloheximide, 137, 362, 408–411, 416, 418
Cyclooctene, 17
Cyclopropylamines, 248, 249, 295
Cyclopropylmethyl rearrangements, 237
p-Cymene, bacterial oxidation of, 438, 439
Cysteine, see Axial ligands, P-450
  see Sequences, primary
Cytochrome b$_5$, 101–108, 142, 222, 223
  antibodies to, 105, 106, 108
  apocytochrome, 107, 467
  changes in product ratios caused by, 413, 421
  distribution and membrane orientation, 102, 103, 126
  as effector, 141
  in electron transfer to P-450, 103–106, 128, 141
  heme substitutions, 107
  hydrophobic tail, 124, 125
  membrane interactions, 122, 128, 133

Cytochrome b$_5$ (cont.)
  obligatory requirement for, 107, 108, 125
  in P-450 affinity column, 107
  in P-450 autoxidation, 219–222, 466
  in P-450 complexes, 107, 139
  properties of, 125, 220
  roles, 102, 217, 412, 417
  spectrum, 102, 103
  structure, 102, 180
Cytochrome b$_5$ reductase, 103–108, 124, 412
Cytochrome c, acetylated, 392, 393
Cytochrome c peroxidase (*Pseudomonas aeruginosa*), 39
Cytochrome c peroxidase (yeast), 33, 35, 47, 65
  Compound I, 31, 37, 223
  protein radical, 37, 518
  reduction by NADH, 39
  spectra, 36, 37
  structure, 31, 65, 509, 511
Cytochrome c, reduction by
  adrenodoxin, 93, 391, 392
  P-450 reductase, 91
Cytochrome P-450 reductase, 91, 92, 350, 358, 366, 367, 481
  antibodies to, 91, 105, 106, 140
  autoxidation, 220
  complex formation with P-450, 98, 100, 122, 126–129, 138–143, 412
  hydrophobic tail, 91, 92, 124, 125
  membrane placement and effects, 125–130, 136
  molecular weight, 92
  one electron-reduced form, 92, 103, 104
  properties, 95–101
  ratio to P-450, 101
  reconstitution, 120–123
  redox potentials, 92
  reduction by NADH, 105
  in reduction of P-450, 92, 96–101, 124, 128
  solubilized form, 91, 92, 124
  structure, 92, 180, 198

DDC, see 3,5-bis(Carbethoxy)-2,6-dimethyl-4-alkyl-1,4-dihydropyridines
DDEP, see 3,5-*bis*(Carbethoxy)-2,6-dimethyl-4-alkyl-1,4-dihydropyridines
Debrisoquine, 530

INDEX

Deoxycholate, 136
11-Deoxycorticosterone, 333, 350, 388, 403, 404
11-Deoxycortisol, 350, 403
Detergents, 121, 133
Dexamethasone, 333, 334, 529
Dianisidine, 484
Diazenes, 292
p-Diazobenzenesulphonate, 124, 125
Dichloroethylene, 240
Diethyldithiocarbamate, 37, 277
2,2-Diethyl-4-pentenamide, 286
Differential scanning microcalorimetry, 134
Diffusion rates for
  cytochrome b_5, 128
  microsomal lipids, 127
  P-450, 127–130
  P-450 complexes, 140
  proteins, 126
Dihydroquinolines, P-450 destruction by, 289
(20R,22R)Dihydroxycholesterol, 397–399, 402
Dilaurylphosphatidylcholine, 49, 120, 121, 139–142
N,N-Dimethylaniline, 32, 43, 48, 107, 227, 247, 248, 484
2,6-Dimethylaniline, 79
Dimethylaniline N-oxides as oxygen donors, 233, 484
Dimethyl-3,3'-dithiobis(propionimidate), 139
Dimyristoylphosphatidylcholine, 123, 127, 133
Dioxygen bond cleavage, see Oxygen activation
Dioxymethylene, see Methylenedioxo
1,6-Diphenyl-1,3,5-hexatriene, 136
Diphenylisobenzofuran, cooxidation of, 53
[1,1-^{2}H]-1,3-Diphenylpropane, 238
Dithionite, reduction of P-450 by, 283, 447
DNA degradation, 479
DNA sequences, 183, 332, 439, 507
11-Dodecynoic acid, 301

Effectors
  of microsomal P-450 enzymes, 141, 143
  of P-450$_{cam}$, 464, 466–468
Electron transfer
  see Ionic strength effects on electron transfer
  mitochondrial shuttle, 93, 393–397

Electron transfer (cont.)
  to P-450 from substrates, 235
    heteroatomic substrates, 245–257, 285, 289, 290, 293, 294
    from radical intermediates, 251, 294
    saturated hydrocarbons, 238
    unsaturated hydrocarbons, 241–243
  to peroxidases, 46, 49, 235, 248, 511
  among reductases, 89–111, 91–93, 390–392
  from reductases to P-450's, 89–111, 128, 389, 406, 448, 449, 453–455, 466, 511
Electrophoresis, see Polyacrylamide gel electrophoresis
Ellipticines, 276, 297
Emulgens, 121, 141
Endopeptidases, 181, 201; see also Peptide maps
Endoplasmic reticulum, 91, 119, 369
  membrane charge, 119
  membrane fluidity, 120
  phospholipid composition, 119–120
  protease treatment, 124
ENDOR, 6
Enterohepatic circulation, 418
Eosin-labeled P-450, 130
Epoxidation
  see Heme adducts
  isotope effects, 239, 240
  mechanism, 56, 65, 240–243, 286
  model systems, 7, 239
  non-P-450 hemeproteins, 53, 54, 56, 59
  olefins, 237, 238–243
  see Oxidation, acetylenes
  see Oxidation, aromatic rings
  see P-450 destruction
  rearrangements, 238, 240, 241
  stereochemistry, 237, 239, 286, 476
Epoxide hydrolase, 80, 132, 140, 198
EPR, 230, 247
  of adrenodoxin reductase, 350
  of bacterial P-450's, 431, 442, 443
  of bleomycin, 479, 480, 482
  of chloroperoxidase, 37
  of horseradish peroxidase, 34, 35
  of 11β-hydroxylase, 350
  of metalloporphyrins, 6, 13, 15, 16
  of microsomal P-450's, 95, 126, 129
  of P-450$_{scc}$, 126, 350, 398, 399, 405, 409
  of spin-labeled lipids, 127, 400
  of spin-trapped radicals, 246

Estradiol
  biosynthesis, 251–253, 364–367, 413–415
  metabolism, 227
Estrogens
  biosynthesis, 364–365, 413–415
  receptor, 316
Ethanol
  as inducer, 163
  as substrate, 227
7-Ethoxycoumarin, 4, 80, 82, 105, 136, 137, 219, 238, 247
7-Ethoxyresorufin, 530
Ethylbenzene, oxidation stereochemistry, 236, 237
Etyhlene, 18, 285, 298
Ethylhydroperoxide, 43, 48
Ethylisocyanide, 79
Ethylmorphine, 105, 227, 421
1-Ethynylpyrene, 297
EXAFS, 6, 13

FAD, 89, 90, 92, 389
Farnesol, bacterial metabolism, 441
Fatty acids
  bacterial hydroxylation, 301, 432
  desaturation, role of cytochrome b_5, 102, 105
  ω and ω-1 hydroxylations, 301, 421
  ω-2 hydroxylations, 301, 432
  as inducers, 433
  β-oxidation, 302, 417
  see also Phospholipids
  spin-labeled, 127
Fenchone, binding to P-450$_{cam}$, 449
Ferredoxins, 389, 391, 418
Ferricyanide, 92, 284, 389, 392, 448
Ferrocyanide, 35, 42, 44–46
Ferryl intermediate
  see appropriate hemeprotein, Compound I
  see Metalloporphyrins
  see Oxygen, activated
Flavonoids as effectors, 143
Flavoprotein, mitochondrial, reduction of, 389, 390
Fluorescence
  -activated cell sorter, 330, 336
  delayed, 130
  energy transfer, 204
  polarization, 136
  recovery after bleaching, 127

Fluoride, as heme ligand, 43
Fluroxene, 286, 287
FMN, 89, 92

Gene regulation, 315, 316, 337–339
Geraniol metabolism, 440, 441
Glucocorticoids, 333, 334, 351, 352, 357, 364
Glucose repression of induction, 435, 437–440
Glutathione
  P-450 activation by, 420
  P-450 protection by, 289
Glutethimide, 297
α-Glycerolphosphate, 400
Glycosylation, 175, 176, 200, 403
Green pigments, see Heme adducts
Guaiacol, oxidation by HRP, 46

Halocarbons, 218, 295, 296, 300
Halothane, 296
Hematin, 48, 54, 62, 227, 230
Hematinic acid, 234
Heme
  accessibility, 204, 232, 511
  adducts, 240, 242, 246, 285–295
  alkylation
    regiochemistry, 83, 84
    stereochemistry, 83, 84, 286–288, 513
  coordination, 274, 280, 283, 284, 301
  degradation, 14, 231, 234, 273, 277, 279, 295–298
  ligands, 186, 189–191, 232, 511
  orientation in hemeproteins, 125, 510, 511
  replacement by analogues in hemeproteins, 107, 465, 466
  thiolate ligand, 6, 189–191, 223, 224, 226, 256, 274, 275, 509
Hemoglobin, 219, 290, 509
  autoxidation, 222
  catalysis of oxygenations, 56, 223, 248
  reaction with peroxides, 47, 55, 518
Hepatocytes, in induction studies, 333, 337
Hexadecane, 435
15-Hexadecynoic acid, 302
3-Hexene, 285
Hexobarbital, 221
Histidine residues, 65, 78, 186, 390
HMG-CoA reductase, 417
Homology, see Sequence

Horseradish peroxidase
  axial ligand, 6, 30, 34, 36
  catalytic cycle, 4, 29, 32, 45
  catalytic residues, 43, 44, 65
  Compound I, 4, 6, 13, 30, 33, 34, 37, 44, 223
  Compound II, 6, 33, 35, 44, 228
  EPR, 35
  hydroperoxide specificity, 34
  inactivation, 42
  iron-oxygen bond length, 34, 35
  kinetics, 35, 42–46
  ligand complexes, 43
  molecular weight, 30
  NMR, 6
  peroxide bond cleavage mechanism, 14
  pH dependence, 42–45
  redox potential, 35, 52
  spectra, 34, 35, 36
  substrate binding, 43
  substrate oxidation, 32, 43, 44–47, 49, 50
  substrate oxygenation, 33, 51, 248
Hybridization, Northern, 317, 324, 332, 360
Hydrazines
  complexes with hemeproteins, 284
  heme destruction by, 290, 293
  radical formation from, 290
Hydrocarbons, P-450 induction in bacteria, 434, 435
Hydrocarbons, polycyclic aromatic, 78–81
  see Aryl hydrocarbon hydroxylation in cultured cells
  cell lines resistant to, 330, 336
  cell mutation by photoactivation of, 336
  metabolism by cultured cells, 327, 332
Hydrocortisone, 334
Hydrodisulfide protein adducts, 277
Hydrogen peroxide, 301
  as product, 4, 108, 220–222, 227, 234, 463
  as substrate, 35, 227, 231, 464
Hydroperoxides, 295
  alkyl, 20, 43, 48, 228, 230, 234
  fatty acid, 52–55, 226, 227, 230, 247
  see Oxygen donors
  sterol, 227, 228, 250, 301, 400
Hydrophobic protein regions
  cytochrome b_5, 125
  P-450, 125, 274
  P-450 reductase, 125

Hydrophobicity profile, 202–204
  homology, 203
  prediction of tertiary structure, 203, 204
  substrate binding site, 204
N-Hydroxy-2-acetylaminofluorene, 230, 247
4-Hydroxy-4-androstene-3,17-dione, 300
19-Hydroxy-4-androstene-3,17-dione, 413, 414
5-Hydroxycamphor, 436
7α-hydroxy-4-cholesten-3-one, 533
7α-Hydroxycholesterol, 420
20α-Hydroxycholesterol, 399, 401
(22R)Hydroxycholesterol, 397–399, 401, 402
25-Hydroxycholesterol, 417
18-Hydroxycorticosterone, 404
Hydroxylamines, 223, 230, 246
25-Hydroxylanosterol, 417
11β-Hydroxylase, see P-450, biosynthetic forms
17α-Hydroxylase/C$_{17-20}$ lyase, see P-450, biosynthetic forms
Hydroxylation
  carbon, 4, 46, 227, 230, 235–238, 471
  isotope effects, 4, 230, 236–238, 473
  ^{18}O-studies, 229
  see Oxidation, aromatic rings
  rearrangements, 4, 5, 237, 238
  regiochemistry, 79, 232, 404
  specificity, 230, 238, 400, 404
  stereochemistry, 78, 237
    camphor, 4, 237, 471–473, 476, 477, 519
    cyclohexenes, 4, 236, 237
    ethyl benzene, 235, 237
    fatty acids, 236
    norbornane, 4, 236, 477
    octane, 236
    steroids, 235, 249, 251, 252, 431, 476, 477
Hydroxynorcocaine, 246, 247
3-Hydroxy-1-phenyl pyrazole, 50
17α-Hydroxyprogesterone, 136, 388, 412
11β-Hydroxyprogesterone, 334
9-Hydroxystearate, 432, 433
7-Hydroxytestosterone, 530
16α-Hydroxytestosterone, 415
Hypercortisolism, 276
Hypochlorite, in metalloporphyrin oxidations, 17
Hypoxia, effect on P-450$_{scc}$, 409

Imidazoles, 62, 79, 224, 275, 276, 301
4'-Imidazolylyacetophenone, 17
Immunochemical relatedness, 167–168
Immunoisolation, 354
Immunoprecipitation, 316
Immunoquantitation, 528
In vitro translation, 316, 317
Inducers, 163, 315–340
  arochlor 1254, 528, 529
  chlorinated biphenyls, 318
  1,1-di(*p*-chlorophenyl)-2,2-dichloroethylene, 170
  cholestyramine, 202
  clofibrate, 164, 371, 526
  ethanol, 163, 170
  glucocorticoids, 333, 334
  imidazole, 170
  isosafrole, 163, 527, 529
  isozyme expression, 528–532
  3-MC, 163, 316–332, 527, 529
  PCN, 332–334, 528, 529
  phenobarbital, 170, 332, 334–337, 528
  pregnancy, 162, 371
  sex differences, 529, 531, 532
  TAO, 163
  TCDD, 318–332,
  TCPOBOP, 335
Inhibitors, 273–303
  irreversible, 273, 285
  lipophilicity of, 275, 276, 281
  mechanism-based, 277, 285–302, 513
  reversible, 273, 274, 283, 297
    amines, 283, 297, 299, 301, 413
    oxygen functionalities, 275
Inositol-1,4-diphosphate, 411
Intestine, 162, 346
Iodide, reaction with HRP, 33, 44–46, 51
Iodobenzene, oxidation of, 244
Iodosobenzene, *see* Oxygen donors
Ionic strength effects
  on electron transfer, 94, 95, 143, 220, 390, 392, 393, 395–397
  on solubilization of Ah receptor complexes, 329
Iron–sulfur proteins, 90, 92, 217, 389, 434, 443
Isocapraldehyde, 398
Isoniazid, 284
2-Isopropyl-4-pentenamide, 285, 286, 297
Isosafrole, 280, 420
Isothiocyanates, 277

Isotope effects
  inactivation of P-450 by acetylenes, 288
  independence from phospholipid composition, 137
  intramolecular, 4, 10, 237, 238, 243, 247, 248, 252, 473–476
  intrinsic, 4, 81, 238, 247
  kinetic (intermolecular),
    π-bond oxidations, 239, 240, 242–244, 288
    dealkylation reactions, 247, 248, 282
    hydrocarbon hydroxylations, 230, 236–238, 252, 473–476
  NADPH versus NADPD, 389, 392
Isozyme expression, *see* Inducers
Isozyme-specific monooxygenase activities, 530, 531

Jerusalem artichoke, 433
Juvenile hormone biosynthesis, 302

5-Ketocamphor, 436
Ketoconazole, 301
Ketolactonase I, 436
Kidney, 162, 163, 346, 356, 368–371, 415, 416, 418, 419,

Lactoperoxidase, 30, 33, 35, 38, 223
  protein radical, 38, 223
Lanosterol demethylation, 253–254, 301, 388, 416, 417, 420
Lauric acid hydroxylation, 227, 228, 236, 301, 302
  bacterial, 432–434
  clofibrate induction, 164, 371, 526, 528
Leukotriene B_4 hydroxylation, 302
Limonene, 438, 440
Linalool metabolism, 440
Linear free energy relationships
  HRP oxidation of substituted sulfides, 51
  P-450 activation by cumene hydroperoxides, 41, 470
  P-450 binding of alkylbenzenes, 79
  P-450 inactivation by cyclopropylamines, 249
  P-450 oxidation of substituted benzenes, 244, 245
  P-450 oxidation of substituted sulfides, 248
Linredoxin, 440
Linredoxin reductase, 440

INDEX

Lipid peroxidation
  effect of NADPH, 227
  inhibition of, 62, 227
  role of P-450, 58, 227
Lipoamide, 467
Lipoic acid, 225, 467
Lipophilicity, see Active sites
  see Inhibitors
Liposomes, 121, 123, 125, 132, 135
Luminol, oxidation by HRP, 46
Lung, 162, 163, 371
Lysine modification
  acylation by chloramphenicol metabolite, 278
  in LM_2, 143

Magnetic circular dichroism, 6, 37, 140, 141
Mechanism-based inhibitors, see Inhibitors
Megaredoxin, 431, 432
Megaredoxin reductase, 431, 432
Membrane
  see Endoplastic reticulum
  fluidity, 120, 135
  permeability, 301
  in regulation of P-450$_{scc}$, 137, 136
  in substrate binding, 135–138, 405
  topology, 204
Menadione, 418
Menthene, 438
Mersalyl, inhibition of P-450 reductase, 97
Messenger RNA, 316, 317, 322, 324, 327, 329, 332, 334, 354, 357, 359–363
Metabolic switching, 4, 81, 82
Metallacyclobutane intermediates
  metalloporphyrin models, 17, 18
  olefin and acetylene metabolism, 242, 243, 287
Metalloporphyrins
  activated oxygen complexes, 11–13, 16, 17, 21, 35, 37, 468
    iron–oxygen bond distance, 13
    oxygen exchange, 13, 14
  aryl hydroxylation, 10
  axial ligand transfer, 16
  carbon hydroxylations, 7–10, 15, 465, 472
  chiral oxidations by, 7
  frontier orbitals, 11, 36
  model complexes, 21, 275, 284, 292, 294, 296

Metalloporphyrins (cont.)
  μ-nitrido complexes, 12
  NMR, 7, 11–13, 284
  olefin
    activation, 19, 20
    epoxidation, 7–10, 13, 17, 19, 20, 66, 476
  oxidation potential, 11
  oxidations, 12, 47
  oxidative halogenations, 8, 9
  μ-oxo complexes, 7, 12
  P-450 models, 6–8, 223–225, 232, 243, 287, 472
  peroxide bond cleavage by, 14, 15, 21, 65, 66, 470
  porphyrin radical cations, 7, 9, 10–13, 37
Methamphetamine, 227
4-Methoxybenzoate, 443
4-Methoxy-3-hydroxybenzoate, 443
p-Methylanisole, 10
3-Methylcholanthrene as inducer, 77, 315–332, 529
  of mitochondrial P-450 enzymes, 419
Methylenecyclohexane, 237
Methylenedioxo, 280, 281
2-Methyl-1-heptene, 285
p-Methylthioanisole, 232
Methylvinylmaleimide, 234
Metyrapone, 58, 275, 276, 403, 404
MI Complexes, 283
Micelles, 120, 121; see also Liposomes
Miconazole, 301
Mineral corticoids, 351, 364
Mitochondrial P-450's, 353–357, 388
  induction by xenobiotics, 419
  P-450$_{scc}$, see Cholesterol, side chain cleavage
  P-450$_{11\beta}$, see P-450, biosynthetic forms
  regulation of electron transport to, 389–397, 407
Monoamine oxidase, 249
Monoclonal antibodies, 167, 168, 531, 532
Mössbauer spectroscopy, 6, 11–13, 35, 219, 284, 455, 479, 481
Myeloperoxidase, 30, 33, 35, 38, 39
Myoglobin, 509
  autoxidation, 222
  catalysis of oxygenations, 56, 223, 230
  ligands, 274, 291, 509
  prosthetic heme destruction, 290, 291
  reaction with peroxides, 14, 35, 41, 518

Myoglobin (*cont.*)
  inactivation, 48
  rate, 40
  spectra, 40
  structure of oxidized myoglobin, 40
  substrate oxidation, 47

NADH synergism, 103–108
  pH profile of, 107–108
NADH cytochrome b_5 reductase, *see* Cytochrome b_5 reductase
NADPH cytochrome P-450 reductase, *see* Cytochrome P-450 reductase
β-Naphthoflavone, 420, 527
Neuraminidase, 403
NIH shift, 10, 228, 243, 245
Nitrenes, 284, 297, 464
Nitrite, oxidation by HRP, 44–46
p-Nitroanisole, 221, 228
Nitrogen
  dealkylation, 10, 32, 247, 248
  oxidation, 245
Nitroso complexes, *see* MI Complexes
NMR, protein, 6, 79, 232
Nonane, bacterial P-450 induction, 435
Non-ionic detergents, 121
Norandrostenedione, 366, 413
Norbornadiene, 238
Norbornane, 4, 236, 238, 465, 472, 477
Norcamphor, 451, 478
Norcarane, 472
19-Nortestosterone, 366
N-oxides, *see* Oxygen donors
Nuclease sensitivity of hemoglobin genes, 338, 339

17-Octadecynoic acid, 302
Octane, 434
[1-^{3}H,^{2}H,^{1}H]-Octane, 236
Octanol–water partition coefficient, 135
Octene, 83, 236
trans-[1-^{2}H]-1-Octene, 239, 286
n-Octylglucoside, 121, 122, 139
1-Octyne, 83, 288
Olefins
  see Epoxidation
  P-450 destruction by, 285–287, 293
Oleic acid, epoxidation stereochemistry, 239
Oligosaccharides, N-linked, 175, 176
Ovary, 346, 351, 363, 364 413–415

β-Oxidation, 302, 417
Oxidation, acetylenes, 241–243
  hydrogen rearrangement, 241–243
  isotope effect, 242, 288
Oxidation, aromatic rings, 78, 80
  epoxide intermediates 243–245
  isotope effects, 243–244
  see NIH shift
  substituent effects, 244, 245
19-Oxoandrostenedione, 413, 414
Oxygen
  binding to P-450, 218, 273, 402, 407
  low tension of, 218
  rate of consumption, 4
Oxygen, activated, 465, 468
  electronic structure of, 3, 4, 9, 13, 36, 37, 217, 223, 224
  exchange with oxygen from the medium, 232, 233, 465
  see Metalloporphyrins, activated oxygen complexes
  reaction selectivity, 8, 52, 230, 238, 242, 472, 477
  redox potential, 52
Oxygen activation, 219
  iron cycloperoxide, 225, 226
  oxygen acylation, 14, 15, 224, 267–468, 519
  oxygen–oxygen bond cleavage,
    metalloporphyrins, 12–14, 54–55, 65
    myoglobin, 40, 41, 55, 56
    P-450, 217, 222–226, 229, 256, 468–471, 519, 520
    peroxidases, 14, 43, 65–67, 226, 518
Oxygen donors, 3, 223, 473
  iodosobenzene, 7, 231–234, 251, 278, 465, 473, 481
  N-oxides, 10, 233
  peracids, 41, 42, 228–230, 247, 265, 473
  perchlorate, 231
  periodate, 231, 432
  peroxides, 14, 35, 41, 226–231, 251, 278, 280, 400, 473
Oxygen, origin of in products formed by
  metalloporphyrins, 54, 55
  myoglobin, 56
  P-450 (alternate oxygen donors), 52, 64, 228–233
  P-450 (reducing cofactor), 225, 252, 286, 414, 434, 467, 476
PGH synthase, 51, 53, 54

INDEX

551

Oxygen "rebound" mechanism, 9, 465
Oxygen-dealkylation, mechanism of, 4, 80, 238, 247

P-420, 46, 135
P-450
  aggregates, 122, 129, 130, 138–142, 166
  see Amino acid composition
  autoxidation of, 219–222, 234, 463–466
  see Axial ligands
  destruction, 290, 296
    acetylenes, 83, 242, 243, 287, 288, 295, 298, 513
    cyclopropylamines, 295
    diazoalkanes, 294
    halocarbons, 277, 278, 296
    heterocycles, 288–290, 292–294
    olefins, 78, 83, 240–243, 285–287, 513
    partition ratio of, 287
    peroxides, 62, 231, 295, 296
    sulfur compounds, 232, 277, 278
  dioxygen complex, 90, 94, 103, 107, 108, 218, 221, 249, 402, 462–464, 479, 481
  distribution, 163–165, 346, 388
  equivalent forms, 162, 526–528, 531, 532
  integration of exogenous enzyme into membrane, 128
  kinetics, 225, 228, 395, 463; see also Catalytic rates
  membrane interactions, 131, 136
  molecular weights, 161–166, 196–198, 431, 432, 442, 443
  nomenclature, 161, 162, 166, 275, 525–528
  orientation in the membrane, 91, 124–126, 204
  as peroxidase, 41, 49, 59, 60, 62, 227, 230
  peroxide complexes, 41, 42, 228–229, 400, 465, 468–471, 481
  as peroxygenase, 58, 62, 67, 400, 467
  proteolytic treatment, 201, 403
  see Reconstitution of P-450 in vesicles
  reduced-CO complex, 96, 129, 163–166, 217, 275, 281, 527
  regio- and stereospecificity, 387
  sex differences, 528–532
  see Spin state
  strain variants, 170, 201–202, 421, 527
P-450, bacterial, 171, 179, 429–520

P-450$_{cam}$, 165, 237, 435–438
  catalytic residues, 466, 467
  crystal structure, 9, 83, 505–520
  DNA sequence, 439, 507
  effectors, 464, 466–468
  Mn-PPIX-reconstituted, 465, 466
  oxygen activation, 218–221, 225, 232
  primary structure, 173, 175–180, 190
  reaction with peroxides, 42, 65, 228–230
  redox potentials, 95
  reduction, 95, 96, 349, 443–456
  secondary structure, 202
  tryptophan fluorescence, 466
  tyrosine role in substrate binding, 478
P-450$_{cym}$, 438–440
P-450$_{8-lin}$, 440
P-450$_{10-lin}$, 440
P-450$_{meg}$, 165, 173, 430
P-450$_{non}$, 434
P-450$_{npd}$, 443
P-450$_{oct}$, 434
P-450$_{Rhizobium}$, 165, 173, 441, 442
P-450$_{\omega-2}$, 432
P-450, biosynthetic forms, 177, 179, 297, 343, 345, 532–534
  see Aromatase
  cholesterol 7α-hydroxylase, 176, 249, 308, 346, 417–420, 533
  cholesterol side chain cleavage
    aggregation, 139
    distribution and location, 346, 349, 351, 352, 388
    electron transport to, 92, 93, 349, 350, 392–397
    enzyme biogenesis, 92, 175, 353–357, 359–362
    inhibition, 254, 297–299, 349
    mechanism, 218, 249–251, 348, 397–403, 476, 477
    membrane interactions and orientation, 122, 123, 133, 136, 405–407
    primary structure, 175–180, 191
    regulation, 137, 138, 352–357, 359–364, 407–411
    role, 348
  see Lanosterol demethylase
  sterol 11β- and/or 18-hydroxylase
    electron transport to, 392–397, 401
    enzyme biogenesis, 359–362

P-450, biosynthetic forms (*cont.*)
  inhibition, 135, 275
  location, 346
  oxidative degradation, 234
  strructure and mechanism, 173, 350, 388, 403–407
  regulation, 353–357, 410
  sterol 12α-hydroxylase, 420, 533
  sterol 21-hydroxylase, 173, 180, 345, 350, 351, 360, 388, 401, 412
  sterol 25-hydroxylase: *see* Vitamin D
  sterol 17α-hydroxylase/C$_{17-20}$ lyase
    degradation, 234
    distribution, 346, 351, 363, 388
    inhibition, 301, 412
    reactions and mechanism, 351, 358, 401, 411–413
    regulation, 357, 358, 362, 363, 413
    structure, 173, 175–177
  see Vitamin D
P-450, bovine, 164, 177, 179; *see also* P-450 Biosynthetic forms
P-450, fish, 165, 177
P-450, guinea pig, 165, 177
P-450, human, 165, 173
P-450, insects, 237, 302
P-450, mouse, 164, 177
  form 1, 175–180, 188, 346, 350, 351, 357, 358, 363
  form 3, 175, 189, 316–318, 323–326, 329
P-450, pig, 165
P-450, plants, 51, 165, 173, 301, 441
P-450, rabbit, 163, 173, 177, 178
P-450$_{LM2}$
  active site, 460
  chemical modification, 143, 191
  see Inducers
  membrane interactions and topology, 124, 125, 129, 130, 132, 133, 137, 204
  oxygen activation, 218–221, 463, 464
  reactions with peroxides, 41, 42, 47, 48, 227, 228, 231
  reconstitution, 123–125, 130–135, 142, 143, 176–180
  redox potentials, 95
  reduction of, 101, 449
  stoichiometry, 233, 234
  structure, 173, 174, 181
  substrate binding, 78, 79
  substrate oxidation mechanisms, 237, 472, 477, 478

P-450$_{LM3}$, 122–124, 137, 166, 173, 180, 182, 201, 219, 421, 432
P-450$_{LM4}$
  active site topology, 79
  autoxidation, 221, 464
  see Inducers
  membrane interaction, 124, 125, 131–134
  peroxidase activity, 47
  primary structure, 173, 176–180, 183, 202
  spin state, 166, 219
  substrates, 419, 478
P-450, rat, 163, 173, 177, 178, 526
P-450$_b$, 77, 83, 174, 176–180, 184, 201, 202
P-450$_c$, 77, 79, 81, 174, 176–180, 185, 316, 317
P-450$_d$, 166, 176–180, 186, 316, 317, 420
P-450$_e$, 176–180, 187, 201
P-450$_p$, 332, 333, 340
P-450, yeast, 165, 254
Papain, *see* Peptide maps
Parathion, 232, 277, 278
Parathyroid hormone, 416
PCB, 318
Pentadecanoic acid, 433
Pentafluoroiodosylbenzene, 10
Peptide hormones, 363–364
Peptide maps, 168–171, 201
Perfluorohexane, 234
Pericyclocamphanone, 471
Periodate, 231
Peroxidases,
  see individual peroxidases
  catalytic cycle, 32
  inactivation, 42
  porphyrin radical cations, 36, 37
  reaction with 5-phenyl-4-pentenyl hydroperoxide, 47, 48, 49, 64
  reduction by impurities, 37, 38
  spectra, 36
  structures, 30, 32
  substrate oxidation, 47
  substrate oxygenation, 51
Peroxygenase, 227–231, 255, 400
pH effects
  on autoxidation, 220, 222
  on electron transfer, 95, 220, 390, 392
  modulation by b$_5$, 413
  on NADH synergism, 107, 108
  on substrate binding, 400
Phase transfer catalysis, 16, 17

Phenanthrene, 80, 228
Phenazine methosulfate, 468
Phenelzine, 290
Phenidone, oxidation by HRP, 49
Phenobarbital, P-450 induction by, 77, 170, 227, 334–337, 433, 528
Phenol(s), 47, 50, 222, 227, 230
 as preservative, 37
Phenylacetylene, 287, 288
Phenylalanine residues, 462, 512, 515
Phenylbutazone
 in cooxidation reactions, 56
 4-hydroperoxide product, 50
 oxidation by HRP, 32, 50, 52
 oxidation by P-450, 52
trans-1-Phenylbutene, 240
p-Phenylenediamine, 418
Phenylhydrazine, 290–292
Phenylhydrazone, 290
Phenylimidazole, 276
1-Phenyloctane, 78, 79
5-Phenyl-4-pentenyl-1-hydroperoxide, 47
2-Phenylperacetic acid, 14, 47, 229, 468, 469, 483
1-Phenylpropene, 17
Phosphatidic acid, 122, 133
Phosphatidylcholine, 119, 122, 127, 133, 405, 406, 431
Phosphatidylethanolamine, 119, 122, 133, 406
Phosphatidylglycerol, 119, 122, 138, 400
Phosphatidylinositol, 119, 122, 133, 138, 406, 410
Phosphatidylserine, 119, 122, 133, 138
Phospholipases, 135, 153
Phospholipid effects, 131–144
 on aggregation, 138–140, 141–144
 charge effects, 142–143
 on complexation, 138–142
 on conformation, 135–136
 on electron transfer, 140–143
 on exchange rate, 134
 on hydroxylation activity, 406, 431
 on 11β-hydroxylase, 406
 on P-450$_{scc}$, 134, 405, 406, 410
 on renaturation, 132, 133
 on spin state changes, 131, 132
 on substrate binding, 134–136, 398
 temperature dependence, 136
Phospholipids, 119, 122, 127, 133, 138
 bilayers, 120, 121

Phospholipids (*cont.*)
 in P-450 reconstitution, 120, 121
 spin-labeled, 136
"Picket fence" porphyrin, 6, 7
Picrotoxin, *see* Barbiturates, receptors for neurological effects
α-Pinene, 438
β-Pinene, 237, 438
Placenta, 346, 365–367, 413–415
Plasmids, 317, 437, 440
Polyacrylamide gel electrophoresis, 139, 161–166, 168, 196, 316, 326, 525
Polyamines, as P-450 effectors, 143
Poly(A$^+$)RNA, 316, 317, 323
Poly-L--lysine, stimulation of substrate binding, 138
Polysomes, 317, 332
Posttranslational modifications, 175–176
 see also Glycosylation
 N-terminus modifications, 180
Pregnenolone, 249, 348, 351, 398, 399, 408
Pregnenolone 16α-carbonitrile, 315, 332–334
Pressure, effect on P-450$_{cam}$, 457–459
Primary structures
 amino acid compositions, 171–175, 196–199
 molecular weights, 161–166, 196–198
 number of residues, 161–166
 prediction of tertiary structure, 202
 sequence homology, 185–189, 192–195, 531, 532
Progesterone, 137, 202, 227, 333, 351, 388, 412, 421, 431
Promoters, 435, 438
Propene, 83, 286
Propentdyopent products of heme degradation, 234
Propoxyphene, 227
Propyne, 83, 286
Prostacyclin synthase, 249, 254, 255
Prostaglandin H synthase, 33, 37, 38
 cooxidation, 53, 54
 mechanism, 47, 53
 structure, 31
 substrate oxygenation, 51
Prostaglandins
 effects on P-450$_{scc}$, 355
 metabolism, 231, 254, 302, 371, 372, 421
Proteases, 124
Protein alkylation, 277, 278

Protein kinase, 408, 411
Proteolytic cleavage, 201, 204, 403
  see also Peptide maps
Pseudomonas citronellis, 441
Pseudomonas incognita, 440
Pseudomonas putida, 89, 435
Putidaredoxin, 89, 90, 93–95, 225, 432, 438, 446, 448, 463–466
Putidaredoxin reductase, 89, 90, 93–95, 438, 463
Pyridine(s), 275, 301
Pyrimidines, 301
Pyrogallol, oxidation by hemoproteins, 46

Quadricyclane, 238
Quinone(s), 273

Radical
  alkoxy, 48, 54, 228, 251
  alkyl, 8, 50, 54, 55, 246, 289, 290
  hydroxyl, 40, 56, 226, 231, 464, 481, 519
  nitroxy, 230, 246
  P-450$_{cam}$, 230
  peroxyl, 50, 54–56, 61
  phenoxy, 50, 56
  protein, 37, 38, 41, 223, 230, 56
  spin trapping of, 246, 289, 290
  trimethylsilyl, 298
Radical cation
  aminopyrine, 230, 231
  hydrocarbon, 238
  nitrogen, 49, 245, 246, 248
  phenidone, 49
  porphyrin, 3, 6, 11–13, 34, 36, 38, 223
  sulfur, 245, 248
Radical intermediates, 4, 8, 230, 233, 237, 238, 241–243, 280, 287, 289, 469
Radius of gyration, 456
Rearrangements, oxidative
  acetylenes, 241–243, 288
  see NIH shift
  olefins, 4, 237, 238, 240, 241
  see Phenylacetylene
  see Quadricyclane
Receptors
  see Ah receptor
  non-P-450 receptors
    GABA, 336
    steroid hormone, 333, 334, 338, 339
  PCN, 315, 333–334
  phenobarbital, 315, 335–336
  TCDD, 318–323

Reconstitution of P-450 in vesicles, 120–124
  distribution of proteins, 124–145
  heme orientation, 125
  isozyme dependence, 124–145
  properties, 123–124
  protection of proteins, 124
  protein diffusion rates, 126–130
  protein mobility, 126–130
  rotational correlation times, 129
  rotational relaxation rates, 129
Redox potentials, 248
  "activated" P-450-oxygen complex, 52
  adrenodoxin, 390, 394
  axial ligand dependence, 217–218, 274
  cytochrome b$_5$, 102
  ferredoxins, 391
  hemoglobin, 482
  microsomal P-450, 95, 96–100, 447
  mitochondrial flavoprotein, 389
  P-450 reductase, 92
  P-450$_{cam}$, 95–96, 447, 482
  P-450$_{LM2}$, 95
  P-450$_{scc}$, 395
  pH effect on, 107–108, 220
  putidaredoxin, 447
  spin-state dependence, 98–100, 446–455
  substrate binding, effect of, 98–100, 218, 395, 453
  tryptophan oxygenase, 482
Reduction kinetics
  dependence on b$_5$, 101–108
  dependence on pH, 107–108
  dependence on substrate, 98–100, 412
  dependence on temperature, 98
  of microsomal P-450, 95–100
  of P-450$_{cam}$, 95
  of P-450$_{LM2}$, 101
  rate limiting step, 104–108
  in reconstituted systems, 100–101, 106, 108, 124
Renodoxin, 419
Renodoxin reductase, 419
Resonance Raman, 12
Resorcinol, 46
Rhizobium japonicum, 441
Rhodococus rhodochrous, 434, 438
mRNA, see Messenger RNA
Rotational diffusion, see Diffusion rates
Rubredoxin, 467

INDEX

Safrole, 297
Saturation transfer EPR, 129
Second derivative spectroscopy, 133, 459–462
Secondary structure prediction, 202
Sequences, primary, 182–195, 438, 507
  alignments, 182–195, 403, 514, 515
  see Amino terminal region
  carboxy-terminus, 180–182
  conserved regions, 185, 186, 403, 514–516
  cysteines, 180, 185, 189, 190, 195, 403, 514, 515
  heme ligands, 185, 186, 191
  homology, 185–189, 192–195, 514–516, 531–532
  in secondary structure predictions, 202
Singlet oxygen, 235, 336, 463
SKF 525-A, 366, 413
Sodium, regulation of, 404
Solvent cage, 54
Spectrum, 296
  absence of substrate-binding spectral changes, 418
  absolute, 218, 431
  difference, 274, 398, 409
  dioxygen complex, 218, 462, 463
  peroxide complex, 228, 229, 400
  Reverse Type I, 228, 447
  Type I, 79, 135, 136, 399, 404, 447
  Type II, 275, 274, 292, 301, 399
  see also P-450, reduced-CO complex
Sphingomyelin, 119
Spin labels, 127, 129, 136; see also Radical, spin trapping of
Spin state, dependence on
  cytochrome b_5, 179
  cytochrome P-450 reductase, 179
  ionic strength, 220, 461
  pH, 108, 220
  phospholipids, 131, 132, 140
  pressure, 457, 458
  substrate binding, 219, 394, 399, 401, 403, 412, 413, 420, 446, 453
  temperature, 132, 398, 412, 446
Spin states, 161–166, 399, 401, 402
  of different isozymes, 163–165
  electronics of, 220, 224
  relation to electron transfer, 99, 100, 394, 446–456
  relaxation rates, 449

Spin states (cont.)
  spectral relationships, 274, 398
  tyrosine solvent exposure, relation to, 460
Spironolactone, 279, 333, 334
Squalene oxidation, 416, 417
Steapsin, 125
Stereoselectivity, 387
Steroid hydroxylases, bacterial, 430
Steroidogenesis, 387
  acute regulation, 352, 387, 388
  chronic regulation, 353, 389
  inducing protein, 362, 363
  mitochondrial, 353–357
  precursors, 356
Sterol carrier proteins, 410, 411
Stilbene, 7, 239
Stoichiometry, 103–107, 220, 221, 233, 234, 405
Stopped-flow spectroscopy, 43, 96, 218, 219, 391, 463, 473
Streptomyces certicillus, 479
Styrene epoxidation, 225, 285
  isotope effects, 239, 240
  by metalloporphyrins, 7, 19
  by non-P-450 hemoproteins, 56
  rearrangements, 241
Suberimidate crosslinking, 140
Substrate, 273
  binding, 77–84, 135–138, 395, 405, 478
  lipophilicity, 79, 135, 275, 276, 281
  reorientation in active site, 4, 81, 82
Succinate, repression of induction by, 435, 437
Suicide substrate, see Inhiibitors, mechanism-based
Sulfur
  dealkylation, 245, 247
  oxidation, 78, 232, 245, 247, 248, 277
  P-450 destruction by, 277, 278
Superoxide, 480
  as autoxidation product, 10, 220–222, 402, 405, 463, 481
  as reactive species, 91, 225
  as reductant, 231
Sydnones, 290, 294
Synergists, 280, 282

Taurodeoxycholate, 533
TCDD, 316, 318–325, 362
  binding proteins, 322, 323, 326–328

TCPOBOP, 335
Terpin hydrate, 438
Testis, 164, 346, 352, 363, 364, 412
Testosterone, 227, 421, 528, 530
3,4,5,6-Tetrachlorocyclohexene, 237
2,3,7,8-Tetrachlorodibenzo-*p*-dioxin, *see* TCDD
Tetracyanoethylene oxide, 18
3,3,6,6-Tetradeuterocyclohexene, 4, 5, 237
exo-Tetradeuteronorbornane, 4, 5, 236, 238
Tetramesitylporphyrin, 7, 13
Tetramethylcyclohexanone, 459
Tetraphenylporphyrin, 6, 7, 11, 12, 17, 20, 21
  halogen substituted, 10
Thioanisole(s), 232, 248
Thiocamphor, binding to P-450$_{cam}$, 461
Thioredoxin, 420
Thioureas, 277
Thromboxane synthase, 256
Thyroid peroxidase, 31, 33
Thyroxine, 33
Toluene oxidation, 41, 228, 230, 274, 470
Tosylimines, oxygen activation by, 233
Transcription, 324, 329, 335, 337, 339, 340, 363
2,4,6-Tri-*tert*-butylphenol, 14
Trichlorethylene, 240
Triiodothyronine, 33
Triton N-101, 121, 132
Triton X-100, 121, 139
Trypsin
  cleavage of cytochrome P-450 reductase, 125
  cleavage of P-450$_{scc}$ into domains, 403
  microsomes treated with, 124, 226
  P-450's treated with, 124, 168–171, 201

Tryptophan residues, 37, 448, 460, 466
Tween 20, 132
Tyrosine
  as axial ligand, 399, 460
  chemical modification, 191
  cooxidation with, 56
  iodination of, 33
  local motions of, 133, 459–462
  oxidation of, 40, 44
  in substrate binding to P-450$_{cam}$, 478, 512
Tyrosine aminotransferase, 333, 334

Uncoupling, 4, 405, 472
10-Undecynoic acid, 301, 302
10-Undecynyl sulfate, 302

Vectors
  Okayama-Berg, 317
  pKT 240, 437
  pUC, 438
Vinca rosea, 441
Vinyl fluoride, 286
Vitamin D$_3$, 415–419
  1α-hydroxylase, 346, 368–370, 388, 419, 534
  25-hydroxylase, 346, 368–370, 388, 418, 420, 421, 533, 534
  metabolism, 368–371

Warfarin, 78, 228, 244, 530
Water
  as catalytic product, 32, 34, 234
  as heme ligand, 191, 233, 399, 512

X-ray crystallography, 9, 284, 290, 505–520

UNIVERSITY OF DELAWARE LIBRARY

Please return this book as soon as you have finished with it. In order to avoid a fine it must be returned by the latest date stamped below.

JAN 1 1987
JAN 1 2 1987
MAY 1 6 1987
MAY 1 3 1987
DEC 3 0 1987
NOV 2 4 1987
JAN 1 2 1988
APR 1 2 1989
FEB 2 0 1991
JUN 1 4 1991

JAN 1 7 1992
JUL 2 8 1992

MAR 0 3 1993

MAR 2 1 2001